Hallucinations

Jan Dirk Blom · Iris E.C. Sommer
Editors

Hallucinations

Research and Practice

Springer

Editors
Jan Dirk Blom, M.D., Ph.D.
Parnassia Bavo Group and
University of Groningen
Kiwistraat 43
2552 DH, The Hague
The Netherlands
jd.blom@parnassia.nl

Iris E.C. Sommer, M.D., Ph.D.
University Medical Center Utrecht and
Rudolf Magnus Institute of Neuroscience
Heidelberglaan 100
3584 CX, Utrecht
The Netherlands
i.sommer@umcutrecht.nl

ISBN 978-1-4614-0958-8 e-ISBN 978-1-4614-0959-5
DOI 10.1007/978-1-4614-0959-5
Springer New York Dordrecht Heidelberg London

Library of Congress Control Number: 2011942898

Printed on acid-free paper

Springer is part of Springer Science+Business Media (www.springer.com)

Foreword

When the editors approached me to write a foreword for this book, I was naturally flattered. They were generous enough to say that they had found some of my publications on the topic nearly 20 years ago to have been a useful starting point for some of their own investigations (Nayani and David 1996).[1] However, it wasn't long before vanity got the better of me and I started to react against being cast far too prematurely as a grand old man. More importantly, I also started to worry that, while I have maintained an interest in the topic, I wasn't confident that I had kept up with all of the latest developments. I should not have been so concerned. This volume is itself the perfect antidote. It reminds us of how diverse and engaging the topic of hallucinations continues to be as well as paying tribute to the truly ancient body of literature that has grown up around the effort to understand it. It is indeed humbling to realise that most seemingly new contributions to the study of hallucinations echo previous thinking.

The scientific study of hallucinations is, however, relatively youthful – perhaps a mere 150 years old, beginning with Esquirol and others. The nineteenth and twentieth centuries were a 'Golden Age' as far as descriptive psychopathology goes, but of course the neuroscientific contribution to psychopathology is more recent, beginning with electroencephalography but now fuelled by functional magnetic resonance imaging. Indeed those of us in the 1990s who had the opportunity to use this 'toy' thought that hallucinations were an obvious target and that they would be explained once and for all. All we needed to do was show that the sensory processing areas of the brain 'lit up' in response to an hallucination in the same way as they did to an external stimulus and the riddle was solved. Such hubris! In fact this early work did lend support to the idea that hallucinations were indeed sensory phenomena to some extent (David et al. 1996) but, like most research, raised new questions about, for example, the relationships between hallucinations, mental imagery and perception; between the generation and reception of images; and, most challenging of all, the nature of belief and reality. In fact, the study of hallucinations was, for me, a salutary

[1] This foreword is dedicated to the memory of Dr. Tony Nayani (1962–2009).

lesson in the limits of neuroimaging and the need to draw on philosophy, the social sciences as well as biomedical science. This book exemplifies this approach and is a tangible demonstration of the benefits of multidisciplinary discourse.

At risk of labouring this point, here is an anecdote. I, like many others, was inspired by Oliver Sacks' observations on musical hallucinations. A review of personal cases and the literature led to a comprehensive review on the topic (Keshavan et al. 1992). It was clear that hallucinations of music and those of voices – a core symptom of schizophrenia – were entirely different. The link with hearing impairment and tinnitus in the former was almost universal as was the lack of associated psychopathology. At last, we had confirmed a clear basis for diagnostic classification. That was until my mentor Alwyn Lishman drew my attention to a case described of an old lady who heard the hymn *The Old Rugged Cross* emanating from her vagina. This led to a pause and consideration that the origin of such a hallucination probably went beyond the scope of the oto-naso-laryngologist.

Another service that this volume provides is to put the spotlight on the variety and modalities of hallucinations. A comparison of these and identification of similarities and differences is long overdue. Why is it that auditory verbal hallucinations (AVHs) are such an important part of psychotic disorders – pointing specifically to language systems and dialogic construction of the Self? Why on the other hand are visual hallucinations the hallmark of 'organic' disorders – pointing to perhaps release phenomena and more diffuse consequences of cerebral perturbation? And why do AVHs in schizophrenia usually respond well to antipsychotic medication while the apparently same phenomena are unresponsive in people with borderline personality disorder? This volume extends to all these modalities and distinctions and will provide much stimulus for integration.

There are other recent developments which have broadened the context for hallucination research and to which Dutch scientists and commentators have made a distinct contribution. In particular, we now know that there are many people who experience hallucinations, of a sort that were once held to be entirely pathological, as part of their daily lives and that these are benign or even positively valued. This prompts the obvious question – what makes, for example, a hallucinated voice a scourge for one person, and a guardian angel for another? Asking voice hearers themselves is a good place to start.

Finally, the editors and their well-chosen contributors demonstrate that these conundrums are not purely theoretical. There is now an array of treatment interventions from psychotherapies to brain stimulation which may be offered to those for whom hallucinations are less than positive, with the added value of informing theoretical advances by a process of reverse translation.

London, UK Anthony S. David

References

David, A., Woodruff, P.W.R., Howard, R., Mellers, J.D.C., Brammer, M., Bullmore, E., Wright, I., Andrew, C., Williams, S.C.R. (1996). Auditory hallucinations inhibit exogenous activation of auditory association cortex. *NeuroReport*, 7, 932–936.

Keshavan, M.S., David, A.S., Steingard, S., Lishman, W.A. (1992). Musical hallucinations: A review. *Neuropsychiatry, Neuropsychology & Behavioral Neurology*, 5, 211–223.

Nayani, T.H., David, A.S. (1996). The auditory hallucination: A phenomenological survey. *Psychological Medicine,* 26, 177–189.

Contents

Part IV Treatment

Contributors

André Aleman, Ph.D. BCN Research School and University Medical Centre, Groningen, The Netherlands
a.aleman@med.umcg.nl

Paul Allen, M.Sc., Ph.D. Department of Psychosis Studies, Institute of Psychiatry, King's College, London, UK
p.allen@kcl.ac.uk

Johanna C. Badcock, Ph.D. School of Psychiatry and Clinical Neurosciences, University of Western Australia, Crawley, Australia

Centre for Clinical Research in Neuropsychiatry, Graylands Hospital, Claremont, Australia
Jo.Badcock@health.wa.gov.au

Ralf-Peter Behrendt, M.R.C.Psych. Elderly Mental Health Team, Princess Elizabeth Hospital, St Martin, Guernsey
rp.behrendt@btinternet.com

German E. Berrios Robinson College, University of Cambridge, Cambridge, UK
geb11@cam.ac.uk

Christine Blanke Voices Clinic, Psychiatry Division, University Medical Center Utrecht, Utrecht, The Netherlands
stienblanke@gmail.com

Olaf Blanke, M.D., Ph.D. Laboratory of Cognitive Neuroscience, Brain-Mind Institute, Ecole Polytechnique Fédérale de Lausanne, Lausanne, Switzerland
olaf.blanke@epfl.ch

Jan Dirk Blom, M.D., Ph.D. Parnassia Bavo Academy, Parnassia Bavo Group and University of Groningen, Groningen, The Hague, The Netherlands
jd.blom@parnassia.nl

Peter Brugger, Ph.D. Department of Neurology, University Hospital Zurich, Zurich, Switzerland
peter.brugger@usz.ch

Sergio Canavero, M.D., F.M.G.E.M.S. Turin Advanced Neuromodulation Group, Turin, Italy
sercan@inwind.it

James Allan Cheyne, Ph.D. Department of Psychology, University of Waterloo, Waterloo, ON, Canada
acheyne@uwaterloo.ca

Daniel Collerton, M.A., M.Sc. Northumberland, Tyne and Wear NHS Foundation Trust, UK

Newcastle University, Newcastle Upon Tyne, UK
daniel.collerton@ncl.ac.uk

Vicka Corey, Ph.D. Harvard Medical School, Boston, MA, USA

The Laboratory for Integrative Psychiatry, Division of Alcohol and Drug Abuse, McLean Hospital, Belmont, MA, USA
vicka@mclean.harvard.edu

Kirstin Daalman, M.Sc. Psychiatry Division, Rudolf Magnus Institute of Neuroscience, University Medical Center Utrecht, Utrecht, The Netherlands
K.Daalman@umcutrecht.nl

Z. Jeff Daskalakis, M.D., Ph.D., F.R.C.P. Department of Psychiatry, University of Toronto, Toronto, ON, Canada

Brain Stimulation and Research Program, Centre for Addiction and Mental Health, Toronto, ON, Canada

Schizophrenia Program, Centre for Addiction and Mental Health, Toronto, ON, Canada
jeff_daskalakis@camh.net

Kelly M.J. Diederen, M.Sc., Ph.D. Psychiatry Division, Rudolf Magnus Institute of Neuroscience, University Medical Center Utrecht, Utrecht, The Netherlands
K.M.J.Diederen@umcutrecht.nl

Rob Dudley, B.A., Ph.D., D.Clin Psy., Cert CBT Institute of Neuroscience, Newcastle University, Newcastle Upon Tyne, UK

Early Intervention in Psychosis Service, Northumberland Tyne and Wear NHS Foundation Trust, UK
r.e.j.dudley@newcastle.ac.uk

Sandra Escher, Ph.D. Association Living with Voices, Gravenvoeren, Belgium board member of Intervoice, editor of de 'Klankspiegel' Resonance magasine member of Intar (international network towards recovery)
sandra@hearing-voices.com

Judith M. Ford, M.A., M.D., Ph.D. Brain Imaging and EEG Laboratory, VA Medical Center, Department of Psychiatry, University of California San Francisco, San Francisco, CA, USA
Judith.Ford@ucsf.edu

Rutger Goekoop, M.D., Ph.D. Parnassia Bavo Group, The Hague, The Netherlands
r.goekoop@parnassia.nl

John H. Halpern, M.D. Harvard Medical School, Boston, MA, USA
The Laboratory for Integrative Psychiatry, Division of Alcohol and Drug Abuse, McLean Hospital, Belmont, MA, USA
john_halpern@hms.harvard.edu

Cor B.M. Hoffer, M.Sc., Ph.D. Parnassia Bavo Group, Rotterdam, The Netherlands
c.hoffer@parnassiabavogroep.nl

Donald D. Hoffman, Ph.D. Professor, Department of Cognitive Science, School of Social Sciences, Donald Bren School of Information and Computer Sciences, University of California, USA
ddhoff@uci.edu

Kenneth Hugdahl, Ph.D. Department of Biological and Medical Psychology, University of Bergen, Bergen, Norway
Division of Psychiatry, Haukeland University Hospital, Bergen, Norway
Hugdahl@psybp.uib.no

Roi Cohen Kadosh, B.A., Ph.D. Department of Experimental Psychology, University of Oxford, Oxford, UK
roi.cohenkadosh@psy.ox.ac.uk

David G. Kingdon, M.D., Ph.D., FRCPsych. Department of Psychiatry, University of Southampton, Southampton, UK
D.Kingdon@soton.ac.uk

Robyn Langdon, Ph.D. ARC Centre of Excellence in Cognition and its Disorders, Macquarie Centre for Cognitive Science, Macquarie University, Sydney, Australia
robyn.langdon@mq.edu.au

Jasper Looijestijn, M.D. Parnassia Bavo Group, The Hague, The Netherlands
J.Looijestijn@parnassia.nl

Ivana S. Marková Department of Psychiatry, University of Hull, Hull, UK
ismarkova@fastmail.co.uk

Gemma Modinos, Ph.D. Department of Psychosis Studies,
Institute of Psychiatry, King's College, London, UK
gemma.modinos@kcl.ac.uk

Urs Peter Mosimann, M.D., Ph.D. Department of Old Age Psychiatry,
University Hospital of Psychiatry, University of Bern, Bern, Switzerland
urs.mosimann@gef.be.ch

Torsten Passie, M.D., M.A. Department for Psychiatry, Social Psychiatry,
and Psychotherapy, Hannover Medical School, Hannover, Germany
dr.passie@qmx.de

Marius Romme, M.D., Ph.D. Emeritus Prof. of Psychiatry,
Maastricht University, Visiting Prof. Birmingham City University
www.hearing-voices.com

Oliver W. Sacks, M.D. Columbia University Medical Center, New York, NY, USA
mail@oliversacks.com

Steven Scholtus, M.Sc. Voices Clinic, Psychiatry Division,
University Medical Center Utrecht, Utrecht, The Netherlands
steven.scholtus@gmail.com

Anna Sforza, Ph.D. Laboratory of Cognitive Neuroscience, Brain-Mind Institute,
Ecole Polytechnique Fédérale de Lausanne, Lausanne, Switzerland
annalaura.sforza@gmail.com

Christina W. Slotema, M.D., Ph.D. Parnassia Bavo Group, The Hague,
The Netherlands
c.slotema@psyq.nl

Iris E.C. Sommer, M.D., Ph.D. Department of Psychiatry, University Medical
Center Utrecht and Rudolf Magnus Institute of Neuroscience,
Utrecht, The Netherlands
i.sommer@umcutrecht.nl

Richard J. Stevenson, D.Phil. Department of Psychology, Macquarie University,
Sydney, Australia
dick.stevenson@mq.edu.au

Devin Blair Terhune, M.Sc., Ph.D. Department of Experimental Psychology,
University of Oxford, Oxford, UK
devin.terhune@psy.ox.ac.uk

Mark van der Gaag, Ph.D. University and EMGO Institute, Amsterdam,
The Netherlands

Psychosis Research, Parnassia Bavo Group, The Hague, The Netherlands
m.vander.gaag@vu.nl

Willemijn A. van Gastel, M.Sc. Psychiatry Division, Rudolf Magnus Institute of Neuroscience, University Medical Center Utrecht, Utrecht, The Netherlands
W.vanGastel@umcutrecht.nl

Remko van Lutterveld, M.Sc. Department of Psychiatry, University Medical Center, Utrecht, The Netherlands and Rudolf Magnus Institute of Neuroscience, Utrecht, The Netherlands
R.vanlutterveld@umcutrecht.nl

Ans Vercammen, Ph.D. Neuroscience Research Australia, Randwick, Sydney, Australia
a.vercammen@neura.edu.au

Chapter 1
General Introduction

Iris E.C. Sommer and Jan Dirk Blom

Hallucinations are fascinating phenomena. The mere possibility of perceiving things that are not there is the stuff that campfire tales are made of. It is one thing to be in a dream state, to be asleep, and to conjure up people, scenes, and landscapes that do not actually exist. But it is quite another to hallucinate: to be wide awake and yet hear that ethereal music, see those costumed figures strolling by, smell the roses that used to grow in your grandfather's garden, feel his hand upon your shoulder, sense his presence somewhere near – and to be the only one able to experience all that. How strange, how fascinating, and how absolutely mind-boggling that would be. And how frightening perhaps, since not all hallucinations involve a walk in the park with loved ones. As clinical psychiatrists, we have come in contact with a great many people who are plagued by voices and visions which compete for priority with what we call "reality." "The voices have completely shaken up my life," as Steven Scholtus – a long-time voice hearer and field expert at the Voices Clinic in Utrecht – writes in this book. "A normal way of living, with a full-time job, a family, the raising of children, is no longer within my reach."

Imagine what that is like.

Or imagine what Christine Blanke, another field expert at the Voices Clinic, has had to endure since she started to hear voices. As she recalls in Chap. 8, "Perhaps for being so busy with other things, I hardly paid any attention to the gentle voices inside my head. Then other people came into my head, conveying evil messages.

I.E.C. Sommer, M.D., Ph.D. (✉)
Department of Psychiatry, University Medical Center Utrecht and Rudolf Magnus Institute of Neuroscience, Heidelberglaan 100, 3584 CX, Utrecht, The Netherlands
e-mail: i.sommer@umcutrecht.nl

J.D. Blom, M.D., Ph.D.
Parnassia Bavo Academy, Parnassia Bavo Group and University of Groningen, Groningen, Kiwistraat 43, 2552 DH, The Hague, The Netherlands
e-mail: jd.blom@parnassia.nl

J.D. Blom and I.E.C. Sommer (eds.), *Hallucinations: Research and Practice*,
DOI 10.1007/978-1-4614-0959-5_1, © Springer Science+Business Media, LLC 2012

I also started to see blood in the streets, and arms, legs, and heads that had apparently been chopped off." It is not difficult to picture the impact of such horrifying hallucinations on the very foundations of one's existence. And yet there are millions of people who experience hallucinations without being overly bothered by them, and who may even find comfort and solace in such experiences. We know that many of them are as intrigued as we are about the nature of their idiosyncratic experiences, whether they attribute them to psychological, neurobiological, or metaphysical sources.

The fact that a person can see or hear things that remain imperceptible to others touches upon a number of fundamental issues. One of those issues is whether there is such a thing as a general – shared – reality. Many philosophers tell us that such a reality does not exist. What we call reality is always perceived by an observer, and is therefore subjective (Schopenhauer 1958). In that sense, the world that we perceive is a creation of our own mind, one that is based only in part on the input provided by our sense organs. Kant even maintained that we are fundamentally incapable of discerning what the real world is like. The mind imposes concepts ("categories") on the information gathered by our senses, thereby determining the way we experience the world (Kant 1933). Since those concepts are applied without our being able to keep track of the way it is done, we have no access to the world beyond the realm of our consciousness.

Plato illustrated this line of thought in his *Allegory of the Cave*, a dialogue between his brother Glaucon and his teacher Socrates. In the Allegory, Socrates describes a group of people who have spent their lives chained to the wall of a cave, facing a blank wall. As they watch the shadows cast onto that wall by the people and things passing between them and a fire which is burning behind them, they begin to interpret those shadows. According to Socrates, those interpretations are as close as we will ever get to grasping what the real world looks like. He then explains how one of the prisoners is freed, and leaves the cave in search of the sun, the world, nature, and all its creatures. That prisoner realizes that the shadows on the wall are quite different from the real world, which he can now perceive in its true shapes and colors. When he returns to the cave, his fellow prisoners do not believe a word of what he says, preferring to stick to their drab, two-dimensional version of reality (Plato 1974). Perhaps some people suffering from complex or compound hallucinations feel like that returning prisoner, as they try to convince their treating physician of the reality of their perceptions.

From a neuroscientific point of view, these philosophical musings make perfect sense, especially when we conceptualize consciousness as a brain function involved in the creation of representations of the external world (Behrendt 2010). Animal experiments have shown that the hippocampus is able to provide rapid representations of our surroundings based on sensory input signals and memory patterns of situations previously experienced (Kahn et al. 2008). As we all have varying experiences, our memories are different and – to some extent – unique. As a consequence, the concepts we apply to the information from our sense organs also tend to be unique, thus providing us with our own idiosyncratic shadows on the wall, i.e., our subjective representations of reality. The study of hallucinations touches upon such neuroscientific topics as perception and consciousness and allows us to explore them in depth.

Another question involves the source of hallucinations. When, as the Bible tells us, the Virgin Mary heard an angel speak to her, was she in fact hallucinating, in the sense that her brain was conjuring up a being that was not there? Socrates, one of the most famous voice hearers, was convinced that a *daimon*, or benevolent spirit, always warned him when he was on the verge of making a grave mistake. Joan of Arc claimed that she was guided by the voices of various saints and even God himself and that they helped her to take command of an army and defeat the British. Although unconfirmed by scholars, it has been maintained that René Descartes also heard voices. Clinical lore holds that they seemed to come from behind, and that they were so realistic that he thought he was actually being followed (Winslow 1840). Did all these historical figures experience hallucinations? Or could it be that gods, angels, demons, *djinns* – and even an evil genius – are readily perceptible to some individuals but not to others?

While the spiritual and religious among us may be more interested in the possible metaphysical sources of such experiences, it is the task of neuroscientists to explore the brain and its functions in order to ascertain their origin. Often it seems as if the two possibilities cancel each other out. By demonstrating brain activity on the fMRI scan of a test person who is experiencing auditory hallucinations, it may seem as if their intracerebral source can be proved. But obviously, a BOLD signal on a scan does not count as a valid argument against the possibility that those hallucinated voices have an external origin. If that same person were to perceive a voice from an actual person (or from God?), then similar speech perception areas may be expected to light up on the scan. The apparent contrast between such spiritual and neurobiological explanations has somehow influenced popular opinion, which holds that hallucinations occurring in the context of psychosis or borderline personality disorder derive from brain dysfunction, while those experienced by people without a psychiatric diagnosis may well stem from metaphysical sources. It is not up to us to give a verdict in this matter, but as clinicians and neuroscientists, we would like to stress the role of the brain in all types of perceptual experience. The involvement of speech *production* areas, for example, as established by our own research group, is hardly commensurable with an external source of verbal auditory hallucinations (Sommer et al. 2008).

A final reason why over the years hallucinations have continued to fascinate so many people is the fact that they compel us to think about the brain in ways that go beyond our current neurobiological discourse. Many fundamental brain mechanisms have been discovered by means of the detailed examination of individuals suffering from pathological conditions. Thus, the function of Wernicke's area in deciphering of speech sounds was discovered by performing an autopsy on a patient with severe sensory aphasia. Similarly, the role of the fusiform gyrus in face recognition was unraveled following the examination of patients with prosopagnosia (i.e., the specific inability to recognize faces while the ability to recognize other objects remains relatively intact). Hallucinations have the potential to shed light on a great number of brain mechanisms currently unknown to us and to provide new insights into perception, consciousness, and many other fundamental brain functions.

For all the above reasons, the brain mechanisms underlying hallucinations have proved to be a popular topic within neuroscience. The technical possibilities currently available to visualize cerebral processes have significantly accelerated our attempts to unravel the neurobiology of hallucinations. But at least as much can be learned from individuals who themselves experience hallucinations. Detailed clinical descriptions and first-person accounts are still a treasure trove for researchers in the field. This book focuses on both aspects. Part I presents various basic approaches to hallucinations which range from the philosophical to the conceptual and neuroscientific. Part II consists of detailed descriptions of the phenomena themselves and the various ways in which they are experienced by patients and by healthy individuals. Part III offers a comprehensive update of findings in the fields of structural and functional neuroimaging, electrophysiology, psychopharmacology, and cognition, while Part IV provides an overview of state-of-the-art treatment methods.

It has been an honor and a pleasure to prepare this book in collaboration with so many experts in the field of hallucinations research. It is gratefully dedicated to the many patients and healthy hallucinators who have inspired them and us.

References

Behrendt, R.P. (2010). Contribution of hippocampal region CA3 to consciousness and schizophrenic hallucinations. *Neuroscience and Biobehavioral Reviews*, 34, 1121–1136.

Kahn, M.C., Siegel, J.J., Jechura, T.J., Bingman, V.P. (2008). Response properties of avian hippocampal formation cells in an environment with unstable goal locations. *Behavioural Brain Research*, 191, 153–163.

Kant, I. (1933). *Critique of pure reason. Second edition*. Translated by N.K. Smith. London: Macmillan.

Plato (1974). *The republic*. Translated by G.M.A. Grube. Indianapolis, IN: Hackett Publishing Company.

Schopenhauer, A. (1958). *The world as will and representation. Volume II*. Translated by E.F.J. Payne. New York, NY: Dover Publications.

Sommer, I.E.C., Diederen, K.M.J., Blom, J.-D., Willems, A., Kushan, L., Slotema, K., Boks, M.P.M., Daalman, K., Hoek, H.W., Neggers, S.F.W., Kahn, R.S. (2008). Auditory verbal hallucinations predominantly activate the *right* inferior frontal area. *Brain*, 131, 3169–3177.

Winslow, F. (1840). *The anatomy of suicide*. London: H. Renshaw.

Part I
Conceptual Issues

Chapter 2
The Construction of Visual Reality

Donald D. Hoffman

2.1 Illusions: What and Why

Many people, when viewing a windmill in the distance, report that the blades sometimes seem to rotate in the wrong direction. This is an example of a visual illusion. The standard account of such illusions says that each is an incorrect perception seen by most people when they view a specific stimulus. Illusions are rare, but the situations that trigger one person to see an illusion are likely to trigger others to see a similar illusion. Hallucinations, by contrast, are incorrect perceptions that are seen by few people and that occur in the absence of an appropriate stimulus. A person with delirium tremens, for instance, might see a spider that no one else sees.

This standard account of visual illusions naturally raises the question as to why our perceptions should be fallible. What is wrong with our visual system that allows false perceptions to occur?

To answer this question, we must understand visual perception as a biological system that has been shaped by natural selection. Each organ of the body has been shaped by natural selection to contribute in specific ways to the fitness of the person. The visual system can be considered as one of the many organs of the body that makes its specific contribution to the fitness of the whole organism.

This still leaves the puzzling question as to why our perceptions are fallible. The standard account of perceptual evolution is that more accurate perceptions are more fit. For instance, the textbook *Vision Science* states that "Evolutionarily speaking, visual perception is useful only if it is reasonably accurate… Indeed, vision is useful precisely because it is so accurate. By and large, *what you see is what you get*. When this is true, we have what is called *veridical perception* … This is almost always the

D.D. Hoffman, Ph.D. (✉)
Professor, Department of Cognitive Science, School of Social Sciences,
Donald Bren School of Information and Computer Sciences
University of California, Irvine 93697, USA
e-mail: ddhoff@uci.edu

J.D. Blom and I.E.C. Sommer (eds.), *Hallucinations: Research and Practice*,
DOI 10.1007/978-1-4614-0959-5_2, © Springer Science+Business Media, LLC 2012

case with vision …" (Palmer 1999, p. 6). Geisler and Diehl (2003) say, "In general, (perceptual) estimates that are nearer the truth have greater utility than those that are wide of the mark." Knill et al. (1996, p. 6) say, "Visual perception … involves the evolution of an organism's visual system to match the structure of the world and the coding scheme provided by nature."

The idea here is that in the struggle for survival, those of our predecessors that saw the world more truly had a fitness advantage over those that saw less truly. Predecessors with truer perceptions had a better chance of becoming our ancestors. Over many generations, this shaped *Homo sapiens* to have more accurate perceptions. We are the offspring of those who saw more truly, and in consequence our perceptions are usually veridical.

From this evolutionary perspective, one answer to the question as to why our perceptions are fallible is simply that evolution is not yet done with us. We are a species in process, not a species that is the end product of an evolutionary great chain of being.

While this last answer is, as far as it goes, correct, it is far from a complete account of why perception is fallible and visual illusions occur. A more complete account requires us to understand that (1) vision is a constructive process and (2) evolution has shaped this constructive process not to deliver truth but to guide adaptive behavior. When these points are understood, we find that we must redefine the notion of illusion. We also find that illusions are an unavoidable feature of perception and cannot be eradicated by further evolution.

2.2 Vision as Construction

Roughly half of the brain's cortex is engaged in vision. Billions of neurons and trillions of synapses are engaged when we simply open our eyes and look around. This is, for many of us, a surprise. We think of visual perception as being a simple process of taking a picture. There is an objective physical world that exists whether or not we look, and vision is just a camera that takes a picture of this preexisting world. We can call this the camera theory of vision. Most of us, to the extent that we think about vision at all, assume that the camera theory of vision is true.

That billions of neurons are involved in vision is a surprise for the camera theory. So much computational power is not necessary to take a picture. Cameras existed long before computers.

The eye is, of course, like a camera. It has a lens that focuses an image on the retina at the back of the eye. But this is just the starting point of the visual system. From there, billions of neurons are engaged in cortical and subcortical processing. Why all this processing power?

The story that has emerged from research in cognitive neuroscience is that vision is a constructive process. When we open our eyes, our visual system constructs in a fraction of a second all the shapes, depths, colors, motions, textures, and objects that we see. The computational power required for such construction is massive, but the

Fig. 2.1 Neon color spreading. The green glowing worm on the *right side* of the figure is a perceptual illusion

construction is done so quickly that we are fooled into thinking that there is no construction at all, that we are simply taking a snapshot of the world as it is.

Why does the visual system bother to do all this construction? Why does it not simply take a picture and be done? That would certainly require less computation, and would reduce the chance of illusions.

The visual system constructs our visual worlds because it must. The starting point of vision is the photoreceptor mosaic in the retina of the eye. Each eye has roughly 120 million photoreceptors, and the activation of each photoreceptor is proportional to the number of photons it catches. One can think of the retina as starting with an array of 120 million numbers, describing the number of photons caught by each photoreceptor. There are no colors, shapes, objects, textures, motions, or depths. There is only a description that says something like, "This photoreceptor caught 5 photons, this one caught 12, this one caught...." From this array of 120 million numbers, the visual system must proceed to construct all the colors, shapes, objects, and depths that constitute our visual world.

This point is painfully clear to computer scientists building robotic vision systems. The input to such a system is an array of numbers from a video camera. If the computer is going to see anything more than just this meaningless array of numbers, then it must have sophisticated programs that set about to construct visual worlds from the video input. Writing such programs has proved exceedingly difficult and has led to great respect for the constructive powers of biological vision systems.

For any image given to the computer, there are always an infinite number of visual worlds that could be constructed and that are compatible with that image. For instance, an infinite number of 3D constructions are always, in principle, compatible with any given 2D image: An ellipse in an image could be the projection of a circle seen at an angle or the projection of any one of an infinite number of different ellipses at different angles. This makes the construction process nontrivial.

A concrete example of visual construction is illustrated in Fig. 2.1. On the left is a collection of green lines. On the right is the same collection of green lines but with black lines attached. Notice that on the right, the green appears to fill in the space between the lines to create a glowing green worm with fairly sharp edges. The glowing green and the sharp edges are all constructed by your visual system, an effect known as neon color spreading (Redies and Spillmann 1981). You can check that you are constructing the neon worm: If you cover the black lines the worm disappears.

Neon color spreading is often used as an example of a visual illusion. It fits the standard definition of an illusion. Most observers see the color spreading when they look at such a figure, and the perceived green spreading where there is no green ink is, most would agree, an incorrect perception. So here we see a case where the constructive power of the visual system leads to a visual illusion. Indeed, each standard visual illusion is in fact a case where we catch the visual system in the act of a construction that is apparently incorrect (for hundreds of illusions and their explanations, see Hoffman (1998) and Seckel (2009)). Illusion and construction are inextricably linked.

Now, the standard view of visual constructions is that they are, in the normal case, *re*constructions. There is an objective physical world with depths, shapes, and colors, and the constructions of the visual system are, in the normal case, fairly accurate reconstructions of the true physical properties. According to this standard view, the reason that visual constructions are usually accurate reconstructions is due to evolution by natural selection. The more accurately an organism's visual system reconstructs the objective physical properties of its environment, the more fit is the organism and the better its chances of surviving long enough to reproduce.

So, the standard view of visual illusions is that they are the result of visual constructions that are not accurate reconstructions of the objective physical world. Evolution by natural selection has made sure that such incorrect constructions are rare.

2.3 Perceptual Evolution

One problem with the standard view of visual illusions is that it is based on an incorrect understanding of evolution by natural selection. As we noted earlier, Geisler and Diehl (2003) say, "In general, (perceptual) estimates that are nearer the truth have greater utility than those that are wide of the mark." Most vision researchers agree that truer perceptions have greater utility and therefore contribute to greater fitness of the organism.

But this assumption, though perhaps plausible, is in fact incorrect. Truth and utility are distinct concepts, and conflating them is a fundamental error. Utility depends on the organism and the world. One cannot assign a utility to the true state of the world unless one specifies an organism. For instance, being 5,000 ft below sea level has high utility for a benthic fish, but is fatal for a person. The same objective feature of the world has radically different utility for people and fish. Mathematically we can write that utility, u, is a function from the objective world, W, and an organism, O, into the real numbers, $\boldsymbol{R}$.

$$u : W \times O \rightarrow R \tag{2.1}$$

So utility and truth are related as shown in (2.1) and therefore are not the same concepts. Now, it might still be the case that although utility and truth are distinct, nevertheless, it happens to be an empirical fact that truer perceptions have greater utility. But this needs to be demonstrated. It cannot simply be assumed to be true.

One way to test this assumption is through the mathematical theory of evolution, known as evolutionary game theory (Maynard Smith 1974; Nowak 2006). Using computer simulations, one can create a wide variety of objective worlds and of organisms with different kinds of perceptual systems. The organisms can compete with each other in evolutionary games, and one can determine whether the organisms that see more truly are in fact the ones that tend to outcompete other organisms and have more offspring.

Results of such simulations have recently been published by Mark et al. (2010). They simulate a variety of worlds with varying numbers of resources and allow organisms to compete. Some see the whole truth, others part of the truth, and still others none of the truth. The organisms in the simulations that see none of the truth have perceptions that are tuned to utilities rather than to the objective structure of the world. For instance, a particular world might have several territories, each having a resource, such as food or water or salt, that can vary in quantity from 0 to 100. The utility of the resource is varied across simulations. Sometimes utility might be a monotonic function of the quantity of the resource, and other times it might be a Gaussian or some other nonmonotonic function.

What Mark et al. find is that true perceptions are not, in general, more fit. In most cases of interest, an organism that sees none of the truth, but instead sees abstract symbols related to utility, drives the truth perceivers to swift extinction. Natural selection does not usually favor true perceptions. It generally drives them to extinction.

One reason is that perceptual information does not come free. There are costs in time and energy for each bit of information that perception reports about the environment. For every calorie an organism spends on perception, it must kill something and eat it to get the calorie. As a result, natural selection pressures perception to be quick and cheap. Getting a detailed description of the truth is too expensive in time and energy. It is also not usually relevant, since utility, not truth, is what perception needs to report.

2.4 Interface Theory of Perception

Simulations using evolutionary game theory show that perceptual systems that report the whole truth or just part of the truth are not as fit as those that report utility (Mark et al. 2010). How shall we understand these fitter perceptual systems? Are there intuitions that can help us understand why they are more fit?

The key idea is that perception serves to guide adaptive behavior. Guiding adaptive behavior is not the same as constructing veridical perceptions. An example of the difference is the windows desktop of a personal computer (Hoffman 1998, 2009). The desktop interface is not there to present a veridical report of the diodes, resistors, magnetic fields, voltages, and software inside the PC. Instead, it is there to allow the user to be ignorant of all this, and yet still interact effectively with the PC to get work done.

If the icon for a file is orange, rectangular, and in the center of the display, this does not mean, of course, that the file itself is orange, rectangular, and in the center of the PC box. The color of the icon is not the true color of the file; files have no colors. The rectangular shape of the icon is not the true shape of the file. The position of the icon on the screen is not the true position of the file in the computer. No property of the icon on the screen is veridical. But this does not mean that the windows interface is useless, or misleading, or an illusion. It means that the purpose of the windows interface is to guide useful interactions with the PC while allowing the user to be free of the burden of knowing its complex details.

So, with the windows interface example, we see that reporting the truth is not the only way to be helpful, and that in fact reporting the truth can be an impediment to progress rather than a help. Perception can be useful even though it is not veridical. Indeed, perception is useful, in part, precisely because it is not veridical and does not burden us with complex details about objective reality. Instead, perception has been shaped by natural selection to be a quick and relatively inexpensive guide to adaptive behavior.

The view of visual perception that emerges from this evolutionary understanding can be summarized as follows: Perceived space and time are simply the desktop of the perceptual interface of *Homo sapiens*. Objects, with their colors, shapes, textures, and motions, are simply the icons of our space-time desktop. Space, time, objects, colors, shapes, and motions are not intended to be approximations to the truth. They are simply a species-specific interface that has been shaped by natural selection to guide adaptive behaviors that increase the chance of having kids.

One objection that often comes to mind at this point is the following: If that bus hurdling down the road is just an icon of your perceptual interface, why do you not step in front of the bus? After you are dead, and your interface theory with you, we will know that perception is not just an interface and that it is indeed a report of the truth.

The reason not to step in front of the bus is the same reason one would not carelessly drag a file icon to the trashcan. Even though the shape and color of the file icon do not resemble anything about the true file, nevertheless if one drags the icon to the trash one could lose the file and many hours of work. We know not to take the icons literally. Their colors and shapes are not literally correct. But we also know to take the icons seriously.

Our perceptions operate the same way. They have been shaped by natural selection to guide adaptive behavior. We had better take them seriously. Those of our predecessors who did not take them seriously were at a selective disadvantage compared to those who did take them seriously. If you see a cliff, do not step over. If you see a spider, back away. If you see a moving bus, do not step in front of it. Take your perceptions seriously. But this does not logically require that you take them to be literally true.

Another objection that often comes to mind regards consensus. If a bus is hurdling down the road, any normal observer will agree that they indeed see a bus hurdling down the road. So, since we all agree about the bus, since there is consensus, does that not mean that we are all seeing the same truth?

But consensus does not logically imply that we are all seeing the truth. It simply implies that we have similar perceptual interfaces and that the rules of visual construction that we use are similar. Just because an icon appears as orange and rectangular on different desktops and to different users does not mean that orange and rectangular are the true color and shape of the file. It just means that the various desktops have similar conventions that they observe.

2.5 Biological Examples

It is one thing to argue from simulations of evolutionary game theory, and from analogies with computer interfaces, that visual perception is simply a species-specific user interface that has been shaped by natural selection to guide adaptive behavior and to hide the complexities of the truth. It is quite another thing to present concrete evidence that this is how perception really works in living biological systems.

Such concrete evidence is abundant. Some of the most salient examples are seen in the phenomena of mimicry, camouflage, supernormal stimuli, and ecological traps. Each of these phenomena can be understood as resulting from natural selection shaping perception to be a quick and inexpensive guide to adaptive behavior rather than a veridical report.

Many dragonflies, for instance, lay their eggs in water. For millions of years, their visual systems have guided them to bodies of water appropriate for oviposition. This is an impressive feat and might suggest that their visual systems have evolved to report the truth about water. Experiments reveal instead that they have evolved a quick and cheap perceptual trick (Horvath et al. 1998). Water slightly polarizes the light that reflects from it, and dragonfly visual systems have evolved to detect this polarization. Unfortunately for the dragonfly, *Homo sapiens* have recently discovered uses for crude oil and asphalt, and these substances polarize light to an even greater degree than does water. Dragonflies find pools of oil even more attractive than bodies of water, and end up dying in large numbers. They also are attracted to asphalt roads. Pools of oil and asphalt roads are now ecological traps for these dragonflies. Apparently their visual system evolved a quick trick to find water: Find something that polarizes light, the more polarization the better. In the environment in which they evolved, this trick was a useful guide to behavior and allowed them to avoid constructing a complex understanding of the truth.

Mimicry and camouflage can be understood as arms races between organisms in which one organism exploits vulnerability in the perceptual interface of a second and in which the second organism in turn sometimes evolves its perceptual interface to remedy that particular vulnerability. Since perception has not evolved to report truth, but is instead a quick and cheap interface that has evolved to guide adaptive behavior, there will always be a myriad of vulnerabilities that can be exploited.

Hence, we find an endless and entertaining variety in the strategies of mimicry and camouflage.

2.6 A New Theory of Illusions and Hallucinations

Vision has evolved to guide adaptive behavior, not to report truth. Our perceptions of space, time, objects, colors, textures, motion, and shapes are useful because they are not true, just as the icons of a computer desktop are useful because they are not true, but simply serve as guides to useful behavior.

Given that none of our perceptions are true, then we must revise the standard definition of illusions, which says that each illusion is an incorrect perception seen by most people when they view a specific stimulus. The key to a new theory of illusions is to think about the evolutionary purpose served by perceptual systems: They have evolved to be guides to adaptive behavior.

This suggests the following new definition: An illusion is a perception, experienced by most people in a specific context, that is not an adaptive guide to behavior.

The windmill illusion, for instance, in which one misperceives the movement of the blades, is an illusion because such a perception is not an adaptive guide. One could be injured by a blade whose movement is misperceived (although, fortunately, the windmill illusion usually disappears if one gets close to the windmill). Similarly, the neon color spreading shown in Fig. 2.1 is an illusion because it is not an adaptive guide and leads the observer to see a surface with certain chromatic properties when it is not adaptive to do so.

We must also revise the standard definition of hallucination, which says that hallucinations are incorrect perceptions that are seen by few people and that occur in the absence of an appropriate stimulus. An evolutionary framework suggests the following new definition: A hallucination is a perception experienced by few people that occurs in the absence of an appropriate context and that is not an adaptive guide to behavior.

The key move in the new definitions of illusion and hallucination is to replace the central role of incorrect perception in the old definitions with the new central role of guiding adaptive behaviors. Our perceptual constructions have been shaped by evolution to be cheap and quick guides to adaptive behaviors in the niches that constituted our environment of evolutionary adaptiveness. Occasionally a situation arises that triggers in most members of the species perceptual constructions that are not adaptive guides to behavior. These are illusions. And occasionally a perceptual system of a member of the species engages in an idiosyncratic perceptual construction that is not an adaptive guide to behavior. This is a hallucination.

The new definitions of illusion and hallucination incorporate an evolutionary understanding of normal perception. They alert us that, when we try to understand the nature and provenance of illusions and hallucinations, it is important to consider how our perceptual systems evolved to serve as guides to adaptive behavior in our environment of evolutionary adaptiveness.

References

Geisler, W.S., Diehl, R.L. (2003). A Bayesian approach to the evolution of perceptual and cognitive systems. *Cognitive Science*, 27, 379–402.

Hoffman, D.D. (1998). *Visual intelligence: How we create what we see*. New York, NY: W.W. Norton.

Hoffman, D.D. (2009). *The interface theory of perception: Natural selection drives true perception to swift extinction*. In: *Object categorization: Computer and human vision perspectives*. Edited by Dickinson, S., Tarr, M., Leonardis, A., and Schiele, B. Cambridge: Cambridge University Press, pp. 148–165.

Horvath, G., Bernath, B., Molnar, G. (1998). Dragonflies find crude oil more attractive than water: Multiple-choice experiments on dragonfly polarotaxis. *Naturwissenschaften*, 85, 292–297.

Knill, D.C., Kersten, D., Yuille, A. (1996). *Introduction: A Bayesian formulation of visual perception*. In: *Perception as Bayesian inference*. Edited by Knill, D. and Richards, W. Cambridge: Cambridge University Press, pp. 1–21.

Mark, J., Marion, B., Hoffman, D.D. (2010). Natural selection and veridical perception. *Journal of Theoretical Biology*, 266, 504–515.

Maynard Smith, J. (1974). The theory of games and the evolution of animal conflicts. *Journal of Theoretical Biology*, 47, 209–221.

Nowak, M.A. (2006). *Evolutionary dynamics: Exploring the equations of life*. Cambridge, MA: Belknap Harvard University Press.

Palmer, S. (1999). *Vision science: Photons to phenomenology*. Cambridge, MA: MIT Press.

Redies, C., Spillmann, L. (1981). The neon color effect in the Ehrenstein illusion. *Perception*, 10, 667–681.

Seckel, A. (2009). Optical illusions: The science of visual perception. Richmond Hill: Firefly Books.

Chapter 3
Consciousness, Memory, and Hallucinations*

Ralf-Peter Behrendt

**"... the 'world' of the hallucinator is no less real to the participant during a hallucinatory phase than is the 'world' of the sane person when awake. To understand the nature of hallucinations it is not sufficient to simply determine the conditions under which non-real mental events (e.g. images, thoughts) somehow become invested with reality. This mistake is made by virtually all investigators of hallucinations in the recent past ... It makes no sense to regard a hallucination as a unique and generally pathological instance of subjective-turned-objective phenomenon, and to enquire into the reason for this, if, according to Kant and Schopenhauer, normal perception is achieved in exactly the same way, ..."*

(Cutting 1997, p. 83)

3.1 Philosophical Idealism

Everyday experience makes us believe that we observe an objective reality that is independent of our presence. The world is experienced as being observable to everybody; hence, it is believed to be observer-independent. Common sense suggests that events and objects are part of an objective reality that exists independently from our mind. Conversely, Kant (1781) and Schopenhauer (1844) recognized that the world around us is a "dreamlike creation." According to philosophical idealism, objects and events are creations of the mind and conform to the rules of the mind. The things that we see, hear, or feel around us are not part of the physical world and do not conform to rules of the physical world. Objects and events that populate our world do not exist without our awareness of them. Objects in the world are conditioned by the subject and only exist for the subject; there being "no object without subject" (Schopenhauer

R.-P. Behrendt, M.R.C.Psych. (✉)
Elderly Mental Health Team, Princess Elizabeth Hospital, St Martin, Guernsey
e-mail: rp.behrendt@btinternet.com

J.D. Blom and I.E.C. Sommer (eds.), *Hallucinations: Research and Practice*,
DOI 10.1007/978-1-4614-0959-5_3, © Springer Science+Business Media, LLC 2012

1844). Objects and events incorporate meaning and fundamentally depend on observation, including observation of constancy or change over time. Although everything that we perceive around us is part of a dreamlike creation, there has to be a material world that is independent of our mind, because the consciously experienced world changes in accordance with learned behavior. Kant and Schopenhauer did not deny the existence of a physical world but pointed out that we have no possibility of knowing what the real world is like. There is a world "in itself," which exists independently of the subject and which, according to Schopenhauer (1844), is wholly undifferentiated; it is only at the level of subjective representation that differentiation occurs. The mind, according to Kant, imposes its "categories" (concepts) upon sensory data emanating from the material world (the world "in itself"). Certain ideas and knowledge are "a priori" (basic) conditions for our experience; they rule our experience of the world. "Categories" of object, event, and causation are examples of such "a priori" knowledge; they are basic to our experience of particular instances of objects, events, and causations. Kant (1781) argued that we automatically apply "a priori" concepts to every observation and have no choice in this. Thus, we see the world in terms of our concepts but have no genuine access to the material world that lies beyond the realm of our conscious experiences.

Gestalt psychologists agreed that perception cannot be broken down into patterns of sensory stimulation and that it is not a derivative of sensory stimulation from the external world (Koehler 1940). Instead, they maintained that perceptual experience is an active achievement of the nervous system. Sensory stimulation serves to link the internally created image of the world to the physical world, ensuring that our internally generated stream of consciousness is adaptive. Sensory information derived from the physical world is not a necessary condition for perceptual experience. The richness and detail of conscious experience in dreaming suggests that even wakeful perception does not have to derive from sensory information to be as complex as it is. Similarly to wakeful perception, objects and sceneries that we see in dreams are substantial and seem to surround us. Dream images are real to the dreamer, and even grotesque violations of logic do not provoke questioning of their reality. Only when we wake up and start interacting with the external world do we understand that those sceneries were a fantasy and must have been produced in our mind. Similarly, images, smells, and sounds that surround us in the state of wakefulness are not identical to what is out there in the physical world. We are prevented from gaining insight into the dreamlike nature of the perceived world in the wakeful state for as long as the phenomena into which the conscious stream differentiates are compatible with behavioral interaction with the external physical world.

True hallucinations, as opposed to pseudohallucinations, are experientially identical to normal conscious perception. These hallucinations appear real to the hallucinator. Patients with acute psychosis typically have no insight into the unreality of their perceptual experiences; they react to their hallucinations as if these were normal perceptions (Jaspers 1946; Cutting 1997). Pseudohallucinations, by contrast, are experienced as "unreal"; they share certain characteristics with mental imagery and are accompanied by preserved insight (Jaspers 1946). From a perspective of "realism," according to which the world around us is an objective reality, true hallucinations are defined as "false perceptions" (Hamilton 1974). What is considered to distinguish hallucinations from normal perception is that hallucinations arise in the absence of a corresponding external object or event. Hallucinations, unlike normal perceptions, are thought to come from within the person's mind (Hamilton 1974). The "realist" approach to hallucinations suggests that hallucinations and normal perception are generated by fundamentally different mechanisms. Hallucinations are suspected to derive from "inner" mental phenomena, such as mental imagery or inner speech. The notion that hallucinations are a derivative of "inner" mental imagery requires the postulation of a process of "external misattribution" – a process that would explain how mental imagery acquires vivid perceptual qualities and becomes alienated from the "self." Cognitive models of hallucinations not only provide unsatisfactory accounts of misattribution, they also fail to explain obvious differences in content and grammatical form between verbal auditory hallucinations in patients diagnosed with schizophrenia (third-person or, less typically, second-person verbal auditory hallucinations) and mental imagery (which may involve first-person inner speech) (Behrendt 1998).

The nature and phenomenology of hallucinations can be explained more fruitfully within a framework that accepts that, similarly to hallucinations and dream imagery, normal conscious awareness of the world during wakefulness is a fundamentally subjective and dreamlike experience (Behrendt and Young 2004). Philosophical idealism predicts that normal perception, hallucinations, and dream imagery are principally manifestations of the same physiological process. Hallucinations and dream imagery differ from normal wakeful conscious experience only with regard to the extent to which they are *constrained* by sensory information from the external physical world. Hallucinations are similar to dreaming in that a lack of sensory constraints on the physiological mechanisms of conscious experience makes these forms of conscious experience maladaptive for interaction with the physical world. Attentional mechanisms modulate the content of conscious experience, whether or not conscious experience is externally constrained by sensory input. In hallucinations, attentional mechanisms are in a position to determine the content of conscious experience without regard for external sensory stimulation. Clinical observation suggests that content and context of verbal auditory hallucinations are crucially dependent on psychological factors relating to personality, psychological conflicts, and social concerns. Hallucinatory experiences that accompany acute psychotic states or chronically persist in patients diagnosed with schizophrenia are characteristically interpersonal in form and content, featuring derogatory voices, verbal commands, or third-person comments on one's actions, often reflecting patients' social anxieties and preoccupations (Linn 1977; Nayani and David 1996; Birchwood et al. 2000). At times of increased social stress and anxiety, patients prone to hallucinations increase their attention to social cues, and it is in the focus of attention where voices talking about or to the patient would emerge. Recognizing the role of attentional pressures attributable to social sensitivities suffered by patients diagnosed with schizophrenia, we can arrive at an explanation of the phenomenology of verbal auditory hallucinations in terms of grammatical form, content, and circumstances of occurrence. Hallucinations experienced by neurological patients have a different phenomenology (Manford and Andermann 1998); however, attentional factors likely play a role in determining form and content of these experiences, too (Behrendt and Young 2004).

3.2 Resonance in Thalamocortical Networks

The thalamus and cerebral cortex constitute a unified oscillatory system (Llinás and Ribary 1993). Projection neurons in specific and nonspecific thalamic nuclei and inhibitory neurons in the adjacent reticular thalamic nucleus form neuronal circuits with interneurons and pyramidal neurons in the cerebral cortex. Neurons connected in thalamocortical circuits have intrinsic resonance rhythmicity that is released by cholinergic input. In the depolarized state, these neurons exhibit subthreshold oscillations of membrane potential around 40 Hz, predisposing them to fire at γ rhythms in response to synaptic excitation (Steriade et al. 1993). Rhythmic discharges from thalamic or cortical neurons can entrain oscillatory activity in connected neurons. Synchronized firing of several neurons will elicit temporally overlapping excitatory postsynaptic potentials in other cells and increase their chance of firing, too. γ Activities in interconnected thalamocortical circuits globally synchronize to form "large functional states" (Llinás and Ribary 1993). Llinás and Paré (1991) pointed out that most of the connectivity in thalamocortical circuits is geared to the generation of internal functional modes, only a minor part of thalamocortical connectivity being devoted to the transfer of sensory information. Thus, consciousness can be viewed as a "closed-loop property" of the thalamocortical system (see Fig. 3.1). Dream imagery associated with paradoxical sleep differs from conscious perception during wakefulness in that, during paradoxical sleep, sensory input exerts only a weak influence over intrinsic thalamocortical resonance. Synchronized thalamocortical γ activity and conscious experience are generated during both wakefulness and paradoxical sleep, but during paradoxical sleep, the external world is mostly excluded from conscious experience (Llinás and Paré 1991; Llinás and Ribary 1993).

Reverberating activity in large assemblies of thalamocortical circuits produces γ (40 Hz) oscillations in magnetic or electrical field potentials recorded over the neocortex (Ribary et al. 1991). Neocortical γ oscillations characterize states of increased attention and alertness (Herculano-Houzel et al. 1999), accompany paradoxical sleep (Llinás and Ribary 1993), and can be recorded in association with hallucinations (Spencer et al. 2009). During wakefulness, sensory stimulation can reset and enhance γ oscillatory activity recorded from the neocortex (Ribary et al. 1991). Such resetting is not observed during paradoxical sleep when random bursts of neocortical 40-Hz oscillations occur in a manner unrelated to sensory stimulation (Llinás and Ribary 1993).

Electroencephalographic activation involves the release of acetylcholine from cholinergic fibers into the thalamus where acetylcholine acts on muscarinic receptors to induce delayed and prolonged membrane depolarization in thalamic projection neurons, thus enabling γ discharge activity (Curro Dossi et al. 1991; Munk et al. 1996). Cholinergic activation during arousal especially affects intralaminar (nonspecific) thalamic nuclei (Steriade et al. 1993). Intralaminar thalamic nuclei project to superficial layers of all neocortical areas in a spatially continuous manner. Neurons in these nuclei have a particularly strong intrinsic 40-Hz rhythmicity that may entrain oscillatory discharge activities in cortical neurons. By distributing

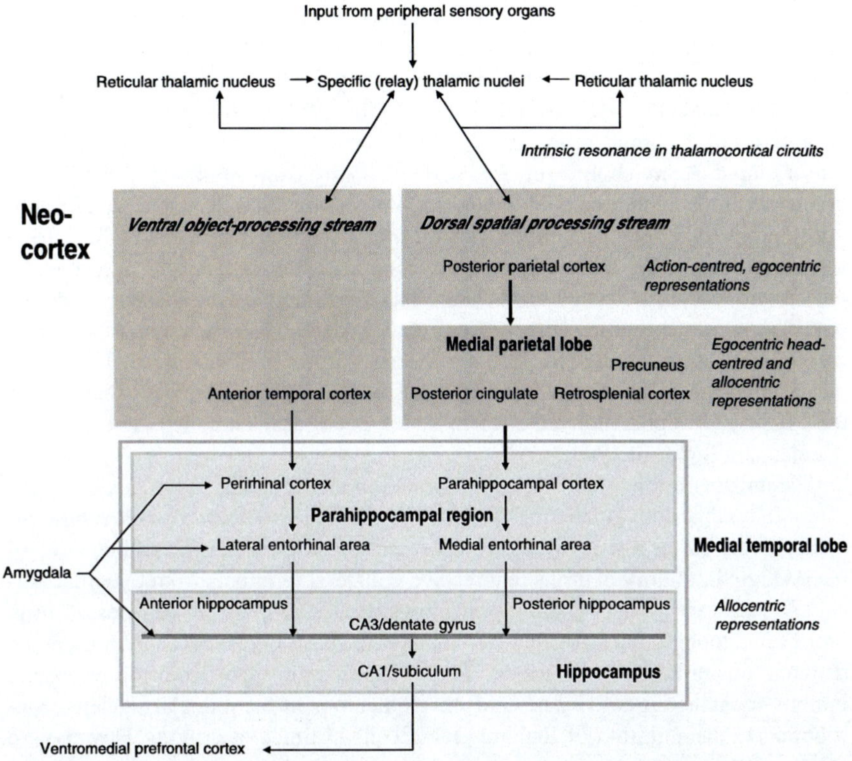

Fig. 3.1 Peripheral sensory information constrains intrinsic resonance processes in the thalamocortical system. Information processed in dorsal parts of the thalamoneocortical system, representing stimuli in action-oriented frames of reference, influences the formation of activity patterns in posterior (*dorsal*) parts of CA3 via medial parietal cortices, parahippocampal cortex, and medial entorhinal cortex. Information processed in ventral parts of the thalamoneocortical system (ventral object-processing stream) influences activity patterns in anterior (*ventral*) parts of CA3 via perirhinal and lateral entorhinal cortices – regions that are concurrently modulated by input from the basolateral amygdala. Arbitrary association patterns rapidly forming at θ rhythms in CA3 may have the information content of discrete epochs of unitary conscious experience and are temporarily stored as event (i.e., episodic) memories. Activity patterns formed in CA3 influence processes in CA1 that may serve to classify the location or situation presently occupied by the subject within its wider spatial or social environment (Behrendt 2010). The medial prefrontal cortex may encode emotional or habitual behavior modes that can be engaged in response to such contextual information

γ rhythms over the neocortex, intralaminar thalamic nuclei can facilitate the synchronization of γ reverberations in specific thalamocortical circuits that are activated by sensory input and attentional mechanisms. It has been suggested that conscious experience is based on coherent 40-Hz coactivation of specific and nonspecific thalamocortical circuits. The content of consciousness would lie in specific thalamocortical circuits, whereas nonspecific thalamocortical circuits may ensure the temporary binding of activated specific thalamocortical circuits toward the creation of a unitary conscious experience (Llinás and Paré 1991; Llinás and Ribary 1993).

Cholinergic, noradrenergic, and serotonergic systems ascending from the brainstem to the thalamus globally facilitate or inhibit fast oscillatory and resonance capabilities of thalamic "relay" cells and modulate their responsiveness to afferent sensory input. Acetylcholine released in the thalamus from terminals of cholinergic brainstem nuclei induces membrane potential depolarization, characterized by subthreshold γ oscillations, and enhances both spontaneous and stimulus-evoked firing activity of relay cells. Effects of cholinergic arousal are mediated by activation of muscarinic receptors located on thalamic relay cells and γ-aminobutyric acidergic (GABAergic) interneurons (Francesconi et al. 1988; McCormick and Pape 1988). By mediating a reduction in the release of GABA in specific thalamic nuclei, activation of muscarinic receptors on interneurons plays an important role in increasing the efficacy of signal transmission in states of arousal and increased attention (Carden and Bickford 1999).

The impact of peripheral sensory information on resonance in the thalamocortical system is partly regulated by the reticular thalamic nucleus. The reticular thalamic nucleus forms a sheet along the outer surface of the thalamus and consists of GABAergic inhibitory neurons that receive collateral terminals from thalamocortical and reentrant corticothalamic projections. Reticular thalamic neurons, in turn, project in a topographically organized manner to specific ("relay") and nonspecific thalamic nuclei (Llinás and Ribary 1993). Cholinergic input from the brainstem inhibits spontaneous activity of GABAergic neurons in the reticular nucleus, contributing to disinhibition of thalamic relay cells at times of arousal. However, in response to certain patterns of sensory stimulation, reticular thalamic neurons can inhibit activity in specific thalamic nuclei during arousal (Villa 1990; Murphy et al. 1994). Stimulus-specific inhibition of thalamic relay cells may be mediated by activation of presynaptic nicotinic receptors on GABAergic terminals from reticular thalamic neurons. Reticular thalamic neurons densely express nicotinic receptors (particularly those with the α_7 subunit). Thus, while inhibition of GABAergic neurons in the thalamus mediated by muscarinic receptor activation may contribute to the global increase of relay cell activity during arousal, nicotinic facilitation of GABAergic transmission from the reticular thalamic nucleus may, at the same time, improve the signal-to-noise ratio of thalamic activity (Lena and Changeux 1997).

Dysfunction of the reticular thalamic nucleus may lead to loss of sensory-specific inhibition in specific thalamic nuclei. This may manifest particularly at times of arousal when thalamic relay cells exhibit increased spontaneous activity. Then, random activity may predominate over stimulus-specific inhibition, and relay cells may become recruited into thalamocortical reverberations without receiving adequate sensory input. Some abnormalities identified in studies of patients diagnosed with schizophrenia or animal models of schizophrenia, including dopaminergic and α_7 nicotinic receptor dysfunctions, may predispose to hallucinations at times of increased stress and anxiety by disrupting the balance between intrinsic activity of the thalamocortical system and constraints imposed by sensory input to the thalamus (Behrendt 2006; Behrendt and Young 2004). Peripheral sensory impairment may constitute another cause for excessive noise in specific thalamic nuclei predisposing to pathological activation of thalamocortical circuits. This mechanism may contribute to musical hallucinations (see Chap. 11), the Charles Bonnet syndrome (see Chap. 6), late paraphrenia, and schizophrenia. Dysfunction of serotonergic raphe nuclei, as may be the case in peduncular hallucinosis, may also cause global disinhibition in specific thalamic nuclei (Behrendt and Young 2004).

3.3 Allocentric Representations and Episodic Memory

Synchronized thalamocortical γ oscillations likely play a role in the generation of conscious experience; however, it remains difficult to understand how widely distributed oscillatory processes can give rise to unified conscious experience of an external world ("binding problem"). The hippocampus, receiving major inputs from the entorhinal cortex, is in a unique position to rapidly integrate stimulus-related and contextual information processed in parietal and temporal association cortices. Superficial layer II of the entorhinal cortex projects, via the perforant path, to the granule cell layer of the dentate gyrus. Granule cells, in turn, send mossy fibers to hippocampal regions CA3 and CA2. Pyramidal neurons of CA3 (cornu ammonis region 3) are extensively interconnected via recurrent axon collaterals. CA3 is thought to form a single "autoassociation network" that displays "attractor dynamics" (Rolls 2007; for a full explanation see Chap. 4). The high degree of internal connectivity and effective synaptic plasticity enable the rapid formation of associations among individual elements of a patterned input reflecting sensory details processed in neocortical areas. The autoassociation network of CA3 stores "arbitrary associations" between object and place information as event memories for durations of seconds to minutes (Kesner 2007; Rolls 2007).

CA3 of the dorsal (posterior) hippocampus preferentially encodes spatial contextual information. Spatial information about the location of objects encoded in the parietal cortex enters the dorsal dentate-CA3 network via the medial entorhinal cortex (Kesner 2007). The posterior parietal cortex encodes sensory information in retinocentric and other egocentric reference frames for the purpose of guiding particular types of motor acts, such as saccadic eye movements, reaching, or grasping ("action-oriented spatial representations") (Colby and Goldberg 1999; Andersen and Buneo 2002). Regions in the medial parietal cortex integrate various feature representations activated in the posterior parietal cortex in order to encode, in egocentric head-centered coordinates, representations of locations or landmarks in the visible environment. Regions in the medial parietal cortex, in turn, are interconnected, via parahippocampal and medial entorhinal cortices, with the dorsal (posterior) hippocampus, which ties various visual feature units together, using "attractor dynamics" (see Chap. 4) to form an allocentric representation of the environment (i.e., a representation that appears fixed to the external environment) (Byrne et al. 2007).

The ventral (anterior) hippocampus is preferentially involved in the acquisition and retrieval of nonspatial memories (Ross and Eichenbaum 2006). The anterior lateral temporal cortex, which forms part of the "ventral visual stream," forwards nonspatial visual information about objects via the perirhinal cortex and lateral entorhinal cortex to the anterior (ventral) hippocampus. The anterior lateral temporal cortex, which, in humans, also processes visual and nonvisual information about social and emotional cues as well as semantic components of language, acts as a "connectional hub" to the anterior hippocampus by way of perirhinal and entorhinal cortices (Kahn et al. 2008). CA3 of the anterior (ventral) hippocampus encodes event memories consisting of arbitrary associations between object/reward information and the "space out there" (allocentric space) (Rolls et al. 2005; Rolls and Xiang 2005). Information about rewards and punishers, represented in the amygdala and orbitofrontal cortex, reaches the anterior hippocampus via the perirhinal and lateral entorhinal cortices. Glutamatergic projections from the basolateral amygdala to superficial layers of the entorhinal cortex promote γ oscillations in the entorhinal cortex by rhythmically depolarizing pyramidal neurons, thereby enabling the integration of neocortical inputs in emotionally arousing situations and promoting the spread of neocortical activity to the hippocampus (Bauer et al. 2007).

During exploratory locomotion and paradoxical sleep, neurons in superficial layers of the entorhinal cortex and their projection targets in the hippocampus discharge synchronously at γ frequencies in relation to the phase of the θ cycle (Chrobak and Buzsáki 1998). Temporal convergence of information-bearing neocortical input to the hippocampus and local θ oscillations, sustained by cholinergic input from the medial septum, results in the encoding of event (episodic) memories. Synchronous γ firing in subsets of CA3 pyramidal neurons that are tuned to the θ rhythm is necessary for the temporary storage of information. Thus, hippocampal θ oscillations, which are selectively present in behavioral states of exploration and attentiveness (but also in paradoxical sleep), ensure the continuous gathering of information about the environment (Buzsáki 1996; Vertes 2005). VanRullen and Koch (2003) argued that θ oscillations, acting as a carrier for γ oscillations, provide the context for conscious memory recall. θ Oscillations may also be important for conscious perception, if we accept that conscious perception is closely intertwined with conscious recognition and event memory retrieval. During paradoxical sleep, θ and γ oscillations are highly synchronized between dentate gyrus and CA3, possibly reflecting the recombination of event memory fragments (Montgomery et al. 2008). This would be consistent with the suggestion that reproduced sequences of patterned ensemble firing in the hippocampus during paradoxical sleep represent reactivated episodic memory traces that form the content of dream states (Louie and Wilson 2001).

Acetylcholine in the thalamus may cooperate with acetylcholine in the hippocampus in enabling neural processes that underlie conscious perception. Cholinergic mechanisms enhance memory encoding by increasing θ oscillations in the hippocampal formation (Hasselmo 2006). In addition, cholinergic input from the medial septum facilitates the encoding of new information in CA3 via activation of nicotinic receptors. Activation of nicotinic receptors in CA3 enhances excitatory synaptic input from the entorhinal cortex and dentate gyrus, while activation of (presynaptic) muscarinic receptors in CA3 suppresses excitatory transmission at recurrent connections between pyramidal cells (Hasselmo 2006). Thus, acetylcholine enables the encoding of new information in part by activating muscarinic receptors and suppressing feedback excitation within the autoassociation network of CA3. Conversely, lower levels of acetylcholine would encourage CA3 to settle on a previously learned event code (Burgess et al. 2001).

Rapid-event memory formation in the autoassociation network of CA3, integrating nonspatial information about landmarks or objects with information about their spatial context, may underlie externalized conscious experience (Behrendt 2010). Identifying consciousness with event memory formation in CA3 is consistent with Zeman's (2001, p. 1281) conclusion that "rather than guiding action from moment to moment, consciousness serves its biological purpose on a longer, more reflective, time scale." The world of our experience – subjective and dreamlike as it may be – appears to be "fixed" to the external physical world. The world of phenomenal space and time seems to be "out there" and contains objects that seem to be "out there," even though this world exists only "in our mind." What we conceptualize as "self" stems from others' attitudes and intentions toward us; it is our reflection in the social situation, which is essentially an aspect of the world that we see around us. Thus, consciousness, including evolutionarily more recent self-awareness, is allocentric. Even if the world of our experience is not constrained by peripheral sensory stimulation, as is the case in dream imagery or hallucinations, it is still externalized, that is, it still appears to be fixed to an external world. The highest-level encoder of allocentric information is the hippocampus, and if we take into account the ability of CA3 to form activity patterns that *bind* information from diverse cortical sensory processing areas into a coherent whole, then the hypothesis arises that consciousness is an emergent property of CA3 attractor dynamics and that dysfunctional regulation of neuronal activity and "parasitic attractor" dynamics in CA3 could be responsible for hallucinatory experiences.

3.4 Schizophrenic Psychosis and Hippocampal Hyperactivity

Schizophrenia is associated with GABAergic hypofunction in the hippocampus, especially in regions CA3 and CA2. GABAergic hypofunction in patients diagnosed with schizophrenia predominantly affects fast-spiking interneurons (basket cells), which contain parvalbumin and synapse onto perisomatic aspects of hippocampal pyramidal cells (Zhang and Reynolds 2002; Benes et al. 2007; Lisman et al. 2008). Decreased functionality of fast-spiking interneurons leads to a reduction of GABA-mediated inhibitory postsynaptic potentials in pyramidal cells and, hence, disinhibition of these cells. Deficient GABAergic transmission, in turn, may be a consequence of NMDA-receptor hypofunction in the hippocampus. Fast-spiking interneurons use NMDA receptors to "sense" the level of activity in surrounding pyramidal neurons (by responding to extracellular glutamate) in order to adjust the synthesis and release of GABA. Malfunctioning NMDA receptors would be relatively insensitive to glutamate levels in the surrounding milieu, which would lead to downregulation of GABA synthesis (Coyle 2006; Lisman et al. 2008). NMDA-receptor hypofunction, resulting in deficient GABA release from fast-spiking interneurons and, hence, disinhibition of hippocampal pyramidal cells, has been implicated in schizophrenia. Similarly, NMDA-receptor antagonists may induce psychotic experiences, such as hallucinations, in healthy subjects by reducing the output of fast-spiking GABAergic interneurons and thereby increasing pyramidal cell activity (Lisman et al. 2008).

The decrease in GABAergic tone in the hippocampus may be an indirect consequence of elevated excitatory input from the basolateral amygdala (Gisabella et al. 2009). The basolateral nucleus of the amygdala provides strong excitatory input directly to CA3 and CA2. In a rat model of schizophrenia, excessive glutamatergic input from the basolateral amygdala to regions CA3 and CA2 causes excitotoxic reductions in GABAergic interneuron density (especially affecting fast-firing basket cells), mirroring postmortem findings of reduced GABAergic interneuron functionality in CA3 and CA2 in patients diagnosed with schizophrenia (Berretta et al. 2004). Reductions in GABA-mediated inhibitory postsynaptic potentials in pyramidal neurons of CA3 occur in association with an increase in hippocampal "long-term potentiation" (Gisabella et al. 2009). Thus, increased responsiveness of the amygdala may contribute to excessive formation of memories for the environmental context of aversive stimulation and, by implication, dysfunctional conscious experiences in the form of hallucinations. Psychotic episodes in patients diagnosed with schizophrenia are often preceded or accompanied by syndromatic or subsyndromatic social phobia characterized by intense apprehensions, generalized across all social encounters, about being criticized, negatively evaluated, or rejected by others (Michail and Birchwood 2009). In neuroimaging studies, patients with social phobia show hyperresponsivity of the amygdala, especially on the left, to others' angry or contemptuous emotional expressions (Stein et al. 2002; Phan et al. 2006).

Electrical stimulation of the human medial temporal lobe, especially in the region of the amygdala and hippocampus, can elicit complex visual hallucinations (see Chap. 6) and sometimes auditory hallucinations (Vignal et al. 2007; see also Chaps. 8–10). Direct stimulation of the amygdala can elicit feelings of fear and anxiety, as well as complex hallucinations. Brain lesions in or near the amygdala are associated with schizophrenia-like psychoses (Fudge and Emiliano 2003). Although hippocampal volume is often reduced in patients diagnosed with schizophrenia, metabolic activity in the hippocampus, as measured by neuroimaging, is often increased, indicating an excess of excitatory neuronal activity. Hyperactivity of the left hippocampus and parahippocampal gyrus is associated with a *tendency*, in these patients, to experience hallucinations and delusions (Liddle et al. 2000; Heckers 2001). *Actual experience* of auditory (Shergill et al. 2000) or visual hallucinations is accompanied by hippocampal activation as demonstrated in fMRI studies, along with activations in higher-order neocortical sensory processing areas (Oertel et al. 2007). These findings agree with the hypothesis that hippocampal hyperactivity might underlie hallucinations and other positive symptoms in patients diagnosed with schizophrenia (Liddle et al. 2000; Heckers 2001). Antipsychotic treatment with risperidone was shown to reduce regional cerebral metabolism in the left hippocampus in association with a gradual reduction in hallucinations and delusions, "consistent with the hypothesis that reduction in aberrant hippocampal firing is a prerequisite for subsequent resolution of reality distortion" (Liddle et al. 2000, p. 406).

Patients diagnosed with schizophrenia, particularly those with "disorganized schizophrenia" (characterized by fleeting and fragmentary hallucinations and prominent thought disorder), are impaired in Gestalt tasks that require perceptual grouping of visual stimuli. In rats, blocking neural activity in one hippocampus induced coactivation, at γ frequencies, of pyramidal cells in the contralateral hippocampus that initially fired independently (Olypher et al. 2006). As a result, the ability of rats to subgroup distal spatial stimuli and segregate them from irrelevant local stimuli was impaired, preventing these rats from effectively avoiding regions of the room where footshock was administered. Coactivation, at γ frequencies, of initially uncoupled hippocampal pyramidal neurons amounts to a failure to segregate cell assemblies encoding unrelated representations. Disorganization of spike timing between cells would provide the basis on which "a pathological steady and stable state of activity" can emerge through γ synchronization: a "parasitic attractor" "that does not reflect reality" but a hallucination instead (Olypher et al. 2006, p. 166).

3.5 Conclusions

Normal perception, dreaming, and hallucinations are equivalent because even normal perception in wakefulness is fundamentally a state of hallucinations, one however that is constrained by external physical reality. Although Kant (1781) was not the first idealist, he can be credited with first recognizing idealism as the appropriate philosophical framework for understanding the nature of the perceived world and its phenomena and, at the same time, recognizing the embeddedness of the phenomenal world in a shared physical world that lies beyond subjective experience. The adaptive state of wakefulness depends on sensory information about changes taking place in the physical world, but we do not see, hear, feel, or smell physical reality itself. Instead, physical reality constrains the internal and fundamentally subjective process of conscious experience.

Activation of thalamic relay cells during arousal is normally balanced by sensory-specific and attention-specific inhibitory input from neurons in the reticular thalamic nucleus. The reticular thalamic nucleus, in turn, is under cholinergic control by the mesencephalic reticular formation and basal forebrain nuclei. In patients diagnosed with schizophrenia, deficient nicotinic activation of reticular thalamic neurons during arousal may lead to a loss of specific inhibition and random activity in specific thalamic nuclei. This would mask sensory input to the thalamus and weaken its impact on thalamocortical self-organization, resulting in impaired γ response synchronization to sensory stimulation. Thalamic relay cells could be recruited into "large functional states" involved in conscious experience without regard for the actual pattern of sensory input. Inhibition of the reticular thalamic nucleus and disinhibition in specific thalamic nuclei may also result from dopaminergic hyperactivity, as in schizophrenia, or exogenous NMDA-receptor antagonists, such as phencyclidine (Behrendt and Young 2004).

Patterns of resonant thalamocortical activity, representing sensory information in egocentric or action-centered frameworks, provide the dynamic substrate from which the hippocampus rapidly constructs allocentric environmental representations that serve mnemonic functions and self-localization. The CA3 autoassociation network, forming unitary activity patterns at θ intervals, integrates information about objects and their spatial context into allocentric representations of the environment. Unitary conscious experiences, referring to *discrete* epochs of conscious experience (VanRullen and Koch 2003), may be an emergent property of unitary activity patterns formed through attractor dynamics in CA3 (Behrendt 2010). Regarding the stream of consciousness as a sequence of higher-order symbols that characterize unique states in CA3 formed through attractor dynamics agrees not only with Kant's idealism but also Chalmers' (1996) argument that consciousness is imbued with nonphysical properties (and therefore cannot make a difference to the trajectory of behavior). Hallucinations may not differ from normal conscious perception in terms of their intricate relationship with episodic memory formation and recall.

Increased excitability in CA3 may cause hallucinations in through the formation of parasitic attractors. Alternatively, pathophysiological processes that predispose to patterns of thalamocortical activity that are underconstrained by peripheral sensory input could be responsible for hallucinations. Ultimately, much of the thalamocortical system, processing external sensory input, interacts with hippocampal region CA3 in producing view-dependent allocentric representations that manifest as discrete epochs of conscious experience. The precise may be phenomenology of hallucinations. The presence of visual hallucinations, alongside auditory hallucinations, in patients diagnosed with disorganized schizophrenia may indicate pathologically increased hippocampal activity, due to GABAergic hypofunction, whereas hallucinations occurring exclusively in the auditory modality in patients diagnosed with paranoid schizophrenia may indicate underconstrained thalamocortical that, of abnormal reticular thalamic nucleus function, is excessively sensitive to attentional.

References

Andersen, R.A., Buneo, C.A. (2002). Intentional maps in posterior parietal cortex. *Annual Review of Neuroscience*, 25, 189–220.

Bauer, E.P., Paz, R., Paré, D. (2007). Gamma oscillations coordinate amygdalo-rhinal interactions during learning. *Journal of Neuroscience*, 27, 9369–9379.

Behrendt, R.P. (1998). Underconstrained perception: A theoretical approach to the nature and function of verbal hallucinations. *Comprehensive Psychiatry*, 39, 236–248.

Behrendt, R.P. (2006). Dysregulation of thalamic sensory transmission in schizophrenia: neurochemical vulnerability to hallucinations. *Journal of Psychopharmacology*, 20, 356–372.

Behrendt, R.P. (2010). Contribution of hippocampal region CA3 to consciousness and schizophrenic hallucinations. *Neuroscience and Biobehavioral Reviews*, 34, 1121–1136.

Behrendt, R.P., Young, C. (2004). Hallucinations in schizophrenia, sensory impairment, and brain disease: a unifying model. *Behavioral and Brain Sciences*, 27, 771–787.

Benes, F.M., Lim, B., Matzilevich, D., Walsh, J.P., Subburaju, S., Minns, M. (2007). Regulation of the GABA cell phenotype in hippocampus of schizophrenics and bipolars. *Proceedings of the National Academy of Sciences USA*, 104, 10164–10169.

Berretta, S., Lange, N., Bhattacharyya, S., Sebro, R., Garces, J., Benes, F.M. (2004). Long-term effects of amygdala GABA receptor blockade on specific subpopulations of hippocampal interneurons. *Hippocampus*, 14, 876–894.

Birchwood, M., Meaden, A., Trower, P., Gilbert, P., Plaistow, J. (2000). The power and omnipotence of voices: Subordination and entrapment by voices and significant others. *Psychological Medicine*, 30, 337–344.

Burgess, N., Becker, S., King, J. A., O'Keefe, J. (2001). Memory for events and their spatial context: Models and experiments. *Philosophical Transactions of the Royal Society of London. Series B, Biological Sciences*, 356, 1493–1503.

Buzsáki, G. (1996). The hippocampo-neocortical dialogue. *Cerebral Cortex*, 6, 81–92.

Byrne, P., Becker, S., Burgess, N. (2007). Remembering the past and imagining the future: a neural model of spatial memory and imagery. *Psychological Review*, 114, 340–375.

Carden, W.B., Bickford, M.E. (1999). Location of muscarinic type 2 receptors within the synaptic circuitry of the cat visual thalamus. *Journal of Comparative Neurology*, 410, 431–443.

Chalmers, D.J. (1996). *The conscious mind*. Oxford: Oxford University Press.

Chrobak, J.J., Buzsáki, G. (1998). Gamma oscillations in the entorhinal cortex of the freely behaving rat. *Journal of Neuroscience*, 18, 388–398.

Colby, C.L., Goldberg, M.E. (1999). Space and attention in parietal cortex. *Annual Review of Neuroscience*, 22, 319–349.

Coyle, J.T. (2006). Glutamate and schizophrenia: beyond the dopamine hypothesis. *Cellular and Molecular Neurobiology*, 26, 365–384.

Curro Dossi, R., Paré, D., Steriade, M. (1991). Short-lasting nicotinic and long-lasting muscarinic depolarizing responses of thalamocortical neurons to stimulation of mesopontine cholinergic nuclei. *Journal of Neurophysiology*, 65, 393–406.

Cutting, J. (1997). *Principles of psychopathology: Two worlds - two minds - two hemispheres*. Oxford: Oxford University Press.

Francesconi, W., Muller, C.M., Singer, W. (1988). Cholinergic mechanisms in the reticular control of transmission in the cat lateral geniculate nucleus. *Journal of Neurophysiology*, 59, 1690–1718.

Fudge, J.L., Emiliano, A.B. (2003). The extended amygdala and the dopamine system: another piece of the dopamine puzzle. *Journal of Neuropsychiatry and Clinical Neurosciences*, 15, 306–316.

Gisabella, B., Cunningham, M.G., Bolshakov, V.Y., Benes, F.M. (2009). Amygdala-dependent regulation of electrical properties of hippocampal interneurons in a model of schizophrenia. *Biological Psychiatry*, 65, 464–472.

Hamilton, M. (1974). *Fish's clinical psychopathology*. Bristol: John Wright & Sons.

Hasselmo, M.E. (2006). The role of acetylcholine in learning and memory. *Current Opinion in Neurobiology*, 16, 710–715.

Heckers, S. (2001). Neuroimaging studies of the hippocampus in schizophrenia. *Hippocampus*, 11, 520–528.

Herculano-Houzel, S., Munk, M.H., Neuenschwander, S., Singer, W. (1999). Precisely synchronized oscillatory firing patterns require electroencephalographic activation. *Journal of Neuroscience*, 19, 3992–4010.

Jaspers, K. (1946, English edition 1963). *General psychopathology. 7th edition*. Manchester: Manchester University Press.

Kahn, I., Andrews-Hanna, J.R., Vincent, J.L., Snyder, A.Z., Buckner, R.L. (2008). Distinct cortical anatomy linked to subregions of the medial temporal lobe revealed by intrinsic functional connectivity. *Journal of Neurophysiology*, 100, 129–139.

Kant, I. (1781/2003). *Critique of pure reason*. Translated by Meiklejohn, J.M.D. Mineola, N.Y.: Dover.

Kesner, R.P. (2007). Behavioral functions of the CA3 subregion of the hippocampus. *Learning & Memory*, 14, 771–781.

Koehler, W. (1940). *Dynamics in psychology*. New York, NY: Liveright.

Lena, C., Changeux, J.P. (1997). Role of Ca^{2+} ions in nicotinic facilitation of GABA release in mouse thalamus. *Journal of Neuroscience*, 17, 576–585.

Liddle, P.F., Lane, C.J., Ngan, E.T. (2000). Immediate effects of risperidone on cortico-striato-thalamic loops and the hippocampus. *British Journal of Psychiatry*, 177, 402–407.

Linn, E.L. (1977). Verbal auditory hallucinations: Mind, self, and society. *Journal of Nervous and Mental Disease*, 164, 8–17.

Llinás, R.R., Paré, D. (1991). Of dreaming and wakefulness. *Neuroscience*, 44, 521–535.

Llinás, R.R., Ribary, U. (1993). Coherent 40-Hz oscillation characterizes dream state in humans. *Proceedings of the National Academy of Sciences USA*, 90, 2078–2081.

Lisman, J.E., Coyle, J.T., Green, R.W., Javitt, D.C., Benes, F.M., Heckers, S., Grace, A.A. (2008). Circuit-based framework for understanding neurotransmitter and risk gene interactions in schizophrenia. *Trends in Neurosciences*, 31, 234–242.

Louie, K., Wilson, M.A. (2001). Temporally structured replay of awake hippocampal ensemble activity during rapid eye movement sleep. *Neuron*, 29, 145–156.

Manford, M., Andermann, F. (1998). Complex visual hallucinations. Clinical and neurobiological insights. *Brain*, 121, 1819–1840.

McCormick, D.A., Pape, H.C. (1988). Acetylcholine inhibits identified interneurons in the cat lateral geniculate nucleus. *Nature*, 334, 246–248.

Michail, M., Birchwood, M. (2009). Social anxiety disorder in first-episode psychosis: incidence, phenomenology and relationship with paranoia. *British Journal of Psychiatry*, 195, 234–241.

Montgomery, S.M., Sirota, A., Buzsáki, G. (2008). Theta and gamma coordination of hippocampal networks during waking and rapid eye movement sleep. *Journal of Neuroscience*, 28, 6731–6741.

Munk, M.H., Roelfsema, P.R., Konig, P., Engel, A.K., Singer, W. (1996). Role of reticular activation in the modulation of intracortical synchronization. *Science*, 272, 271–274.

Murphy, P.C., Uhlrich, D.J., Tamamaki, N., Sherman S.M. (1994). Brain-stem modulation of the response properties of cells in the cat's perigeniculate nucleus. *Visual Neuroscience*, 11, 781–791.

Nayani, T.H., David, A.S. (1996) The auditory hallucination: A phenomenological survey. *Psychological Medicine*, 26, 177–189.

Oertel, V., Rotarska-Jagiela, A., Van de Ven, V.G., Haenschel, C., Maurer, K., Linden, D.E. (2007). Visual hallucinations in schizophrenia investigated with functional magnetic resonance imaging. *Psychiatry Research*, 156, 269–273.

Olypher, A.V., Klement, D., Fenton, A.A. (2006). Cognitive disorganization in hippocampus: a physiological model of the disorganization in psychosis. *Journal of Neuroscience*, 26, 158–168.

Phan, K.L., Fitzgerald, D.A., Nathan, P.J., Tancer, M.E. (2006). Association between amygdala hyperactivity to harsh faces and severity of social anxiety in generalized social phobia. *Biological Psychiatry*, 59, 424–429.

Ribary, U., Ioannides, A.A., Singh, K.D., Hasson, R., Bolton, J.P., Lado, F., Mogilner, A., Llinás, R. (1991). Magnetic field tomography of coherent thalamocortical 40-Hz oscillations in humans. *Proceedings of the National Academy of Sciences USA*, 88, 11037–11041.

Rolls, E.T. (2007). An attractor network in the hippocampus: theory and neurophysiology. *Learning & Memory*, 14, 714–731.

Rolls, E.T., Xiang, J.Z. (2005). Reward-spatial view representations and learning in the primate hippocampus. *Journal of Neuroscience*, 25, 6167–6174.

Rolls, E.T., Xiang, J., Franco, L. (2005). Object, space, and object-space representations in the primate hippocampus. *Journal of Neurophysiology*, 94, 833–844.

Ross, R.S., Eichenbaum, H. (2006). Dynamics of hippocampal and cortical activation during consolidation of a nonspatial memory. *Journal of Neuroscience*, 26, 4852–4859.

Schopenhauer, A. (1844/1966). *The world as will and representation. Two volumes.* Translated by Payne, E.F.J. New York, NY: Dover.
Shergill, S.S., Brammer, M.J., Williams, S.C., Murray, R.M., McGuire, P.K. (2000). Mapping auditory hallucinations in schizophrenia using functional magnetic resonance imaging. *Archives of General Psychiatry*, 57, 1033–1038.
Spencer, K.M., Niznikiewicz, M.A., Nestor, P.G., Shenton, M.E., McCarley, R.W. (2009). Left auditory cortex gamma synchronization and auditory hallucination symptoms in schizophrenia. *BMC Neuroscience*, 10, 85.
Stein, M.B., Goldin, P.R., Sareen, J., Zorrilla, L.T., Brown, G.G. (2002). Increased amygdala activation to angry and contemptuous faces in generalized social phobia. *Archives of General Psychiatry*, 59, 1027–1034.
Steriade, M., Curro Dossi, R., Contreras, D. (1993). Electrophysiological properties of intralaminar thalamocortical cells discharging rhythmic (approximately 40 HZ) spike-bursts at approximately 1000 HZ during waking and rapid eye movement sleep. *Neuroscience*, 56, 1–9.
VanRullen, R., Koch, C. (2003). Is perception discrete or continuous? *Trends in Cognitive Sciences*, 7, 207–213.
Vertes, R.P. (2005). Hippocampal theta rhythm: a tag for short-term memory. *Hippocampus*, 15, 923–935.
Vignal, J.P., Maillard, L., McGonigal, A., Chauvel, P. (2007). The dreamy state: hallucinations of autobiographic memory evoked by temporal lobe stimulations and seizures. *Brain*, 130, 88–99.
Villa, A.E. (1990). Physiological differentiation within the auditory part of the thalamic reticular nucleus of the cat. *Brain Research and Brain Research Reviews*, 15, 25–40.
Zeman, A. (2001). Consciousness. *Brain*, 124, 1263–1289.
Zhang, Z.J., Reynolds, G.P. (2002). A selective decrease in the relative density of parvalbumin-immunoreactive neurons in the hippocampus in schizophrenia. *Schizophrenia Research*, 55, 1–10.

Chapter 4
A Network Model of Hallucinations

Rutger Goekoop and Jasper Looijestijn

4.1 Introduction

The study of hallucinations is complicated by the huge number of factors that determine the occurrence and phenomenological characteristics of these phenomena. Until recently, this has made it impossible to develop a "global view" of the events that govern their existence. Recent breakthroughs in network science allow for a graphical representation and modeling of large numbers of interacting factors (Barabasi 2003; Barabasi et al. 2011), which may bring such a global view within reach. In this chapter, we will summarize a number of theoretical issues that are required for a basic understanding of network models. Finally, a network model of hallucinations is presented that aims to integrate a substantial number of clinical and research findings pertaining to the origin and phenomenological expression of hallucinations.

4.2 General Network Models

By the mid-twentieth century, network science began to take shape as a separate discipline, thanks to Paul Erdős (1913–1996) and many other brilliant physicists and mathematicians. In the 1950s, biological organisms were generally considered to be too complex to be described in terms of mathematical formulas. That all changed during the 1960s, when computers became available that allowed for complex

R. Goekoop, M.D., Ph.D. (✉) • J. Looijestijn, M.D.
Parnassia Bavo Group, The Hague, The Netherlands
e-mail: r.goekoop@parnassia.nl; J.Looijestijn@parnassia.nl

J.D. Blom and I.E.C. Sommer (eds.), *Hallucinations: Research and Practice*,
DOI 10.1007/978-1-4614-0959-5_4, © Springer Science+Business Media, LLC 2012

simulations of anything ranging from molecules to cells, organs, individuals, and markets. Classical network theory was born, which earned a serious reputation when it produced successful explanations and descriptions of complex phenomena such as the crystallization of atoms, phase transitions in matter, and navigation (e.g., the traveling-salesman problem). And yet it would take until the 1990s before a number of important discoveries would allow for a revolution in network science to take place, the consequences of which are only beginning to be felt in modern medicine and the current neurosciences.

A network is a mathematical concept that describes interactions between agents that can be identified separately in space (Watts and Strogatz 1998). These agents may themselves be in a certain state, which can be transferred from one agent to another in the course of time. It was Albert Einstein (1879–1955) who first remarked that all natural phenomena can be described in terms of events (states) that take place in space and time (Russell 2011). Since the addition of "scale" as a final descriptor, the central thesis has become that *states can interact with one another on different spatial and temporal scales*. This type of representation is so general that it allows most natural phenomena to be described in terms of networks.

The graphic representation of a network is called a network graph (see Box 4.1). Network graphs contain "nodes" and "links," which together determine network *structure*. States traveling between the nodes along the links in the course of time reflect network *function*. Classical network theory was based on the assumption that nodes were randomly connected to other nodes. Biological systems turn out to violate this rule completely, and are best represented by networks in which many nodes have relatively few connections, whereas the remaining nodes have many connections. Those richly connected nodes are called "hubs." Hubs connect many different nodes within the network, thus forming clusters of tightly interconnected nodes that are called "modules." Hubs interconnect the modules, which themselves can serve as nodes to form superclusters at ever higher levels of spatial organization. Viewed this way, life can be characterized as an endless variation of multimodular-hierarchic network structures which collectively display a so-called small-world topology (see Box 4.1).

Box 4.1 Small-World Network Structures

Nearly all biological systems, including collections of interacting genes, proteins, organelles, cells, tissues, organs, and individuals, can be described in terms of a "small-world" network topology. Such a network structure is characterized by the presence of hubs (see main text) that form clusters, which themselves are connected by hubs into ever larger networks, thus creating a multimodular-hierarchic structure (see Fig. 4.1). In such structures, states can

(continued)

Box 4.1 (continued)

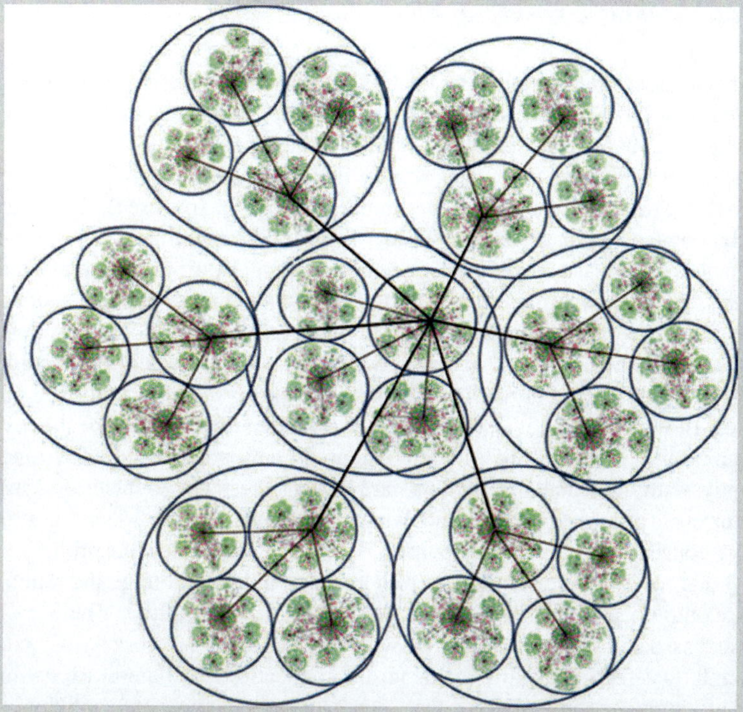

Fig. 4.1 Network graph showing an example of a "small-world" network structure. Hubs (i.e., high-degree nodes) connect clusters of lesser-degree nodes into superclusters (*marked by circles*) and so on. This hierarchical network structure is called "scale-free" or "fractal-like" since a similar structure can be found in nature at all spatial-scale levels of organization, including networks of genes, proteins, metabolites, organelles, neurons, brain areas, social networks, and markets

travel from one node to any other node in the network along very short routes. It turns out that every human being is part of various communities and hierarchies and connected to any other human being through an average of only six degrees of acquaintance (or six degrees of separation). In other words, the average "path length" of small-world networks is low. Because of the short distances between the nodes, such networks are called "small-world" networks (see Fig. 4.1), after the expression "It's a small world after all" (uttered, e.g., after meeting a total stranger who turns out to be an acquaintance of your best friend). In social systems, small-world networks promote the dissipation of information across a group of cooperating individuals. In neural networks, they help to optimize the transfer of information at minimal "costs" (i.e., connections).

4.3 Neural Network Models

4.3.1 Attractor Network Models

Neural networks are special types of biological networks. Like most biological systems, they have a multimodular, hierarchic network structure that displays a small-world topology (Bassett et al. 2006; Meunier et al. 2010). This network layout allows for maximum efficiency of information transfer and processing at minimal costs (expressed in terms of the number of links being used). The study of hallucinations by means of network theory involves the examination of (pathological) changes in neural network structure and function at various spatial and temporal scales, varying from micrometers to decimeters, and milliseconds to years. Such studies are nearly impossible to perform in living human beings. To overcome this obstacle, computer models of neural networks have been created that incorporate findings from different fields of study, including postmortem, in vitro, and animal research. In such models, it is possible to manipulate any parameter of interest and to examine the effects of such manipulations on network structure and function. Currently, "attractor network models" are among the most sophisticated models used for such purposes (Brunel and Wang 2001; see also Fig. 4.3). An attractor network consists of a network of neurons (e.g., pyramidal cells in the primary visual cortex) that receive dendritic inputs from an external location (e.g., the retina) and produce output via axonal connections that travel to another attractor network (e.g., the visual association cortex). Neurons within this attractor network have excitatory collaterals that feed back to their dendritic connections (collateral excitation) as well as to inhibitory interneurons that – on average – suppress the activity in networks lying outside of the attractor network (collateral inhibition) (Brunel and Wang 2001). Thus the excitatory and inhibitory collaterals form positive and negative feedback loops, which operate via N-methyl-D-aspartic acid (NMDA) and gamma-aminobutyric acid (GABA) receptors, respectively. From this basic layout, the brain is able to generate the multitude of visual representations of its environment that are required to fulfill the needs of its owner.

4.3.2 The Attractor State

When dendritic input enters the attractor network, neural activity reinforces itself by means of its positive feedback loop, thus aiding itself to persist even after the original stimulus has ceased (Rolls and Deco 2011; Chumbley et al. 2008). Meanwhile, the negative feedback loop suppresses the activity of neurons located (on average) outside of the attractor network. Collectively, the positive and negative feedback loops regulate the "persistence" of network activity within the attractor network (Brunel and Wang 2001). It has been shown that the application of an external stimulus to an attractor network can induce sustained and stimulus-specific neural activity within the network (Wu et al. 2008). Dendritic input pertaining to a particular stimulus selects a subset of neurons that together enter a "preferential state" that is specific to that particular stimulus. Such preferential states are called "attractor states" (see Box 4.2). It turns out that attractor states are among the best candidates for the neural correlates of subjective experience as currently known to neuroscientists. Although definitive proof is still required, specific smells, tastes, sounds, and visual experiences are thought to be encoded by stimulus-specific attractor states. Additionally, thoughts, feelings, rule-based decision making, and memory states have all been linked to activity within specific attractor networks (Braun and Mattia 2010).

Attractor networks can assume two different stable states. One of those states is the attractor state itself, which can develop when the network has received dendritic input. This "active state" (or "persistent state") is characterized by high-frequency firing rates of neurons within the network, which are difficult to disrupt by alternative dendritic input (distracting stimuli). Another stable state develops when attractor networks are not receiving any dendritic input. Under such circumstances, attractor networks enter a so-called resting state, characterized by low-frequency, random-spiking activity as a result of spontaneous depolarizations of neurons within the attractor network (Eliasmith 2007; Rolls 2010). In the resting state, attractor networks are freely roaming their state spaces. They may switch between active and resting states, depending on the amount of "energy" applied to them. Such energy typically takes the form of novel dendritic input or of spontaneous noise fluctuations generated within the network itself.

Box 4.2 Network Function and State Space

Each node within a network can be in a certain *state* (often 0 or 1, but states may vary along continuous scales). Those states are transferred from one node to another along the links in the course of time. The temporal evolution of those states in the network is referred to as network *function*. Just as network *structure* is usually represented by a graph, network function can be depicted as a "state space" (Wuensche 2011). State space is a Euclidian space, characterized by a multidimensional coordinate system with perpendicular axes that represent the states of the nodes or clusters present within the network. In time, the network "travels" through its state space and leaves behind a "state-space trajectory," reflecting its past behavior (see Fig. 4.2a). By recording changes in the states of nodes at different moments in time, the behavior of the entire network across infinite time can be described using just a single image.

After a certain period of time, each network will enter a state that it has encountered before. For instance, a simple network with $n=3$ nodes (or clusters) may cycle from state-space coordinates (0 0 0) to (1 1 1) to (1 1 0) and back to (0 0 0). Thus a loop of a certain size has been made in state space from one particular (macro) state back onto itself, through a series of intermediate states. That loop is called a "state cycle," and the point, line,

(continued)

Box 4.2 (continued)

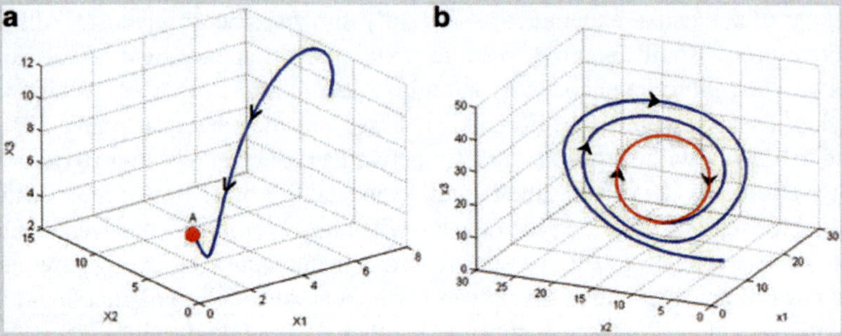

Fig. 4.2 State space trajectories, see box text (Reproduced from Eliasmith (2007). with permission)

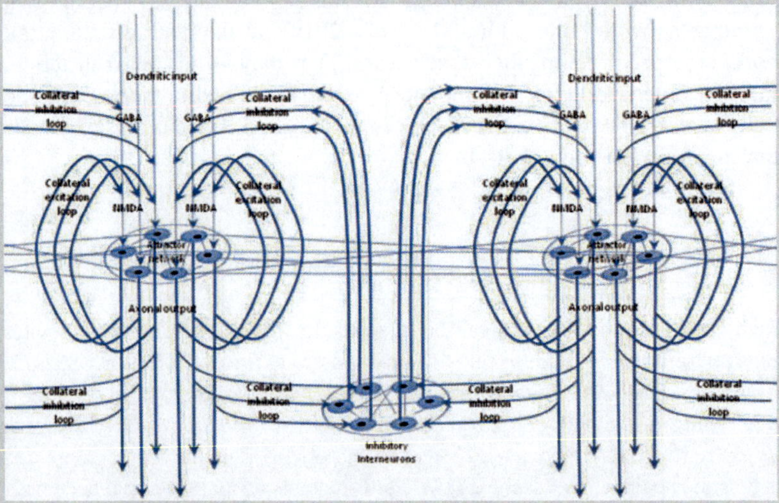

Fig. 4.3 The network structure of an attractor network, showing two attractor networks that are connected into a larger network. Dendritic input selects a subpopulation of neurons that forms the attractor network for that particular stimulus. Excitatory (NMDA-related) output loops back onto the dendritic input connections, causing a self-perpetuation of attractor activity. Excitatory output also connects to inhibitory (GABAergic) interneurons, which loop back to the dendritic input connections of all surrounding neurons. Thus attractor networks compete for activity by promoting the persistence of their own attractor states and suppressing activity in neighboring attractor networks

surface, or volume in state space that has been circled is called the "attractor" (see Fig. 4.2b). The term "attractor" was chosen because irrespective of the initial state of a network at any point in time, it will eventually enter a cycle of states that orbit the attractor at hand (see Fig. 4.3).

4.3.3 *Signal-to-Noise Processing*

Any influence that has the potential to affect attractor states can be referred to as "input" or "energy." With respect to a particular attractor network, two different sources of input can be identified. "Extrinsic input" refers to dendritic input originating from networks outside the attractor network. Such input constitutes either high- or low-frequency neural activity, corresponding to the active or resting states of upstream attractor networks. "Intrinsic input" is generated by spontaneous spiking activity of the neurons within the attractor network itself (Brunel and Wang 2001; Chumbley et al. 2008). Both sources of input can provide the energy needed to generate new attractor states or to disrupt existing ones. Attractor states constantly reinforce their own existence (as well as the "signal" they confer) through positive feedback while suppressing their surrounding states through collateral inhibition (see Fig. 4.4). Hence, attractor states are in a constant state of mutual competition (Rabinovich et al. 2001). As a result, robust signals are selected out of a sea of noise, which is called "signal-to-noise processing" (Rolls et al. 2008). This process allows organisms to separate distracting stimuli (noise) from meaningful signals, which is a major precondition for survival. Malfunctions of signal-to-noise processing have long been considered a candidate mechanism for the mediation of hallucinations and delusions.

4.3.4 *Bottom-Up and Top-Down Processing of Information*

Recent studies have shown that the human brain as a whole has a scale-free (i.e., fractal-like), multimodular, hierarchic network structure that displays small-world characteristics (Bassett et al. 2006; Meunier et al. 2010). Such a network structure enables the extraction of information from sensory input at an unprecedented level. At a small spatial scale, the primary visual system contains stimulus-dedicated neural networks that respond selectively to isolated perceptual stimuli such as direction, speed, color, movement, and texture (Horton and Adams 2005). Repeated encounters with similar isolated stimuli (e.g., yellow color, furry texture, vertical black stripes) will cause the networks that correspond with those percepts to strengthen their mutual connections to form a global ("tiger") network and a corresponding attractor state. This takes place according to the principle of "neurons that fire together, wire together," a process that involves the long-term potentiation of signal transduction at the level of individual synapses (Malenka and Bear 2004). Thus higher-order "associative" stimulus representations are generated from lower-level network activity. Such representations are formed in most association cortices (Bullmore and Sporns 2009). The combined states of those associative networks are in turn sampled by means of dendritic connections that converge onto an attractor network of hub neurons located in even higher-order association cortices (A and B in Fig. 4.4) (Zamora-Lopez et al. 2010; Kaiser et al. 2010). This process of associating

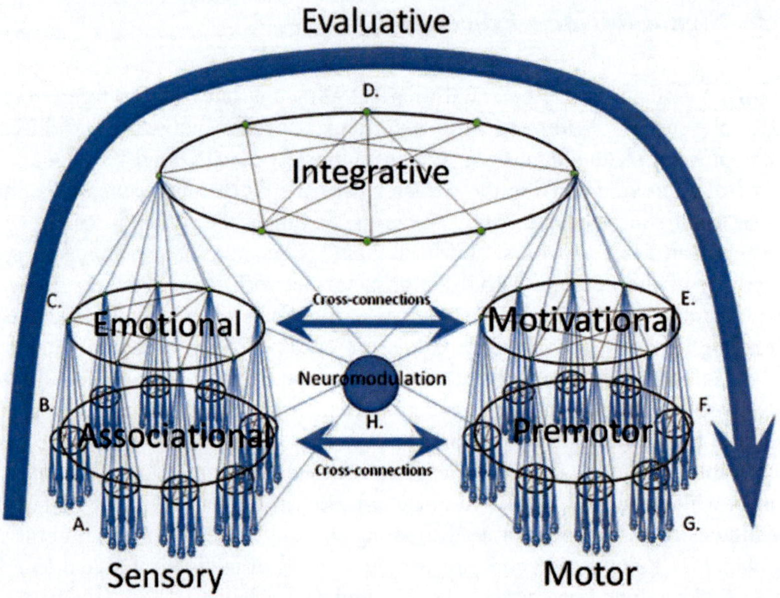

Fig. 4.4 A simplified model of the scale-free (i.e., "fractal-like") structure of the human brain and the flow of information through its neuronal architecture. As a result of sensory stimulation, stimulus-specific neural columns (*A*) enter attractor states that are associated into larger-scale networks through the process of long-term potentiation. Those states are integrated by means of dendritic connections that converge onto hubs, and they are transferred to higher-order stimulus-specific networks in sensory association cortices (*B*). The higher-order representations from multiple sensory modalities (not shown) are further integrated and transferred to emotional centers for evaluation (*C*). Combined sensory and emotional information is transferred to even higher-level integrative networks (such as the limbic system and prefrontal cortex) (*D*). Subsequently, attractor states of higher-order centers affect lower-level (sensory), motivational (*E*), and premotor (*F*) areas, and finally primary motor cortex (*G*) to elicit responses in individual muscle fibers. Thus the general flow of information involves a stimulus-evaluation-response loop, or "sensorimotor" loop. Cross-connections exist between the various levels of information processing to create stimulus–response loops of varying lengths. This allows for fast reflexive responses based on minimal sensory information in stressful situations, and iteratively processed, slower responses in more quiet situations. Information transfer in all loops is biased by neuromodulatory systems, including the noradrenergic, cholinergic, dopaminergic, and serotonergic systems (H). See main text for further details

and integrating information creates hierarchies of increasingly complex multimodal representations that are passed on to the limbic system and prefrontal cortex for further information processing (C and D in Fig. 4.4) (Freeman 1999).

At the highest level of information processing, the state spaces of all lower-level domains are bound together by a network of large-scale hubs to create a single multimodal representation of the organism's environment (Varela et al. 2001). Such a representation includes information from multiple sensory, emotional, motivational, premotor, and linguistic domains (including thoughts and feelings), and is processed in the medial prefrontal cortex, which is one of the largest hub regions of the human brain (D in Fig. 4.4) (Bullmore and Sporns 2009). This multimodal form of processing is often referred to as "cognition" and is thought to be experienced consciously as the "sense of self" (Tononi 2005). The states of these high-level areas are subsequently transferred to motivational and premotor areas (E and F in Fig. 4.4), from which action-related information determines the states of lower-level motor areas and eventually pyramidal cells (G in Fig. 4.4), which elicit stimulus-directed contractions in individual muscle fibers. Thus a sensorimotor loop is completed in which lower-level attractor states may dominate higher-level states (referred to as "bottom-up processing of information"), while higher-level states may dominate lower-level states (i.e., "top-down processing"). Higher-level states have their own relative autonomy and may exert control on lower-level processing stages by inhibiting brain activity in these areas. The failure of top-down control and the subsequent disinhibition of lower-level perceptive areas has been associated with the occurrence of hallucinations (see below).

4.3.5 Neuromodulation

Neuromodulatory neurotransmitter systems (such as the dopaminergic, noradrenergic, cholinergic, and serotonergic systems) can interfere in basic neural signaling by adjusting the degree of NMDA-related excitation or GABAergic inhibition within the attractor networks (Rolls et al. 2008; Loh et al. 2007). This may affect signal-to-noise ratios within neural networks along the entire length of the sensorimotor loop within the human brain (see Fig. 4.3) (Beck and Kastner 2009). Thus neuromodulatory systems are involved in biasing the competition between attractor states, allowing for a balancing of top-down and bottom-up control routes (Beck and Kastner 2009). As an example, the presence of a high-energy (i.e., "clear") tiger attractor within the visual cortex of an organism can induce specialized attractor states of the amygdala (signal). Such states in turn induce activity within the noradrenergic system, which connects back to the visual cortex. Here noradrenalin increases the signal-to-noise ratios of attractor networks by increasing both NMDA-receptor activity and GABAergic inhibition (Hu et al. 2007; Tully and Bolshakov 2010). As a result, noise is suppressed, while the attractor states are sustained for longer periods of time. This process is known as "attentional biasing" (Browning et al. 2010). In a behavioral sense, such shifts of balance in neural activity render organisms more prone to certain stimuli and actions in specific situations. In stressful situations, this translates into a heightened state of alertness, focus, and attention with respect to the sensory and evaluative parts of information processing, and into an increase in motivation and dexterity with respect to the motor and executive parts. This allows organisms to better spot minute stimuli, such as subtle clues as to the presence of a predator, and to quickly mobilize evasive or aggressive action (Bishop 2008). In contrast, when clear stimuli are lacking (e.g., when a tiger is hiding between the yellow leaves of a bush on a dark autumn night), the signal-to-noise ratios are shifted in favor of signal detection at the cost of noise suppression. That process, called "neural adaptation," is thought to be the combined effect of receptor sensitization and neuromodulation on the signal-to-noise ratios of lower-level perceptive networks (Clifford et al. 2007). As the potential costs of false-negative decisions in the case of predators and other threats may be high (to the extent of impending death), the network is biased slightly toward false-positive decisions (Dolgov and McBeath 2005).

4.4 Hallucinations in Network Terms

Since attractors represent specific subjective experiences, hallucinations are thought to reflect the presence of attractor states in the absence of an external source (Loh et al. 2007; Blom 2010). In what follows, we will discuss some of the many factors that can be held responsible for the production of such "false-positive" attractor states.

4.4.1 A Network Model of Sensory Deprivation

In some famous experiments of sensory deprivation, healthy subjects were brought into a state of almost complete disconnection from the external world by submerging them in flotation tanks containing salty water at body temperature. They were deprived of any patterned sources of light and sound and were instructed to float around weightlessly to avoid the corrective influences of gravity. Within 8 h, all of those subjects were hallucinating (Lilly 1956; Walters et al. 1964). The emergence of hallucinations in normal subjects under such extreme circumstances can be explained by means of the dynamics of attractor networks. Due to the absence of external stimuli, attractor networks enter their resting states. The absence of external stimuli prevents the use of collateral inhibition as a mechanism to suppress noise generated by extrinsic or intrinsic input to the attractor networks. Meanwhile, neural adaptation to the low stimulus intensities further reduces the number of GABAergic inhibitory currents and decreases the firing thresholds of neurons within attractor networks. As a result, these neurons become disinhibited, and attractor networks will switch more easily from their resting states to their active states (Behrendt and Young 2004). Collectively, those factors create a supersensitive network in which only a small amount of energy is required to trigger a false-positive attractor state. In the absence of any external input, such energy is provided by noise, either from the random-spiking activity of neurons within the attractor network itself (i.e., "intrinsic input") or from the dendritic input by other (disinhibited) attractor networks (i.e., "extrinsic input") (Deco et al. 2009). Depending on the types of network in which the attractors occur, sensory deprivation may yield different types of hallucination. Thus visual networks may generate images (see Chap. 6), whereas auditory networks may generate voices (see Chaps. 8–10), music (see Chap. 11), and other sounds.

Sensory deprivation can be aggravated by diminished sensory functioning (e.g., deafness or blindness) as well as social deprivation (e.g., loneliness in the elderly, which may entail paranoid delusions and hallucinations) (Pierre 2010; Teunisse et al. 1996). Stress and anxiety states during sensory or social deprivation may further enhance a network's sensitivity through the noradrenergic enhancement of NMDA signaling. The reason why many hallucinations and delusions involve situations of a frightening or threatening nature would seem to be that perceptive networks are biased toward the false-positive detection of predators and other threats. The increased sensitivity of attractor networks under sensory or socially deprived and stressful circumstances may be sufficient to turn that bias into a persistent attractor state (Dolgov and McBeath 2005; Van Os 2009).

4.4.2 A Network Model of Hallucinations

In more pathological cases, hallucinations can occur under normal circumstances. In such cases, individual differences in the structure and/or function of attractor networks can be held responsible for their mediation. As a general rule, such differences ultimately affect the levels of excitation, inhibition, and/or neuromodulation that govern the activity of attractor networks. With respect to the nature of such changes, various important clues have emerged from studies in patients diagnosed with schizophrenia (Rolls et al. 2008). One of the most solid findings is a decrease in the activity of the NMDA receptor, which is a primary predisposing factor for psychosis (see also Chap. 3). In addition, deficits in GABAergic inhibitory function have been established in patients diagnosed with schizophrenia, as well as elevated ratios of the activity of D2- versus D1-dopamine receptors (the receptor subtypes against which most antipsychotic agents are targeted). Finally, electroencephalography (EEG) and functional magnetic resonance imaging (fMRI) studies show decreased signal-to-noise ratios in patients diagnosed with schizophrenia.

Collectively, those findings allow for the following conceptualization of hallucinations. A decrease in NMDA conductance and self-reinforcing collateral excitation produces labile attractor states that can be disrupted more easily by noise.[1] Meanwhile, reduced GABAergic activity reduces noise suppression, making it easier for networks to enter attractor states. Increased D2- versus D1-receptor activity amplifies the effects caused by the NMDA and GABA deficiencies, as shown by simulation studies that report shallower basins of attraction in the energy landscape in both resting and active states (see Box 4.3) (Rolls and Deco 2011). Under such conditions, a small amount of energy in the form of intrinsic or extrinsic input (see above) is capable of inducing false-positive attractor states, which are easily

[1] Such labile attractor states may well explain some of the cognitive symptoms in patients diagnosed with schizophrenia, including distractability, loss of concentration, and impaired working memory.

disrupted, and alternate with resting-state activity. The corresponding network activity is characterized by fast alternations between high-frequency (active) and low-frequency (resting) states (Loh et al. 2007; Winterer and Weinberger 2004). That process is analogous to a ping-pong ball that moves erratically from one valley in an energy landscape to another (see Box 4.3). The erratic switching between high- and low-frequency active and resting states, respectively, may well explain the huge variability ("noise") in the signals of EEG and fMRI recordings of brain activity in patients diagnosed with schizophrenia. In a behavioral sense, that may translate into the subjective experience of hallucinations (intrinsic noise) or of illusions perceived in synchrony with external stimuli (extrinsic noise) such as auditory pareidolias (e.g., hearing voices in the hissing of a radiator tube).

The above explanations form a general model of alterations in network function that can explain the final common pathway in the mediation of various types of misperception. The key parameters in this model are the excitation, inhibition, and modulation of attractor states. Those parameters can be affected by an array of different processes. Chemical substances such as psychoactive drugs and medication can interfere directly with the neurochemistry that underlies the functioning of attractor networks (Skosnik et al. 2006; Ehrlichman et al. 2009; see also Chap. 22). Such substances alter the connections, weights, and/or firing thresholds of neural networks that eventually affect attractor persistence and formation. With the aid of computer simulations, it is possible to identify drug targets that predict the effects of psychoactive drugs on subjective experiences and to test their efficacy as part of future – actual – treatments. Thus it has been shown that NMDA agonists and D1-receptor agonists are capable of improving attractor stability (thereby having the potential to reduce negative symptoms), whereas D2-receptor antagonists are capable of reducing the formation of false-positive attractors (thereby reducing hallucinations and other positive symptoms) (Rolls et al. 2008). Apart from such neurochemical changes at microlevel, macrolevel structural changes can alter the function of attractor networks and produce hallucinations. Such changes, referred to as "connectivity changes," will be discussed below.

Box 4.3 The Concept of Information

The resting state of an attractor network defines the largest volume of state space that the network can occupy. In that state, the network is freely roaming its state space. Dendritic input leads to a reduction of the total number of possible states of the network, causing the network to occupy a more confined volume of state space (i.e., the attractor). As a corollary, we may think of perception (or any other form of "measurement") as a deterministic process characterized by the "pruning of possibilities." That reduction of potential states is what defines the concept of "in-formation" in information theory: the number of possible states that is lost as a result of external input is equivalent

(continued)

Box 4.3 (continued)

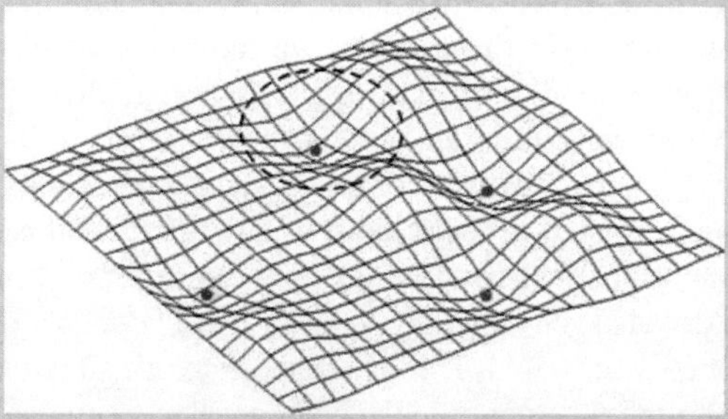

Fig. 4.5 Example of an "energy landscape." Signals or noise can provide the energy necessary to push networks from one stable attractor state (low-energy valley) into another. Those valleys are the active and resting states of attractor networks. A single network may harbor different low-energy attractor states, depending on the input to the network. This is represented by multiple valleys in the energy landscape (as depicted by the circle) (Reproduced from Eliasmith (2007) with permission)

to the amount of information gained as a result of the input (Heylighen and Joslyn 2001). The constraint that a particular stimulus context imposes on the occupied volume of state space of an attractor network can be linked directly to the degree of order (attractor state) or disorder (free-roaming state) that is allowed to exist within the state space. This has allowed researchers to connect the concept of information directly to changes in network entropy (Heylighen and Joslyn 2001). Since entropy itself is related to the amount of free energy contained in the network, "information transfer" can be understood as a complex flow of energy through network systems (Friston 2010). This signifies how fundamental the act of information transfer is in both physical and biological terms. The two stable states of each attractor network (the active and resting states) can be seen as two valleys in a landscape, where each valley represents a low-energy state (see Fig. 4.5). A ping-pong ball may leave one valley and enter a next, depending on the amount of energy applied to it. Likewise, an attractor network may switch between its resting and active states, depending on the amount of energy (signal or noise) applied to it. Simulation studies have shown that low levels of NMDA and GABA (e.g., in patients diagnosed with schizophrenia) increase the shallowness of the energy valleys. As a consequence, less energy (noise) is needed to induce false-positive attractor states (e.g., hallucinations). Shallow basins of attraction also mean that attractor states can disrupt more easily, thus producing a disorganized phenotype for hallucinations.

4.4.3 Structure-Function Relationships

A confusing aspect of network neuroscience is the distinction between structural and functional connectivity. In this chapter, we refer to network *structure* as that part of a network that can be described using a network graph, and to network *function* as that part of a network that can be described in terms of its state space. The brain has a distinct physical structure of anatomical connections that support the exchange of states (i.e., function) between spatially distributed nodes and clusters. It is thought that the functional connections between different brain areas (e.g., as measured by EEG or fMRI) closely follow the structural connections, but this is not always the case (Deco et al. 2011). The difference between structural and functional connections in the brain is analogous to the difference between the World Wide Web (which is a network of copper wires and glass-fiber cables that connects computers worldwide) and the Internet (which is a network of interconnected .HTML pages). This leads to a difference between structural connectivity (for which network graphs and state spaces can be created) and functional connectivity (for which different graphs and state spaces can be created) (Bullmore and Sporns 2009). In discussing the effects of changes in network structure upon network function, we will therefore commence with discussing changes in structural connectivity and then move on to changes in functional connectivity in patients diagnosed with schizophrenia.

4.4.4 Structural Connectivity

In studies of gray-matter density among patients diagnosed with schizophrenia, the most consistent finding is a loss of gray-matter density in the middle and bilateral superior temporal gyri (Nenadic et al. 2010; García-Martí et al. 2008; Glahn et al. 2008; Hulshoff Pol and Kahn 2008). Such alterations have varying correlations with the severity of hallucinations (Nenadic et al. 2010; García-Martí et al. 2008). In white-matter connectivity studies, a loss of connectivity has been found between frontal regions and medial and superior temporal areas in patients with verbal auditory hallucinations (Kubicki et al. 2007; Shergill et al. 2007). Neuroscientists are now able to combine such findings and to incorporate them into network topologies. As we saw, the human brain can be said to possess small-world characteristics, with "small-worldness" being defined as the ratio of average clustering and average path length (see Box 4.1). In network terms, a loss of gray matter can be considered a loss of network nodes (i.e., attractor networks) that reduces the possibilities for local clustering. Likewise, a loss of anatomical connectivity can be considered a loss of network links. Such a loss may affect the efficiency with which information travels through the anatomical network. In patients diagnosed with schizophrenia, reduced small-worldness has repeatedly been observed in graphs showing the structural connectivity of their brains. They show a longer path length, a reduced probability of frontal high-degree hubs, and the emergence of novel, nonfrontal hubs (Bassett et al. 2008; Van den Heuvel et al. 2010). Those network abnormalities correlate with the severity of positive as well as negative symptoms (Rotarska-Jagiela et al. 2009).

The small-world, fractal-like topology of the human brain makes it ideally fit for extracting relevant information from the external world, ranging from the tiniest bits to a global, integrated whole. Small-world networks are characterized by nonrandomness of connectivity patterns, resulting in clear hub-like structures. Small-worldness is associated with a highly efficient transfer of information (because information travels over relatively small distances), an optimal balance between functional segregation and integration (allowing for constant recombinations of subnetworks to accomplish varying tasks and the formulation of creative solutions to novel problems), and unprecedented levels of integration of information, allowing for abstract representations of the environment. A disruption of small-worldness can therefore be expected to lead to a reduction in these abilities. Specifically, the increase in average path length can produce a general slowing down of information transfer, which may explain some of the cognitive and negative symptoms found in patients diagnosed with schizophrenia. A loss of clusters will reduce the number of attractor states that are available to a patient, reducing the efficacy of stimulus discrimination. Phenotypically, a loss of stimulus discrimination may well underlie the phenomenon of "jumping to conclusions," i.e., the drawing of conclusions on the basis of little or no evidence. Since clusters represent attractor states and alternative explanations, only few alternative explanations for a given event are available for consideration. This may feed paranoid interpretations of experienced events.[2] A loss of clusters may also reduce possibilities for the recombination of attractor states to generate creative solutions to novel problems, causing patients to persist in certain (maladaptive) coping strategies. Additionally, a loss of clusters reduces the number of possibilities for collateral inhibition by alternative (i.e., competing) attractor states, which may promote the formation of false-positive attractor states. This may induce hallucinations in a way similar to that in sensory deprivation. Increased randomness of anatomical connections (a possible result of the excessive pruning of synapses (Hoffman and McGlashan 2001)) may lead to the experience of bizarre delusions (e.g., the conviction that a thermonuclear power plant resides within one's belly).

[2] Jumping to conclusions (for instance, "My grandmother came by the other day." "My car broke down the same day." "She sabotaged it.") may well result from an overly sensitive form of "pattern completion," i.e., attractor networks entering their full attractor states as a result of minimal dendritic input. Such oversensitivity may be due to a hyperconnectivity between neurons within a single attractor network, leading to an "overrecruitment" of neurons receiving dendritic input. A likely cause of such hyperconnectivity is increased D2-versus-D1-receptor activity, which is capable of enhancing long-term potentiation and the formation of associative ties between networks that would otherwise remain separated. As a result of such an increased connectivity between attractor networks, smaller clusters with a relatively high degree of specialization will be replaced by larger clusters with a more generalistic function. That may well explain the overall decrease in clustering, as well as the emergence of novel high-degree hubs, found in patients diagnosed with schizophrenia.

Finally, a loss of small-worldness may reduce a patient's capacities for integration, abstraction, and overview (Van den Heuvel et al. 2010). Specifically, a reduced number of high-level frontal hubs may entail a reduction of the capacity to integrate high-level abstract information, allowing patients to miss out on the "global picture," to disregard corrective clues, and to take expressions literally (i.e., concretisms). The dissolution of higher-order (prefrontal) clusters reduces the amount of top-down control, causing a disinhibition of lower-level sensory attractor networks that may result in the production of hallucinations (Bassett et al. 2008).

4.4.5 Functional Connectivity

Functional connectivity can be examined during active, resting, and sleeping states. Generally speaking, functional connectivity studies of patients diagnosed with schizophrenia show changes in network topology that are similar to those in anatomical connectivity. Only recently have more sophisticated analyses of network topology (such as small-worldness) been performed in functional-connectivity studies. These studies show that healthy human brain function, like the physical network structure that supports it, is characterized by a small-world structure (Bullmore and Sporns 2009; Van den Heuvel et al. 2008). Analogously, a decrease in small-worldness is observed in functional connectivity studies of patients diagnosed with schizophrenia, which shows a loss of high-degree (frontal) hubs and reduced local clustering (Liu et al. 2008; Demirci et al. 2009; Yu et al. 2011; Guye et al. 2010; Rubinov et al. 2009; Lynall et al. 2010). Those deficits correlate negatively with behavioral performances on verbal-fluency tasks (Lynall et al. 2010) and positively with the duration of illness (Liu et al. 2008). Patients also show reduced functional connectivity between the prefrontal cortex and the temporal lobe (Wolf et al. 2007; Pettersson-Yeo et al. 2011). The severity of positive symptoms correlates positively with the degree of functional disconnectivity of fronto-temporal and auditory networks (Vercammen et al. 2010; Gavrilescu et al. 2010; Rotarska-Jagiela et al. 2010), which is in keeping with the assumption that a loss of prefrontal function results in a disinhibition of attractor states in the temporal lobe and the production of verbal auditory hallucinations (Allen et al. 2008). Indeed, functional MRI studies have reliably shown an increase of activity in auditory networks accompanying the subjective experience of such hallucinations (Stephan et al. 2009).

When the brain as a whole enters its resting state (i.e., eyes closed but awake), characteristic regions are activated which are associated with mind-wandering and musing (Deco et al. 2011). Patients diagnosed with schizophrenia show an increase of functional connectivity between various regions of their resting-state network that is associated with the severity of positive symptoms (Garrity et al. 2007; Whitfield-Gabrieli et al. 2009). Such overactivity of the resting-state network may be due to a low NMDA-receptor conductance, which prevents high-frequency (i.e., active-state) activity to take over from low-frequency resting-state activity. This may prevent patients from switching between active and resting states, thus forcing them to remain in the resting state for prolonged periods of time. As a result, patients tend to be more focused on their inner experiences and to be less open to external corrections of their behavior (Bagby et al. 2005). Such perceptual and/or social disconnection may lead to positive symptoms through the process described above for sensory deprivation. Many patients indeed report that they experience an increase in hallucinatory experiences when they are alone and engaged in a resting state (see also Chaps. 8, 11, and 28). In such situations, a switch to a more active state (by talking to other people, e.g., or whistling) may reduce the occurrence of hallucinations.

During sleep, the functional architecture of the human brain moves closer toward a veritable small-world structure (Ferri et al. 2008). Hence a loss of small-worldness can have profound effects on the quality of sleep. Patients diagnosed with schizophrenia often suffer from a lack of deep sleep, which may have debilitating effects on their daily functioning (Sarkar et al. 2010). Such sleep disturbances occur regularly in psychotic patients without any form of psychoactive medication, which may well point in the direction of a fundamental inability to switch between the sleep-stage equivalents of active and resting attractor states (i.e., rapid eye movement (REM) sleep and deep sleep). During REM sleep, the free-roaming state of the default network is thought to induce attractor states that constitute the source of dreams (Ghosh et al. 2008; Sämann et al. 2011). Low NMDA and GABA conductance may induce inconsistent dreams and an inability to enter a deep-sleep stage. As a result, patients may have the experience that they have not slept at all, whereas the nursing staff report that they have been asleep all night. Impaired switching between REM sleep and the waking resting state may well explain the dynamics of hypnagogic and hypnopompic hallucinations occurring during the transitional periods between waking and sleeping (Ben-Aaron 2003), while abnormally persistent attractors generated during the REM phase can be experienced consciously during the waking state as hallucinations.

4.5 An Integrated View

Network theory offers a comprehensive view of the pathogenesis of hallucinations by examining the mutual relationships between networks of biochemical agents, structural and functional connectivity, subjective symptoms, and social functioning. Figure 4.6 shows an example of various coupled graphs for patients experiencing hallucinations. Ultimately, the whole network can be described dynamically in terms of its various state spaces. When such models become increasingly detailed, computer simulations may well be able to predict which brain areas should be targeted to reduce specific symptom dimensions – such as the intensity, severity, frequency, and duration of hallucinations – and help to reduce the personal and social consequences of such symptoms.

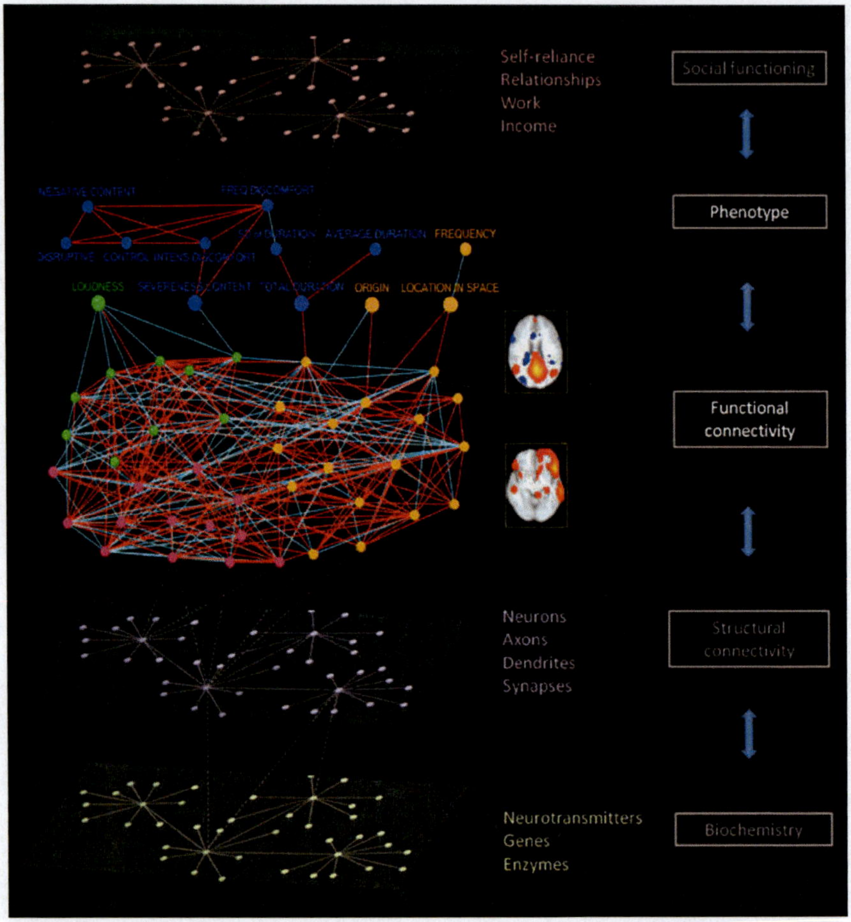

Fig. 4.6 Coupled network graphs of functional connectivity in relation to a network structure of symptoms of patients who experience verbal auditory hallucinations (Looijestijn et al., data in preparation). Measured data are shown in color. Functional connectivity: network of neurophysiological data (independent components of functional magnetic resonance signal). Phenotype: subjective symptoms as rated with the aid of a questionnaire (PSYRATS) and registered hallucination timings during magnetic resonance imaging. Hypothetical additional levels of organization are shown in gray and can be added to the model at later stages. Based on knowledge of the relationships between the various factors that relate to the origin and phenomenological expression of hallucinations, therapeutic intervention foci can be predicted at each level of network organization (e.g., medication, transcranial magnetic stimulation, psychotherapy, work, etc.)

4.6 Conclusion

In this chapter, we reviewed a number of biochemical and neurophysiological changes that accompany the experience of hallucinations. Collectively, those changes constitute a multitude of factors that give rise to a colorful palette of subjective experiences.

Network theory offers a way to integrate those findings in an unprecedented way, by offering a common terminology and methodology to describe relationships between events on different spatial and temporal scales. The specification of network models that incorporate the key players in the production of hallucinations allows researchers to make predictions with respect to specific targets for treatment. At a small spatial scale, new molecular targets are already being identified. At a larger spatial scale, network simulations may allow for a more specific targeting of brain areas using techniques that alter brain function, such as transcranial magnetic stimulation (see Chap. 25), electrocortical stimulation, and deep brain stimulation. With the progression of knowledge and the use of integrative science, it is hoped that a bigger arsenal of treatment options will become available for hallucinations and other psychotic symptoms in the foreseeable future.

References

Allen, P., Larøi, F., McGuire, P.K., Aleman, A. (2008). The hallucinating brain: a review of structural and functional neuroimaging studies of hallucinations. *Neuroscience and Biobehavioral Reviews*, 32, 175–191.

Bagby, R.M., Marshall, M.B., Georgiades, S. (2005). Dimensional personality traits and the prediction of DSM-IV personality disorder symptom counts in a nonclinical sample. *Journal of Personality Disorders*, 19, 53–67.

Barabasi, A.L. (2003). *Linked. How everything is connected to everything else and what it means for business, science, and everyday life*. New York, NY: Plume Books.

Barabasi, A.L., Gulbahce, N., Loscalzo, J. (2011). Network medicine: a network-based approach to human disease. *Nature Reviews. Genetics*, 12, 56–68.

Bassett, D.S., Bullmore, E., Verchinski, B.A., Mattay, V.S., Weinberger, D.R., Meyer-Lindenberg, A. (2008). Hierarchical organization of human cortical networks in health and schizophrenia. *Journal of Neuroscience*, 28, 9239–9248.

Bassett, D.S., Meyer-Lindenberg, A., Achard, S., Duke, T., Bullmore, E. (2006). Adaptive reconfiguration of fractal small-world human brain functional networks. *Proceedings of the National Academy of Sciences USA*, 103, 19518–19523.

Beck, D.M., Kastner, S. (2009). Top-down and bottom-up mechanisms in biasing competition in the human brain. *Vision Research*, 49, 1154–1165.

Behrendt, R.P., Young, C. (2004). Hallucinations in schizophrenia, sensory impairment, and brain disease: A unifying model. *Behavioral and Brain Sciences*, 27, 771–787.

Ben-Aaron, M. (2003). The function of 'brain waves': A cybernetic model of electroencephalography. *Medical Hypotheses*, 61, 593–596.

Bishop, S.J. (2008). Neural mechanisms underlying selective attention to threat. *Annals of the New York Academy of Sciences*, 1129, 141–152.

Blom, J.D. (2010). *A dictionary of hallucinations*. New York, NY: Springer.

Braun, J., Mattia, M. (2010). Attractors and noise: Twin drivers of decisions and multistability. *Neuroimage*, 52, 740–751.

Browning, M., Holmes, E.A., Harmer, C.J. (2010). The modification of attentional bias to emotional information: A review of the techniques, mechanisms, and relevance to emotional disorders. *Cognitive, Affective and Behavioral Neuroscience*, 10, 8–20.

Brunel, N., Wang, X.J. (2001). Effects of neuromodulation in a cortical network model of object working memory dominated by recurrent inhibition. *Journal of Computational Neuroscience*, 11, 63–85.

Bullmore, E., Sporns, O. (2009). Complex brain networks: graph theoretical analysis of structural and functional systems. *Nature Reviews. Neuroscience*, 10, 186–198.

Chumbley, J.R., Dolan, R.J., Friston, K.J. (2008). Attractor models of working memory and their modulation by reward. *Biological Cybernetics*, 98, 11–18.

Clifford, C.W.G, Webster, M.A., Stanley, G.B., Stocker, A.A., Kohn, A., Sharpee, T.O., Schwartz, O. (2007). Visual adaptation: neural, psychological and computational aspects. *Vision Research*, 47, 3125–3131.

Deco, G., Jirsa, V.K., McIntosh, A.R. (2011). Emerging concepts for the dynamical organization of resting-state activity in the brain. *Nature Reviews. Neuroscience*, 12, 43–56.

Deco, G., Rolls, E.T., Romo, R. (2009). Stochastic dynamics as a principle of brain function. *Progress in Neurobiology*, 88, 1–16.

Demirci, O., Stevens, M.C., Andreasen, N.C., Michael, A., Liu, J., White, T., Pearlson, G.D., Clark, V.P., Calhoun, V.D. (2009). Investigation of relationships between fMRI brain networks in the spectral domain using ICA and Granger causality reveals distinct differences between schizophrenia patients and healthy controls. *NeuroImage*, 46, 419–431.

Dolgov, I., McBeath, M.K. (2005). A signal-detection-theory representation of normal and hallucinatory perception. *Behavioral and Brain Sciences*, 28, 761–762.

Eliasmith, C. (2007). *Attractor network. Second edition*. Scholarpedia, 2, 1380.

Ehrlichman, R.S., Gandal, M.J., Maxwell, C.R., Lazarewicz, M.T., Finkel, L.H., Contreras, D., Turetsky, B.I., Siegel, S.J. (2009). N-methyl-d-aspartic acid receptor antagonist-induced frequency oscillations in mice recreate pattern of electrophysiological deficits in schizophrenia. *Neuroscience*, 158, 705–712.

Ferri, R., Rundo, F., Bruni, O., Terzano, M.G., Stam, C.J. (2008). The functional connectivity of different EEG bands moves towards small-world network organization during sleep. *Clinical Neurophysiology*, 119, 2026–2036.

Freeman, W.J. (1999). *How brains make up their minds*. New York, NY: Columbia University Press.

Friston, K. (2010). The free-energy principle: a unified brain theory? *Nature Reviews. Neuroscience*, 11, 127–138.

García-Martí, G., Aguilar, E.J., Lull, J.J., Martí-Bonmatí, L., Escartí, M.J., Manjón, J.V., Moratal, D., Robles, S., Sanjuán, J. (2008). Schizophrenia with auditory hallucinations: a voxel-based morphometry study. *Progress in Neuro-psychopharmacology and Biological Psychiatry*, 32, 72–80.

Garrity, A.G., Pearlson, G.D., McKiernan, K., Lloyd, D., Kiehl, K.A., Calhoun, V.D. (2007). Aberrant "default mode" functional connectivity in schizophrenia. *American Journal of Psychiatry*, 164, 450–457.

Gavrilescu, M., Rossell, S., Stuart, G.W., Shea, T.L., Innes-Brown, H., Henshall, K., McKay, C., Sergejew, A.A., Copolov, D., Egan, D.F. (2010). Reduced connectivity of the auditory cortex in patients with auditory hallucinations: a resting state functional magnetic resonance imaging study. *Psychological Medicine*, 40, 1149–1158.

Ghosh, A., Rho, Y., McIntosh, A.R., Kotter, R., Jirsa, V.K. (2008). Noise during rest enables the exploration of the brain's dynamic repertoire. *Public Library of Science Computational Biology*, 4, e1000196.

Glahn, D.C., Laird, A.R., Ellison-Wright, I., Thelen, S.M., Robinson, J.L., Lancaster, J.L., Bullmore, E., Fox, P.T. (2008). Meta-analysis of gray matter anomalies in schizophrenia: application of anatomic likelihood estimation and network analysis. *Biological Psychiatry*, 64, 774–781.

Guye, M., Bettus, G., Bartolomei,.F, Cozzone, P.J. (2010). Graph theoretical analysis of structural and functional connectivity MRI in normal and pathological brain networks. *MAGMA*, 23, 409–421.

Heylighen, F., Joslyn, C. (2001). *Cybernetics and second-order cybernetics*. New York, NY: Academic Press.

Hoffman, R.E., McGlashan, T.H. (2001). Neural network models of schizophrenia. *Neuroscientist*, 7, 441–454.

Horton, J.C., Adams, D.L. (2005). The cortical column: a structure without a function. *Philosophical Transactions of the Royal Society of London. Series B, Biological Sciences*, 360, 837–862.

Hu, H., Real, E., Takamiya, K., Kang, M.-G., Ledoux, J., Huganir, R.L., Malinow, R. (2007). Emotion enhances learning via norepinephrine regulation of AMPA-receptor trafficking. *Cell*, 131, 160–173.

Hulshoff Pol, H.E., Kahn, R.S. (2008). What happens after the first episode? A review of progressive brain changes in chronically ill patients with schizophrenia. *Schizophrenia Bulletin*, 34, 354–366.

Kaiser, M., Hilgetag, C.C., Kotter, R. (2010). Hierarchy and dynamics of neural networks. *Frontiers in Neuroinformatics*, 4, 112.

Kubicki, M., McCarley, R., Westin, C.F., Park, H.J., Maier, S., Klinikis, R., Jolesz, F.A., Shenton, M.E. (2007). A review of diffusion tensor imaging studies in schizophrenia. *Journal of Psychiatric Research*, 41, 15–30.

Lilly, J.C. (1956). Mental effects of reduction of ordinary levels of physical stimuli on intact, healthy persons. *Psychiatric Research Reports*, 5, 1–9.

Loh, M., Rolls, E.T., Deco, G. (2007). A dynamical systems hypothesis of schizophrenia. *PLoS Computational Biology*, 3, e228.

Lynall, M.E., Bassett, D.S., Kerwin, R., McKenna, P.J., Kitzbichler, M., Muller, U., Bullmore, E. (2010). Functional connectivity and brain networks in schizophrenia. *Journal of Neuroscience*, 30, 9477–9487.

Liu, Y., Liang, M., Zhou, Y., He, Y., Hao, Y., Song, M., Yu, C., Liu, H., Liu, Z., Jiang, T. (2008). Disrupted small-world networks in schizophrenia. *Brain*, 131, 945–961.

Malenka, R.C., Bear, M.F. (2004). LTP and LTD: an embarrassment of riches. *Neuron*, 44, 5–21.

Meunier, D., Lambiotte, R., Bullmore, E.T. (2010). Modular and hierarchically modular organization of brain networks. *Frontiers in Neuroscience*, 4, 200.

Nenadic, I., Smesny, S., Schlosser, R.G., Sauer, H., Gaser, C. (2010). Auditory hallucinations and brain structure in schizophrenia: voxel-based morphometric study. *British Journal of Psychiatry*, 196, 412–413.

Pettersson-Yeo, W., Allen, P., Benetti, S., McGuire, P., Mechelli, A. (2011). Dysconnectivity in schizophrenia: Where are we now? *Neuroscience and Biobehavioral Reviews*, 35, 1110–1124.

Pierre, J.M. (2010). Hallucinations in nonpsychotic disorders: toward a differential diagnosis of "hearing voices". *Harvard Review of Psychiatry*, 18, 22–35.

Rabinovich, M., Volkovskii, A., Lecanda, P., Huerta, R., Abarbanel, H.D., Laurent, G. (2001). Dynamical encoding by networks of competing neuron groups: winnerless competition. *Physical Review Letters*, 87, 068102.

Rolls, E.T. (2010). Attractor networks. *Wiley Interdisciplinary Reviews: Cognitive Science*, 1, 119–134.

Rolls, E.T., Deco, G. (2011). A computational neuroscience approach to schizophrenia and its onset. *Neuroscience and Biobehavioral Reviews*, 35, 1644–1653.

Rolls, E.T., Loh, M., Deco, G., Winterer, G. (2008). Computational models of schizophrenia and dopamine modulation in the prefrontal cortex. *Nature Reviews. Neuroscience*, 9, 696–709.

Rotarska-Jagiela, A., Oertel-Knoechel, V., DeMartino, F., Van de Ven, V., Formisano, E., Roebroeck, A., Rami, A., Schoenmeyer, R., Haenschel, C., Hendler, T., Maurer, K., Vogeley, K., Linden, D.E.J. (2009). Anatomical brain connectivity and positive symptoms of schizophrenia: a diffusion tensor imaging study. *Psychiatry Research*, 174, 9–16.

Rotarska-Jagiela, A., Van de Ven, V., Oertel-Knöchel, V., Uhlhaas, P.J., Vogeley, K., Linden, D.E. (2010). Resting-state functional network correlates of psychotic symptoms in schizophrenia. *Schizophrenia Research*, 117, 21–30.

Rubinov, M., Knock, S.A., Stam, C.J., Micheloyannis, S., Harris, A.W., Williams, L.M., Breakspear, M. (2009). Small-world properties of nonlinear brain activity in schizophrenia. *Human Brain Mapping*, 30, 403–416.

Russell, B. (2011). *History of Western philosophy and its connection with political and social circumstances from the earliest times to the present day*. London: Routledge, pp. 861–863.

Sämann, P.G., Wehrle, R., Hoehn, D., Spoormaker, V.I., Peters, H., Tully, C., Holsboer, F., Czisch, M. (2011). Development of the brain's default mode network from wakefulness to slow wave sleep. *Cerebral Cortex*, doi: 10.1093/cercor/bhq295.

Sarkar, S., Katshu, M.Z., Nizamie, S.H, Praharaj, S.K. (2010). Slow wave sleep deficits as a trait marker in patients with schizophrenia. *Schizophrenia Research*, 124, 127–133.

Shergill, S.S., Kanaan, R.A., Chitnis, X.A., O'Daly, O., Jones, D.K., Frangou, S., Williams, S.C.R., Howard, R.J., Barker, G.J., Murray, R.M., McGuire, P. (2007). A diffusion tensor imaging study of fasciculi in schizophrenia. *American Journal of Psychiatry*, 164, 467–473.

Skosnik, P.D., Krishnan, G.P., Aydt, E.E., Kuhlenshmidt, H.A., O'Donnell, B.F. (2006). Psychophysiological evidence of altered neural synchronization in cannabis use: relationship to schizotypy. *American Journal of Psychiatry*, 163, 1798–1805.

Stephan, K.E., Friston, K.J., Frith, C.D. Dysconnection in schizophrenia: from abnormal synaptic plasticity to failures of self-monitoring. *Schizophrenia Bulletin*, 35, 509–527.

Teunisse, R.J., Cruysberg, J.R., Hoefnagels, W.H., Verbeek, A.L., Zitman, F.G. (1996). Visual hallucinations in psychologically normal people: Charles Bonnet's syndrome. *Lancet*, 347, 794–797.

Tononi, G. (2005). Consciousness, information integration, and the brain. *Progress in Brain Research*, 150, 109–126.

Tully, K., Bolshakov, V.Y. (2010). Emotional enhancement of memory: how norepinephrine enables synaptic plasticity. *Molecular Brain*, 3, 15.

Van den Heuvel, M.P., Mandl, R.C., Stam, C.J., Kahn, R.S., Hulshoff Pol, H.E. (2010). Aberrant frontal and temporal complex network structure in schizophrenia: a graph theoretical analysis. *Journal of Neuroscience*, 30, 15915–15926.

Van den Heuvel, M.P., Stam, C.J., Boersma, M., Hulshoff Pol, H.E. (2008). Small-world and scale-free organization of voxel-based resting-state functional connectivity in the human brain. *NeuroImage*, 43, 528–539.

Van Os, J. (2009). A salience dysregulation syndrome. *British Journal of Psychiatry*, 194, 101–103.

Varela, F., Lachaux, J.P., Rodriguez, E., Martinerie, J. (2001). The brainweb: phase synchronization and large-scale integration. *Nature Reviews. Neuroscience*, 2, 229–239.

Vercammen, A., Knegtering, H., Den Boer, J.A., Liemburg, E.J., Aleman, A. (2010). Auditory hallucinations in schizophrenia are associated with reduced functional connectivity of the temporo-parietal area. *Biological Psychiatry*, 67, 912–918.

Walters, C., Parsons, O.A., Shurley, J.T. (1964). Male–female differences in underwater sensory isolation. *British Journal of Psychiatry*, 110, 290–295.

Watts, D.J., Strogatz, S.H. (1998). Collective dynamics of 'small-world' networks. *Nature*, 393, 440–442.

Whitfield-Gabrieli, S., Thermenos, H.W., Milanovic, S., Tsuang, M.T., Faraone, S.V., McCarley R.W., Shenton M.E., Green A.I., Nieto-Castanon A., Laviolette P.S., Wojcik J., Gabrieli J.D.E., Seidman L.J. (2009). Hyperactivity and hyperconnectivity of the default network in schizophrenia and in first-degree relatives of persons with schizophrenia. *Proceedings of the National Academy of Sciences USA*, 106, 1279–1284.

Winterer, G., Weinberger, D.R. (2004). Genes, dopamine and cortical signal-to-noise ratio in schizophrenia. *Trends in Neurosciences*, 27, 683–690.

Wolf, D.H., Gur, R.C., Valdez, J.N., Loughead, J., Elliott, M.A., Gur, R.E., Ragland, J.D. (2007). Alterations of fronto-temporal connectivity during word encoding in schizophrenia. *Psychiatry Research*, 154, 221–232.

Wu, S., Hamaguchi, K., Amari, S. (2008). Dynamics and computation of continuous attractors. *Neural Computation*, 20, 994–1025.

Wuensche, A. (2011). DDLab - Tools for researching cellular automata, random Boolean networks, multi-value discrete dynamical networks, and beyond. 2011:http://www.ddlab.com/.

Yu, Q., Sui, J., Rachakonda, S., He, H., Pearlson, G., Calhoun, V.D. (2011). Altered small-world brain networks in temporal lobe in patients with schizophrenia performing an auditory oddball task. *Frontiers in Systems Neuroscience*, 5, 7.

Zamora-Lopez, G., Zhou, C., Kurths, J. (2010). Cortical hubs form a module for multisensory integration on top of the hierarchy of cortical networks. *Frontiers in Neuroinformatics*, 4, 1.

Chapter 5
The Construction of Hallucination: History and Epistemology

German E. Berrios and Ivana S. Marková

5.1 Introduction

Whether due to epistemological opaqueness or flawed conceptualization, little is known about the phenomena currently called "hallucinations." Given that only the second option is susceptible to remedial work, it is proposed in this chapter that a reanalysis be undertaken of the historical process that gave rise to the "received" view. Unpacking the reasons and assumptions informing such a process should allow for a fresh approach and hopefully for the development of management routines that might help those who want to be rid of hallucinatory experiences.

5.2 The Received View

The received view has given rise to much unproductive correlational research: To the nineteenth century surveyal and neuropathological work, the twentieth century added electroencephalography (EEG), brain cartography, electrode stimulation, sensory deprivation studies, neurochemical, neuroimaging, and genetic studies. Regardless of the technique employed, this research limited itself to linking proxy variables representing putative "changes" in the brain to proxy variables representing

---

G.E. Berrios (✉)
Robinson College, University of Cambridge, Cambridge, UK
e-mail: geb11@cam.ac.uk

I.S. Marková
Department of Psychiatry, University of Hull, Hull, UK
e-mail: ismarkova@fastmail.co.uk

J.D. Blom and I.E.C. Sommer (eds.), *Hallucinations: Research and Practice*,
DOI 10.1007/978-1-4614-0959-5_5, © Springer Science+Business Media, LLC 2012

aspects of hallucinatory experiences. Often enough, the correlations thereby obtained have been furtively converted into "cause-effect" accounts.[1]

In all likelihood, this unproductiveness has resulted from the epistemological incoherence that characterizes the received view. Throughout this time, the latter has been masked by the sauntering plausibility of the received view and by costly academic investment. The said incoherence is but a reflection of tensions and contradictions generated by identifiable primary and secondary antinomies: the former built into the received view at the time of its construction; the latter added to alleviate the tensions caused by the primary antinomies.

5.2.1 *Primary and Secondary Antinomies*

The primary antinomies relate to four epistemological issues: (1) representation versus nonrepresentation,[2] (2) perception versus nonperception,[3] (3) unitary versus multiple, and (4) "organic" versus "psychiatric."[4] Esquirol (1817) gambled on the view that hallucinatory experiences were perceptual events, tokens of a unitary concept, best accommodated by a representational epistemology, and explained by either organic or psychological accounts. The definitional instability that has affected the concept of hallucination ever since can be explained by the choice of contradictory antinomies. This came to a head in 1855 when an (inconclusive) debate on the nature of hallucinations erupted at the *Association Médico-Psychologique* in Paris (Ey 1935) (see below).

To relieve these tensions, secondary antinomies were added during the second half of the nineteenth century: (a) true versus pseudo (or pale) hallucinations,[5] (b) hallucinations with and without insight, (c) hallucinations in the sane and in the insane,[6] (d) voice hallucinations versus auditory and other sense modality

[1]The literature on hallucinations in the main European vernaculars is absolutely enormous, and it would be unnecessary and inimical to list only some of them. Suffice it to say that the best-quality historical and conceptual work is still to be found in French (e.g., Quercy 1930; Paulus 1941; Ey 1973; Lanteri Laura 1991).

[2]The term representation is used here in its philosophical sense (Bernheimer 1961; Cummins 1989; Sterelny 1990; Ibarra and Mormann 2000; Clapin 2002, etc.).

[3]Perception is used here in both its philosophical and psychological sense. The literature on this theme is also vast (e.g., Merleau-Ponty 1945; Allport 1955; Hamlyn 1957, 1961; Armstrong 1961; Gibson 1966; Carterette and Friedman 1974; Dicker 1980; Yolton 1984, 1996; Fish 2009).

[4]This antinomy concerns the issue of whether hallucinations result from changes in the brain or from psychogenic mechanisms. Traditionally it has been traced back to the contrast that was established between the hallucinations affecting the German patient Nicolai and the French patient Berbiguier (for references and details, see Berrios 1990).

[5]For a full history of this concept, see Berrios and Dening (1996).

[6]On hallucinations with and without insight and on the debate on hallucinations in the same, see Gurney (1885) and Parish (1897).

hallucinations, (e) relevant versus nonrelevant hallucinatory content, (f) dream versus nondream view of the nature of hallucinations,[7] and (g) strict versus broad definition of hallucination (with the broad view including premonitions and visionarism; synesthesia; eidetic imagery; palinacusis and palinopsia; hypnagogic states; obsessional, negative, reflex, and extracampine hallucinations; oneiric states and image-related mental automatism; autoscopy; Charles Bonnet syndrome; formication and Ekbom states; écho de la pensée; unilateral hallucinations).[8]

The fact that there is little conceptual debate on the nature of hallucinations is less due to the coherence of the received view than to the fact that it has been adopted hook, line, and sinker by biological psychiatry which remains the predominant approach. In other words, hallucinations (regardless of sense modality) are considered as direct expressions of perturbations of brain regions related to the perceptual system.[9] To understand the infertility of the received view, the moment and circumstances of its construction must be analyzed.

5.2.2 *The Construction of the Received View*

Reports of phenomena redolent of hallucinatory experiences can be found in the literature of the ages. In earlier times, these reports seem to have been (a) socially and culturally integrated, particularly within certain religious contexts, and (b) considered to carry portentous messages concerning the life of the individual or its ancestors (Berrios 1996). The fact that on occasions hallucinatory experiences may have also been considered as manifestations of lunacy, madness, or insanity does not affect the appropriateness of the above claim. For reasons which need further historical elucidation, during the eighteenth century, hallucinatory experiences began to be included in the nosological listings of the time.[10] These attempts at medicalizing some hallucinatory experiences were *not* driven by the belief that they might constitute "symptoms of madness" (the notion of "mental symptom" is a nineteenth-century construct) but expressed the different belief that on occasions hallucinatory experiences might constitute independent "diseases." When thus considered, the predominant view was that hallucinatory experiences resulted not from pathological changes in the faculty of perception but of imagination.

---

[7]Maury (1865) remains the best classical source on this issue.

[8]On each of these concepts, the reader will find a raft of references. Ey (1973) remains an important source in this regard.

[9]For a variety of presentations of this view, see Mourgue (1932), West (1962), and Aleman and Larøi (2008), etc.

[10]Reports of "hallucinatory experiences" can be frequently found in eighteenth-century nosologies sometimes as "hallucination" and sometimes under a different name (e.g., Boissier de Sauvages 1772). In all cases, each is considered as a separate disease and certainly not as "symptoms" considered as part of a "disease." This is because the concept of mental symptoms was only constructed during the nineteenth century (Berrios and Markova 2006; Markova and Berrios 2009)

The socioeconomic and epistemological context of the early nineteenth century encouraged the adoption by medicine of the so-called anatomo-clinical model of disease. The latter model, together with factors such as associationism (Locke's psychology),[11] representationalism (part of Cartesian dualism), and romantic philanthropy (including mental hospital building, penal reform, and abolitionism) participated in the medicalization of madness.

Both the epistemological and psychological versions of associationism redefined understanding as "analysis" and encouraged the breaking up of wholes into their constituent "units." Representationalism, in turn, defined knowledge as the organized accumulation of internal and private representations of reality. Although there were various versions of representationalism, according to the most popular, the knower was sure of the existence of the representation itself (e.g., in Descartes), but knowledge of the quality of the relationship between representation and reality itself was inferred rather than captured by intuition. Very conveniently to later empirical research into hallucinations, the new physiology of the senses was also based on the epistemology of representationalism.

The philanthropic movement contributed to insanity legislation and to the building of mental hospitals. The latter acted as repositories for mental patients with similar complaints and allowed both long-term observation of their complaints and the possibility of undertaking correlational autopsies. Lastly, the anatomo-clinical model of disease encouraged the belief that madness was a composite of symptoms and signs, and that the latter resulted from disturbances in specific brain sites (Berrios and Freeman 1991).

5.2.3 Background Factors

"Alienism" is the name of the discipline and trade that resulted from the medicalization of madness. By the 1880s, such a discipline included a body of medical and paramedical staff specialized in madness, rites of passage (examinations), specialized databases (journals and textbooks), venues (hospitals, institutes, societal headquarters), and organized communicational exchanges (associations and congresses) (Berrios and Freeman 1991). By the turn of the century, alienism became psychiatry. The alienist differentiated himself from other medical practitioners and other social agents by developing a specialized language (psychopathology), a theoretical syntagma, and supporting rhetorical devices. The descriptive language of madness needed to be constructed anew, and the syntagma combined borrowings from the medical sciences and from the philosophy of mind, psychology, and the burgeoning human sciences.

---

[11]For an analysis of the relationship between associationism and psychiatry, and full references, see Berrios (1988).

This Janus-like conceptual profile allowed alienists to participate both in the world of medicine and in the social management of madness. To implement the latter, alienists carved for themselves a role in the social, legal, ethical, and political world of the nineteenth century. Their double conceptual profile was based both on the natural sciences and on the developing human sciences. Thus, the epistemology of psychiatry has been hybrid from its inception. In the event, the objects of inquiry that alienists constructed for themselves were hybrid too (Berrios and Marková 2002). However, alienists tried from the start to create the impression that there was no difference between their objects of inquiry and those of the rest of medicine. This may have resulted from the fact that some were genuinely unaware of the hybrid nature of mental symptoms and disorders; others, however, may have done so for self-serving reasons. Whatever the explanation, the historical fact is that the concept of "hallucination" was constructed against this ambivalent context.

Esquirol (1817) both coined the generic name "hallucination" and tooled a unitary concept to embrace, irrespective of sense modality, all hallucinatory experiences: "If a man has the inner *conviction* of truly experiencing a sensation for which there is no external object, he is in a hallucinated state, he is a visionary." And: "Hallucinations of eyesight have been called visions but this term is appropriate only for that sensory modality. One cannot talk about 'auditory visions', 'taste visions', or 'olfactory visions'? ... The latter phenomena, however, share with vision the same mechanisms and are seen in the same diseases. A generic term is needed for all. I *propose* the word hallucination." (our Italics). There is no space in this chapter to unpack Esquirol's reasons for expressing this view. The issue to remember is that before 1817 there were only "hallucinatory experiences" resulting from a pathology of imagination, and after this date, and irrespective of sense modality, they were reconceptualized as tokens of the same underlying phenomenon.

5.3 The 1855 Debate

In 1855, a debate broke out at the *Société Médico-Psychologique* on the nature of hallucinations. In the session of the 26th of February, Alfred Maury challenged remarks by Delasiauve and Moreau on a putative association between hallucinations and mystic states. The untidy debate that ensued lingered on until April 1856. The philosophical aspects of the matter in hand encouraged the participation of nonclinicians such as Maury, Bouchez, Peisse, and Garnier, particularly in relation to the questions of whether (a) hallucinations might be seen in the normal; (b) sensation, image, and hallucination formed a continuum; and (c) hallucinations, dreams, and ecstatic trances were similar states. Nonmedical participants also wanted to know whether hallucinations might have a "psychological" origin. By the end of May 1856, the debate had ended inconclusively in spite of efforts by Baillarger, Michéa, and Parchappe to draw some useful conclusions (Berrios 1996).

5.3.1 The Aftermath

By the end of the nineteenth century, opposite factions in the hallucination debate had made their positions clear: On the one hand, there were the alienists (by then, called psychiatrists or experts in psychological medicine) who supported the received view that hallucinations were a pathological and unitary disturbance of regions of the brain associated with perception; and on the other, there was a growing group of nonmedical academics and amateurs interested in psychical research who believed that hallucinations were frequent in the general population and were increasingly calling into question the received view by suggesting that hallucinations may have a psychological origin and on occasions carry a communicational meaning of sorts (Gurney 1885). Indeed, the interesting results of the Census of Hallucinations undertaken by the Society for Psychical Research during the 1880s seemed to support their view (Parish 1897). At the turn of the century, Freud threw yet another spanner in the wheels of the received view by introducing the concept of "negative hallucination" (first put forward by Bernheim) (Duparc 1992). To this day, the debate on hallucinations has been conducted along the tramlines set by these three approaches.

5.4 The Epistemology of Hallucinations

Fully to show the epistemological flaws of the received view would require a detailed analysis of all the primary and secondary antinomies. Lack of space precludes such an undertaking in this chapter; instead, only few of the polarities in question will be explored.

5.4.1 Representation Versus Nonrepresentation

The deepest antinomy embedded in the received view concerns the view that the hallucinatory experience is a sensation acting as a representation *manqué* (in the sense that in real or veridical perception, sensations are always vouchsafed by real external objects). The representational hypothesis of hallucinations has been particularly successful because (a) it identifies an easy explanatory locus (in other words, it makes the study of the "sensation manqué" the central theme in hallucinations research), and (b) it shares the same epistemological assumptions with sensory neurophysiology and neuropsychology, and hence, empirical research by means of the latter is bound to provide some support for the received view.

The representationist view of hallucinations provides reality (the noumena) with a permanent alibi by putting the blame on either of two components of the presentation: (1) the relationship between reality and its representation, or (2) the relationship between the subject and his inner representation. Hence, a hallucinatory experience occurs either because (a) a representation presents itself to awareness without being vouchsafed by reality or (b) there is a failure in the subject's cognitive and emotional management of the representation.

Overemphasizing the explanatory power of the sensation manqué begs the question as to why hallucinations consisting of representations of nonperceptual material not vouchsafed by reality (e.g., hallucinatory emotions or hallucinatory volitions) do not seem to be reported in the clinical literature. This dearth of reports may of course be due to the fact that such phenomena do not "occur in nature." But it could also be that (if the representationalism view is correct) they do exist, but because the current definition of hallucination also operates under a perceptual constraint (i.e., that to be called "hallucination," a false representation must relate to material acquired by a sensory modality), then suspicious representations of a nonperceptual nature are made to fall outside the purview of the concept of hallucination.

In general terms, representationalism works better in association with the view that external reality is independent and fully constituted (realism in its many varieties) than in relation to constructivist epistemologies according to which man's cognition participates actively in the conformation of reality. For example, while for Kant phenomena are accessible to modulation by means of conceptions of understanding (i.e., categories; see also Chap. 1), noumena are cognitively unreachable.

According to the received view on hallucinations, representations (in general) are ontologically and epistemologically vouchsafed by reality itself and not by the subject who entertains them. This makes representations ideal loci for explanation in the sense that hallucinations can be described as representations manqué. This is because if the subject were to be allowed a major role in the vouchsafing (construction) of his own representations, then these would lose plausibility and power as potential loci of explanation. The naïve representationalism of the received view also excludes psychogenetic and cultural configurators from playing a central role in the construction of hallucinations.

To circumvent the rigidity of the received view, it may be necessary to return to nonrepresentational models according to which the relationship of the subject and his surrounding reality is conceived as direct, continuous, and unmediated. This purview was developed, for example, by the Scottish Philosophers of Common Sense to correct the epistemological difficulties posed by passive Lockean representationalism (Grave 1960; Yolton 1984). Following this lead, nonrepresentational models of hallucinations, such as Gibson's (1966), must be reclaimed.

5.4.2 *Perception Versus Nonperception*

The second antinomy built into the epistemology of the received view concerns the nature of the relationship between hallucinations and perception. The origin of such a "relationship" is likely to have been the observation that reports of hallucinatory experiences seem to indicate that the "content" of experience was an image, a sensation, or a perception. The issue here is why it has been assumed that those reports entail *simpliciter* that the experiences in question are actually related to (or "caused" by) a pathological change in the perceptual system. The standard answer to this question is that there is plenty of clinical evidence that tinkering with the neuroanatomy, neurophysiology, and neuropharmacology of the perceptual system may trigger "perceptual" experiences which can be considered as similar to reported hallucinations. The evidential force of this reply, however, depends upon the assumption that such triggered perceptual experiences are identical to the hallucinatory experiences reported by subjects diagnosed with schizophrenia, mania, melancholia, obsessional disorder, hysteria, etc. This assumption is groundless and related to the belief that "organic" and "psychiatric" hallucinations are the same phenomenon (see below).

The perceptual nature of psychiatric hallucinations is called into question by the fact that the statement "S is experiencing an image" cannot be meaningfully differentiated from the statement "S believes that he is experiencing an image" (i.e., that S is having a perceptual delusion). A short digression into the historical background may throw light on this issue. The meaning of "perception" (*catalepsis* in Greek and *perceptio* in Latin) has changed in the West. Originally used to refer to the action of grabbing, collecting, and bringing things into oneself, it was later used in a metaphorical sense to refer to learning, i.e., to carrying information from the external world into the mind. At this stage, the nature of this information was *not* sensorial. The association between perception and sensation only starts with Descartes who uses the term to refer to the action of obtaining information about the world via the collection of sensations. From then on, perception was to carry two epistemological meanings: (a) a *general* meaning of getting to know the world in general (e.g., via intuition – which obviates intermediaries and representations), and (b) a *sensorial* meaning according to which such knowledge is exclusively obtained via sense-data (Locke's primary ideas, Condillac, etc.) (Hamlyn 1961).

The general and sensorial meanings of the term "perception" are often confused, and this may (partly) explain the problem at hand. While it would be perfectly in order to accept that hallucinations are related to the "general" meaning of perception (i.e., gaining information about the world, *simpliciter*), it might not be correct to state that hallucinations are causally or explanatorily related to "sensory" perception. A working compromise would be to state that (a) organic hallucinations may be the result of an impairment of the sensory perceptual system and (b) psychiatric hallucinations are related to a dysfunction of the general system of perceptual apprehension of the world, which would include intellectual, emotional, and volitional mechanisms as well as intuition. This compromise assumes that these two types of hallucination are different phenomena (see Sect. 5.4.4).

5.4.3 Unitary Versus Multiple

The third antinomy concerns the question of whether (a) all hallucinations are to be considered as the same phenomenon or (b) hallucinations differ (i.e., are different phenomena) according to the sensory modality with which they are associated. A variation of (b) would be to consider hallucinations associated with public senses (vision, audition, and olfaction) as different from those associated with nonpublic senses (touch, taste, and coenesthesis). This is based on the observation that both groups seem to require different identificatory ascertainment (Berrios 1982).

The original unitary view built into Esquirol's definition has rarely been challenged. And yet there are interesting epistemological differences that show up the instability of the received view. According to the official epistemology of the latter and the current view of mental symptoms (e.g., DSM IV), hallucinations are self-contained objects whose recognition should not be based on context or history but on intrinsic attributes alone. Thus if a subject declares that he can see his grandmother sitting in a chair in front of him, the correct identification of the phenomenon ought to be carried out by ascertaining that the chair is empty and there is no one sitting on it. This method of consensual ascertainment can be accomplished vis-à-vis hallucinations predicated of objects sitting in the public domain. But it is otherwise in relation to nondistant senses. For example, if the subject were to report an itch in his right hand or a strange sensation inside his abdomen, the consensual ascertainment test no longer applies (see also Chap. 13). Indeed, the epistemology of the received view cannot differentiate for certain between a real and a hallucinated itch. In psychiatric practice, this epistemological discrepancy is concealed by the fact that hallucinations (like the rest of mental symptoms) are rarely if ever diagnosed on the bases of intrinsic attributes alone but against a clinical context constituted by other mental symptoms, history, assessment of personality, etc.

There is also the interesting issue raised by the received view and (DSM) definition of hallucination according to which "hallucination is a sensory perception that has the compelling sense of reality of a true perception but that occurs without external stimulation of the relevant sensory organ." This is interesting because hallucinations rarely occur against an absolute vacuum. The everyday world is populated by many objects, and in most cases hallucinatory experiences occur against a rich backdrop of stimuli. When the definition claims that the hallucination "occurs without external stimulation of the relevant sensory organ," it is likely to mean that a hallucination of a grandmother occurs when there is no grandmother sitting on the chair. But the subject's retinas are at that very moment being stimulated by the chair on which grandmother is supposed to be sitting and by a myriad of other background visual stimuli.

This analysis brings hallucinations very close to the definition of illusion because in both cases stimuli are being perceived in a distorted manner. Whether a reported perception is declared a hallucination or an illusion would depend, therefore, on the degree of distortion of the stimulus. A coat hanging behind a door "seen" as a man hiding is considered to be an illusion simply because their shapes are grossly similar. A chair "seen" as a grandmother is called a hallucination because their shapes are vastly different, and the claim that one is an illusion of the other would be considered as implausible. And yet, the question here is one of grades of differentiation between shapes, concepts, sounds, smells, etc., rather than there being *no* external stimulation.

5.4.4 "Organic" Versus "Psychiatric" Types

The issue here is whether, irrespective of putative cause, (a) *all* hallucinations are the same phenomenon, or (b) hallucinations seen in the context of neurological disease or experimental manipulation of the brain are radically different from the "hallucinations" observed in relation to psychiatric disorders. When Esquirol put together his original definition, rather wisely, he left the door open to the possibility that they might be different phenomena. Throughout the nineteenth century, alienists backed either poles of this antinomy, and this led to periods of predominance of one or the other. When the "neuropsychiatric view" of hallucinations predominated, alienists tended to be unitarians, i.e., it was believed that hallucinations were the same, regardless of putative cause. This view predominates at the moment and is sponsored by what can be called the third wave of neuropsychiatric or biological psychiatry. The problem with the unitarian view is that it may be misleading in regard to the nature and etiology of "psychiatric hallucinations." Hallucinatory experiences in patients diagnosed with obsessive-compulsive disorder, schizophrenia, melancholia, etc., may be fundamentally different, and this needs to be taken into account. For example, the musical hallucinations of funereal marches heard by a patient with melancholia and Cotard syndrome differ in every respect from the musical hallucinations of elderly subjects with tinnitus and hearing impairment (Berrios 1990a).

5.5 Toward a New Epistemology of Hallucinatory Experiences

5.5.1 The Cambridge Model

According to the Cambridge model (see Fig. 5.1), subjective mental symptoms are "hybrid" objects, i.e., entities constituted by both organic (neurobiological signaling) and semantic (an admixture of cultural, social, individual configurators conferring personal meaning to an experience) components. Little as yet is known about the respective roles of the organic components and semantic configurators (or indeed about their interrelationships), but this is liable to vary in regard to different mental symptoms. It is likely that the process of semantic or cultural configuration modifies both the gain and the specificity of the subjective experience associated with the neurobiological signal. The process can be said to start when signals issued out by dysfunctional or distressed brain addresses enter the patient's awareness and generate primitive experiences whose prelinguistic and preconceptual status makes them ineffable. In order to be communicated, such experiences need to be configured into mental and speech acts. Thus, from the start, mental symptoms are "hybrid objects," i.e., dense combinations of information issued out by both the cultural configurators and the proxies representing the neurobiological signal. Four pathways by means

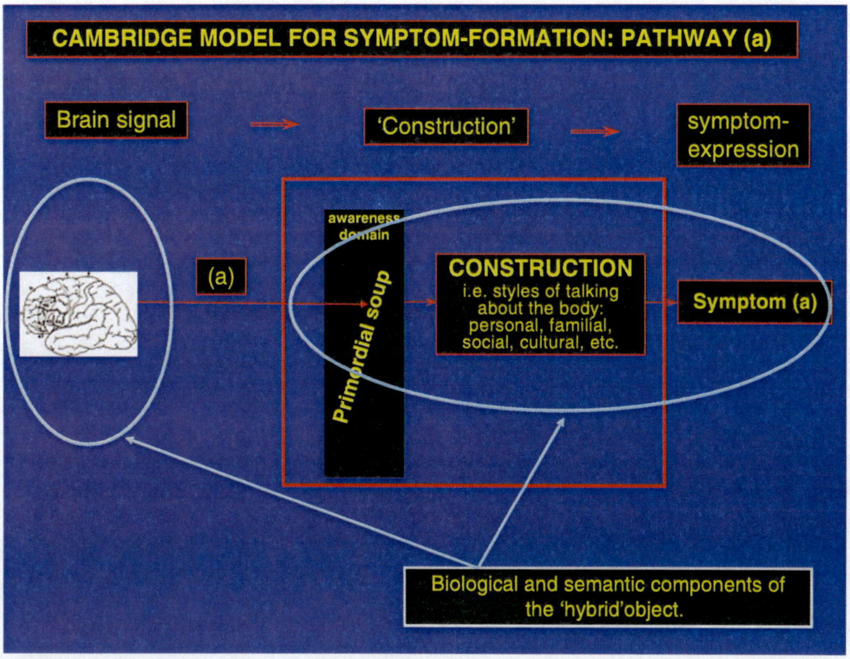

Fig. 5.1 Cambridge model of symptom formation (pathway a)

of which mental symptoms can be constructed have been proposed elsewhere (Berrios and Marková 2002, 2006).

According to pathway (a) patients construct subjective mental symptoms (e.g., hallucinations) on the basis of particular changes they experience in their mental state. Such changes in their awareness while mediated by neurobiological signaling may be in response to either internal (primary biological alteration) or external (perceived distress) stressors. The resulting inchoate, preconceptual and prelinguistic experiences have been called the "primordial soup." To integrate such experiences into the concert of his mind, the patient needs to provide them with meaning and articulate them into words. This configuratory process is achieved by means of personal, familial, and cultural templates (semantic components). Only then is the patient ready to communicate his experience via an utterance or behavioral gesture. It is important to remark that at this stage the "content" of the configured mental symptom may have little to do with the functional signature of the brain site that originated the signal in the first place. This means that the same biological signal may be configured into different mental symptoms, or that different brain signals may be configured into the same mental symptom.

Once the mental symptom has been configured, the subject may decide to complain, i.e., to report it to another person (e.g., a clinician). This dialogical interaction adds another stage to the process of configuration, for it entails conceptual haggling

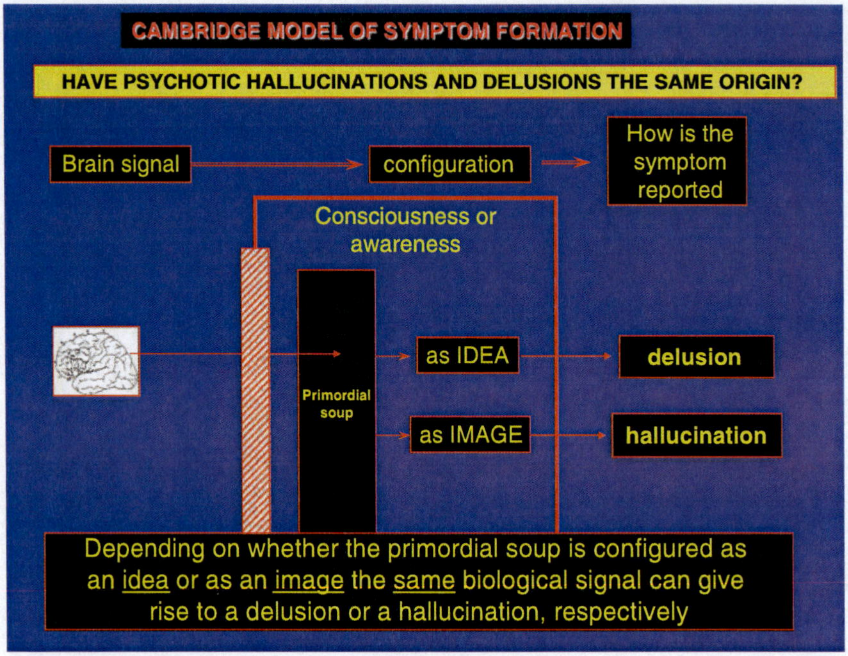

Fig. 5.2 Same origin of psychotic hallucinations and delusions

and negotiation. Clinicians do not behave like "phenomenological" amanuenses but as active negotiators. Against the context of a shared or different culture and influenced by the clinician's (diagnostic) view, each of these configured experiences is reshaped into what will become the final "mental symptom" (as recorded in the case notes).

There is no space in this chapter to deal with pathways (b) and (c). However, pathway (d) is interesting for it allows for the reconfiguration of a mental symptom already formed. Thus, mental symptoms configured via pathway (a) may undergo a secondary configuration based on specific representations and tropes that are of special importance to the individual. In this case, the neurobiological kernel built into the original mental symptom no longer plays any role as content and meaning are fully determined by the cultural configurators. It could be said that mental symptoms generated by pathway (d) correspond to those generically called "psychogenic" (as per dissociation, rerepresentation, etc.).

Against this background, one could then envisage a situation when the primordial soup (whatever its neuropsychological origin in the brain) could be putatively configured as an image or an idea or belief. If the former, the declaration by the subject would be deemed to be a hallucination; if the latter, a delusion (see Fig. 5.2). It is clear from this example that nonperceptual neurobiological signals can give rise to hallucinations and that perceptual signals may be configured as mental symptoms other than hallucinations.

5.5.2 *Hybrid Objects*

The concept of "hybrid object" includes components originating in the natural and social worlds. The proportional contribution of these two types of components and the manner of their combination and interaction require further research, and it is likely that such proportionality should be evaluated in relation to each hybrid object. In the case of "biological" objects (e.g., orchids or horses), describing and classifying are considered as overlapping operations in that both are meant to capture and release information about the objects classified. This is not the case in regard to ideal or abstract objects (e.g., virtues, revolutions, categories). Hybrid objects (particularly mental symptoms) are sui generis in this regard: (a) Their biological component may not in fact be informative at all, and (b) in spite of all the ongoing neuroscientific research, little is known about the "biological basis" of mental symptoms.

Although hybrid objects (e.g., hallucinations) may have biological and semantic components, their proportionality is yet to be determined, and hence any models developed to capture and classify them must take this into account. In general, however, it is likely that when it comes to psychiatric hallucinations, the semantic (cultural) component will be predominant to the extent that there will be little point in bothering with an analysis of their neurobiological kernel.

Comparing "hybrid objects" with physical and abstract objects throws further light on their nature, novelty, and usefulness. First of all, hybrid objects should not be considered as a mere combination of the other two. They represent the creative and configurative action of moral agents and hence are imbued with the emotional, volitional, and cognitive force that only persons can generate when confronted with a complex and (often) painful and perplexing experience (primordial soup). As dynamic responses, hybrid objects are fully consonant with personality and mental state. They are the expression of the manner in which beliefs, cultural codes, and views of the world get knitted together in response to a strange experience (see Figs. 5.3 and 5.4).

5.5.3 *Consequences and Inferences*

Although the Cambridge model still operates within a representational epistemology, it is able to do justice to (a) the existence of putative brain signaling constituting the kernel of mental symptoms (as dictated by the neuropsychiatric principle that mental disorders are disorders of the brain) and (b) to the operation of psychogenic and cultural factors. It offers a viable alternative to the received view of hallucinations, for it takes cultural configurators seriously. Instead of considering cultural modulation to be an "end-of-cascade" event with culture just coloring the "content" of the hallucination, the Cambridge model proposes that the cultural modulation occurs at the beginning of the cascade and hence it is drastic enough to

	CONCRETE	ABSTRACT	HYBRID
examples	orchids, dogs, gold	numbers, virtues, sacraments	works of art, maps mental symptoms
time	yes	no	yes
space	yes	no	yes
reducible without residuum	yes	no	no
epistemological approach	explanation	understanding	understanding
classification	Biological taxonomy	Artificial object	Artificial object
physical proxy variables	yes	no	yes
need for proxy variables	no	yes	yes
ontological origin	nature, evolution, god	man-made	man-made
relation to language	independent	dependent	dependent
semantic envelope	no	yes	yes
causally active	yes	yes?	yes

Fig. 5.3 Hybrid objects differentiated

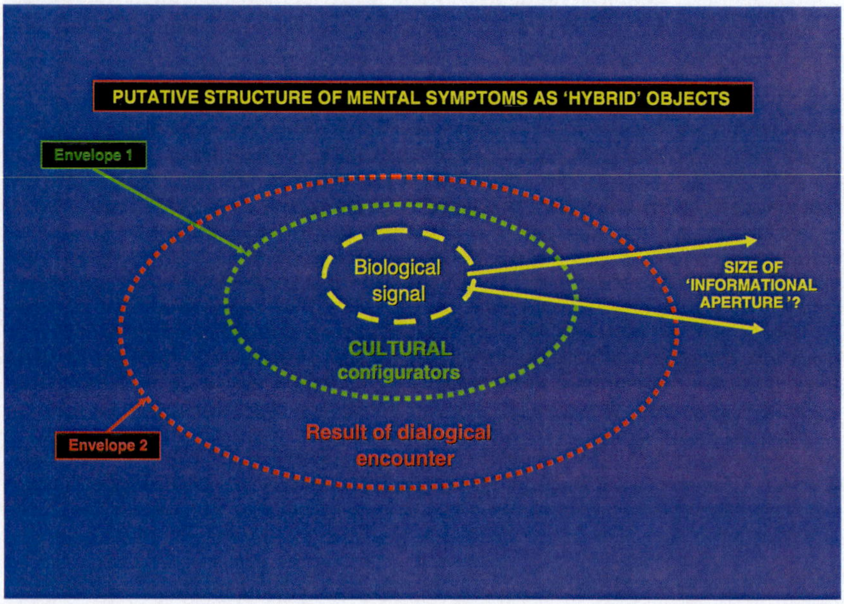

Fig. 5.4 Structure of hybrid objects

attenuate (and on occasions obliterate) the informational load of the biological signal.

As mentioned above, the Cambridge model allows for delusions and hallucinations (and for other mental symptoms) to share the same brain origin but to be differentiated in their form, meaning, content, and clinical import by the cultural configurators. In this respect, the biological signal would only provide the original trigger or force for a symptom to appear. What the symptom becomes must be considered as the total responsibility of the cultural and semantic configurators. This hybrid quality of the mental symptoms has important consequences for their brain localization and therapeutic management. But that is another story.

5.6 Conclusions

The current concept of hallucination was constructed during the early nineteenth century and carries epistemological assumptions belonging to this period. The received view has been useful to catalyze research but is proving infertile with each generation repeating the correlational research paradigm of the last except for making use of different research techniques. The same representational epistemology governs both conceptual and empirical research into hallucinations, and it is imposing severe constraints on both.

Deconstruction of the structure of the received view shows that it is supported by a raft of primary and secondary antinomies. The former were built into the concept at the moment of its inception; the latter were added throughout the nineteenth century to alleviate the tensions created by the original assumptions. The representational structure of the received view identifies the "representation" as the weakest point in the explanatory chain, both in its relationship to reality itself and to its management by the mind. Given the unknowable nature of reality, this is a pyrrhic victory. The fact that the nature of the representation itself cannot be fully ascertained means that declarations of entertaining an image in the mind's eye are conflated with declarations of believing that such an image is being entertained.

The relationship between hallucinations and perceptions is based on the fact that the images are part of the narrative of hallucinations. There is no reason to believe that this legitimizes the inference that (particularly psychiatric) hallucinations result from a pathology of the perceptual system. Likewise, hallucinations are unlikely to be a unitary phenomenon. Those linked to the distance and nondistance sense modalities are crucially differentiable, for consensual ascertainment is no longer available for the second group. The belief that hallucinations are the same phenomenon irrespective of etiology must also be revised. Organic hallucinations, i.e., those seen in relation to brain disease or experimental manipulation of the brain, may be related to perception in ways that psychiatric hallucinations are not.

An alternative hybrid model of hallucinations is proposed according to which what determines whether the experience is reported as an image or as a thought does not depend upon the nature of the brain signal that triggers the experience but upon the cultural configurators that modulate its form and content.

Arguments have been offered to believe that a radical revision of the epistemology of hallucinations is long overdue. The new epistemology of hallucinations must first be formulated on the conceptual drawing board and should offer a clear alternative to the received view. This alternative view should escape the constraints imposed by the representationist, perceptual, and unitary antinomies. Many human beings entertain hallucinatory experiences, some report them, and some may want to be rid of them. For the sake of the latter group, psychiatrists should try to break new grounds in regard to the understanding of these extraordinary and little-understood phenomena.

References

Aleman, A., Larøi, F. (2008). *Hallucinations. The science of idiosyncratic perception*. Washington, DC: American Psychological Association.

Allport, F.H. (1955). *Theories of perception*. New York, NY: Wiley.

Armstrong, D.M. (1961). *Perception and the physical world*. London: Routledge & Kegan Paul.

Bernheimer, R. (1961). *The nature of representation*. New York, NY: New York University Press.

Berrios, G.E. (1982). Tactile Hallucinations. *Journal of Neurology, Neurosurgery and Psychiatry*, 45, 285–293.

Berrios, G.E. (1988). *Historical background to abnormal psychology*. In *Adult abnormal psychology*. Edited by E. Miller and P. Cooper. London: Churchill Livingstone, pp. 26–51

Berrios, G.E. (1990). Introduction to Tamburini's A theory of hallucinations. Classic text N°1. *History of Psychiatry*, 1, 145–156.

Berrios, G.E. (1990a). Musical hallucinations: a historical and clinical study. *British Journal of Psychiatry*, 156, 188–194.

Berrios, G.E. (1996). *The history of mental symptoms*. Cambridge: Cambridge University Press.

Berrios, G.E., Freeman, H. (1991). *150 years of British psychiatry. Volume 1*. London: Gaskell.

Berrios, G.E., Dening, T. (1996). Pseudohallucinations: A conceptual history. *Psychological Medicine*, 26, 753–763.

Berrios, G.E., Marková, I.S. (2002). *Conceptual issues*. In *Biological Psychiatry*. Edited by H. D'haenen, J.A. den Boer, and P. Willner. Chichester: John Wiley & Sons, pp. 9–39.

Berrios, G.E., Marková, I.S. (2006). *Symptoms - historical perspectives and effect on diagnosis*. In *Psychosomatic Medicine*. Edited by M. Blumenfield and J.J. Strain. Philadelphia, PA: Lippincott Williams & Wilkins, pp. 27–38.

Boissier de Sauvages (1772). *Nosologie méthodique. Tome 7*. Lyon: Bruyset, pp. 43–190.

Carterette, E.C., Friedman, M.P., eds. (1974). *Handbook of perception. Volume 1: Historical and philosophical roots of perception*. New York, NY: Academic Press.

Clapin, H., ed. (2002). *Philosophy of mental representation*. Oxford: Clarendon Press.

Cummins, R. (1989). *Meaning and mental representation*. Cambridge, MA: MIT Press.

Dicker, G. (1980). *Perceptual knowledge*. Dordrecht: Reidel.

Duparc, F. (1992). Nouveaux développements sur l'hallucination négative et la représentation. *Revue Française de Psychanalyse*, 61, 101–121.

Esquirol, E. (1817). *Hallucinations*. In *Dictionnaire des Sciences Médicales, par une Société de Médecins et de Chirurgiens*. Paris: Panckoucke, pp. 64–71.

Ey, H. (1935). La discussion de 1855 à la Société Médico-Psychologique sur l'hallucination et l'état actuel du problème de l'activité hallucinatoire. *Annales Médico-psychologiques*, 93, 584–612.

Ey, H. (1973). *Traité des hallucinations. Tome 1 et 2*. Paris: Masson.
Fish, W. (2009). *Perception, hallucination and illusion*. Oxford: Oxford University Press.
Gibson, J.J. (1966). *The senses considered as perceptual systems*. London: Allen & Unwin.
Grave, S.A. (1960). *The Scottish philosophy of common sense*. Oxford: Clarendon Press.
Gurney, E. (1885). Hallucinations. *Mind*, 10, 161–199.
Hamlyn, D.W. (1957). *The psychology of perception*. London: Routledge & Kegan Paul.
Hamlyn, D.W. (1961). *Sensation and perception. A history of the philosophy of perception*. London: Routledge & Kegan Paul.
Ibarra, A., Mormann, T., eds. (2000). *Variedades de la representación en la ciencia y la filosofía*. Barcelona: Ariel.
Lanteri-Laura, G. (1991). *Les hallucinations*. Paris: Masson.
Marková, I.S., Berrios, G.E. (2009). The epistemology of mental symptoms. *Psychopathology*, 42, 343–349.
Maury, L.F.A. (1865). *Le sommeil et les rêves*. Paris: Didier.
Merleau-Ponty, M. (1945). *Phénoménologie de la perception*. Paris: Gallimard.
Mourgue, R. (1932). *Neurobiologie de l'hallucination*. Bruxelles: Lamertin.
Parish, E. (1897). *Hallucinations and illusions*. London: Walter Scott.
Paulus, J. (1941). *Le problème de l'hallucination et l'évolution de la psychologie d'Esquirol à Pierre Janet*. Paris: Les Belles Lettres.
Quercy, P. (1930). *Études sur l'hallucination. Tome 1 et 2*. Paris: Alcan.
Sterelny, K. (1990). *The representational theory of mind*. Oxford: Blackwell.
West, L.J., ed. (1962). *Hallucinations*. New York, NY: Grune & Stratton.
Yolton, J.W. (1984). *Perceptual acquaintance from Descartes to Reid*. Oxford: Blackwell.
Yolton, J.W. (1996). *Perception and reality. A history from Descartes to Kant*. Ithaca, NY: Cornell University Press.

Part II
Hallucinatory Phenomena

Chapter 6
Visual Hallucinations

Daniel Collerton, Rob Dudley, and Urs Peter Mosimann

6.1 Introduction

Visual experiences of 'things that are not there' have, in different ways, been recorded throughout history. Thus night-time visual hallucinations of a figure accompanied by a sense of terror and paralysis currently classified as a hypnagogic or hypnopompic hallucination in the context of a parasomnia were previously construed as supernatural night hags, incubi or nightmares in Western cultures (Blom 2010), with equivalent kanashibari reported in Eastern cultures (Goswami et al. 2010; see also Chaps. 17 and 18). Over the same time span, people of many cultures have shown a marked willingness to induce hallucinations as a means of reaching spirit worlds through the use of a range of plants and venoms (Perry and Laws 2010). Even in these early periods, however, not all explanations were supernatural. In one of the most famous hallucinations in literature, Macbeth's vision of a dagger,

D. Collerton, M.A., M.Sc. (✉)
Northumberland, Tyne and Wear NHS Foundation Trust, UK

Newcastle University, Newcastle Upon Tyne, UK
e-mail: daniel.collerton@ncl.ac.uk

R. Dudley, B.A., Ph.D., D.Clin Psy, Cert CBT
Institute of Neuroscience, Newcastle University, Newcastle Upon Tyne, UK

Early Intervention in Psychosis Service, Northumberland Tyne and Wear
NHS Foundation Trust, UK
e-mail: r.e.j.dudley@newcastle.ac.uk

U.P. Mosimann, M.D., Ph.D.
Department of Old Age Psychiatry, University Hospital of Psychiatry,
University of Bern, Bern, Switzerland
e-mail: urs.mosimann@gef.be.ch

J.D. Blom and I.E.C. Sommer (eds.), *Hallucinations: Research and Practice*,
DOI 10.1007/978-1-4614-0959-5_6, © Springer Science+Business Media, LLC 2012

Fig. 6.1 Charles Bonnet (1720–1792)

Shakespeare proposes a cortical explanation for hallucinations – 'or art thou but, A dagger of the mind, a false creation, Proceeding from the heat-oppressed brain?' (Macbeth, Act II Scene 1).

The shift from supernatural to scientific explanations became decisive with the European Enlightenment. In the eighteenth century, Charles Bonnet (1720–1792; see Fig. 6.1), a Swiss philosopher and scientist, reported the complex visual hallucinations experienced by his grandfather, Charles Lullin. '[He] sees from time to time, in front of him, figures of men, of women, of birds, of carriages, of buildings, et cetera. He sees these figures make various movements: getting closer, going away, fleeing, diminishing or increasing in size, appearing or disappearing; he sees the buildings rise in front of his eyes and a display of all the outside construction material.' An unequivocal physical (in this case bilateral cataracts) rather than a supernatural explanation was given together with a proposed neurological basis which would still stand scrutiny today: 'All of this appears to have a seat in that part of the brain involved with sight. It is not difficult to imagine physical causes, strong enough to shake sensitive bundles of fibres that will produce in the mind, the picture of various objects with as much veracity as if the objects themselves had stimulated the fibres' (Hedges 2007). Even in those still seeking supernatural explanations, for example, Sidgwick's mammoth survey of psychic experience in late Victorian Britain (Blom 2010), a scientific approach gradually took hold.

Table 6.1 Relative population frequency of common associates of formed (complex) visual hallucinations (Adapted from Collerton et al. 2005)

State	Relative frequency of visual hallucinations
Sleep-related hallucinations	100
Delirium	77
Nonpathological daytime hallucinations	33
Neurodegenerative disorders	28
Psychosis	22
Eye disease	20
Others (epilepsy, migraine, stroke, peduncular hallucinosis, narcolepsy, etc.)	5
Bereavement	1

Interest in clinical visual hallucinations slowly grew through the nineteenth and twentieth centuries. Charles Bonnet syndrome was named in the 1930s by the Swiss neurologist Georges de Morsier (1894–1982) (O'Farrell et al. 2010), and there was a gradual recognition of visual hallucinations in relatively rare neurological disorders, such as narcolepsy and peduncular hallucinosis, but also as core features of a much more common, albeit previously undiscovered, disorder – dementia with Lewy bodies. In psychosis, the historic focus on auditory hallucinations (and delusional beliefs) as core diagnostic characteristics overshadowed research on other modalities, with early interest in models of visual hallucinations in patients diagnosed with schizophrenia (Horowitz 1975) lapsing for a while.

By the start of the twenty-first century, however, four disorders – dementia, psychosis, delirium and eye disease – had become recognised as common associates of visual hallucinations (see Table 6.1). The hallucinatory consequences of physical or sensory deprivation and the flashbacks associated with psychological trauma also attracted increasing attention. Paralleling this clinical interest, but developing more slowly, was the recognition of visual hallucinations in nonpathological states, particularly on the borders of sleep (D'Agostino et al. 2010, see also Chap. 17) or following bereavements (Rees 1971; Alroe and McIntyre 1983).

However, research remained fragmented until the beginning of the twenty-first century, with most studies limited to a single disorder or mental state. Several factors have combined within the last decade to push these disparate fields together (Collerton and Mosimann 2010).

Firstly, developments in the understanding of normal visual processes in the latter decades of the twentieth century allowed the relationship of visual input and visual perception to be more clearly understood. The recognition that all visual perception is based upon generated internal models of the visual environment, rather than being a projection of visual input on the brain, provided a theoretical framework for integrating a range of hallucinations into disturbances within a single, albeit distributed and complex, system. Secondly, systematic comparisons of visual hallucinations in separate disorders highlighted not only consistent similarities across these in, for example, phenomenology, but also regular differences in, as an instance, frequency. This data provided a focus for what models have to explain and a way of testing those that are developed. Finally, improvements in the accessibility of the scientific literature and in communication have made it much easier for research groups working in different fields to influence each other.

6.2 Defining Visual Hallucinations

The move away from visual perception as a simple analysis of visual input to the conception of subjective reality as an internal model which is checked by visual input is a double-edged sword for hallucination researchers. On the plus side, it easily allows the existence of hallucinations if the balance between model and input is tilted too far from the input. On the other hand though, it means that, conceptually, all perception is at least partially hallucinatory. The conundrum of defining veridical as opposed to hallucinatory perception remains unsolved. Within a constructive model of visual perception, there is no sharp boundary around any visual phenomenon. Perhaps as a consequence, many visual disturbances commonly co-exist with hallucinations in the same person, hence the pragmatic, though problematic, definition of hallucinations as seeing something that other people do not see. In experimental practice, this can be tested fairly reliably by comparing a stimulus with the self-report of an associated perception, but more difficulty arises in naturalistic settings when what is actually present in the visual environment is not easily measurable.

6.3 The Measurement of Visual Hallucinations

Setting aside these conceptual issues, there is no difficulty in describing an astonishing variety of visual experiences which occur in the absence of environmental equivalents (see Table 6.2). The reader with a special interest into the phenomenology of hallucinations may find further information in Blom's extensive catalogue (2010), with a briefer survey in ffytche et al. (2010). What is rather more challenging is to record and classify these experiences in a systematic, valid and reliable manner. Without privileged access to another person's perception and a means of relating this to the visual environment, distinguishing true hallucinations (where there is no environmental equivalent for the perception) from illusionary misperceptions (in which there is a partial equivalent) to visual re-experiences (in which veridical perceptions recur beyond the initial stimulus) can prove very difficult (see also Chap. 2). At present, there is no validated instrument for capturing all types of hallucinatory experiences, though reliable specific measures of, for example, complex object hallucinations do exist (Mosimann et al. 2008).

Table 6.2 Varieties of visual hallucinations (Adapted from ffytche et al. 2010; Blom 2010)

Unformed	Simple	Visual hallucinations with lowest degree of complexity. They include flashes, dots, spots or phosphenes (Greek phōs (light); phainein (to shine)), visual experience of a transient flash or spark of light
Intermediate (not clearly formed or unformed)	Geometric	Hallucinations characterised by lines, geometrical shapes, circles, patterns
	Teichopsia	Greek: teichos (wall) and opsis (seeing), usually seen as zigzag lines in one hemifield in the context of migraine
Formed	Complex	Phenomenologically rich and well-organised visual hallucinations, including people, animals or objects. May be lilliputian (small)
	Panoramic, scenic or landscape	The whole visual input is replaced by hallucinatory precepts
	Text	Words and figures
Partial (one aspect of the perception is hallucinatory)	Colour hallucination/illusion	
	Hyperchromatopsia	Greek; Huper (to exceed certain boundaries), chrōma (colour) and opsis (seeing). Colours are perceived as exceptionally vivid and brilliant
	Face hallucination/illusion	
	Facial intermetamorphosis	Latin: inter (between); metamorphoun (to change one's shape) Misperception of an individual with the belief that the person has been transformed
	Prosopometamorphopsia	Greek: prosōpon (face, expression, part, mask); metamorphoun (to change the form); opsis (seeing), i.e. seeing faces in altered forms
Visual re-experiences (perception is re-experienced after the initial stimulus)	Visual perseveration	
	Trailing phenomenon	
	Palinopsia	Palin (again) opsis (seeing)
	Illusory visual spread	
	Positive afterimages	
	PTSD flashbacks	
	Flashbulb memories	
	Memory hallucination	
	Re-perceptive hallucination	

6.4 The Phenomenology of Visual Hallucinations

Though hallucinations come in many forms, it is not yet clear whether they cluster into a smaller number of core types, and if they do, where the edges lie between different types (ffytche 2005; Collerton and Mosimann 2010). Existing classificatory schemes use the content of the hallucination. Thus formed (sometimes called complex) hallucinations of figures or objects are distinguished from unformed (or simple) hallucinations of unformed dots, shapes or flashes. Beyond the problem of how to classify intermediate forms (tessellations, for example), it is not clear that content is the only potential divider since different hallucinations are also associated with variations in duration, distinctiveness, distress and other associated sensory, cognitive and behavioural phenomena. As an instance, hallucinations in other modalities are variably present in the same person (uncommon in eye disease, but common in Lewy body disorders, psychosis and delirium), but multimodal hallucinations are rare, even in these cases (see also Chap. 18). In hallucination-prone individuals, both internal (alertness and fatigue) and external (lighting and visual environment) factors may influence occurrence. Even in highly hallucinating individuals, however, veridical perception tends to be the rule with individual hallucinations being brief and episodic.

Comparison of visual hallucinations across different normal states and clinical disorders has the potential to test different ways of dividing up hallucinatory experiences (Collerton et al. 2005). However, differences in methodology (resulting partially from the conceptual and measurement problems mentioned previously) make it hard at present to know the contribution of reporting biases to these apparent differences (Collerton and Mosimann 2010). When different disorders are assessed within a single study, results are much more similar than comparisons of different studies might suggest. Bearing this caveat in mind, we will briefly review the evidence base for hallucinations in the most common clinical and nonpathological states. More systematic surveys can be found in Collerton et al. (2005), Aleman and Larøi (2008), ffytche et al. (2010) and Larøi and Aleman (2010).

Just about everyone experiences a visual hallucination at some time (Ohayon 2000), usually for a nonclinical reason. Prominent amongst these nonpathological states are the hypnagogic and hypnopompic hallucinations which occur on the borders of sleep, bereavement hallucinations and hallucinations associated with deprivation; both extreme physical deprivation and prolonged visual deprivation. In total, nonpathological hallucinations, mainly sleep related, are roughly as common as pathological ones (Collerton et al. 2005).

Though there is relatively little comparative study of nonpathological states, sleep- and bereavement-related hallucinations tend to be of formed figures (c.f. the night hags mentioned earlier), and are often fleeting and indistinct (D'Agostino et al. 2010). Similarly, reports of hallucinated figures are reported in physical extremis (Ryan 1995). In contrast, sensory deprivation is more commonly associated with unformed hallucinations (Meppelink et al. 2010), suggesting more of a commonality with eye disease in which unformed hallucinations also predominate.

Hallucinogens are a group of substances that have the potential to alter consciousness and to evoke phenomena such as hallucinations, illusions and other sensory distortions (Nichols 2004, see also Chap. 22). As an example, Albert Hofmann (1906–2008), the discoverer and inadvertent first user of lysergic acid diethylamide (LSD), described his initial experiences as follows: 'At home I lay down and sank into a not unpleasant intoxicated-like condition, characterized by an extremely stimulated imagination. In a dreamlike state, with eyes closed (I found

Table 6.3 Examples of the visual effects of hallucinogenic substances adapted from Blom 2010

Substance	Chemical properties	Possible visual experience
Serotonin receptor agonists		
Lysergic acid diethylamide (LSD)	Ergoline alkaloid, partial 5-hydroxytryptamine (5-HT2) agonist	Out-of-body experience, animation of extracorporeal world, sparkling, phosphenes, teleopsia, metamorphosis
Mescaline	Phenethylamine alkaloid from peyote cacti; 5-HT agonist	More vivid perception, kaleidoscopic hallucinations, kinesthetic hallucination
Psilocybin	Tryptamine alkaloid from mushrooms	Distorted vision, geometric complex visual hallucinations, complex scenes or forms
Bufotenine	Tryptamine alkaloid, from toad skin and eggs	Simple geometric visual hallucinations, increased intensity of colours
Acetylcholine receptor antagonists		
Atropa belladonna; Datura; Herba apollinaris; Mandragora officinarum	Tropane alkaloid, acetylcholine receptor antagonist, night shade plant	Vivid dreams, some illusions and complex hallucinations, but also anticholinergic side effects
Cannabinoids		
Tetrahydrocannabinol (THC); Dronabinol	Anandamide neurotransmitter, cannabinoid receptor agonists	Subtle changes of sensory acuity. In high doses it may be associated with illusions and hallucinations, possibly in conjunction with thought disorder

the daylight to be unpleasantly glaring), I perceived an uninterrupted stream of fantastic pictures, extraordinary shapes with intense, kaleidoscopic play of colours.' The number of potentially hallucinogenic plants is large, and the list of their visual effects is long (see Table 6.3). Systemic effects, for example mydriasis, blurred vision, tachycardia, vertigo, dry throat, constipation and urinary retention with anticholinergic compounds limit usage. Dosage with plant compounds is challenging, and in overdose, psychedelics, particularly those with anticholinergic properties, can easily lead to delirium, stupor or death. The shift to synthetic hallucinogens, typified by LSD, initially appeared to open up the possibility of direct research into the pharmacological mechanisms of hallucinations, but discovery of the long-term disadvantages of hallucinogens, including visual flashbacks, visual snow and hallucinogen-induced persistent perception disorder (HPPD) in predisposed adults, combined to produce an abrupt stop to initial enthusiasm. Now, indirect studies based on case reports, or treatment studies for existing hallucinations, tend to predominate.

Hallucinations in eye disease are highly variable within and across people (ffytche and Howard 1999; ffytche 2009; ffytche 2010) and range from flashes, dots and other unformed experiences, to perceptual distortions and re-experiences, through to hallucinations of figures and objects, and ending with perceptions of whole landscapes. As an instance, A.B. – a person with macular degeneration – described her hallucination as '…a huge, vast canyon. I had to step over the edge to get to my seat. I have seen large churches with gravestones inside and everything was larger than life, but doors appeared shorter. The rooms just carried on, and I have to walk through things to get around – like a bath or a wall.'

There is an inverse relationship between complexity and frequency (Collerton and Mosimann 2010). Thus simple dots and flashes are almost invariable, whilst complex scenes are rare. The nomenclature of simple to complex hallucinations and the phenomenology of hallucinations implies a continuum in these experiences, but it is not clear whether this is so.

Turning to neurodegenerative disorders, the strong association of figure and object hallucinations with dementia with Lewy bodies initially obscured the range of other visual experiences that patients report. To illustrate, P.J. – a person with dementia with Lewy bodies – reported his experiences as follows: 'Every night I would see a man and a young child standing in the corner of the room staring at me… it was really queer. They would move but not come any closer to me and didn't say anything… they both had on old-fashioned clothing, like Victorian style with cloaks on.' Though still the archetypal hallucination is of a figure, less formed hallucinations and perceptual distortions are very common (Mosimann et al. 2006). As with other disorders, hallucinations are associated with increased distress and disability and poorer outcome (Mosimann and Collerton 2010).

People with psychosis report a broad range of odd or anomalous experiences both as triggers for delusional interpretations and as psychotic phenomena in their own right (Garety et al. 2001). As an example, K.F. – a person with psychosis – reported that 'I saw a man hanging from a noose in a tree – the man is dead.' Even before the person has made the transition to a psychosis, there may be evidence of subtle anomalous visual experiences which are not necessarily disturbing. People often report odd or unusual experiences such as greater intensity of colours and sounds (Bell et al. 2006; Freeman and Fowler 2009) more akin to optical distortions and illusions than hallucinations.

Where people have made the transition to overt psychosis, the prevalence of visual hallucinations is high (Bracha et al. 1989) and associated with greater levels of distress and impairment (Mueser et al. 1990). Most visions are of figures. The remaining visions are either unformed or of animals or objects (Gauntlett-Gilbert and Kuipers 2003, 2005). Visions are not usually accompanied by voices, though the majority hear voices at other times. Fused visual and auditory experiences in which, for example, the person sees lips moving on an otherwise normal, real face at the same time as hearing a voice are relatively rare (Hoffman and Varanko 2006).

The extent to which psychosis and hallucinations are independent is under review. The classic belief that hallucinations are, by definition, psychotic symptoms (though not necessarily as part of a psychotic disorder) has been countered by the perspective that they can be nonpathological, though usually infrequent, phenomena which become part of a psychotic syndrome through cognitive and emotional biases towards distress. The physiological overlap between visual hallucinations in psychosis and other disorders, for example, the role of cholinergic function (Hussain 1971; Patel et al. 2010), is starting to be investigated.

The final member of the big four hallucinatory disorders is delirium, an extremely common but poorly understood disorder (Fearing and Inouye 2009). In contrast to other disorders, there is little systematic evidence on the form of visual hallucinations in delirium, in large part because of the practical difficulties in doing research with actively delirious participants (a problem also found with acutely hallucinating people with psychosis and dementing disorders). However, it does seem that the phenomenology of hallucinations in delirium is more akin to that seen in dementing illnesses, whilst the interpretation and emotional reaction is more akin to that seen in psychosis.

6.5 Mechanisms of Visual Hallucinations

Any model of how people can see things that are not there has to be based on how people see things that are there. Early conceptualisations of vision emphasised a bottom-up abstraction of perception from an increasingly abstract and global analysis of visual input. Thus, perceptions of objects were abstracted from shapes which were built from lines and angles which were in turn recognised from patterns of excitation in primary visual cortex.

In contrast, current generative models see perception as an internal, sparse, functional, predictive, dynamic representation of the visual input that the brain would receive if that model were correct (Heerkeren et al. 2008). Hence, what is seen is the least unlikely match between predicted and actual visual input (Friston 2002). This moves the emphasis in vision away from input from the eye as the determinant of perception towards input as the means of checking or limiting the perceptual representation. This shift in model is associated with a parallel development in neuroscience towards investigating a distributed visual processing system which incorporates not only posterior visual areas, but also frontal attentional cortex, brainstem and thalamic regulatory mechanisms, as well as interconnecting tracts.

However, as might be expected from the continuing discussion on the existence of distinct visual hallucinatory syndromes, there is no established consensus on the mechanisms within this distributed system which may underlie these visual experiences. Different groups working on the pathogenesis of visual hallucinations variably stress the importance of disturbances in frontal, visual, regulatory and connective systems (Horowitz 1975; Manford and Andermann 1998; Pelaez 2000; Lee et al. 2003; Collerton and Perry 2004; Behrendt and Young 2004; Collerton et al. 2005; ffytche 2008).

The causes of visual hallucinations are not yet firmly established. In eye disease, for example, the extent of visual loss is a consistent risk factor, but other factors such as age, cognitive status and gender are inconsistent (ffytche 2009; Graham et al. 2011). Specifically, the factors which account for the variations in hallucinatory phenomenology from one person to another are very poorly understood, and it remains unclear why similar levels of visual impairment do not always lead to visual hallucinations. Neuroimaging indicates that those areas active in hallucinations are also those that are active in sense perceptions of similar stimuli. Conversely, if the ability to perceive or image something is lost through damage to the brain visual systems, so is the ability to hallucinate that image. Clear eye pathology suggesting a primary role for loss of visual input supported by the relatively high frequency of unformed hallucinations in both eye disease and visual deprivation, and the similarity in content of the hallucinations seen in eye disease and following serotonin receptor agonists, places the emphasis on posterior, primarily visual, areas (ffytche 2007, 2010).

Cholinergic, and to a lesser extent, dopaminergic function is a key neuropathological concomitant of visual hallucinations in dementia with Lewy bodies. Unlike eye disease, dementia with Lewy bodies has a distributed pathology with consistent damage to frontal and secondary visual cortices. Putting this together with distributed models of veridical visual perception which highlight similar brain areas, with the cognitive characteristics of dementia with Lewy bodies and with evidence that cholinergic manipulation both induces and abolishes figure and object hallucinations heightens the emphasis on interactive models of visual hallucinations (Collerton et al. 2005). Unlike previous models which emphasised single areas of dysfunction, albeit within complex systems (Manford and Andermann 1998), these interacting models propose a number of interacting impairments, none of which in itself is sufficient to cause hallucinations. Thus the Perception and Attention Deficit (PAD) model (Collerton et al. 2005) stipulates combined dysfunctions in attentional and perceptual processes, and Diedrich et al.'s (2005, 2009) Activation, Input and Modulation (AIM) disturbance model proposes simultaneous shifts in external perception and internal image generation.

Subsequent research has confirmed that co-existing impairments in central top-down attentional/generational processes and in bottom-up perceptual processes do lead to a greatly increased chance of visual hallucinations (Ramírez-Ruiz et al. 2007; Ozer et al. 2007; Barnes and Boubert 2008; Imamura et al. 2008; Meppelink et al. 2008). However, just as in veridical perception, the nature of the dysfunctional interaction has remained elusive (Bronnick et al. 2011). Imaging has identified consistent dysfunction in posterior and ventral visual areas, but equivalent findings in more frontal attentional cortex have been less reliable (Diederich et al. 2009; Sanchez-Castaneda et al. 2010).

Research into Lewy body disorders also led to a renewed interest in the dream intrusion hypothesis which was initially proposed as an explanation for psychotic and sleep-associated hallucinations (D'Agostino et al. 2010), although this time it was the association with sleep disturbance and rapid eye movement sleep behaviour disorder rather than phenomenological similarities between dreams and hallucinations which sparked interest (Arnulf et al. 2000; Barnes et al. 2010). As in other areas, however, the association between disturbed dreams and disturbed perception appears more to reflect the fortuitous disruption of anatomically close but separable systems (Collerton and Perry 2011).

Current models of psychotic symptoms emphasise the importance of anomalous visual experiences both in and of themselves and as key drivers in the development of delusional beliefs (Garety et al. 2001). However, in many cases, there is a notable absence of peripheral sensory or central perceptual impairments, suggesting either a heightened role for the influence of generative processes on the experience of visual hallucinations (perhaps in the context of image memories from the previous traumatic experiences which are common in psychosis) or for an anomalous interpretation of mundane experiences. For example, some people report seeing shadows in the corner of the eye. Normally dismissed as an insignificant transient illusion, the person with psychosis may interpret it as a ghost or vampire.

Despite the difference between age of onset of hallucinations in psychosis (usually in late adolescence or early adulthood) and other disorders (usually well into later life), transition to a full psychosis with distressing voices, visual hallucinations and delusions is associated with the same factors which increase the risk of hallucinations in other disorders. Thus, many people report that the onset of psychosis is associated with sleep disturbance, stress, social withdrawal and isolation, as well as with substance misuse. Whilst the use of hallucinogens is not a frequent factor, the sustained use of strong cannabis or amphetamine use seems to be frequent in the emergence of psychotic symptoms including visual hallucinations. Imaging (Oertel et al. 2007), though less extensive than in eye disease or dementia, also seems to indicate a relationship between the content of the hallucination and the activity of relevant brain areas.

Moving away from the mechanisms generating the hallucinatory experience, wider research in psychosis, aimed at understanding the psychological mechanisms which drive abnormal perceptions and interpretations, and the distress which flows from these, has led to a focus not only on the hallucinatory mechanisms themselves but also on an interest in the emotional and behavioural impact of these experiences.

Analogies from the experience of auditory hallucinations suggest that it is not the hallucination per se which drives distress (Andrews et al. 2007; Birchwood 2003; Collerton and Dudley 2004), but the interpretation of the hallucination as a threat (Dudley et al. in press) and subsequent behavioural response to that interpretation which is key.

The mechanisms underlying hallucinations in delirium are very poorly understood, but what evidence there is does suggest similar risk factors to those seen in dementia (Brown et al. 2009).

6.6 Visual Hallucinations: An Integrative Model?

There still remain many areas of less than perfect knowledge where visual hallucinations are concerned, but it now seems accepted that hallucinations are not phenomena that stand separate from normal vision, nor from other visual disturbances. Both conceptually and in practice, there are no clear dividing lines between veridical perceptions and hallucinations.

Looking across disorders and states, the beginnings of more integrative, multidimensional models are emerging (Aleman and Larøi 2008; see also Chaps. 2 and 3). Disturbances in several parts of the distributed visual system can all cause a transitory

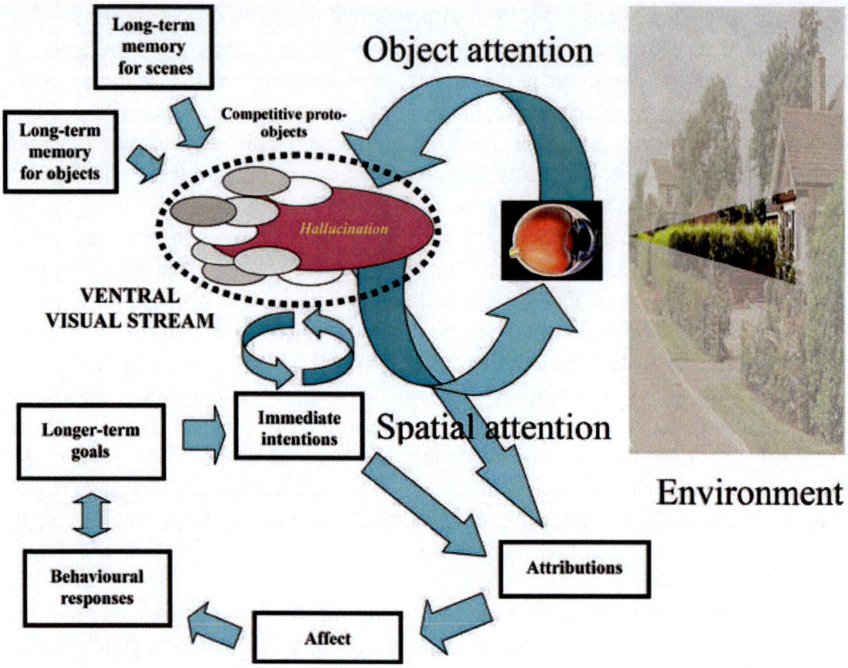

Fig. 6.2 An integrative model of hallucinations

disturbance in perception such that a perception including a hallucinatory element is a better predictor of visual input than is a wholly veridical perception.

From eye disease and sensory deprivation comes evidence on the effects of reducing or eliminating visual input; from neurodegenerative disorders and delirium comes data on the effects of distributed disturbances in brain function, particularly on attentional and top-down processes; from psychosis come models of emotional disturbance, psychological development and maintenance mechanisms. Putting these together, a tentative model can be sketched (see Fig. 6.2). In this model, spatial and object attentions shape sensory input which then interacts with expectancies biased by long-term memories of visual contexts and immediate intentions to increase the chance of a hallucinatory perception. This perception is then not disconfirmed since sensory input is, in itself, impaired. The content of the hallucination and the intentional context in which it occurs then trigger emotional and behavioural reactions which may lead to longer-term goals (for example, to avoid similar experiences in the future), which then feed into emotional reactions, if they do recur.

A number of consequences flow from this model.

Firstly, to test it is not trivial. Not only are many components of the model underspecified and hard to measure, but evaluation of it depends upon the accurate simultaneous measurement of dynamic perception, processing and stimulus during the occurrence of relatively rare, unpredictable, episodic phenomena. At present, our measurement of the static factors which increase the chance of hallucinations is much better than our measurement of the dynamic factors which lead to a specific hallucination. We do not understand enough of the temporal course of visual hallucinations and their relationship with thought, affect and behaviour. Serial investigations, including functional imaging of people at risk of hallucinations whilst they are and are not hallucinating would allow hypotheses to be tested, but as yet, the practical problems of capturing an unpredictable and transient experience with a very large machine have only rarely been surmounted (ffytche et al. 1998; Oertel et al. 2007).

Secondly, it suggests that there may well be a number of potential avenues for developing effective treatments (Larøi and Aleman 2010). Hallucinations appear to be inherently unstable – veridical perception is the default position even in those people who do hallucinate. Thus, targeting specific cognitive or biological risk factors either for the occurrence of hallucinations or for their disabling emotional consequences may be effective; though it is fair to say that, as yet, no truly effective interventions have been discovered. The co-existence of multiple impairments in eye function, sensory processing and cognition in the population group most at risk of hallucinations – i.e. the elderly – together with the limitations of medication use in older people (see also Chap. 24) does suggest that optimism should be tempered by caution.

Looking to the future, most of the tools that are needed for progress are at our hands already. Despite our imperfect definitions and measures of hallucinations, their consequences and the neural systems underlying them, we can do these things well enough to move forward. Systematically using the tools that we do have across hallucinations in different clinical and natural states will undoubtedly refine our existing understandings and open up new areas of investigation.

References

Aleman, A., Larøi, F. (2008). *Hallucinations. The science of idiosyncratic perception*. Washington, DC: American Psychological Association.

Alroe, C.J., McIntyre, J.N. (1983). Visual hallucinations. The Charles Bonnet syndrome and bereavement. *Medical Journal of Australia*, 2, 674–675.

Andrews, C., Collerton, D., Mosimann, U., Dudley, R. (2007). Emotional experiences and complex visual hallucinations. *PSIGE Newsletter*, 97, 51–53.

Arnulf, I., Bonnet, A.-M., Damier, P., Bejjani, B.-P., Seilhean, D., Derenne, J.-P., Agid, Y. (2000). Hallucinations, REM sleep, and Parkinson's disease. A medical hypothesis. *Neurology*, 55, 281–288.

Barnes, J., Boubert, L. (2008). Executive functions are impaired in patients with Parkinson's disease with visual hallucinations. *Journal of Neurology, Neurosurgery, and Psychiatry*, 79, 190–192.

Barnes, J., Connelly, V., Wiggs, L., Boubert, L., Maravic, K. (2010). Sleep patterns in Parkinson's disease patients with visual hallucinations. *International Journal of Neuroscience*, 120, 564–569.

Behrendt, R.-P., Young, C. (2004). Hallucinations in schizophrenia, sensory impairment, and brain disease: a unifying model. *Behavioural and Brain Sciences*, 27, 771–787.

Bell, V., Halligan, P.W., Ellis, H.D. (2006). The Cardiff Anomalous Perceptions Scale (CAPS): A new validated measure of anomalous perceptual experience. *Schizophrenia Bulletin*, 32, 366–377.

Birchwood, M. (2003). Pathways to emotional dysfunction in first-episode psychosis *British Journal of Psychiatry*, 182, 373–375.

Blom, J.D. (2010). *A dictionary of hallucinations*. New York, NY: Springer.

Bracha, H.S., Wolkowitz, O.M., Lohr, J.B., Karson, C.N., Bigelow, L.B. (1989). High prevalence of visual hallucinations in research subjects with chronic schizophrenia. *American Journal of Psychiatry*, 146, 526–528.

Bronnick, K., Emre, M., Tekin, S., Haugen, S.B., Aarsland, D. (2011). Cognitive correlates of visual hallucinations in dementia associated with Parkinson's disease. *Movement Disorders*, 26, 824–829.

Brown, L.J.E., McGrory, S., McLaren, L., Starr, J.M., Deary, I.J., MacLullich, A.M.J. (2009). Cognitive visual perceptual deficits in patients with delirium. *Journal of Neurology, Neurosurgery, and Psychiatry*, 80, 594–599.

Collerton, D., Mosimann, U.P. (2010). *Visual hallucinations*. In *Wiley interdisciplinary reviews: Cognitive science*. Edited by M. Petersen and L. Nadel. New York, NY: John Wiley & Sons, pp. 781–786.

Collerton, D., Perry, E. (2011). Dreaming and hallucinations - continuity or discontinuity? Perspectives from dementia with Lewy bodies. *Consciousness and Cognition*, 20, 1016–1020.

Collerton, D., Perry, E., McKeith, I. (2005). Why people see things that are not there. A novel Perception and Attention Deficit Model for recurrent complex visual hallucinations. *Behavioural and Brain Sciences*, 28, 737–794.

Collerton, D., Dudley, R. (2004). A cognitive behavioural framework for the treatment of distressing visual hallucinations in older people. *Behavioural and Cognitive Psychotherapy*, 32, 443–455.

Collerton, D., Perry, E. (2004). Thalamocortical dysfunction and complex visual hallucinations in brain disease - are the primary disturbances in the cerebral cortex? *Behavioural and Brain Sciences*, 27, 789–790.

D'Agostino, A., Limosani, I., Scarone, S. (2010). *Dreaming as a physiological process: connecting states of consciousness*. In *New horizons in the neuroscience of consciousness*. Edited by E. Perry, D. Collerton, H. Ashton, and F. Lebeau. Amsterdam: John Benjamins, pp. 239–248.

Diederich, N.J., Goetz, C.G., Stebbins, G.T. (2005). Repeated visual hallucinations in Parkinson's disease as disturbed external/internal perceptions: focused review and a new integrative model. *Movement Disorders*, 20, 130–140.

Diederich, N.J., Fénelon, G., Stebbins, G., Goetz, C.G.A.F. (2009). Hallucinations in Parkinson disease. *Nature Reviews. Neurology*, 5, 331–342.

Dudley, R., Wood, M., Spencer, H., Mosimann, U.P., Collerton, D. (In press). Identifying specific interpretations and use of safety behaviours in people with distressing and visual hallucinations. Behavioural and Cognitive Psychotherapy.

Fearing, M.A., Inouye, S.K. (2009). *Delirium*. In *The American psychiatric publishing textbook of geriatric psychiatry*. Edited by D.G. Blazer and D.C. Steffens. Washington, DC: American Psychiatric Publishing, pp. 229–241.

ffytche, D.H. (2010). *The visual unconscious. Perspectives from the Charles Bonnet syndrome*. In *New horizons in the neuroscience of consciousness*. Edited by E. Perry, D. Collerton, H. Ashton, and F. Lebeau. Amsterdam: John Benjamins, pp. 215–226.

ffytche, D.H. (2005). Two visual hallucinatory syndromes. *Behavioural and Brain Sciences*, 28, 763–764.

ffytche, D.H. (2009). Visual hallucinations in eye disease. *Current Opinions in Neurology*, 22, 28–35.

ffytche, D.H. (2008). The hodology of hallucinations. *Cortex*, 44, 1067–1083.

ffytche, D.H. (2007). Visual hallucinatory syndromes: past, present and future. *Dialogues in Clinical Neuroscience*, 9, 173–189.

ffytche, D.H., Howard, R.J. (1999). The perceptual consequences of visual loss: 'positive' pathologies of vision. *Brain*, 122, 1247–1260.

ffytche, D.H., Howard, R.J., Brammer, M.J., David, A., Woodruff, P., Williams, S. (1998). The anatomy of conscious vision: An fMRI study of visual hallucinations. *Nature Neuroscience*, 1, 738–742.

ffytche, D.H., Blom, J.D., Catani, M. (2010). Disorders of visual perception. *Journal of Neurology, Neurosurgery, and Psychiatry*, 81, 1280–1297.

Freeman, D., Fowler, D. (2009). Routes to psychotic symptoms: trauma, anxiety and psychosis like experiences. *Psychiatry Research*, 169, 107–112.

Friston, K. (2002). Functional integration and inference in the brain. *Progress in Neurobiology*, 68, 113–143.

Frucht, S.J., Bernsohn, L. (2002). Visual hallucinations in PD. *Neurology*, 59, 1965.

Garety, P.A., Kuipers, E., Fowler, D., Freeman, D., Bebbington, P.E. (2001). A cognitive model of the positive symptoms of psychosis. *Psychological Medicine*, 31, 189–195.

Gauntlett-Gilbert, J., Kuipers, E. (2003). Phenomenology of visual hallucinations in psychiatric conditions. *Journal of Nervous and Mental Disorders*, 191, 203–205.

Gauntlett-Gilbert, J., Kuipers, E. (2005) Visual hallucinations in psychiatric conditions: appraisals and their relationship to distress. *British Journal of Clinical Psychology*, 44, 77–87.

Goswami, M., Pandi-Perumal, S.R., Thorpy, M.J. (2010). *Narcolepsy*. New York, NY: Springer.

Graham, G., Dean, J., Mosimann, U.P., Colbourn, C., Dudley, R., Clarke, M., Collerton, D. (2011). Specific attentional impairments and complex visual hallucinations in eye disease. *International Journal of Geriatric Psychiatry*, 26, 263–267.

Hedges, T.R. (2007). Charles Bonnet, his life, and his syndrome. *Survey of Ophthalmology*, 52, 111–114.

Heerkeren, H.R., Marrett, S., Ungerleider, L.G. (2008). The neural systems that mediate human perceptual decision making. *Nature Reviews. Neuroscience*, 9, 467–469.

Hoffman, R.E., Varanko, M. (2006). "Seeing voices": fused visual/auditory verbal hallucinations reported by three persons with schizophrenia-spectrum disorder. *Acta Psychiatrica Scandinavica*, 114, 290–292.

Horowitz, M.J. (1975). *Hallucinations: an information processing approach*. In: *Hallucinations. Behavior, experience and theory*. Edited by R.K. Siegel and L.J. West. New York, NY: John Wiley & Sons.

Hussain, M.Z. (1971). Benzhexol-induced hallucinosis. *Canadian Medical Association Journal*, 104, 420–421.

Imamura, K., Wada-Isoe, K., Kitayama, M., Nakashima, K.A.F., Imamura, K., Wada-Isoe, K., Kitayama, M., Nakashima, K. (2008). Executive dysfunction in non-demented Parkinson's disease patients with hallucinations. *Acta Neurologica Scandinavica*, 117, 255–259.

Larøi, F., Aleman, A., eds. (2010). *Hallucinations. A guide to treatment and management*. Oxford: Oxford University Press.

Lee, K.-H., Williams, L.M., Breakspear, M., Gordon, E. (2003). Synchronous gamma activity: a review and contribution to an integrative neuroscience model of schizophrenia. *Brain Research Reviews*, 41, 57–78.

Manford, M., Andermann, F. (1998). Complex visual hallucinations: clinical and neurobiological insights. *Brain*, 121, 1819–1840.

Meppelink, A.M., Koerts, J., Borg, M., Leenders, K.L., Van Laar, T.A.F. (2008). Visual object recognition and attention in Parkinson's disease patients with visual hallucinations. *Movement Disorders*, 23, 1906–1912.

Meppelink, A.M., De Jong, B.M., Van der Hoeven, J.H., Van Laar, T. (2010). Lasting visual hallucinations in visual deprivation; fMRI correlates and the influence of rTMS. *Journal of Neurology, Neurosurgery, and Psychiatry*, 81, 1295–1296.

Mosimann, U., Rowan, E., Partington, C., Collerton, D., Littlewood, E., O'Brien, J.T., Burn, D.J., McKeith, I.G. (2006). Characteristics of visual hallucinations in Parkinson's disease dementia and dementia with Lewy bodies. *American Journal of Geriatric Psychiatry*, 14, 154–160.

Mosimann, U.P., Collerton, D. (2010). *Hallucinations in the context of dementing illnesses*. In *Hallucinations. A guide to treatment and management*. Edited by F. Larøi and A. Aleman. Oxford: Oxford University Press, pp. 323–350.

Mosimann, U.P., Collerton, D., Dudley, R., Meyer, T.D., Graham, G., Dean, J.L., Bearn, D., Killen, A., Dickinson, L., Clarke, M.P., McKeith, I.G. (2008). A semi-structured interview to assess visual hallucinations in older people. *International Journal of Geriatric Psychiatry*, 23, 712–718.

Mueser, K.T., Bellack, A.S., Brady, E.U. (1990). Hallucinations in schizophrenia. *Acta Psychiatrica Scandinavica*, 82, 26–29.

Nichols, D.E. (2004). Hallucinogens. *Pharmacology and Therapeutics*, 101, 131–181.

Oertel, V., Rotarska-Jagiela, A., Van de Ven, V.G., Haenschel, C., Maurer, K., Linden, D.E. (2007). Visual hallucinations in schizophrenia investigated with functional magnetic resonance imaging. *Psychiatry Research*, 156, 269–273.

O'Farrell, L., Lewis, S., McKenzie, A., Jones, L. (2010). Charles Bonnet syndrome: A review of the literature. *Journal of Visual Impairment and Blindness*, 104, 261–274.

Ohayon, M.M. (2000). Prevalence of hallucinations and their pathological associations in the general population. *Psychiatry Research*, 97, 153–164.

Ozer, F., Merai, H., Hanoglu, L., Ozturk, O., Aydemir, T., Cetin, S., Atmaca, B., Tiras, R. (2007). Cognitive impairment patterns in Parkinson's disease with visual hallucinations. *Journal of Clinical Neuroscience*, 14, 742–746.

Patel, S.S., Attard, A., Jacobsen, P., Shergill, S. (2010). Acetylcholinesterase inhibitors (AChEI's) for the treatment of visual hallucinations in schizophrenia: a case report. *BMC Psychiatry*, 10, 68.

Pelaez, J.R. (2000). Towards a neural network based therapy for hallucinatory disorders. *Neural Networks*, 13, 1047–1061.

Perry, E., Collerton, D., Ashton, H., Lebeau, F., eds. (2010). *New frontiers in consciousness research*. Amsterdam: John Benjamins.

Perry, E.K., Laws, V. (2010). *Plants of the Gods and shamanic journeys*. In *New horizons in the neuroscience of consciousness*. Edited by E. Perry, D. Collerton, H. Ashton, and F. Lebeau. Amsterdam: John Benjamins, pp. 239–248.

Ramírez-Ruiz, B., Junqué, C., Martí, M.J., Valldeoriola, F., Tolosa, E. (2006). Neuropsychological deficits in Parkinson's disease patients with visual hallucinations. *Movement Disorders*, 21, 1483–1487.

Ramírez-Ruiz, B., Junqué, C., Martí, M.-J., Valldeoriola, F., Tolosa, E. (2007) Cognitive changes in Parkinson's disease patients with visual hallucinations. *Dementia and Geriatric Cognitive Disorders*, 23, 281–288.

Rees, W.D. (1971). The hallucinations of widowhood. *British Medical Journal*, 4, 37–41.

Ryan, C. (1995). *The one that got away*. London: Century.

Sanchez-Castaneda, C., Rene, R., Ramírez-Ruiz, B., Campdelacreu, J., Gascon, J., Falcon, C., Calopa, M., Jauma, S., Juncadella, M., Junqué, C.A.F. (2010). Frontal and associative visual areas related to visual hallucinations in dementia with Lewy bodies and Parkinson's disease with dementia. *Movement Disorders*, 25, 615–622.

Chapter 7
Synaesthesias

Devin Blair Terhune and Roi Cohen Kadosh

7.1 Introduction

Most people believe that everyone else experiences the world in a similar way to them. However, a small percentage of the population, individuals with synaesthesia, experiences various aspects of their environment in profoundly different ways from the rest of us. *Synaesthesia* (literally meaning 'joined sensation') is an unusual neurological condition in which a stimulus, such as the number *5*, the note *C*, or the feel of sandpaper, reliably and involuntarily elicits an *ancillary*, atypical experience, such as the colour red or the emotion of excitement, alongside the typical response (Ward and Mattingley 2006). For individuals with synaesthesia, numbers may trigger colours, tastes may elicit tactile perceptions, and months of the year may be distributed in space. A number of historical figures, such as Pythagoras (see Cytowic and Eagleman 2009), suspected of having synaesthesia, but the first documented case dates to 1812 (Jewanski et al. 2009), with the first systematic study of the phenomenon occurring later in the century (Galton 1883). In the parlance of contemporary research, the stimulus that elicits the synaesthetic response is referred to as the *inducer*, whereas the ancillary event, which may be affective, cognitive, or perceptual, is termed the *concurrent* (Grossenbacher and Lovelace 2001). In what follows, we summarize the methods by which synaesthesia is typically studied, its principal characteristics and relationship to other phenomena such as hallucinations, and how recent research has informed our understanding of its origins and mechanisms.

D.B. Terhune, M.Sc., Ph.D. (✉) • R. Cohen Kadosh, B.A., Ph.D.
Department of Experimental Psychology, University of Oxford, Oxford, UK
e-mail: devin.terhune@psy.ox.ac.uk; roi.cohenkadosh@psy.ox.ac.uk

J.D. Blom and I.E.C. Sommer (eds.), *Hallucinations: Research and Practice*,
DOI 10.1007/978-1-4614-0959-5_7, © Springer Science+Business Media, LLC 2012

7.2 Measuring Synaesthesia

Before turning to its principal characteristics, it is imperative to outline how synaesthesia is typically measured and authenticated. There is broad consensus amongst researchers that an individual's synaesthesia is genuine if it is involuntary, or *automatic*, and if the inducer-concurrent pairs are *consistent* over time (Ward and Mattingley 2006).

The criterion of automaticity can be easily measured using different selective attention tasks in which inducers are alternately paired with stimulus dimensions that are either congruent or incongruent with an individual's concurrents for the respective inducers. For example, if a synaesthete experiences the colour *red* for the letter *E*, one can measure how quickly the synaesthete is able to identify the actual colour of the letter when presenting the letter *E* coloured in red (congruent condition) as compared to the letter *E* coloured in blue (incongruent condition). It has been repeatedly demonstrated that synaesthetes exhibit interference effects, as indicated by slower response times or lower accuracy rates for incongruent than congruent inducer-concurrent pairs, whereas control participants do not (Dixon et al. 2000). Cumulatively, these results suggest that synaesthesia is resistant to control.

The criterion of consistency requires assessing the reliability of participants' inducer-concurrent pairs. This is typically done by indexing an individual's self-reported concurrents (e.g. colours) for a set of inducers (e.g. numbers) at two time points separated by a few weeks or months. Insofar as synaesthetes reliably display high consistency, this feature is widely regarded as a marker of genuine synaesthesia. As a result, consistency is typically used to verify synaesthesia and as an inclusion criterion for experimental research (Simner 2011).

Despite the clear utility and widespread use of these criteria, they are not without their limitations. First, non-synaesthetes who are trained to associate graphemes with colours display interference effects akin to those observed by some synaesthetes (Cohen Kadosh et al. 2005; Meier and Rothen 2009). Grapheme-colour synaesthesia (and corresponding interference effects) can also be induced in highly suggestible non-synaesthetes with a posthypnotic suggestion (Cohen Kadosh et al. 2009). Accordingly, interference effects alone may be insufficient to verify an individual's synaesthesia. Simner (2011) has similarly argued that inducer-concurrent consistency may be an erroneous identifying characteristic of synaesthesia whose use will invariably lead to the specious exclusion of real synaesthetes. It is unclear whether this argument will gain traction amongst researchers; it is likely that consistency testing will remain the gold standard for the identification of synaesthetes for some time.

7.3 Principal Characteristics

Although still uncommon, synaesthesia is not as rare as once believed. Recent research estimates the prevalence of synaesthesia in the general population to be approximately 4% with equivalent frequency of occurrence for men and women (Simner et al. 2006b). Synaesthesias include a diverse set of experiences that although notably similar may not represent a uniform set of phenomena with shared characteristics.

7.3.1 Types of Synaesthesia

There are a wide variety of forms of atypical binding that can be classified as synaesthesia. To date, more than one hundred different forms have been identified (Cytowic and Eagleman 2009). In the most representative study to date, the most common form of synaesthesia (2.8%) was coloured days, in which days of the week elicit colour concurrents (Simner et al. 2006b). By contrast, grapheme-colour synaesthesia, the mostly commonly studied form of this condition, had a prevalence of 1.4% in two independent samples. Another study found the prevalence of mirror-touch synaesthesia, in which viewing another person being touched produces corresponding tactile hallucinations, to be 1.6% (Banissy et al. 2009).

A critical question is how the different forms of synaesthesia are related to one another. It is well known that having one form increases the likelihood of possessing another. For instance, as many as 50% of synaesthetes have two or more forms of synaesthesia (Cytowic and Eagleman 2009). Indeed, having another form of synaesthesia is often used as an argument that a non-prototypical form is indeed a true form of synaesthesia (Simner et al. 2006b). However, whether synaesthesias represent a unitary or a diverse set of similar phenomena remains poorly understood. Eagleman (2010) has recently presented data that bear on this question. Applying factor analysis to the forms of synaesthesia exhibited by a large sample of synaesthetes, he found that 21 different forms fell into six higher-order factors, each comprised of multiple forms: coloured sequence (e.g. number → colour), non-visual concurrents (e.g. sound → smell), coloured flavour, coloured music, coloured affect (conceptual) (e.g. emotion → colour), and coloured affect (physical) (e.g. touch → colour). An individual with any of the four forms within the coloured-sequence cluster has a high probability of having another form from within this cluster, but only a chance probability of having forms from other clusters. Notably, spatial-sequence synaesthesia, such as experiencing numbers as localized in space, was only weakly related to the other forms of synaesthesia and thereby appears to represent a distinct condition. Mapping the similarities and differences between these clusters will be an important area for future research.

7.3.2 Unity and Diversity

The principal feature that unites all forms of synaesthesia is the atypical binding of two normally distinct experiences, in which either an environmental (e.g. a sound) or endogenous (e.g. an aural image) stimulus produces an atypical ancillary experience (e.g. a colour photism [image or percept]). Most synaesthetes do not experience *conscious* bidirectional synaesthesia (e.g. digit ↔ colour), although a minority does (Cohen Kadosh et al. 2007; Cytowic and Eagleman 2009). Some grapheme-colour synaesthetes do exhibit bidirectional synaesthesia *implicitly*, that is, colours are associated with particular numerical values, associations that influence behavioural response patterns even though the association does not breach conscious awareness (Gevers et al. 2010). One persistent finding is that the experience of colour, as a photism, is the most frequent synaesthetic concurrent, present in approximately 77% of synaesthetes (Simner et al. 2006b). A further piece of evidence for the unity of synaesthesia is the aforementioned finding that many synaesthetes reliably exhibit other forms of synaesthesia. Finally, it is notable that synaesthetes' inducer-concurrent pairs are often remarkably consistent (Ward and Mattingley 2006). Concordance in inducer-concurrent pair consistency across different forms of synaesthesia would provide further support for the claim that synaesthesias are different expressions of the same underlying phenomenon.

There are also a number of noteworthy similarities in the inducer-concurrent mappings of synaesthetes. For example, *0*, *O*, *1*, and *I* are commonly white, and *A* is often red (Barnett et al. 2008; Rich et al. 2005; Simner et al. 2005), although there is variegation in other grapheme-colour pairs across studies. Similar consistencies are present in sound-colour and in grapheme-luminance pairs. Recent data by Eagleman (2010) suggest that mappings across synaesthetes may be determined by low-level features of the stimuli (e.g. shape) and occur through associative mechanisms during development. Specifically, for many synaesthetes, it appears that novel graphemes are assigned similar colours as similarly shaped graphemes. In some cases, mappings are likely due to semantic associations (*A* is associated with *apples*, which are often associated with the colour red), but such an explanation can only account for a small number of mappings. Further, similarly shaped graphemes (e.g. *S* and *5*) have been found to elicit different concurrents, which may in part be determined by their semantic context (Dixon et al. 2006).

Despite the shared commonalities between and amongst different forms of synaesthesia, there exist equally important differences. The most salient divergence from uniformity amongst synaesthetes is perhaps Eagleman's (2010) finding of different higher-order clusters of synaesthesia. This set of results calls into question the idea that synaesthesia is a uniform condition. We anticipate that this will be the subject of a considerable amount of research in the future.

Diversity in the expression of synaesthesia is also present amongst individuals experiencing the same form. For example, grapheme-colour synaesthetes have been found to vary with regard to whether they experience concurrents spatially localized to the stimulus (*projectors*) or as endogenous representations (*associators*) (Dixon et al. 2004). These two subtypes display distinct behavioural response patterns (Dixon et al. 2004; Ward et al. 2007) as well as differences in grey and white matter volumes and connectivity in regions related to sensory processing and memory (Rouw and Scholte 2007, 2010). However, synaesthetes' associator-projector status may be unreliable (Edquist et al. 2006); it is somewhat unclear how individuals are best stratified on this dimension (see Van Leeuwen et al. 2010; Ward et al. 2007), and the two subtypes do not always display behavioural and neurophysiological differences (Gebuis et al. 2009; Rouw and Scholte 2007; Weiss and Fink 2009). Grapheme-colour synaesthetes may also vary in the extent to which low-level features of an inducer (e.g. colour contrast with background) influence concurrents, pointing to possible *higher* (concurrents independent of stimulus features) and *lower* (concurrents influenced by stimulus features) subtypes of synaesthesia (Ramachandran and Hubbard 2001). The associator-projector and lower-higher dimensions appear to be orthogonal (Ward et al. 2007). It has also been shown that mirror-touch synaesthetes may be comprised of one subtype that experiences mirror-touch perceptions as looking through a mirror, and another that experiences these perceptions as though they were in the same position as the individual being touched (Banissy et al. 2009).

Ignoring heterogeneity amongst synaesthetes has important consequences. It may result in the failure to uncover important individual differences amongst synaesthetes as well as the failure to replicate findings across laboratories (Dixon and Smilek 2005). Notwithstanding these concerns, whether the aforementioned subgroups reflect discrete subtypes or positions on continua in an otherwise uniform population is an issue that can be empirically resolved and which warrants greater attention (Cohen Kadosh and Terhune 2011).

7.3.3 Relationship with Cross-Modal Mapping

Some findings presented in the preceding section point to commonalities between synaesthetes and non-synaesthetes. One explanation for this concordance is that these similarities reflect a latent continuum in which synaesthetes are positioned at an extreme end of a normal distribution. For instance, individuals without synaesthesia display colour-pitch, pitch-size, and letter-colour associations that closely match those found in synaesthetes (Ward et al. 2006). There are also similarities between the implicit representation of numbers in space in non-synaesthetes and their explicit representation in space amongst number-space synaesthetes (Cohen Kadosh and Henik 2007). However, synaesthetes' experiences of atypical binding are of much greater magnitude and exhibit greater specificity and consistency than the cross-modal mappings of non-synaesthetes. Perhaps the most fundamental difference is that synaesthetes are *conscious* of their condition. Further quantitative evidence that synaesthetes are a discrete group is garnered when the consistency of grapheme-colour pairs is plotted. Critically, synaesthetes and controls are nearly wholly distinct from one another, resulting in a bimodal distribution (Barnett et al. 2008; Ward and Simner 2005). This challenges a continuum account of synaesthesia and suggests that the synaesthesia phenotype represents a discrete category. Further studies are needed to resolve these conflicting views. It might be that some types of synaesthesia exhibit continua-like properties, whereas others reflect discrete categories of experience.

7.3.4 Correlates

Synaesthesia may confer abilities on an individual that have a positive or deleterious impact on cognition. These effects may be specific to an individual's form of synaesthesia or be independent of it, stemming from common underlying mechanisms.

The cognitive correlate of synaesthesia that has attracted the greatest attention is memory. Insofar as many synaesthetes have access to two channels by which a string of information can be encoded (e.g. a string of letters and a string of colours), they presumably have access to a greater number of retrieval cues and thereby may be expected to exhibit superior memory. There are a number of case studies of synaesthetes with extraordinary memory abilities (see for example Luria 1968), but few investigations of groups. Group studies point to heightened mnemonic abilities amongst synaesthetes, although there have been some mixed results (Rothen and Meier 2009, 2010; Yaro and Ward 2007). Moreover, synaesthetes' performances are well within the normal range. Superior memory for graphemes and colours may underlie greater grapheme-colour pair consistency in this population (Yaro and Ward 2007) and may contribute to the greater *range* of colours that synaesthetes can describe (Simner et al. 2005), which may, in turn, facilitate greater colour discrimination.

Experiencing the world in a profoundly different way is likely to impact how synaesthetes interact with others. Banissy and Ward (2007) found that mirror-touch synaesthetes exhibit greater empathy than control participants. This effect presumably arises from the tactile 'hallucinations' these synaesthetes experience whilst viewing other individuals being touched, although it is critical to ensure that this effect is restricted to this particular form of synaesthesia. A related finding is that more grapheme-colour synaesthetes than controls report *Mitempfindung*, a neurological condition in which tactile stimulation produces sensation in a different location from the stimulation (Burrack et al. 2006).

Less is known about the deleterious effects of synaesthesia on cognition. There is empirical evidence that synaesthetes display difficulties in mathematics and in mislocating sensory stimuli to the contralateral side of the body (*allochiria*, see also Chap. 13) relative to non-synaesthetes (Cytowic and Eagleman 2009; Rich et al. 2005; Ward et al. 2009). One hypothesis is that number-form synaesthetes may utilize inflexible spatial strategies during computations (Ward et al. 2009). Cytowic and Eagleman (2009) note how a number of these and other deficits observed in this population are associated with parietal lobe functioning and thus may indirectly implicate the parietal cortex in synaesthesia.

This brief review illustrates how poorly understood the cognitive profile of the synaesthete is and how ripe for investigation this subject matter is. It will be important for future researchers to dissociate the relationships between particular cognitive (dis)abilities and different forms of synaesthesia.

7.3.5 Relationship to Other Hallucinatory Phenomena

Insofar as the concurrents of some synaesthetes qualify as hallucinations, it is worth considering the relationship between synaesthesia and other hallucinatory phenomena. Some colour-taste synaesthetes experience the respective taste localized in the mouth, grapheme-colour projector synaesthetes experience colour photisms localized in space, and mirror-touch synaesthetes experience tactile hallucinations resembling touch. These experiences can be regarded as hallucinations because they constitute perceptual experiences that do not have a corresponding stimulus in the environment. At the same time, it needs to be noted that they are only produced when exposed to an inducer (e.g. a taste, a number, etc.) and are thus qualitatively different from other hallucinations. By analogy, projector photisms would be comparable to visual hallucinations that occur *only*, and *reliably*, when another specific stimulus is presented. In addition, the majority of synaesthetes know that concurrents are produced by their unique cognitive apparatus and are not actually present in the environment. In this regard, synaesthetic hallucinations do not meet conservative criteria that require that an individual believes that her or his hallucination is a veridical perception.

It is plausible that synaesthesia may share mechanisms with multisensory hallucinations. A recent case study, using multiple electrophysiological and neuroimaging techniques in a patient diagnosed with schizophrenia who experienced auditory-visual hallucinations, found that he displayed reduced cortical thickness in multiple regions (e.g. the superior temporal gyrus, see Jardri et al. 2009), whereas synaesthetes display the converse pattern in other regions (see Sect. 7.4.2 below). Whilst the activation of the fusiform gyrus during multisensory hallucinations parallels its activation during synaesthesia (see Sect. 7.4.2), the former appears to serve a binding/integration function, whereas the latter is associated with grapheme and colour processing. Cytowic and Eagleman (2009) note how synaesthesia may be experienced in the context of release hallucinations. Similarly, cross-modal percepts akin to synaesthesia are sometimes reported during seizures in individuals with temporal lobe epilepsy and in a minority of individuals with concussions; although it is unclear whether these phenomena occur through similar mechanisms as synaesthesia (Cytowic and Eagleman 2009). The oft-cited fact that LSD and other psychedelic drugs (see also Chap. 22) can produce synaesthesia-like experiences (Nichols 2004) has received virtually no empirical attention, but it is important to note that many of the hallmark phenomenological properties of synaesthesia such as automaticity and consistency do not appear to be present in psychedelic hallucinations (see also Cytowic and Eagleman 2009). It is clear that many hallucinations arise from a disruption of monitoring functions in which an endogenous representation is misattributed to the environment or an environmental stimulus is misinterpreted. One potentially fruitful avenue for research in this area would be to consider whether projector synaesthesias and other forms with hallucinatory properties represent instances in which concurrents are misattributed to one's environment because of a source-monitoring deficit. At present, how synaesthesias and hallucinations may or may not be related has received such little attention that it is difficult to provide any firm conclusions regarding this relationship.

7.4 Mechanisms

A considerable amount of research on synaesthesia has been devoted to its mechanisms. In this section, we summarize work on the genetic and developmental origins of synaesthesia as well as its cognitive and neural mechanisms.

7.4.1 Genetic and Developmental Origins

When considering the genetic and developmental contributions to the incidence and phenomenology of synaesthesia it is imperative to distinguish between three types of variables: (1) those that determine whether one has synaesthesia; (2) those that determine one's form of synaesthesia; and (3) those that determine the characteristics of one's synaesthesia, such as inducer-concurrent pairs and localization of concurrents. Some evidence has been yielded for the variables underlying each of these features.

Early research by Galton (1883) suggested that synaesthesia runs in families. More recent research corroborates his results and indicates that synaesthesia is a heritable condition (Barnett et al. 2008; Baron-Cohen et al. 1996; Ward and Simner 2005). Familial studies have shown that around 40% of synaesthetes have a relative with synaesthesia (Barnett et al. 2008; Rich et al. 2005; Ward and Simner 2005). In contrast, case studies of monozygotic twin pairs have produced inconsistent results: in one case, both twins had synaesthesia (Hancock 2006), whereas in two other cases, only one twin had synaesthesia (Smilek et al. 2002, 2005). Smilek et al. (2005) speculated that discordance amongst monozygotic twins may result from genetic mutations or changes in brain morphology during development. The results of a whole-genome analysis of auditory-visual synaesthetes and their families yielded evidence for a link with chromosome 2q and suggestive links with other chromosomes, indicating that synaesthesia is a complex condition with genetic heterogeneity (Asher et al. 2009).

Data bearing on the heritability of the features of synaesthesia indicate that its phenomenological expression is not wholly constrained by the synaesthesia of one's relatives. Amongst a group of synaesthetes and family members with synaesthesia, only grapheme-colour synaesthesia was found in 73% of the families. However, the remaining families had members with different forms of synaesthesia, suggesting that a specific form is not inherited (Barnett et al. 2008; Ward et al. 2005), although it is plausible that one may inherit a predisposition for a particular higher-order cluster (Eagleman 2010, see Sect. 7.3.1). Barnett and colleagues also found that grapheme-colour pairs differed across family members and were comparable to those observed between unrelated synaesthetes. Finally, synaesthetes' associator-projector status did not appear to be inherited. Cumulatively, these results indicate that synaesthesia is heritable but that its form and phenomenological features are determined by other factors.

Synaesthesia is typically reported as having been experienced since early childhood and thus is generally considered a congenital condition. However, environmental influences during development appear to contribute to the phenomenology of synaesthesia. The clearest evidence is seen with inducer-concurrent pairs. As described above (see Sect. 7.3.2), low-level features and semantic properties of graphemes as well as syllable stress position of words (Simner et al. 2006a) appear to determine colour concurrents. Similar effects may be at play in sound-colour synaesthesia (Ward et al. 2006). However, patterns of inducer-concurrent pairs exhibit considerable diversity and often appear to be idiosyncratically determined by features of an individuals' unique environment, such as through explicit exposure to coloured graphemes during childhood (see also Simner et al. 2009; Witthoft and Winawer 2006).

7.4.2 *Cognitive and Neural Mechanisms*

The mechanisms underlying synaesthesia have been a consistent source of debate and motivation for research. One point of contention amongst synaesthesia researchers is the extent to which synaesthesia requires attention. Synaesthesia was initially widely regarded as a sensory-perceptual phenomenon that occurred through low-level mechanisms and could be induced by exposure to stimulus properties (Ramachandran and Hubbard 2001). In contrast, recent research has emphasized the importance of conceptual representations for synaesthesia (Simner 2011) and indicates that attention is indeed necessary for synaesthesia (Rich and Mattingley 2010; Ward et al. 2010). As described above (see Sect. 7.3.2), the extent to which sensory or conceptual properties of inducers trigger synaesthetic experiences may vary across synaesthetes. Behavioural differences amongst grapheme-colour synaesthetes may emerge because of differential shifting of attention from one spatial reference frame to another, which will produce different attentional demands across synaesthetes (Ward et al. 2007).

There is broad consensus that synaesthesia results from the enhanced communication, or crosstalk, between cortical regions governing processing of the inducer and concurrent. However, the mechanism supporting crosstalk is the source of fervent debate. To date, two distinct, albeit not wholly competing, neurological accounts have been advanced to account for crosstalk.

The two-stage model of Hubbard (2007) proposes that excess structural connectivity between particular brain regions gives rise to their cross-activation during experience of an inducer but that binding processes in the parietal cortex produce a coherent synaesthetic experience (see Fig. 7.1). Functional neuroimaging studies have implicated the fusiform gyrus, including V4/V8, which is known to support colour, letter, and word processing, in various forms of synaesthesia (Barnett et al. 2008; Brang et al. 2010; Nunn et al. 2002; Rouw and Scholte 2007; Sperling et al. 2006). Grapheme-colour synaesthetes also appear to exhibit larger cortical volume than controls in the left and right anterior fusiform gyrus (Jancke et al. 2009) and the left intraparietal sulcus and right fusiform gyrus grey matter (Weiss and Fink 2009). Further support for the role of the parietal cortex in synaesthesia is gleaned from two studies that demonstrated that transcranial magnetic stimulation (see also Chap. 25) applied to the right parietal occipital junction transiently disrupted grapheme-colour synaesthesia (Esterman et al. 2006; Muggleton et al. 2007). Finally, increased structural connectivity in the left superior parietal cortex, the motor cortex, and the right inferior temporal cortices (adjacent to the fusiform gyrus) has been observed in grapheme-colour synaesthetes (Rouw and Scholte 2007, 2010). Cumulatively,

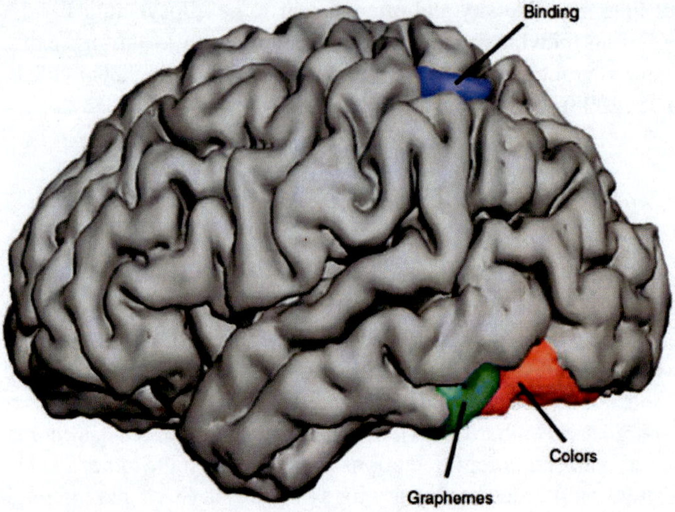

Fig. 7.1 The cortical regions implicated in the two-stage model of grapheme-colour synaesthesia (Hubbard 2007). Crosstalk between V4 (*red*) and the visual word form area in the fusiform gyrus (*green*), produced by excess anatomical connectivity, gives rise to the experience of colour during the presentation of graphemes. Grapheme-colour binding is further enabled by the intraparietal sulcus (*blue*) (Reprinted by permission from Macmillan Publishers Ltd: Nature Neuroscience (Hubbard 2007). Copyright 2007)

these results support Hubbard's (2007) model, but it should be noted that evidence supporting a *causal* link between hyperconnectivity and synaesthesia has not yet been found.

A second theory of synaesthesia (see Fig. 7.2) asserts that synaesthesia results from cortical disinhibition, in which functional neural pathways that are normally inhibited in the general population are disinhibited in synaesthetes (Cohen Kadosh and Henik 2007; Grossenbacher and Lovelace 2001). Insofar as neural pathways strengthen with repeated use, disinhibition could also give rise to hyperconnectivity, that is, excess connectivity may be a *consequence* rather than a *cause* of synaesthesia (Cohen Kadosh and Walsh 2008). The finding that synaesthesia can be transiently induced by suggestion further indicates that hyperconnectivity is not *necessary* for the occurrence of synaesthesia because the suggestion is unlikely to produce excess cortical connections so rapidly (Cohen Kadosh et al. 2009). A recent magnetoencephalography study of grapheme-colour synaesthesia found that the activation differences associated with synaesthesia in the posterior temporal lobe (grapheme region) preceded corresponding differences in V4 (colour region) by approximately 5 ms (Brang et al. 2010). This finding has been interpreted as challenging a disinhibition account, but the prior activation of the grapheme region is arguably consistent with feedback disinhibition.

A number of caveats regarding results bearing on the neural basis of synaesthesia are worth mentioning. First, many of the results concern grapheme-colour

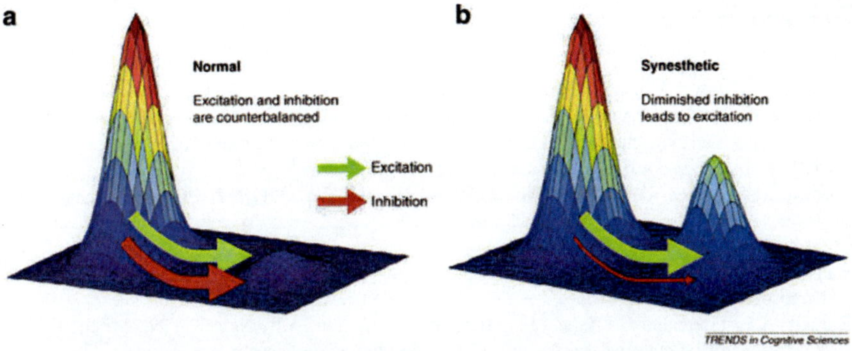

Fig. 7.2 Diagram of a proposed mechanism in the disinhibition model of synaesthesia. (**a**) When levels of excitation (*red*) and inhibition (*green*) are equivalent, crosstalk is disrupted. (**b**) Diminished inhibition from the region processing the inducer gives rise to the activation of the concurrent-selective region (Reprinted from Eagleman and Goodale (2009). With permission from Elsevier)

synaesthesia, and it is unclear whether they generalize to other forms of synaesthesia. As an example, a recent model of mirror-touch synaesthesia does not invoke a parietal binding mechanism but instead asserts parietal involvement in this form of synaesthesia through its modulation of processes supporting bodily representation and disembodiment (Banissy et al. 2009). Second, V4 activations are not always observed across studies, and some studies implicate left-hemisphere parietal regions whereas others observe right-hemisphere effects (see Hubbard 2007). Finally, it is imperative that caution be exerted when assuming that all forms of synaesthesia occur through uniform mechanisms (Cohen Kadosh and Walsh 2008). For instance, within-modality forms may be the product of excess structural connectivity facilitating increased cross-activation, whereas cross-modality forms may be produced more often through disinhibition.

7.5 Conclusions

In the preceding pages, we have reviewed recent research on the characteristics and mechanisms of synaesthesia. It is abundantly clear that considerable knowledge has been gained and theoretical accounts are becoming increasingly sophisticated. However, it is also evident that many questions remain unanswered and many fascinating avenues remain unexplored. We are confident that further investigation of the characteristics and mechanisms of synaesthesia will yield insights into the nature of other anomalous experiences as well as subjective experience itself.

Acknowledgements DBT is supported by the Cogito Foundation. RCK is supported by the Wellcome Trust (WT88378).

References

Asher, J.E., Lamb, J.A., Brocklebank, D., Cazier, J.-B., Maestrini, E., Addis, L., Sen, M., Baron-Cohen, S., Monaco, A.P. (2009). A whole-genome scan and fine-mapping linkage study of auditory-visual synesthesia reveals evidence of linkage to chromosomes 2q24, 5q33, 6p12, and 12p12. *American Journal of Human Genetics*, 84, 279–285.

Banissy, M.J., Cohen Kadosh, R., Maus, G.W., Walsh, V., Ward, J. (2009). Prevalence, characteristics and a neurocognitive model of mirror-touch synaesthesia. *Experimental Brain Research*, 198, 261–272.

Banissy, M.J., Ward, J. (2007). Mirror-touch synesthesia is linked with empathy. *Nature Neuroscience*, 10, 815–816.

Barnett, K.J., Finucane, C., Asher, J.E., Bargary, G., Corvin, A.P., Newell, F.N., Mitchell, K.J. (2008). Familial patterns and the origins of individual differences in synaesthesia. *Cognition*, 106, 871–893.

Baron-Cohen, S., Burt, L., Smith-Laittan, F., Harrison, J., Bolton, P. (1996). Synaesthesia: prevalence and familiality. *Perception*, 25, 1073–1079.

Brang, D., Hubbard, E.M., Coulson, S., Huang, M., Ramachandran, V.S. (2010). Magnetoencephalography reveals early activation of V4 in grapheme-color synesthesia. *NeuroImage*, 53, 268–274.

Burrack, A., Knoch, D., Brugger, P. (2006). Mitempfindung in synaesthetes: co-incidence or meaningful association? *Cortex*, 42, 151–154.

Cohen Kadosh, R., Cohen Kadosh, K., Henik, A. (2007). The neuronal correlate of bidirectional synesthesia: a combined event-related potential and functional magnetic resonance imaging study. *Journal of Cognitive Neuroscience*, 19, 2050–2059.

Cohen Kadosh, R., Henik, A. (2007). Can synaesthesia research inform cognitive science? *Trends in Cognitive Science*, 11, 177–184.

Cohen Kadosh, R., Henik, A., Catena, A., Walsh, V., Fuentes, L.J. (2009). Induced cross-modal synaesthetic experience without abnormal neuronal connections. *Psychological Science*, 20, 258–265.

Cohen Kadosh, R., Sagiv, N., Linden, D.E., Robertson, L.C., Elinger, G., Henik, A. (2005). When blue is larger than red: Colors influence numerical cognition in synesthesia. *Journal of Cognitive Neuroscience*, 17, 1766–1773.

Cohen Kadosh, R., Terhune, D.B. (2011). Redefining synaesthesia? *British Journal of Psychology* doi: 10.1111/j.2044-8295.2010.02003.x.

Cohen Kadosh, R., Walsh, V. (2008). Synaesthesia and cortical connections: Cause or correlation? *Trends in Neurosciences*, 31, 549–550.

Cytowic, R.E., Eagleman, D.M. (2009). *Wednesday is indigo blue: Discovering the brain of synaesthesia*. Cambridge, MA: MIT Press.

Dixon, M.J., Smilek, D. (2005). The importance of individual differences in grapheme-color synesthesia. *Neuron*, 45, 821–823.

Dixon, M.J., Smilek, D., Cudahy, C., Merikle, P.M. (2000). Five plus two equals yellow. *Nature*, 406, 365.

Dixon, M.J., Smilek, D., Duffy, P.L., Zanna, M.P., Merikle, P.M. (2006). The role of meaning in grapheme-colour synaesthesia. *Cortex*, 42, 243–252.

Dixon, M.J., Smilek, D., Merikle, P.M. (2004). Not all synaesthetes are created equal: projector versus associator synaesthetes. *Cognitive, Affective and Behavioral Neuroscience*, 4, 335–343.

Eagleman, D.M. (2010). What has large-scale analysis taught us? Paper presented at the annual meeting of the UK Synaesthesia Association, Brighton, UK.

Eagleman, D.M., Goodale, M.A. (2009). Why color Synesthesia involves more than Color. *Trends in Cognitive Sciences*, 13, 288-292.

Edquist, J., Rich, A.N., Brinkman, C., Mattingley, J.B. (2006). Do synaesthetic colours act as unique features in visual search? *Cortex*, 42, 222–231.

Esterman, M., Verstynen, T., Ivry, R.B., Robertson, L.C. (2006). Coming unbound: Disrupting automatic integration of synesthetic color and graphemes by transcranial magnetic stimulation of the right parietal lobe. *Journal of Cognitive Neuroscience*, 18, 1570–1576.

Galton, F. (1883). *Enquiries into human faculty and its development*. London: Macmillian and Co.

Gebuis, T., Nijboer, T.C., Van der Smagt, M.J. (2009). Multiple dimensions in bi-directional synesthesia. *European Journal of Neuroscience*, 29, 1703–1710.

Gevers, W., Imbo, I., Cohen Kadosh, R., Fias, W., Hartsuiker, R.J. (2010). Bidirectionality in synesthesia: evidence from a multiplication verification task. *Experimental Psychology*, 57, 178–184.

Grossenbacher, P.G., Lovelace, C.T. (2001). Mechanisms of synesthesia: Cognitive and physiological constraints. *Trends in Cognitive Sciences*, 5, 36–41.

Hancock, P. (2006). Monozygotic twins' colour-number association: A case study. *Cortex*, 42, 147–150.

Hubbard, E.M. (2007). Neurophysiology of synesthesia. *Current Psychiatry Reports*, 9, 193–199.

Jancke, L., Beeli, G., Eulig, C., Hanggi, J. (2009). The neuroanatomy of grapheme-color synesthesia. *European Journal of Neuroscience*, 29, 1287–1293.

Jardri, R., Pins, D., Bubrovszky, M., Lucas, B., Lethuc, V., Delmaire, C., Vantyghem, V., Despretz, P., Thomas, P. (2009). Neural functional organization of hallucinations in schizophrenia: multisensory dissolution of pathological emergence in consciousness. *Consciousness and Cognition*, 18, 449–457.

Jewanski, J., Day, S.A., Ward, J. (2009). A colorful albino: The first documented case of synaesthesia, by Georg Tobias Ludwig Sachs in 1812. *Journal of the History of the Neurosciences*, 18, 293–303.

Luria, A.R. (1968). *The mind of a mnemonist*. Cambridge, MA: Harvard University Press.

Meier, B., Rothen, N. (2009). Training grapheme-colour associations produces a synaesthetic Stroop effect, but not a conditioned synaesthetic response. *Neuropsychologia*, 47, 1208–1211.

Muggleton, N., Tsakanikos, E., Walsh, V., Ward, J. (2007). Disruption of synaesthesia following TMS of the right posterior parietal cortex. *Neuropsychologia*, 45, 1582–1585.

Nichols, D.E. (2004). Hallucinogens. *Pharmacology and Therapeutics*, 101, 131–181.

Nunn, J.A., Gregory, L.J., Brammer, M., Williams, S.C., Parslow, D.M., Morgan, M.J., Morris, R.G., Bullmore, E.T., Baron-Cohen, S., Gray, J.A. (2002). Functional magnetic resonance imaging of synesthesia: activation of V4/V8 by spoken words. *Nature Neuroscience*, 5, 371–375.

Ramachandran, V.S., Hubbard, E.M. (2001). Synaesthesia - A window into perception, thought and language. *Journal of Consciousness Studies*, 8, 3–34.

Rich, A.N., Bradshaw, J.L., Mattingley, J.B. (2005). A systematic, large-scale study of synaesthesia: implications for the role of early experience in lexical-colour associations. *Cognition*, 98, 53–84.

Rich, A.N., Mattingley, J.B. (2010). Out of sight, out of mind: the attentional blink can eliminate synaesthetic colours. *Cognition*, 114, 320–328.

Rothen, N., Meier, B. (2009). Do synesthetes have a general advantage in visual search and episodic memory? A case for group studies. *Public Library of Science One*, 4, e5037.

Rothen, N., Meier, B. (2010). Grapheme-colour synaesthesia yields an ordinary rather than extraordinary memory advantage: evidence from a group study. *Memory*, 18, 258–264.

Rouw, R., Scholte, H.S. (2007). Increased structural connectivity in grapheme-color synesthesia. *Nature Neuroscience*, 10, 792–797.

Rouw, R., Scholte, H.S. (2010). Neural basis of individual differences in synesthetic experiences. *Journal of Neuroscience*, 30, 6205–6213.

Simner, J. (2011). Defining synaesthesia. *British Journal of Psychology* doi: 10.1348/000712610X528305.

Simner, J., Glover, L., Mowat, A. (2006a). Linguistic determinants of word colouring in grapheme-colour synaesthesia. *Cortex*, 42, 281–289.

Simner, J., Harrold, J., Creed, H., Monro, L., Foulkes, L. (2009). Early detection of markers for synaesthesia in childhood populations. *Brain*, 132, 57–64.

Simner, J., Mulvenna, C., Sagiv, N., Tsakanikos, E., Witherby, S.A., Fraser, C., Scott, K., Ward, J. (2006b). Synaesthesia: The prevalence of atypical cross-modal experiences. *Perception*, 35, 1024–1033.

Simner, J., Ward, J., Lanz, M., Jansari, A., Noonan, K., Glover, L., Oakley, D. (2005). Non-random associations of graphemes to colours in synaesthetic and non-synaesthetic populations. *Cognitive Neuropsychology*, 22, 1069–1085.

Smilek, D., Dixon, M.J., Merikle, P.M. (2005). Synaesthesia: Discordant male monozygotic twins. *Neurocase*, 11, 363–370.

Smilek, D., Moffatt, B.A., Pasternak, J., White, B.N., Dixon, M.J., Merikle, P.M. (2002). Synaesthesia: A case study of discordant monozygotic twins. *Neurocase*, 8, 338–342.

Sperling, J.M., Prvulovic, D., Linden, D.E., Singer, W., Stirn, A. (2006). Neuronal correlates of colour-graphemic synaesthesia: a fMRI study. *Cortex*, 42, 295–303.

Van Leeuwen, T.M., Petersson, K.M., Hagoort, P. (2010). Synaesthetic colour in the brain: beyond colour areas. A functional magnetic resonance imaging study of synaesthetes and matched controls. *Public Library of Science One*, 5, e12074.

Ward, J., Huckstep, B., Tsakanikos, E. (2006). Sound-colour synaesthesia: To what extent does it use cross-modal mechanisms common to us all? *Cortex*, 42, 264–280.

Ward, J., Jonas, C., Dienes, Z., Seth, A. (2010). Grapheme-colour synaesthesia improves detection of embedded shapes, but without pre-attentive 'pop-out' of synaesthetic colour. *Proceedings. Biological Sciences*, 277, 1021–1026.

Ward, J., Li, R., Salih, S., Sagiv, N. (2007). Varieties of grapheme-colour synaesthesia: a new theory of phenomenological and behavioural differences. *Consciousness and Cognition*, 16, 913–931.

Ward, J., Mattingley, J.B. (2006). Synaesthesia: an overview of contemporary findings and controversies. *Cortex* 42, 129–136.

Ward, J., Sagiv, N., Butterworth, B. (2009). The impact of visuo-spatial number forms on simple arithmetic. *Cortex*, 45, 1261–1265.

Ward, J., Simner, J. (2005). Is synaesthesia an X-linked dominant trait with lethality in males? *Perception*, 34, 611–623.

Ward, J., Simner, J., Auyeung, V. (2005). A comparison of lexical-gustatory and grapheme-colour synaesthesia. *Cognitive Neuropsychology*, 22, 28–41.

Weiss, P.H., Fink, G.R. (2009). Grapheme-colour synaesthetes show increased grey matter volumes of parietal and fusiform cortex. *Brain*, 132, 65–70.

Witthoft, N., Winawer, J. (2006). Synesthetic colors determined by having colored refrigerator magnets in childhood. *Cortex*, 42, 175–183.

Yaro, C., Ward, J. (2007). Searching for Shereshevskii: What is superior about the memory of synaesthetes? *Quarterly Journal of Experimental Psychology*, 60, 681–695.

Chapter 8
Auditory Verbal Hallucinations, First-Person Accounts

Steven Scholtus and Christine Blanke

8.1 What It Is Like to Hear Voices

Steven Scholtus

All the time that I have been hearing voices, altogether some 25 years now, many people have asked me what it is like. It is not as complex as it seems. I hear voices in two different ways: within my head and outside my head. When I hear voices inside my head, it is as if someone is talking to me. There are male and female voices, loud and clear, easily distinguishable from my own thoughts, and sounding very realistic. That is why it is so easy to believe in telepathy. I can communicate with these voices, and they react to my thoughts. It is difficult to ignore them (Fig. 8.1).

Voices outside my head are easier to handle. They sound as if they are coming from a radio, placed somewhere in a high corner of the room. These voices sound very realistic, too, but the distance appears to be larger, which makes them less direct.

I have been hearing voices since 1986. At first I thought I was having a paranormal experience (many people who hear voices develop an interest in the occult). I was convinced that people were watching me, and that they wanted me dead for some reason. I heard voices all day long and developed a psychosis. I did not seek any mental health care, and the psychosis remitted spontaneously after 10 months. The years afterward, I was on the edge of a new psychotic episode, but I managed to finish my education (history) at an academic level. Then the second psychosis hit me, less severe than the first one, and yet it wreaked more havoc. This time I did seek help and was diagnosed with schizophrenia.

S. Scholtus, M.Sc. (✉) • C. Blanke
Voices Clinic, Psychiatry Division, University Medical Center Utrecht, Utrecht, The Netherlands
e-mail: steven.scholtus@gmail.com; stienblanke@gmail.com

J.D. Blom and I.E.C. Sommer (eds.), *Hallucinations: Research and Practice*,
DOI 10.1007/978-1-4614-0959-5_8, © Springer Science+Business Media, LLC 2012

Fig. 8.1 Steven Scholtus

Over the years I have developed various different explanations for the origins of my voices. In the beginning I thought that they stemmed from paranormal sources such as spirits or 'entities'. Later I learned about the neurobiological processes that govern them.

Learning to cope with voices has been a central theme in my life. This was, and still is, a process with ups and downs. An important aspect was learning to accept the presence of the voices. I assume that I will be hearing them all my life. My magic word for coping with them is *distraction*. I just try to continue whatever it is that I was doing before the voices started. I know that I should not listen to them too much, and that I should not let them upset me: no fighting against the voices, nor becoming attached to them, even at times when their messages are quite joyful.

I am convinced that the hearing of voices is a state of mind. I can 'set' my consciousness in a way that makes me very vulnerable to them. When I do so, I am always able to hear them. It is also easier to hear them when I am tired or a bit drowsy. In the early mornings and near the end of the day, when I am usually more drowsy, it costs much more effort not to give in to the hearing of voices. I have two options to avoid that: I can then either try to focus on some activity – such as playing the drums – or, alternatively, try to take a nap. Any intermediate states of mind, in which I am neither awake and focussed nor fully asleep, are the most dangerous to me, as they render me too vulnerable to the voices.

The voices have completely shaken up my life. A normal way of living, with a full-time job, a family, the raising of children, is no longer within my reach. And yet the voices have not only brought me bad things. I have learned who my real friends are, I have learned to really enjoy social contact, and I have met inspiring people. I would rather have been without the voices, but even in their presence I am able to enjoy my life.

8.2 A Delicate Balance

Christine Blanke

I heard a child crying. There was no one around, no parent or grandparent. I went looking for that child. But there was no child. So I went home, where I heard the child crying again. I looked everywhere, I didn't understand. When I went outside a second time, I heard my name called out in a soft and friendly tone. I saw no one who knew me, and again I heard that crying child. I saw no one who could have called out, which frightened me. I started running and kept yelling, 'Blue! Blue!' The crying child and the lady who had called my name were nowhere to be found (Fig. 8.2).

At the time I did not know that it was possible to hear voices or to see things that aren't there. Therefore, those voices represented real people to me. Or perhaps deceased people, I thought, whose spirits might be dwelling between heaven and earth. Perhaps they wanted to protect me, but I was not sure about that. Sometimes I wish that I could go back to that time, when I only heard a crying child and a friendly lady.

I started to work as a trainee. I made friends and took notice of all the injustice in this world. Perhaps for being so busy with other things, I hardly paid any attention to the gentle voices inside my head. Then other people came into my head, conveying evil messages. I also started to see blood in the streets, and arms, legs, and heads that had apparently been chopped off.

Whenever I hear voices I look behind me, mostly to the right. At times the voices have gotten very close to me, becoming massive. At other times they came into my body, and I felt their massiveness within me. I was in their power, but only when they were so close to me.

They will kill me! I look behind me. There is a gentleman walking behind me. I am positive that he is going to kill me. He is putting his hand into his pocket. I am sure that he will be taking out a gun. But no, it's his handkerchief (Fig. 8.3).

They tell me that I am ill. That I have schizophrenia, and what that is all about. When I look behind me, they call me suspicious, which means that I am ill. I do behave like someone who is ill. However, I never took the things they say too seriously. That may be the reason that I manage to live happily with my own reality.

Fig. 8.2 Christine Blanke

Fig. 8.3 *Droedel 53*. Red ink on paper (Copyright 2007 Christine Blanke)

It may be deviant. If I would find my own reality abnormal it would frighten me, and I would have a problem. Whenever I accept my deviant reality, I am fine.

I still hear them, they haven't changed at all. And yet they are not inside my intimate zone in the upper right corner anymore. Why do I care? They are a part of me. Or maybe the voices do belong to real people after all. In the shopping district you can hear the weirdest things, but not everything applies to me. That is what I have had to learn.

Voices are not thoughts! Thoughts have a vibrant quality to them. Rather monotonous. Thoughts may be compulsory too, and they may be in my way, but they are not voices. My voices sound unfamiliar to me. A male voice, a child that I do not know. I don't know any of these voices, and yet they sound like real people. The difference with real people of flesh and blood is that the voices utter only short phrases, using a lot of repetitions. My thoughts are audible inside my head, whereas the voices are audible in my ear.

Pills do help. I am convinced that they do. But when you are not ill, you don't need to take any pills. I have noticed that pills can expand my world, that they allow me to improve the way I make contact with other people. And yet I have always been sure that the pills I took were only a placebo. If you think hard enough that such tasteless candies can make you feel better, they will.

I create the world with my thoughts. I see and make up all kinds of beautiful things. That makes me happy. If the voices go on saying that I will be killed, then that is part of my world. It will not scare me anymore, or make me sad. I ignore them, or make jokes in order to take away the heaviness. Sometimes I can walk the streets full of giggles, because I was to be murdered again, and yet it didn't happen…again.

They think that I will fall.

But I won't.

Balance can be so delicate.

Chapter 9
Auditory Verbal Hallucinations

Kelly M.J. Diederen and Iris E.C. Sommer

9.1 Introduction

In 1971, in a village in the Philippines, a woman was found unconscious and fevered in a field where she had been plucking fruit. The woman later reported, "As I busied myself picking fresh fruit I felt as though a gush of strong wind passed by. Then all of a sudden I heard human voices crying, pleading, and asking not to be shot. Some were cursing. Then there was silence. A few minutes later the voices came again, agonizing groans of men about to die, writhing in pain. I started to run but I could not move my legs. I tried to shout but I couldn't. Then the world started to turn round; I did not know what happened next." According to the villagers, the woman's experiences were caused by spirits called *bahoy,* which haunt places where violent deaths have happened in the past (Jocano 1971).

While auditory verbal hallucinations (AVH) or "voices" are frequently attributed to possession by a spirit in non-Western societies (see also Chap. 18), in modern Western societies they are generally considered an aspect of disease (Al-Issa 1995). This contrasts with earlier Western accounts, in which powerful men were supposedly being guided by gods speaking to them. The Greek philosopher Socrates (470–399 BC), for instance, was reportedly aided by a voice to make important decisions, and in the nineteenth century Joan of Arc (1412–1431), the "Iron Maiden," was declared a Saint by the Catholic Church because she had heard the voice of God telling her how to liberate France from English domination.

AVH can occur in a wide variety of individuals, including patients with a neurological or neurodegenerative disease, patients with a psychiatric disorder, and

K.M.J. Diederen, M.Sc.. Ph.D. (✉) • I.E.C. Sommer, M.D., Ph.D.
Psychiatry Division, Rudolf Magnus Institute of Neuroscience,
University Medical Center Utrecht, Utrecht, The Netherlands
e-mail: K.M.J.Diederen@umcutrecht.nl; i.sommer@umcutrecht.nl

J.D. Blom and I.E.C. Sommer (eds.), *Hallucinations: Research and Practice*,
DOI 10.1007/978-1-4614-0959-5_9,

healthy individuals in the general population (Aleman and Larøi 2008; Asaad and Shapiro 1986). Moreover, they can be induced by illicit substances such as cannabis, amphetamines, and cocaine, as well as by prescribed drugs and alcohol (Asaad and Shapiro 1986) – although it should be noted here that substance-induced hallucinations tend to occur most frequently in the visual modality. They have also been reported in association with progressive deafness (Slade and Bentall 1988). But irrespective of the context in which they occur, AVH often co-occur with hallucinations in any of the other sensory modalities, as well as with delusions and formal thought disorder (Asaad and Shapiro 1986; Sommer et al. 2010).

Over time, various different approaches have evolved to study AVH. Thus, a number of studies have investigated the prevalence rates of AVH in different subgroups and in the general population. Other studies have focused on distinguishing different subtypes of AVH, based on specific characteristics such as frequency, intensity, and content. In addition, specific characteristics of AVH have been compared among different groups of individuals to elucidate whether they should be considered phenomenologically similar or not. Finally, various theories have been proposed to account for the mediation of AVH in an etiological and pathophysiological sense. Each of these approaches will be discussed below.

9.2 Patients with a Neurological or Neurodegenerative Disorder

AVH tend to occur in the context of a number of neurological disorders, including epilepsy, brainstem pathology, brain tumors, cerebrovascular infarctions, migraine, and delirium (Asaad and Shapiro 1986; Brasic 1998; Brasic and Perry 1997). They can also occur in the context of neurodegenerative diseases such as Lewy body dementia, Parkinson's disease, and Alzheimer's disease (Bassiony and Lyketsos 2003; Inzelberg et al. 1998; McKeith et al. 1992).

Although various studies investigated the prevalence rates of hallucinations in patients with neurodegenerative disorders, only a handful of them focused exclusively on AVH. For instance, Inzelberg et al. (1998) reported that 37% of a group of patients with Parkinson's disease experienced hallucinations, that 29% of their sample experienced only visual hallucinations, and that 8% experienced visual as well as auditory verbal hallucinations. In agreement with this, Fénelon et al. (2000) showed that hallucinations were present in 39.8% of patients diagnosed with Parkinson's disease, while hallucinations in the auditory modality were experienced by 9.7%. Interestingly, cognitive impairment was more common among the hallucinating patients.

In an early study, Wolff and Curran (1935) found auditory hallucinations to occur in 41.5% of patients diagnosed with Alzheimer's disease. Almost 70 years later, Bassiony and Lyketsos (2003) reviewed all prior studies on Alzheimer's disease, and showed that the prevalence rates ranged from 4% to 76% for all types of hallucinations, and from 1% to 29% for auditory hallucinations. In addition, the authors reported that hallucinations tended to persist over time, to recur throughout the disease process, and to be associated with negative consequences such as functional impairment and aggression.

From those studies, it can be concluded that although auditory hallucinations are relatively common in Parkinson's and Alzheimer's disease, visual hallucinations tend to occur more frequently in those patient groups. This is in sharp contrast with patients diagnosed with psychiatric disorders, in whom hallucinations of the auditory modality tend to be the most prevalent ones (Baethge et al. 2005).

9.3 Patients with a Psychiatric Disorder

AVH frequently occur in the context of bipolar disorder, major depressive disorder, borderline or schizotypal personality disorder, posttraumatic stress disorder, and dissociative identity disorder (Ross et al. 1990; Siegel 1984; Skaf et al. 2002). However, their prevalence is highest in patients diagnosed with schizophrenia, as defined by the DSM-IV-TR (American Psychiatric Association 2000).

A number of studies have investigated the prevalence rates of auditory hallucinations in psychiatric patients. For instance, the International Pilot Study of Schizophrenia (WHO 1973) recorded AVH in 74% of patients diagnosed with schizophrenia. In agreement with this, Sartorius et al. (1986) reported them to occur in 70% of their cases. However, Slade and Bentall (1988) reported a somewhat lower prevalence rate (60%).

In patients diagnosed with bipolar disorder, the frequency of hallucinations was established in 22.9% in patients with a mixed episode, 11.2% in those with a manic episode, and 10.5% in those with a depressive episode. Among the bipolar group presenting with hallucinations, 56.9% heard voices (Baethge et al. 2005). Among the patients diagnosed with a unipolar mood disorder, 5.9% reported hallucinations, and 40.6% of the latter group AVH (Baethge et al. 2005). Reviewing studies published between 1922 and 2007, Goodwin and Geddes (2007) showed that auditory hallucinations had been recorded in 18% of all patients diagnosed with bipolar disorder. Finally, Kingdon et al. (2010) found that 50% of the patients diagnosed with borderline personality disorder, 66% of those diagnosed with schizophrenia, and 90% of those with both diagnoses experienced auditory hallucinations. Among those, the auditory hallucinations were reported most frequently.

Some studies found sex differences associated with the prevalence rates for hallucinations in psychiatric patients. For instance, Marneros (1984) reported a significantly higher prevalence of auditory hallucinations among women (25%) than among men (15%) diagnosed with schizophrenia. Interestingly, the prevalence rates of hallucinations in their sample are much lower than generally reported (Sartorius et al. 1986; Slade and Bentall 1988; Wing 1974). This might be due to the fact that the authors only included patients who were hospitalized for the first time. Rector and Seeman (1992) showed that while 54% of the male participants in their study experienced auditory hallucinations, 78% of the female patients experienced them.

Cetingok et al. (1990) reported that hallucinations were more frequent among married Turkish women diagnosed with schizophrenia than among unmarried Turkish women, Turkish men, and Americans of either sex. However, another study of the prevalence rates of AVH in individuals diagnosed with schizophrenia did not report any sex differences (Mueser et al. 1990). As regards mood disorders, various studies showed a higher prevalence rate for hallucinations in women than men (Baethge et al. 2005; Bowman and Raymond 1931).

9.4 The General Population

Although AVH are often associated with pathological conditions, they also occur in healthy individuals in the general population (Sidgwick et al. 1894; Sommer et al. 2008a; Tien 1991; Van Os et al. 2000; see also Chap. 28). More than a century ago, Henry Sidgwick (1838–1900) et al. (1894) were the first to study hallucinations in the general population. In their sample of 17,000 individuals who were primarily of British descent, 9.9% claimed having experienced visual, tactile, or auditory hallucinations. AVH were reported by 3.6% of all respondents. Over a century later, Tien (1991) reported that visual, tactile, or auditory hallucinations occurred in 13% of all healthy individuals in the USA. The frequency of AVH varied with age, ranging from 1.5% to 3.2%. Roughly similar rates for AVH were found in New Zealand (3.4%) by Caspi et al. (2005) and in the Netherlands (2–4%) by Escher et al. (2005). A much higher prevalence rate (16%) was reported in a French study by Verdoux et al. (1998), which is in keeping with the findings of Léon Marillier (1862–1901), who collaborated with Sidgwick et al. in the nineteenth-century Census of Hallucinations, and reported a steeping prevalence rate of 20% for hallucinations (visual, auditory, or tactile) among the French (Sidgwick et al. 1894). The reported differences in prevalence rates are probably at least partially due to differences in study design and demographic characteristics of the cohorts under study (Linscott and Van Os 2010; Beavan et al. 2011).

While the prevalence rates of hallucinations are rather similar across population groups in Western countries, there are striking differences to be found among specific subgroups of those populations. For instance, 14% to 71% of college students in the USA report having experienced AVH at least once in their lives (Barrett and Etheridge 1992; Bentall and Slade 1985; Posey and Losch 1983), which is substantially more often than in the general population (Caspi et al. 2005; Escher et al. 2005; Sidgwick et al. 1894; Tien 1991; Verdoux et al. 1998). In concordance with the prevalence rates recorded in individuals with a psychiatric disorder, Young et al. (1986) found that female students had a significantly greater propensity to hallucinate than men. Likewise, in Tien's sample (Tien 1991), women reported more auditory and olfactory hallucinations, whereas visual hallucinations were reported slightly more frequently by men.

Prevalence rates for hallucinations would also seem to depend on ethnic and cultural differences. Jocano (1971), for instance, reported that 13.3% of the individuals in a village in the Philippines experienced supernatural experiences which to us would qualify as auditory hallucinations. In addition, Johns et al. (2002) found a higher prevalence rate for hallucinatory experiences in subjects of the general Western population as compared to (originally) non-Western individuals living in the UK. Interestingly, subjects belonging to ethnic minorities reported fewer hallucinations when they were born abroad and had migrated later in life, as compared to those who were born in the UK. The exception was a Caribbean subgroup, where hallucinations were reported 2.5-fold more often than by Caucasian respondents.

9.4.1 A Continuum Hypothesis

As AVH can be experienced by psychiatric and neurological patients as well as by healthy individuals in the general population, a number of authors have argued that so-called clinical and nonclinical hallucinations are points on a continuum that do not differ qualitatively from each other (Aleman and Larøi 2008; Linscott and Van Os 2010; Strauss 1969; Van Os et al. 2000, 2009). As a corollary, they assume that AVH in clinical and nonclinical groups result from the same underlying mechanisms, and that the need for treatment depends primarily on the percipients' reaction to them rather than on the presence of AVH themselves. At present, it is insufficiently clear, however, whether AVH in clinical and nonclinical individuals can indeed be considered identical (David 2010; Lawrie et al. 2010; Linscott and Van Os 2010). Comparing the phenomenological characteristics of AVH in different groups may well help to shed light on this matter (David 2010; see also Chap. 23), although it should be borne in mind that this issue is ultimately a matter of conceptualization.

9.5 Phenomenology

A striking aspect of AVH is their variegated nature. While one person may hear a single voice giving friendly advice approximately once per hour, someone else may continuously hear multiple voices gossiping about him. This variable nature of AVH has been recognized for a long time and has led to the conception of numerous subclassifications. Starting from the assumption that phenomenological differences reflect differences in the underlying neurobiological mechanisms, those subclassifications may well be of aid in elucidating the neural underpinnings of AVH (Blom and Sommer 2010). In addition, it may well be that phenomenologically different AVH require different types of treatment. For instance, frequent AVH with a predominantly negative content often call for pharmacotherapeutic interventions, whereas less frequent AVH with a benevolent character may well respond to psychotherapy (see, e.g., Chap. 27 on the Coping-With-Voices Protocol).

9.5.1 *Classifications*

Although early classification studies generally define subgroups of hallucinations based on observation and clinical experience, contemporary studies predominantly use data-driven approaches to identify independent clusters of AVH characteristics. For instance, Haddock et al. (1999) identified three hallucination factors, comprising emotional characteristics (i.e., distress and negative content), physical characteristics (i.e., frequency, loudness, etc.), and a cognitive interpretation factor (i.e., beliefs about the voices, control, etc.). Stephane et al. (2003) reported two clusters of which the first included components such as control strategies, self attribution and repetitive content, and the second systematized content, high linguistic complexity, and a number of other components. Singh et al. (2003) identified two factors, which they called "reality of hallucinatory perception" and "immersion in the hallucination." Finally, Hayashi et al. (2004) identified four independent factors, consisting of "the intractable nature of the experience" (comprising negative voice content, negative patient responses, and uncontrollability of the voices), "delusional reality distortion," "influence," and "externality" (which was composed of perception of external or internal voices and their origins). Variations among the nature and number of factors identified are striking. They may well result from a number of factors, including the ethnicity of the participants, and the use of dissimilar interview scales (Carter et al. 1995; Chadwick and Birchwood 1994; Haddock et al. 1999). For a detailed overview of the classifications rendered by these studies, see Table 9.1.

9.6 Comparisons

9.6.1 *Comparisons 1: Within-Subjects*

Multiple studies have compared the characteristics of AVH among different groups of individuals and across different disease states. For instance, Larkin (1979) studied inpatients diagnosed with schizophrenia during the acute phases of their illness and during phases of remission. While during the acute phases, the hallucinatory content was reported to be predominantly threatening and isolating, and during phases of remission, it tended to be supportive directed at social interaction. Moreover, during phases of remission, patients were better able to exert control over their auditory hallucinations. Nayani and David (1996) showed that, over time, psychotic patients are likely to undergo a gradual shift from experiencing their voices in the extracorporeal world to experiencing them inside the head. Meanwhile, the complexity of those hallucinations tends to increase, in the sense of voices being added, dialogues becoming more extensive, and the relation between the experient and his voices becoming more intimate. Interestingly, the patients' levels of distress and their coping skills both improved over time.

Table 9.1 Overview of factor structures for auditory hallucinations

Study	Country	Studied sample	Interview scales	Analysis	Identified factors		
Haddock et al. (1999)	UK	71 Patients (schizophrenia/ schizoaffective disorder)	Psychotic symptom	Principal component analysis	(1) Emotional characteristics	(2) Physical characteristics	(3) Cognitive interpretation
			Rating scales (auditory hallucination subscale)		Distress	Loudness	Disruption
					Negative content	Frequency	Control
						Location	Beliefs about the voices
Stephane et al. (2003)	USA	30 Patients (schizophrenia/ schizoaffective disorder/ psychotic depression)	Semi-structured interview (unnamed)	Hierarchical cluster analysis	(1) Cluster 1	(2) Cluster 2	
					Control strategies	Systematized content	
					Self attribution	High linguistic complexity	
					Repetitive content	Conversation	
					Other hallucinations	Inner space location	
					Clear acoustics	Multiple voices	
					Low linguistic complexity	Attribution of voices to others	
					Words	Nosognosia	
					Outer space location	Episodic occurrence	
						Spontaneous occurrence	
						Linguistic complexity	
						Sentences	

(continued)

Table 9.1 (continued)

Study	Country	Studied sample	Interview scales	Analysis	Identified factors			
Singh et al. (2003)	India	75 Patients (schizophrenia)	Phenomenology of Hallucinations Scale	Principal component analysis	(1) Reality of hallucinatory perception	(2) Immersion in the hallucination		
					Reality (current)	Frequency		
					Reality (past)	Duration		
					Sensory intensity	Overt behavior		
						Control		
						Time		
						Content-affect		
Hayashi et al. (2004)	Japan	214 Patients (schizophrenia/ schizoaffective disorder)	Matsuzawa Assessment Schedule for Auditory Hallucinations	Principal component analysis	(1) Intractability	(2) Delusion	(3) Influence	(4) Externality
					Unpleasant feelings	Generalization of delusion	Ego disturbance	Voices speaking to patient
					Hostility of voices	Conviction	Audible thoughts	Conversation among voices
					Malevolent content	Unusualness of origin	Influence of voices	Outside location of origin
					Distress	Delusional explanation	Preoccupation	Perception through ears
					Recognition of failure coping	Illness	Imperative content	Voices from present figures
					Controllability of voices	Identification of origins		Outside
						Actualness		

9.6.2 Comparisons 2: Different Patient Groups

Lowe (1973) compared the characteristics of hallucinations in patients diagnosed with manic-depressive disorder, schizophrenia, and organic psychosis (i.e., resulting from a known physical disorder) and proposed to use them as discriminatory indicators for differential diagnosis because numerous differences could be found between them. By contrast, Junginger and Frame (1985) compared the AVH in patients diagnosed with schizophrenia and those with an affective disorder, and reported no significant differences at all. Cutting (1987) showed that while AVH occurred in 34% of patients diagnosed with schizophrenia, they occurred in only 18% of those with an organic psychosis. Moreover, unknown voices in the second or third person occurred in 22% of the group diagnosed with schizophrenia and in no more than 4% in the organic-psychosis group. Mitchell and Vierkant (1991) compared hallucinations among cocaine abusers and patients diagnosed with schizophrenia and showed that while command hallucinations were found in both groups, those in the schizophrenia group more frequently involved the harming or killing of others. Gonzalez et al. (2006) compared auditory hallucinations among groups of psychotic patients with persistent and sporadically occurring hallucinations. Pleasurable hallucinations were more frequent in the persistent hallucinators. Finally, Kingdon et al. (2010) observed that patients diagnosed with borderline personality disorder scored higher on the negative content of the voices and distress associated with them than patients diagnosed with schizophrenia (see also Chap. 10).

9.6.3 Comparisons 3: Psychiatric Patients and Healthy Individuals

The first study comparing AVH in psychotic and nonpsychotic voice hearers (Leudar et al. 1997) focused on pragmatic properties of AVH, such as the familiarity of the voices, the type of action demanded by those voices, and the degree of dialogical engagement of voices and voice hearers. Honig et al. (1998) reported that the form of hallucinatory experiences was not significantly different among patients diagnosed with schizophrenia or dissociative identity disorder on the one hand and healthy voice hearers on the other. However, in contrast to the patient groups, healthy voice hearers perceived their voices as predominantly positive: They were not alarmed or upset by their voices and felt in control of the experience. Daalman et al. (2011) studied the phenomenological characteristics of AVH in a substantial sample of psychotic and nonpsychotic individuals. Differences between the groups included the emotional valence of their content, the frequency of AVH, and the control subjects experienced over their AVH. An additional difference was that the onset of AVH tended to be at a younger age in the healthy subjects, which might well be due to a – as yet unknown – difference in the underlying mechanisms of origin. Other characteristics of the AVH, such as experienced location and loudness, perceived reality, the number of voices, and personification (i.e., attribution to one or more actual persons), were similar in both groups.

9.6.4 *Comparisons 4: Differences in Cultural and Ethnical Background*

Other variables which may influence the characteristics of hallucinations are the individuals' cultural or ethnical background. For instance, Kent and Wahass (1996) found that while AVH tended to have religious and superstitious connotations in patients from Saudi Arabia, verbal instructions and a running commentary were more common in patients from the UK. The frequency and loudness of the AVH did not differ among the groups. Gecici et al. (2010) observed that auditory hallucinations were more frequent in patients diagnosed with schizophrenia from central Turkey than in those from western Turkey. In addition, voices ordering, conversing, commenting, and threatening, as well as voices from prophets and goblins, were more frequent in the group from central Turkey. To disentangle cultural and racial influences, Suhail and Cochrane (2002) compared Pakistani patients living in their home country with a sample of Pakistani immigrants in the UK, as well as to Caucasian patients of British origin. The patients living in Pakistan reported visual hallucinations of spirits or ghosts significantly more often than the two groups living in the UK. Moreover, auditory hallucinations were significantly less frequent in the Pakistan-dwelling group as compared to the other two groups.

9.7 Theoretical Frameworks

Yet another approach involves hypothesis-based studies of the theoretical frameworks proposed to account for them. Although numerous theories have been proposed (Aleman and Larøi 2008), most studies focus on either of four models.

The most influential model proposes that AVH are due to a failure to recognize self-generated inner speech, which entails the false belief that they stem from an external agent (Frith 1991; Frith and Done 1989). In support of this theory, patients diagnosed with schizophrenia were shown to have difficulties in identifying their own actions and thoughts, and to commonly misattribute self-generated behavior to an external source (Waters et al. 2010, see also Chap. 26). Further support comes from observed abnormalities in brain regions implicated in self-processing (Allen et al. 2007b). However, the tendency to misidentify inner speech appears to be related to positive symptoms in general rather than to AVH per se (Allen et al. 2007a).

A second model suggests that AVH may well result from aberrant activation of the primary auditory cortex (Lennox et al. 1999). Support for this hypothesis comes from a number of neuroimaging studies that show activation of the primary auditory cortex during the experience of AVH (Dierks et al. 1999; Jardri et al. 2010).

In addition, the perception of external verbal stimuli was shown to compete for the same neural resources as AVH (Dierks et al. 1999; Hubl et al. 2007; Jardri et al. 2010). In support of this hypothesis, Hunter et al. (2006) found increased spontaneous activity (i.e., in the absence of specific external stimulation) in the auditory cortex of healthy individuals.

A third model proposes that AVH result from the spontaneous recollection of memories, leading to the auditory reexperience of previously encoded information (Copolov et al. 2003; Waters et al. 2006). This hypothesis offers an explanation for the strong association between trauma and AVH (Read et al. 2005). Support for it comes from studies reporting the (de)activation in the hippocampus and parahippocampal gyri (involved in memory processes) during and preceding the onset of AVH (Copolov et al. 2003; Diederen et al. 2010; Hoffman et al. 2008; Silbersweig et al. 1995).

Finally, AVH have been hypothesized to result from the release of language activity in the right hemisphere, which is normally inhibited in the healthy brain (Sommer and Diederen 2009). Support for this theory comes from studies on language lateralization, which consistently show a decreased language lateralization in patients diagnosed with schizophrenia (Li et al. 2009; Sommer et al. 2001) and other psychiatric patients experiencing AVH (Sommer et al. 2007). In addition, neuroimaging studies investigating brain activity during AVH in patients with psychosis showed activation of the right homologues of the language areas (Diederen et al. 2010; Jardri et al. 2010; Sommer et al. 2008b).

Thus, at present, empirical support for all four models is present yet limited. Most likely not one but several, and perhaps all of the proposed mechanisms, play a role in their mediation.

9.8 To Conclude

This chapter provides an overview of the phenomenological characteristics and prevalence rates of AVH, and of the neurobiological mechanisms possibly underlying them. Studies of the prevalence rates of AVH show rather consistent results in healthy individuals as well as in individuals with psychiatric or neurological disorders. In contrast, investigations into the phenomenology of AVH yield highly heterogeneous results. This variability can perhaps be partially explained by the use of dissimilar interview scales and differences in the groups of participants under study. As a corollary, future studies may well benefit from the use of standardized methodologies. In addition, efforts to divide AVH into different types have not proved helpful, if only because the studies at hand have yielded different subclassifications. Investigations into the phenomenology of AVH may well benefit from focusing on specific symptoms rather than symptom clusters. Finally, hypothesis-based studies have shown that dysfunctions in a number of cognitive domains, including language, memory, self-monitoring, and auditory perception, are involved in the mediation of AVH. However, it is as yet unclear how dysfunctions in these domains interact with each other, and how they contribute to the mediation of AVH. Therefore, a major challenge for future studies lies in integrating phenomenological and cognitive factors associated with AVH in order to obtain a more comprehensive understanding of their origin and expression.

References

Aleman, A., Larøi, F. (2008). *Hallucinations. The science of idiosyncratic perception.* Washington, DC: American Psychological Association.

Al-Issa, I. (1995). The illusion of reality or the reality of illusion. Hallucinations and culture. *British Journal of Psychiatry*, 166, 368–373.

Allen, P., Aleman, A., McGuire, P.K. (2007a). Inner speech models of auditory verbal hallucinations: Evidence from behavioural and neuroimaging studies. *International Review of Psychiatry*, 19, 409–417.

Allen, P., Amaro, E., Fu, C.H., Williams, S.C., Brammer, M.J., Johns, L.C., McGuire, P.K. (2007b). Neural correlates of the misattribution of speech in schizophrenia. *British Journal of Psychiatry*, 190, 162–169.

American Psychiatric Association. (2000). *Diagnostic and statistical manual of mental disorders. Fourth edition, text revision.* Washington, DC: American Psychiatric Association.

Asaad, G., Shapiro, B. (1986). Hallucinations: Theoretical and clinical overview. *American Journal of Psychiatry*, 143, 1088–1097.

Baethge, C., Baldessarini, R.J., Freudenthal, K., Streeruwitz, A., Bauer, M., Bschor, T. (2005). Hallucinations in bipolar disorder: Characteristics and comparison to unipolar depression and schizophrenia. *Bipolar Disorders*, 7, 136–145.

Barrett, T.R., Etheridge, J.B. (1992). Verbal hallucinations in normals, I: People who hear 'voices'. *Applied Cognitive Psychology*, 6, 379–387.

Bassiony, M.M., Lyketsos, C.G. (2003). Delusions and hallucinations in Alzheimer's disease: Review of the brain decade. *Psychosomatics*, 44, 388–401.

Beavan, V., Read, K., Cartwright, C. (2011). The prevalence of voice-hearers in the general population: A literature review. *Journal of Mental Health*, 20, 281–229.

Bentall, R.P., Slade, P.D. (1985). Reality testing and auditory hallucinations: A signal detection analysis. *British Journal of Clinical Psychology*, 24, 159–169.

Blom, J.D., Sommer, I.E. (2010). Auditory hallucinations: Nomenclature and classification. *Cognitive and Behavioral Neurology*, 23, 55–62.

Bowman, K.M., Raymond, A.F. (1931). A statistical study of hallucinations in the manic-depressive psychoses. *American Journal of Psychiatry*, 88, 299–309.

Brasic, J.R. (1998). Hallucinations. *Perceptual and Motor Skills*, 86, 851–877.

Brasic, J.R., Perry, R. (1997). Unilateral auditory hallucinations in a boy with ipsilateral conductive hearing loss. *Journal of Neurology, Neurosurgery, and Psychiatry*, 62, 302.

Carter, D.M., Mackinnon, A., Howard, S., Zeegers, T., Copolov, D.L. (1995). The development and reliability of the Mental Health Research Institute Unusual Perceptions Schedule (MUPS): An instrument to record auditory hallucinatory experience. *Schizophrenia Research*, 16, 157–165.

Caspi, A., Moffitt, T.E., Cannon, M., McClay, J., Murray, R., Harrington, H., Taylor, A., Arseneault, L., Williams, B., Braithwaite, A., Poulton, R., Craig, I.W. (2005). Moderation of the effect of adolescent-onset cannabis use on adult psychosis by a functional polymorphism in the catechol-O-methyltransferase gene: Longitudinal evidence of a gene X environment interaction. *Biological Psychiatry*, 57, 1117–1127.

Cetingok, M., Chu, C.C., Park, D.B. (1990). The effect of culture on the sex differences in schizophrenia. *International Journal of Social Psychiatry*, 36, 272–279.

Chadwick, P., Birchwood, M. (1994). The omnipotence of voices. A cognitive approach to auditory hallucinations. *British Journal of Psychiatry*, 164, 190–201.

Copolov, D.L., Seal, M.L., Maruff, P., Ulusoy, R., Wong, M.T., Tochon-Danguy, H.J., Egan, G.F. (2003). Cortical activation associated with the experience of auditory hallucinations and perception of human speech in schizophrenia: A PET correlation study. *Psychiatry Research*, 122, 139–152.

Cutting, J. (1987). The phenomenology of acute organic psychosis. Comparison with acute schizophrenia. *British Journal of Psychiatry*, 151, 324–332.

Daalman, K., Boks, M.P., Diederen, K.M., de Weijer, A.D., Blom, J.D., Kahn, R.S., Sommer, I.E. (2011). The same or different? A phenomenological comparison of auditory verbal hallucinations in healthy and psychotic individuals. *Journal of Clinical Psychiatry*, 72, 320–325.

David, A.S. (2010). Why we need more debate on whether psychotic symptoms lie on a continuum with normality. *Psychological Medicine*, 40, 1935–1942.

Diederen, K.M.J., Neggers, S.F.W., Daalman, K., Blom, J.D., Goekoop, R., Kahn, R.S., Sommer, I.E.C. (2010). Deactivation of the parahippocampal gyrus preceding auditory hallucinations in schizophrenia. *American Journal of Psychiatry*, 167, 427–435.

Dierks, T., Linden, D.E.J., Jandl, M., Formisano, E., Goebel, R., Lanfermann, H., Singer, W. (1999). Activation of Heschl's gyrus during auditory hallucinations. *Neuron*, 22, 615–621.

Escher, S., Hayward, M., Kingdon, D., Pembroke, L., Thomas, P. (2005). Learning to live with voices. *Mental Health Today*, 18–20.

Fénelon, G., Mahieux, F., Huon, R., Ziegler, M. (2000). Hallucinations in Parkinson's disease: prevalence, phenomenology and risk factors. *Brain*, 123, 733–745.

Frith, C. (1991). Positron emission tomography studies of frontal lobe function: Relevance to psychiatric disease. *Ciba Foundation Symposium*, 163, 181–191.

Frith, C., Done, J. (1989). Positive symptoms of schizophrenia. *British Journal of Psychiatry*, 154, 569–570.

Gecici, O., Kuloglu, M., Guler, O., Ozbulut, O., Kurt, E., Onen, S., Ekinci, O., Yesilbas, D., Caykoylu, A., Emül, M., Alatas, G., Albayrak, Y. (2010). Phenomenology of delusions and hallucinations in patients with schizophrenia. *Bulletin of Clinical Psychopharmacology*, 20, 204–212.

Gonzalez, J.C., Aguilar, E.J., Berenguer, V., Leal, C., Sanjuan, J. (2006). Persistent auditory hallucinations. *Psychopathology*, 39, 120–125.

Goodwin, G.M., Geddes, J.R. (2007). What is the heartland of psychiatry? *British Journal of Psychiatry*, 191, 189–191.

Haddock, G., McCarron, J., Tarrier, N., Faragher, E.B. (1999). Scales to measure dimensions of hallucinations and delusions: The Psychotic Symptom Rating Scales (PSYRATS). *Psychological Medicine*, 29, 879–889.

Hayashi, N., Igarashi, Y., Suda, K., Nakagawa, S. (2004). Phenomenological features of auditory hallucinations and their symptomatological relevance. *Psychiatry and Clinical Neuroscience*, 58, 651–659.

Hoffman, R.E., Anderson, A.W., Varanko, M., Gore, J.C., Hampson, M. (2008). Time course of regional brain activation associated with onset of auditory/verbal hallucinations. *British Journal of Psychiatry*, 193, 424–425.

Honig, A., Romme, M.A., Ensink, B.J., Escher, S.D., Pennings, M.H., De Vries, M.W. (1998). Auditory hallucinations: A comparison between patients and nonpatients. *Journal of Nervous and Mental Disease*, 186, 646–651.

Hubl, D., Koenig, T., Strik, W.K., Garcia, L.M., Dierks, T. (2007). Competition for neuronal resources: How hallucinations make themselves heard. *British Journal of Psychiatry*, 190, 57–62.

Hunter, M.D., Eickhoff, S.B., Miller, T.W., Farrow, T.F., Wilkinson, I.D., Woodruff, P.W. (2006). Neural activity in speech-sensitive auditory cortex during silence. *Proceedings of the National Academy of Sciences U S A*, 103, 189–194.

Inzelberg, R., Kipervasser, S., Korczyn, A.D. (1998). Auditory hallucinations in Parkinson's disease. *Journal of Neurology, Neurosurgery, and Psychiatry*, 64, 533–535.

Jardri, R., Pouchet, A., Pins, D., Thomas, P. (2010). Cortical activations during auditory verbal hallucinations in schizophrenia: A coordinate-based meta-analysis. *American Journal of Psychiatry*, 168, 73–81.

Jocano, F.L. (1971). Varieties of supernatural experiences among Filipino peasants: Hallucination or idiom of cultural cognition? *Transcultural Psychiatry*, 8, 43–45.

Johns, L.C., Nazroo, J.Y., Bebbington, P., Kuipers, E. (2002). Occurrence of hallucinatory experiences in a community sample and ethnic variations. *British Journal of Psychiatry*, 180, 174–178.

Junginger, J., Frame, C.L. (1985). Self-report of the frequency and phenomenology of verbal hallucinations. *Journal of Nervous and Mental Disease*, 173, 149–155.

Kent, G., Wahass, S. (1996). The content and characteristics of auditory hallucinations in Saudi Arabia and the UK: A cross-cultural comparison. *Acta Psychiatrica Scandinavica,* 94, 433–437.

Kingdon, D.G., Ashcroft, K., Bhandari, B., Gleeson, S., Warikoo, N., Symons, M., Taylor, L., Lucas, E., Mahendra, R., Ghosh, S., Mason, A., Badrakalimuthu, R., Hepworth, C., Read, J., Mehta, R. (2010). Schizophrenia and borderline personality disorder: Similarities and differences in the experience of auditory hallucinations, paranoia, and childhood trauma. *Journal of Nervous and Mental Disease*, 198, 399–403.

Larkin, A.R. (1979). The form and content of schizophrenic hallucinations. *American Journal of Psychiatry*, 136, 940–943.

Lawrie, S.M., Hall, J., McIntosh, A.M., Owens, D.G., Johnstone, E.C. (2010). The 'continuum of psychosis': Scientifically unproven and clinically impractical. *British Journal of Psychiatry*, 197, 423–425.

Lennox, B.R., Park, S.B., Jones, P.B., Morris, P.G. (1999). Spatial and temporal mapping of neural activity associated with auditory hallucinations. *Lancet*, 353, 644.

Leudar, I., Thomas, P., McNally, D., Glinski, A. (1997). What voices can do with words: Pragmatics of verbal hallucinations. *Psychological Medicine*, 27, 885–898.

Li, X., Branch, C.A., DeLisi, L.E. (2009). Language pathway abnormalities in schizophrenia: A review of fMRI and other imaging studies. *Current Opinion in Psychiatry*, 22, 131–139.

Linscott, R.J., Van Os, J. (2010). Systematic reviews of categorical versus continuum models in psychosis: Evidence for discontinuous subpopulations underlying a psychometric continuum. Implications for DSM-V, DSM-VI, and DSM-VII. *Annual Review of Clinical Psychology*, 6, 391–419.

Lowe, G.R. (1973). The phenomenology of hallucinations as an aid to differential diagnosis. *British Journal of Psychiatry*, 123, 621–633.

Marneros, A. (1984). Frequency of occurrence of Schneider's first rank symptoms in schizophrenia. *European Archives of Psychiatry and Neurological Sciences*, 234, 78–82.

McKeith, I.G., Perry, R.H., Fairbairn, A.F., Jabeen, S., Perry, E.K. (1992). Operational criteria for senile dementia of Lewy body type (SDLT). *Psychological Medicine*, 22, 911–922.

Mitchell, J., Vierkant, A.D. (1991). Delusions and hallucinations of cocaine abusers and paranoid schizophrenics: A comparative study. *Journal of Psychology*, 125, 301–310.

Mueser, K.T., Bellack, A.S., Brady, E.U. (1990). Hallucinations in schizophrenia. *Acta Psychiatrica Scandinavica*, 82, 26–29.

Nayani, T.H., David, A.S. (1996). The auditory hallucination: A phenomenological survey. *Psychological Medicine*, 26, 177–189.

Posey, T.B., Losch, M.E. (1983). Auditory hallucinations of hearing voices in 375 normal subjects. *Imagination, Cognition and Personality*, 2, 99–113.

Read, J., Van Os, J., Morrison, A.P., Ross, C.A. (2005). Childhood trauma, psychosis and schizophrenia: a literature review with theoretical and clinical implications. *Acta Psychiatrica Scandinavica*, 112, 330–350.

Rector, N.A., Seeman, M.V. (1992). Auditory hallucinations in women and men. *Schizophrenia Research*, 7, 233–236.

Ross, C.A., Miller, S.D., Reagor, P., Bjornson, L., Fraser, G.A., Anderson, G. (1990). Schneiderian symptoms in multiple personality disorder and schizophrenia. *Comprehensive Psychiatry*, 31, 111–118.

Sartorius, N., Jablensky, A., Korten, A., Ernberg, G., Anker, M., Cooper, J.E., Day, R. (1986). Early manifestations and first-contact incidence of schizophrenia in different cultures. A preliminary report on the initial evaluation phase of the WHO Collaborative Study on determinants of outcome of severe mental disorders. *Psychological Medicine*, 16, 909–928.

Siegel, R.K. (1984). Hostage hallucinations. Visual imagery induced by isolation and life-threatening stress. *Journal of Nervous and Mental Disease*, 172, 264–272.

Sidgwick, H., Johnson, A., Myers, F.W.H., Podmore, F., Sidgwick, E. (1894). *Report on the Census of Hallucinations*. In *Proceedings of the Society for Psychical Research Vol. XXVI*. London: Kegan Paul, Trench, Trübner & Co, pp. 25–422.

Silbersweig, D.A., Stern, E., Frith, C., Cahill, C., Holmes, A., Grootoonk, S., Seaward, J., McKenna, P., Chua, S.E., Schnorr, L., Jones, T., Frackowiak, R.S.J. (1995). A functional neuroanatomy of hallucinations in schizophrenia. *Nature*, 378, 176–179.

Singh, G., Sharan, P., Kulhara, P. (2003). Phenomenology of hallucinations: A factor analytic approach. *Psychiatry and Clinical Neuroscience*, 57, 333–336.

Skaf, C.R., Yamada, A., Garrido, G.E., Buchpiguel, C.A., Akamine, S., Castro, C.C., Busatto, G.F. (2002). Psychotic symptoms in major depressive disorder are associated with reduced regional cerebral blood flow in the subgenual anterior cingulate cortex: A voxel-based single photon emission computed tomography (SPECT) study. *Journal of Affective Disorders*, 68, 295–305.

Slade, P.D., Bentall, R.P. (1988). *Sensory deception: A scientific analysis of hallucination*. Baltimore, MA: Johns Hopkins University Press.

Sommer, I.E., Daalman, K., Rietkerk, T., Diederen, K.M., Bakker, S., Wijkstra, J., Boks, M.P. (2008a). Healthy individuals with auditory verbal hallucinations; who are they? Psychiatric assessments of a selected sample of 103 subjects. *Schizophrenia Bulletin*, 36, 633–641.

Sommer, I.E., Derwort, A.M., Daalman, K., De Weijer, A.D., Liddle, P.F., Boks, M.P. (2010). Formal thought disorder in non-clinical individuals with auditory verbal hallucinations. *Schizophrenia Research*, 118, 140–145.

Sommer, I.E., Diederen, K.M. (2009). Language production in the non-dominant hemisphere as a potential source of auditory verbal hallucinations. *Brain*, 32, e124.

Sommer, I.E., Diederen, K.M., Blom, J.D., Willems, A., Kushan, L., Slotema, K. Boks, M.P.M., Daalman, K., Hoek, H.W., Neggers, S.F.W., Kahn, R.S. (2008b). Auditory verbal hallucinations predominantly activate the *right* inferior frontal area. *Brain*, 131, 3169–3177.

Sommer, I.E., Ramsey, N.F., Kahn, R.S. (2001). Language lateralization in schizophrenia, an fMRI study. *Schizophrenia Research*, 52, 57–67.

Sommer, I.E., Van de Veer, A.J., Wijkstra, J., Boks, M.P., Kahn, R.S. (2007). Comparing language lateralization in psychotic mania and psychotic depression to schizophrenia; A functional MRI study. *Schizophrenia Research*, 89, 364–365.

Stephane, M., Thuras, P., Nasrallah, H., Georgopoulos, A.P. (2003). The internal structure of the phenomenology of auditory verbal hallucinations. *Schizophrenia Research*, 61, 185–193.

Strauss, J.S. (1969). Hallucinations and delusions as points on continua function. Rating scale evidence. *Archives of General Psychiatry*, 21, 581–586.

Suhail, K., Cochrane, R. (2002). Effect of culture and environment on the phenomenology of delusions and hallucinations. *International Journal of Social Psychiatry*, 48, 126–138.

Tien, A.Y. (1991). Distributions of hallucinations in the population. *Social Psychiatry and Psychiatric Epidemiology*, 26, 287–292.

Van Os, J., Hanssen, M., Bijl, R.V., Ravelli, A. (2000). Strauss (1969) revisited: A psychosis continuum in the general population? *Schizophrenia Research*, 45, 11–20.

Van Os, J., Linscott, R.J., Myin-Germeys, I., Delespaul, P., Krabbendam, L. (2009). A systematic review and meta-analysis of the psychosis continuum: Evidence for a psychosis proneness-persistence-impairment model of psychotic disorder. *Psychological Medicine*, 39, 179–195.

Verdoux, H., Maurice-Tison, S., Gay, B., Van Os, J., Salamon, R., Bourgeois, M.L. (1998). A survey of delusional ideation in primary-care patients. *Psychological Medicine*, 28, 127–134.

Waters, F., Woodward, T., Allen, P., Aleman, A., Sommer, I. (2010). Self-recognition deficits in schizophrenia patients with auditory hallucinations: A meta-analysis of the literature. *Schizophrenia Bulletin*, doi: 10.1093/schbul/sbq144.

Waters, F.A., Badcock, J.C., Michie, P.T., Maybery, M.T. (2006). Auditory hallucinations in schizophrenia: intrusive thoughts and forgotten memories. *Cognitive Neuropsychiatry*, 11, 65–83.

Wing J.K. (1974). Psychiatry in the Soviet Union. *Br Med J,* 1(5905), 433–436.

Wolff, H.G., Curran, D. (1935). Nature of delirium and allied states: The dysergastic reaction. *Archives of Neurology and Psychiatry*, 33, 1175–1215.

World Health Organization (1973). *The international pilot study of schizophrenia. Volume 1. Results of the initial evaluation phase*. Geneva: World Health Organization.

Young, H.F., Bentall, R.P., Slade, P.D., Dewey, M.E. (1986). Disposition towards hallucination, gender and EPQ scores: A brief report. *Personality and Individual Differences*, 7, 247–249.

Chapter 10
Auditory Verbal Hallucinations in Patients with Borderline Personality Disorder

Christina W. Slotema and David G. Kingdon

10.1 Introduction

In 1938, the expression 'borderline' was coined to designate the overlap area of neurotic and psychotic symptoms (Stern 1938). The term is somewhat confusing and unsatisfactory – if only because patients diagnosed with schizophrenia or affective disorder can also experience symptoms belonging to this nosological twilight zone – but for the sake of clarity we will stick to the established term 'borderline personality disorder' (BPD) in this chapter. In contrast to the name, the concomitant concept does have significant clinical utility, being operationalized in terms of a combination of affective dysregulation, impulsive-behavorial dyscontrol, cognitive-perceptual symptoms (including suspiciousness, referential thinking, paranoid ideation, illusions, derealization, depersonalization, and 'hallucination-like symptoms'), as well as persistent personal invalidation (Skodol et al. 2002).

Since the 1940s, transient psychotic episodes are recognized as possible manifestations of BPD (Hoch and Polatin 1949), and during the 1980s, they were added to the diagnostic criteria for BPD in the Diagnostic and Statistical Manual of Mental Disorders (DSM-III-R, APA 1987). Nevertheless, the current DSM criteria for BPD (DSM-IV-TR, APA 2000) fail to include auditory verbal hallucinations (AVH) and other types of hallucination, even though two studies (with admittedly small sample sizes) demonstrated their occurrence in 21% and 54% of the cases, respectively (George and Soloff 1986; Chopra and Beatson 1986). Moreover, clinical experience indicates that in these patients, the occurrence of AVH can result in suicidal or self-injurious behaviour due to their imperative nature or the burden they cause.

C.W. Slotema, M.D., Ph.D. (✉)
Parnassia Bavo Group, The Hague, The Netherlands
e-mail: c.slotema@psyq.nl

D.G. Kingdon, M.D., Ph.D., F.R.C.Psych.
Department of Psychiatry, University of Southampton, Southampton, UK
e-mail: D.Kingdon@soton.ac.uk

J.D. Blom and I.E.C. Sommer (eds.), *Hallucinations: Research and Practice*,
DOI 10.1007/978-1-4614-0959-5_10, © Springer Science+Business Media, LLC 2012

Only few studies have systematically assessed the phenomenology and severity of AVH and other psychotic features occurring in the context of BPD. The reason for this is unclear, although it should be noted that the execution of such studies is complicated by the fact that a substantial number of patients diagnosed with BPD have a comorbid axis-I disorder associated with an increased risk for psychosis (i.e. mood disorder in 97% of the cases, and substance abuse in 62%; see Zanarini et al. 2004). Moreover, the interpretation of the results of earlier studies tends to be hampered by the use of such terms as 'psychotic' and 'psychotic-like' to designate cognitive rather than perceptual symptoms in patients diagnosed with BPD (Zanarini et al. 1990).

As the diagnostic criteria of BPD fail to account for the occurrence of longer-lasting hallucinations, clinicians and researchers often find themselves struggling for words when confronted with AVH experienced by patients thus diagnosed. Because of the contradiction in terms of a borderline patient experiencing AVH, those hallucinations tend to be explained in four different ways. First, they are conceptualized as 'pseudohallucinations'. The term as well as the concept were originally introduced by Friedrich Hagen (1814–1888) (Hagen 1868), and elaborated further by Victor Kandinsky (1849–1889) (Kandinsky 1885) and many others, which eventually resulted in an impressive number of terms and connotations (for an overview see Blom 2010) that all differ somewhat from Hagen's original version. As a result, true consensus is lacking, although there is some agreement that pseudohallucinations might be perceptions that are experienced inside the head, with preserved insight into their nature (Van der Zwaard and Polak 1999). This is consistent with a number of publications in which psychotic features occurring in the context of BPD are described as transient, affecting no more than one or two areas of life (such as work and family), atypical (possibly reality-based or totally fantastic in content) or not genuinely psychotic (Soloff 1979; Zanarini et al. 1990; Skodol et al. 2002), and in which they are described in terms of 'quasipsychotic thought' or 'hallucination-like symptoms'. Secondly, AVH experienced by patients diagnosed with BPD may be considered quite similar to those occurring in individuals without a psychiatric or neurological diagnosis. Daalman et al. (2011) compared the characteristics and ensuing distress of AVH experienced by individuals without a diagnosis with those in patients with a psychotic disorder, and found differences in the frequency of AVH and the emotional valence of their content, with higher scores for the patients diagnosed with a psychotic disorder. In addition, control over the voices was found to be higher in individuals without a diagnosis. In the third place, AVH occurring in the context of BPD are considered to lie on a continuum with those experienced by individuals without a diagnosis or diagnosed with schizophrenia, with the BPD group holding some sort of middle ground. And fourth, AVH are conceptualized as occurring across different psychiatric disorders, including BPD. So far, it is unknown whether the voices experienced by patients diagnosed with BPD comply with any of those explanations. We are aware of only two studies that investigated the phenomenology and ensuing distress of AVH in the context of BPD (Kingdon et al. 2010; Slotema et al. submitted). In this chapter, we present the results of those two studies, and discuss the consequences of their findings for the diagnosis and treatment of BPD patients.

10.2 Methods and Results

10.2.1 Methods

Details of the two studies are described in Kingdon et al. (2010) and Slotema et al. (submitted). Participants were included when they experienced AVH, and were diagnosed with BPD or schizophrenia. The Kingdon study added a third diagnostic group, consisting of patients diagnosed with BPD *and* schizophrenia, whereas the Slotema study also included patients with a schizoaffective disorder in the schizophrenia group. The Slotema study was matched for age and gender, and included only female patients. In both studies, the phenomenological characteristics of AVH and the ensuing distress were quantified on a five-point scale with the aid of the AVH-related items of the Psychotic Symptom Rating Scales (PSYRATS, Haddock et al. 1999). In the Slotema study, four additional questions were asked:

1. Do the voices comment on the thoughts and behaviour of the patient?
2. Does the patient experience thoughts being heard out loud?
3. Do the voices converse with other hallucinated voices?
4. Do the voices have an imperative character?

The Kingdon study used three additional questionnaires, i.e. the Beck Depression Inventory II (BDI, Beck et al. 1996), the Beck Anxiety Inventory (BAI, Beck et al. 1988), and the Childhood Trauma Questionnaire (CTQ, Bernstein 1998).

10.2.2 Results

Forty-eight patients diagnosed with BPD, 86 with schizophrenia or schizoaffective disorder, and 17 with BPD *and* schizophrenia were included in the two studies. Their demographic data are presented in Table 10.1.

The patients diagnosed with BPD experienced AVH for a mean duration of 17 years, the majority of them in a frequency of at least once per hour, and with a duration of various minutes or more per episode. More than half of them believed that the AVH originated from an internal source, and experienced them accordingly inside the head. The content of the voices was negative in the majority of the cases, the ensuing distress and disruption of life were high, and control over the voices was low. The mean scores on the items *frequency*, *duration*, *negative content*, and *controllability* are presented in Fig. 10.1.

No significant differences among the three groups of patients were found regarding the mean length of time during which AVH had been experienced, the frequency, duration, loudness, and perceived location of AVH (i.e. whether they were experienced inside or outside the head) or their perceived source (i.e. whether they were believed to originate from an internal or an external source). No significant group differences were found for their controllability (which, incidentally, tended to be low).

Table 10.1 Demographic information

	BPD	Schizophrenia	Both	*p* value
Kingdon et al. (2010)				
AVH present, *n* (%)	15 (46)	35 (59)	17 (90)	0.002
Age, mean (Sd)	32.9 (10.0)	40.2 (10.7)	32.5 (10.4)	0.002
Gender, female, *n* (%)	30 (91)	19 (32)	14 (74)	<0.001
n, % outpatients	21 (67)	16 (27)	12 (63)	<0.001
Slotema et al. (submitted)				
Schizoaffective disorder, *n* (%)	0	15 (29)		
AVH present, *n* (%)	33 (100)	51 (100)	–	ns
Age, mean (Sd)	33 (10.4)	37 (9.8)	–	ns
Gender, female, *n* (%)	33 (100)	51 (100)	–	ns
n, % outpatients	33 (100)	25 (93)	–	ns

BPD borderline personality disorder, *n* number of patients, *ns* not significant, *sd* standard deviation

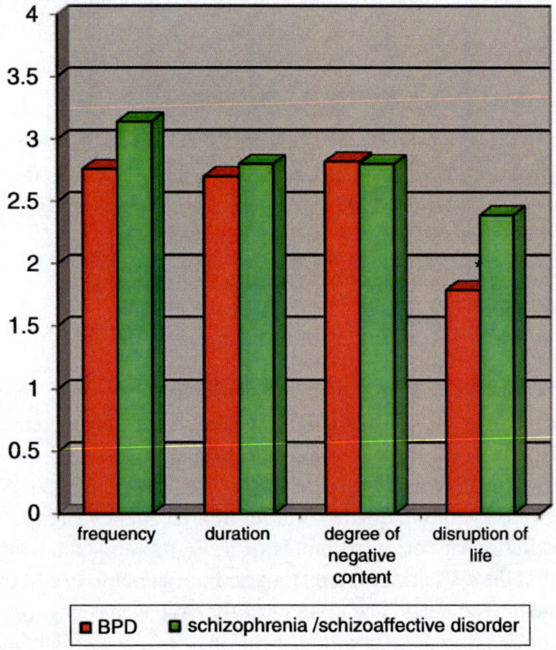

Fig. 10.1 Mean scores on the auditory-verbal-hallucinations-related items of the Psychotic Symptom Rating Scales (Adapted from Slotema et al. submitted)

The only differences were higher scores on the items *negative content* and *ensuing distress* for the BPD group in the Kingdon study, and a higher score on the item *disruption of life* for the schizophrenia group in the Slotema study. No differences were found in the proportion of patients experiencing AVH-related Schneiderian criteria (i.e. voices commenting on the thoughts and behaviour of the patient, thoughts being heard out loud, and voices conversing with other hallucinated voices) or in the proportion of patients experiencing voices with an imperative character.

In addition, the BPD group showed higher scores for depression and anxiety. Severe emotional abuse and severe sexual abuse were significantly more frequent in patients diagnosed with BPD. Logistic regression using the CTQ category found emotional abuse to be the key predictor of allocation-to-diagnostic-group, with sexual abuse only adding marginally to the variance.

10.3 Discussion

In this chapter, we present the first two studies in which the phenomenology of auditory verbal hallucinations (AVH), as well as the distress they may yield, were systematically investigated in patients diagnosed with borderline personality disorder (BPD). The results were compared with those in patients diagnosed with schizophrenia and schizoaffective disorder. In summary, the two studies failed to find any significant differences in the frequency, duration, perceived location, and loudness of AVH, or in the beliefs about their source. Patients diagnosed with BPD had a long history of experiencing AVH. The ensuing burden was high, and did not differ from that in the other diagnostic groups, except for the items 'distress' and 'negative content' (which were higher in the BPD group as found by Kingdon et al. (2010)), and a higher score for 'disruption of life' in the schizophrenia group as found by Slotema et al. (submitted). Severe emotional abuse and severe sexual abuse were reported significantly more often by patients diagnosed with BPD. In conclusion, the phenomenological characteristics of AVH and the ensuing distress in patients diagnosed with BPD do not differ from those in patients diagnosed with schizophrenia, as indicated foremost by the high frequency and long duration of AVH in both groups.

An interesting question might be whether these results can be explained by assuming that the patients diagnosed with BPD should in fact also have been diagnosed with schizophrenia. However, this possibility was ruled out by the strict inclusion criteria and the use of standardized interviews by trained psychiatrists. Evidence of the contrary was also found in the long histories of experiencing AVH in the BPD group. After so many years, it would be highly unlikely if those individuals would go on to develop schizophrenia after all.

Our results imply that the AVH experienced by patients diagnosed with BPD and by those with schizophrenia show many similarities, and only few differences. Therefore, we argue that terms such as 'pseudohallucination' and 'hallucination-like symptom' fail to do justice to the hearing of voices in the context of BPD. Using those terms – or terms of similar meaning – brings the risk of repudiating the AVH experienced by those patients, and of undertreating them, whereas our own two studies indicate that they fulfil all the criteria of hallucinations proper, and therefore deserve to be named and treated as such.

Although a direct comparison has never been performed, AVH occurring in the context of BPD appear to differ from those in individuals without a diagnosis, in the sense that the latter have lower scores on the items 'frequency' and 'duration', as well as on the scores indicating 'burden', as compared to patients diagnosed with a psychotic disorder (Daalman et al. 2011).

Another important question is whether our current diagnostic criteria for BPD and schizophrenia constitute an accurate reflection of clinical reality (Blom and Van Praag in press). Perhaps BPD and schizophrenia should be conceptualized as continuous disorders with a broad variety of clinical presentations, with 'pure schizophrenia' lying on one end of the spectrum, and 'pure BPD' on the other. The group diagnosed with schizophrenia as well as BPD is particularly challenging in clinical and scientific terms, but is usually excluded from treatment trials (Turkington et al. 2006). However, while our studies indicate that AVH occurring in the context of BPD or in the context of schizophrenia are quite similar, we would rather conceptualize AVH as nonspecific for any particular kind of psychiatric disorder, thus endorsing previous studies that have consistently demonstrated that hallucinations are found across different psychiatric disorders without having any diagnostically predictive value (Pierre 2010; Mott et al. 1965; Goodwin et al. 1971; Nayani and David 1996; Copolov et al. 2004).

As our results indicate that patients diagnosed with BPD can suffer from genuine AVH, it is important to consider all treatment strategies for this indication. Unfortunately, studies reporting on the treatment of AVH in patients diagnosed with BPD are rare. Those reporting on the effects of antipsychotic agents, and especially those using questionnaires to quantify those effects, are virtually nonexistent. Cognitive-behavioural therapy for psychotic symptoms (Kingdon and Turkington 2005, see also Chaps. 27 and 28) is used at a modest scale for patients diagnosed with BPD, but needs to be assessed alongside other psychological interventions such as dialectical behaviour therapy.

Our advice is that more attention be given to the occurrence of genuine AVH in borderline patients, as well as to the burden they may cause. In every patient diagnosed with BPD, therefore, caregivers should explore whether any AVH are experienced. Secondly, we argue that more research is needed to assess the prevalence of AVH and other psychotic symptoms in larger patient samples. Further research should preferentially focus on the phenomenological characteristics of other psychotic phenomena occurring in the context of BPD, and on the distress they may cause. In the third place, it would be interesting to learn whether the symptoms occurring in the contexts of BPD and psychotic disorders, which show so many similarities at a phenomenological level, also share a similar neurophysiology. But most importantly, the use of antipsychotic agents and psychotherapy should be explored in randomized controlled trials with questionnaires assessing their effects on the severity of psychotic features.

10.4 Conclusion

The phenomenological characteristics of auditory verbal hallucinations (AVH) do not differ significantly among patients diagnosed with borderline personality disorder (BPD) and those with schizophrenia. Both groups tend to have long histories of experiencing AVH, and to experience high levels of distress due to the voices they hear.

Therefore, in the treatment of BPD, more attention should be paid to AVH. This implies that those AVH should not be given a different name, as this may lead to misdiagnosis and consequently to insufficient medical care. Antipsychotic agents and cognitive behavioural therapy may well be proper treatment options, although this has not yet been properly investigated. More studies are needed to investigate other psychotic features that may occur in the context of BPD, and to assess the neurophysiological basis and therapeutic options for AVH.

References

American Psychiatric Association (1987). *Diagnostic and statistical manual of mental disorders. Third edition – revised.* Washington, DC: American Psychiatric Association.

American Psychiatric Association (2000). *Diagnostic and statistical manual of mental disorders. Fourth edition, text revision.* Washington, DC: American Psychiatric Association.

Beck, A.T., Steer, R.A., Brown, G.K. (1996). *Beck Depression Inventory. Second edition.* San Antonio, CA: The Psychological Corporation.

Beck, A.T., Epstein, N., Steer, R.A., Brown, G.K. (1988). An inventory for measuring clinical anxiety: psychometric properties. *Journal of Consulting and Clinical Psychology*, 56, 893–897.

Bernstein, D.F.L. (1998). *The childhood trauma questionnaire: A retrospective self-report.* San Antonio, CA: The Psychological Corporation.

Blom, J.D. (2010). *A dictionary of hallucinations.* New York, NY: Springer.

Blom, J.D., Van Praag, H.M. Schizophrenia: it's broken and it can't be fixed. A conceptual analysis at the centenary of Bleuler's *Dementia praecox oder Gruppe der Schizophrenien. Israeli Journal of Psychiatry* (in press).

Chopra, H.D., Beatson, J.A. (1986). Psychotic symptoms in borderline personality disorder. *American Journal of Psychiatry*, 143, 1605–1607.

Copolov, D. Trauer, T., Mackinnon, A. (2004). On the non-significance of internal versus external auditory hallucinations. *Schizophrenia Research*, 69, 1–6.

Daalman, K., Boks, M.P., Diederen, K.M., De Weijer, A.D., Blom, J.D., Kahn, R.S., Sommer, I.E. (2011). The same or different? A phenomenological comparison of auditory verbal hallucinations in healthy and psychotic individuals. *Journal of Clinical Psychiatry*, 72, 320–325.

George, A., Soloff, P.H. (1986). Schizotypal symptoms in patients with borderline personality disorders. *American Journal of Psychiatry*, 143, 212–215.

Goodwin, D.W., Alderson, P., Rosenthal, R. (1971). Clinical significance of hallucinations in psychiatric disorders A study of 116 hallucinatory patients. *Archives of General Psychiatry*, 24, 76–80.

Haddock, G., McCarron, J., Tarrier, N., Faragher, E.B. (1999). Scales to measure dimensions of hallucinations and delusions: the psychotic symptom rating scales. *Psychological Medicine*, 29, 879–889.

Hagen, F.W. (1868). Zur Theorie der Halluzinationen. *Allgemeine Zeitschrift für Psychiatrie*, 25, 1–113.

Hoch, P., Polatin, P. (1949). Pseudoneurotic forms of schizophrenia. *Psychiatric Quarterly*, 23, 248–276.

Kandinsky, V. (1885). *Kritische und klinische Betrachtungen im Gebiete der Sinnestäuschungen.* Berlin: Verlag von Friedländer und Sohn.

Kingdon, D.G., Ashcroft, K., Bhandari, B., Gleeson, S., Warikoo, N., Symons, M., Taylor, L., Lucas, E., Mahendra, R., Ghosh, S., Mason, A., Badrakalimuthu, R., Hepworth, C., Read, J., Mehta, R. (2010). Schizophrenia and borderline personality disorder. Similarities and differences in the experience of auditory hallucinations, paranoia and childhood trauma. *Journal of Nervous and Mental Disease*, 198, 399–403.

Kingdon, D.G., Turkington, D. (2005). *Cognitive therapy of schizophrenia.* New York, NY: The Guilford Press.

Mott, R.H., Small, I.F., Anderson, J.M. (1965). Comparative study of hallucinations. *Archives of General Psychiatry*, 12, 595–601.

Nayani, T.H., David, A.S. (1996). The auditory hallucination: a phenomenological survey. *Psychological Medicine*, 26, 177–189.

Pierre, J.M. (2010). Hallucinations in nonpsychotic disorders: Toward a differential diagnosis of "hearing voices". *Harvard Review of Psychiatry*, 18, 22–35.

Skodol, A.E., Gunderson, J.G., Pfohl, B., Widiger, T.A., Livesley, W.J., Siever, L.J. (2002). The borderline diagnosis I: psychopathology, comorbidity, and personality structure. *Biological Psychiatry*, 51, 936–950.

Slotema, C.W., Daalman, K., Blom, J.D., Diederen, K.M., Hoek, H.W., Sommer, I.E. Auditory verbal hallucinations in patients with borderline personality disorder are similar to those in schizophrenia (submitted).

Soloff, P.H. (1979). Physical restraint and the nonpsychotic patient: Clinical and legal perspectives. *Journal of Clinical Psychiatry*, 40, 302–305.

Stern, A. (1938). Psychoanalytic investigation of and therapy in the borderline group of neuroses. *Psychoanalytic Quarterly*, 7, 467–489.

Turkington, D., Kingdon, D., Rathod, S., Hammond, K., Pelton, J., Mehta, R. (2006). Outcomes of an effectiveness trial of cognitive-behavioural intervention by mental health nurses in schizophrenia. *British Journal of Psychiatry*, 189, 36–40.

Van der Zwaard, R., Polak, M.A. (1999). Pseudohallucinaties. Een literatuuronderzoek naar de waarde van het concept en de relaties met aanverwante symptomatologie. *Tijdschrift voor Psychiatrie*, 41, 25–35.

Zanarini, M.C., Frankenburg, F.R., Hennen, J., Reich, D.B., Silk, K.R. (2004). Axis I comorbidity in patients with borderline personality disorder: 6-year follow-up and prediction of time to remission. *American Journal of Psychiatry*, 161, 2108–2114.

Zanarini, M.C., Gunderson, J.G., Frankenburg, F.R. (1990). Cognitive features of borderline personality disorder. *American Journal of Psychiatry*, 147, 57–63.

Chapter 11
Musical Hallucinations

Oliver W. Sacks and Jan Dirk Blom

11.1 Introduction

In 1846, the French alienist Jules Baillarger (1806–1891) published the first known medical account of musical hallucinations (Baillarger 1846). Since then no more than 150 studies on the subject have been published in the international literature, most of them involving case reports. Although the past century saw ten large epidemiological surveys of hallucinations in the general population, yielding prevalence rates for auditory hallucinations of around 4% (for an overview see Blom 2012), none of them discriminated between various types of auditory hallucination. As a consequence, it is not possible to extract any prevalence figures for musical hallucinations from them (Cole et al. 2002). This state of affairs would seem to reflect – and perhaps also sustain – the long-held opinion that musical hallucinations are rare, and for the most part irrelevant from a clinical point of view (Fig. 11.1).

And yet, working over the past 50 years with elderly patients living in hospitals and nursing homes, and in neurology clinics for patients with migraine, epilepsy, movement disorders, sleep disorders, and other conditions, I have been given many accounts of musical hallucinations from my patients (*the first person in this chapter will refer to one of the authors, O.S.*). After 1985, when I published case histories of two such patients in my book *The Man Who Mistook His Wife For A Hat*, (Sacks 1985) many readers wrote to me about their own musical hallucinations. I have thus had contact, in person or via correspondence, with hundreds of patients with musical hallucinations, and certain trends emerge from their accounts.

O.W. Sacks, M.D. (✉)
Columbia University Medical Center, New York, NY, USA
e-mail: mail@oliversacks.com

J.D. Blom, M.D., Ph.D.
Parnassia Bavo Academy, Parnassia Bavo Group and University of Groningen, Groningen
Kiwistraat 43, 2552 DH, The Hague, The Netherlands
e-mail: jd.blom@parnassia.nl

J.D. Blom and I.E.C. Sommer (eds.), *Hallucinations: Research and Practice*,
DOI 10.1007/978-1-4614-0959-5_11,

Fig. 11.1 *Musical hallucinations II* (Artwork by Marten Blom, copyright 2011)

11.2 Phenomenological Characteristics

Although there is a wide range in the quality of individual musical hallucinations – sometimes they are soft, sometimes disturbingly loud; sometimes simple, sometimes complex – there are certain characteristics common to all of them. First and foremost, they are perceptual in quality and seem to emanate from an external source – in this way, they are distinct from all forms of imagery (even "earworms," the often annoying, repetitious musical imagery which most of us are prone to from time to time; see Sacks 2006). People with musical hallucinations will often search for an external source – a radio, a neighbor's television, and a band in the street – and only when they fail to find any such external source do they realize that the source must be in themselves. Thus they may speak of a tape recorder, a jukebox, or an iPod in the brain, something autonomous with a will of its own, something not felt to be a controllable, integral part of the self. That there should be something like this in one's head arouses bewilderment and, not infrequently, fear. The fear is that one is going mad, or perhaps that the phantom music may be a sign of a tumor, a stroke, or a dementia. Such fears often inhibit people from acknowledging that they have hallucinations; perhaps this is the reason why musical hallucinations have long been considered rare – although it is now realized that this is far from the case. It may require direct but gentle and discreet questioning to elicit the fact that hallucinations are occurring, and it may afford immense relief to patients (even if their hallucinations cannot be cured) to be assured, after careful questioning and examination and testing if appropriate, that there is no evidence of mental illness or progressive cerebral pathology.

Hallucinations share the subjective character (and much of the physiological character too) of perception, and they can intrude upon and even overwhelm perception; musical hallucinations (like tinnitus) can be so loud as to make it impossible to hear someone speak; imagery never rivals perception in this way.

Musical hallucinations usually appear suddenly, out of the blue; but sometimes, they may follow a tinnitus or an external noise (like the drone of a plane engine or a lawn mower), the hearing of real music, or verbal or other associations to a particular piece of music. Some people with musical hallucinations experience them virtually nonstop, while others hallucinate only intermittently. The hallucinated music is mostly familiar (though not always liked; thus, one of my patients hallucinated Nazi songs from his youth, which terrified him). It may be vocal or instrumental, classical or popular, but in almost all cases, it is music which was heard in the patient's early years such as children's songs, folk songs, hymns, arias, and advertising tunes (Warner and Aziz 2005; Sacks 2007; Coebergh et al. 2009). Very occasionally, patients may hear "meaningless phrases and patterns," as one of my correspondents, a gifted musician, put it.

Hallucinated music can be very detailed, so that every note in a piece, every instrument in an orchestra, is distinctly "heard." Such detail and accuracy may be astonishing, no least to the hallucinator, who may be relatively "unmusical" and scarcely able, normally, to hold a simple tune in his head, let alone an elaborate chorale or instrumental composition. Typically though, only a few bars, a segment, or a single theme is hallucinated – which may then be repeated, as with a stuck record, again and again. One patient of mine heard part of *O Come, All Ye Faithful* nineteen and a half times in 10 min (her husband timed this) and was tormented by never hearing the entire hymn. Hallucinatory music can wax slowly in intensity, and then slowly wane, but it may also come on suddenly full blast in midbar and then stop with equal suddenness (like a switch turned on and off, patients often comment).

Musical hallucinations may continue in their own way, irrespective of whether one attends to them consciously. Some patients may sing along with them; others ignore them – it makes no difference. And having their own autonomous mechanism, musical hallucinations can continue, pursue their own course, even if one is listening to or playing something else. Thus, Gordon B., an Australian violinist described in *Musicophilia* (Sacks 2007), sometimes hallucinates a piece of music while he is actually performing an entirely different piece at a concert.

The natural history of musical hallucination is usually one of spreading. A familiar tune, an old song, may start the process; this is likely to be joined, over the ensuing weeks, by another song, and then another, until a whole repertoire of hallucinatory music has been built up. And this repertoire itself tends to change – one tune will drop out, another will replace it. There is usually very little conscious control of musical hallucinations. One cannot voluntarily start or stop the hallucinations coming, though some people may be able, on occasion, to replace them with other hallucinated music. Thus one man who said he had "an intracranial jukebox" found that he could switch at will from one "record" to another, provided there was some similarity of style or rhythm, though he could not turn on or turn off the "jukebox" as a whole.

The assumption that musical hallucinations are rare is no longer tenable (Evers and Ellger 2004). Although they still seem to be rare in children, one boy I have seen, Michael, has had them since the age of 5 or 6. His music is nonstop and overwhelming, and often prevents him from focusing on anything else.

11.3 Classification and Differential Diagnosis

Musical hallucinations are conceptualized as auditory hallucinations characterized by songs, tunes, melodies, harmonics, rhythms, and/or timbres (Berrios 1990; Blom 2010). They can be perceived either unilaterally or bilaterally, and within the head or as if coming from the environment. Whether they should be classified as verbal or nonverbal auditory hallucinations depends on the presence or absence of any vocal elements. When they occur in the absence of any associated pathology other than hearing impairment, they are called idiopathic musical hallucinations (Pasquini and Cole 1997). When occurring in the presence of such pathology, they are called symptomatic musical hallucinations. Table 11.1 lists these types of musical hallucination, as well as various phenomena from which they should be distinguished.

Musical hallucinations should be distinguished, for example, from music "heard in" the background noise of engines, refrigerators, computer ventilators, and so on, a phenomenon known as auditory illusion or auditory pareidolia. The term musical palinacusis is reserved for the continued perception – or reperception – of music after some actual piece of music has stopped. A final phenomenon from which they should be distinguished is musical tinnitus (Vernon and Sanders 2001). Although the two may be indistinguishable in clinical practice, the latter term refers to musical

Table 11.1 Glossary of musical hallucinations and related phenomena (Adapted from Blom 2010)

Phenomenon	Characterization
Musical hallucination (musical hallucinosis, musical ear syndrome, auditory Charles Bonnet syndrome, Oliver Sacks' syndrome)	An auditory hallucination characterized by songs, tunes, melodies, harmonics, rhythms, and/or timbres
Idiopathic musical hallucination	A musical hallucination occurring in the absence of any type of associated pathology other than hearing loss
Symptomatic musical hallucination	A musical hallucination occurring in the presence of any type of associated pathology other than hearing loss
Auditory illusion (auditory pareidolia)	An auditory percept resulting from a misrepresentation or misinterpretation of random auditory stimuli, which may or may not have a musical quality
Musical illusion	A paradoxical auditory illusion created with the aid of actual musical sounds
Musical tinnitus	A musical hallucination with evolution from "ringing in the ears"
Musical palinacusis	The continued perception or reperception of an actual piece of music
Musical imagery	An imagined or remembered piece of music, having no perceptual characteristics
Earworm	A tune going round in the head, having no perceptual characteristics

hallucinations which evolved out of tinnitus or "ringing in the ears." The music experienced by Gordon B., who had suffered for more than 20 years from a tonal type of tinnitus before it changed into "the most horrific grinding," and then, a few weeks later, into a nonstop flow of musical phrases and patterns, constitutes an apt example of musical tinnitus.

11.4 Pathophysiology

Positron emission tomography (PET) and functional magnetic resonance imaging (fMRI) have shown that musical hallucination, like actual musical perception, is associated with the activation of an extensive network involving many areas of the brain – auditory areas, motor cortex, visual areas, basal ganglia, brainstem, pons, tegmentum, cerebellum, hippocampi, amygdala, and, according to some, even the peripheral auditory system (Gordon 1999). Apparently, any of those regions – and probably many additional ones – are capable of mediating musical hallucinations (Stewart et al. 2006). The factors responsible for this would seem to be even more diverse. The musical network can be stimulated directly, on occasion, as by an irritable focus as in temporal lobe epilepsy, but what seems to occur in most cases of musical hallucination is a "release" of activity in the musical network when inhibition or constraints which are normally operative are weakened (Keshavan et al. 1992). The commonest cause of such a release is auditory deprivation or deafness. In this way, the musical hallucinations of the elderly deaf are analogous to the visual hallucinations of the blind (Charles Bonnet syndrome (CBS), see also Chap. 6). In both cases, it is as if the networks normally employed in musical or visual perception escape into a spontaneous and autonomous activity of their own if there is no longer adequate perception to constrain them. We therefore speak of such hallucinations as "release hallucinations."

But although the musical hallucinations of deafness and CBS hallucinations may be akin physiologically, they have striking differences phenomenologically, which reflect the very different nature of our visual worlds and our musical worlds – differences evident in the ways we perceive, recollect, or imagine them. We are not given an already-made, preassembled visual world; we have to construct our (own) visual world as best we can. This construction entails analysis and synthesis at many functional levels in the brain, starting with orientation receptors, perception of lines and angles and orientation, in the occipital cortex (see Chap. 2). At higher levels, in the inferotemporal cortex, the "elements" of visual perception are of a more complex sort, appropriate for the putting together and recognition of natural scenes, objects, animal and plant forms, and faces. But all of these visual objects have to be broken down and reassembled, whereas music is already made and assembled for us.

Visual hallucinations of the CBS type commonly include faces, but typically they are not familiar or recognizable. They may sometimes be deformed or grotesque, with exaggerated eyes or teeth. The hallucinations of CBS often contain elements that change kaleidoscopically; they have the quality of bric-a-brac, of bits of images, continually taken apart and reassembled, by the early visual system, unconnected with the hippocampi or amygdala, unconnected to memory, intention, or emotion.

Musical hallucinations are quite different. With music, although there are separate functional systems for perceiving pitch, timbre, rhythm, etc., the musical networks of the brain work as a whole, and a piece of music cannot be significantly altered (in melodic contour or tempo or rhythm) without losing its musical identity – without, indeed, its ceasing to be music. We apprehend a piece of music as a whole. Whatever the initial process of musical perception and memory may be, once a piece of music is known, it is retained not as an assemblage of elements, but as a completed procedure or performance; music is performed by the mind/brain, whenever it is recollected; and this is also so when it surfaces spontaneously, as a hallucination.

Table 11.2 provides a list of risk factors for the mediation of musical hallucinations extracted from the literature. It should be noted, however, that some of those risk factors have been corroborated more thoroughly than others. Hearing impairment, tinnitus, and advanced age appear to be the most robust risk factors presently known. Female sex, on the other hand, may well be a spurious finding due to a historical underrepresentation of male hallucinators. Cerebral pathology in all its many forms is associated with an increased risk for musical hallucinosis, although here it should be noted that – the other way round – the prevalence of musical hallucinations among such cases is remarkably low, especially if we consider the vastness of the brain networks involved and the a priori chance of damage to those networks in the case of cerebral pathology.

Table 11.2 also features a number of psychiatric disorders. As suggested by some authors, musical hallucinations may well be more prevalent in the context of psychosis (Baba and Hamada 1999) or other psychiatric conditions (Hermesh et al. 2004). In fact, the literature boasts more than 30 such cases (Evers 2006). However, given the relatively high prevalence of those disorders it is as yet unclear whether the occurrence of musical hallucinations in the context of psychiatric disease is any higher than in the general population (Evers and Ellger 2004; Tuerlings et al. 2009). Our own clinical experience speaks against the assumption that psychosis predisposes to musical hallucinations, as does the largest study to date carried out among psychiatric patients. That study identified 5 individuals out of 128 (3.9%) as being familiar with musical hallucinations (Schakenraad et al. 2006), 3 of them being of advanced age and only 1 of them being diagnosed with a psychotic disorder. Finally, Table 11.2 features a number of drugs and illicit substances reportedly associated with musical hallucinations, as well as a group of miscellaneous issues, ranging from sensory deprivation, Hashimoto's encephalopathy, and cochlear implantation to electroconvulsive treatment, Lyme disease, and Behçet's disease, which are all in need of further investigation (Fig. 11.2).

Table 11.2 Risk factors for musical hallucinations, as derived from the literature

Hearing impairment
Tinnitus
Older age
Female sex (possibly)
Cerebral pathology
Epilepsy
Brain tumor
Stroke
Hemorrhage
Meningitis
Alzheimer's disease
Neurosyphilis
Localized atrophy
Traumatic lesion
Psychiatric disorder
Schizophrenia spectrum disorder
Bipolar disorder
Psychotic depression
Depression
Obsessive-compulsive disorder
Adaptation impairment
Personality disorder
ADHD
Cocaine dependence
Intoxication
Alcohol
Antidepressants
Opioids
Antibiotics
Beta-blockers
Quinine
Salicylates
Miscellaneous
Behçet's disease
Hashimoto's encephalopathy
Lyme disease
Electroconvulsive treatment
Cochlear implantation
Sensory deprivation

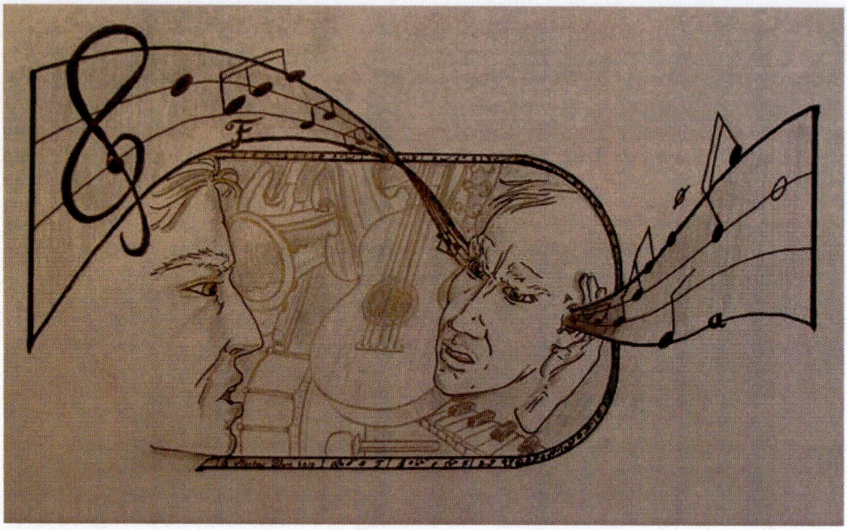

Fig. 11.2 *Musical hallucinations I* (Artwork by Marten Blom, copyright 2011)

11.5 Diagnosis and Treatment

Diagnosis hinges on proper history-taking, and, if necessary, auxiliary investigations such as blood testing, audiological evaluation, electroencephalography, and an MRI scan. If present, any underlying pathology should be treated first. The symptomatic treatment of musical hallucinations tends to consist of anticonvulsant, antidepressant, or antipsychotic drugs. In the literature, a few reports can be found on successful treatments with the anticonvulsants gabapentin (Holroyd and Sabeen 2008; Sacks 2007), carbamazepine (Gertz et al. 1996), and valproic acid (Coebergh et al. 2009), and with the antidepressant clomipramine (Matsui et al. 2003); however, none of these drugs have yet shown a broad usefulness, and perhaps the majority of cases are refractory to treatment. Nonpharmacological treatment may consist of simple reassurance (i.e., that musical hallucinations are certainly no sign of dementia or psychiatric illness), a hearing aid (for the hearing impaired and probably being more useful in cases of auditory pareidolia than in hallucinations proper), and the advice to engage in other activities that may divert attention. Most patients will discover for themselves that getting absorbed in such activities – conversation, physical activity, playing bridge, sometimes listening to real music – may drive their hallucinations into the background. But this may not always work, and some people may have musical hallucinations which are loud and intrusive, sometimes maddeningly so. At the current time, there are few if any reliable therapeutic approaches. With the passage of months or years, most people with musical hallucinations find ways of living with them, accommodating to them, so that what was initially felt as so alien and intrusive may become tolerable or even companionable. One lady, Sheryl C., whom

I followed for several years, became resigned to living with her musical hallucinations, and said, "It's *my* music, now. It's like I have a circuit in my head. I think I am landed with it forever." I asked another woman, June B., whether, after more than a decade, her hallucinated music had become important in her life in either a positive or a negative way. "If it went away," I asked, "would you be pleased or would you miss it?"

"Miss it," she answered at once. "I would miss the music. You see, it is now part of me."

References

Baba, A., Hamada, H. (1999). Musical hallucinations in schizophrenia. *Psychopathology*, 32, 242–251.

Baillarger, J. (1846). Des hallucinations, des causes qui les produisent et des maladies qui les caractérisent. *Mémoires de l'Académie Royale de Médecine*, 12, 273–475.

Berrios, G.E. (1990). Musical hallucinations. A historical and clinical study. *British Journal of Psychiatry*, 156, 188–194.

Blom, J.D. (2010). *A dictionary of hallucinations*. New York, NY: Springer.

Blom, J.D. (2012). *Hallucinations in psychiatric disorders*. In: *The neuroscience of hallucinations*. Jardri, R., Thomas, P., Cachia, A., & Pins, D., eds. New York, NY: Springer.

Coebergh, J.A.F., Shaya, M., Koopman, J., Blom, J.D. (2009). Muzikale hallucinaties. *Nederlands Tijdschrift voor Geneeskunde*, 153, B313.

Cole, M.G., Dowson, L., Dendukuri, N., Belzile, E. (2002). The prevalence and phenomenology of auditory hallucinations among elderly subjects attending an audiology clinic. *International Journal of Geriatric Psychiatry*, 17, 444–452.

Evers, S. (2006). Musical hallucinations. *Current Psychiatry Reports*, 8, 205–210.

Evers, S., Ellger, T. (2004). The clinical spectrum of musical hallucinations. *Journal of the Neurological Sciences*, 227, 55–65.

Gertz, H.J., Göhringer, K., Schimmelpfennig, C. (1996). Erfolgreiche Carbamazepintherapie bei 2 Fällen von Musikhalluzinose. *Nervenarzt*, 67, 387–389.

Gordon, A.G. (1999). Do musical hallucinations always arise from the inner ear? *Medical Hypotheses*, 53, 89–90.

Hermesh, H., Konas, S., Shiloh, R., Dar, R., Marom, S., Weizman, A., Gross-Isseroff, R. (2004). Musical hallucinations: prevalence in psychotic and nonpsychotic outpatients. *Journal of Clinical Psychiatry*, 65, 191–197.

Holroyd, S., Sabeen, S. (2008). Successful treatment of hallucinations associated with sensory impairment using gabapentin. *Journal of Neuropsychiatry and Clinical Neurosciences*, 20, 364–366.

Keshavan, M.S., Davis, A.S., Steingard, S., Lishman, W.A. (1992). Musical hallucinations: a review and synthesis. *Neuropsychiatry, Neuropsychology, and Behavioral Neurology*, 5, 211–223.

Matsui, T., Matsunaga, H., Ohya, K., Iwasaki, Y., Koshimune, K., Miyata, A., and Kiriike, N. (2003). Clinical features in two cases with musical obsessions who successfully responded to clomipramine. *Psychiatry and Clinical Neurosciences*, 57, 47–51.

Pasquini, F., Cole, M.G. (1997). Idiopathic musical hallucinations in the elderly. *Journal of Geriatric Psychiatry and Neurology*, 10, 11–14.

Sacks, O. (1985). The man who mistook his wife for a hat, and other clinical tales. New York, NY: Summit Books.

Sacks, O. (2006). The power of music. *Brain*, 129, 2528–2532.

Sacks, O. (2007). *Musicophilia. Tales of music and the brain*. New York, NY: Alfred A. Knopf.

Schakenraad, S.M., Teunisse, R.J., Olde Rikkert, M.G.M. (2006). Musical hallucinations in psychiatric patients. *International Journal of Geriatric Psychiatry*, 21, 394–397.

Stewart, L., von Kriegstein, K., Warren, J.D., Griffiths, T.D. (2006). Music and the brain: disorders of musical listening. *Brain*, 129, 2533–2553.

Tuerlings, J.H.A.M., Wijnen, H., Boerman, R., Verwey, B. (2009). Muziek als ongenode gast: de auditieve variant van syndroom van Charles Bonnet. *Tijdschrift voor Psychiatrie*, 51, 853–858.

Vernon, J.A. and Sanders, B.T. (2001). *Tinnitus. Questions and answers*. Boston, MA: Allyn and Bacon.

Warner, N., Aziz, V. (2005). Hymns and arias: Musical hallucinations in older people in Wales. *International Journal of Geriatric Psychiatry*, 20, 658–660.

Chapter 12
Olfactory and Gustatory Hallucinations

Richard J. Stevenson and Robyn Langdon

12.1 Introduction

This chapter examines hallucinations in the chemical senses. Almost all of the literature, which is not abundant, concerns olfaction, with little coverage of taste and flavor – that is the integration of taste, smell, and somatosensation during eating (Rozin 1982). For this reason, most of the chapter will focus on olfaction.

Olfactory hallucinations (OHs) are smell experiences which occur in the absence of an appropriate stimulus (Greenberg 1995). OHs can occur in healthy people, as a primary symptom in the smell disorder phantosmia (see Table 12.1 for definitions of common chemosensory disorders) and as a secondary symptom in a range of disease states, notably psychotic disorders, temporal lobe epilepsy (TLE), brain injury, and migraine. Unlike hallucinations in other modalities, it does not appear possible to classify OHs into simple and complex forms. Most OHs involve the experience of odor objects (e.g., coffee) – there being no obvious parallel to seeing spots of light (simple visual hallucination) or angels (complex visual hallucination). The OH can also differ in other ways, in time course (brief to persistent) and location – being perceived as originating in the environment, arising from the self, but rarely from the mouth. The latter is surprising given that olfaction is a significant input into flavor perception via retronasal delivery (Rozin 1982). However, flavor hallucinations may be misreported as gustatory ones, just as flavor is typically misidentified as

R.J. Stevenson, D.Phil (✉)
Department of Psychology, Macquarie University, Sydney, Australia
e-mail: dick.stevenson@mq.edu.au

R. Langdon, Ph.D.
ARC Centre of Excellence in Cognition and its Disorders,
Macquarie Centre for Cognitive Science, Macquarie University, Sydney, Australia
e-mail: robyn.langdon@mq.edu.au

J.D. Blom and I.E.C. Sommer (eds.), *Hallucinations: Research and Practice*,
DOI 10.1007/978-1-4614-0959-5_12, © Springer Science+Business Media, LLC 2012

Table 12.1 Common chemosensory disorders and their definitions

Sensory modality	Name of disorder	Definition
Taste		
	Ageusia	Inability to perceive or detect tastes
	Hypogeusia	Reduced ability to perceive or detect tastes
	Dysgeusia	Altered taste perception, even with tastes absent
Smell		
	Anosmia	Inability to perceive or detect odors
	Hyposmia	Reduced ability to perceive or detect odors
	Parosmia (cacosmia, troposmia)	Distortion of smell, usually unpleasant
	Phantosmia	Long-lasting olfactory hallucination

"taste" – but this possibility has not been explored. The term gustatory hallucination (GH) refers to the experience of taste sensations, namely sweet, sour, bitter, umami (meaty), and salty, in the absence of an appropriate stimulus (Brasic 1998).

This chapter opens with a brief look at GHs, followed by a more extensive survey of OHs, examining their incidence, phenomenology, and correlates in healthy people, and then in conditions where they feature as primary or secondary symptoms. The latter half of the chapter diverges from previous reviews (Greenberg 1995, Stevenson 2009a; Velakoulis 2006) focusing on current and emerging theories of OHs.

12.2 Gustatory Hallucinations (GHs)

GHs occur in healthy people (Ohayon 2000), but they can also be a secondary symptom in brain tumors, TLE, migraine, drug misuse, and psychotic disorders (Brasic 1998). Detailed reports of GHs are rare, and of those known to the authors, involve experience of diffuse sensations of taste (unspecified but often unpleasant) either in the mouth or throat (see Chen et al. 2003; Hausser-Hauw and Bancaud 1987). Cases of highly localized GHs have also been reported (Daly (1958) describes a TLE patient who experienced a taste experienced between two teeth). The incidence of GHs has not been well established; although the TLE literature would suggest a figure far lower than for OHs, with Chen et al.'s (2003) data suggesting 0.5% and Hausser-Hauw and Bancaud (1987) 4.0%. A further factor that makes the study of GHs especially problematic, including establishing their incidence, is the confusion between a GH and a flavor hallucination. A "pure" GH involves the experience of just a taste (i.e., sweet, sour, salty, bitter, or umami), while a flavor hallucination involves orally located food- or drink-like experiences, which include gustatory, olfactory, and somatosensory components – that is the senses that compose flavor under normal perceptual conditions. The etiology of GHs has not been investigated. However, it would seem that at least for TLE, many of the brain structures known to be involved in OHs may be abnormally active here as well (see Hausser-Hauw and Bancaud 1987).

12.3 Olfactory Hallucinations (OHs)

12.3.1 OHs in Healthy People

OHs are relatively common in healthy adults. Ohayon (2000), in a large-scale survey of hallucination frequency, found that for daytime hallucinations, OHs were the most frequent, second only to GHs. OHs were reported to have occurred at least once in 8.6% of the sample, with 0.9% having experienced multiple OHs in the preceding week. OHs may be more likely to occur in those healthy adults who score highly on measures of psychosis proneness (Kwapil et al. 1996).

12.3.2 OHs as a Primary Symptom

Phantosmia is a condition in which the primary presenting symptom is an OH. Leopold (2002) noted that incidence data is unavailable but observed that phantosmia may be more frequent in women, with a sudden mid-life onset. OHs in phantosmia may be unpleasant and enduring, lasting hours or days, and can be triggered by actual olfactory events. Both of these latter features seem to distinguish phantosmia from other types of OH discussed below, which tend to have no obvious trigger and whose time course is more akin to normal smelling. There are currently no established treatments for phantosmia, and the disease shows a relapsing and remitting presentation (Frasnelli et al. 2004).

A further condition in which OHs may feature prominently is olfactory reference syndrome. Here, patients may visit multiple medical specialists in the belief that they are emitting a foul body odor (Pryse-Phillips 1971). The key feature of this disorder is the unshakeable belief that one smells foul – even when tests reveal no underlying pathology. This may suggest a delusional condition rather than the presence of OHs, but this issue remains unresolved (see Phillips et al. 2006).

12.3.3 OHs as a Secondary Symptom

OHs have been documented as occurring in many conditions as a secondary symptom (Greenberg 1995). However, as symptom frequency varies between conditions, as with the literature devoted to it, we present independent sections for psychotic disorders, temporal lobe epilepsy (TLE), migraine, brain tumors, and injuries, along with a miscellanea section for rarer presentations.

12.3.3.1 Psychotic Disorders

Estimates of the incidence of OHs in the context of psychotic disorders vary from 1% to 35% (Kopala et al. 1994). Drawing from all such studies produces an incidence rate of around 14% (Stevenson et al. 2011) and one comparable to the 13% incidence rate obtained from the WHO-10 data set (Jablensky et al. 1992) and the 17% incidence rate estimated from Australian data (Langdon et al. 2011). While the WHO-10 data set indicates cross-cultural variation in this figure (with no obvious geographical pattern; Langdon et al. 2011), these incidence rates suggest that OHs are unusual but not rare in the context of psychotic disorders.

OHs in psychotic disorders may be more common in females (Kopala et al. 1994) and in patients who experience referential or control delusions (Langdon et al. 2011) as well as tactile hallucinations (Stevenson et al. 2011; Langdon et al. 2011). However, contrary to clinical lore (see Kwapil et al. 1996), the presence of OHs does not seem to be associated with worse symptomology or disease course (Stevenson et al. 2011). Early reports on OHs suggested principally negative experiences, with smoky, decay, body, and animal odors predominant (Bromberg and Schilder 1934; Rubert et al. 1961). We interviewed 51 people diagnosed with schizophrenia and a history of OHs to determine their phenomenology (Stevenson et al. 2011). Affectively negative experiences were dominant (51% of all reported OHs), but a significant minority were either neutral (19%) or positive (30%). Qualitatively, the OHs covered a broad range of smells, with some participants experiencing the same OH repeatedly. Although most OHs were localized as stemming from the environment, a minority were reported as emanating from the self, echoing other reports (e.g., Bromberg and Schilder 1934). Our study did not, however, reveal any bizarre smells, as have been observed before and which may be better characterized as delusions (e.g., "devil's breath" – see Greenberg 1995). As for time course, we found that OHs generally mimicked normal olfactory experience. The parallel with normal olfactory experience, as noted by other workers, included reports of engaging in appropriate behavior after experiencing the OH (Rubert et al. 1961). Insight varied and tended to be more common if the OH was unpleasant (Stevenson et al. 2011).

12.3.3.2 Temporal Lobe Epilepsy (TLE)

The presence of an OH prior to a seizure (or a migraine attack – see below) is referred to as an aura rather than a hallucination. Phenomenologically, epileptic olfactory auras closely resemble OHs reported in other conditions. Olfactory auras appear to be largely restricted to TLE (West and Doty 1995), and they (like other prodromal cues) tend to signal an approaching seizure. Incidence estimates of OHs in epilepsy vary from 1% to 30% (West and Doty 1995; Chen et al. 2003), with a mean across studies of around 9% – making them relatively rare. The actual incidence in TLE may be somewhat higher because early studies may have included patients whose seizure focus fell outside of the temporal lobe. Olfactory auras are brief; they may be accompanied by other types of aura (epigastric auras and nausea may predominate; see Chen et al. 2003); they can be repetitive and foul smelling, although pleasant-smelling auras have also been reported (Efron 1956). Insight tends to be rapid because the odors that are experienced seem to stem from the external environment and are sufficiently realistic – especially if unpleasant – to prompt people to actively search for their cause (Efron 1956).

12.3.3.3 Migraine

The literature on OHs – auras – in the context of migraine is less well developed and it is not possible to estimate any incidence rates. Based upon case reports, auras in migraine seem to have the signal quality that characterizes OHs in TLE, occurring mainly as part of the migraine prodrome (Crosley and Dhamoon 1983). In the nine cases described in the literature, most were unpleasant and environmentally localized (Diamond et al. 1985). Time course varied between 5 min and 2 h, but whether this represents a distinct feature or a sampling bias (i.e., a long-lasting OH is likely to be both more notable and to prompt verification) is not currently known (Diamond et al. 1985). In several of the reported cases, patients had actively attempted to identify the odor, leading them to recognize the experience as hallucinatory (Crosley and Dhamoon 1983).

12.3.3.4 Brain Tumors and Injuries

One of the first papers to link the characteristic clinical features of TLE with OHs (Jackson and Stewart 1899) concluded that the epilepsy, the other symptoms (diplopia, facial weakness, hemiplegia), and the OHs probably resulted from a tumor in the temporal lobe. Tumors can present with symptoms that may initially resemble TLE, and in Chen et al.'s (2003) detailed case series of 217 patients, of whom 12 had olfactory auras, 3 of these were revealed to have brain tumors (astrocytoma and glioblastoma; the rate of brain tumors without olfactory auras was not reported). This type of observation has led some authors to conclude that olfactory auras signal tumors (Acharya et al. 1998). Not only can a variety of tumors (e.g., gliomas, meningiomas; see Greenberg 1995) in the temporal lobe produce seizures and OHs but so can tumors at more distant sites that indirectly impact on olfactory centers. Faris and Terrence (1989) describe a case with extensive OHs resulting from a cyst in the third ventricle, which they suggest exerted a mass effect on the medial dorsal nucleus of the thalamus (MDNT) – the thalamo-cortical connection of the olfactory system. While tumors and other masses adjacent to olfactory structures may be associated with the presence of OHs, direct injury to similar regions of the brain via aneurysm, stroke, or hematoma can also result in OHs (Toone 1978). The relatively few reports make any general remarks about the phenomenology uninformative, especially as the time courses for the OHs identified in these reports varied from minutes to days.

12.3.3.5 Miscellanea

OHs have also been reported following cancer chemotherapy (Nesse et al. 1983), in chronic cocaine users (Brasic 1998), and in acute alcohol withdrawal (Bromberg and Schilder 1934). Several reports have identified OHs as occurring when a person was reexposed to cues relating to an earlier traumatic incident – such as smelling petrol while traveling in a car (following an earlier car accident) and smoke (following an earlier experience of fire; Burstein 1987). OHs have also been reported in Parkinson's disease (Sandyk 1981), depression (Brasic 1998), Alzheimer's disease (Greenberg 1995), and in Charles Bonnet syndrome (Alroe and McIntyre 1983), which is normally characterized by visual hallucinations (see Chap. 6).

12.4 Cause

It seems likely that several different causes contribute to the occurrence of OHs. This is suggested by the diverse conditions in which OHs are found (Greenberg 1995) and also by emerging evidence that there may be variations in cause within a particular disease state, notably in psychotic disorders (Stevenson et al. 2011). Moreover, different disease states may share common causes. For these reasons, the following section is organized around putative causes, not conditions, and for each class of cause its potential contribution to different conditions is outlined. Three primary causes are presented, peripheral (receptors) and central (brain), with a predominantly physiological focus and psychological explanations. These are not discrete and the overlaps are highlighted. Finally, it should be borne in mind that it is only recently that efforts have been made to understand the cause of OHs, and so what follows should be considered a preliminary survey.

12.4.1 Peripheral Contributions

There are two basic types of peripheral cause. The first concerns some form of olfactory receptor abnormality, resulting from an infectious agent, mechanical or chemical damage, or genetic disposition, which affects the ability to sense volatile chemicals (Smith and Duncan 1992). Such damage could produce three forms of abnormality that might lead a person to experience an OH or regard an experience as an OH – a distinction that will become apparent below. If the damage is extensive, such as sheering of the olfactory nerve (with the cribriform plate acting as a "guillotine"), the ability to detect an odorant may be impaired or even lost (Smith and Duncan 1992). This may result in release hallucinations whereby reduced peripheral input leads to cortical hyperexcitability, producing activations of olfactory sensory memories – and thus OHs (Donnelly et al. 1989). However, if the damage is less severe, or if there has been damage followed by abnormal neuronal regrowth, this may lead to misperceptions – parosmia – which are typically unpleasant (Leopold 2002). In this case, the pattern of neural activity generated by the receptors for a particular odorant becomes deviant, and so the resulting percept does not match external cues (e.g., lemon detergent may be perceived as smelling of feces). The unusual and alarming nature of the misperception (or parosmia) may then lead a person to falsely conclude that they have experienced a hallucination (Greenberg 1995).

A characteristic feature of olfaction is rapid adaptation, so that the system is optimized to detect new environmental odorants (Wilson and Stevenson 2006). Adaptation has both a peripheral and a central component, and either or both of these can be compromised. Adaptation failures might lead to misperceptions, if, for example, spontaneous recovery occurred to an odor to which a person had adapted. If this occurred, it would *seem* as if an odor had spontaneously appeared in the environment, and again, this might be regarded, incorrectly, as a hallucination.

A case for such peripheral contributions can be made in psychotic disorders, brain injury, phantosmia, and TLE. For psychotic disorders, there is a large literature demonstrating abnormalities in olfaction, and these abnormalities are independent of antipsychotic medication (Moberg et al. 1999). People diagnosed with schizophrenia are more likely to have elevated detection thresholds, impaired odor identification ability, restricted affective response to odors, as well as poorer discriminative and odor memory capacity (Moberg et al. 1999). At least some of these impairments can be observed in healthy first-degree relatives of patients, and one report has obtained evidence of small cranio-facial abnormalities that may contribute to olfactory impairments (Moberg et al. 1999). In particular, the presence of elevated detection thresholds suggests some peripheral abnormality, although these are probably of insufficient magnitude to produce release hallucinations. Only three studies have examined whether people diagnosed with schizophrenia who experience OHs have peripheral deficits. Kerekovic (1972) conducted a detailed otorhinological examination of 39 patients diagnosed with schizophrenia and experiencing OHs, comparing them to 155 who did not report OHs. The OH group did not differ in nasal pathology, the presence of anosmia, hyposmia, or nasal allergies, from the remainder, suggesting that a peripheral cause of their OHs was unlikely. A similar conclusion was reached in two further studies in individuals diagnosed with schizophrenia by Stedman and Clair (1998) and Kopala et al. (1994), who found no difference in odor identification ability between those with or without OHs. Impaired ability to detect odors is known to impact on identification ability, and so this test is sensitive to peripheral damage.

While these findings point toward a more central cause for OHs in individuals diagnosed with schizophrenia, it would be premature to abandon peripheral accounts just yet. First, it may be that only a subgroup of OHs in those individuals result from a peripheral cause. This possibility was suggested in a recent study, which used an exploratory cluster analysis of participants' hallucinatory phenomenology (Stevenson et al. 2011). This analysis revealed that a subset of participants scored higher on a "sensory dysfunction" score (i.e., presence of sinusitis, prior head injury, smoking history, smoky OHs, length of OHs), all of which are associates of parosmia (Leopold 2002). Second, a further study (Stevenson and Langdon in press) suggests that there may be impairments in adaptation, which may well have a peripheral basis.

Peripheral damage to the olfactory epithelium may also be a contributory factor in TLE, brain injury, and phantosmia. For TLE, several studies have found that preoperative otorhinological workouts show significant differences in olfactory thresholds (both increased and decreased sensitivity), indicative of peripheral abnormality (West and Doty 1995). For brain injury, as noted above, especially following shearing injury, not only will anosmia occur, but during recovery there may be significant distortions of smell, which are frequently unpleasant (parosmia), and which may be misinterpreted as OHs (Greenberg 1995). For phantosmia, the evidence for peripheral involvement is more direct (Donnelly et al. 1989), and it has been suggested that it results from a peripheral deafferentation syndrome following trauma of some kind (Frasnelli et al. 2004). Nonetheless, some patients with phantosmia do not appear to have abnormal detection thresholds, and at least one study has suggested that there may be abnormalities in levels of the neurotransmitter gamma-aminobutyric acid (GABA) at several central olfactory sites, including the amygdala (Levy and Henkin 2004). This highlights the problem in clearly disambiguating peripheral from central causes because the olfactory system contains extensive feedforward and feedbackward circuits (Wilson and Stevenson 2006). Abnormalities in one will inevitably lead to abnormalities in the other, making determination of the initial locus difficult.

There may be an additional peripheral cause, although its inclusion in this section could be disputed. As noted earlier, the presence of tactile and olfactory hallucinations are associated in individuals diagnosed with schizophrenia (Greenberg 1995). In our recent study of OHs, a cluster analysis also revealed a group characterized by highly frequent tactile hallucinations (Stevenson et al. 2011). Nasal somatosensation is important in olfaction, both for magnitude constancy (i.e., bigger sniffs deliver more odorant but are not accompanied by equally large increments in perceptual magnitude) and for localization (i.e., is the odorant in the mouth or at the nose; Wilson and Stevenson 2006). The act of sniffing invokes neural activity in all parts of the olfactory system even if odorless air is used (Sobel et al. 1998). It may be that a tactile hallucination of nasal airflow may activate olfactory pathways sufficiently to generate an OH in some people diagnosed with schizophrenia, which might constitute a form of synesthesia.

12.4.2 Central Contributions

While cortical release represents one account of hallucinations, another is cortical irritation. The essence of this latter theory is that some form of irritation results in excess neural activity, which results in hallucinatory experience. This theory owes much to the brain stimulation experiments of Wilder Penfield (1891–1976), but in fact these and more recent studies find little evidence that stimulation of olfactory brain regions can generate any olfactory experience (Penfield and Perot 1963). Nonetheless, olfactory experiences were noted by one patient of Gloor's (1990) with amygdala stimulation and in four out of five non-epileptic participants receiving stimulation of the thalamus (Nashold and Wilson 1969). So while induced electrical activity can generate illusory olfactory percepts, it does not appear to be common.

The amygdala has been one site that has attracted attention as a potential neural locus responsible for the mediation of OHs. The amygdala is involved in processing affective reactions to odors (Royet et al. 2000). Amygdala abnormalities are also known to occur in association with psychotic disorders (Shenton et al. 2001) and TLE (West and Doty 1995), and these abnormalities may in turn be associated with the generation of OHs in the following way. Rather than experiencing an odor memory per se, cortical irritation (generated by diseased tissue) may lead to amygdala activity and hence negative affect, which is then misattributed to the olfactory system. It is possible to generate variants of this type of model. For example, if particular parts of the amygdala were selectively associated with olfaction, abnormal activity at that location could be automatically rendered "olfactory" without any need for misattribution. Whatever, a general prediction would be of greater amygdala dysfunction in olfactory hallucinators with TLE or psychotic disorder. So far, the only evidence pertinent to this hypothesis – one advanced by a number of authors (e.g., Chen et al. 2003; Acharya et al. 1998) – comes from a study comparing the neuropsychological profile of people diagnosed with schizophrenia who experience OHs with those who experience auditory verbal hallucinations (but no OHs) and normal matched controls (Arguedes et al. submitted). We found a small impairment in facial recognition of negative emotions, consistent with amygdala dysfunction, in just the patients who had experienced OHs. However, the most notable impairment was in tests sensitive to orbitofrontal cortex processing. While this structure functions as secondary olfactory cortex, it has not previously been implicated as a putative neural correlate of OHs. Interestingly, it too has a major role in olfactory hedonics (Wilson and Stevenson 2006).

12.4.3 Psychological Contributions

The presence of OHs in otherwise healthy individuals would suggest that psychological accounts are important to consider as well. In this regard, olfactory stimuli often present a special problem relative to those detected by the other senses in that olfactory experiences can be difficult to independently verify (e.g., if you smell pizza, is it *really* there, especially if there are no other cues to its presence?). A further problem concerns misidentification. Human ability to name odors, again in the absence of other contextual cues, is demonstrably poor, and it may be that misidentified innocuous household odors lead to an erroneous report of a hallucination (Stevenson and Case 2005). This could apply with equal force to any of the other conditions discussed in this chapter.

Explanations of visual and auditory hallucinations, especially in psychotic disorders, have also had recourse to psychological accounts (Slade 1994). One possibility concerns the misattribution of mental imagery or involuntary memories to the external milieu – a source-monitoring failure (Bentall et al. 1991). The applicability of this type of account to olfaction is potentially problematic because there is active debate as to whether we can experience olfactory sensation in the absence of any appropriate stimulation (Stevenson and Case 2005). This has led some observers, including ourselves, to question the relevance of this type of account to OHs. However, recent evidence suggests that this conclusion may be premature. Using a novel odor-source-monitoring task, in which participants either imagined or smelled several odors, revealed that participants diagnosed with schizophrenia and experiencing OHs were far less efficient at distinguishing real from imagined odors on a later source-discrimination task than either participants diagnosed with schizophrenia and experiencing auditory verbal hallucinations or healthy controls (Arguedes et al. submitted). These results, albeit on a small sample, suggest that the ability to track "real" and "imagined" stimuli may be compromised in people diagnosed with schizophrenia who experience OHs. Moreover, it raises the interesting possibility that olfactory imagery in this group might differ in some qualitative way relative to normal participants, namely that people who experience OHs may under certain conditions be able to experience odor images.

12.5 Discussion and Conclusions

Hallucinations in the chemical senses are unusual. While OHs have received rather limited attention, even less is known about gustatory and flavor hallucinations. OHs can be observed in healthy normal people, possibly due to the difficulty in verifying the source of an odorant, and they can occur as an independent disease – phantosmia – which may stem initially from damage to the periphery of the olfactory system. OHs can also occur as secondary symptom in several conditions, notably in psychotic disorders, TLE, and migraine, as well as with brain tumors and injuries. These conditions all affect some part of the olfactory system – either centrally and/or peripherally. OHs also occur, albeit more rarely, in post-traumatic stress disorder (PTSD), drug abuse, depression, Parkinson's disease, and Alzheimer's disease.

That the causes of OHs in particular have not been well investigated is to be lamented for two reasons. First, its study may shed light on the genesis of hallucinations in other modalities, especially if conventional theories turn out to be inapplicable to olfaction. Indeed, even if conventional theories are correct – as some recent data on source monitoring may suggest – this provides a new avenue of theory verification, as it may for others. The second reason for lamenting the lack of study of OHs is because of what it may tell us about routine aspects of olfactory perception. Olfaction is emerging as a particularly interesting system to study in regard to consciousness, as its divergent neural organization and unusual psychological properties challenge the generality of a number of contemporary theories of consciousness

(Stevenson 2009b). However, relative to the visual and auditory systems, we know little about olfaction. Many of its basic cognitive capacities (e.g., for imagery and the nature of short-term memory) are unresolved; the neural basis for these and other processes, especially attention, are not understood; and even apparently basic questions such as the "stimulus problem" (i.e., which features of the stimulus are associated with particular types of sensation) remains unsolved. Studying OHs can provide a new approach to addressing many of these questions.

References

Acharya, V., Acharya, J., Luders, H. (1998). Olfactory epileptic auras. *Neurology*, 51, 56–61.

Arguedes, D., Langdon, R., Stevenson, R.J. Neurocognitive processes associated with olfactory hallucinations in schizophrenia (submitted).

Alroe, C.J., McIntyre, J.N.M. (1983). Visual hallucinations: The Charles Bonnet syndrome and bereavement. *Medical Journal of Australia*, 2, 674–675.

Bentall, R., Baker, G., Havers, S. (1991). Reality monitoring and psychotic hallucinations. *British Journal of Clinical Psychology*, 30, 213–222.

Brasic, J.R. (1998). Hallucinations. *Perceptual and Motor Skills*, 86, 851–877.

Bromberg, W., Schilder, P. (1934). Olfactory imagination and olfactory hallucinations. *Archives of Neurology and Psychiatry*, 32, 467–492.

Burstein, A. (1987). Olfactory hallucinations. *Hospital and Community Psychiatry*, 38, 80.

Chen, C., Shih, Y., Yen, D., Lirng, J., Guo, Y., Yu, H., Yiu, C. (2003). Olfactory auras in patients with temporal lobe epilepsy. *Epilepsia*, 44, 257–260.

Crosley, C.J., Dhamoon, S. (1983). Migrainous olfactory aura in a family. *Archives of Neurology*, 40, 459.

Daly, D. (1958). Uncinate fits. *Neurology*, 8, 250–260.

Diamond, S., Freitag, F.G., Prager, J., Gandi, S. (1985). Olfactory aura and migraine. *New England Journal of Medicine*, 312, 1390–1391.

Donnelly, J., Cain, W., Scott, A. (1989). Parosmia among patients with olfactory complaints. *Chemical Senses*, 14, 695.

Efron, R. (1956). The effect of olfactory stimuli in arresting uncinate fits. *Brain*, 79, 267–281.

Faris, A.A., Terrence, C.F. (1989). Limbic system symptomatology associated with colloid cyst of the third ventricle. *Journal of Neurology*, 236, 60–61.

Frasnelli, J., Landis, B., Heilmann, S., Hauswald, B., Huttenbrink, K., Lacroix, J., Leopold, D., Hummel, T. (2004). Clinical presentation of qualitative olfactory dysfunction. *European Archives of Otorhinolaryngology*, 261, 411–415.

Gloor, P. (1990). Experiential phenomenon of temporal lobe epilepsy. Facts and hypotheses. *Brain*, 113, 1673–1694.

Greenberg, M.S. (1995). *Olfactory hallucinations*. In *Science of Olfaction*. Edited by M.J. Serby and K. Chobor. New York, NY: Springer, pp. 467–499.

Hausser-Hauw, C., Bancaud, J. (1987). Gustatory hallucinations in epileptic seizures. *Brain*, 110, 339–359.

Jackson, H., Stewart, P. (1899). Epileptic attacks with a warning of a crude sensation of smell and with the intellectual aura (Dreamy state) in a patient who had symptoms pointing to gross organic disease of the right temporo-sphenoidal lobe. *Brain*, 22, 534–549.

Jablensky, A., Sartorius, N., Ernberg, G., Anker, M., Korten, A., Cooper, J.E., Day, R., Bertelsen, A. (1992). Schizophrenia: Manifestations and incidence and course in different cultures: A world health organization ten country study. *Psychological Medicine*, 20, 1–97.

Kerekovic, M. (1972). The relationship between objective disorders of smell and olfactory hallucinations. *Acta Oto-Rhino-Laryngologica Belgica*, 26, 518–523.

Kopala, L.C., Good, K.P., Honer, W.G. (1994). Olfactory hallucinations and olfactory identification ability in patients with schizophrenia and other psychiatric disorders. *Schizophrenia Research*, 12, 205–211.

Kwapil, T.R., Chapman, J.P., Chapman, L.J., Miller, M.B. (1996). Deviant olfactory experiences as indicators of risk for psychosis. *Schizophrenia Bulletin*, 22, 371–382.

Langdon, R., McGuire, J., Stevenson, R.J., Catts, S.V. (2011). Clinical correlates of olfactory hallucinations in schizophrenia. *British Journal of Clinical Psychology*, 50, 145–163.

Leopold, D. (2002). Distortion of olfactory perception: Diagnosis and treatment. *Chemical Senses*, 27, 611–615.

Levy, L., Henkin, R. (2004). Brain GABA levels are decreased in patients with phantageusia and phantosmia demonstrated by magnetic resonance spectroscopy. *Journal of Computer Assisted Tomography*, 28, 721–727.

Moberg, P.J., Agrin, R., Gur, R.E., Gur, R.C., Turetsky, B.I., Doty, R.L. (1999). Olfactory dysfunction in schizophrenia: A qualitative and quantitative review. *Neuropsychopharmacology*, 21, 325–340.

Nashold, B.S., Wilson, W.P. (1969). Olfactory hallucinations evoked from stimulation of human thalamus. *Confinia Neurologica*, 32, 298–306.

Nesse, R.M., Carli, T., Curtis, G.C., Kleinman, P.D. (1983). Pseudohallucinations in cancer chemotherapy patients. *American Journal of Psychiatry*, 140, 483–485.

Ohayon, M.M. (2000). Prevalence of hallucinations and their pathological associations in the general population. *Psychiatric Research*, 97, 153–164.

Penfield, W., Perot, P. (1963). The brain's record of auditory and visual experience. *Brain*, 86, 595–696.

Phillips, K., Gunderson, C., Gruber, U., Castle, D. (2006). *Delusions of body malodour: the olfactory reference syndrome.* In *Olfaction and the brain.* Edited by W. Brewer, D. Castle, and C. Pantelis. Melbourne: Cambridge University Press, pp. 334–353.

Pryse-Phillips, W. (1971). An olfactory reference syndrome. *Acta Psychiatrica Scandinavica*, 47, 484–509.

Royet, J., Zald, D., Versace, M., Costes, N., Lavenne, F., Koenig, O., Gervais, R. (2000). Emotional responses to pleasant and unpleasant olfactory, visual, and auditory stimuli: a positron emission tomography study. *Journal of Neuroscience*, 20, 7752–7759.

Rozin, P. (1982). "Taste-smell confusions" and the duality of the olfactory sense. *Perception & Psychophysics*, 31, 397–401.

Rubert, S.L., Hollender, M.H., Mehrhof, E.G. (1961). Olfactory hallucinations. *Archives of General Psychiatry*, 5, 313–318.

Sandyk, R. (1981). Olfactory hallucinations in Parkinson's disease. *South African Medical Journal*, 60, 950.

Shenton, M., Dickey, C., Frumin, M., McCarley, R. (2001). A review of MRI findings in schizophrenia. *Schizophrenia Research*, 49, 1–52.

Slade, P.D. (1994). *Models of hallucination: From theory to practice*. In *Neuropsychology of schizophrenia.* Edited by A. David and J. Cutting. Hove: Psychology Press, pp. 245–253.

Sobel, N., Prabhakaran, V., Zhao, Z., Desmond, J.E., Glover, G.H., Sullivan, E.V., Gabrieli, J.E.D. (1998). Sniffing and smelling: Separate subsystems in the human olfactory cortex. *Nature*, 392, 282–286.

Stedman, T.J., Clair, A.L. (1998). Neuropsychological, neurological and symptom correlate of impaired olfactory identification in schizophrenia. *Schizophrenia Research*, 32, 23–30.

Stevenson, R.J. (2009a). *Olfactory hallucinations*. In *Encyclopedia of neuroscience*. Edited by M.D. Binder, N. Hirokawa, and U. Windhorst. Berlin: Springer, pp. 2989–2992.

Stevenson, R.J. (2009b). Phenomenal and access consciousness in olfaction. *Consciousness and Cognition*, 18, 1004–1017.

Stevenson, R.J., Case, T.I. (2005). Olfactory imagery: A review. *Psychonomic Bulletin & Review*, 12, 244–265.

Stevenson, R.J., Langdon, R., McGuire, J. (2011). Olfactory hallucinations in schizophrenia and schizoaffective disorder: A phenomenological survey. *Psychiatry Research*, 185, 321–327.

Stevenson, R.J., Langdon, R. Olfactory function in olfactory and auditory-verbal hallucinators with schizophrenia, and normal controls (in press).

Smith D., Duncan, H. (1992). *Primary olfactory disorders: Anosmia, hyposmia and dysosmia.* In *Science of olfaction.* Edited by M.J. Serby and K. Chobor. New York, NY: Springer, pp. 430–466.

Toone, B.K. (1978). Psychomotor seizures, arterio-venous malformation and the olfactory reference syndrome. *Acta Psychiatrica Scandinavica*, 58, 61–66.

Velakoulis, D. (2006). *Olfactory hallucinations.* In *Olfaction and the brain.* Edited by W. Brewer, D. Castle, and C. Pantelis. Melbourne: Cambridge University Press, pp. 322–333.

West, S.E., Doty, R.L. (1995). Influence of epilepsy and temporal lobe resection on olfactory function. *Epilepsia*, 36, 531–542.

Wilson, D., Stevenson, R.J. (2006) *Learning to smell: Olfactory perception from neurobiology to behavior.* Baltimore, MA: Johns Hopkins University Press.

Chapter 13
Hallucinations of Bodily Sensation

Jan Dirk Blom and Iris E.C. Sommer

13.1 Introduction

A young man complains of a scraping sensation inside the head, as if a plant with pointed leaves were going up and down against the inner surface of his skull. Another person complains of needle pricks in the back. Because he repeatedly feels them while sitting alone in the waiting room, he figures that there must be "some sort of James Bond elevator" rising and falling behind him, carrying a person who sticks the needle in before he silently disappears. A third man complains of a pressure in the corners of the mouth, and the sensation of his toes turning into paws. He has searched the internet for possible explanations, and the one that seems most plausible to him is that he is turning into a werewolf. A patient in her 30s keeps visiting doctors because she is convinced that the left side of her body is much larger than the other side. Not only her extremities, but even the left side of her brain appears to have grown, so that they no longer match the other half. Another young woman feels snakes moving inside her belly, and yet another one cannot feel her body at all. In bed, in the dark, she is unable to tell whether she is lying on her back or otherwise, and needs tactile information to verify her position. An elderly woman suffers from continuous orgasmic sensations which seriously incapacitate her daily functioning. Among a group of 176 medical staff members of a hospital in Springfield, MA, 75% experienced vibrations of their pager or cellular phone while

J.D. Blom, M.D., Ph.D. (✉)
Parnassia Bavo Academy, Parnassia Bavo Group and University of Groningen, Groningen, Kiwistraat 43, 2552 DH, The Hague, The Netherlands
e-mail: jd.blom@parnassia.nl

I.E.C. Sommer, M.D., Ph.D.
Department of Psychiatry, Rudolf Magnus Institute of Neuroscience, University Medical Center Utrecht, Heidelberglaan 100, 3584 CX, Utrecht, The Netherlands
e-mail: i.sommer@umcutrecht.nl

J.D. Blom and I.E.C. Sommer (eds.), *Hallucinations: Research and Practice*,
DOI 10.1007/978-1-4614-0959-5_13, © Springer Science+Business Media, LLC 2012

the devices are not vibrating (Rothberg et al. 2010). What do experiences such as these have in common? As varied as they may be, they all represent some aberration of corporeal sensation. In this chapter, these phenomena are subsumed under the heading of "bodily hallucinations," an umbrella term used to designate a group of widely divergent corporeal percepts, phenomenologically rich in nature, yet poorly described and often equally poorly understood.

13.2 The Psychopathology of Bodily Sensation

Bodily hallucinations are also known as hallucinations of bodily sensation and body sensation hallucinations. They are hallucinatory percepts which have bearing on the body itself, as opposed to visual, auditory, olfactory, and gustatory hallucinations, which all constitute percepts projected into the extracorporeal world. Although the latter are sometimes designated as "hallucinations of the distance senses" (Berrios 1982), bodily hallucinations defy characterization as mere hallucinations "of the proximity senses" or "of intracorporeal sensation." They are also more variegated than hallucinations of the "distance senses," if only because the latter are all unimodal in nature. Table 13.1 provides a classification of the various types of bodily hallucination and the sensory modalities involved. The exact delineation of these phenomena may be somewhat arbitrary, and different classifications may well yield different categories (for a well-considered alternative, see Gibson 1966), but what this overview illustrates is that they are definitely multimodal in nature, and that they differ in many ways from hallucinations of the so-called distance senses.

Before we proceed to discuss these types of hallucination, a few introductory remarks are in place. First of all that, historically speaking, bodily sensation has always been subordinated to the other senses, with touch only reluctantly being designated as a "fifth sense" by Aristotle (Aristotle 1968), and pain being classified for centuries as a "bodily passion" rather than a percept (Starobinski 1990). The exact source of this conceptual uneasiness is unknown, but it would seem to be connected with varying philosophical and biomedical insights into the nature of the senses – referring to the sense organs themselves, as well as to the mental constructs based on perceptual information – and associated concepts of "reality" (Rodaway 1994). It was only near the end of the eighteenth century that bodily sensation became an object of study in its own right, with notions such as coenesthesis (the sensation of physical existence) and *Gemeingefühl* ("common sensation") reigning supreme (Jütte 2005). But this Golden Age of bodily sensation lasted no longer than a century. Late nineteenth- and early twentieth-century discoveries of modality-specific neural substrates such as the pain spots (Von Frey 1894) and nociception (Sherrington 1906) gave the deathblow to the coordinating concepts of olden days (Boring 1942). Ever since, the focus of biomedical research has been primarily on constituent parts of bodily sensation rather than on bodily sensation as a whole.

Secondly, it should be noted that some types of hallucination listed in Table 13.1 would seem to pose irreconcilable conceptual and methodological problems, in the

Table 13.1 Classification of bodily hallucinations and concomitant sensory modalities

Sensory modality	Type of hallucination
1. Exteroceptive modality	Tactile (haptic) hallucination
2. Interoceptive modality	Somatic hallucination, visceral hallucination
3. Proprioceptive modality	Proprioceptive hallucination
4. Kinesthetic modality	Kinesthetic hallucination
5. Vestibular modality	Vestibular hallucination
6. Coenesthetic modality	Coenesthetic hallucination
7. Pain modality	Hallucinated pain, central pain
8. Sexual modality	Sexual hallucination
9. Temperature modality	Thermal (thermic) hallucination

sense that they appear to stretch the definition of the term "hallucination" beyond its rightful limits. Pain, vertigo, and sexual arousal are cases in point of subjective sensations that can only counterintuitively be pictured as hallucinations. Below we will explain why the concept of hallucinated pain arose in the first place, but the very existence of concepts such as these begs the question of whether "imagined" or "unreal" subjective experiences can ever be said to exist. As observed by the American philosopher Saul Kripke, "For a sensation to be felt as pain is for it to be pain" (Kripke 1980). This simple statement pretty much sums up the central problem of the bodily hallucination concept and explains many of the difficulties we meet in clinical practice, notably in trying to differentiate between bodily hallucinations and bodily percepts due to unexplained somatic conditions.

That being said, we will now turn to the major classes of bodily hallucination, outline the phenomenological characteristics of their various members, and summarize the findings regarding their neurobiological correlates.

13.3 Tactile Hallucinations

Tactile hallucinations, also known as tactile phantasmata, haptic hallucinations, superficial hallucinations, touch hallucinations, and hallucinations of touch (Blom 2010; Sims 2003), are corporeal percepts seemingly evoked by a stimulus from outside the body. Examples are the needle pricks experienced by the young man described above, and the phantom vibrations experienced by medical staff members, a condition recently introduced into the literature as "phantom vibration syndrome" (Rothberg et al. 2010). The group of tactile hallucinations comprises many other variants, an overview of which is given in Table 13.2.

Among them are formicative hallucinations, characterized by the sensation of ants or other bugs crawling upon or beneath the skin (Bers and Conrad 1954). When attributed to the use of cocaine, these hallucinations are referred to as Magnan's sign, Magnan-Saury's sign, or "cocaine bugs," and in the context of amphetamine use as "crank bugs" (Siegel 1978). Hygric hallucinations, mimicking the sensation

Table 13.2 Classification of tactile hallucinations (Adapted from Blom 2010)

Type of tactile hallucination	Characterization
Allachesthesia (allesthesia, atopognosis)	A horizontal or diagonal displacement of localization for touch, confined to the ipsilateral side of the body
Allochiria	A displacement of localization for touch to the contralateral side of the body
Auditory-tactile synesthesia	A tactile hallucination triggered by a regular auditory percept
Extracampine hallucination	A hallucination experienced outside the regular field of perception (in the case of tactile hallucinations: e.g., in the hair or the brain)
Formicative hallucination (formication, dermatozoic hallucination, insect hallucination)	The hallucinated sensation of ants or other bugs crawling upon or beneath the skin
Hygric hallucination	A tactile hallucination mimicking the sensation of wetness and fluids upon the skin
Mirror-touch synesthesia	A tactile hallucination triggered by observing touch to another person
Phantom alloesthesia	A tactile hallucination experienced beneath the stump of an amputated limb, following touch to the contralateral, remaining limb
Phantom vibration syndrome	A tactile hallucination mimicking the vibrations of an electronic device such as a pager or a cellular phone
Stereognosic hallucination (stereognostic hallucination)	A tactile hallucination characterized by the sensation of touching one or more solid objects
Spontaneous stereognosic sensation (spontaneous stereognostic sensation)	A stereognosic hallucination mimicking the sensation of a solid object held in the palm of one's hand
Synchiria (diplohaptia)	A bilateral tactile sensation produced by unilateral tactile stimulation
Tactile polyesthesia	A tactile phenomenon in which a single tactile stimulus is at first perceived and localized properly, and then perceived again in one or more different parts of the body
Visual-tactile synesthesia	A tactile hallucination which is triggered by a regular visual percept

of wetness and fluids upon the skin (Jaspers 1997), are considered quite rare. The same holds true for stereognosic (or stereognostic) hallucinations, characterized by the sensation of one or more solid objects touching the skin, and a variant called spontaneous stereognosic sensation, which mimics the sensation of a solid object held in the palm of one's hand (Critchley 1953).

Allachesthesia and allochiria are hallucinations characterized by a "confusion of sides" rather than a spontaneous tactile sensation (Halligan et al. 1992). In allachesthesia, stimuli are experienced upon actual touch, affecting the ipsilateral side of the body, although in a different place, whereas in allochiria they are experienced at the contralateral side (Meador et al. 1991). Phenomenologically, both types of percept are related to synchiria – in which unilateral tactile stimulation produces a bilateral tactile percept – and tactile polyesthesia, in which a single tactile stimulus is at first perceived and localized properly, and then perceived again in one or more different parts of the body (Critchley 1953). Perhaps these phenomena are also related to extracampine hallucinations (i.e., tactile hallucinations experienced outside the regular field of tactile perception; see Bleuler 1903; Chan and Rossor 2003) and tactile hallucinations triggered by a regular percept in any of the other sensory modalities, such as a sound (auditory-tactile synesthesia) or a specific color (visual-tactile synesthesia) (Ward 2007, see also Chap. 7). Yet another variant of this class of phenomena is mirror-touch synesthesia, in which a tactile hallucination is triggered by observing touch to another person (Banissy et al. 2009).

The neurobiological correlates of these phenomena have not been fully elucidated, but the cross-activation hypothesis is traditionally considered an attractive explanation (Parish 1897; Harrison 2001). It suggests that perceptual information may feed (or "cross over") from one sensory modality into another or, when the information remains within the tactile modality, from one representational center into another. Competing models attribute the mediation of such secondary hallucinatory phenomena to (1) disinhibition of adjacent brain areas, (2) an increase in feedback connections between successive stages of the sensory hierarchy, and (3) excess activity between those successive stages, for example, as a result of a disinhibition of feedback connections (Ramachandran and Hubbard 2001).

13.4 Somatic Hallucinations

Somatic (i.e., somesthetic, visceral) hallucinations are experienced within the body rather than at its surface. The woman described above, with the sensation of snakes moving about in her belly, as well as the young man with the scraping sensation inside the skull, provide apt examples of this class of phenomena. Although it is very likely that the phenomena at hand were hallucinations proper, one can never rule out with certainty whether they may rather be sense perceptions caused by some unknown somatic condition. The same holds true for other types of somatic hallucination, an overview of which is given in Table 13.3. This basic uncertainty is incidental to this group of phenomena, and practically insoluble given the state of the art of present-day medicine, but in clinical practice we use the rule of thumb that we are dealing with a somatic hallucination when the percept at hand is inexplicable in physiological terms, especially so when it is provided with a delusional explanation by the patient (Berrios 1982).

Somatic hallucinations are notoriously hard to classify, but traditionally the following phenomena are distinguished. Audioalgesic synesthesias are painful sensations triggered by a regular auditory percept (Jacome and Gumnit 1979). Not unlike the types of synesthesia discussed above, they tend to be attributed to intermodal cross-activation. Hallucinated headache is a painful sensation experienced within the head, occurring in the absence of any demonstrable organic cause. It may be considered similar to tension headache (i.e., psychogenic headache), but there is a

Table 13.3 Classification of somatic hallucinations (Adapted from Blom 2010)

Type of somatic hallucination	Characterization
Audioalgesic synesthesia	A painful bodily sensation triggered by a regular auditory percept
Hallucinated headache	A somatic hallucination characterized by pain in the head, attributed to psychic "emptiness"
Hallucinated pain	A painful somatic hallucination attributed to psychic "emptiness"
Painful somatosensory hallucination	A painful somatic hallucination
Paresthesia	A sensation of "pins and needles," usually located in or beneath the skin
Somatosensory aura	A somatic hallucination experienced within a clearly defined region of the body, attributed to epilepsy

subtle conceptual difference here that is worth noting. The concept of hallucinated headache evolved out of the psychoanalytic dictum that any type of hallucination can be caused by hunger – either actual hunger or affective hunger (Forrer 1962). Rather than attributing hallucinated headache to muscular or peripheral nervous stimulation, Forrer conceptualized it as a means to fill a subjectively experienced "void" or "emptiness." Elaborating on that theme, he expanded the hallucinated headache concept to arrive at hallucinated pain (Forrer 1968). For this type of pain, he designed a diagnostic triad, comprising (1) description by the patient in vague, uncertain, inappropriate, or bizarre terms, often violating neuroanatomical boundaries; (2) contingent circumstances such as actual or psychological emptiness, hunger, or thirst; and (3) improvement or disappearance of the pain with oral activity such as eating or drinking. A fundamental objection against Forrer's concept of hallucinated pain is the above-mentioned argument of Kripke's that "for a sensation to be felt as pain is for it to be pain." However, the concept finds some support from the experience of dentists and surgeons that physical pain is not seldom indicated by patients moments before a tooth or other body part is actually being touched, and of hypnotists who find that pain can be evoked as well as alleviated through suggestion. Although clinical experiences such as these may not suffice to legitimize the concept of hallucinated pain, they do confirm that the dynamics of painful sensations involve more than tissue damage and nociception alone. Stripped from its psychoanalytic overtones, the concept is known today as "painful somatosensory hallucination," and operationalized in terms of a spontaneously occurring pain without adequate external stimuli (Bär et al. 2002). The neural correlates of that type of pain are as yet unclear, but an fMRI experiment by Bär et al. (2002) suggests that – in their patient, at least – a parietal area of the posterior medial wall may be involved in its mediation. All in all, it has evolved from a psychoanalytic concept to a neuroscientific one that comes close to the concepts of central pain (which is associated with damage to specific areas of the brain, brainstem, or spinal cord; see Canavero and Bonicalzi (2007) as well as Chap. 14) and phantom pain (i.e., pain experienced in an amputated, aplasic, or anesthetic limb; see Brugger et al. 2000, as well as Chap. 16, this volume).

Somatic hallucinations can also occur in the context of paroxysmal neurological disorders such as epilepsy or migraine. In such cases, they are called somatosensory aurae. Theoretically, an aura may mimic any type of bodily sensation, but in clinical practice, it tends to present chiefly in the form of paresthesias (Lüders et al. 1998). These tingling somatic sensations in or beneath the skin ("pins and needles") also occur in the context of peripheral nerve compression and a variety of other conditions, but among the somatosensory aurae they are reportedly the most prevalent. Other well-known manifestations of somatosensory aurae are numbness, pain, rising epigastric sensations, thermal hallucinations, splitting of the body image (i.e., the subjective sensation of one's body being split in two; see Podoll and Robinson 2002), and other body schema illusions.

13.5 Sexual Hallucinations

The term sexual hallucination refers to a variety of hallucinated percepts that are sexually charged, such as visual hallucinations with a sexually explicit or pornographic quality, auditory hallucinations delivering a message with a sexual connotation, and bodily hallucinations experienced in an erogenic zone. It is the latter type of hallucination that concerns us here. When experienced in the penis or vulva, such hallucinations are referred to as genital hallucinations. It should be noted, however, that hallucinations experienced in the mammae, buttock, anus, mouth, and in fact any other erogenic zone can also qualify as sexual hallucinations, depending on their ability to arouse sexual feelings. When occurring during or after anesthesia, the term nitrous oxide hallucination applies (Bennett 1980; Balasubramaniam and Park 2003). Persistent sexual arousal syndrome (PSAS) is a condition occurring mostly in women between ages 40 and 65, characterized by a persistent feeling of genital arousal – or orgasmic sensation – that is incongruous with the patient's affect (Leiblum and Nathan 2001). An example is the description above of the elderly woman whose continuous orgasmic sensations had nothing to do with hyperarousal or hypersexuality, and were experienced by her as ego-dystonic. The pathophysiology of PSAS is largely unknown, although a few case reports exist about associations with the use of trazodone and other drugs, and with pelvic arteriovenous malformations.

13.6 The Coenesthesiopathies

The coenesthesiopathies comprise a group of somatic hallucinations that alter one's sense of physical existence. The young man described above whose subjective physical changes made him think he was turning into a werewolf, and the woman who was unable to feel her body at all, suffered from coenesthesiopathy (for a detailed account, see Blom et al. 2010). More specifically, the man was diagnosed with clinical lycanthropy, and the woman with acenesthesia (see below).

Table 13.4 Classification of coenesthesiopathies (Adapted from Blom et al. 2010)

Type of coenesthesiopathy	Characterization
Coenesthesiopathy (cenesthopathy)	A pathological alteration in the sense of bodily existence, caused by aberrant bodily sensations
Acoenesthesiopathy (acenesthesia, total asomatognosia)	A total absence of the sense of bodily existence
Hypercoenesthesiopathy	A hypertrophic alteration in the sense of bodily existence, caused by aberrant bodily sensations
Hypocoenesthesiopathy	A hypotrophic alteration in the sense of bodily existence, caused by aberrant bodily sensations
Paracoenesthesiopathy	A qualitative alteration in the sense of bodily existence, caused by aberrant bodily sensations
Clinical lycanthropy	The delusional conviction of having changed into a wolf or having the potential of becoming a wolf

The concept of coenesthesiopathy proceeds from the late eighteenth-century notion of coenesthesis (coenesthesia, cenesthesis, or in German, *Gemeingefühl*), which refers to the "common sensation" or "common general sensibility" arising from the sum of all bodily sense impressions. As explained by Tanzi (1909), "The united and incessant exercise of the sensory functions is the perennial source, not only of all special information that is supplied to us regarding the external world and our body, but also of a general and indistinct, but often very active, consciousness that enables us from moment to moment to recognize the functional intactness of the body in all its parts, including those which, owing to being isolated and in a normal condition, never make themselves felt." Thus, coenesthesis stands for a strictly intimate, personal feeling of physical existence (Schiller 1984), characterized in general terms as "the sense of feeling well." The notion that this "sense of feeling well" can be compromised by all sorts of bodily discomfort was recognized right away, but interestingly the concomitant concept of coenesthesiopathy was introduced only a century later when the general scientific interest in coenesthesis was already on the wane (Deny and Camus 1905a, b; Dupré and Camus 1907). Table 13.4 provides an overview of this group of disorders, which were found to be so prevalent in psychosis (i.e., 18% in a group of 223 individuals with a clinical diagnosis of schizophrenia) that they led Huber (1957, 1992) to conceptualize a special subgroup of schizophrenia, featuring in the International Classification of Diseases as cenesthopathic schizophrenia (WHO 1992; Jenkins and Röhricht 2007).

If we follow the classification devised by Deny and Camus, the woman who was unable to feel her body should be diagnosed with hypocoenesthesiopathy (i.e., a strongly diminished sense of bodily existence) or even acoenesthesiopathy (a total absence of the sense of bodily existence, see Meador et al. 2000). The man who thought he was turning into a werewolf, on the other hand, experienced a hypertrophic alteration in the sense of bodily existence, diagnosable as a hypercoenesthesiopathy

Fig. 13.1 Lycanthropy, artist's impression (Copyright 2011 by Marten Blom)

or paracoenesthesiopathy. And yet, because of the werewolf theme, he also fits the criteria for clinical lycanthropy, a rare delusional conviction that should be distinguished from lycanthropy (see Fig. 13.1) as described in mythological accounts referring to the actual transformation of humans into wolves and back again (Moselhy 1999; Garlipp et al. 2004).

Today the coenesthesiopathies would seem to be underreported, underdiagnosed, and undertreated (Blom et al. 2010). And yet one should be cautious to establish those diagnoses too soon. As demonstrated by the following case vignette, the differential diagnosis includes a number of somatic disorders, some of which may require urgent treatment.

A man diagnosed with chronic schizophrenia had repeatedly complained that the size of his feet had increased to the extent that the wearing of shoes had become a painful affair. He also complained that his rings had become too narrow for his fingers, and that his hat had become too small for his head. He interpreted these signs as messages from God that he was not allowed to wear any shoes, jewelry, or hats. While participating in an MRI study for entirely different reasons, it turned out that he suffered from a large pituitary tumor. That tumor had produced growth hormone responsible for an actual increase of the size of the body parts he had mentioned.

13.7 Discussion

As the neuronal circuits within the parietal lobe and thalamus play an important role in the mediation of bodily percepts and the representation of the body scheme, it would seem there – and in the interconnecting white-matter tracts – that we should localize the neurobiological correlates of hallucinations of bodily sensation. Imaging studies indicate that this is at least partly correct. Arzy et al. (2006) found damage to the right premotor cortex in a patient with acoenesthesiopathy, and Bär et al. (2002) found transient activation of the precuneus and supplementary motor cortex in a patient with painful somatosensory hallucinations. Functional imaging studies also indicate that bodily hallucinations and artificially induced bodily sensations have shared correlates in primary and secondary somatosensory cortical areas (Bär et al. 2002; Shergill et al. 2001). This may well be the reason why the two types of percept can be experienced as equally "real" (Sommer et al. 2005). Another, and only recently "rediscovered" structure that would seem to play a major role in the mediation of bodily sensations and corporeal awareness is the insula, also referred to as the "fifth lobe of the brain" (Brooks et al. 2005; Craig 2009). Featuring a somatotopic organization, this structure may well hold the key to our future understanding of the pathophysiology of bodily hallucinations.

All things considered, however, we must conclude that our understanding of bodily hallucinations is still very limited. Phenomenologically, these hallucinated percepts would seem to lie on a continuum with hyperschematia or left size distortion (i.e., the conviction that the left side of the body is larger than the right one, as described by the woman above), body schema illusions (Kelly 1992), body dysmorphic disorder, phantom sensations, macrostereognosia (i.e., the sensation that an object held in the palm of one's hand, for example, is significantly more bulky and massive than it actually is; see Halpern 1945), macro- and microsomatognosia (i.e., the sensation that the body, or a part thereof, is significantly larger or smaller than it actually is; see Podoll and Robinson 2000), the Isakower phenomenon (a sleep-related group of phenomena that includes the sensation of a shrinking or swelling of the hands, the sensation of having something dry, crumpled or sandy on the skin or in the mouth, and the sensation of floating; see Isakower 1938 and Dann 1992), and numerous other symptoms and syndromes described in the literature, as well as bodily percepts due to unknown somatic conditions. It may be tempting to attribute our lack of understanding of these phenomena to the limited availability – as well as the modest size – of neurobiological studies in this area, but an equally grave problem would seem to be at the level of conceptualization. As noted above, the conceptual distinction between sense perceptions and hallucinated percepts allows for a practicable operationalization in clinical and neuroscientific terms as long as the "distance senses" are involved but fails to do so in the case of many types of bodily sensation. The notion that any type of percept can be either "real" or "imagined" may well be true, but it quickly loses its meaning when our judgment must be made without the possibility of the external verification of sensory input signals.

13.8 Conclusion

The umbrella term "bodily hallucination" refers to a group of widely divergent and only rudimentarily understood corporeal sensations. In clinical practice it may be helpful to let the diagnosis of bodily hallucination depend on the co-occurrence of a delusional explanation by the patient, but even this procedure is not airtight. Future studies seeking to enhance our understanding of these phenomena will have to address – as always, in biomedicine – neurobiological as well as conceptual issues.

References

Aristotle (1968). *De anima. Books II and III.* Translated by Hamlyn, D.W. Oxford: Clarendon Press.

Arzy, S., Overney, L.S., Landis, Th., Blanke, O. (2006). Neural mechanisms of embodiment: Asomatognosia due to premotor cortex damage. *Archives of Neurology*, 63, 1022–1025.

Balasubramaniam, B., Park, G.R. (2003). Sexual hallucinations during and after sedation and anaesthesia. *Anaesthesia*, 58, 549–553.

Banissy, M.J., Cohen Kadosh, R., Maus, G.W., Walsh, V., Ward, J. (2009). Prevalence, characteristics and a neurocognitive model of mirror-touch synaesthesia. *Experimental Brain Research*, 198, 261–272.

Bär, K.-J., Gaser, Chr., Nenadic, I., Sauer, H. (2002). Transient activation of a somatosensory area in painful hallucinations shown by fMRI. *NeuroReport*, 13, 1–4.

Bennett, C.R. (1980). Nitrous oxide "hallucinations". *Journal of the American Dental Association*, 101, 595–597.

Berrios, G.E. (1982). Tactile hallucinations: conceptual and historical aspects. *Journal of Neurology, Neurosurgery, and Psychiatry*, 45, 285–293.

Bers, N., Conrad, K. (1954). Die chronische taktile Halluzinose. *Fortschritte der Neurologie und Psychiatrie*, 22, 254–270.

Bleuler, E. (1903). Extracampine Hallucinationen. *Psychiatrisch-neurologische Wochenschrift*, 25, 261–264.

Blom, J.D. (2010). *A dictionary of hallucinations*. New York, NY: Springer.

Blom, J.D., Neven, A., Aouaj, Y., Jonker, B. Hoek, H.W. (2010). De coenesthesiopathieën. *Tijdschrift voor Psychiatrie*, 52, 695–704.

Boring, E.G. (1942). *Sensation and perception in the history of experimental psychology*. New York, NY: Appleton-Century-Crofts.

Brooks, J.C.W., Zambreanu, L., Godinez, A., Craig, A.D., Tracey, I. (2005). Somatotopic organisation of the human insula to painful heat studied with high resolution functional imaging. *NeuroImage*, 27, 201–209.

Brugger, P., Kollias, S.S., Müri, R.M., Crelier, G., Hepp-Reymond, M.-C., Regard, M. (2000). Beyond re-membering: phantom sensations of congenitally absent limbs. *Proceedings of the National Academy of Sciences USA*, 97, 6167–6172.

Canavero, S., Bonicalzi, V. (2007). *Central pain syndrome. Pathophysiology, diagnosis and management*. Cambridge: Cambridge University Press.

Chan, D., Rossor, M. (2003). "-but who is that on the other side of you?" Extracampine hallucinations revisited. *Lancet*, 360, 2064–2066.

Craig, A.D. (2009). How do you feel -- now? The anterior insula and human awareness. *Nature Reviews. Neuroscience*, 10, 59–70.

Critchley, M. (1953). *The parietal lobes*. London: Edward Arnold & Co.

Dann, O.T. (1992). The Isakower phenomenon revisited: a case study. *International Journal of Psycho-Analysis*, 73, 481–491.

Deny, G., Camus, P. (1905a). Sur une forme d'hypochondrie aberrante due à la perte de la conscience du corps. *Revue Neurologique*, 9, 461–467.

Deny, G., Camus, P. (1905b). Sur un cas de délire métabolique de la personnalité lié à des troubles de la cœnesthésie. *Archives de Neurologie*, 20, 257–268.

Dupré, E., Camus, P. (1907). Les cénesthopathies. *L'Encéphale*, 2, 616–631.

Forrer, G.R. (1962). Hallucinated headache. *Psychosomatics*, 3, 120–128.

Forrer, G.R. (1968). The hallucinated pain syndrome: A new diagnostic triad. *Headache: The Journal of Head and Face Pain*, 8, 71–75.

Garlipp, P., Gödecke-Koch, T., Dietrich, D.E., Haltenhof, H. (2004). Lycanthropy - psychopathological and psychodynamical aspects. *Acta Psychiatrica Scandinavica*, 109, 19–22.

Gibson, J.J. (1966). *The senses considered as perceptual systems*. Boston, MA: Houghton Mifflin Company.

Halligan, P.W., Marshall, J.C., Wade, D.T. (1992). Left on the right: allochiria in a case of left visuo-spatial neglect. *Journal of Neurology, Neurosurgery and Psychiatry*, 55, 717–719.

Halpern, L. (1945). Macrostereognosia. An unusual phenomenon in a case of thalamic syndrome. *Journal of Nervous and Mental Disease*, 102, 260–264.

Harrison, J. (2001). *Synaesthesia. The strangest thing*. Oxford: Oxford University Press.

Huber, G. (1957). Die coenästhetische Schizophrenie. *Fortschritte der Neurologie und Psychiatrie*, 25, 491–520.

Huber, G. (1992). 'Cenesthetic schizophrenia'- a subtype of schizophrenic disease. *Neurology, Psychiatry and Brain Research*, 1, 54–60.

Isakower, O. (1938). A contribution to the patho-psychology of phenomena associated with falling asleep. *International Journal of Psycho-Analysis*, 19, 331–345.

Jacome, D.E., Gumnit, R.J. (1979). Audioalgesic and audiovisuoalgesic synesthesias: Epileptic manifestation. *Neurology*, 29, 1050–1053.

Jaspers, K. (1997). *General psychopathology. Volume 1*. Translated by Hoenig, J. and Hamilton, M.W. Baltimore, MA: Johns Hopkins University Press.

Jenkins, G., Röhricht, F. (2007). From cenesthesias to cenesthopathic schizophrenia: A historical and phenomenological review. *Psychopathology*, 40, 361–368.

Jütte, R. (2005). *A history of the senses. From Antiquity to cyberspace*. Cambridge: Polity Press.

Kelly, C. (1992). *Status and investigation of body image delusions*. In: *Delusions and hallucinations in old age*. Edited by Katona, C. and Levy, R. London: Gaskell, pp. 115–135.

Kripke, S. (1980). *Naming and necessity*. Cambridge, MA: Harvard University Press.

Leiblum, S.R., Nathan, S.G. (2001). Persistent sexual arousal syndrome: A newly discovered pattern of female sexuality. *Journal of Sex and Marital Therapy*, 27, 365–380.

Lüders, H., Acharya, J., Baumgartner, C., Banbadis, S., Bleasel, A., Burgess, R., Dinner, D.S., Ebner, A., Foldvary, N., Geller, E., Hamer, H., Holthausen, H., Kotagal, P., Morris, H., Meencke, H.J., Noachtar, S., Rosenow, F., Sakamotot, A., Steinhoff, B.J., Tuxhorn, I., Wyllie, E. (1998). Semiological seizure classification. *Epilepsia*, 39, 1006–1013.

Meador, K.J., Allen, M.E., Adams, R.J., Loring, D.W. (1991). Allochiria vs allesthesia. Is there a misperception? *Archives of Neurology*, 48, 546–549.

Meador, K.J., Loring, D.W., Feinberg, T.E., Lee, G.P., Nichols, M.E. (2000). Anosognosia and asomatognosia during intracarotid amobarbital inactivation. *Neurology*, 55, 816–820.

Moselhy, H.F. (1999). Lycanthropy: New evidence of its origin. *Psychopathology*, 32, 173–176.

Parish, E. (1897). *Hallucinations and illusions. A study of the fallacies of perception*. London: Walter Scott.

Podoll, K., Robinson, D. (2000). Macrosomatognosia and microsomatognosia in migraine art. *Acta Neurologica Scandinavica*, 101, 413–416.

Podoll, K., Robinson, D. (2002). Splitting of the body image as somesthetic aura symptom in migraine. *Cephalalgia*, 22, 62–65.

Ramachandran, V.S., Hubbard, E.M. (2001). Synaesthesia - A window into perception, thought and language. *Journal of Consciousness Studies*, 8, 3–34.

Rodaway, P. (1994). *Sensuous geographies. Body, sense and place.* London: Routledge.
Rothberg, M.B., Arora, A., Hermann, J., Kleppel, R., St Marie, P., Visintainer, P. (2010). Phantom vibration syndrome among medical staff: a cross sectional survey. *British Medical Journal*, 341, bmj.c6914.
Schiller, F. (1984). Coenesthesis. *Bulletin of the History of Medicine*, 58, 496–515.
Shergill, S.S., Cameron, L.A., Brammer, M.J., Williams, S.C.R., Murray, R.M., McGuire, P.K. (2001). Modality specific correlates of auditory and somatic hallucinations. *Journal of Neurology, Neurosurgery and Psychiatry*, 71, 688–690.
Sherrington, C.S. (1906). *The integrative action of the nervous system.* New York, NY: Scribners.
Siegel, R.K. (1978). Cocaine hallucinations. *American Journal of Psychiatry*, 135, 309–314.
Sims, A. (2003). *Symptoms in the mind. An introduction to descriptive psychopathology. Third edition.* London: Saunders.
Sommer, I.E.C., Selten, J.P., Sierink, H.D., Bloemkolk, D. (2005). Bizarre lichamelijke klachten als uiting van een schizofrene stoornis. *Nederlands Tijdschrift voor Geneeskunde*, 149, 217–220.
Starobinski, J. (1990). A short history of bodily sensation. *Psychological Medicine*, 20, 23–33.
Tanzi, E. (1909). *A text-book of mental diseases.* Translated by W. Ford Robertson and T.C. Mackenzie. London: Rebman Limited.
Von Frey, M. (1894). Beiträge zur Physiologie des Schmerzsinnes. *Berichte über die Verhandlungen der Königlich Sächsischen Gesellschaft der Wissenschaften zu Leipzig*, 46, 185–196, 283–296.
Ward, J. (2007). Acquired auditory-tactile synaesthesia. *Annals of Neurology*, 62, 429–430.
World Health Organization (1992). *The ICD-10 classification of mental and behavioral disorders. Clinical descriptions and diagnostic guidelines.* Geneva: World health Organization.

Chapter 14
Hallucinatory Pain: Central Pain

Sergio Canavero

14.1 Introduction and Historical Note

Pain from physical trauma is a common experience. Pricking a finger, for example, engenders acute, temporary pain, classified by physiologists as nociceptive pain since specialized receptors (i.e., nociceptors) are involved in their mediation. The mere possibility that pain may also arise "spontaneously," that is to say in the absence of a direct noxious stimulus, has been considered an impossibility until the latter half of the nineteenth century. In fact such "spontaneous" pains are very much akin to visual and auditory hallucinations, as they stem from a focal disturbance of the brain.

In 1521 the French surgeon Ambroise Paré (c. 1510–1590) described with great clarity phantom limbs (see also Chap. 16), and in 1866 the US neurologist Silas Weir Mitchell (1829–1914) published his first account of phantom limbs in the *Atlantic Monthly* as an anonymously written short story, *The Case Of George Dellow*, wherein the protagonist has both legs amputated during the American Civil War and upon awakening is suddenly aware of a "sharp cramp" in the left leg, which, however, is missing. In 1888 "Frau R.," having agonized with "terrible pains" for 2 years, committed suicide; not even opium could relieve her pains, which followed cerebral infarction. By studying this case in 1891, Ludwig Edinger (1855–1918), working in Frankfurt am Main, published his seminal paper in which he hypothesized the existence of "centrally arising pains" (Edinger 1891). The concept that pain could be released independently of peripheral nociceptive stimulation was born.

Actually, patients suffering from central pain (CP) had already been described by other authors, both in western and eastern sources, at least since the early 1800s; nonetheless, nobody actually intuited that brain damage alone could trigger spontaneous, unrelenting pain. In 1938 George Riddoch (1888–1947) published his three-part

S. Canavero, M.D., F.M.G.E.M.S. (✉)
Turin Advanced Neuromodulation Group, Turin, Italy
e-mail: sercan@inwind.it

J.D. Blom and I.E.C. Sommer (eds.), *Hallucinations: Research and Practice*,
DOI 10.1007/978-1-4614-0959-5_14, © Springer Science+Business Media, LLC 2012

review of CP in *The Lancet*, thereby establishing it as a definite nosological entity (Riddoch 1938). By that time, it was clear that CP could be caused by brain damage at various levels (cortex, subcortex, thalamus, brainstem), but also by spinal cord injury (so-called paraplegia pain), as amply noted during World War I.

In 1906 Jules Dejerine (1849–1917) and Gustave Roussy (1874–1948) described the so-called thalamic syndrome, in which pain was experienced among other symptoms and signs due to thalamic stroke (Dejerine and Roussy 1906). A few years later, a scholarly paper by Henry Head (1861–1940) and Gordon Holmes (1876–1965) defined the sensory deficits that characterize CP (Head and Holmes 1911). Unfortunately, an exclusive focus on the thalamus sidelined the cortex and other brain areas, and for the best part of the twentieth century, CP was attributed mainly to thalamic stroke. Only with the advent of computerized tomography (CT) in 1973 and magnetic resonance imaging (MRI) in 1981 it became clear that pure thalamic lesions are not the primary triggers for CP (Canavero and Bonicalzi 2011). Finally, a series of epidemiological studies found CP to be a major, under-recognized nosological entity.

14.2 Definition and Epidemiology

CP is an all-encompassing term which refers to pain, dysesthesias, paresthesias, and – as recently realized – pruritus (Canavero et al. 1997; for an explanation of all these terms see Table 14.1) initiated by a central nervous system (CNS) lesion impinging on or interfering with the spinothalamic tract en route to the parietal somatosensory areas, i.e., the path responsible for the conduction of nociceptive and thermal stimuli. The anatomical basis of CP was highlighted in a major work by Cassinari and Pagni (1969) and was later confirmed by electrophysiological studies employing laser-evoked potentials, which specifically explored the pain- and temperature-conducting pathways in the CNS (Canavero and Bonicalzi 2011).

Before the term "central pain" gained general acceptance, various other terms were proposed, including thalamic pain, pseudothalamic pain, and insular pain. CP after spinal injury was priorly referred to as paraplegia pain or remote pain (as opposed to end-zone pain), among many other terms. All those terms have now been discarded in favor of *central pain of brain or brainstem origin* (as seen in medullary stroke, i.e., Wallenberg's syndrome) and *central pain of cord origin* (Canavero and Bonicalzi 2011).

Long considered to be a neurological rarity, we now know that no less than 9–10% of all CNS strokes (both ischemic and hemorrhagic, see Andersen et al. 1995 and Bowsher 2001); no less than 20–25% of all spinal cord injuries (SCI), including syringomyelia (Siddall et al. 1999); almost 20% of all cases of multiple sclerosis (MS) (Solaro et al. 2004); perhaps 2% of all cancer cases (Gonzales et al. 2003); and an undefined number of other neurological conditions coincide with CP, thus running in the millions worldwide (see further series in Canavero and Bonicalzi 2011). In the USA alone, no less than 600,000 patients suffer from CP, which is ten times more than previously estimated. Males are generally overrepresented, in older age in cases

Table 14.1 Types of central pain

Type of pain	Characterization
Allodynia	Pain due to a stimulus which does not normally evoke pain
Central pain	A group of spontaneous and/or evoked anomalous bodily sensations, painful or nonpainful in nature, which are projected into a body part congruent with a clearly imaged lesion of the central nervous system
Dysesthesia	An unpleasant and abnormal bodily sensation, whether spontaneous or evoked
Evoked pain	An umbrella term for allodynia, hyperalgesia, and hyperpathia
Hyperalgesia	An increased response to a tactile stimulus which normally is painful
Hyperpathia	An abnormally painful reaction to a tactile stimulus which lasts beyond the stimulus duration
Paresthesia	An abnormal bodily sensation of "pins and needles," whether spontaneous or evoked

of stroke, in younger age after SCI, the only exception being MS, which predominantly affects females. An often unrecognized cause of CP is iatrogenic lesions: CP can be a complication of neurosurgical procedures both on the spine and the brain, ranging from disc ablation at dorsal levels to tumor excision in the parietal lobe, brainstem, or cord. Unfortunately, the lack of awareness of this possibility tends to delay effective treatment. On the other hand, the possibility that neurosurgical pain-relieving procedures (e.g., thalamotomies, mesencephalotomies, cordotomies, and so on) inadvertently entail a new type of pain, i.e., CP, is a well-recognized paradox (Cassinari and Pagni 1969).

Roughly 1% of all epileptic patients suffer painful fits at least once in their lives: we consider this a central-pain-allied condition (CPAC) since there is no actual damage to the parietal projection of the spinothalamic tract. Nonetheless, the end cells of that pathway are likely to be involved (Canavero and Bonicalzi 2011). For a long time, dysesthetic pain as reported by some patients with Parkinson's disease was considered a form of CP, but this has been refuted (Canavero 2009a, b).

14.3 Clinical Features and Diagnostic Evaluation

CP has three components (Tasker 2001): (1) a continuous, spontaneous pain (which may be described as aching, burning, pricking, lacerating, or cramping); dysesthesias; paresthesias; pruritus; or combinations thereof; more than one kind of pain is experienced in more than 90% of the patients, while in cord CP and MS-associated CP, dysesthesias are often reported (Beric 1999); (2) a spontaneous, intermittent, generally lancinating pain, which is experienced by about 10–20% of the cases; and (3) evoked pain, experienced by about two thirds of the patients (rarely constituting the sole presenting symptom). In the latter cases, a nonpainful or only mildly painful stimulus is felt as painful or burning (i.e., allodynia) or very painful (hyperalgesia, or hyperpathia if it lasts beyond the moments of stimulation; see also Table 14.1). Evoked pain, just as in peripheral neuropathic pain, can be elicited by mechanical (static or dynamic) or thermal (cold more than heat) stimuli. In cases described as hyperpathic, the pain is usually unbearable and evokes violent emotional and defensive reactions; characterized by late onset and poor localization, it generally irradiates from the stimulated point to the entire half of the body and persists for an unusually long time after the stimulation has ceased. Evoked pain and intermittent shooting pains can hinder the patient's daily activities because innocuous maneuvers can trigger fits of intolerable pain. Patients may have to wear a glove when the hand is affected. All those anomalous sensations are referred inside a larger area of sensory loss or hypesthesia on the side of the body contralateral to the cerebral or spinal damage. Pain may be felt superficially, deep, or both. Symptoms may be focal (hand/arm, hemichest, foot and/or leg) or affect half of the body – as generally seen after thalamic and cortical lesions – or one side of the face, ipsilaterally, and the rest of the hemisoma, contralaterally (Wallenberg's syndrome) following brainstem injury. CP after SCI (both complete and incomplete) may affect the entire body region below the level of injury, but usually it is more intense in the sacral dermatomes, buttocks, genitalia, and feet. Signs of dystrophy in affected areas may be observed in some CP patients (Canavero and Bonicalzi 2011). The intensity of the pain varies from a mild, unpleasant tingling to one of the most agonizing torments known to humans.

CP greatly impairs one's quality of life, interfering with sleep patterns and driving people to suicide. Its highly unpleasant quality tends to disable patients even when the intensity is low (Canavero and Bonicalzi 2011; Tasker 2001).

The diagnosis of CP is rather straightforward when a patient complains of pain or other abnormal sensations following CNS injury. The neurological examination usually reveals areas of hypoanesthesia to thermal stimuli and pinches. Those areas must be assessed clinically with cold (e.g., an ice cube) and warm stimuli as well as with pinpricks. Frequently, evoked pain will be elicited. In the clinical setting, pain scales may be employed, although they are more useful in research settings.

Due to its clinical features, which can be misinterpreted as peripheral neuropathic pains (e.g., diabetic neuropathy), all patients must be assessed neuroradiologically and neurophysiologically to exclude any sensory neuropathies. MRI is the instrument of choice in the assessment of CP; a lesion (whether it be an ischemic or hemorrhagic lesion, a tumor, a demyelinating plaque, etc.) is usually seen along the spinothalamic pathway. In doubtful cases, laser-evoked potentials (LEPs) may confirm damage to the (sensory) C/A delta fibers. Shoulder-hand pain, which frequently accompanies stroke, is generally considered nociceptive, but a neuropathic component cannot be excluded (Canavero and Bonicalzi 2011). Pain caused by muscle cramping or dystonia due to an abnormal tone, posture, or muscle excitability is often seen after CNS damage and must be differentiated from CP.

14.4 Natural History

CP generally starts days, weeks, or months after a CNS insult but may present suddenly or take one or more years to develop; when the delay is in the year range after SCI, a syrinx may be found on MRI (Beric 1999). Once set in, CP remains with the patient for a lifetime in the vast majority of the cases. It may fluctuate during the day, depending on such factors as stress, weather changes, and effort. Rest and distraction may lessen CP. Unlike brain CP, which usually does not change significantly over time, except in degree, cord CP may change remarkably over the years: it may increase in severity for several years and even change in distribution and quality, sometimes dramatically (Canavero and Bonicalzi 2011; Beric 1999).

Careful studies reveal that CP may completely and suddenly regress after removal of the inciting lesion in some patients (Canavero and Bonicalzi 2001; Kim et al. 2006; Schott 2001). Likewise, a further stroke along the parietothalamic axis may abolish the pain (Canavero and Bonicalzi 2011; Canavero 1994). These observations have pathophysiological consequences that will be discussed below.

14.5 The Genesis of Central Pain

Scores of different theories have been proposed to explain CP. For a long time, the thalamus has played a starring role in them. Animal studies over the last 20 years have not added any insights to the field for the simple reason that the brain structure and neurochemistry of animals differ significantly from those in humans (Canavero and Bonicalzi 2011). In the end, a theory in biomedicine is considered a good one if it leads to therapeutic advances. In that sense, most theories have failed.

The highly popular view that central and peripheral neuropathic pain can both be understood within the framework of so-called deafferentation pain (Tasker 2001) has never made it into international classifications, and although sharing similar pain descriptors, differences between the two types of neuropathic pain are so numerous that a common genesis is out of the question. For instance, while brachial plexus avulsion pain responds well to standard dorsal root entry zone (DREZ) lesions, since the pain generator is a hyperactive focus in the spinal dorsal horns, CP of cord origin does not; likewise, thalamic stimulation and thalamic surgical lesions are far more effective for peripheral neuropathic pain than for CP (Canavero and Bonicalzi 2011). Subhypnotic propofol, an intravenous anesthetic with gamma-aminobutyric-acidergic (GABA-ergic) properties, can relieve CP but only rarely peripheral neuropathic pain (Canavero et al. 1995).

It has become clear that CP must be understood within a framework that includes both the thalamus and the sensory (somatosensory area I/motor area I) cortex (Canavero 1994; Canavero et al. 1995). In particular, it is a derangement of the oscillatory pattern (and hence of the content of the information transmitted) inside the sensory corticothalamocortical loop that best explains the pathophysiology of CP (dynamic reverberation theory of central pain, Canavero 1994). That derangement is possible because of the establishment of an "attractor state" (see Chap. 4, this volume; Canavero and Bonicalzi 2011, Chap. 26) in the cortical energy landscape, exploiting the highly recurrent nature of cortical microcircuitry. Layers V and VI of primary somatosensory cortex (SI) responsible for the massive output to the thalamus cannot "escape," unless supraphysiologic inhibition is applied pharmacologically or by way of electrical stimulation.

In-depth recordings in patients' brains point to various anomalies in the brainstem, thalamus, and cortex (Canavero and Bonicalzi 2011; Tasker 2001). In particular, hyperactivity of the bursting type within the thalamus, both in sensory and aspecific (central lateral) nuclei, has been considered indicative of a local dysrhythmia (Sarnthein et al. 2006; Lenz and Dougherty 1997). This has been refuted, however, as bursting is a modality of transmission of the normal brain (Radhakrishnan et al. 1999; see Canavero and Bonicalzi 2011 for further contrary evidence). Thalamic dysrhythmia as a mechanism for central pain is also unlikely because of two simple observations: (1) central-lateral (CL) thalamotomies are poorly effective for CP, and (2) the descending corticothalamic arm is facilitatory, not inhibitory (Canavero and Bonicalzi 2011). On the other hand, a disturbance in neural patterns of activity applies to CP as well as to many other neurologic disorders (Montgomery 2010).

Another popular theory emphasizes neuroplastic changes in the CNS following "deafferentation" (Canavero and Bonicalzi 2011; Tasker 2001). "Entrenched" neuroplasticity, spanning from the changes in brain maps to the sensitization of central structures, has been put forth to explain chronicity – and so purported neurodegenerative changes as evidenced by MR spectroscopy studies – but a sudden resolution of CP after the extirpation of inciting lesions (Pagni and Canavero 1995) or a further stroke (Canavero 1994) clearly suggests that plasticity (or neurodegeneration) cannot underlie constant pain, as both would then be reversible under such circumstances (Schott 2001). Moreover, careful studies by Tasker's group reveal major differences between what occurs in the human patient and the experimental animal in terms of plastic changes (Kiss et al. 1994).

Importantly, CP can be canceled by lesions interrupting the thalamocortical loop in the subparietal corona radiata and internal capsule. Careful analysis of such cases reveals that it is the much larger descending arm of the thalamocortical loop that is key to maintaining CP over time (Canavero and Bonicalzi 2011; Canavero 1994). A role in feeding the loop comes from hyperactive reticular cells along the spinotruncothalamic axis, whose destruction may help allay CP (Bowsher 2001; Finnerup et al. 2003). This has been highlighted by studies dealing with mesencephalic reticulotomy (Amano et al. 1992) and extended DREZ-tomies (Falci et al. 2002). The difference between subparietal lesions and the latter ones lies in the rate of disappearance of CP, i.e., suddenly versus gradually.

The possibility of hyperexcitation of the spinothalamic pathway by the reticulothalamic system, which is in turn modulated by the medial lemniscus, is supported by clinical data (Kim 2007). Neuroimaging studies using techniques such as single photon emission computed tomography (SPECT), positron emission tomography (PET), and functional magnetic resonance imaging (fMRI) have provided valuable data in this area of research. Unfortunately, however, most of those studies focused on evoked pains, which, as stated, are not key to CP. Studies investigating the spontaneous component with drug dissection, although few, have confirmed the role of the sensory cortex as well as the thalamus (Pagni and Canavero 1995; Canavero et al. 1993; Hirato et al. 1993; Canavero et al. 1996; see also Figs. 14.1, 14.2, 14.3, 14.4, and 14.5), while evoked pains elicit different – and wider – patterns of neural activity, particularly in prefrontal areas (Ducreux et al. 2006).

Fig. 14.1 Superposition of computerized tomography (*CT*)/positron emission tomography (*PET*) images in a case of brain central pain. Note that the regional oxygen extraction ratio is increased in the cerebral cortex around the central sulcus (sensory cortex I/motor cortex I) (From Hirato et al. (1993). Reproduced with permission)

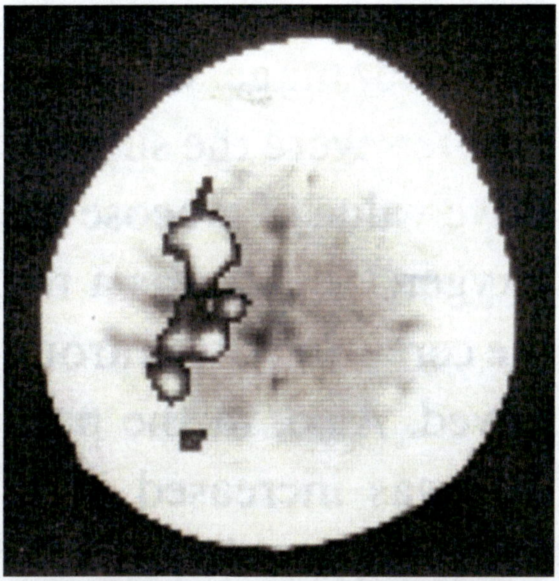

Fig. 14.2 Regional cerebral blood flow (*rCBF*) decreases at primary somatosensory cortex (*SI*) and thalamic levels in a case of cord central pain (From Canavero and Bonicalzi 2011)

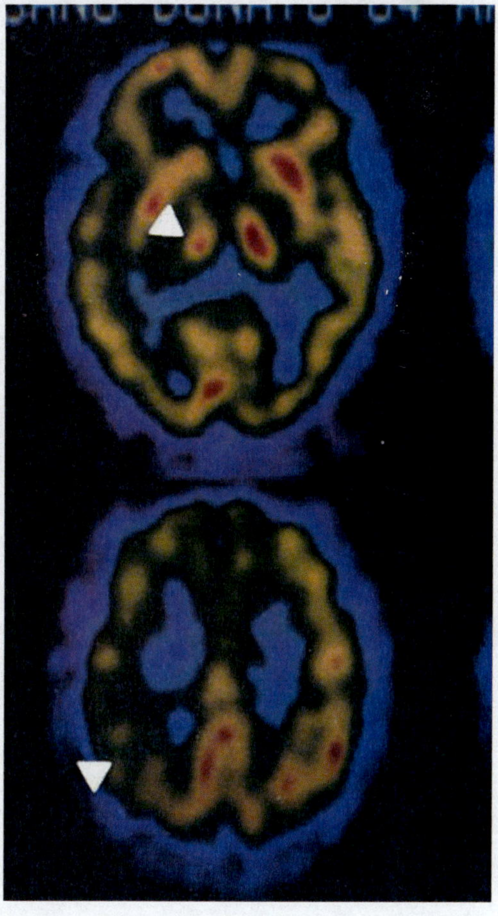

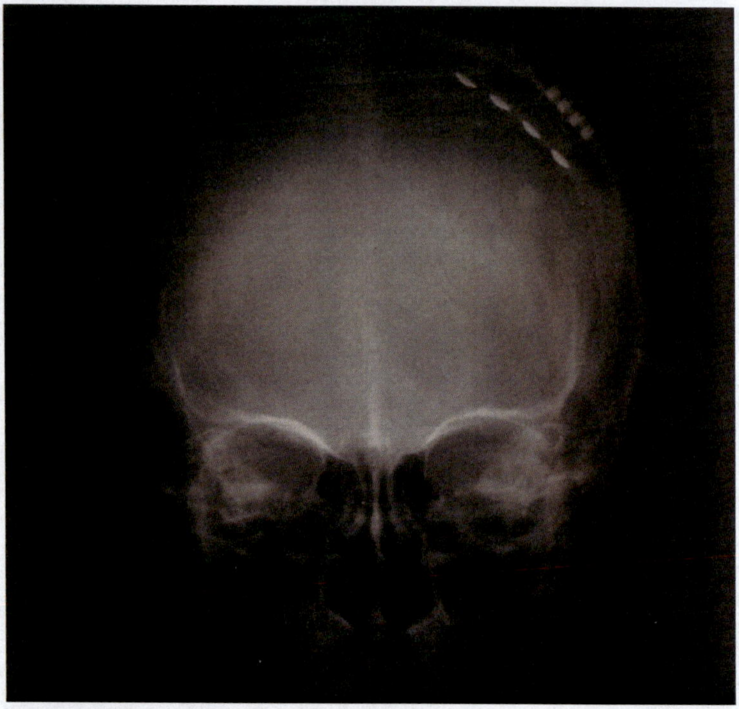

Fig. 14.3 Skull X-rays of an extradural cortical stimulating (*ECS*) paddle in place

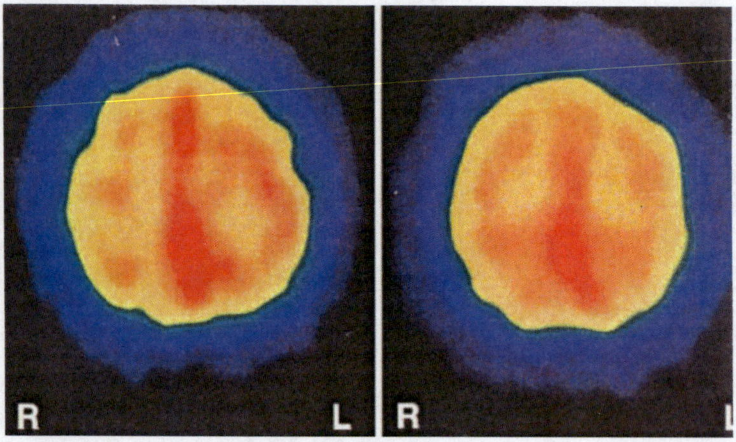

Fig. 14.4 Regional cerebral blood flow (*rCBF*) renormalization at the primary somatosensory cortex (*SI*) level following extradural cortical stimulation (*ECS*) (From Canavero and Bonicalzi 1995)

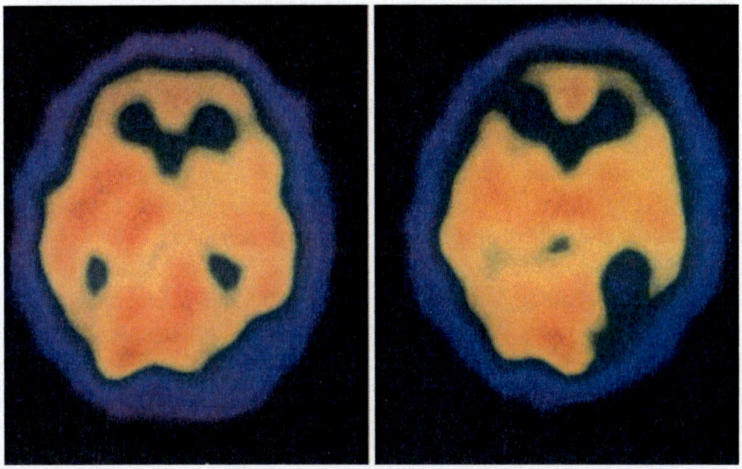

Fig. 14.5 Renormalization of thalamic hypoperfusion following a bolus of propofol at subhypnotic doses in a case of brain central pain (From Canavero et al. 1995)

A recent study espoused the view that "the evidence of blood flow, stimulation, and lesion studies forcefully make the case that (thalamic primary sensory nucleus) Vc and sensorimotor cortex are involved in [central post-stroke pain]" (Kim et al. 2007).

Data from Tasker's group also support the view that the generator of ongoing CP can be shifted to the healthy hemisphere (Parrent et al. 1992), thus explaining cases of CP after massive destruction of one thalamus or sensory cortex. Clinical observations likewise support the possibility of a unilateral CNS lesion triggering bilateral CP (Canavero 1996; Kim 1998, 1999).

It has been proposed that a certain number of people may be genetically predisposed to develop CP. In fact a lesion anywhere along the spinothalamocortical pathway only triggers CP in a minority of patients, while sparing most others (Tasker 2001). Thus spinothalamocortical damage is necessary, but not sufficient, for CP to arise (Canavero and Bonicalzi 2011; Finnerup et al. 2003; Ducreux et al. 2006). In particular, there is a suggestion that sparing of the lemniscal system, along with damage to the spinothalamic system, may trigger dysesthetic pain (Beric 1999). Neuropharmacological data with GABA agonists point to a specific derangement of GABA transmission at the basis of CP (Canavero and Bonicalzi 1998). It is surmised that CNS damage in these cases triggers an acute GABA loss, which then unbalances the oscillatory pattern along the thalamocortical loop responsible for conscious sensory processing, starting in the somatosensory cortex (Canavero and Bonicalzi 2011).

In this context, anomalies seen in the cingular and temporal areas are unspecific findings present in most chronic pain patients without them constituting the primary cause of CP. According to Craig (1998), CP is subtended by a hyperactive spinothalamocingular pathway after selective damage to a specific spinothalamoinsular pathway, i.e., the posterior part of the ventral medial nucleus (VMpo). However, this thermosensory disinhibition hypothesis has been disproven on clinical and neurophysiological grounds (see Canavero and Bonicalzi 2011 for discussion and references), and its anatomic foundations have been refuted (Jones 2007). The insula, unlike what some authors currently think, is not paramount to the genesis of central pain: primary somatosensory cortex (SI) is central to pain awareness even in acute pain (Nir et al. 2008), and it is the only area which topographically feeds back to the sensory thalamus along the path that cancels central pain.

On the other hand, the different components of evoked pain are most likely due to different pathophysiological mechanisms, i.e., thermal versus mechanical allodynia (Canavero and Bonicalzi 2011; Ducreux et al. 2006; Greenspan et al. 2004). Thus central sensitization along spared – thermally or mechanically coded – pathways may well play a major role in the genesis of those types of pain (Canavero and Bonicalzi 2011; Tasker 2001; Greenspan et al. 2004).

The sympathetic nervous system plays no role whatsoever in the genesis or sustenance of CP (Canavero and Bonicalzi 2011; Rondinelli et al. 2008; Bonicalzi and Canavero 2000) for the simple reason that sympathetic-nociceptor interactions do not exist (Campero et al. 2010).

14.6 Therapy

CP is best understood as a "cancer of the spirit" which nibbles away each day at the patient's quality of life until severe disruption in daily living inexorably sets in. Therefore, it is of the utmost importance that valuable time not be lost trying a smorgasbord of – often useless – drugs. If a trial of oral pharmacotherapy produces no benefit within 6 months, neuromodulation must be the next therapeutic step without any further delay.

14.6.1 Drug Therapy

Various effective agents have been identified for the purpose of reducing CP, but only few are suitable for oral use. Lamotrigine and amitriptyline are the only oral drugs that have a proven benefit for brain CP, whereas they appear to be less effective for CP of cord origin. Although not submitted to any formal studies, mexiletine appears to be a valuable drug, whereas gabapentin and pregabalin are effective in only few patients (Canavero and Bonicalzi 2011). Many other drugs, including opioids, have been tried, with most patients failing to benefit or only mildly so (Canavero and Bonicalzi 2003). Cannabinoids have not advanced therapy to any meaningful extent, and a major drawback is that they can trigger psychosis. It should be noted anyway that all these drugs have side effects that limit their use, above all in special patient populations such as the elderly and those with spinal injuries (who are vulnerable to the anticholinergic effects of amitriptyline), while some of them can lead to rare but major toxicities (e.g., Stevens-Johnson syndrome during lamotrigine consumption). Topiramate, levetiracetam, reboxetine, valproate, duloxetine, and the selective serotonin reuptake inhibitors (SSRIs) – among others – are generally ineffective (Canavero and Bonicalzi 2011), whereas oral NMDA (N-methyl-D-aspartate) blockers such as dextromethorphan are modestly effective yet poorly tolerated.

14.6.2 Electrical Neuromodulation

As far as medication-resistant brain CP is concerned, the technique of choice is extradural cortical stimulation (ECS) of the primary motor or sensory cortex contralateral to the pain (Canavero 2009b). This is a minimally invasive neurosurgical technique in which a stimulating paddle is inserted through two burr holes or a small craniotomy and placed on the dura overlying the appropriate area to cover the painful region. More than half of the patients experience a pain reduction greater than 40% after 4 years of treatment (Canavero 2009b). No mortality or permanent morbidity (including the kindling of an epileptic syndrome) has been reported in hundreds of reported cases since the technique's introduction in 1989 (Canavero 2009b). Both spontaneous and evoked components are altered favorably. Patients may be selected for ECS on the basis of pharmacological dissection with GABA agonists (propofol, barbiturates) and trials of transcranial magnetic stimulation (TMS) (see also Chap. 25) or transcranial direct current stimulation (tDCS), in which stimuli are applied noninvasively (Canavero 2009b). The latter techniques per se may give relief to some patients, but the therapeutic benefits tend to be short-lived, even when repeated cyclically. Neuroimaging studies and neuroengineering modeling clearly show that cortical stimulation (CS) can affect the underlying sensory cortex and the corticothalamic loop; the contention that CS works by altering the affective component of pain is not supported (Canavero 2009b). Deep brain stimulation (DBS), in which one or two electrodes are inserted into the sensory thalamus or mesencephalon (periaqueductal-gray-matter (PAG) and periventricular-gray-matter (PVG) areas), is not effective in the great majority of the patients. Surgical complications include intracranial hemorrhage (up to 10%), stroke (up to 2%), and death (up to 4.4%). DBS has a long history, having been introduced for the treatment of CP during the 1960s. Experience over the years has not borne out the initial results (Hamani et al. 2006; Rasche et al. 2006). Spinal cord stimulation (SCS), in which a paddle is extradurally directed at cord segments, plays a limited role in no more than a few patients suffering from brain CP. For CP of cord origin with at least partially preserved lemniscal sensibility, SCS is the primary technique although its effects often abate within 1 year (Canavero and Bonicalzi 2011). In treatment-resistant cases and cases with complete loss of sensibility, and in which SCS is totally ineffective, the choice is between ECS and DBS. Unfortunately, not enough patients have accrued over time to evaluate the effects of ECS, while DBS is often ineffective. Transcutaneous electrical nerve stimulation (TENS), although the least invasive of all the electrical neuromodulatory procedures, is of scarce benefit in the vast majority of patients. Moreover, it must be applied several times a day, thus hindering activities of daily living. If elected, it is best added on to an oral drug therapy (Canavero and Bonicalzi 2011). Some patients have been submitted to electroconvulsive therapy (ECT; see also Chap. 24), however, with mixed results. As a consequence, this is considered a technique of last resort reserved for highly refractory cases (Canavero and Bonicalzi 2011).

14.6.3 Chemical Neuromodulation

Various drugs can be infused into the subarachnoid space in order to control CP. However, no evidence-based recommendations are possible; only anecdotal evidence exists, and the few controlled studies lack the necessary power and follow-up data. A GABA-ergic agent like baclofen or midazolam can be infused with the aid of an implanted pump, best in combination with clonidine, an adrenergic agent (Canavero and Bonicalzi 2011). Opioids are only rarely effective and, in the long term, have major endocrinologic and immunologic toxicity. Ziconotide, a recently approved drug, appears insignificantly effective on CP and must be considered unsafe.

14.6.4 Neuroablation

Ever since the introduction of stereotactic surgery at the end of the 1940s, neurosurgical ablative techniques such as thalamotomies and mesencephalotomies have been offered to distraught patients, but results have not held up over time in most studies, in particular as regards the spontaneous components of CP (Canavero and Bonicalzi 2011; Tasker 2001). Furthermore, mortality and permanent morbidity often offset an initially positive result. Cingulotomy has proven of no benefit to the sensory components of CP. Surgery on the cord, including cordectomies and DREZ coagulations, may relieve evoked and paroxysmal components in some patients in the long run, but with unacceptable morbidity (Canavero and Bonicalzi 2011; Tasker 2001). The only ablative technique that makes sense in the light of the discussed pathophysiology is a small stereotactic lesion deep in the corona radiata/internal capsule in order to interrupt the descending arm of the corticothalamic loop: this has been confirmed in a patient with central post-stroke pain (CPSP), who was totally and immediately relieved by subparietal leucotomy/capsulotomy (Koszewski et al. 2003; see also Figs. 14.6 and 14.7). Such a selective lesion can also be produced by stereotactic radioneurosurgery or high-intensity focused ultrasound (HIFU) and DTI imaging of the descending arm in the individual patient.

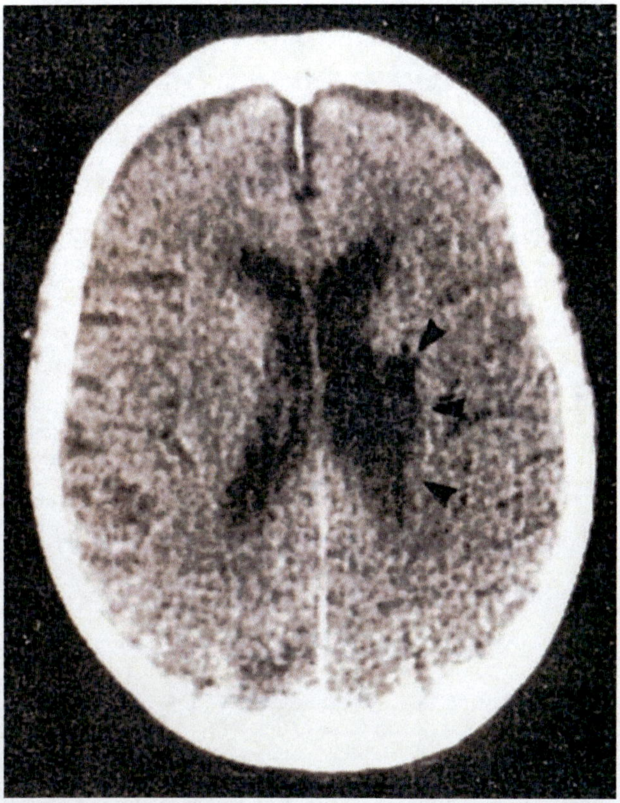

Fig. 14.6 Stroke in the subparietal corona radiata abolishing brain central pain (From Canavero and Bonicalzi 2011)

14.7 Conclusion

Central pain can easily be described as "hallucinatory pain," being couched as a disturbance of sensory information transfer along an uninhibited reverberating corticothalamocortical sensory channel. This generator alters the widely distributed small-world architecture of the brain (see Chap. 4, this volume, and Canavero and Bonicalzi 2011, Chap. 26), which can be neuroimaged, for instance, as an alteration of brain rhythms in multiple brain areas. These by themselves are not responsible for the sustenance of the pain but may contribute to the different qualities thereof. Incidentally, phantom pain can be abolished by a lesion along the same axis (Canavero 1994), disclosing a common mechanism underlying such hallucinatory pains.

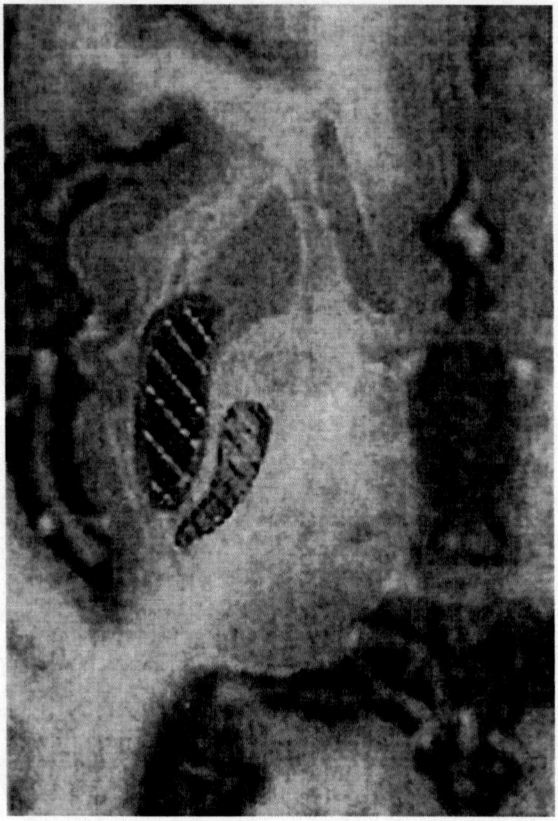

Fig. 14.7 Surgical stereotactic lesion interrupting the descending arm of the corticothalamocortical reverberatory loop abolishing brain central pain (From Canavero and Bonicalzi 2011)

References

Amano, K., Kawamura, H., Tanikawa, T., Kawabatake, H., Iseki, H., Taira, T. (1992) Stereotactic mesencephalotomy for pain relief. A plea for stereotactic surgery. *Stereotactic and Functional Neurosurgery*, 59, 25–32.

Andersen, G., Vestergaard, K., Ingeman-Nielsen, M., Jensen, T.S. (1995). Incidence of central post-stroke pain. *Pain*, 61, 187–193.

Beric, A. (1999). *Spinal cord damage: injury.* In: *Textbook of pain. Fourth edition.* Edited by Wall, P.D. and Melzack, R. Edinburgh: Churchill-Livingstone, pp. 915–927.

Bonicalzi, V., Canavero, S. (2000). Sympathetic pain again? *Lancet*, 360, 1426–1427.

Bowsher, D. (2001). Stroke and central poststroke pain in an elderly population. *Journal of Pain*, 2, 258–261.

Campero, M., Bostock, H., Baumann, T.K., Ochoa, J.L. (2010). A search of C nociceptors by sympathetic fibers in complex regional pain syndrome. *Clinical Neurophysiology*, 121, 1072–1079.

Canavero, S. (1994). Dynamic reverberation: a unified mechanism of central and phantom pain. *Medical Hypotheses*, 42, 203–207.

Canavero, S. (1996). Bilateral central pain. *Acta Neurologica Belgica*, 96, 135–136.

Canavero, S. (2009a). Central pain and Parkinson disease. *Archives of Neurology*, 66, 282–283.
Canavero, S., ed. (2009b) *Textbook of therapeutic cortical stimulation.* New York, NY: Nova Science.
Canavero, S., Bonicalzi, V. (1998). The neurochemistry of central pain: evidence from clinical studies, hypothesis and therapeutic implications. *Pain*, 74, 109–114.
Canavero, S., Bonicalzi, V. (2001). Reversible central pain. *Neurological Sciences*, 22, 271–273.
Canavero, S., Bonicalzi, V. (2003). Chronic neuropathic pain. *New England Journal of Medicine*, 348, 2688–2689.
Canavero, S., Bonicalzi, V. (2011). *Central pain syndrome: Pathophysiology, diagnosis and management. Second edition.* Cambridge: Cambridge University Press.
Canavero, S., Bonicalzi, V., Castellano, G. (1996). Two in one: the genesis of central pain. *Pain*, 64, 394–395.
Canavero, S., Bonicalzi, V., Massa-Micon, B. (1997). Central neurogenic pruritus: a literature review. *Acta Neurologica Belgica*, 97, 244–247.
Canavero, S., Bonicalzi, V., Pagni, C.A. (1995). Propofol analgesia in central pain: preliminary clinical observations. *Journal of Neurology*, 242, 561–567.
Canavero, S., Pagni, C.A., Castellano, G., Bonicalzi, V., Belló, M., Duca, S., Podio, V. (1993). The role of cortex in central pain syndromes: preliminary results of a long-term technetium-99 hexamethylpropyleneamineoxime single photon emission computed tomography study. *Neurosurgery*, 32, 185–191.
Cassinari, V., Pagni, C.A. (1969). *Central pain: a neurosurgical survey*. Cambridge, MA: Harvard University Press.
Craig, A.D. (1998). A new version of the thalamic disinhibition hypothesis of central pain. *Pain Forum*, 7, 1–14.
Dejerine, J., Roussy, G. (1906). Le syndrome thalamique. *Revue Neurologique*, 14, 521–532.
Ducreux, D., Attal, N., Parker, F., Bouhassira, D. (2006). Mechanisms of central neuropathic pain: a combined psychophysical and fMRI study in syringomyelia. *Brain*, 129, 963–976.
Edinger, L. (1891). Giebt es central entstehende Schmerzen? *Deutsche Zeitschrift für Nervenheilkunde*, 1, 262–282.
Falci, S., Best, L., Bayles, R., Lammertse, D., Starnes, C. (2002). Dorsal root entry zone microcoagulation for spinal cord injury-related central pain: operative intramedullary electrophysiological guidance and clinical outcome. *Journal of Neurosurgery (Spine 2)*, 97, 193–200.
Finnerup, N.B., Johannesen, I.L., Fuglsang-Frederiksen, A., Bach, F.W., Jensen, T.S. (2003). Sensory function in spinal cord injury patients with and without central pain. *Brain*, 126, 57–70.
Gonzales, G.R., Tuttle, S.L., Thaler, H.T., Manfredi, P.L. (2003). Central pain in cancer patients. *Journal of Pain*, 4, 351–354.
Greenspan, J.D., Ohara, S., Sarlani, E., Lenz, F.A. (2004). Allodynia in patients with post-stroke central pain (CPSP) studied by statistical quantitative sensory testing within individuals. *Pain*, 109, 357–366.
Hamani, C., Schwalb, J.M., Rezai, A.R., Dostrovsky, J.O., Davis, K.D., Lozano, A.M. (2006). Deep brain stimulation for chronic neuropathic pain: long-term outcome and the incidence of insertional effect. *Pain*, 125, 188–196.
Head, H., Holmes, G. (1911). Researches into sensory disturbances from cerebral lesions. *Brain*, 34,102–254.
Hirato, M., Horikoshi, S., Kawashima, Y., Satake, K., Shibasaki, T., Ohye, C. (1993). The possible role of the cerebral cortex adjacent to the central sulcus for the genesis of central (thalamic) pain - a metabolic study. *Acta Neurochirurgica (Suppl.)*, 58, 141–144.
Jones, E.G. (2007). *The thalamus. Second edition.* New York, NY: Cambridge University Press.
Kim, J.H., Greenspan, J.D., Coghill, R.C., Ohara, S., Lenz, F.A. (2007). Lesions limited to the human thalamic principal somatosensory nucleus (ventral caudal) are associated with loss of cold sensations and central pain. *Journal of Neuroscience*, 27, 4995–5005.
Kim, J.S. (1998). Delayed-onset ipsilateral sensory symptoms in patients with central post-stroke pain. *European Neurology*, 40, 201–206.

Kim, J.S. (1999). Aggravation of poststroke sensory symptoms after a second stroke on the opposite side. *European Neurology*, 42, 200–204.
Kim, J.S. (2007). Medial medullary infarct aggravates central poststroke pain caused by previous lateral medullary infarct. *European Journal of Neurology*, 58, 41–43.
Kim, L.J., Klopfenstein, J.D., Zabramski, J.M., Sonntag, V.K.H., Spetzler, R.F. (2006). Analysis of pain resolution after surgical resection of intramedullary spinal cord cavernous malformations. *Neurosurgery*, 58, 106–111.
Kiss, Z.H.T., Dostrovsky, J.O., Tasker, R.R. (1994). Plasticity in human somatosensory thalamus as a result of deafferentation. *Stereotactic and Functional Neurosurgery*, 62, 153–163.
Koszewski, W., Jarosz, J., Pernak-De Gast, J. (2003). Stereotactic posterior capsulo-lentiform deafferentation as an effective treatment in central poststroke pain. A new surgical method for intractable central pain control? *Pain Clinic*, 15, 115–123.
Lenz, F.A., Dougherty, P.M. (1997). *Pain processing in the human thalamus*. In: *Thalamus. Volume II. Experimental and clinical aspects*. Edited by Steriade, M., Jones, E.G., and McCormick, D.A. Amsterdam: Elsevier, pp. 617–652.
Montgomery, E.B. Jr. (2010). *Deep brain stimulation programming. Principles and practice*. New York, NY: Oxford University Press.
Nir, R.R., Lev, R., Moont, R., Gronovsky, Y., Sprecher, E., Yarnitsky, D. (2008). Neurophysiology of the cortical pain network: revisiting the role of S1 in subjective pain perception via standardized low-resolution brain electromagnetic tomography (sLORETA). *Journal of Pain*, 9, 1058–1069.
Pagni, C.A., Canavero, S. (1995). Functional thalamic depression in a case of reversible central pain due to a spinal intramedullary cyst. *Journal of Neurosurgery*, 83, 163–165.
Parrent, A.G., Lozano, A.M., Dostrovsky, J.O., Tasker, R.R. (1992). Central pain in the absence of functional sensory thalamus. *Stereotactic and Functional Neurosurgery*, 59, 9–14.
Radhakrishnan, V., Tsoukatos, J., Davis, K.D., Tasker, R.R., Lozano, A.M., Dostrovsky, J.O. (1999). A comparison of the burst activity of lateral thalamic neurons in chronic pain and non-pain patients. *Pain*, 80, 567–575.
Rasche, D., Rinaldi, P.C., Young, R.F., Tronnier, V.M. (2006). Deep brain stimulation for the treatment of various chronic pain syndromes. *Neurosurgical Focus*, 21, E8.
Riddoch, G. (1938). The clinical features of central pain. Lumleian lecture. *Lancet*, 234, 1093–1098, 1150–1156, 1205–1209.
Rondinelli, R.D. et al., eds. (2008). *Guides to the evaluation of permanent impairment. Sixth edition.* Chicago, IL: American Medical Association, pp. 450–451.
Sarnthein, J., Stern, J., Aufenberg, C., Rousson, V., Jeanmonod, D. (2006). Increased EEG power and slowed dominant frequency in patients with neurogenic pain. *Brain*, 129, 55–64.
Schott, G.D. (2001). Delayed onset and resolution of pain. Some observations and implications. *Brain*, 124, 1067–1076.
Siddall, P.J., Taylor, D.A., McClelland, J.M., Rutkowski, S.B., Cousins, M.J. (1999). Pain report and the relationship of pain to physical factors in the first 6 months following spinal cord injury. *Pain*, 81, 187–197.
Solaro, C., Brichetto, G., Amato, M.P., and the PaISM Study Group. (2004). The prevalence of pain in multiple sclerosis. A multicenter cross-sectional study. *Neurology*, 63, 919–921.
Tasker, R.R. (2001). *Central pain states*. In: *Bonica's management of pain. Third edition.* Edited by Loeser, J.D. Philadelphia, PA: Lippincott Williams & Wilkins, pp. 433–457.

Chapter 15
Autoscopic Phenomena: Clinical and Experimental Perspectives

Anna Sforza and Olaf Blanke

15.1 Introduction

Autoscopic phenomena (from the Greek *autos* (self) and *skopeo* (looking at)) encompass a wide range of experiences involving the visual illusory reduplication of one's own body in extrapersonal space. They have been described in psychiatric disorders such as dissociation, schizophrenia, depression, and anxiety, as well as in various neurological disorders. Those neurological disorders include generalized diseases such as cerebral infection, meningitis, encephalitis, intoxications and generalized epilepsy, as well as focal diseases such as focal epilepsy (Devinsky et al. 1989), traumatic brain damage (Todd and Dewhurst 1955), migraine (Lippman 1952), vascular brain damage (Kölmel 1985), and neoplasia (Todd and Dewhurst 1955). Autoscopic phenomena have also been described in the healthy population, and they have widely inspired the literature (i.e., Fjodor Dostoevsky (1821–1881), Vladimir Nabokov (1899–1977), E.T.A. Hoffmann (1776–1822), Edgar Allan Poe (1809–1849), and Guy de Maupassant (1850–1893)) and the visual arts (for an overview see Blanke 2005).

15.1.1 Classification and Phenomenology

The scientific systematization of autoscopic phenomena (AP) began later than that of visual and non-visual illusions related to single body parts such as disconnection, dislocation, illusory movement, and the illusory reduplication of body parts (Brugger et al. 1997; Hécaen and de Ajuriaguerra 1952; Ramachandran and Hirstein 1998;

A. Sforza, Ph.D. (✉) • O. Blanke, M.D., Ph.D.
Laboratory of Cognitive Neuroscience, Brain-Mind Institute,
Ecole Polytechnique Fédérale de Lausanne, Lausanne, Switzerland
e-mail: annalaura.sforza@gmail.com; olaf.blanke@epfl.ch

J.D. Blom and I.E.C. Sommer (eds.), *Hallucinations: Research and Practice*,
DOI 10.1007/978-1-4614-0959-5_15, © Springer Science+Business Media, LLC 2012

Heydrich et al. 2010). The initial classification proposed by Devinsky et al. (1989) was extended by Brugger et al. (1997, 2002; see also Chap. 16), and subsequently by Blanke et al. (2004, 2008). It distinguishes three main forms of AP: out-of-body experiences (OBEs), heautoscopy (HAS), and autoscopic hallucinations (AHs), which differ with respect to phenomenological characteristics such as (1) self-location (i.e., the apparent location of the self within or outside one's body) and (2) the experienced visuo-spatial perspective. Autoscopy (i.e., the seeing of one's own body) is generally common to all three forms of AP.

15.1.1.1 OBEs: Out-of-Body Experiences

During an OBE, the experient has the subjective feeling of being awake, and typically experiences the "self," or center of awareness, as being located outside the physical body, at a somewhat elevated level. It is from this elevated extrapersonal location that the body and the world are experienced (Blackmore 1982; Devinsky et al. 1989; Brugger 2002; Blanke et al. 2004). Most OBE experients see their own body as lying on the ground or in bed, and the experience tends to be described as vivid and realistic. Thus an OBE is characterized by an abnormal, extracorporeal self-location (disembodiment), a distanced and elevated first-person perspective, and the seeing of one's own body (i.e., autoscopy) from the vantage point of this elevated perspective (see Fig. 15.1 for a graphic depiction of these OBE features).

15.1.1.2 AHs: Autoscopic Hallucinations

During an AH, people are seeing a double (doppelgänger) of themselves in extrapersonal space. Compared to OBEs, AHs are seen from the habitual first-person perspective, while the "self," or center of awareness, remains inside the physical body (normal self-location). Experientially, AHs are set apart from the other AP in the sense that the subject tends to realize the hallucinatory nature of the experience. AHs are usually very brief, and they are often accompanied by other visual hallucinations or illusions, as well as by visual field deficits (Blanke and Mohr 2005). Often only one's own face or the upper part of the trunk is perceived (Blanke and Mohr 2005; Zamboni et al. 2005, see also Fig. 15.1). Recently, however, Bolognini et al. (2011) described an atypical case of AH in which the patient reported seeing a profile view of the autoscopic body. What made this case even more interesting was the autoscopic body's tendency to react to motor activity. As long as the patient was standing still, the hallucination was limited to the head and upper trunk. Passive displacement of the patient's arms and active walking, however, were followed by an "extension" of the autoscopic body with arms and legs. As we shall explain in Sect. 15.1.2, these findings suggest that not only sensory mechanisms but also sensorimotor and proprioceptive mechanisms would seem to play a role in the mediation of AHs.

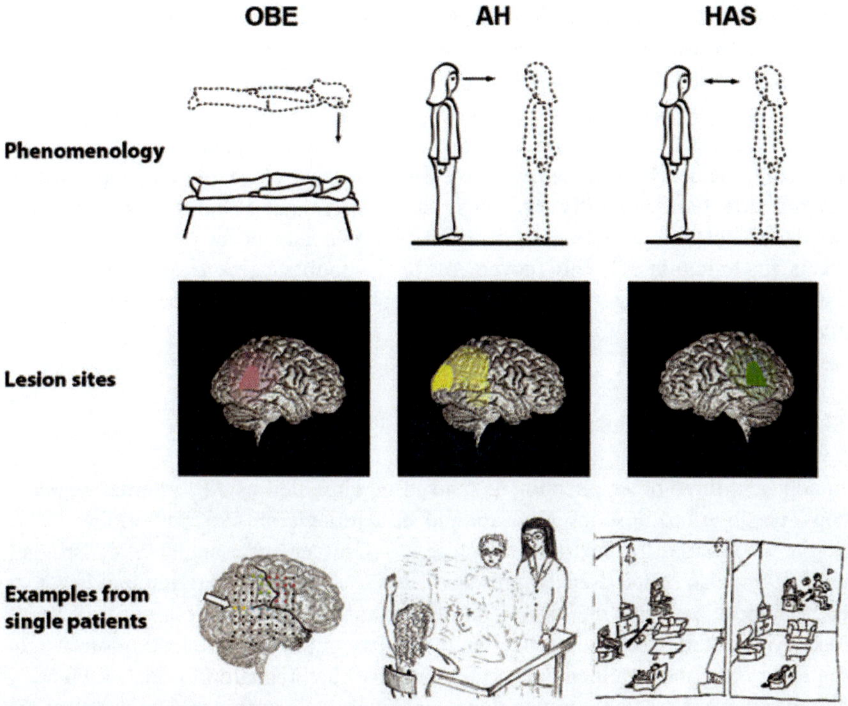

Fig. 15.1 Summary of the phenomenological and neural aspects of autoscopic phenomena: *Phenomenology.* In out-of-body experiences (*OBEs*), self-localization and the first-person perspective are at the position of the illusory body, that is, a position that is somewhat elevated with respect to the physical body; during an autoscopic hallucination (*AH*), the person sees an illusory copy of his own body from the vantage point of the habitual first-person perspective, thus maintaining a normal self-localization; heautoscopy (*HAS*) is an intermediate form in which self-localization and the first-person perspective are either at the position of the physical body, at the position of the illusory body, or at both positions simultaneously or alternatingly (Lopez et al. 2008). *Lesion overlap of brain damage.* Patients with OBEs show damage in the right temporo-parietal cortex; patients with AHs in the right parieto-occipital or right temporo-occipital cortex; patients with HAS in the left temporo-parietal cortex (Blanke and Metzinger 2009). *Examples from single-case studies.* Stimulation of the electrode sites indicated by the arrow (on the right angular gyrus) induced OBEs in an epileptic patient with implanted subdural electrodes (Blanke et al. 2002); graphical representation of an AH by Zamboni et al. (2005), based on their patient's description: visual features and movements appear in a mirror fashion; only the upper trunk and the face were seen; graphical illustration of experienced HAS (as drawn by the patient himself): during an initial episode he experienced being elevated with his living room chair into the air. In an immediately following moment, he experienced a "second" body, which continued to be elevated, but left the patient's body from the elevated position in the chair (Blanke et al. 2004)

15.1.1.3 HAS: Heautoscopy

A third form of AP is HAS, which can be conceptualized as lying on a continuum between AHs and OBEs. The person experiencing HAS is also seeing a double of himself/herself in extrapersonal space. However, it is difficult for him/her to decide whether the self is located within the physical or the autoscopic body (Blanke et al. 2004, see also Fig. 15.1). In addition, the subject often reports seeing the physical and autoscopic bodies alternatingly or even simultaneously from different first-person perspectives (physical body, double's body). Moreover, the double may or may not mirror the subject's exact physical characteristics. And yet, it tends to be recognized as "being one self." This feeling of self-identification is usually accompanied by marked alterations in bodily awareness, such as an unusual lightness of the body, vestibular illusions, or feelings of detachment. Frequently, heautoscopic echopraxia, that is, the imitation of bodily movements by the double, helps to create the illusion that it is the doppelgänger that "contains the real mind" (Lukianowicz 1958; Brugger et al. 1994).

15.1.1.4 Other Forms of AP

In addition, three other phenomena tend to be classified as AP: internal heautoscopy, negative heautoscopy (Hécaen and de Ajuriaguerra 1952; Lhermitte 1939; Sollier 1903), and the feeling of a presence (Grüsser and Landis 1991; Brugger et al. 1996; also called "sensed presence," see Chap. 17). During internal heautoscopy, subjects perceive one or several of their inner organs. During negative heautoscopy, they fail to see their reflection in a mirror or any other reflecting surface. In the latter case, the confinement of the (negative) hallucination to one's own body and the frequent association with depersonalization are considered evidence of the close relationship of the phenomenon with positive autoscopic experiences. Negative heautoscopy and internal heautoscopy are considered extremely rare. Only few case reports exist, and even fewer discussing their neuropsychological characteristics or their association with focal brain damage (see also Brugger et al. 1997 and Blanke et al. 2008).

The feeling of a presence (FP), described by Jaspers 1913 as *leibhafte Bewusstheit*, is defined as the conviction that there is another person close by without the percipient being able to actually perceive that person. The phenomenon has received much attention over time, and different authors have used many different names when referring to it. Thus, FP has been called *hallucination du compagnon* (Lhermitte 1939), idea of a presence (Critchley 1955), hallucination of presence, false proximate awareness, false bodily awareness, intruder hallucination, somaesthetic phantom double, somaesthetic doppelgänger, heautoscopy without optical image, phantom impressions, *Anwesenheit*, and concrete awareness (see Blom 2010).

15.1.2 Multisensory and Anatomical Mechanisms

It has been proposed that the phenomenological differences between the three main forms of autoscopy should be attributed to systematic variations in underlying brain mechanisms. Most authors conceptualize autoscopic phenomena generically as disorders of own-body perception and/or corporeal awareness and attribute them to multisensory and/or sensorimotor disintegration. Blanke and Mohr (2005) offer a detailed review of the literature outlining the contribution of the various sensory modalities that may be involved in mediating them. Multisensory mechanisms in the higher-level integrative brain areas of the extrastriate visual system would seem to be primarily involved in the case of AH, which could be mainly described as a visuo-tactile or "specular" hallucination or pseudohallucination. Indeed those phenomena are often lateralized and associated with the side of the hemianopia, without any concomitant vestibular hallucinations or body-schema disturbances (Blanke and Mohr 2005).

Proprioceptive and kinesthetic processing are considered particularly relevant for the mediation of HAS (especially in comparison with AHs; see Sollier 1903). In 1903, Sollier already postulated that heautoscopy is essentially a proprioceptive-kinesthetic disturbance associated with a strong psychological affinity of the physical body for the autoscopic body. The aforementioned phenomenon of echopraxia, found mainly in HAS, is often mentioned to sustain the alleged role of proprioception and kinesthesia in such disturbances. However, we propose that the three main forms of AP can be accounted for more systematically by reference to the vestibular system, which conveys sensations of the body's orientation in three-dimensional space.

Elaborating on a proposal by Grüsser and Landis (1991), Blanke et al. (2004) have suggested that systematic differences in vestibular function must underlie AHs, HAS, and OBEs. That assumption was based on clinical data pointing toward a strong association of OBEs with gravitational, otolithic, and vestibular disturbances (i.e., feelings of elevation and floatation) and the presence of a more variable type of vestibular dysfunction in patients with HAS, often characterized by rotational components, as well as the absence of any dysfunctioning of this system in AHs.

Support for this clinical hypothesis comes from a detailed analysis of the brain areas that were found to be damaged most frequently in patients with AP. Early studies stressed the involvement of posterior regions within the temporal, parietal, and occipital lobes (Brugger et al. 1997; Devinsky et al. 1989). More recently, however, Blanke et al. (2002, 2008) showed that OBEs and HAS are primarily associated with structural damage or electrophysiological hyperactivity at the temporo-parietal junction (TPJ; see also De Ridder et al. 2007 and Ionta et al. 2011), an area associated with vestibular processing, multisensory integration, and the perception of human bodies and body parts (Blanke and Arzy 2005; Lopez et al. 2008), while AHs are primarily associated with damage to the parieto-occipital cortex (Brandt et al. 2005). The latter finding would seem to be in line with the phenomenology of AHs, which are predominantly visual in nature, and do not involve any dramatic alterations in perspective and embodiment.

In summary, lesion data suggest that OBEs are associated primarily with damage to the right TPJ, HAS with damage to the left TPJ, and AHs with damage to the right parieto-occipital cortex (Blanke and Mohr 2005). Thus, AP in general would seem to derive from a double failure in multisensory integration processes: those related to proprioceptive, tactile, and visual information regarding one's own body (disintegration in personal space) and an additional vestibular dysfunction leading to a disintegration of personal (vestibular) space and extrapersonal (visual) space (especially in the case of OBEs, but also in HAS). The question to be answered in the forthcoming sections is how these clinical and neurobiological findings can be translated to the research laboratory and the study of self-consciousness.

15.2 Experimental Approaches: Rubber-Hand Illusion, Full-Body Illusion, and Face Illusion

15.2.1 The Rubber-Hand Illusion: A Method Designed to Manipulate Self-attribution

The rubber-hand illusion (RHI) is a perceptual phenomenon in which normal individuals are made to experience an artificial limb (usually a hand) as if it were a part of their own body, due to a short period of synchronous stroking of their hand, which is hidden out of view, and a rubber hand lying next to it. The RHI was described for the first time in a seminal paper by Botvinick and Cohen (1998). It provided the first experimental means to manipulate the sense of ownership for a hand in healthy subjects. Until then, knowledge about the neural mechanisms underlying the sense of bodily ownership depended almost exclusively on clinical reports of patients with somatoparaphrenia, a condition due to temporo-parietal damage, which, in right-handed individuals, leads to the misattribution of their left hand or foot to another person (Giummarra et al. 2008).

The first explanation for the RHI was that of a three-way interaction between vision, touch, and proprioception: vision of the touch to the rubber hand dominates over touch experienced by the real hand, resulting in a mislocalization of the tactile percept toward the spatial location of the visual percept. This sense of referred touch to the rubber hand, also named "illusory touch," influences the felt position (proprioception) of one's own hand, which is perceived closer to the rubber hand. This is reflected in a measurable proprioceptive drift toward the rubber hand, which has been used as an objective, behavioral proxy of the illusion for a great part of the subsequent studies exploring hand ownership (Tsakiris and Haggard 2005, see also Fig. 15.2a).

Various authors have demonstrated that the RHI can only be elicited when synchronous (rather than alternating) brushstrokes are applied to the real and fake hands, and when the rubber hand is placed in a position congruent with the subject's own hand (Botvinick and Cohen 1998; Armel and Ramachandran 2003; Tsakiris and Haggard 2005). More recently, it was demonstrated that the illusion of ownership and proprioceptive displacement can also be evoked with the aid of a three-dimensional projection of an arm, as long as the appropriate synchronous visuo-tactile stimulation is provided in a virtual environment (Slater et al. 2008; Sanchez-Vives et al. 2010).

Subsequent studies exploring the brain activity during the experience of the RHI revealed various regions possibly related to hand perception, cross-modal integration, and conflict processing (Ehrsson et al. 2004; Tsakiris et al. 2007). Those studies

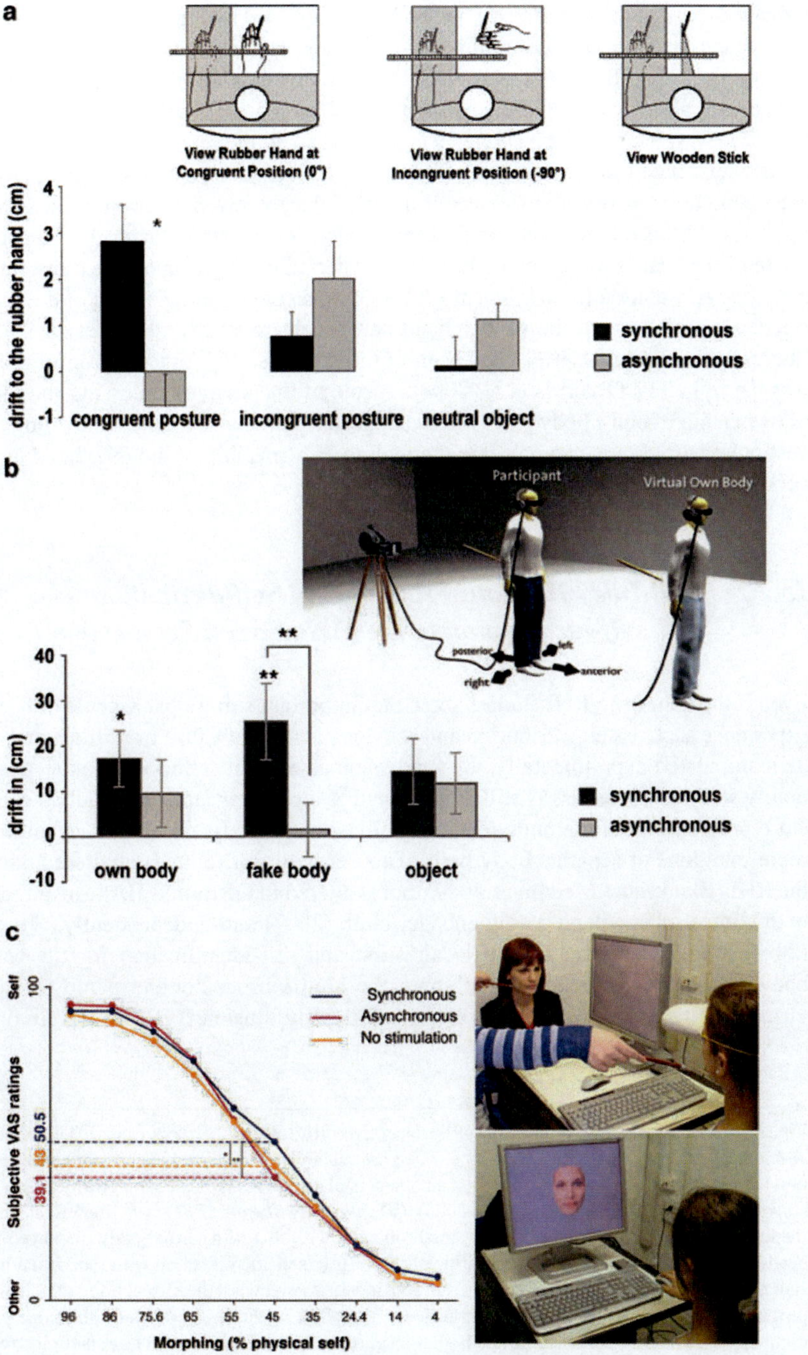

Fig. 15.2 Set-up and principal findings regarding the rubber-hand, full-body, and face-related illusions. **(a)** *Rubber-hand illusion (RHI):* stroking synchronously the subject's hand and a rubber

indicated that the feeling of body ownership is mediated by brain activity in regions involved in the integration of multisensory signals, such as bilateral premotor cortices and the cerebellum (Ehrsson et al. 2004), as well as the right posterior insula (Tsakiris et al. 2007), a region consistently implicated in self-attribution (Farrer and Frith 2002), self-processing (Vogeley et al. 2004), and the representation of an egocentric reference frame (Fink et al. 2003). The left inferior posterior parietal lobule was found to be involved in the recalibration of the perceived position of one's own limb toward the rubber hand. In fact this region is activated during the induction phase of the RHI (Ehrsson et al. 2004) and, when targeted by repetitive transcranial magnetic stimulation (rTMS, see also Chap. 25), a consistent reduction of the proprioceptive drift toward the rubber hand can be observed (Kammers et al. 2008). Interestingly, a similar effect has been described when the target region of rTMS was the right TPJ (Tsakiris et al. 2008). A role of this area in processing the sense of ownership of one's body parts is also suggested by recent evidence indicating that vestibular stimulation can increase the feeling of ownership of the fake hand in the context of the RHI (Lopez et al. 2010).

15.2.2 *Full-Body Illusion: From Global Self-attribution and Self-localization to the First-Person Perspective*

Even if the data from RHI studies show that important sub-global aspects of bodily experience such as self-attribution and self-localization of a fake or virtual hand can be manipulated experimentally, the fundamental sense of selfhood associated with bodily self-consciousness is still experienced as a singular feature, namely a coherent representation of the undivided, spatially situated body, rather than as multiple representations of separate body parts. This sense of self remains unaltered during the RHI (Blanke and Metzinger 2009), but is altered in OBEs and HAS. Inspired by both clinical phenomena, Lenggenhager et al. (2007) and, independently, Ehrsson (2007) induced changes in self-localization and self-identification for the entire body in healthy subjects by combining the multisensory conflict paradigm with virtual-reality technology, thus creating a full-body illusion (FBI). In the study by

Fig. 15.2 (continued) hand induced a proprioceptive drift toward the rubber hand. The effect is absent when the stroking is asynchronous, when the rubber hand is placed in an incongruent position with respect to the real hand, or when an object is placed in front of the subject which does not resemble a hand (Tsakiris and Haggard 2005). (**b**) *Full-body illusion* (*FBI*): a similar effect can be induced by stroking the subject's back synchronously with that of a virtual body displayed in a head-mounted display (*HMD*). The manipulation leads to self-identification with the virtual body (but not with an object) and induces a drift in self-localization toward the virtual body which occupies an extracorporeal position (Lenggenhager et al. 2007). (**c**) *Enfacement effect:* the same paradigm applied to faces, demonstrating a higher level of self-identification with morphed pictures of the subject and a partner's face when synchronous stroking is delivered to their cheeks and while they are sitting looking at each other in a mirror fashion (Sforza et al. 2010)

Lenggenhager et al. (2007), subjects observed the synchronous or asynchronous (i.e., delayed) stroking of their own backs, projected on head-mounted displays in front of them. Only after synchronous stimulation did they observe changes in self-localization (i.e., mislocalization of the position of their own body, in the sense of its being closer to their virtual body) in analogy with the proprioceptive drift that was found in the RHI experiments (see Fig. 15.2b). In this experimental situation, subjects also reported experiencing illusory touch on the virtual body, and they were keener to identify with it (irrespective whether the virtual body was their own filmed body or a fake one). This was not the case when an object, rather than a body, was projected and stroked synchronously with their own bodies. This result would seem to imply the influence of top-down cognitive constraints on multisensory integration, self-localization, and self-identification (see also Tsakiris and Haggard 2005 and Armel and Ramachandran 2003).

While comparing their results with the clinical manifestations during OBEs, Lenggenhager et al. (2007) observed how the illusion was neither associated with overt disembodiment nor with quantitative changes in the first-person perspective (see also Blanke and Metzinger 2009). They argued that only some aspects of OBE had been elicited in this experiment, and that an additional conflict between visual and vestibular information may be required to induce a more complete OBE (see Ionta et al. 2011). Thus they concluded that the FBI would rather seem to resemble HAS, in which patients tend to report seeing themselves from behind, and to identify with the illusory body (Blanke and Mohr 2005).

Ehrsson (2007) complemented these results by showing physiological changes related to embodiment of the virtual body as a consequence of multisensory integration: he reported significantly greater skin-conductance responses in reaction to threats against the virtual body when the latter was stroked in a synchronous, (illusory) manner. Recently even more aspects of the FBI have been elucidated. Thus Aspell et al. (2009) reported systematic alterations in the mapping of tactile stimuli during the FBI, as quantified through repeated reaction-time and accuracy measures. They found a greater visual capture of touch, in terms of a mislocalization of tactile stimuli toward the virtual body (measured through the visuo-tactile cross-modal congruency effect; see Pavani et al. 2000) in conditions associated with stronger changes in self-location and self-identification with the visually perceived body (synchronous condition). More recent data have shown that during the FBI, pain thresholds tend to be altered (Hänsel et al. 2011), a feature which has also been found in association with spontaneous OBEs. Moreover, muscle vibrations applied to the legs of subjects involved in FBI experiments were found to decrease the classical dependency of self-identification and congruency effects on the synchrony of stroking (Palluel et al. 2011). Finally, Petkova and Ehrsson (2008) demonstrated how critical visual information from the first-person perspective is for the experience of owning a body. Subjects in their experiments were wearing a head-mounted display (HMD) connected to a camera placed on a mannequin's body. Thus they were made to perceive the mannequin's body where they expected to see their own. Such a manipulation of the first-person perspective, in combination with the application of correlated multisensory information from the body (synchronous visuo-tactile stimulation), turned out to be sufficient to trigger the illusion that someone else's body – or an artificial body, for that matter – was their own, irrespective of the concomitant execution of synchronous or asynchronous voluntary movements (i.e., the shaking of hands), and even when the subjects were directly facing their own real body (body-swapping illusion). These data on the first-person perspective were recently extended by Ionta et al. (2011), who used robotic stroking during functional magnetic resonance imaging (fMRI), and thus were able to induce experimental changes in the experienced self-location and direction of the first-person perspective during FBIs. These changes were reflected in the increased activity of the bilateral TPJs. The authors corroborated their findings by comparing the fMRI results with a lesion analysis of a large group of patients experiencing OBEs, whose brain damage was also localized at TPJ (although mainly on the right side). As regards the neural underpinnings of the FBI, Lenggenhager et al. (2011) recorded high-density electroencephalograms (EEGs) during the FBI and described differential alpha-band power modulations in bilateral sensorimotor cortices following synchronous and asynchronous stimulation of the virtual body, as well as a correlation of the alpha power in the medial prefrontal cortex, a brain region associated with a large variety of self-related cognitions, with the degree of drift in self-location.

Collectively, these studies indicate that self-identification (or the feeling of ownership of an entire body), self-location, and the first-person perspective are the results of a consistent pattern of spatially and temporally congruent visuo-tactile and sensorimotor signals that are integrated with additional visuo-vestibular mechanisms related to perspectival coding in a distributed network.

15.2.3 From Embodiment to "Enfacement": Challenging the Sense of Identity

Obviously our face is crucial for our sense of identity and constitutes our most distinctive external feature: when we look into a mirror we immediately know that the reflected image is ours. The face is also a special body part because, while we can inspect other body parts directly, we can only see the face when it is reflected in a mirror or some other reflecting surface. The self-identification of seen faces has been demonstrated only in a limited number of species: they include humans and apes, and recently it has also been possible to demonstrate signs of self-recognition in dolphins, elephants, magpies, and pigeons (Keenan et al. 2003; Plotnick et al. 2006; Prior et al. 2008). It develops in children around the age of two, in the same period of language development characterized by the first use of the pronouns "me" and "you" (Keenan et al. 2003). Moreover, for what concerns AP, in several cases of AH, patients reported to see only the face or the upper part of the trunk of the autoscopic body (Zamboni et al. 2005; Hécaen and de Ajuriaguerra 1952), suggesting a potentially different involvement of the upper and front-facing own body in bodily self-consciousness, likely due to different multisensory mechanisms.

Despite the relevance of self-face recognition for self-consciousness (Gallup 1982), experimental research in healthy subjects in this area has been sparse, and self-face recognition studies have focused primarily on visual mechanisms related to the perception and retrieval of one's own facial characteristics (Keenan et al. 2000). Recent research explored the impact of multisensory stimulation on self-face recognition, an ability that has been shown to be largely impenetrable in cases of brain damage. Indeed, even when transient difficulties in recognizing one's own face were reported by normal subjects (e.g., when looking at old pictures; see Brédart and Young 2004), persisting deficits in own-face recognition (not extending to other familiar faces, as in cases of prosopagnosia) have only rarely been reported and almost exclusively in patients with severe neurological disorders (such as the mirrored self-misidentification sign, i.e., the inability to recognize one's own reflected image, as shown by some patients at an early stage of dementia; see Breen et al. 2001) or psychiatric disorders (Kircher et al. 2007). Recent studies have extended procedures from the RHI and FBI experiments to the study of the perception of the own face and manipulated self-identification with one's own face experimentally (Sforza et al. 2010; Tsakiris 2008; Paladino et al. 2010). By using conflicting visuo-tactile information, Sforza et al. (2010) induced a modulation of self-face recognition associated with higher self-identification when participants were asked to rate morphed pictures. In particular, touching the faces of subjects who were viewing synchronous touches on a partner's face was shown to induce illusory self-identification or "enfacement" (Sforza et al. 2010), in the sense that the partner's facial features became incorporated into the representation of the participant's own face. As confirmed by psychophysical measurements, people participating in the experiment judged that morphed images of themselves and their partner contained more "self" than "other" only after synchronous, but not asynchronous, stroking (see Fig. 15.2c). Moreover, the incorporation of other's facial features into the self-face representation depended on the facial attractiveness of the other as well as on the participants' empathic abilities. Paladino et al. (2010) observed that such visuo-tactile facial stimulation not only leads to facial self-other merging (resemblance or bodily self-consciousness) but also extends its effect to the domain of social perception. In line with previous studies on bodily self-consciousness, these findings suggest that the visual capture of facial touch updates central representations of one's own physical facial appearance, maybe thus allowing us to update any slow physical changes related to the passage of time (Sforza et al. 2010; Arzy et al. 2009). Evidence that self-face representation might well be more malleable than previously believed also comes from a recent study exploring mirror gazing under specific light conditions (Caputo 2010). Thus, after a period of own-face observation in a mirror (in a sparsely illuminated room), individuals reported perceiving their own face as deformed, unknown or strange, and even as belonging to another person. Similar effects have been described by Barnier et al. (2008) with the aid of hypnosis. Collectively, these self-face effects are reminiscent of mirrored self-misidentification syndromes (see Breen et al. 2001).

15.3 Conclusion

Understanding the foundation of our sense of self has been a major ambition for science, philosophy, and religion since time immemorial. Only recently, however, has cognitive neuroscience started an attempt to clarify the psychological and cerebral mechanisms underlying the sense of self. The studies reviewed so far focused on an elementary sense of awareness that is strongly linked to bodily processing and body representation. Studies of hallucinations involving one's own body perception on the one hand, and experimental studies seeking to replicate these phenomena in healthy subjects on the other, suggest that the integration of visual and multisensory bodily signals from the entire body is important for three major aspects of bodily self-consciousness, namely self-location (i.e., the volume in space where humans experience the location of the self: "Where I experience to be"), first-person perspective (i.e., the directedness of conscious experience: "From which vantage point I experience to perceive the world"), and self-identification (i.e., the degree to which humans identify with their own bodies: "What I experience as my body") (Blanke and Metzinger 2009). Recently, a multisensory integration model has been proposed for the sense of identity as regards the representation of one's own face, challenging its relative stability and stressing the continuous need to update our facial representation through multisensory inputs (Sforza et al. 2010; Tsakiris et al. 2008).

Future studies will need to aim at integrating own-body and own-face representations to allow us to understand to which extent the sense of self in cases such as these has any neural and functional mechanisms in common. In this sense, the various phenomenological features of autoscopic phenomena offer a tantalizing model to be tested within the context of the paradigms described. Can multisensory conflicts applied to the bodies as well as the faces of patients with out-of-body experiences, heautoscopy, and autoscopic hallucinations modulate their hallucinatory experiences in such a way that they can be found to differ for each type of autoscopic phenomenon? Bolognini et al. (2011) provided the first important example of this approach by testing the effects of sensory and motor stimulation in a patient experiencing prolonged autoscopic phenomena. As those phenomena are rare, and mostly of a paroxysmal nature, we believe that, especially, laboratory work – based on such clinical observations – will allow for a scientific understanding of self-consciousness in health as well as in disease.

References

Armel, K.C., Ramachandran, V.S. (2003). Projecting sensations to external objects: Evidence from skin conductance response. *Proceedings. Biological Sciences*, 270, 1499–1506.

Arzy, S., Collette. S., Ionta. S., Fornari, E., Blanke, O. (2009). Subjective mental time: the functional architecture of projecting the self to past and future. *European Journal of Neuroscience*, 30, 2009–2017.

Aspell, J.E., Lenggenhager, B., Blanke, O. (2009). Keeping in touch with one's self: multisensory mechanisms of self-consciousness. *PLoS One*, 4, e6488.

Barnier, A.J., Cox, R.E., O'Connor, A., Coltheart, M., Langdon, R., Breen, N., Turner, M. (2008). Developing hypnotic analogues of clinical delusions: Mirrored-self misidentification. *Cognitive Neuropsychiatry*, 13, 406–430.

Berlucchi, G., Aglioti, S. (1997). The body in the brain: neural bases of corporeal awareness. *Trends in Neurosciences*, 20, 560–564.

Blackmore, S.J. (1982). *Beyond the body. An investigation of out-of-body experiences*. London: Heinemann.

Blanke, O., Ortigue, S., Landis, T., Seeck, M. (2002). Stimulating illusory own-body perceptions. *Nature*, 419, 269–270.

Blanke, O., Landis, T., Spinelli, L., Seeck, M. (2004). Out-of-body experience and autoscopy of neurological origin. *Brain*, 127, 243–258.

Blanke, O. (2005). Perception and experience of the self in autoscopic phenomena and self-portraiture. *Swiss Archives of Neurology and Psychiatry*, 156, 173–188.

Blanke O., Arzy S. (2005). The out-of body experience. Disturbed self-processing at the temporo-parietal junction. *Neuroscientist*, 11, 16–24.

Blanke, O., Mohr, C. (2005). Out-of-body experience, heautoscopy, and autoscopic hallucination of neurological origin. Implications for neurocognitive mechanisms of corporeal awareness and self-consciousness. *Brain Research Reviews*, 50, 184–199.

Blanke, O., Arzy, S., Landis, T. (2008). *Illusory reduplications of the human body and self*. In *Handbook of clinical neurology. Volume 88*. Edited by M. Aminoff, F. Boller, and D. Swaab. Amsterdam: Elsevier, pp. 429–458.

Blanke, O., Metzinger, T. (2009). Full-body illusions and minimal phenomenal selfhood. *Trends in Cognitive Sciences*, 13, 7–13.

Blom, J.D. (2010). *A dictionary of hallucinations*. New York, NY: Springer.

Bolognini, N., Làdavas, E., Farnè, A. (2011). Spatial perspective and coordinate systems in autoscopy: a case report of a "fantome de profil" in occipital brain damage. *Journal of Cognitive Neuroscience*, 23, 1741–1751.

Botvinick, M., Cohen, J. (1998). Rubber hands 'feel' touch that eyes see. *Nature*, 391, 756.

Brandt, C., Brechtelsbauer, D., Bien, C.G., Reiners, K. (2005). Out-of-body experience as possible seizure symptom in a patient with a right parietal lesion. *Nervenarzt*, 76, 61–62.

Brédart, S., Young, A.W. (2004). Self-recognition in everyday life. *Cognitive Neuropsychiatry*, 9, 183–197.

Breen, N., Caine, D., Coltheart, M. (2001). Mirrored-self misidentification: Two cases of focal onset dementia. *Neurocase*, 7, 239–254.

Brugger, P. (2002). Reflective mirrors: Perspective-taking in autoscopic phenomena. *Cognitive Neuropsychiatry*, 7, 179–194.

Brugger, P., Agosti, A., Regard, M., Wieser, H.-G., Landis, T. (1994). Heautoscopy, epilepsy, and suicide. *Journal of Neurology, Neurosurgery, and Psychiatry*, 57, 838–839.

Brugger, P., Blanke, O., Regard, M., Bradford, D.T., Landis, T. (2006). Polyopic heautoscopy: Case report and review of the literature. *Cortex*, 42, 666–674.

Brugger, P., Regard, M., Landis, T. (1996). Unilaterally felt "presences": the neuropsychiatry of one's invisible doppelgänger. *Neuropsychiatry, Neuropsychology, and Behavioral Neurology*, 9, 114–122.

Brugger, P., Regard, M., Landis, T. (1997). Illusory reduplication of one's own body: Phenomenology and classification of autoscopic phenomena. *Cognitive Neuropsychiatry*, 2, 19–38.

Caputo, G.B. (2010). Strange-face-in-the-mirror illusion. *Perception*, 39, 1007–1008.

Critchley M. (1955). The idea of a presence. Acta Psychiatrica et Neurologica Scandinavica, 30, 155–168.

De Ridder, D., Van Laere, K., Dupont, P., Menovsky, T., Van de Heyning, P. (2007). Visualizing out-of-body experience in the brain. *New England Journal of Medicine*, 357, 1829–1833.

Devinsky, O., Feldmann, E., Burrowes, K., Bromfield, E. (1989). Autoscopic phenomena with seizures. *Archives of Neurology*, 46, 1080–1088.

Ehrsson, H.H. (2007). The experimental induction of out-of-body experiences. *Science*, 317, 1048.

Ehrsson, H.H., Spence, C., Passingham, R.E. (2004). That's my hand! Activity in premotor cortex reflects feeling of ownership of a limb. *Science*, 305, 875–877.

Farrer, C., Frith, C.D. (2002). Experiencing oneself vs another person as being the cause of an action: The neural correlates of the experience of agency. *NeuroImage*, 15, 596–603.

Fink, G.R., Marshall, J.C., Weiss, P.H., Stephan, T., Grefkes, C., Shah, N.J., Zilles, K., Dieterich, M. (2003). Performing allocentric visuospatial judgments with induced distortion of the egocentric reference frame: An fMRI study with clinical implications. *NeuroImage*, 20, 1505–1517.

Gallup, G.G., Jr. (1982). Self-awareness and the emergence of mind in primates. *American Journal of Primatology*, 2, 237–248.

Giummarra, M.J., Gibson, S.J., Georgiou-Karistianis, N., Bradshaw, J.L. (2008). Mechanisms underlying embodiment, disembodiment and loss of embodiment. *Neuroscience and Biobehavioral Reviews*, 32, 143–160.

Grüsser, O.-J., Landis, T. (1991). *The splitting of "I" and "me": heautoscopy and related phenomena.* In V*isual agnosias and other disturbances of visual perception and cognition.* Edited by O.-J. Grüsser and T. Landis. London: MacMillan, pp. 297–303.

Hänsel, A., Lenggenhager, B., Von Känel, R., Curatolo M., Blanke, O. (2011). Outside my body, I feel less pain: Pain perception changes during illusory own body perception. *European Journal of Pain*, 15, 874–879.

Hécaen, H., de Ajuriaguerra, J. (1952). *Méconnaissances et hallucinations corporelles.* Paris: Masson, pp. 310–343.

Heydrich, L., Dieguez, S., Grunwald, T., Seeck, M., Blanke, O. (2010). Illusory own body perceptions: case reports and relevance for bodily self-consciousness. Consciousness and Cognition, 19, 702–710.

Ionta, S., Heydrich, L., Lenggenhager, B., Mouthon, M., Fornari, E., Chapuis, D., Gassert, R., Blanke, O. (2011). Multisensory mechanisms in temporo-parietal cortex support self-location and first-person perspective. *Neuron*, 70, 363–374.

Jaspers, K. (1913). Über leibhaftige Bewusstheiten (Bewussheitstaüschungen). Ein psychopathologisches Elementarsymptom. *Zeitschrift für Psychopathologie*, 2, 150–161.

Kammers, M.P., Verhagen, L., Dijkerman, H.C., Hogendoorn, H., De Vignemont, F., Schutter, D.J. (2008). Is this hand for real? Attenuation of the rubber hand illusion by transcranial magnetic stimulation over the inferior parietal lobule. *Journal of Cognitive Neuroscience*, 10, 1–10.

Keenan, J.P., Gallup, G.G., Falk, D., Jr. (2003). *The face in the mirror: The search for the origins of consciousness.* New York, NY: Harper Collins.

Keenan, J.P., Wheeler, M.A., Gallup, G.G., Jr, Pascual-Leone, A. (2000) Self-recognition and the right prefrontal cortex. *Trends in Cognitive Sciences*, 4, 338–344.

Kircher, T.T., Seiferth, N.Y., Plewnia, C., Baar, S., Schwabe, R. (2007). Self-face recognition in schizophrenia. *Schizophrenia Research*, 94, 264–272.

Kölmel, H.W. (1985). Complex visual hallucinations in the hemianopic field. *Journal of Neurology, Neurosurgery, and Psychiatry*, 48, 29–38.

Lenggenhager, B., Halje, P., Blanke, O. (2011). Alpha band oscillations correlate with illusory self-location induced by virtual reality. *European Journal of Neuroscience*, 33, 1935–1943.

Lenggenhager, B., Tadi, T., Metzinger, T., Blanke, O. (2007). Video ergo sum: Manipulating bodily self-consciousness. *Science*, 317, 1096–1099.

Lhermitte, J. (1939). *L'image de notre corps.* Paris: Editions de la Nouvelle Revue Critique.

Lippman, C.W. (1952). Certain hallucinations peculiar to migraine. *Journal of Nervous and Mental Disease*, 116, 346–351.

Lopez, C., Halje, P., Blanke, O. (2008). Body ownership and embodiment: Vestibular and multisensory mechanisms. *Neurophysiologie Clinique*, 38, 149–161.

Lopez, C., Lenggenhager, B., Blanke, O. (2010). How vestibular stimulation interacts with illusory hand ownership. *Consciousness and Cognition*, 19, 33–47.

Lukianowicz, N. (1958). Autoscopic phenomena. *Archives of Neurology and Psychiatry*, 80, 199–220.

Paladino, M.P., Mazzurega, M., Pavani, F., Schubert, T.W. (2010). Synchronous multisensory stimulation blurs self-other boundaries. *Psychological Science*, 21, 1202–1207.

Palluel, E., Aspell, J.E., Blanke, O. (2011). Leg muscle vibration modulates bodily selfconsciousness: Integration of proprioceptive, visual, and tactile signals. *Journal of Neurophysiology*, 105, 2239–2247.

Pavani, F., Spence, C., Driver, J. (2000). Visual capture of touch: out-of-the-body experiences with rubber gloves. *Psychological Science*, 11, 353–359.

Petkova, V.I., Ehrsson, H.H. (2008). If I were you: Perceptual illusion of body swapping. *PLoS ONE*, 3, 3832.

Plotnik, J.M., De Waal, F.B., Reiss, D. (2006). Self-recognition in an Asian elephant. *Proceedings of the National Academy of Sciences U S A*, 103, 17053–17057.

Prior, H., Schwarz, A., Güntürkün, O. (2008). Mirror-induced behavior in the magpie (Pica pica): Evidence of self-recognition. *PLoS Biology*, 6, e202.

Ramachandran, V.S., Hirstein, W. (1998). The perception of phantom limbs. *Brain*, 121, 1603–1630.

Sanchez-Vives, M.V., Spanlang, B., Frisoli, A., Bergamasco, M., Slater, M. (2010). Virtual hand illusion induced by visuomotor correlations. *PLoS One*, 5, e10381.

Sforza, A., Bufalari, I., Haggard, P., Aglioti, S.M. (2010). My face in yours: visuo-tactile facial stimulation influences sense of identity. *Social Neuroscience*, 5, 148–162.

Slater, M., Perez-Marcos, D., Ehrsson, H.H., Sanchez-Vives, M.V. (2008) Towards a digital body: The virtual arm illusion. *Frontiers in Human Neuroscience*, 2, 6.

Sollier, P. (1903). *Les phénomènes d'autoscopie*. Paris: Félix Alcan.

Todd, J., Dewhurst, K. (1955). The double: its psycho-pathology and psychophysiology. *Journal of Nervous and Mental Disease*, 122, 47–55.

Tsakiris, M. (2008). Looking for myself: Current multisensory input alters self-face recognition. *PLoS ONE*, 3, e4040.

Tsakiris, M., Costantini, M., Haggard, P. (2008). The role of the right temporo-parietal junction in maintaining a coherent sense of one's body. *Neuropsychologia*, 46, 3014–3018.

Tsakiris, M., Haggard, P. (2005).The rubber hand illusion revisited: Visuotactile integration and self-attribution. *Journal of Experimental Psychology. Human Perception and Performance*, 31, 80–91.

Tsakiris, M., Hesse, M.D., Boy, C., Haggard, P., Fink, G.R. (2007). Neural signatures of body ownership: A sensory network for bodily self-consciousness. *Cerebral Cortex*, 17, 2235-2244.

Vogeley, K., May, M., Ritzl, A., Falkai, P., Zilles, K., Fink, G.R. (2004). Neural correlates of first-person perspective as one constituent of human self-consciousness. *Journal of Cognitive Neuroscience*, 16, 817–827.

Zamboni, G., Budriesi, C., Nichelli, P. (2005). "Seeing oneself": A case of autoscopy. *Neurocase*, 11, 212–215.

Chapter 16
Phantom Limb, Phantom Body, Phantom Self: A Phenomenology of "Body Hallucinations"

Peter Brugger

16.1 Introduction

Behavioral neurology has a strong focus on deficits: It deals with functional breakdowns in the language system, with the many ways memories can be lost, and with the various impairments of action and perception. Hallucinations are intriguing examples of perceptual impairments. They are unique in that something is added to, rather than falling apart or falling entirely away from our perceptual experience. In the domain of the perception of our body, such productive symptoms of experience can take many faces, collectively being subsumed as "phantom experiences." These can be defined as experiences of bodily functions spatially dissociated from the objective body on the level of "flesh and blood." In the various types of phantom-limb experience, physically or phenomenally lost body parts continue to live on in corporeal awareness. But a phantom may also replace one lateral half of a person's body, most typically after a stroke in the hemisphere controlling that side has rendered it hemiplegic (i.e., paralyzed and devoid of any sensory functions). Finally, the complex experience of oneself as a phantom is known as a "hallucination of the double." This chapter is a guide to phantom experiences on all these levels, from single-limb phantoms to hemibody phantoms, and finally to phantoms of the entire body and self.

16.2 Phantom Body Parts

Phantom body parts are most frequently experienced after the physical loss of a limb. In a first section, we will review the history and phenomenology of amputation phantoms and emphasize their meaningfulness for theories of neural plasticity

---

P. Brugger, Ph.D. (✉)
Department of Neurology, University Hospital Zurich, Zurich, Switzerland
e-mail: peter.brugger@usz.ch

J.D. Blom and I.E.C. Sommer (eds.), *Hallucinations: Research and Practice*,
DOI 10.1007/978-1-4614-0959-5_16, © Springer Science+Business Media, LLC 2012

and reorganization. The question of innate components of the shape and functionality of our canonical, four-limbed body will be addressed in two subsequent sections; the first dealing with phantom experiences in people born without limbs and the second with a condition arguably consisting of the absence of an innate representation of one or more limbs. A further section will discuss phantom body parts that are experienced despite the fact that the relevant parts are still present, at least physically. One subsection will focus on the experience of such "supernumerary phantoms" in patients with spinal cord injury, another on patients with brain damage. Yet another one will deal with the experimental evocation of phantom body parts in healthy volunteers.

16.2.1 Phantom Body Parts After Amputation

Reports about phantom limbs are probably as old as survival from the loss of an arm or a leg. Already in the tenth century, soldiers who had lost an extremity in combat described that they continued to feel its presence. Folk psychology took these accounts as evidence for a divine resurrection of the limb, in analogy with the purported resurrection of the entire body after death (Price and Twombly 1976). It is a sad fact that medical knowledge about phantom limbs is also primarily based on what we learned from war experiences (Solonen 1962; Weeks et al. 2010). This was already the case in the sixteenth century, when Ambroise Paré (1510–1590), a barber by training, had advanced to be the most successful military surgeon. The Frenchman, whose diverse contributions to medicine are compiled in a work of ten volumes, stated in 1551 that "the patients, long after the amputation is made, say they still feel pain in the amputated part. Of this, they complain strongly, a thing worthy of wonder and almost incredible to people who have not experienced this." (Paré 1840/41, p. 221). As Paré's reports remained largely descriptive, they did not stimulate much dispute among professionals, and the topic of the phenomenal persistence of lost limbs vanished from the medical literature for more than three centuries. It resurfaced in the late nineteenth century in the work of the American neurologist Silas Weir Mitchell (1829–1914). Mitchell was confronted with the "sensory ghost" of lost limbs in the course of the American Civil War, and it is to him that we owe the term "phantom limb."

In conventional textbooks, much emphasis is given to *painful* phantom sensations. This may be a consequence of the severity of phantom-limb pain which, if present, is mentioned spontaneously by an amputee, while painless phantom-limb phenomena such as posture, weight, or movement sensations are not given comparable priority (see Fig. 16.1 for an illustration of both painful and painless phantom phenomena). Furthermore, phantom-limb pain is largely resistant to treatment. This holds for physiotherapeutic manipulation of the stump to a broad range of pharmacological interventions and invasive techniques involving thalamic stimulation or removal of cortical tissue (see Chap. 14). Not one single method has proven effective for all amputees, and once they yield relief, this tends to be only

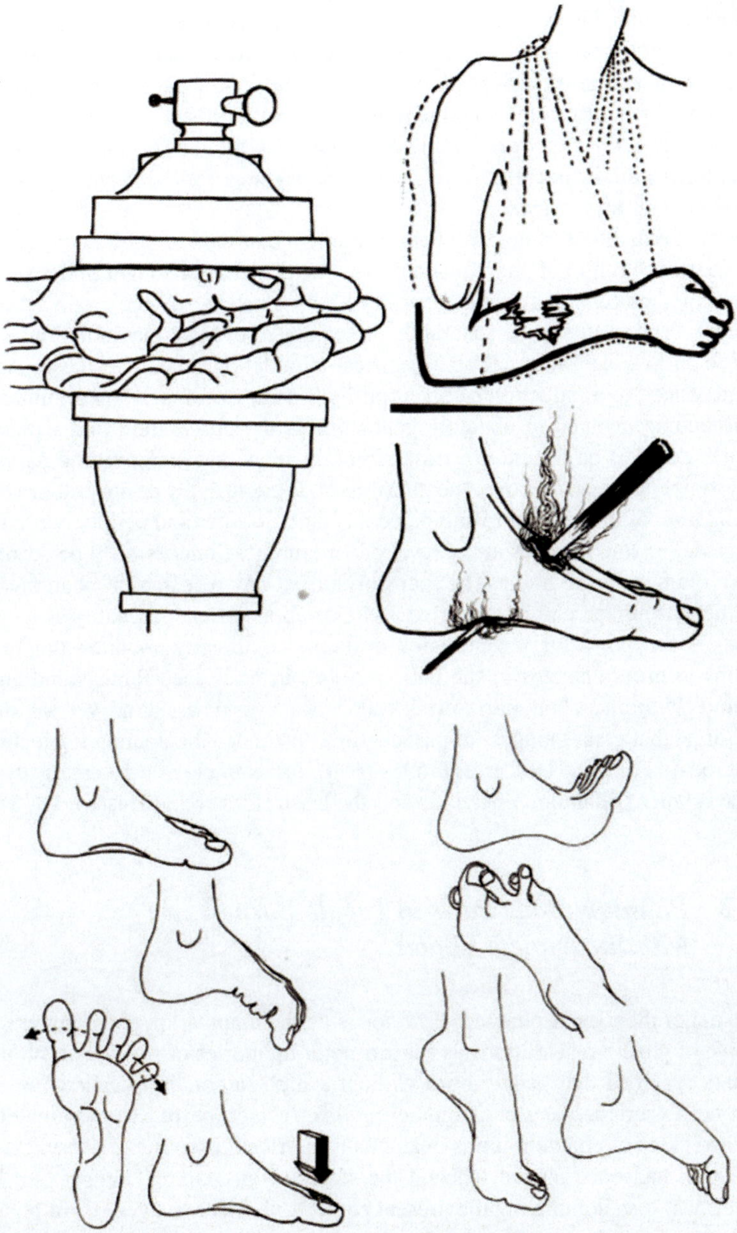

Fig. 16.1 Painful (*top*) and painless (*bottom*) phantom sensations. Pain can be clenching, burning, and/or stabbing, and even preamputation pain may persist. Non-painful phantoms mainly represent posture, weight, and movement. They can also comprise temperature and distinct impressions about paraphernalia, like for instance, rings or wristwatches (not shown) (Reproduced, with kind permission of the editors of *Acta Orthopaedica*, from Solonen 1962)

transient. As will be evident from the remaining sections of this chapter, any overemphasis on phantom *pain* does not adequately reflect the phenomenological richness of phantom experiences. The narrow focus on pain in a lost limb is comparable to a unilateral treatment of, say, color in a discussion of visual hallucinations. Color may predominate in single reports, but brightness, contrast, texture, form, and the occurrence of specific visual cues such as faces or places are of equal general importance.

During the first half of the twentieth century, one prominent theoretical question was whether the origin of phantom-limb sensations had to be sought in the amputation stump ("peripheral theories") or rather at the level of the brain ("central theories"). Propagators of peripheral theories pointed out that manipulations of the stump (e.g., massage, thermal and electrical stimulation, local anesthesia) would sometimes lead to a reduction of phantom-limb pain. Conversely, an argument for central factors involved in phantom generation is the observation that a phantom limb may cease to be felt after a stroke in brain areas that represent the particular limb. From today's perspective, the question of an exclusively peripheral or central determination of the phantom-limb percept is moot. Center and periphery are interconnected, i.e., any changes at the level of the stump will necessarily be accompanied by changes in the brain. The focus of current-day research on phantom-limb phenomena is almost entirely directed at the cerebral cortex – arguably as a consequence of the powerful technologies available to directly monitor the brain's reactions to gross changes in the periphery of our body (see Ramachandran and Hirstein (1998) and Giummarra et al. (2007) for overviews). And yet we should never forget that great thoughts of past centuries preceded the dominating technologies of today: For René Descartes (1596–1650), for example, it was already evident that the origin of phantom sensations was the brain (Finger and Hustwit 2003).

16.2.2 Phantom Body Parts in People Born with Incomplete Bodies

As the major theories of phantom limb focus on the amputation phantom, and thus conceive of phantom sensations as sensorimotor memories of previous limb use, it was long assumed that people born without a limb (a condition called "amelia") would not experience any phantom sensations. In fact, despite approximately 60 published cases of phantom limbs in amelia (see Price (2006) for an overview), this assumption had once almost reached the status of an axiom (Brugger and Funk 2007). It was also dominating the present author's views on the genesis of phantom experiences when he first met A.Z., a then 44-year-old woman born without legs and forearms who claimed vivid phantom experiences of all missing body parts. This view was thoroughly shaken after an extensive research collaboration initiated to substantiate these claims. A variety of methods, from clinical observation to behavioral testing and functional neuroimaging, showed that the phantom sensations reported by A.Z. were more than just wishful thinking, and reflected a cortical representation of body parts that had never physically developed (see Table 16.1).

Table 16.1 Methods used to substantiate claims about phantoms of congenitally absent limbs in A.Z., a person born without forearms and legs

Method	Main findings	Reference
Clinical observation: A.Z. was asked to fold her phantom arms, and clasp her phantom fingers (a) in a habitual way, and (b) reversed, i.e., the other arm/thumb on top	Reverse folding/clasping elicited a spontaneous reaction of postural discomfort and made A.Z. emphasize that she would never fold her arms /clasp her hands that way	Brugger and Funk (2007)
Clinical observation: A.Z. was asked to mimic finger postures as displayed on drawings of hands	A.Z.'s vividness ratings for the imagined postures reflected the degree of anatomical awkwardness of the postures	Hilti and Brugger (2010)
Behavioral testing: A.Z. was shown photographs of hands and feet in varying degrees of rotation. She provided laterality decisions of these body parts ("a left or a right limb?")	A.Z.'s reaction times of correct laterality decisions were a function of the rotation angle of the displayed limb, i.e., she showed the same constraints in this task as normally limbed observers do	Brugger et al. (2000), Funk and Brugger (2008)
Behavioral testing: A.Z. was shown pairs of photographs of a limb in two different postures eliciting an apparent motion percept	A.Z.'s apparent motion trajectories were a function of the flash rate; like for normally limbed observers, apparent motion percepts were influenced by anatomical joint constraints	Funk et al. (2005)
Behavioral testing: A.Z. had to sort labels of body parts and activities based on whether she considered them related or not. Sorting was achieved in a hierarchical splitting paradigm	Structure dimensional analysis revealed that some of the sorting characteristics by A.Z. reflected the presence of her phantoms, while others reflected peculiarities in her daily activities (e.g., writing with her mouth)	Bläsing et al. (2010)
Functional magnetic resonance imaging: repeated opening/closing of a phantom hand	Bilateral activation of premotor and parietal cortex; no activation of the primary sensorimotor area of hand representation	Brugger et al. (2000)
Transcranial magnetic stimulation: single-pulse magnetic stimuli were applied over areas anterior and posterior of the vertex, bilaterally. Verbal reports of phantom sensations were recorded verbatim	Phantom-hand sensations were elicited by contralateral stimulation. They were mostly reported when there was a simultaneous motor-evoked response from the deltoid muscle in the stump. Phantom movements were experienced also after stimulations posterior to the central sulcus and without accompanying motor-evoked potentials	Brugger et al. (2000)

Control experiments in an aplasic person born with a body similar to that of A.Z., but who had never experienced any phantom sensations, showed that these representations were specific to the presence of the phantom percept (Bläsing et al. 2010; Funk and Brugger 2008; Funk et al. 2005).

How can phantom sensations of limbs that have never been physically developed be explained? Do we have to assume that a four-limbed body as a canonical form of the human body is hardwired in our genes (Abramson and Feibel 1981)? As we have already proposed in our first report on the work with A.Z. (Brugger et al. 2000), the assumption of a genetic basis of "body schema" as a central representation of the body is not warranted by the present data. What must be considered are theories of the tight interplay between action observation and action execution (Gallese and Goldman 1998). It is now firmly established that a "mirror-neuron" system codes for both the motor execution and the visual observation of limb movements. Thus, in the absence of a physical substrate for the execution of an action, i.e., in the (congenital) absence of limbs, an individual can still activate networks mediating visuomotor limb representations while observing conspecifics moving their extremities. We admit that we are unable to explain why only a minority of aplasic people (up to 20% according to Melzack et al. (1997)) experience phantom limbs. However, given the very large heterogeneity in appearance and etiology of congenital limb deficiencies, it does not seem too farfetched to assume large inter-individual differences in matching observation and (imaginary) execution of parts of the body. Contemporary authors agree that prospective studies of the mirror-neuron system in persons born without limbs are badly needed (Price 2006; Giummarra et al. 2007; Wood and Stuart 2009). Equally important is a consideration of intrauterine life events. In many amelic persons, the ultimate reason for a (non-teratogenically caused) limb deficiency remains unknown. While fetal limb movements have been studied to infer possible factors responsible for the genesis of congenital phantoms (Gallagher et al. 1998), the problem of matching the "incarnation" of a limb (i.e., its realization on the physical level) to its "animation" (i.e., the phenomenal experience of owning that limb) has not yet been sufficiently explored by current-day neuroembryology.

16.2.3 "Wannabe Amputees" and the Concept of "Negative Phantom Limbs"

The previous section dealt with the failed incarnation of a limb that is still animated in the sense of being experienced as a phantom. In the present section, we will very briefly address the question of whether the converse case can be observed: Is it possible that incarnation of a limb has taken place flawlessly, but animation has failed? Our answer is yes, and the condition we have in mind is called "body-integrity identity disorder" (BIID). People with BIID complain about being "overcomplete." They suffer from having four limbs and desire to get rid of one or several of them to match their physical appearance with the idea they have about how their body should have been shaped. Typically, this desire for amputation has been around "as long they can think," quite similar to the phantom awareness in people with congenitally missing limbs. We have contrasted these two phantom-limb phenomena: the one designating an "animation without incarnation," the other, BIID, an "incarnation without animation" (Hilti and Brugger 2010). Unpublished data on the dreams of people with BIID who have realized their desire (i.e., succeeded in having the non-animated limb removed) indicate that the time they need to dream themselves amputated is much shorter than that for persons with traumatic amputations. This may indicate that their amputation desire was based on a central representation of their body, which diverged from the canonical four-limb format right from the beginning of their lives. Viewed as an instance of a "negative hallucination" in the domain of corporeal awareness, BIID needs to be considered in any comprehensive work on hallucinations.

16.2.4 Supernumerary Phantom Body Parts

A "supernumerary" phantom is the phenomenal experience of a body part in the absence of its peripheral loss (or congenital absence) but *in addition to* the respective physical limb. In former times, the condition was described as an "illusory reduplication of body parts" (Weinstein et al. 1954) and thought to occur only in cases of psychosis or organic brain damage. We propose here to label "supernumerary" also those phantom limbs experienced by patients with a dissected spinal cord or by healthy participants subject to procedures that elicit a phantom-limb percept.

16.2.4.1 Supernumerary Phantom Body Parts After Spinal Cord Injury

Spinal cord injury (SCI) produces a disconnection between brain and peripheral body parts. Thus a complete section of spinal nerve fibers leads to a "deafferentation" (no sensory information reaches the cortex) and a "deefferentation" (no central motor commands reach the peripheral organ). The reports about phantom limbs in this state of "functional amputation" are too numerous to be cited here (for some references, see Ettlin et al. (1980) and Drysdale et al. (2009)). In a neuroimaging approach, we have shown that the vividness of phantom-foot movements in patients with paraplegia correlates highly with the activation volumes of relevant cortical regions during motor imagery tasks (Alkadhi et al. 2005). A recent observation in a patient with an incomplete tetraplegia (paralysis of all four extremities) described spontaneous bilateral phantom-arm sensations that occurred exclusively in a reclined position (Curt et al. 2011). However, subjected to a "rubber-hand illusion" paradigm (Botvinick and Cohen 1998; see Sect. 16.2.4.3), a sudden supernumerary right-hand phantom occurred also in a sitting position, and remained as long as the rubber hand remained in view (see Fig. 16.2). This observation supports theories of

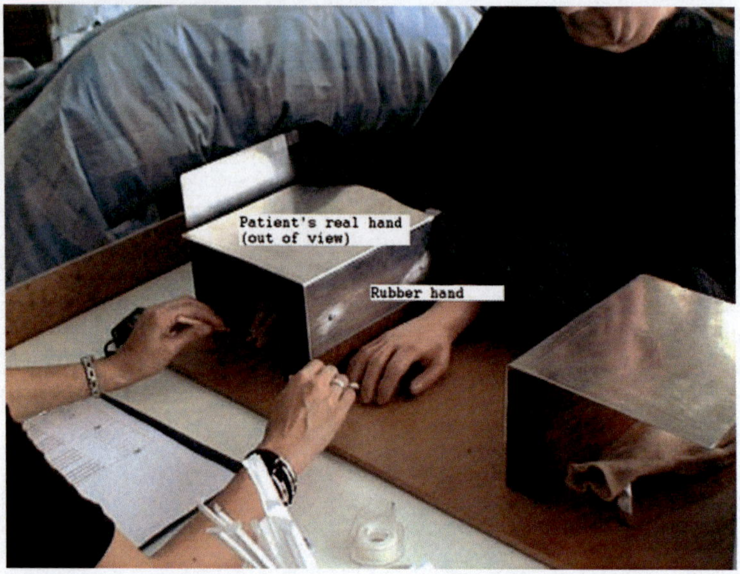

Fig. 16.2 The setup for inducing the "rubber-hand illusion" (*RHI*) in a patient with incomplete quadriplegia after a ski accident (Curt et al. 2011). While in healthy volunteers the RHI leads to a displacement of the unseen hand, touched in synchrony with observed touches to a rubber hand, in this patient the procedure resulted in inducing a supernumerary phantom-hand percept. See text for details (Reproduced, with kind permission of the editors of *Spinal Cord*, from Curt et al. 2011)

the rubber-hand illusion as reflecting a mismatch between visual, tactile, and proprioceptive information (Tsakiris and Haggard 2005), and illustrates the potential usefulness of illusion paradigms in clinical settings.

16.2.4.2 Supernumerary Phantom Body Parts After Brain Damage

Supernumerary phantoms are also observed after a deafferentation and/or deefferentation at subcortical or cortical levels. A recent study found the prevalence of supernumerary phantoms in patients with a stroke to be higher than 50% (Antoniello et al. 2010). Although phenomenologically similar, supernumerary phantoms after stroke and phantoms after the loss of a limb differ in important respects. First, pain is frequently associated with the latter but is only exceptionally an attribute of the former. Second, as "real" as amputation phantoms may feel, they are always recognized as illusory percepts. Reality monitoring is less perfect in the case of supernumerary phantoms after brain damage, the presence of which is often commented on in delusional ways. Thus patients sometimes refer to the phantom as a "spare limb" or claim that they "grew it for protection" (Weinstein et al. 1954, p. 50). Alternatively, hostile actions have been ascribed to supernumerary, but not to amputation phantoms (Hari et al. 1998). Third, amputation phantoms are never visualized by mentally healthy amputees, whereas some patients with supernumerary phantoms have reported a visual, in addition to a somesthetic, awareness of their "extra" limb (Critchley 1953; Khateb et al. 2009). This visual component constitutes an important argument for the conceptual similarity between phantoms of single limbs and the "phantom double," which will be the subject of Sect. 16.4.

16.2.4.3 Supernumerary Phantom Body Parts in Healthy Participants

A rather painful method to induce supernumerary phantom limbs in healthy subjects involves the transient deafferentation of a limb by intravenous regional anesthesia or pressure-cuff ischemia of one limb. Brasil-Neto et al. (1992) used this method combined with a measurement of the functional reorganization of the corticospinal tract. They found a very rapid adaptation of the nervous system to this experimental, transient, and reversible "amputation" of a limb. "Vibratory myesthetic illusions" were described by Lackner (1988) to evoke phantom sensations of various body parts. Biceps vibration, for instance, fools the brain in believing that the forearm is extended, and as a result the subjects (as long as blindfolded!) feel the hand and forearm in extracorporeal space. One intriguing modification of this type of biceps vibration requires the subject to touch her nose with the fingers during the vibration period. A conflict arises: The hand is felt some 30 cm away from the face, yet the sensation of contact with the nose is maintained. As a consequence, subjects report a phantom nose, up to 30 cm long! (Lackner 1988). Apparently, the brain "fills in" the gap between the perceived hand location and the spatial position of the tip of the nose. This so-called Pinocchio illusion offers a tool for the investigation of short-term brain plasticity but may also aid in understanding individual differences in the susceptibility to experience spontaneous perceptual aberrations regarding one's own body (Burrack and Brugger 2005). Finally, the rubber-hand illusion paradigm, first described by Botvinick and Cohen (1998), and mentioned already in Sect. 16.2.4.1, produced an unanticipated interest by the neuroscience community (see Tsakiris and Haggard (2005) for an overview). Briefly, the observation of touch delivered to a rubber hand while one's unseen real hand is touched simultaneously (see Fig. 16.2) leads to the "incorporation" of the rubber hand, or, in other words, to a dislocation of one's own hand toward the rubber hand. This dislocation of a body part into extracorporeal space can be quantified by assessing the proprioceptive drift (Tsakiris and Haggard 2005) or the decrease in hand temperature (Moseley et al. 2008).

16.3 Hemiplegic Twins and Invisible Phantom Bodies

In the presence of brain damage, a supernumerary phantom limb is often no longer recognized as belonging to the own self. Either its actions seem alien (and often threatening, see Hari et al. 1998) or altered sensations are interpreted as a sign that

Fig. 16.3 Drawing by a 57-year-old, right-handed woman, showing herself and a phantom hemibody to her left side. The patient had suffered a large intracerebral bleeding in the right hemisphere, which produced a dense hemiplegia for which she remained anosognosic. The phantom hemibody was initially experienced as comforting, but then as an intruder. The patient attempted to expel the phantom from her bed by biting it – the bite marks on her left upper arm required medical treatment

the limb in fact belongs to another person. Likewise, hemiplegic patients, paralyzed along an entire half side of their body, sometimes fail to recognize the paralysis (a condition called "anosognosia") but conceive of their numb hemibody as another person ("personification anosognosia" or "somatoparaphrenia"; see Vallar and Ronchi (2009) for an overview; see also Fig. 16.3). This disownership of large parts of one's body and its simultaneous attribution to someone else's form a conceptual link between the phantom limb and the various shades of "phantom body" that will be discussed in Sect. 16.4. Common to phantom limb is the state of deafferentation, common to phantom body the conviction that another person is nearby.

The perhaps most obvious difference between the "hemiplegic twin" (Critchley 1953) and the phantom double is that the former is not usually visualized, but just "felt to be around." However, not all types of phantom body usually subsumed under the heading of "autoscopic phenomena" (see Chap. 15) manifest themselves in the visual domain. There is one type of phantom body (see Table 16.2) that is not visualized at all, even if it is sometimes felt to be "just at the fringe of vision." There is

Table 16.2 Four main types of autoscopic phenomena. Sketches adapted from Blanke et al. (2004), original taxonomy after Brugger et al. (1997)

Autoscopic phenomenon	Invisible phantom double	Autoscopy	Heautoscopy	Out-of-body experience
	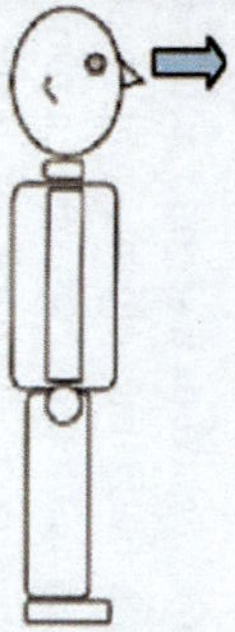		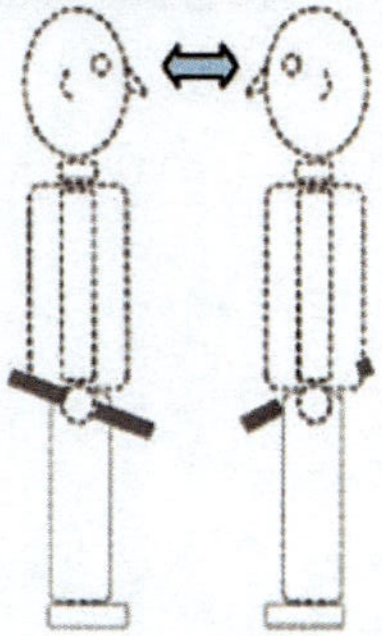	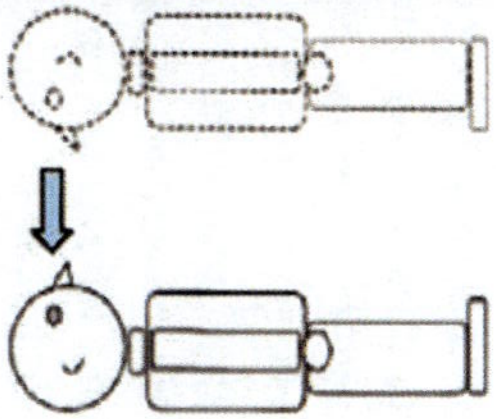
Sensory modality involved	Only somesthetic	Only visual	Somesthetic and visual	Somesthetic and visual
Core brain region affected (in case of brain damage)	Parietal	Occipital	Parieto-occipital	Parieto-temporal
Experience and localization of self	One self; within body	One self; within body	Two selves; primarily within body	One self; completely out of body
Reports about an experimental induction (key references)	Arzy et al. (2006)	Kitamura (1938) and Schilder (1935)	–	Stratton (1899), Ehrsson (2007), Lenggenhager et al. (2007), and Altschuler and Ramachandran (2007)

a vast and most heterogeneous literature on this invisible phantom (see Geiger 2009), and the bewildering terminological variety gives testimony to the prescientific state of its understanding. It has been referred to as an "invisible *doppelgänger*," a "felt presence," or a "shadow person" (Arzy et al. 2006), and explorers and mountaineers know it as the "third man" (see Brugger et al. (1997) for a more comprehensive list of terms). Unilateral forms and auditory variants of the phenomenon have been described, and famous writers' accounts have put it into most uncanny contexts (see Brugger et al. 1994 and 1997 for the literature). To be sure, with the invisible phantom body, we reach the borderlands between tales of the supernatural and the scientific study of hallucinations. What sense is it anyway, in which we hallucinate, if we feel the presence of another being close by, at a precise distance from our body, yet without reliance on visual, auditory, tactile, or olfactory cues? We argue that it is the sense of having a body, although some authors are reluctant to consider it a perceptual phenomenon at all (e.g., Thompson 1982). In any case, all visualization characterizing the phantom experiences to which we now turn is rather secondary to any changes in corporeal awareness.

16.4 Visible Phantoms: Autoscopic Phenomena

Autoscopic phenomena involve an illusory duplication of the own body. As the motif of the double, or *doppelgänger*, they are well known to students of the literary arts. There is a vast literature on the folklore and ethnology of the subject (see Brugger et al. (1997) for overview), and a considerable interest by philosophers, who feel that whole-body phantoms, whether observed in a clinical context or induced experimentally, may help us gain significant insights into the nature of the human self (Metzinger 2009).

Four main types of autoscopic phenomena need to be distinguished. We already mentioned the invisible phantom double in the previous section; the duplication may not be immediately apparent in this type, as the "sensed presence" (as it is called, see Chap. 17) is not usually experienced as a second self. However, the fact that "the other" typically imitates own body movements and sometimes "shares one's thoughts" makes clear that the invisible companion is a projection of one's own somesthetic impressions into extracorporeal space. An autoscopic hallucination is a unimodal, visual hallucination of one's own body, frequently only of the upper body or face. The perceived image is usually brightly colored, its duration very brief, and the defining feature is a mirror reversal, i.e., things carried in the right hand are displayed in the autoscopic double's left hand (Brugger 2002). This absence of a perspective transformation indicates the absence of parietal lobe involvement (Blanke et al. 2005) and constitutes the main difference with heautoscopy, another type of autoscopic reduplication, extensively researched by the Austrian neuropsychiatrist Erich Menninger-Lerchenthal (1898–1966), who wrote two monographs on the topic (Menninger-Lerchenthal 1935, 1946). Heautoscopy literally means "seeing one's self" and thus underlines the presence of two selves, one seeing the other. This duality can lead to complex interactions between a person and her double; although supporting doubles have been reported, hostile interactions are the rule (Brugger 2007; Brugger et al. 1994) and make heautoscopy so attractive as a literary motif. Finally, the core feature of out-of-body experiences is the apparent separation of the self from what is considered the "empty body," usually observed from an extracorporeal vantage point (see Table 16.2). The recent neuroscientific literature has seen an upsurge of contributions that link this experience to certain parts of the brain (Blanke et al. 2002) or to distinct cerebral pathologies (Blanke et al. 2004). A topical issue is the experimental evocation of out-of-body states; among the methods introduced are mirror techniques (Altschuler and Ramachandran 2007; see Stratton (1899) for an early attempt) or full-body variants of the rubber-hand illusion (Ehrsson 2007; Lenggenhager et al. 2007). These methods allow the study of hallucinations of corporeal awareness under controlled conditions. We have to be aware, however, that experimentally induced phantom experiences, whether on the level of single limbs or the entire body, are phenomenally worlds apart from what they attempt to mimic.

16.5 "Phantomology": The Study of Body Hallucinations

Phantom experiences, the productive symptoms of corporeal awareness, form a heterogeneous class of hallucinations of one's own body. Borrowing a term from the Polish science-fiction writer Stanislas Lem (1921–2006), we could define "phantomology" (Lem 1964) as the science of the virtual reality of the body. Phantoms are "virtually real" in the sense of knowingly imaginary, yet phenomenally as convincing as the experience of our physical body. They help us recognize that it is by no means self-evident to "own a body" and make us aware that the concept of the self is largely grounded in basic sensorimotor processes (see also Chap. 13). One of the basic tenets of phantomology is that there is a continuum of phantom experiences – from phantom limb to phantom body. Phantomology can thus profit from circumstances not usually given in other fields of hallucination research: Most phantom limbs are experienced by persons who are healthy, wide-awake, and sober. In addition, they tend to be experienced almost continuously. This makes a laboratory approach feasible and has helped develop methods to investigate those phantoms that have hitherto been enshrouded in mysticism or regarded as too rare to be accessible to scientific explorations. As pointed out by Blanke and Metzinger (2009) and by Brugger (in press), the demystification of current-day concepts of "the self" will be among the foremost future prospects of phantomology. To this end, existing gaps between cognitive neurology and the philosophy of mind will have to be bridged.

References

Abramson, A.S., Feibel, A. (1981). The phantom phenomenon: Its use and disuse. *Bulletin of the New York Academy of Medicine*, 57, 99–112.

Alkadhi, H., Brugger, P., Hotz Boendermaker, S., Crelier, G., Curt, A., Hepp-Reymond, M.-C., Kollias, S.S. (2005). What disconnection tells about motor imagery. Evidence from paraplegic patients. *Cerebral Cortex*, 15, 131–140.

Altschuler, E.L., Ramachandran, V.S. (2007). Last but not least - A simple method to stand outside oneself. *Perception*, 36, 632–634.

Antoniello, D., Kluger, B.M., Sahlein, D.H., Heilman, K.M. (2010). Phantom limb after stroke: An underreported phenomenon. *Cortex*, 46, 1114–1122.

Arzy, S., Seeck, M., Ortigue, S., Spinelli, L., Blanke, O. (2006). Induction of an illusory shadow person. *Nature*, 443, 287.

Blanke, O., Landis, T., Spinelli, L., Seeck, M. (2004). Out-of-body experience and autoscopy of neurological origin. *Brain*, 127, 243–258.

Blanke, O., Mohr, C., Michel, C.M., Pascual-Leone, A., Brugger, P., Seeck, M., Landis, T., Thut, G. (2005). Linking out-of-body experiences and self-processing to mental own-body imagery at the temporoparietal junction. *Journal of Neuroscience*, 25, 550–557.

Blanke, O., Metzinger, T. (2009). Full-body illusions and minimal phenomenal selfhood. *Trends in Cognitive Sciences*, 13, 7–13.

Blanke, O., Ortigue, S., Landis, T., Seeck, M. (2002). Stimulating illusory own-body perceptions. *Nature*, 419, 269–270.

Bläsing, B., Schack, T., Brugger, P. (2010). The functional architecture of the human body: Assessing body representation by sorting body parts and activities. *Experimental Brain Research*, 203, 119–129.

Botvinick, M., Cohen, J. (1998). Rubber hands "feel" touch that eyes see. *Nature*, 391, 756.

Brasil-Neto, J.P., Cohen, L.G., Pascual-Leone, A., Jabir, F.K., Wall, R.T., Hallett, M. (1992). *Neurology*, 42, 1302–1306.

Brugger, P. (2007). Hostile interactions between body and self. *Dialogues in Clinical Neuroscience*, 9, 210–213.

Brugger, P. *Autoscopic phenomena*. In *Encyclopedia of the mind*. Edited by H. Pashler. Thousand Oaks, CA: Sage Publications (in press).

Brugger, P. (2002). Reflective mirrors: Perspective transformation in autoscopic phenomena. *Cognitive Neuropsychiatry*, 7, 179–194.

Brugger, P., Agosti, R., Regard, M., Wieser, H.-G., Landis, T. (1994). Heautoscopy, epilepsy and suicide. *Journal of Neurology, Neurosurgery, and Psychiatry*, 57, 838–839.

Brugger, P., Funk, M. (2007). *Out on a limb. Neglect and confabulation in the study of aplasic phantoms*. In *Tall tales about the brain. Things we think we know about the mind, but ain't so*. Edited by S. Della Sala. London: Wiley, pp. 348–368.

Brugger, P., Kollias, S.S., Müri, R.M., Crelier, G., Hepp-Reymond, M.-C., Regard, M. (2000). Beyond re-membering: Phantom sensations of congenitally absent limbs. *Proceedings of the National Academy of Sciences USA*, 97, 6167–6172.

Brugger, P., Regard, M., Landis, T. (1997). Illusory reduplication of one's own body: Phenomenology and classification of autoscopic phenomena. *Cognitive Neuropsychiatry*, 2, 19–38.

Burrack, A., Brugger, P. (2005). Individual differences in susceptibility to experimentally induced phantom sensations. *Body Image*, 2/3, 307–313.

Critchley, M. (1953). *The parietal lobes*. London: Hafner.

Curt, A., Yengue, C.N., Hilti, L.M., Brugger, P. (2011). Supernumerary phantom limbs in spinal cord injury. *Spinal Cord*, 49, 588–595.

Drysdale, D.G., Shem, K., Walbom, A., Miner, M.D., MacLachlan, M. (2009). Phantom sensations in people with complete spinal cord lesions: a grounded theory perspective. *Disability and Rehabilitation*, 31, 267–276.

Ehrsson, H.H. (2007). The experimental induction of out-of-body experiences. *Science*, 317, 1048.

Ettlin, T.M., Seiler, W., Kaeser, H.E. (1980). Phantom and amputation illusions in paraplegic patients. *European Neurology*, 19, 12–19.

Finger, S., Hustwit, M.P. (2003). Five early accounts of phantom limb in context: Paré, Descartes, Lemos, Bell, and Mitchell. *Neurosurgery*, 52, 675–686.

Funk, M., Brugger, P. (2008). Mental rotation of congenitally absent hands. *Journal of the International Neuropsychological Society*, 14, 81–89.

Funk, M., Shiffrar, M., Brugger, P. (2005). Hand movement observation by individuals born without hands. Phantom limb experience constrains visual limb perception. *Experimental Brain Research*, 341–346.

Gallagher, S., Butterworth, G.E., Lew, A., Cole, J. (1998). Hand-mouth coordination, congenital absence of limb, and evidence for innate body schemas. *Brain and Cognition*, 38, 53–65.

Gallese, V., Goldman, A. (1998). Mirror neurons and the simulation theory of mind-reading. *Trends in Cognitive Sciences*, 2, 493–501.

Geiger, J. (2009). *The third man factor*. Toronto: Penguin.

Giummarra, M.J., Gibson, S.J., Georgiou-Karistianis, N., Bradshaw, J.L. (2007). Central mechanisms in phantom limb perception: The past, present and future. *Brain Research Reviews*, 54, 219–232.

Hari, R., Hänninen, R., Mäkinen, T., Jousmäki, V., Forss, N., Seppä, M., Salonen, O. (1998). Three hands: Fragmentation of human bodily awareness. *Neuroscience Letters*, 240, 131–134.

Hilti, L.M., Brugger, P. (2010). Incarnation and animation: physical versus representational deficits of body integrity. *Experimental Brain Research*, 204, 315–326.

Khateb, A., Simon, S.R., Dieguez, S., Lazeyras, F., Momjian-Mayor, I., Blanke, O., Landis, T., Pegna, A.J., Annoni, J.M. (2009). Seeing the phantom: A functional magnetic resonance imaging study of a supernumerary phantom limb. *Annals of Neurology*, 65, 698–705.

Kitamura, S. (1938). Untersuchungen über das autoskopische Erlebnis in dem Vorstellungsleben I & II. *Tohoku Psychologica Folia*, 6, 1–34 and 169–188.

Lackner, J.R. (1988). Some proprioceptive influences on the perceptual representation of body shape and orientation. *Brain*, 111, 281–297.

Lem, S. (1964/1976). *Summa technologiae*. Frankfurt am Main: Insel.

Lenggenhager, B., Tadi, T., Metzinger, T., Blanke, O. (2007). Video ergo sum: manipulating bodily self-consciousness. *Science*, 317, 1096–1099.

Melzack, R., Israel, R., Lacroix, R., Schultz, G. (1997). Phantom limbs in people with congenital limb deficiency or amputation in early childhood. *Brain*, 120, 1603–1620.

Menninger-Lerchenthal, E. (1935). Das Truggebilde der eigenen Gestalt. Heautoskopie, Doppelgänger. *Abhandlungen aus der Neurologie, Psychiatrie, Psychologie und ihren Grenzgebieten*, 74, 1–196.

Menninger-Lerchenthal, E. (1946). Der eigene Doppelgänger. *Beiheft zur Schweizerischen Zeitschrift für Psychologie und ihre Anwendungen*, 11, 1–95.

Metzinger, T. (2009). Why are out-of-body experiences interesting for philosophers? The theoretical relevance of OBE research. *Cortex*, 45, 256–258.

Moseley, G.L., Olthof, N., Venema, A., Don, S., Wijers, M., Gallace, A., Spence, C. (2008). Psychologically induced cooling of a specific body part caused by the illusory ownership of an artificial counterpart. *Proceedings of the National Academy of Sciences USA*, 105, 13169–13173.

Paré, A. (1840/41). *Oeuvres complètes d' Ambroise Paré*. Vol. II. Edited by J.F. Malgaigne. Paris: Baillère (work originally published in 1551).

Price, D.B., Twombly, N.J. (1976). *The phantom limb phenomenon: a medical, folkloric, and historical study*. Washington, DC: Georgetown University Press.

Price, E.H. (2006). A critical review of congenital phantom limb cases and a developmental theory for the basis of body image. *Consciousness and Cognition*, 15, 310–322.

Ramachandran, V.S., Hirstein, W. (1998). The perception of phantom limbs: The DO Hebb Lecture. *Brain*, 121, 1603–1630.

Schilder, P. (1935). *The image and appearance of the human body*. London: Kegan Paul, Trench, Trübner & Co.

Solonen, K.A. (1962). The phantom phenomenon in amputated Finnish war veterans. *Acta Orthopaedica Scandinavica*, Suppl. 54, 1–37.

Stratton, G.M. (1899). The spatial harmony of touch and sight. *Mind*, 8, 492–505.

Thompson, C. (1982). *Anwesenheit* - psychopathology and clinical associations. *British Journal of Psychiatry*, 141, 628–630.

Tsakiris, M., Haggard, P. (2005). The rubber hand illusion revisited: visuotactile integration and self-attribution. *Journal of Experimental Psychology - Human Perception and Performance*, 31, 80–91.

Vallar, G., Ronchi, R. (2009). Somatoparaphrenia: a body delusion. A review of the neuropsychological literature. *Experimental Brain Research*, 192, 533–551.

Weeks, S.R, Anderson-Barnes, V.C., Tsao, J.W. (2010). Phantom limb pain. Theories and therapies. *The Neurologist*, 16, 277–286.

Weinstein, E.A., Kahn, R.L., Malitz, S., Rozanski, J. (1954). Delusional reduplication of parts of the body. *Brain*, 77, 45–60.

Wood, R., Stuart, S.A.J. (2009). Aplasic phantoms and the mirror neuron system: An enactive, developmental perspective. *Phenomenology and the Cognitive Sciences*, 8, 487–504.

Chapter 17
Sensed Presences

James Allan Cheyne

17.1 Intuitions of "Something There"

> It is as if there were in the human consciousness *a sense of reality, a feeling of objective presence, a perception* of what we may call '*something there*,' more deep and more general than any of the more special and particular 'senses.'
>
> William James, *Varieties of Religious Experience*, 1902, p. 58

It is likely that the reader will have had an experience, in the absence of any tangible evidence, of sensing some unseen presence nearby: along the trail, around the corner, in another room. Such feelings may range from a vague and almost embarrassingly unwarranted suspicion to a feeling of absolute conviction. We associate such feelings most acutely with darkness, strange surroundings, and, very often, being alone. We may experience this when walking at night on a poorly lit street, or in darkly shaded woodlot, or alone in our own homes. For most of us and for the most part these episodes are subtle, fleeting sensations. Under certain conditions, however, such feelings can become protracted delusions accompanied by hallucinatory imagery.

Although the sensed presence has, in itself, no sensory components, it is more than an internal feeling state, being perceptually projected into extrapersonal space (Brugger et al. 1996; Cheyne and Girard 2004) often within precise and reliable coordinates (Girard et al. 2007; see Fig. 17.1). The sensed presence typically has a specific location and distance from the experient, usually within a few meters or less, behind or to one side just out of sight (Arzy et al. 2004; Brugger et al. 1997; Girard et al. 2007). In addition to such spatial specification, the presence is often explicitly understood to be an *intentional* agent with interests, wants, desires, and goals, ranging from altruistic through benign to strongly menacing (Geiger 2008; Cheyne 2001). With rare exceptions, only relatively recently have such strange and

J.A. Cheyne, Ph.D. (✉)
Department of Psychology, University of Waterloo, Waterloo, ON, Canada
e-mail: acheyne@uwaterloo.ca

J.D. Blom and I.E.C. Sommer (eds.), *Hallucinations: Research and Practice*,
DOI 10.1007/978-1-4614-0959-5_17,

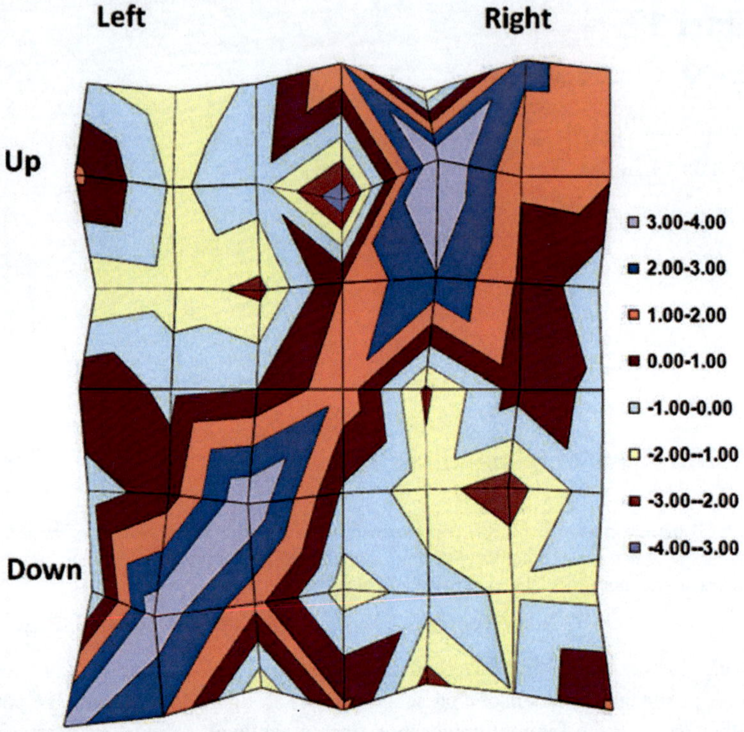

Fig. 17.1 The systematic spatial distribution of hypnagogic experiences of sensed presence and associated hallucinations. The numbers in the legend are residuals reflecting the spatial bias in locating the presence. Higher numbers (Z-scores) and peaks in 3-D surface chart indicate a stronger spatial bias (Adapted from Girard et al. (2007). Reproduced with permission)

apparently mysterious experiences been subjected to very close scrutiny by neuroscience. This chapter reviews a small selection of this recent work and its bearing on two neuropsychological hypotheses.

One dominant neuropsychological view is that the sensed presence is a reduplication of one's own body image (Blanke et al. 2008; Brugger et al. 1997) associated with neural events at the temporoparietal junction (TPJ) and sometimes in frontal motor and premotor areas of the brain resulting in the (dis)integration of bodily self-sensations (Blanke et al. 2003). Consistent with this hypothesis is the observation of striking parallels and transitions between sensed-presence and autoscopic experiences (seeing oneself in extrapersonal space, see Chap. 15), and that the presence sometimes seems to the experient to imitate her movements. Arguing somewhat against this hypothesis is that the presence is seldom recognized as self. In addition, the sensed presence, though itself without sensory qualities, is sometimes accompanied by visual hallucinations, which seldom resemble the experient, may be wearing different clothing, and be of a different age or sex. The body-image hypothesis also fails to explain the intentional nature of the presence or that, in some contexts, the other is experienced as something terrifyingly alien.

An alternative hypothesis is that the sense of presence is the activation of a distinct and evolutionarily functional "sense of the other" (Cheyne and Girard 2007a, b; Cheyne et al. 1999; Picard 2010) implicating structures deep within the temporal lobe specialized for the detection of cues for agency, especially those potentially associated with threat or safety. This hypothesis implies that the feeling itself is simply a dissociation of the normal feeling of the presence of others associated with the normal cues for their physical presence. The sense of the presence of others is pervasive and continuous when one *is* with others, but we rarely explicitly note or remark on it because, after all, there really are others present, and this is the normal feeling when they are there. We sometimes feel it explicitly when in intense relationships and interactions, but mostly it is just a background feeling. When this feeling becomes disconnected from explicit perceptual cues of the presence of others, we may then have the experience of the (disembodied) sensed presence (Cheyne 2009). This hypothesis has little to say, however, about the intriguing parallels between sensed presence and autoscopic hallucinations. Likely, neither hypothesis is adequate to explain all cases, though some combination of the two might come close.

This chapter will review several salient studies of sensed-presence experiences in four contexts selected because they bear upon these hypotheses and provide particular insights into the underlying sources of our experiences of sensed presence, self, and others. These contexts are: neurological and psychiatric conditions, brain stimulation, extreme environments, and sleep paralysis.

17.2 Neurological Cases: The Self-Interrupted

Sensed presences have been reported in neurological cases associated with tumors, lesions, and epileptic foci in occipital, parietal, and temporal lobes (Ardila and Gomez 1988; Brugger et al. 1996; Dening and Berrios 1994; Devinsky et al. 1989; Landtblom 2006). Recent theory and research on neurological cases of sensed presence have focused on paroxysmal somatognosic disorders (i.e., acute distortions of body image, see Chap. 13), often associated with temporoparietal dysfunction (Blanke et al. 2008). In addition, patients diagnosed with schizophrenia frequently demonstrate misattributions of their own actions (see Chap. 26), which also have been associated with the events in temporoparietal cortex (Farrer et al. 2004).

Brugger et al. (1997, see also Chap. 16) developed a useful taxonomy of *autoscopic phenomena* including autoscopic hallucinations, out-of-body experiences (OBEs), and heautoscopy, as well as the sensed presence. The sensed presence is, however, something of an anomaly in this taxonomy in that, on the face of it, it is neither *auto* nor *scopic*. That is, it is not experienced as a self, and it is not seen, in contrast to each of the other phenomena. Autoscopic hallucinations, for example, involve seeing a visual image of oneself in extracorporeal space. Heautoscopy, which also involves autoscopy, usually of a double of oneself, also includes somesthetic and vestibular sensations, which tend to contribute feelings of depersonalization (spatial and psychological detachment from one's body), and ambiguity about the location of one's self. The double can sometimes become transparent

Fig. 17.2 Artist's drawing according to a patient's description of four doubles. The patient's own body is shown as split, with the right side feeling slightly alien. All doubles were located to the right of the patient's body. Although the "double" immediately to the right did not resemble the patient, he was experienced still as belonging to the self. Other figures were also felt to belong to the patient's self, at least in a family sense. The degree of belonging decreased with increasing distance from the patient. An additional figure (not shown) was vaguely sensed about 20 m away (From Brugger et al. (2006). Reproduced with permission)

and "ghost-like" and tends to be referred to in the third person by experients (Brugger et al. 1997). Heautoscopy is, therefore, a potential bridge between autoscopic hallucinations and sensed presence.

Brugger and colleagues present a case that appears to illustrate a continuum from hallucinations of the self through increasingly distant and different identities resolving, over time, into a vague sense of an autonomous presence. A patient with a lesion in the left insula (implicated in interoceptive and vestibular processing) extending into the frontal-temporal cortex experienced hallucinations with three levels of dissociation from self (Brugger et al. 2006). Although the left half of the patient's body felt normal, the right half was emotionally and physically detached, though apparently still clearly interpreted as his own, while immediately to his right, he perceived another man, who, despite physical differences in appearance, still seemed to be a part of himself (see Fig. 17.2). The space further to the right was populated by hallucinated individuals of different ages and sexes, whom he referred to as a "family," though not his own in reality. Eventually all of them faded until their presence was felt in "some hardly discernable sense." Despite the vagueness of the later experiences, he was able still to precisely localize them in the room. Picard (2010) also reports a case of a patient with epilepsy reporting the presence of multiple familiar persons. Although she could neither see nor hear them, she identified them as four family members.

17.3 Sensed Presence by Brain Stimulation: The Other as the Shadow of the Self

Arzy et al. (2004) report a case involving electrical stimulation of the cerebral cortex during presurgical evaluation of a patient with epilepsy. Stimulation applied at the left temporoparietal junction (see Fig. 17.3a) produced the sensation that there was somebody behind her. With further stimulation, the patient described the person as young and of indeterminate sex. During a subsequent stimulation, she was sitting with her arms around her knees and experienced the shadow person as sitting directly behind and embracing her (see Fig. 17.3b–d). She now referred to the presence as male and to the experience as unpleasant. Later during a naming task that required her to hold a card in her right hand, she complained that the presence "wants to take the card" and "doesn't want me to read." Given the importance of the region of cortex stimulated for the integration of bodily sensations, the authors reasonably concluded that the stimulation produced a disruption in the multisensory

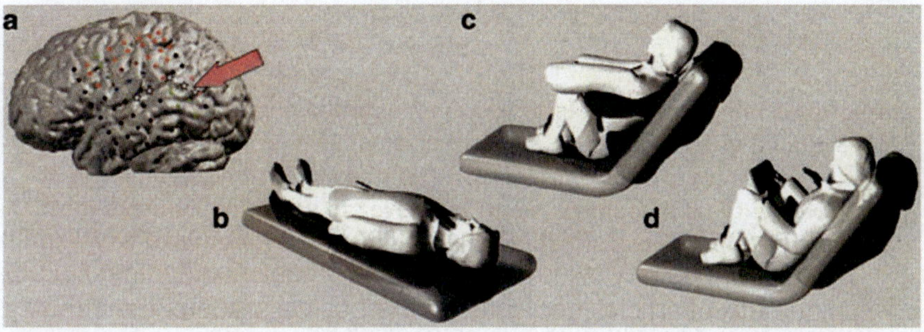

Fig. 17.3 **(a)** Three-dimensional reconstruction (*MRI*) of the left-hemisphere surface of a patient's brain. Subdural electrodes were implanted in the brain of a patient undergoing presurgical evaluation. The locations at which focal electrical stimulation evoked different responses are shown (*red*, motor; *blue*, somatosensory; *green*, language). The *arrow* indicates the site where the "feeling of a presence" could be induced; stars indicate the epileptic focus. **(b–d)** Positions and postures of the patient's body (*white*) and that of the "shadow" person during cortical stimulation. **(b)** The shadow "is behind me, almost at my body, but I do not feel it." **(c)** The shadow clasps the patient in his arms. **(d)** The shadow "wants" to take a book from the patient (From Arzy et al. (2004). Reproduced with permission)

processing of body-self information and that the shadow person was a result of disturbed perception of her own body (i.e., the shadow was the alienated sense of the self).

Brugger and colleagues cite a similar misattribution in Strindberg's self-recorded experiences in the course of probable schizophrenia:

> This unknown man never uttered a word; he seemed to be occupied in writing something behind the wooden partition that separated us. All the same, it was odd that he should push back his chair every time I moved mine. He repeated my every movement in a way that suggested that he wanted to annoy me by imitating me .... When I went to bed the man in the room next to my desk went to bed too.... I could hear him lying there, stretched out parallel to me. I could hear him turning the pages of a book, putting out the lamp, breathing deeply, turning over and falling asleep.

Like Strindberg, the patient in the Arzy study appeared not to draw the inference from the shadow's imitation of her own actions that she might be experiencing her own movements displaced. Perhaps from a first-person perspective, more compelling alternatives were available. For example, the patient also reported a definite impression that the presence was trying to take the card from her, suggesting that there were positive elements of otherness in her experience as in the case of Strindberg. This is not to deny that that she was sensing her own movements, but rather to suggest that she was construing these experiences as though their source were indeed another person. This might be taken as evidence that the senses of self and of others share certain brain mechanisms.

17.4 Illusory Companions in Times of Need

> Who is the third who always walks beside you?
> When I count, there are only two of you and I together
> But when I look ahead up the white road
> There is always another one walking beside you…
>
> T.S. Eliot, *The Wasteland*

Commenting on these lines, Eliot tells us that they "were stimulated by the account of one of the Antarctic expeditions [probably Shackleton's].... It was related that the party of explorers at the extremity of their strength, had the constant conviction that there was *one more member* than could actually be counted." The Shackleton expedition experiences are prototypical of the "companion" version of the sensed presence in a number of ways. First, there were the rigors of a protracted Antarctic expedition gone awry, with members at the limits of their endurance. Second, the events occurred near the end of their ordeal while scaling mountainous terrain on the island of South Georgia. Third, although Eliot describes the experience as a visual hallucination, it was originally described by Shackleton merely as a "curious feeling" of an extra person. Finally, there was a strong conviction concerning the reality of the companion. An illusory companion may be a stranger nearby but out of sight, often just off to one side, or just behind one's shoulder, or perhaps felt to be just a little way along the trail. Alternatively, the companion may be experienced as a familiar person: a friend, a favorite aunt, or sometimes a recently deceased actual companion, any of whom, in the guise of the presence, may sometimes provide, usually implicit, moral support and/or guidance (Geiger 2009).

Common precipitating events for companion presences are the misadventures of sailors, polar explorers, and especially mountaineers, that result in extreme conditions that include isolation, cold, injury, dehydration, hunger, fatigue, darkness, barren landscapes, and fear, as well as, in the case of mountaineering, high-altitude effects – all typically extreme, persistent, and in combination (Suedfeld and Geiger 2008; Suedfeld and Mocellin 1987).

17.4.1 Neurophysiological Effects of Altitude

Perhaps the most common and informative unusual context for companion experiences is extreme mountaineering, especially over 8,000 m (Geiger 2009). This is almost certainly related to the pervasive physiological and neuropsychological effects of high-altitude exposure (Maa 2010; Virtués-Ortega et al. 2004). Human physiology places fairly strict limits on our ability to thrive at high altitudes. A particularly distinctive feature of the stresses imposed by high altitude is the reduction in ambient pressure and resulting poor oxygen exchange in the lungs. Neural cells are greedy consumers of oxygen and hypersensitive to oxygen deficits. In addition, a well-known acute effect is high-altitude cerebral edema, which likely causes mechanical stresses because of the inelastic skull and resulting elevated intracranial pressure (Maa 2010). Although neurophysiological effects experienced by mountaineers in the form of acute mountain sickness have long been known, only recently have associated neuropsychological aspects been subject to detailed study including perceptual, motor, memory, and psychomotor deficits (see Maa 2010 for a review).

Using precise neuroimaging techniques and extreme samples, Paola et al. (2008) studied nine world-class mountaineers using quantitative magnetic resonance imaging (MRI) techniques prior to and following extreme ascents: Mount Everest and K2. These are, in terms of altitude, the most extreme possible test environments at 8,848 and 8,611 m, respectively, at less than 50% of sea level atmospheric pressure (Maa 2010). Paola and colleagues found their subjects, relative to controls, to have baseline (i.e., prior to their ascents for the study) reductions in white matter density/volume in the left pyramidal tract involving frontal cortex (primary and supplementary motor areas). Moreover, in comparison to their own baseline, the mountain climbers were found to have reduced gray matter in the region of the angular gyrus of the parietal lobe following the ascents (i.e., regions centrally implicated in the processing of multisensory body-related information, and hence in bodily self-perception and self-other distinctions). These results are not surprising as the brain areas in question were likely relatively poorly supplied with oxygen while being subjected to the most intense metabolic demands in mountaineering. On the supply side, these areas, called "watershed areas," are more removed from major arteries, and hence tend to be most at risk for reduced available blood oxygen. On the demand side, part of the relevance of the localization on the supply side is revealed in just what mountaineers do when they climb mountains. They engage in extremely challenging motor activity and in considerable way-finding, thus making particular demands of parietal and motor cortex when oxygen availability is considerably reduced in those regions. Keeping track of motor planning, spatial location, whole-body orientation, and relative body-part arrangements are all high-demand activities during mountain climbing, activities in which the parietal angular gyrus is implicated. Brugger et al. (1999) also report a combination of somesthetic hallucinations, neuropsychological impairments, and EEG and MRI abnormalities among high-altitude climbers. Moreover, it is worth noting that the laboratory observations were of the relatively modest residual sequelae of the neurophysiological stresses experienced during climbs that, though extreme, were likely much less acutely stressful than the events that produce companion experiences (see Geiger 2008).

17.5 Night Creatures: The Theme of the Intruder

Arguably the most common and almost certainly the most terrifying context for the sensed presence is the group of hallucinatory experiences associated with sleep paralysis. Sleep paralysis is classified as a REM-related parasomnia or sleep disorder in which someone, when falling asleep or waking, experiences a brief episode of complete paralysis of the major muscles. The paralysis is often, though not always, accompanied by a variety of frightening experiences, the most common of which is the sensed presence (Cheyne and Girard 2007a). There is something profoundly malevolent and evil inherent in this instance of the presence. The uncanny dread and impending doom associated with this presence sometimes generates supernatural themes of "the grim reaper" or "the angel of death" (Cheyne 2001). In the sleep-paralysis context, the sensed-presence experience is often elaborated into explicit visual, auditory, and tactile intruder hallucinations (see Fig. 17.4a). Although – in modern industrialized societies – these presence experiences are often interpreted as human intruders, in numerous cultures around the world, nocturnal visitation by a sensed presence is experienced as an oppressive night spirit with a variety of specific names that equate with ghost, demon, hag, witch, *djinn*, and so on (Adler 2010; Hufford 1982).

Initially, and often exclusively, the experience is merely of a "watcher," i.e., a monitoring presence (see Fig. 17.5). Even simply being stared at can, of course, be quite disconcerting. Staring is seldom a neutral act but rather is viewed as a threatening gesture across cultures and among most primates (Eibl-Eibesfeldt 1989). That the intruder present in our room during sleep paralysis is staring at us is an obvious inference to draw. Yet this seems not a deduction consciously arrived at – one experiences the threat as an event, not as an inference.

Frequently, a complex scenario subsequently unfolds as the presence approaches and climbs onto the bed. Sometimes a presence will simply stop at some point,

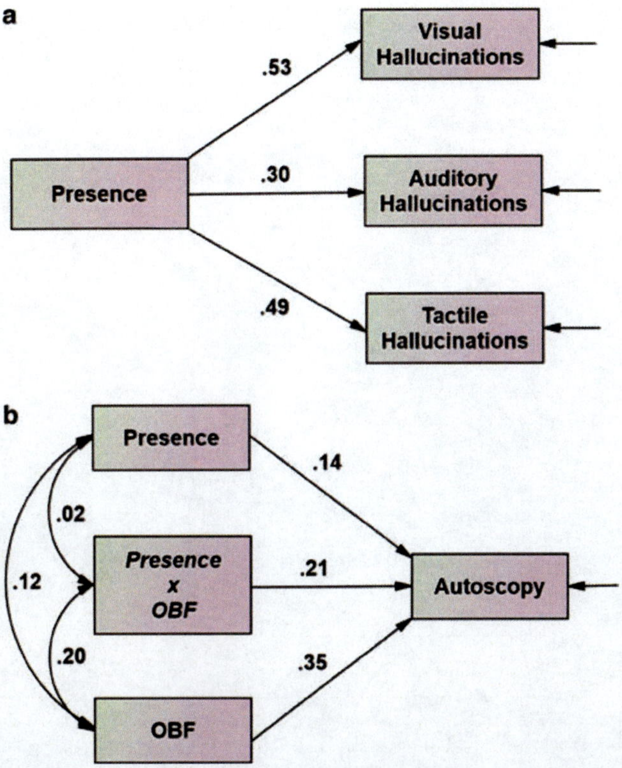

Fig. 17.4 Presence as others and as self: Path analyses of presence experiences during sleep paralysis. **(a)** Path coefficients with presence as predictors of visual, auditory, and tactile intruder hallucinations. **(b)** Path coefficients with presence experiences, vestibular-motor out-of-body feelings, and their interaction predicting autoscopic experiences (Based on data from Cheyne and Girard (2007a, b, 2009). Reproduced with permission)

leaving the anguished experient in a state of considerable suspense. A feeling perhaps preferable to the alternative experience of the mattress being pressed down as one realizes that one's bed is being shared with the presence. Frequently, the presence will be experienced as sitting on one's chest (the incubus experience: see Fig. 17.5), smothering and otherwise physically, and sometimes sexually, assaulting the experient.

17.5.1 Detecting Threatening Agents

Throughout evolutionary history, and into the present, for the large majority of humans, the world has been filled with ambiguous and unseen agents (see also Chap. 18); and often, just those presenting the greatest threat are cryptic, elusive, and, as

Fig. 17.5 Smarra: The watcher and the incubus. The tradition of the presence as watcher or threatening/assaulting intruder is an old one in the Judeo-Christian tradition. The feeling of being watched by an unseen presence is one of the more common experiences of the waking nightmare (Lithograph after Tony Johannot (1845). Paris: Bibliothèque Nationale)

the adage, *Homo homini lupus*, reminds us, often members of our own species (Keeley 1996; LeBlanc 2003). Shape, color, shade, immobility, and stealth conceal these agents from detection. Hence, agency detection must become very sensitive under presumed threat, even at the risk of false positives (see also Chap. 4). Rustling leaves or shadowy movements are at such times sufficient for us to experience the presence of agency, even without bodily signs of the agent. The experience of disembodied threatening agents is therefore intrinsic to our collective and individual evolutionary, historical, and developmental histories. Hence, evolutionary scientists

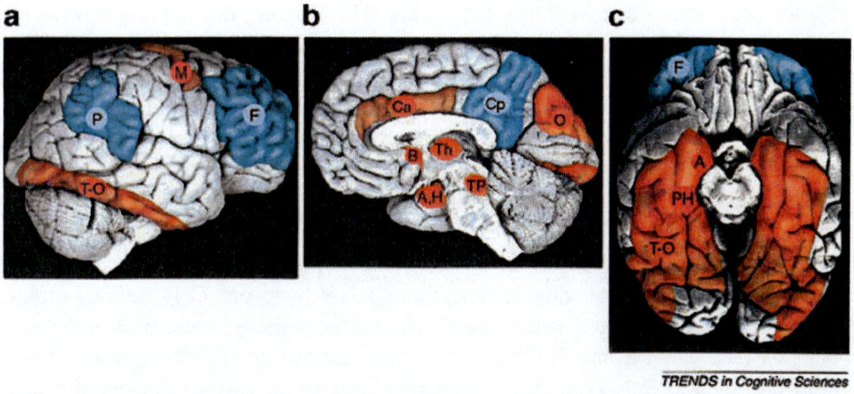

Fig. 17.6 Schematic representation of the relative increases and decreases in neural activity associated with REM sleep. *Red* regions indicate increased, and *blue* regions decreased, regional cerebral blood flow (*rCBF*) during REM sleep. *F* prefrontal cortex, *A* amygdala, *PH* parahippocampal gyrus, *T-O* temporo-occipital extrastriate cortex, *P* parietal supramarginal cortex, *M* motor cortex, *Ca* anterior cingulate gyrus, *Cp* posterior cingulate, *Oc* occipital lateral cortex, *B* basal forebrain, *Th* thalamus, *H* hypothalamus, *TP* pontine tegmentum. (**a**) Lateral (**b**) Medial and (**c**) Ventral (From Schwartz and Maquet (2002). Reproduced with permission)

have argued for evolved special perceptual sensitivities and biases (see Chap. 2), suggesting that experiences of unseen agents such as spirits and gods simply reflect our hypersensitive agent detectors (Barrett 2004; Guthrie 1993). On our view, in the threat simulation mode of sleep paralysis, such agency-detection devices come especially to the fore as criteria are lowered, and agency-detection bias is increased via activation of the amygdala and related structures (Cheyne et al. 1999).

17.5.2 *The Role of REM State*

As a REM phenomenon, the sensed-presence experience during sleep paralysis shares a number of relevant features with the previously discussed contexts, but it also has some unique features. In contrast to the earlier contexts, for example, the sensed presence during sleep paralysis is overwhelmingly associated with intense fear. During REM, and consistent with the previous contexts reviewed, activity of temporoparietal and frontal cortex is suppressed (Schwartz and Maquet 2002). Hence, integration of body sensations and orientation are compromised during REM states. In contrast, the activation of structures associated with the amygdala, including the anterior cingulate cortex, is enhanced (see Fig. 17.6). These areas appear to constitute a threat-activated vigilance system (TAVS, see Cheyne 2003) playing an important role in affective and strategic responses to threat (Whalen 1998) and hence likely to contribute to the sense of threat so particularly associated with the presence of sleep-paralysis episodes. Exacerbating this hyperactivation of the TAVS is REM deactivation of prefrontal cortex, which normally modulates amygdalar activity. Finally, that one is supine, paralyzed, often alone, and in the dark during episodes must further heighten the sense of terror and uncertainty, further activating the TAVS (Cheyne and Girard 2007a, b).

The TAVS hypothesis suggests the further possibility that individual differences in fear-detection thresholds would lead to differential susceptibility to threatening presence experiences. The limited evidence bearing on this hypothesis is, however, not very consistent. A number of studies have implicated heightened stress and anxiety in sleep-paralysis experiences, though few have precisely defined either stress or anxiety or related either specifically to the sensed presence during sleep paralysis (see Simard and Neilson 2005; Solomonova et al. 2008). Simard and Neilson as well as Solomonova and colleagues did assess various aspects of social anxiety and various sleep-paralysis experiences, and reported correlations with sleep-paralysis experiences, but found no more than suggestive evidence that social anxiety is uniquely associated with sensed presence during sleep paralysis.

Although many people do imagine the intruder to be demonic and otherworldly, the threatening presence is not always subject to such supernatural interpretations. Often, people make assessments of the situation in terms of very material sources of danger: prowlers, robbers, and rapists, as well as simply accepting them as hallucinations (Cheyne 2001). It is important to realize, however, that *rational assessments need not detract from the vividness or realism of the experiences*. The compelling nature of the threat during sleep paralysis is such that we have suggested that it constitutes a paroxysmal paranoid delusion (Cheyne and Girard 2007a, b). The resistance of such experiences to alteration by rational assessment is consistent with a primitive modular source, such as the TAVS, that is highly immune to being overridden by higher-order cognitions (see also Chap. 26).

17.6 Reconciling Sensed Presence as Self and as Others

Despite the apparently absolute sense of otherness of the night creature, there is an aspect of the threatening presence that is consistent with the notion that it is somehow linked to the sense of self. The presence is sometimes taken to be a demonic entity intent on "taking over one's body" or "possessing one's soul." People thus tend to regard themselves as locked in a desperate struggle for the very integrity of their innermost self. Yet, how can something so utterly alien and other share any aspect of the self?

A beginning of an answer might be found in the observation that a good deal of human motor and perceptual learning throughout development occurs in the context of coordinated interactions with others. James Mark Baldwin (1861–1934) was one of the first, at the end of the nineteenth century, to consider seriously the parallels between the imitation of others and the repetition of one's own actions ("circular reaction") and hence that whenever we learn something about another person we learn something about ourselves, and whenever we learn something about ourselves we learn something about others. Another early developmental psychologist, Lev Vygotsky (1896–1934), also noted that a child's early skills typically begin as shared capacities with adults, who "fill in" the gaps in the child's performance. Both Baldwin and Vygotsky implied that our own sense of ourselves as intentional agents developed in tandem with, and under the guidance of, the actions and implied intentions of others. With such a social background, we cannot help but understand ourselves as others do, or perhaps more accurately, understand ourselves very much as we understand others in a very embodied sense.

Recent evidence corroborates these early intuitions that the perception and interpretation of implicit and overt bodily self-activity and the actions of others are very closely linked across multiple shared systems. These include observations as diverse as infant facial imitation (Melzoff and Moore 1977), automatic imitation (Stürmer et al. 2000), the social Simon effect (Sebanz et al. 2003), and the existence of mirror neurons responsive to similar own and other actions (Rizzolati and Craighero 2004).[1] Interestingly, given the inherent intentionality of the sensed presence, in both the case of mirror neurons and automatic imitation, the awareness of the implicitly inferred intentions of the partner or model appears to be as or more important than concrete behavioral cues (Liepelt and Brass 2010; Vlainic et al. 2010). Moreover, the parameters during movement imitation map onto the physical topography of the bodies of model and imitator (Chartand and Bargh 1999). The locations of mirror neurons in frontal premotor areas and inferior parietal areas (Iacoboni et al. 1999) clearly correspond to the general areas most often implicated in sensed presence across many of the contexts considered here (Buccino et al. 2001; Chaminade et al. 2005). Thus, the anomalous activation of mirror neurons may sometimes obscure the distinction between self and others. Companion experiences and the presence experiences reported by neurological patients appear to involve these same shared motor and perceptual systems. The observation that the presences in neurological disorders and elicited contexts sometimes imitate the movements of corresponding body parts is consistent with the everyday interactions with others, and therefore should not automatically lead to the inference that an anomalous imitating agent is just a projection of our bodily selves. Companion experiences, in particular, might reflect a bias introduced by the intense need for companionship and assistance. TAVS-activated threatening presences would conversely bias interpretation to an external source of threat.

It is important not to omit, however, to mention that sleep-paralysis experiences also include some very dramatic anomalies of the bodily self (Cheyne and Girard 2009). These bodily self-experiences include vestibular and motor hallucinations such as flying, floating, falling, and spinning (see Chap. 13), as well as out-of-body

[1] The fact that the very primates in which mirror neurons were first observed do not appear to be capable of imitation suggests that the close linkage between the observed actions of self and others precedes the ability for explicit imitation, and may find its roots in the need to coordinate the activity of self and others.

experiences (OBEs), which include both kinesthetic out-of-body dissociative experiences and out-of-body autoscopy, and all of these have always formed a separate factor from the intruder hallucinations. Thus, in the sleep-paralysis context, we have consistently found a clear separation of the bodily self and threatening-other hallucinations. Nonetheless, we recently found tentative evidence that the sensed presence when combined with strong bodily sensations can contribute to hallucinations of self as well as of others (Cheyne and Girard 2009). Within the bodily self-experiences, we have found evidence for a potential path from vestibular-motor hallucinations to out-of-body dissociative experiences to autoscopic experiences. Although the sensed-presence experience itself is not strongly correlated with any of the OBE phenomena, we have found that the combination of sensed presence and out-of-body dissociation does significantly improve the prediction of autoscopic experiences (see Fig. 17.4b).

17.7 Concluding Remarks

The neuropsychological hypotheses concerning sensed presence admittedly remain tentative and crudely formulated, limited not only by our imagination but also by the paucity of empirical data. Moreover, our knowledge of brain functioning and its relation to everyday normal action and experience remains only very roughly worked out despite enormously exciting discoveries coming almost daily. In addition, as empirical phenomena, sensed presences remain elusive and comparatively rare, and hence challenging to observe under controlled conditions. As sensed-presence experiences constitute subjective experiences, investigating them is always limited on the phenomenal side by the challenge for the experients to describe what are, for all their gripping vividness, unfamiliar and fleeting sensations. Yet, as I hope this brief and selective review illustrates, the last decade has seen some fascinating progress in understanding this elusive and mysterious phenomenon from a neuropsychological perspective. More generally, and perhaps more importantly, however, is the potential for this work to help untangle the curiously intertwined oppositions of self and others.

References

Adler, S.R. (2010). *Sleep paralysis: Night-mares, nocebos, and the mind-body connection*. New Brunswick, NJ: Rutgers University Press.

Ardila, A., Gomez, J. (1988). Paroxysmal "feeling of somebody being nearby". *Epilepsia*, 29, 188–189.

Arzy, S., Seeck, M., Ortigue, S., Spinelli, L., Blanke, O. (2004). Induction of an illusory shadow person. *Nature*, 443, 287.

Barrett, J.L. (2004). *Why would anyone believe in god?* Walnut Creek, CA: Altimira.

Blanke, O., Ortigue, S., Coeytaux, A., Martory, M.D., Landis, T. (2003). Hearing of a presence. *Neurocase*, 9, 329–339.

Blanke O., Arzy S., Landis T. (2008). *Illusory perceptions of the human body and self.* In: *Handbook of clinical neurology*. Edited by Goldenberg, G.B.M. Amsterdam: Elsevier, pp. 429–458.

Brugger, P., Blanke, O., Regard, M., Bradford, D.T., Landis, T. (2006). Polyopic heautoscopy: Case report and review of the literature. *Cortex*, 42, 666–674.

Brugger, P., Regard, M., Landis, T. (1997). Illusory reduplication of one's own body: Phenomenology and classification of autoscopic phenomena. *Cognitive Neuropsychiatry*, 2, 19–38.

Brugger, P., Regard, M., Landis, T., Oelz, O. (1999). Hallucinatory experiences in extreme-altitude climbers. *Neuropsychiatry, Neuropsychology, and Behavioral Neurology*, 12, 67–71.

Brugger, P., Regard, M. Landis, T. (1996). Unilaterally felt "presences:" The neuropsychiatry of one's invisible doppelganger. *Neuropsychiatry, Neuropsychology, and Behavioral Neurology*, 9, 114–122.

Buccino, G., Binkofski, F., Fink, G.R., Fadiga, L., Fogassi, L., Gallese, V., Seitz, R.J., Zilles, K., Rizzolatti, G., Freund, H.J. (2001). Action observation activates premotor and parietal areas in a somatotopic manner: an fMRI study. *European Journal of Neuroscience*, 13, 400–404.

Chaminade, T., Meltzoff, A.N., Decety, J. (2005). An fMRI study of imitation: action representation and body schema. *Neuropsychologia*, 43, 115–127.

Chartand, T.L., Bargh, J.A. (1999). The chameleon effect: The perception-motor link and social interaction. *Journal of Personality and Social Psychology*, 76, 893–910.

Cheyne, J.A. (2001). The ominous numinous: Sensed presence and 'other' hallucinations. *Journal of Consciousness Studies*, 8, 133–150.

Cheyne, J.A. (2003). Sleep paralysis and the structure of waking-nightmare hallucinations. *Dreaming*, 13, 163–179.

Cheyne, J.A. (2009). Sensed presences in extreme contexts: A review of "The third man factor: The secret of survival in extreme environments" by John Geiger. *Skeptic*, 15, 68–71.

Cheyne, J.A., Girard, T.A. (2004). Spatial characteristics of hallucinations associated with sleep paralysis. *Cognitive Neuropsychiatry*, 9, 281–300.

Cheyne, J.A., Girard, T.A. (2007a). Paranoid delusions and threatening hallucinations: A prospective study of hypnagogic/hypnopompic hallucinations during sleep paralysis. *Consciousness and Cognition*, 16, 959–974.

Cheyne, J.A., Girard, T.A. (2007b). The nature and varieties of felt presence experiences: A reply to Nielsen. *Consciousness and Cognition*, 16, 984–991.

Cheyne, J.A., Girard, T.A. (2009). The body unbound: Vestibular-motor hallucinations and out-of-body experiences. *Cortex*, 45, 201–215.

Cheyne, J.A., Rueffer, S.D., Newby-Clark, I.R. (1999). Hypnagogic and hypnopompic hallucinations during sleep paralysis: Neurological and cultural construction of the night-mare. *Consciousness and Cognition*, 8, 319–337.

Dening, T.R., Berrios, G.E. (1994). Autoscopic phenomena. *British Journal of Psychiatry*, 165, 808–817.

Devinsky, O., Feldmann, E., Burrowes, K., Bromfield, E. (1989). Autoscopic phenomena with seizures. *Archives of Neurology*, 46, 1080–1088.

Eibl-Eibesfeldt, I. (1989). *Human ethology*. New York, NY: Aldine de Gruyter.

Farrer, C., Franck, N., Frith, C.D., Decety, J., Georgieff, N., d'Amato, T., Jeannerod, M. (2004). Neural correlates of action attribution in schizophrenia. *Psychiatry Research*, 131, 31–44.

Geiger, J. (2009). *The third man factor: the secret to survival in extreme environments.* Toronto: Penguin.

Girard, T.A., Martius, D.L.M.A., Cheyne, J.A. (2007). Mental representation of space: Insights from oblique distribution of hallucinations. *Neuropsychologia*, 45, 1257–1269.

Guthrie, S.E. (1993). *Faces in the clouds: A new theory of religion.* Oxford: Oxford University Press.

Hufford, D.J. (1982). *The terror that comes in the night: An experience-centered study of supernatural assault traditions*. Philadelphia, PA: University of Pennsylvania Press.

Iacoboni, M., Woods, R.P., Brass, M., Bekkering, H., Mazziotta, J.C., Rizzolatti, G. (1999). Cortical mechanisms of human imitation. *Science*, 286, 2526–2528.

Keeley, L.H. (1996). *War before civilization.* Oxford: Oxford University Press.
Landtblom, A.-M. (2006). The "sensed presence": An epileptic aura with religious overtones. *Epilepsy and Behavior*, 9, 186–188.
LeBlanc, S.A. (2003). *Constant battles: Why we fight.* New York, NY: St. Martin's Press.
Liepelt, R., Brass, M. (2010). Top-down modulation of motor priming by belief about animacy. *Journal of Experimental Psychology*, 57, 221–227.
Maa, E.H. (2010). Hypobaric hypoxia cerebral insults: The neurological consequences of going higher. *Neurorehabilitation*, 26, 73–84.
Melzoff, A.N., Moore, M.K. (1977). Imitation of facial and manual gestures by human neonates. *Science*, 7, 74–78.
Paola, M.D., Bozzali, M., Fadda, L., Musicco, M., Sabatini, U., Caltagirone, C. (2008). Reduced oxygen due to high-altitude exposure relates to atrophy in motor-function areas. *European Journal of Neurology*, 15, 1050–1057.
Picard, F. (2010). Epileptic feeling of multiple presences in frontal space. *Cortex*, 46, 1037–1042.
Rizzolati, G., Craighero, L. (2004). The mirror neuron system. *Annual Review of Neuroscience*, 24, 169–192.
Schwartz, S., Maquet, P. (2002). Sleep imaging and the neuropsychological assessment of dreams. *Trends in Cognitive Sciences*, 6, 23–30.
Sebanz, N., Knoblich, G., Prinz, W. (2003). Representing others' actions: just like one's own? *Cognition*, 88, B11-B21.
Simard, V., Nielsen, T.A. (2005). Sleep paralysis-associated sensed presence as a possible manifestation of social anxiety. *Dreaming*, 15, 245–260.
Solomonova, E., Neilsen, T. Stenstrom, P., Simard, V. Frantova, E., & Donderi, D. (2008). Sensed presence as a correlate of sleep paralysis distress, social anxiety and waking state imagery. Consciousness and Cognition, 17, 49–63.
Stürmer, B., Aschersleben, G., Prinz, W. (2000). Correspondence effects with manual gestures and postures: a study of imitation. *Journal of Experimental Psychology. Human Perception and Performance*, 26, 1746–1759.
Suedfeld, P., Geiger, J. (2008). The sensed presence as a coping resource in extreme environments. In: *Miracles: God, science, and psychology in the paranormal. Vol. 3.* Edited by Ellens, J.H. London: Praeger.
Suedfeld, P., Mocellin, J.S.P. (1987). The "sensed presence" in unusual environments. *Environment and Behavior*, 19, 33–52.
Virtués-Ortega, J., Buela-Casal, G., Garrido, E., Alcázar, B. (2004). Neuropsychological functioning associated with high-altitude exposure. *Neuropsychology Review*, 14, 197–224.
Vlainic, E., Liepelt, R., Colzato, L.S., Prinz, W., Hommel, B. (2010). The virtual-co-actor: The social Simon effect does not rely on online feedback from the other. *Frontiers in Psychology*, doi: 10.3389/fpsyg.2010.00208.
Whalen, P.J. (1998). Fear, vigilance, and ambiguity: Initial neuroimaging studies of the human amygdala. *Current Directions in Psychological Science*, 7, 177–188.

Chapter 18
Djinns

Jan Dirk Blom and Cor B.M. Hoffer

18.1 Introduction

"Djinn" and "hallucination" are very different notions. Deriving from widely divergent discourses, the two are hardly commensurable (Fig. 18.1). And yet the present chapter is devoted to djinns, in a book dealing with hallucinations. The reason for including this topic is that individuals with an Islamic background show a marked tendency to attribute any hallucinatory experiences they may have to a djinn, and therefore seek help from religious healers (see Fig. 18.2) before ever consulting a biomedical practitioner. Biomedical practitioners, in turn, and particularly those in Western societies, tend to know preciously little about djinns and Arabic-Islamic healing methods. When their Islamic patients finally show up on their doorstep, they establish diagnoses in conformity with their own psychiatric classifications, prescribe state-of-the-art pharmacological or psychotherapeutic treatments, and then almost invariably have to see with disappointment how these patients fail to recover and silently slip out of their treatment programs. In an increasingly multicultural society such as ours (i.e., the Netherlands),[1] this is an undesirable situation for patients and their families, as well as for the mental health practitioners involved. A second reason for addressing this topic is that hallucinations attributed to djinns would seem to possess quite extraordinary phenomenological characteristics,

[1]In 2010, 907,000 people in the Netherlands (i.e., 6% of the population) were Muslims, with 329,000 of them being of Turkish origin, and 314,000 of Moroccan origin (FORUM 2010).

J.D. Blom, M.D., Ph.D. (✉)
Parnassia Bavo Academy, Parnassia Bavo Group and University of Groningen, Groningen, Kiwistraat 43, 2552 DH, The Hague, The Netherlands
e-mail: jd.blom@parnassia.nl

C.B.M. Hoffer, M.Sc., Ph.D.
Parnassia Bavo Group, Rotterdam, The Netherlands
e-mail: c.hoffer@parnassiabavogroep.nl

J.D. Blom and I.E.C. Sommer (eds.), *Hallucinations: Research and Practice*,
DOI 10.1007/978-1-4614-0959-5_18, © Springer Science+Business Media, LLC 2012

Fig. 18.1 *Djinn*. Oil painting by Marten Blom (Copyright 2010) (Reproduced with permission)

Fig. 18.2 Religious healer at Sidi Harazem, Morocco (Photograph by Cor Hoffer 2007)

setting them apart from the types of hallucination we tend to encounter in Western patients diagnosed with a psychotic disorder. As these characteristics are as yet largely uncharted, and the biomedical literature on the subject is limited (Khalifa and Hardie 2005; Sheikh 2005), this chapter also draws on anthropological and religious sources to arrive at a characterization of this niche phenomenon in the area of hallucinations research.

18.2 What Are Djinns?

Djinn is Arabic for spirit or ghost. It literally means "that which is concealed from people's sight" (Al-Ashqar 2005). Alternatives for "djinn" are *jinn*, *jin*, *jinni*, *cin* (in Turkish), and *shed* (in Hebrew). In the plural, it is *djinns*, *jins*, *jinn*, *dzjenoun*, *djnoun*, *jnoun*, *jenoun*, *jnûn*, *cinler*, and *shedim*. According to traditional Islamic faith, djinns were created by Allah out of smokeless fire (Qur'an 15:27). As such, Muslims generally consider these creatures part and parcel of the living world and believe that they actively participate in the lives and social interactions of humans, as do angels and *Iblis* (i.e., Satan) for that matter. Although often endowed with malign intentions, good djinns are also believed to exist. The latter are considered capable of helping humans to attain valuable goals in life and of coexisting with them in the form of djinn companions (*qareen*).

Being made out of fire, djinns are considered bodiless and hence invisible. And yet they are deemed capable of perceiving human beings (Qur'an 7:27) and of manifesting themselves to humans in a variety of shapes, including a vapor, cloud, cat, dog, snake, vulture, camel, donkey, onager, dragon, or human form. One of the Moroccan patients in our hospital used to see an old, female djinn hovering over the bed, reminiscent of her mother, but shorter, and with her feet bent backward in an anatomically impossible position. Another patient described to us a small, dark cloud that zoomed across the room, jumped off the furniture and walls, and sometimes disappeared through closed doors, whereas a third patient reported snakes and black dogs crawling and jumping up his legs and biting him in the thighs and belly. It is believed that djinns have a life cycle not unlike our own, in the sense that they are born, grow up, breathe, laugh, speak, eat, drink, urinate, defecate, have sexual intercourse, form families, appoint leaders, and so on, the only difference being that they get much older than we do before they die (Sakr 2001). Their preferred meal is said to consist of bones and dung (Al-Asqar 2005). In analogy with human demographics, djinns are classified according to gender, ethnicity, religious affiliation (i.e., Muslim, Jewish, Christian, pagan, or other), social class, political affiliation, moral standards, and so on (Bilu 1979). Some djinns are believed to be nameless, whereas others have names, such as the notorious female djinn Aisha Qandisha in Morocco and Alkarisi in Turkey (Crapanzano 1973; Hoffer 2000). Djinns are deemed to have humid, impure places as their natural habitat, including rivers, wells, cemeteries, caves, ruins, market places, garbage dumps, bathrooms, and the vicinity of humans who neglect their personal hygiene. And yet they are also associated with fire, darkness, and blood. Reportedly, they feel particularly attracted to individuals going through transitional phases such as birth, circumcision, marriage, pregnancy, and the process of dying (Bilu 1979). According to some, the vulnerability to djinns is highest in cases of blood loss, extreme affective states, and negligence of prayer (Ghubash and Eapen 2009).

Like humans, djinns are considered sensitive to offense. It is believed that they can be easily provoked by insults and inadvertently disturbed by simple actions such as turning a stone or stepping over it (Stein 2000), killing a snake indoors, or pouring boiling hot water into the kitchen sink without saying "*bismillah*" ("In the name of Allah"). It is widely held that djinns can be conjured up by magicians and sent out to gather information from remote places with incredible speed – according to some, even from the gates of heaven – or to transport heavy objects from one place to another (Sakr 2001). Religious scholars distinguish many techniques used by djinns to influence humans, including seduction (*nazagha*), temptation (*azalla*), presenting something as attractive (*zayyana*), driving one astray (*aghwā*), beguilement (*fatana*) (Maarouf 2007), and marrying a person who has remained single for too long (El-Zein 2009), although the latter is considered extremely rare (Al-Ashqar 2005). They also distinguish three basic modes of interaction with humans: from a relative distance, for example, by whispering in one's ear (*waswasa*), by striking a person (*Ṣira*), and by entering the body and possessing it (*aslai*) (Maarouf 2007). The following two cases illustrate how versatile and intrusive those perceptual experiences can be.

18.3 Two Cases

Patient A is a 45-year-old Tunisian man who used to work as a chef in various restaurants. Fifteen years ago someone startled him by pushing him in the back while he was working, and at that very moment he had the feeling a djinn had gotten into him. Ever since he has felt inexplicable movements, as if the djinn were relocating itself from one arm to the other, and down to his chest, belly, and legs. After 5 years, he started seeing the djinn too, in the shape of a blond woman in her twenties, dressed in a white robe and surrounded by a soft white light. She moves the way real people do and looks him in the eye when she addresses him. Patient A tends to listen to her intently, believing she has an important message to convey, but he can seldom understand what she says. Sometimes he hears in Dutch, "We will destroy your life" or "You will never get married," but most of the time, he hears a language he does not master. Sometimes the woman changes shape, ending up with multiple feet or a half-animal appearance. The sight of her tends to be accompanied by the smell of rotting cadavers, a sour taste in the mouth, pinches in an arm or a leg, and the sensation of a strong wind. During nights, he often sees the woman sitting on his chest, and feels her exerting pressure. At times, he even feels her having sexual intercourse with him. He always feels wide awake and yet unable to move or scream, and in the morning he often finds that he has ejaculated. Since the time he experienced the djinn, patient A lost his job, got into quarrels and fights, spoiled three marriages, and became socially isolated. He consulted various imams, who advised him to pray, visit the mosque, read Qur'anic texts, and undertake a pilgrimage to Mecca. After 10 years, he consulted a biomedical practitioner who diagnosed him with borderline personality disorder and prescribed antidepressants. Finding no avail, patient A discharged himself after 2 years and, after three more years, turned to our hospital. He was interviewed and offered an EEG and an MRI scan to rule out temporal epilepsy, but backed out once again and never returned.

Patient B is a married, 26-year-old Moroccan woman, a former cashier and mother of a single daughter.[2] Four years ago, she fell down from the lavatory pan at work, dislocated her pink, and lost consciousness for half an hour. Believing that such a curious incident in such an unhygienic place must have been the work of a djinn, she set out to consult physicians and myriad other health professionals and even had herself admitted to a rehabilitation center, initially because of limb dysfunction resulting from the dislocation, but soon thereafter also because of depression, headache, stomachache, and fainting. After 1.5 year, she started hearing a male voice which commanded her to steal, smoke, drown herself, cut herself, and jump off a balcony or in front of a running train. At this point, she consulted an imam, who confirmed that she had been attacked by a djinn and treated her by reading Qur'anic verses out loud. Meanwhile she consulted the outpatient ward of our psychiatric hospital, where she was diagnosed with borderline personality disorder and treated with antidepressants, benzodiazepines, and antipsychotics. Her behavior deteriorated to the extent that she had to be admitted several times a week because of self-mutilation, suicide attempts, and physical aggression toward members of her family. Upon reexamination, it turned out that she did not only hear voices but also saw a dark figure facing her and talking to her. In addition, she turned out to suffer from mood swings and nonepileptic seizures. A neurological examination, blood tests, an EEG, and an MRI scan did not reveal any underlying pathology. Patient B's diagnosis was adjusted to schizoaffective disorder, and she was treated with clozapine and valproic acid. A second imam, working in the service of our hospital, said that he doubted whether she had been stricken by a djinn. Nevertheless, he took time to pray with her and read the Qur'an. Patient B recovered insofar that the seizures, aggression, and self-mutilating behavior came to a halt, and that after 2 years of hospitalization, she was sufficiently stable to be transferred to an open long-stay ward, from where she could regularly visit her daughter and other family members.

---

[2]The case of patient B was described before in Blom et al. (2010).

18.4 Descriptive Psychopathology

In Muslim societies, many people consider their lives so much intertwined with those of djinns that any kind of illness, misfortune, and unconventional behavior can be attributed to them (Drieskens 2008). From this vantage point, it is quite understandable that the two patients described above would see such entities as an explanation for their experiences, but does this also explain their phenomenological characteristics? Our own phenomenological and anthropological studies of the beliefs of Islamic healers and their clients (Hoffer 2000, 2009) as well as the medical examination of some 30 clinical patients – most of them diagnosed with psychotic disorder – indicate that hallucinations experienced by Islamic patients in the Netherlands tend to be multimodal in nature (as opposed to the predominantly unimodal or bimodal hallucinations experienced by the majority of our Dutch patients), and that they tend to add up to *personifications*, i.e., compound hallucinations depicting human or humanoid beings (see Staudenmaier 1912). As we saw, patient A experienced somatic, tactile, visual, verbal auditory, olfactory, and gustatory hallucinations, as well as a possible aura (the sensation of a wind blowing against the body) and an incubus phenomenon, i.e., sleep paralysis combined with the lively and often multimodal sensation of a creature sitting on the chest (see Cheyne 2003, as well as Chap. 17, this volume). This array of hallucinatory phenomena combined to conjure up the image of a living being carrying out coordinated actions. In the case of our patient A, the clinical picture may well have been attributable to temporal epilepsy, but in many other cases, including that of patient B, epilepsy was ruled out with reasonable certainty whereas the hallucinations at hand were similarly elaborate in nature. One male patient described to us a female djinn walking toward him, touching his chest, disappearing into his body, delivering a baby inside, and then leaving, after which he heard the baby crying inside his head. A second man described to us the experience of a small, primordial creature attaching itself to his face, thus preventing him to speak or breathe. The same creature was experienced by him as sitting behind his back, anus to anus, forcing feces into his intestines, which he could only get rid of by forcefully straining back.

18.5 Traditional Islamic Versus Biomedical Interpretations

Thus far we have consistently spoken of "hallucinations" when we referred to the perceptual experiences of our Islamic patients. After all, we designate a percept as hallucinatory when it is experienced by a waking individual in the absence of an appropriate stimulus from the extracorporeal world. But does this approach do sufficient justice to the topic at hand? In anthropology, it is customary to distinguish between "illness" (i.e., a feeling of not being healthy or normal) and "disease" (i.e., a diagnosable pathological condition), with "illness" predominantly reflecting the patient's perspective and "disease," the health practitioner's perspective. In the case of Islamic patients attributing their perceptual experiences to djinns, and biomedical health practitioners treating them with antipsychotic medication, illness and disease have diametrically opposed ontological connotations. While the health practitioners attribute the percepts at hand to some neurophysiological aberration in the brain, the patients interpret them as testimony of the presence of actual living beings in the world "out there." If the health professionals were radical empiricists in the vein of William James (1842–1910), they might be willing to accept that the ultimate stuff of reality is pure experience, and that different perspectives therefore do not only entail different worldviews but also different worlds (James 1996). But a pluralistic universe is not what most of us have in mind when we think about the world "out there." On the contrary, we usually envisage the world as a singular, physical entity with objectifiable characteristics, which is probably the principal reason why it is so hard for us to accept the validity of more than a single explanatory model at a time (see also Chap. 2). While incommensurable explanatory models are a sure way to compromise the working relation between any two parties (Kleinman 1980), one cannot expect health practitioners to abandon their own model in favor of their patients', if only because they owe their professional skills to that model. Nor can we expect patients to simply give up their idiosyncratic beliefs in favor of their practitioners' worldviews. So the question here is how biomedical practitioners can bridge the gap between the Islamic religious model and their own biomedical model when it comes to the interpretation and treatment of perceptual phenomena attributed to djinns.

18.6 Religious Context and Folk Belief

To complicate things further, it should be noted that the term "Islamic religious model" refers to at least two different discourses, namely, the official Islam and the body of folk beliefs prevalent in Muslim societies. The official Islam comprises the Sunnite and Shi'ite faiths propagated by imams, *ûlamâ*, and *mullahs* in mosques and other orthodox institutions. Islamic folk beliefs, on the other hand, are rooted in historical medico-religious traditions such as Islamic-Arabic medicine, prophetic medicine, Sufism, and all sorts of local cultural traditions and folklore (Hoffer 2000), examples of which can be found in abundance on the internet and in the popular media (Drieskens 2008). The differences between the two discourses are of utmost importance for representatives of the official Islam, who consider many of the folk beliefs as superstitions. The majority of nonorthodox Muslims, however, tend to borrow elements from both discourses, mixing them up freely in their daily lives, and applying them whenever they see fit. As a corollary, many of them do not only believe in the existence of djinns but also in magic (*s'hour*), the evil eye (*l'ayne*), hagiolatry, and many other phenomena and practices not sanctioned by the official Islam (Hoffer 2000; Hermans 2007). Moreover, Muslims tend to see little harm in adding elements from the biomedical discourse, as testifies the consultation of a psychiatrist as well as an Islamic healer by patients A and B. In this sense, we may well ask ourselves the same question as Van der Geest (1985), who wrote, "Why all this fuss, one could ask, about integration? In their heads, clients of health care have already achieved an 'integration' of medical traditions."

18.7 Treatment

Traditional Arabic-Islamic treatment methods tend to fuse elements from the official Islam and Islamic folk medicine. Some of them are directed at the expulsion or ousting of djinns, whereas others aim to establish a state of peaceful coexistence between humans and djinns (Crapanzano 1973). Protection against djinns is offered by imams, *fuqaha* (religious teachers), *ashraf* (descendants of a holy family), and other religious healers in the form of prayer, Qur'anic verses (read out loud or recited), lavages, faith healing, the laying on of hands (see Fig. 18.3), magnetism, magical rituals, the induction of trance states, amulets (see Fig. 18.4), talismans (see Fig. 18.5), dietary measures, herbs, and fumigation.

Qur'anic verses are also written down in saffron, dissolved in water, and then offered to the alleged victims of djinns either to drink or to wash themselves (Hoffer 2000; Sengers 2000). Measures of a more sweeping nature include ecstatic dances, pilgrimages, visits to a holy shrine (see Fig. 18.6), animal sacrifices, the cursing of djinns, exorcism, incarceration, and caning (Crapanzano 1973; Hermans 2007; Lebling 2010). Various patients of ours had been locked up for 40 days, some of them chained to a wall. One of them told us that during that time, she underwent a foot whipping every Thursday, 40 strokes at a time, and another patient told us that he was thrown backward into a desert well, with the apparent intention of taking

Fig. 18.3 The laying on of hands by a religious healer in West Morocco (Photograph by Cor Hoffer 2007)

Fig. 18.4 Amulet (*nazarlik*) (Photograph by Cor Hoffer 2000)

Fig. 18.5 Leather talisman with Qur'anic verses (Photograph by Cor Hoffer 2000)

Fig. 18.6 Islamic holy shrine in West Morocco (Photograph by Cor Hoffer 2000)

him by surprise and thus ousting his djinn by frightening the life out of him.[3] As regards incarceration and immobilization, we gained the impression from our patients' accounts that these measures reflected the despair of all involved rather than any malign intentions. Caning, whippings, and beatings – which tend to be frowned upon by biomedical practitioners and Muslim scholars alike – were sometimes justified by our patients through the assertion that pain is thus afflicted to the djinn rather than to the possessed person. Here, too, the motivation to apply such measures would seem to be powerlessness rather than hostility (conform Hanley 2005). As noted by Dein et al. (2008) and Mölsä et al. (2010), today, many religious healers – especially those practicing in Western countries – are prepared to consider biomedical diagnoses first, to the extent that they adopt the idiom and accept the therapeutic techniques employed by psychiatrists, and that they reserve their more drastic techniques for cases where all else has failed, perhaps comparable to the way biomedical practitioners take refuge to electroconvulsive treatment (ECT).

Biomedical practitioners tend to treat their Islamic patients in conformity with biomedical diagnostic algorithms, using methods ranging from psychotherapy to antidepressants, antipsychotics, mood stabilizers, sedatives, anticonvulsants, and ECT (see Chap. 24). But unless they have the proper background information on their patients' attributional habits, their treatments are seldom successful (Hoffer 2009). This appears to be the rule rather than the exception, considering the

[3]This type of treatment, called water shock treatment, was also practiced in eighteenth- and early nineteenth-century European psychiatry (Guislain 1826; Kraepelin 1918).

prevalence of djinn attributions among Islamic patients, their reluctance to discuss them, and the unfamiliarity of many biomedical practitioners with the issue at hand (Dein et al. 2008). Our own experience is that it takes considerable effort and genuine interest to encourage Islamic patients to discuss the true nature of their concerns and to have them shed some light on the therapeutic strategies attempted so far. But when they do so, one is often able to obtain surprisingly detailed accounts. We, too, establish diagnoses in conformity with the Diagnostic and Statistical Manual of Mental Disorders (DSM) and offer biomedical treatments to match. But in addition, we offer our patients the possibility of consulting an imam in the service of our hospital, who establishes his own religious diagnoses, and offers religious treatments in the form of prayer, Qur'an reading, lavages, and sometimes dietary measures. This two-track policy is of a complementary nature, in the sense that both parties encourage the patient to adhere to both types of treatment, thus tolerating the coexistence of the explanatory models involved and offering a treatment program that addresses the biomedical as well as the religious issues at stake. An important benefit of this strategy is the improved compliance of our patients, in addition to an improved patient-physician relationship, which, after all, is the sine qua non for any therapeutic success.

18.8 Concluding Remarks

The prevalence of hallucinations attributed to djinns is unknown, but we estimate that in the Netherlands, some 80% of Islamic patients receiving a – biomedical – diagnosis of psychotic disorder consider djinns as an explanation for their condition (Blom et al. 2010). In conformity with the literature on the subject (Akerele 1987; Saeed et al. 2000; Khalifa and Hardie 2005; Sheikh 2005; Hoffer 2005), we advocate a culturally sensitive approach that involves medical and transcultural history taking, proper medical diagnosis, proper medical treatment, as well as the consultation of a qualified imam or religious healer for the purpose of dealing with any religious issues at stake. In doing so, however, one should take care to avoid conflicting – and potentially harmful – therapeutic approaches. As to future research, we recommend systematic studies of the phenomenological characteristics and neurobiological correlates of hallucinations attributed to djinns, in the vein of relatively recent studies on autoscopy and out-of-body experience (Brugger et al. 1997; Blanke et al. 2004; see also Chaps. 15 and 16). It does not take a Kant or Freud to appreciate the notion that we constantly project our own ideas onto the world we think we are perceiving, but there might well be more to this type of hallucination than the mere idea that we are dealing here with a "culture-bound syndrome."

References

Akerele, O. (1987). Bringing traditional medicine up to date. *Social Science & Medicine*, 24, 177–181.

Al-Ashqar, 'U.S. (2005). *The world of the jinn & devils in the light of the Qur'an and Sunnah.* Translated by Al-Khattab, N. Riyadh: International Islamic Publishing House.

Bilu, Y. (1979). Demonic explanations of disease among Moroccan Jews in Israel. *Culture, Medicine and Psychiatry*, 3, 363–380.

Blanke, O., Landis, T., Spinelli, L., Seeck, M. (2004). Out-of-body experience and autoscopy of neurological origin. *Brain*, 127, 243–258.

Blom, J.D., Eker, H., Basalan, H., Aouaj, Y., Hoek, H.W. (2010). Hallucinaties toegedicht aan djinns. *Nederlands Tijdschrift voor Geneeskunde*, 154, A973.

Brugger, P., Regard, M., Landis, Th. (1997). Illusory reduplication of one's own body: Phenomenology and classification of autoscopic phenomena. *Cognitive Neuropsychiatry*, 2, 19–38.

Cheyne, J.A. (2003). Sleep paralysis and the structure of waking-nightmare hallucinations. *Dreaming*, 13, 163–179.

Crapanzano, V. (1973). *The Hamadsha. A study in Moroccan ethnopsychiatry.* Los Angeles, CA: University of California Press.

Dein, S., Alexander, M., Napier, D. (2008). *Jinn*, psychiatry and contested notions of misfortune among East London Bengladeshis. *Transcultural Psychiatry*, 45, 31–55.

Drieskens, B. (2008). *Living with djinns. Understanding and dealing with the invisible in Cairo.* London: SAQI.

El-Zein, A. (2009). *Islam, Arabs, and the intelligent world of the jinn.* Syracuse, NY: Syracuse University Press.

FORUM (2010). *De positie van moslims in Nederland. Feiten en cijfers. Factbook editie 2010.* Utrecht: FORUM.

Ghubash, R., Eapen, V. (2009). Postpartum mental illness: perspectives from an Arabian Gulf population. *Psychological Reports*, 105, 127–136.

Guislain, J. (1826). *Traité sur l'aliénation mentale et sur les hospices des aliénés. Tome II.* Amsterdam: J. van der Hey et fils, et les héritiers H. Gartman.

Hanley, J. (2005). Possession and jinn. *Journal of the Royal Society of Medicine*, 98, 486.

Hermans, P. (2007). *De wereld van de djinn. Traditionele Marokkaanse geneeswijzen.* Amsterdam: Bulaaq.

Hoffer, C. (2000). *Volksgeloof en religieuze geneeswijzen onder moslims in Nederland. Een historisch-sociologische analyse van religieus-medisch denken en handelen.* Amsterdam: Thela Thesis.

Hoffer, C. (2005). 'Psychose' of 'djinn': verklaringsmodellen en interculturele communicatie in de GGZ. *Patient Care*, 4, 141–146.

Hoffer, C. (2009). *Psychische ziekten en problemen onder allochtone Nederlanders: beleving en hulpzoekgedrag.* Assen: Van Gorcum.

James, W. (1996). *A pluralistic universe. Hibbert lectures at Manchester College on the present situation in philosophy.* Lincoln, NE: University of Nebraska Press.

Khalifa, N., Hardie, T. (2005). Possession and jinn. *Journal of the Royal Society of Medicine*, 98, 351–353.

Kleinman, A. (1980). *Patients and healers in the context of culture. An exploration of the borderland between anthropology, medicine and psychiatry.* Berkeley, CA: University of California Press.

Kraepelin, E. (1918). *Hundert Jahre Psychiatrie. Ein Beitrag zur Geschichte menschlicher Gesittung.* Berlin: Verlag von Julius Springer.

Lebling, R. (2010). *Legends of the fire spirits. Jinn and genies from Arabia to Zanzibar.* London: I.B. Tauris.

Maarouf, M. (2007). *Jinn eviction as a discourse of power. A multidisciplinary approach to Moroccan magical beliefs and practices.* Leiden: Brill.

Mölsä, M.E., Hjelde, K.H., Tiilikainen, M. (2010). Changing conceptions of mental distress among Somalis in Finland. *Transcultural Psychiatry*, 47, 276–300.

Saeed, K., Gater, R., Husain, A., Mubbashar, M. (2000). The prevalence, classification and treatment of mental disorders among attenders of native faith healers in rural Pakistan. *Social Psychiatry and Psychiatric Epidemiology*, 35, 480–485.

Sakr, A.H. (2001). *Al-jinn*. New York, NY: Islamic Book Service.

Sengers, G. (2000). *Vrouwen en demonen. Zar en korangenezing in hedendaags Egypte.* Amsterdam: Het Spinhuis.

Sheikh, A. (2005). Jinn and cross-cultural care. *Journal of the Royal Society of Medicine*, 98, 339–340.

Staudenmaier, L. (1912). *Die Magie als experimentelle Naturwissenschaft.* Leipzig: Akademische Verlagsgesellschaft.

Stein, D. (2000). Views of mental illness in Morocco: Western medicine meets the traditional symbolic. *Canadian Medical Association Journal*, 163, 1468–1470.

Van der Geest, S. (1985). Integration or fatal embrace? The uneasy relationship between indigenous and western medicine. *Curare*, 8, 9–14.

Part III
Research

Chapter 19
Structural Neuroimaging in Psychotic Patients with Auditory Verbal Hallucinations

Paul Allen and Gemma Modinos

19.1 Introduction

Over the last few decades, neuroimaging techniques have allowed us to investigate the structural correlates of symptoms in individuals diagnosed with schizophrenia and other psychoses. Brain scans, using computerised tomography (CT) or magnetic resonance imaging (MRI), are often used clinically to rule out focal lesions such as a tumour or a stroke in patients who present with psychotic symptoms. In a seminal study, Johnstone et al. (1976) were the first to use CT scans to show structural alterations in the brains of patients diagnosed with schizophrenia. Since then, a significant body of research has used neuroimaging to examine neuroanatomy in patients with psychosis (Ellison-Wright and Bullmore 2009; Glahn et al. 2008; Shenton et al. 2001). A smaller number of studies have used structural neuroimaging to examine the brain in patients diagnosed with schizophrenia who experience frequent hallucinations, usually in the auditory modality. In this chapter, a comprehensive overview of studies examining grey- and white-matter changes in patients experiencing hallucinations is provided. The focus will be predominantly on studies of patients diagnosed with schizophrenia or schizophreniform disorders who experience auditory verbal hallucinations (AVHs).

19.2 Hallucinations and Brain Lesions

Braun and colleagues reviewed case studies of post-lesion hallucinations in the visual, auditory, and somatic modalities. They observed that lesions are almost always located in the brain pathway of the sensory modality of the hallucination and

P. Allen, M.Sc., Ph.D. (✉) • G. Modinos, Ph.D.
Department of Psychosis Studies, Institute of Psychiatry, King's College, London, UK
e-mail: p.allen@kcl.ac.uk; gemma.modinos@kcl.ac.uk

J.D. Blom and I.E.C. Sommer (eds.), *Hallucinations: Research and Practice*,
DOI 10.1007/978-1-4614-0959-5_19,

suggested that the lesioned tissue must have contained a predominance of inhibitory over excitatory neurons for the sensory modality in question (Braun et al. 2003). These inhibitory neurons would normally be responsible for 'gauging down' brain circuits containing complex sensory representations. In addition to localised lesions and dysfunction to specific brain regions (topological), dysfunction related to connections between brain regions (hodological) has also been proposed since patients prone to hallucinations have complex, task-specific hodological abnormalities that persist between hallucination episodes (ffytche 2008). Despite the ostensible relationship between hallucinations and lesions in corresponding sensory regions seen in neurological patients, in psychiatric patients who experience hallucinations, no gross brain abnormalities can usually be seen on brain scans at the individual level. For this reason, it has been necessary to combine brain scans of groups of subjects with hallucinations and compare these to healthy individuals and/or patients diagnosed with schizophrenia without hallucinations in order to identify neuroanatomical changes associated with these symptoms.

19.3 Grey-Matter Changes in Patients with Hallucinations

Early region of interest (ROI) studies investigating hallucinations in patients diagnosed with schizophrenia showed that their brains were characterised by volume reductions of the superior temporal gyrus (STG) and lateral ventricles. In particular, hallucination severity was inversely correlated with left STG volume (Barta et al. 1990; Flaum et al. 1995). More recently, Sumich and colleagues assessed positive and negative symptomatology in 25 first-episode patients and its relationship with grey-matter volume (GMV) in the temporal lobe (Sumich et al. 2005). The study used linear regression to establish associations between symptom dimensions and stereological measurements of Heschl's gyrus and the planum temporale. Volume decrease in left Heschl's gyrus (primary auditory cortex) was associated with hallucinations, whilst increased volumes in the left planum temporale were associated with delusions. Onitsuka et al. (2004) reported that increased severity of AVH is correlated with reduced grey-matter volumes (GMVs) in the left anterior STG and middle temporal gyrus (MTG). A summary of research studies using an ROI approach is provided in Table 19.1.

More recently, voxel-based morphometry (VBM), a whole-brain, unbiased, semiautomated technique for characterising regional cerebral differences in structural MRI (Ashburner and Friston 2000; Mechelli et al. 2005), has been applied to investigate structural abnormalities associated with AVH. O'Daly et al. (2007) reported an association between AVH severity and GMV loss in a wider range of temporal regions than those typically included in ROI studies, including the right STG and fusiform gyrus, and the left inferior temporal gyrus. Martí-Bonmatí et al. (2007) reported GMV reduction also in bilateral STG and medial temporal gyri in 21 patients diagnosed with schizophrenia and experiencing persistent hallucinations. Nenadic et al. (2010) studied a relatively large group of 99 participants

Table 19.1 Studies investigating brain structure with a region of interest approach

Authors	Sample	Design	Findings
Barta et al. (1990)	15 SCZ, 15 HC	Correlation severity AVH	Less left STG
Flaum et al. (1995)	166 SCZ	Correlation severity AVH	Less left STG
Onitsuka et al. (2004)	23 SCZ, 28 HC	Correlation severity AVH	Less left STG, MTG
Shin et al. (2005)	25 SCZ (17 AVH, 8 no AVH)	Comparison AVH>no AVH	More temporal GMV and WMV, more frontal GMV
Sumich et al. (2005)	25 SCZ	Correlation severity AVH	Less left Heschl's gyrus
Hubl et al. (2010)	13 SCZAVH, 13 SCZ no AVH, 13 HC	Group comparisons	AVH more volume in the right, but not in the left Heschl's gyrus (WM and GM)

AVH auditory verbal hallucinations, *GMV* grey-matter volume, *HC* healthy controls, *MTG* middle temporal gyrus, *SCZ* patients diagnosed with schizophrenia, *SCZAVH* patients diagnosed with schizophrenia and auditory verbal hallucinations, *SCZ no AVH* patients diagnosed with schizophrenia without auditory verbal hallucinations, *STG* superior temporal gyrus, *WMV* white-matter volume

diagnosed with schizophrenia. Voxel-wise correlations with a score of AVH severity identified areas in the left and right STG (including Heschl's gyrus) and left supramarginal/angular gyrus, although it is not clear whether these correlations represent increased or decreased GMV. Together, these results are consistent with previous ROI studies demonstrating main effects of auditory hallucinations related to modality-specific superior temporal areas including primary and secondary auditory cortices.

Also using VBM techniques, a number of studies have reported structural alterations in non-sensory regions to be associated with AVH. Shapleske et al. (2002) analysed whole-brain grey and white matter in patients with and without hallucinations and age- and gender-matched controls. Relative to non-hallucinating patients, those with AVH showed reduced grey-matter volume in the left insula and adjacent temporal pole. Neckelmann et al. (2006) reported GMV reduction that correlated with hallucination severity in the thalamus and cerebellum as well as the left STG. Martí-Bonmatí et al. (2007) described GMV reductions associated with the severity of AVH in the right precentral, left Rolandic gyrus, left inferior opercular frontal and left superomedial frontal gyri, bilateral orbitomedial frontal, right posterior cingular, bilateral anterior cingular, right parahippocampus, right insula, and right precuneus. The VBM study by Nenadic and colleagues also reports GMV reduction associated with AVH in the left post-central gyrus and left posterior cingulate cortex. Modinos et al. (2009) reported that GMV in the left inferior frontal gyrus was positively correlated with the severity of AVH. Furthermore, hallucination severity influenced the pattern of structural covariance between this region and the left STG and MTG, the right inferior frontal gyrus and hippocampus, and the insula bilaterally. Gaser et al. (2004) studied 85 patients diagnosed with schizophrenia, of which 29 experienced hallucinations. They found severity of hallucinations to be correlated with volume loss in the right dorsolateral prefrontal cortex as well as the left transverse temporal Heschl's and supramarginal gyri. The volume loss in the right prefrontal cortex is of interest given the role that has been ascribed to frontotemporal interactions in volitional auditory perception (Silbersweig and Stern 1998). In light of this evidence, impairments in this network could erase the volitional signature of subjective perceptual awareness that arises from frontotemporal interactions and thus explain why hallucinations are experienced as involuntary. Interestingly, the right prefrontal area described by Gaser et al. (2004) partially overlaps with the homologue of Broca's area. Homotopic brain regions are connected to each other by inhibitory tracts (Karbe et al. 1998), which are the densest of all the interhemispheric pathways (McGuire et al. 1991a, b). Impairment of one region may thus lead to hyperactivation of the homotopic region. Support for this idea comes from a recent functional imaging study that revealed activation in the right homologue of Broca's area in patients whilst experiencing hallucinations (Sommer et al. 2008).

Overall, the imaging literature has primarily focused on how brain structure relates to AVH using indices of its overall severity. A recent study, however, investigated how GMV relates to a specific characteristic of AVH, namely spatial location. Plaze et al. (2009) used VBM to examine the neural substrate of this clinical feature in two subgroups of patients according to the spatial localization of their hallucinations. Patients with only outer-space hallucinations (voices heard outside the head) and patients with only inner-space hallucinations (voices heard inside the head) were compared. Convergent anatomical differences (white-matter volume) were detected between the patient subgroups in the right temporoparietal junction (rTPJ). In comparison to healthy subjects, opposite deviations in white-matter volumes and sulcus displacements were found in patients with inner-space hallucinations and patients with outer-space hallucinations. The authors concluded that spatial localization of auditory hallucinations is associated with the rTPJ anatomy, a key region of the 'where' auditory pathway. The rTPJ plays an important role in spatial awareness and in self-processing and integration of personal and extrapersonal spaces. In this context, the authors speculated that, regarding the spatial localization of AVH, the reported differences in rTPJ anatomy could contribute to the differential attributions of spatial coordinates to hallucinations in an egocentric referential manner (external vs. internal).

Whilst the majority of studies using either ROI or VBM analysis have mainly reported *reductions* in grey matter to be associated with AVH, *increases* in grey- and white-matter volumes have also been documented in hallucinating relative to non-hallucinating patients. Using a semiautomated Talairach atlas–based parcellation method, Shin et al. (2005) compared 17 first-episode patients with AVH to 8 first-episode patients without. Larger temporal grey- and white-matter and frontal grey-matter volumes were found in hallucinating patients. The authors suggest that increased volumes may be due to patient characteristics, as their study comprised of unmedicated first-episode patients. However, the results from this study should be

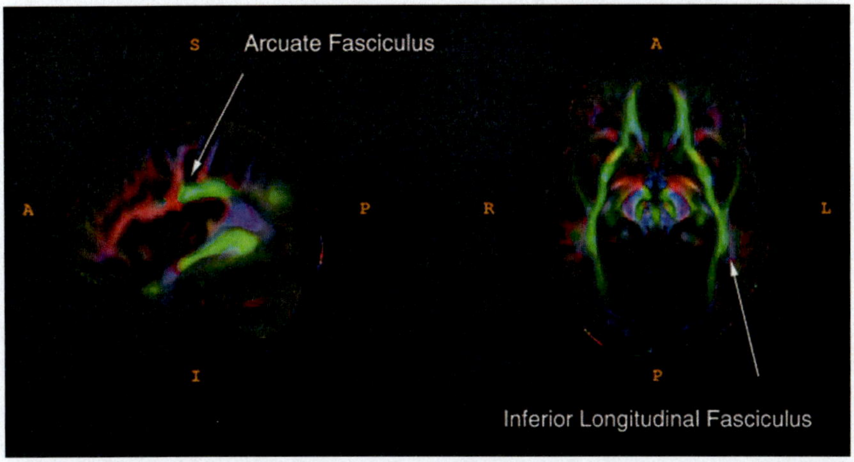

Fig. 19.1 The most consistently reported areas of grey-matter reductions in association with auditory verbal hallucinations in schizophrenia comprise the left superior temporal gyrus (*blue*) and the left middle temporal gyrus (*red*). The images depict the anatomical landmarks of these regions on the sagittal (*left*) and axial (*right*) views. They have been prepared with anatomically predefined masks according to the Talairach Deamon in the WFU_PickAtlas software in SPM5 and overlaid on the standard brain with MRICron

interpreted cautiously due to the small sample size. Hubl et al. (2010) also reported a bigger volume of Heschl's gyrus in the right hemisphere in hallucinating patients, caused by both grey- and white-matter increases, and Modinos et al. (2009) reported a positive correlation between volume in the left IFG and the overall severity of AVH. Such findings may be consistent with theories of increased connectivity between cerebral regions in hallucinating patients. This will be further discussed in the next section on white-matter studies, as increased integrity of frontotemporal white-matter bundles have been reported in hallucinators relative to non-hallucinators (Hubl et al. 2004; Shergill et al. 2007).

Despite the emergence of some clear trends in the findings, such as the apparent alteration in STG volume reported in patients, several studies have also been published that failed to find a relationship between STG volume and hallucinations (Cowell et al. 1996; DeLisi et al. 1994; Havermans et al. 1999; Plaze et al. 2006). In addition, and extending the findings from ROI studies, which are limited to the a priori selection of regions of interest, VBM studies have identified regions that have not been included in any of the main theoretical accounts of AVH (external misattribution of inner speech, modality-specific sensory cortical dysfunction, dysfunctional neural substrates of episodic verbal memory, see Seal et al. 2004; Waters et al. 2010; Northoff and Qin 2011) such as post-central cortical regions (García-Martí et al. 2008; Nenadic et al. 2010).

Overall, grey-matter reduction in the left STG/MTG, including the primary auditory cortex, is the most consistently reported finding in patients with AVH (see Fig. 19.1). This finding traverses methodological boundaries, as has been repeatedly reported both by studies using stereological and studies using VBM data analysis methods. This finding is also consistent with lesion studies which report that the lesion is almost always in the brain region of the sensory modality of the hallucination (Braun et al. 2003). Furthermore, reduced grey-matter volumes in non-sensory regions have also been reported in a number of studies. Amongst these, there appears to be little consistency with regard to non-sensory regions showing significant abnormalities, although GMV anomalies in portions of the prefrontal cortex are reported by five studies. Volume reductions in the prefrontal and cerebellar cortices may be associated with impairments in awareness and volition of internal speech, and the assignation of a self-tag to the experience of AVH, as proposed by a recent review (Northoff and Qin 2011). Finally, the one study investigating a clinical feature of AVH, spatial location, reports that the spatial location of AVH is associated with right TPJ anatomy. A summary of research studies using VBM may be seen in Table 19.2.

19.4 White-Matter Changes in Patients with Hallucinations

White-matter (WM) changes are usually investigated with an MRI technique called diffusion tensor imaging (DTI) that assesses the directionality of water diffusion (anisotropy), which is restricted by boundaries such as white-matter fibres. Considered to reflect the underlying structural integrity (Basser and Pierpaoli 1996), reduced fractional anisotropy (FA) is believed to represent a loss of WM integrity. A recent meta-analysis identified two consistent locations of FA reduction in patients diagnosed with schizophrenia: one in the deep WM of the left frontal lobe and the other in the deep WM of left temporal lobe (Ellison-Wright and Bullmore 2009). The association between AVH and WM integrity has also been examined by a small number of studies. Using magnetic resonance DTI, Hubl et al. (2004) investigated the integrity of WM tracts in the brains of patients diagnosed with schizophrenia with and without AVH and a healthy comparison group. Patients with AVH were shown to have significantly higher FA values relative to both control groups in the lateral temporoparietal section of a major fibre tract known as the arcuate fasciculus. This tract connects language production areas (e.g. Broca's area) with auditory processing and language perception areas. The authors speculate that, during inner speech, the apparently stronger connectivity between such areas in patients with hallucinations may lead to dysfunctional coactivation of regions related to the acoustical processing of external stimuli, which may account for their inability to distinguish self-generated thoughts from external stimulation. Shergill et al. (2007) used DTI in patients diagnosed with schizophrenia to examine the integrity of the major WM fasciculi, which connects the frontal, temporal, and parietal cortices and the corpus callosum. Across all patients, there was reduced FA in regions corresponding to the longitudinal fasciculi bilaterally and in the genu of the corpus callosum. Within the patients group, AVHs were associated with relatively increased FA in the superior longitudinal fasciculi (SLF), which is a part of the arcuate fasciculus, and in the anterior cingulum. Seok et al. (2007) reported

Table 19.2 Studies examining brain structure with voxel-based morphometry

Study	Sample	Design	Findings
Shapleske et al. (2002)	41 SCZAVH, 31 SCZ no AVH, 32 HC	Comparison AVH>no AVH	Less left INS and adjacent temporal lobe
Gaser et al. (2004)	85 SCZ	Correlation severity AVH	Less left transverse temporal gyrus of Heschl, left (inferior) SMG, middle/inferior right PFC
Neckelmann et al. (2006)	12 SCZAVH, 12 HC	Correlation severity AVH	Less left STG (transverse), left THA, bilateral cerebellum
Plaze et al. (2006)	15 SCZAVH	Correlation severity AVH	No correlation GMD and severity AVH
Martí-Bonmatí et al. (2007)	21 SCZAVH, 10 HC	Comparison AVH>HC, correlation severity AVH	Less right precentral, left Rolandic gyrus, left inferior opercular frontal, left superomedial frontal, bilateral orbitomedial frontal, right PCC, bilateral ACC, both medial temporal gyri, both STG, right PHC, right INS and right PreC.
O'Daly et al. (2007)	28 SCZAVH, 32 HC	Correlation severity AVH	Less right STG, fusiform gyrus, left ITG
García-Martí et al. (2008)	18 SCZAVH, 19 HC	Comparison AVH>HC. Correlation severity AVH	Less in bilateral INS, STG, and left AMY. Severity AVH correlated with less left IFG and right post-central gyri
Modinos et al. (2009)	26 SCZAVH	Correlation severity AVH	More left IFG
Plaze et al. (2009)	45 SCZAVH (12 outer, 15 inner), 20 HC	Comparison outer>inner	Less WMV rTPJ. No difference GMV
Nenadic et al. (2010)	99 AVH	Correlation severity AVH	Less left/right STG (including Heschl's gyrus), left SMG/ANG gyrus, left post-central gyrus, left PCC

ACC anterior cingulate cortex, *AMY* amygdala, *ANG* angular gyrus, *AVH* auditory verbal hallucinations, *GMD* grey-matter density, *GMV* grey-matter volume, *HC* healthy controls, *IFG* inferior frontal gyrus, *INS* insula, *PFC* prefrontal cortex, *PHC* parahippocampal gyrus, *PCC* posterior cingulate cortex, *PreC* precuneus, *rTPJ* right temporoparietal junction, *SCZ* patients diagnosed with schizophrenia, *SCZAVH* patients diagnosed with schizophrenia and auditory verbal hallucinations, *SCZ no AVH* patients diagnosed with schizophrenia without auditory verbal hallucinations, SMG supramarginal gyrus, *STG* superior temporal gyrus, *THA* thalamus, *WMV* white-matter volume that, in patients diagnosed with schizophrenia, FA of the WM regions was significantly decreased in the left SLF, whereas WM density was significantly increased in the left inferior longitudinal fasciculus (ILF). The mean FA value of the left frontal part of the SLF was positively correlated with the severity of AVH. Lee et al. (2009) used DTI to investigate altered structural integrity in STG grey and white matter in patients diagnosed with chronic schizophrenia compared with healthy controls. Relative to controls, the patients demonstrated reduced volume bilaterally in STG grey matter but not in white matter. For DTI measures, the patients showed increased mean diffusivity (a scalar measure of the total diffusion of the water molecules within a voxel, which is restricted by structures like membranes and myelin sheets), bilaterally, in STG grey matter, and in left STG white matter. In addition, mean diffusivity in left STG white matter showed a significant correlation with AVH. Lee and colleagues concluded that the increased water diffusivity in left-side STG, which was associated with auditory hallucinations, is due to a disconnection amongst auditory/language processing regions in patients diagnosed with schizophrenia. Ashtari et al. (2007) reported that adolescents with a diagnosis of schizophrenia or schizoaffective disorder and a history of visual hallucinations had lower FA in the left ILF than patients without visual hallucinations. Using a white-matter parcellation technique, Makris et al. (2010) compared the volume of brain fibre systems between patients diagnosed with schizophrenia ($n=88$) and matched healthy controls ($n=40$). White-matter regions of local and distal associative fibre systems were significantly different in the patients, and there were significant positive correlations between volumes (larger) in occipital, cingulate, and sagittal temporal regions and positive symptoms, in particular hallucinations. Table 19.3 shows a summary of research studies using DTI.

Based on the evidence accumulated to date, Whitford et al. (2010) proposed a direct link between WM abnormalities seen in patients diagnosed with schizophrenia and mechanistic accounts of AVH proposed by Frith (Frith et al. 1991, 2000). Specifically, it is proposed that abnormalities in frontal myelination of white-matter fasciculi result in conduction delays in the efference copies initiated by willed actions. These conduction delays cause the resulting corollary discharges to be generated too late to suppress the sensory consequences of the willed actions. The resulting ambiguity as to the origins of these actions represents a phenomenologically and neurophysiologically significant prediction error. On a phenomenological level, the perception of salience in a self-generated action leads to confusion as to its origins and, consequently, passivity experiences and auditory hallucinations.

In summary, a number of studies now show an association between white-matter alterations and hallucinations in patients diagnosed with schizophrenia. Although both increases and decreases in FA are reported, the majority of studies examining white matter report an increase in FA or white-matter volume associated with hallucinations. Furthermore, whilst there is considerable variation in the regions of reported white-matter alterations, the inferior and superior longitudinal fasciculi (or arcuate fasciculus) have been reported in three studies, implicating a disruption amongst cerebral networks supporting language and attentional processes (Fig. 19.2).

Table 19.3 Studies examining white-matter characteristics with diffusion tensor imaging

Authors	Sample	Design	Findings
Hubl et al. (2004)	13 SCZAVH, 13 SCZ no AVH, 13 HC	Comparison SCZAVH > SCZ no AVH + HC, SCZAVH > SCZ no AVH	SCZAVH > SCZ no AVH + HC = FA lateral parts, TP section of the arcuate fasciculus, and in parts of the ACC SCZAVH > SCZ no AVH = FA left hemispheric fibre tracts, including the cingulate bundle
Ashtari et al. (2007)	23 SCZ (9 SCZvisualH, 12 SCZnovisualH), 21 HC	Comparison SCZvisualH > SCZnovisualH	Less FA in left ILF
Seok et al. (2007)	15 SCZAVH, 15 SCZ no AVH, 22 HC	Correlation severity AVH	More mean FA value of the left frontal part of the SLF
Shergill et al. (2007)	33 SCZ, 40 HC	Correlation severity AVH	More FA in SLF and AC
Lee et al. (2009)	21 SCZ, 22 HC	Correlation severity AVH	More mean diffusivity in left STG
Makris et al. (2010)	88 SCZ, 40 HC	Correlation severity AVH	More WM volume in occipital, cingulate, and sagittal temporal regions

AC anterior cingulum, *AVH* auditory verbal hallucinations, *HC* healthy controls, *FA* fractional anisotropy, *TP* temporoparietal, *ILF* inferior longitudinal fasciculus, *SCZ* patients diagnosed with schizophrenia, *SCZAVH* patients diagnosed with schizophrenia and auditory verbal hallucinations, *SCZ no AVH* patients diagnosed with schizophrenia without auditory verbal hallucinations, *SCZvisualH* patients diagnosed with schizophrenia and visual hallucinations, *SCZnovisualH* patients diagnosed with schizophrenia without visual hallucinations, *SLF* superior longitudinal fasciculus, *STG* superior temporal gyrus, *WM* white matter

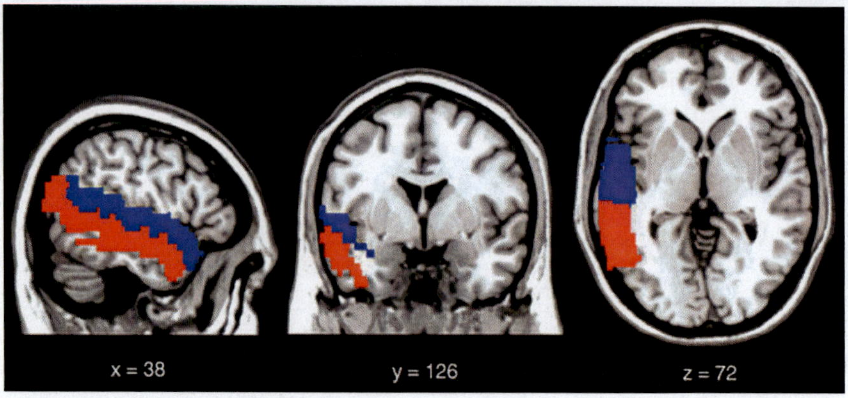

Fig. 19.2 Diffusion tensor imaging studies have most commonly reported white-matter abnormalities in association with auditory verbal hallucinations in schizophrenia which comprise the arcuate fasciculus (*left*) and the inferior longitudinal fasciculus (*right*). The image has been prepared using the IIT2 template (Described by Zhang et al. (2011). Enhanced ICBM diffusion tensor template of the human brain. Neuroimage, 54, 974–984)

19.5 Lateralisation and Gyrification Studies

Reduced cerebral lateralisation of language in patients diagnosed with schizophrenia has been documented in a substantial number of studies (Sommer et al. 2001). More specifically to patients with AVH, four studies have reported altered lateralisation. Shapleske et al. (2001) compared structural brain asymmetry of the planum temporale (PT) and sylvian fissure (SF) of patients with no history of hallucinations ($n=30$) and patients with a strong definitive history of AVH ($n=44$), in addition to 32 matched healthy volunteers. They failed to find differences between the groups on these measures. The only significant finding was a modest correlation between leftward asymmetry of the SF and hallucinations within the prominent hallucinator group. A recent functional imaging study also failed to find widespread lateralisation differences that were specific to AVH. Although patients with AVH did show decreased functional lateralisation, healthy individuals with AVH did not (Diederen et al. 2010).

One study examined cortical folding abnormalities in patients with treatment-resistant AVH. Cachia et al. (2008) used an automated method to extract, label, and measure the sulcus area in the whole cortex. They reported that for both hemispheres, patients diagnosed with schizophrenia had a lower global sulcal index. The local-sulcal index decrease was not homogeneous across the whole cortex and was more significant in the superior temporal sulcus bilaterally, in the left middle frontal sulcus, and in the left SF (Broca's area). It was hypothesised that sulcal abnormalities in language-related areas of the cortex might underlie these patients' particular vulnerability to hallucinations. A limitation of this study, however, was that a non-hallucinating control group was not studied.

19.6 Conclusions

From the six region-of-interest studies on patients diagnosed with schizophrenia and experiencing auditory verbal hallucinations, four have reported volumetric decreases in sensory regions, mainly the superior temporal gyrus. From the ten voxel-based morphometry studies, seven found less grey-matter volume in the superior temporal gyrus. From the six diffusion tensor imaging studies examining white-matter connectivity in patients with auditory verbal hallucinations, five reported an increase in fractional anisotropy or mean diffusivity indices. In terms of the directionality of the results, the earlier lesion studies showed tissue loss in sensory regions. Region-of-interest studies suggest volume reductions in sensory regions. Voxel-based morphometry analyses are largely consistent with lesion and region-of-interest studies, whilst also showing significant effects in non-sensory regions such as areas of the prefrontal cortex, and areas of the neuronal circuitry underlying emotional processes (insula, amygdala, anterior cingulate cortex, parahippocampal gyrus). The majority of connectivity studies reported enhanced integrity of white-matter tracts relevant to language processing (arcuate fasciculus and inferior longitudinal fasciculus), indicating increased coactivation of language-processing regions in auditory verbal hallucinations. Hence, despite some divergence between studies, abnormalities in the auditory cortex and language-related brain regions seem to be the most replicated finding, consistent with evidence from functional neuroimaging studies in auditory verbal hallucinations (Jardri et al. 2011). In addition, there are abnormalities in white-matter integrity, indicating perturbed connectivity, particularly between frontal and temporal regions involved in language and attention processes. This aligns well with a recent comprehensive review of structural and functional MRI studies on brain connectivity in patients diagnosed with schizophrenia, which evidenced that the symptoms attributed to schizophrenia are associated with connectivity reductions, across all stages of the disorder and regardless of the neuroimaging methodology employed (Pettersson-Yeo et al. 2011).

The structural evidence reviewed can be linked to a model put forward from functional neuroimaging studies postulating bottom-up dysfunction through over-activation in secondary (and occasionally primary) sensory cortices that lead to the experience of vivid perceptions in the absence of sensory stimuli (Northoff and Qin 2011). Subsequently, hallucinatory experiences may be augmented by a weakening of top-down control from the ventral anterior cingulate, prefrontal, premotor, and cerebellar cortices, which, through a breakdown in monitoring and volitional assignment, may further lead to the experience of externality (Allen et al. 2008). Finally, structural alterations in regions involved in the experience and regulation of emotion (parahippocampal gyri, cingulate, orbitofrontal cortex, insula) may be implicated in the often affect-laden characteristics of hallucinations. It has been proposed that it is not the presence of hallucinations per se but rather their negative emotional content and the perceived distress that appear to constitute an important difference between more benign positive experiences and psychopathology in healthy populations with hallucinatory predisposition (Sommer et al. 2008). Finally, aberrant structural and functional connectivity between sensory cortices and frontal regions may be central to this dysregulation.

Interestingly, a fourth conceptual framework for auditory verbal hallucinations has been put forward by Northoff and Qin (2011), which proposes that the brain's resting state activity prior to onset of auditory verbal hallucinations enables and predisposes to auditory verbal hallucinations, and thus tentatively provides an explanation for the initial overactivation of sensory regions (bottom-up dysfunction). There is some support for this theory as structural abnormalities have been reported in voxel-based morphometry studies in regions of the default-mode network (including the anterior and posterior cingulate cortex and the prefrontal cortex) (Raichle et al. 2001).

To conclude, for any theoretical framework of auditory verbal hallucinations to be comprehensive, it is important to integrate findings implicating relevant brain regions involved in bottom-up sensory and top-down monitoring processes, self-referential and emotional processing, and dysconnectivity between these networks. Future studies should aim at the integration of imaging data from different modalities (functional, structural, neurochemical), with large-enough samples (including patients with and without hallucinations as well as healthy individuals with and without subclinical hallucinatory experiences), in order to illuminate the mechanisms by which the human brain is capable of generating an auditory verbal hallucination.

References

Allen, P., Laroi, F., McGuire, P.K., Aleman, A. (2008). The hallucinating brain: a review of structural and functional neuroimaging studies of hallucinations. *Neuroscience and Biobehavioral Reviews*, 32, 175–191.

Ashburner, J., Friston, K.J. (2000). Voxel-based morphometry – the methods. *Neuroimage*, 11, 805–821.

Ashtari, M., Cottone, J., Ardekani, B.A., Cervellione, K., Szeszko, P.R., Wu, J., Chen, S., Kumra, S. (2007). Disruption of white matter integrity in the inferior longitudinal fasciculus in adolescents with schizophrenia as revealed by fiber tractography. *Archives of General Psychiatry*, 64, 1270–1280.

Barta, P.E., Pearlson, G.D., Powers, R.E., Richards, S.S., Tune, L.E. (1990). Auditory hallucinations and smaller superior temporal gyral volume in schizophrenia. *American Journal of Psychiatry*, 147, 1457–1462.

Basser, P.J., Pierpaoli, C. (1996). Microstructural and physiological features of tissues elucidated by quantitative-diffusion-tensor MRI. *Journal of Magnetic Resonance, Series B*, 111, 209–219.

Braun, C.M., Dumont, M., Duval, J., Hamel-Hebert, I., Godbout, L. (2003). Brain modules of hallucination: an analysis of multiple patients with brain lesions. *Journal of Psychiatry and Neuroscience*, 28, 432–449.

Cachia, A., Paillere-Martinot, M.L., Galinowski, A., Januel, D., de Beaurepaire, R., Bellivier, F., Artiges, E., Andoh, J., Bartres-Faz, D., Duchesnay, E., Rivière, D., Plaze, M., Manqin, J.F., Martinot, J.L. (2008). Cortical folding abnormalities in schizophrenia patients with resistant auditory hallucinations. *NeuroImage*, 39, 927–935.

Cowell, P.E., Kostianovsky, D.J., Gur, R.C., Turetsky, B.I., Gur, R.E. (1996). Sex differences in neuroanatomical and clinical correlations in schizophrenia. *American Journal of Psychiatry*, 153, 799–805.

DeLisi, L.E., Hoff, A.L., Neale, C., Kushner, M. (1994). Asymmetries in the superior temporal lobe in male and female first-episode schizophrenic patients: measures of the planum temporale and superior temporal gyrus by MRI. *Schizophrenia Research*, 12, 19–28.

Diederen, K.M., De Weijer, A.D., Daalman, K., Blom, J.D., Neggers, S.F., Kahn, R.S., Sommer, I.E. (2010). Decreased language lateralization is characteristic of psychosis, not auditory hallucinations. *Brain*, 133, 3734–3744.

Ellison-Wright, I., Bullmore, E. (2009). Meta-analysis of diffusion tensor imaging studies in schizophrenia. *Schizophrenia Research*, 108, 3–10.

ffytche, D.H. (2008). The hodology of hallucinations. *Cortex*, 44, 1067–1083.

Flaum, M., O'Leary, D.S., Swayze, V.W. 2nd, Miller, D.D., Arndt, S., Andreasen, N.C. (1995). Symptom dimensions and brain morphology in schizophrenia and related psychotic disorders. *Journal of Psychiatric Research*, 29, 261–276.

Frith, C.D., Blakemore, S., Wolpert, D.M. (2000). Explaining the symptoms of schizophrenia: abnormalities in the awareness of action. *Brain Research. Brain Research Reviews*, 31, 357–363.

Frith, C.D., Friston, K., Liddle, P.F., Frackowiak, R.S. (1991). Willed action and the prefrontal cortex in man: a study with PET. *Proceedings. Biological Sciences*, 244, 241–246.

García-Martí, G., Aguilar, E.J., Lull, J.J., Martí-Bonmatí, L., Escartí, M.J., Manjón, J.V., Moratal, D., Robles, M., Sanjuán, J. (2008). Schizophrenia with auditory hallucinations: a voxel-based morphometry study. *Progress in Neuro-psychopharmacology & Biological Psychiatry*, 32, 72–80.

Gaser, C., Nenadic, I., Volz, H.P., Buchel, C., Sauer, H. (2004). Neuroanatomy of "hearing voices": a frontotemporal brain structural abnormality associated with auditory hallucinations in schizophrenia. *Cerebral Cortex*, 14, 91–96.

Glahn, D.C., Laird, A.R., Ellison-Wright, I., Thelen, S.M., Robinson, J.L., Lancaster, J.L, Bullmore, E., Fox, P.T. (2008). Meta-analysis of gray matter anomalies in schizophrenia: application of anatomic likelihood estimation and network analysis. *Biological Psychiatry*, 64, 774–781.

Havermans, R., Honig, A., Vuurman, E.F., Krabbendam, L., Wilmink, J., Lamers, T., Verheecke, C.J., Jolles, J., Romme, M.A., Van Praag, H.M. (1999). A controlled study of temporal lobe structure volumes and P300 responses in schizophrenic patients with persistent auditory hallucinations. *Schizophrenia Research*, 38, 151–158.

Hubl, D., Dougoud-Chauvin, V., Zeller, M., Federspiel, A., Boesch, C., Strik, W., Dierks, T., Koenig, T. (2010). Structural analysis of Heschl's gyrus in schizophrenia patients with auditory hallucinations. *Neuropsychobiology*, 61, 1–9.

Hubl, D., Koenig, T., Strik, W., Federspiel, A., Kreis, R., Boesch, C., Maier, S.E., Schroth, G., Lovblad, K., Dierks, T. (2004). Pathways that make voices: white matter changes in auditory hallucinations. *Archives of General Psychiatry*, 61, 658–668.

Jardri, R., Pouchet, A., Pins, D., Thomas, P. (2011). Cortical activations during auditory verbal hallucinations in schizophrenia: a coordinate-based meta-analysis. *American Journal of Psychiatry*, 168, 73–81.

Johnstone, E.C., Crow, T.J., Frith, C.D., Husband, J., Kreel, L. (1976). Cerebral ventricular size and cognitive impairment in chronic schizophrenia. *Lancet*, 308, 924–926.

Karbe, H., Herholz, K., Weber-Luxenburger, G., Ghaemi, M., Heiss, W.D. (1998). Cerebral networks and functional brain asymmetry: evidence from regional metabolic changes during word repetition. *Brain and Language*, 63, 108–121.

Lee, K., Yoshida, T., Kubicki, M., Bouix, S., Westin, C.F., Kindlmann, G., Niznikiewicz, M., Cohen, A., McCarley, R.W., Shenton, M.E. (2009). Increased diffusivity in superior temporal gyrus in patients with schizophrenia: a Diffusion Tensor Imaging study. *Schizophrenia Research*, 108, 33–40.

Makris, N., Seidman, L.J., Ahern, T., Kennedy, D.N., Caviness, V.S., Tsuang, M.T., Goldstein, J.M. (2010). White matter volume abnormalities and associations with symptomatology in schizophrenia. *Psychiatry Research*, 183, 21–29.

Martí-Bonmatí, L., Lull, J.J., García-Martí, G., Aguilar, E.J., Moratal-Perez, D., Poyatos, C., Robles, M., Sanjuán, J. (2007). Chronic auditory hallucinations in schizophrenic patients: MR analysis of the coincidence between functional and morphologic abnormalities. *Radiology*, 244, 549–556.

McGuire, P.K., Bates, J.F., Goldman-Rakic, P.S. (1991a). Interhemispheric integration: I Symmetry and convergence of the corticocortical connections of the left and the right principal sulcus (PS) and the left and the right supplementary motor area (SMA) in the rhesus monkey. *Cerebral Cortex*, 1, 390–407.

McGuire, P.K., Bates, J.F., Goldman-Rakic, P.S. (1991b). Interhemispheric integration: II Symmetry and convergence of the corticostriatal projections of the left and the right principal sulcus (PS) and the left and the right supplementary motor area (SMA) of the rhesus monkey. *Cerebral Cortex*, 1, 408–417.

Mechelli, A., Friston, K.J., Frackowiak, R.S., Price, C.J. (2005). Structural covariance in the human cortex. *Journal of Neuroscience*, 25, 8303–8310.

Modinos, G., Vercammen, A., Mechelli, A., Knegtering, H., McGuire, P.K., Aleman, A. (2009). Structural covariance in the hallucinating brain: a voxel-based morphometry study. *Journal of Psychiatry & Neuroscience*, 34, 465–469.

Neckelmann, G., Specht, K., Lund, A., Ersland, L., Smievoll, A.I., Neckelmann, D., Hugdahl, K. (2006). MR morphometry analysis of grey matter volume reduction in schizophrenia: association with hallucinations. *International Journal of Neuroscience*, 116, 9–23.

Nenadic, I., Smesny, S., Schlosser, R.G., Sauer, H., Gaser, C. (2010). Auditory hallucinations and brain structure in schizophrenia: voxel-based morphometric study. *British Journal of Psychiatry*, 196, 412–413.

Northoff, G., Qin, P. (2011). How can the brain's resting state activity generate hallucinations? A 'resting state hypothesis' of auditory verbal hallucinations. *Schizophrenia Research*, 127, 202–214.

O'Daly, O.G., Frangou, S., Chitnis, X., Shergill, S.S. (2007). Brain structural changes in schizophrenia patients with persistent hallucinations. *Psychiatry Research*, 156, 15–21.

Onitsuka, T., Shenton, M.E., Salisbury, D.F., Dickey, C.C., Kasai, K., Toner, S.K., Frumin, M., Kikinis, R., Jolesz, F.A., McCarley, R.W. (2004). Middle and inferior temporal gyrus gray matter volume abnormalities in chronic schizophrenia: an MRI study. *American Journal of Psychiatry*, 161, 1603–1611.

Pettersson-Yeo, W., Allen, P., Benetti, S., McGuire, P., Mechelli, A. (2011). Dysconnectivity in schizophrenia: Where are we now? *Neuroscience and Biobehavioral Reviews*, 35, 1110–1124.

Plaze, M., Bartrés-Faz, D., Martinot, J.L., Januel, D., Bellivier, F., De Beaurepaire, R., Chanraud, S., Andoh, J., Lefaucheur, J.P., Artiges, E., Pallier, C., Paillère-Martinot, M.L. (2006). Left superior temporal gyrus activation during sentence perception negatively correlates with auditory hallucination severity in schizophrenia patients. *Schizophrenia Research*, 87, 109–115.

Plaze, M., Paillère-Martinot, M.L., Penttilä, J., Januel, D., de Beaurepaire, R., Bellivier, F., Andoh, J., Galinowski, A., Gallarda, T., Artiges, E., Olié, J.-P., Mangin, J.-F., Martinot, J.-L., Cachia, A. (2009). "Where do auditory hallucinations come from?" -- A brain morphometry study of schizophrenia patients with inner or outer space hallucinations. *Schizophrenia Bulletin*, 37, 212–221.

Raichle, M.E., MacLeod, A.M., Snyder, A.Z., Powers, W.J., Gusnard, D.A., Shulman, G.L. (2001). A default mode of brain function. *Proceedings of the National Academy of Sciences USA*, 98, 676–682.

Seal, M.L., Aleman, A., McGuire, P.K. (2004). Compelling imagery, unanticipated speech and deceptive memory: neurocognitive models of auditory verbal hallucinations in schizophrenia. *Cognitive Neuropsychiatry*, 9, 43–72.

Seok, J.H., Park, H.J., Chun, J.W., Lee, S.K., Cho, H.S., Kwon, J.S., Kim, J.J. (2007). White matter abnormalities associated with auditory hallucinations in schizophrenia: a combined study of voxel-based analyses of diffusion tensor imaging and structural magnetic resonance imaging. *Psychiatry Research*, 156, 93–104.

Shapleske, J., Rossell, S.L., Chitnis, X.A., Suckling, J., Simmons, A., Bullmore, E.T., Woodruff, P.W., David, A.S. (2002). A computational morphometric MRI study of schizophrenia: effects of hallucinations. *Cerebral Cortex*, 12, 1331–1341.

Shapleske, J., Rossell, S.L., Simmons, A., David, A.S., Woodruff, P.W. (2001). Are auditory hallucinations the consequence of abnormal cerebral lateralization? A morphometric MRI study of the sylvian fissure and planum temporale. *Biological Psychiatry*, 49, 685–693.

Shenton, M.E., Dickey, C.C., Frumin, M., McCarley, R.W. (2001). A review of MRI findings in schizophrenia. *Schizophrenia Research*, 49, 1–52.

Shergill, S.S., Kanaan, R.A., Chitnis, X.A., O'Daly, O., Jones, D.K., Frangou, S., Williams, S.C., Howard, R.J., Barker, G.J., Murray, R.M., McGuire, P. (2007). A diffusion tensor imaging study of fasciculi in schizophrenia. *American Journal of Psychiatry*, 164, 467–473.

Shin, S.E., Lee, J.S., Kang, M.H., Kim, C.E., Bae, J.N., Jung, G. (2005). Segmented volumes of cerebrum and cerebellum in first episode schizophrenia with auditory hallucinations. *Psychiatry Research*, 138, 33–42.

Silbersweig, D.A., Stern, E. (1998). Towards a functional neuroanatomy of conscious perception and its modulation by volition: implications of human auditory neuroimaging studies. *Philosophical Transactions of the Royal Society of London. Series B. Biological Sciences*, 353, 1883–1888.

Sommer, I.E., Diederen, K.M., Blom, J.D., Willems, A., Kushan, L., Slotema, K., Boks, M.P., Daalman, K., Hoek, H.W., Neggers, S.F., Kahn, R.S. (2008). Auditory verbal hallucinations predominantly activate the right inferior frontal area. *Brain*, 131, 3169–3177.

Sommer, I.E., Ramsey, N.F., Kahn, R.S. (2001). Language lateralization in schizophrenia, an fMRI study. *Schizophrenia Research*, 52, 57–67.

Sumich, A., Chitnis, X.A., Fannon, D.G., O'Ceallaigh, S., Doku, V.C., Faldrowicz, A., Sharma, T. (2005). Unreality symptoms and volumetric measures of Heschl's gyrus and planum temporal in first-episode psychosis. *Biological Psychiatry*, 57, 947–950.

Waters, F., Woodward, T., Allen, P., Aleman, A., Sommer, I. (2010). Self-recognition deficits in schizophrenia patients with auditory hallucinations: A meta-analysis of the literature. *Schizophrenia Bulletin*, doi: 10.1093/schbul/sbq144.

Whitford, T.J., Ford, J.M., Mathalon, D.H., Kubicki, M., Shenton, M.E. (2010). Schizophrenia, myelination, and delayed corollary discharges: A hypothesis. *Schizophrenia Bulletin*, doi: 10.1093/schbul/sbq105.

Zhang, S., Peng, H., Dawe, R.J., Arfanakis, K. (2011). Enhanced ICBM diffusion tensor template of the human brain. *Neuroimage*, 54, 974–984.

Chapter 20
Functional Neuroimaging of Hallucinations

André Aleman and Ans Vercammen

20.1 Introduction

The first functional neuroimaging studies of hallucinations appeared in the mid-1990s, just after PET and functional MRI began to be widely used as techniques for "capturing the brain in action." Positron emission tomography (PET, see Fig. 20.1) is a nuclear-medicine imaging technique that produces a three-dimensional image of functional processes in the body. The system detects pairs of gamma rays emitted by a positron-emitting radionuclide (tracer), which is introduced into the body on a biologically active molecule. Images of tracer concentration in three-dimensional space within the body are then reconstructed by a computer. If the biologically active molecule chosen for PET is fludeoxyglucose (FDG), an analogue of glucose, the concentrations of tracer then reflect tissue metabolic activity in terms of regional glucose uptake. For brain-activation studies, radiolabeled water is also frequently used (Cherry and Phelps 2002). Functional magnetic resonance imaging (fMRI) is a technique based on changes in local blood oxygenation that accompany neural activation (Huettel et al. 2004). This is referred to as the blood-oxygen-level-dependent (BOLD) signal. The advent of those neuroimaging techniques has spurred attempts to localize brain areas involved in the puzzling experience of hallucination. Sixteen years after the first study (Silbersweig et al. 1995), there are now dozens of studies that have indexed brain activity during the experience of hallucinations. In this chapter, we summarize the main findings of these studies and discuss their implications for our understanding of the neural basis of hallucinations. We will also briefly address the studies' implications for cognitive models of hallucination.

A. Aleman, Ph.D. (✉)
BCN Research School and University Medical Centre, Groningen, The Netherlands
e-mail: a.aleman@med.umcg.nl

A. Vercammen, Ph.D.
Neuroscience Research Australia, Randwick, Sydney, Australia
e-mail: a.vercammen@neura.edu.au

J.D. Blom and I.E.C. Sommer (eds.), *Hallucinations: Research and Practice*,
DOI 10.1007/978-1-4614-0959-5_20,

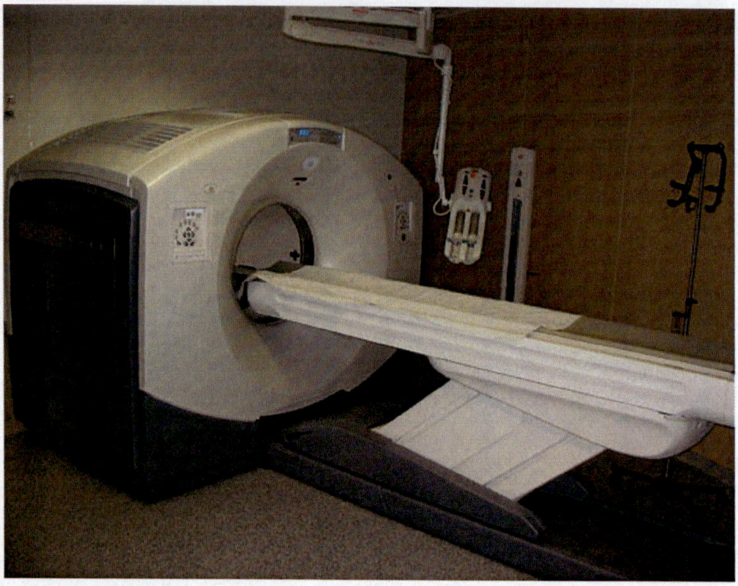

Fig. 20.1 Positron emission tomography (PET) scanner (Photograph courtesy of Wikimedia Commons, 2009. Reproduced with permission)

With regard to the neural basis of hallucinations, four different types of study can be distinguished. First, a significant number of studies have attempted to measure brain activity that directly correlates with the ongoing experience of hallucinations, as indicated by the patient with the aid of button presses. In such designs, a patient typically experiences intermittent hallucinatory episodes, which allows the investigator to contrast scans made during the hallucinatory episodes with those made during nonhallucinatory episodes. A second type of study is the perceptual-interference study in which auditory or visual stimuli are presented to patients who are actively hallucinating, as contrasted to patients without any hallucinations. This procedure allows researchers to determine to which extent hallucinations and external sensory stimuli utilize shared resources in brain areas involved in perception. A third type is the cognitive-activation study in which a particular cognitive task (e.g., speech monitoring) is carried out by groups of patients with and without hallucinations. This allows for the investigation of cognitive processes involved in the disposition toward hallucinations. Studies such as these can also be conducted in healthy individuals with a propensity to hallucinate. Finally, studies can focus on the functional connectivity between various brain regions. Synchronization of spontaneous neural activity across the brain, as observed in the low-frequency spectrum of the BOLD fMRI signal, is considered indicative of functional connectivity between those areas. Alternatively, correlations between time courses of different areas can be calculated. We will discuss these four types of study below. Most emphasis will be on studies of auditory hallucinations in patients diagnosed with schizophrenia (see also Chaps. 8–10), as these are the most prevalent ones in the published literature. Whenever possible, however, we will also discuss findings regarding visual hallucinations, e.g., as experienced by patients with Charles Bonnet syndrome (see also Chap. 6).

20.2 Brain Activity During Hallucinations

The first study to report on the activity of brain areas during episodes of hallucinations was published by McGuire et al. (1993). They obtained patterns of brain activity from 13 patients diagnosed with schizophrenia during periods when they experienced hallucinations and compared them to patterns obtained at a later occasion when the patients did not experience any hallucinations. The neuroimaging method they used was SPECT (single photon emission computed tomography), which is based on the same principles as PET, providing information about regional blood flow in the brain, although yielding images with a somewhat limited spatial resolution. McGuire et al. observed stronger activity in Broca's area in the hallucination condition than in the control condition. Two other areas, i.e., the anterior cingulate and left temporal cortex, also showed increased blood flow in the hallucination condition, although to a lesser extent than Broca's area. Suzuki et al. (1993) investigated a group of five patients and also used a design in which periods of hallucinatory activity were contrasted with those without. Their SPECT scans showed increased activity in left temporal cortex in the hallucination condition. A limitation of these studies is that when patients are scanned during hallucination-free episodes, other symptoms may also have improved, as well as the patients' general level of functioning. Therefore, the established differences in brain activity may not be exclusively attributable to differences in hallucinatory activity. The first study to circumvent this problem was published by Silbersweig et al. (1995), who were the first to use the so-called button-press method. This method involves the pressing of a button by the patients inside the scanner while they are hearing voices or experiencing visual hallucinations, and the release of the button when the hallucinations are subsiding. This allows the researcher to distinguish periods of hallucination-related activity from nonhallucination-related activity. However, as this method requires that patients experience various discrete periods of hallucinations during their stay in the scanner (typically an hour), only small groups can generally be included in such studies. As a consequence, the power of those studies tends to be limited, and their results should be interpreted with some caution. In their study of five patients with auditory hallucinations, Silbersweig et al. (1995) used PET scans to index concomitant brain activity and reported the involvement of a network of regions comprising the bilateral parahippocampal gyri, the right anterior cingulate gyrus, the left orbitofrontal cortex, the thalamus, the putamen, and the caudate nucleus. On the basis of their findings, the authors suggested that hallucinations may well originate from subcortical structures that spread activity toward cortical perceptual areas, which in turn determine the specific content of the percepts at hand.

Publications that appeared after these landmark studies generally confirmed the involvement of higher-order sensory areas (e.g., the superior temporal gyrus in the case of auditory hallucinations) and the involvement of a distributed network comprising multiple cortical and subcortical structures. For example, Lennox et al. (2000) reported activity in the bilateral superior temporal gyri in four patients who were diagnosed with schizophrenia and experienced auditory hallucinations, along with activity in left inferior parietal cortex and the left middle frontal gyrus. A minority of studies found activity in primary auditory cortex (Heschl's gyrus) during auditory hallucinations (Dierks et al. 1999; Van de Ven et al. 2005), whereas the involvement of the superior temporal gyrus was ubiquitous in all those studies. Shergill et al. (2000) argued that the results of button-press studies may perhaps be compromised by activity in those regions involved in the monitoring of voices (i.e., a task requirement). To overcome this drawback, they used a random sampling method in which a large number of individual scans were acquired at unpredictable time intervals in intermittently hallucinating individuals. Immediately after each scan, the subjects were asked whether they had experienced any hallucinations during that session. Thus participants retrospectively reported whether or not they had experienced any hallucinations. With the aid of this procedure, Shergill et al. found activity in the bilateral temporal gyri, which was stronger in the right hemisphere. Other areas included inferior frontal and insular cortex, the anterior cingulate, the right thalamus and inferior colliculus, and left hippocampal and parahippocampal cortex. Using spatial independent component analysis, Van de Ven et al. (2005) observed involvement of the auditory cortex (including Heschl's gyrus) in three out of six patients diagnosed with schizophrenia. The largest study to date was published by Sommer et al. (2008). It presents 24 patients diagnosed with schizophrenia who were investigated with the aid of the button-press method. Group analysis revealed activity in the right-hemisphere homotope of Broca's area, the bilateral insula, the bilateral supramarginal gyri, and the right superior temporal gyrus. The patients also performed a language task (consisting of verb generation) during their stay in the scanner. In contrast to the hallucination-related activity, this task mainly co-occurred with activity in the left hemisphere, notably Broca's area and the left superior temporal gyrus. Interestingly, the lateralization of brain activity during hallucinatory episodes did not correlate with the lateralization of language activation, but rather with the degree of the negative content of hallucinations. The authors also investigated the time course of the BOLD signal associated with hallucinatory activity (Diederen et al. 2010), reporting deactivation of the left parahippocampal gyrus preceding the onset of hallucinations. In addition, significant deactivation preceding the onset of hallucinations was found in the left superior temporal, right inferior frontal, and left middle frontal gyri, as well as in the right insula and the left cerebellum. They interpreted the involvement of the parahippocampus as indicative of memory retrieval and the cortical activity as the subsequent reception of information from the hippocampus by the auditory association areas. In a study among six patients diagnosed with schizophrenia, Hoffman et al. (2008) reported that prehallucination periods co-occurred with activity in the left anterior insula and the right middle temporal gyrus, as well as with deactivation of the anterior cingulate and parahippocampal gyri. Shergill et al. (2004) found increases in activity in the left inferior frontal and right middle temporal gyri 6–9s before the onset of hallucinations and no deactivations. Activity in the bilateral temporal gyri and the left insula coincided with the perception of the hallucination. The authors interpreted this as the generation of inner speech that may precede the engagement of areas implicated in the perception of auditory verbal material. Clearly, more studies are needed to establish the exact time course of hallucinatory activity, as well as the role of subcortical areas such as the thalamus (Silbersweig et al. 1995; Behrendt 1998; Aleman and Larøi 2008).

Jardri et al. (2011) integrated the evidence from ten neuroimaging studies indexing auditory-verbal-hallucination-related activity using PET or fMRI. For their meta-analysis, they used an activation likelihood estimation by combining the foci of activity reported across the various studies. Thus they assessed the combined findings in 68 patients diagnosed with a schizophrenia spectrum disorder and the stereotactic coordinates of 129 significant foci. Those patients who experienced auditory verbal hallucinations (AVH) showed significantly increased activation likelihoods in a bilateral neural network that included Broca's area, the anterior insula, the precentral gyrus, the frontal operculum, the middle and superior temporal gyri, the inferior parietal lobule, and the hippocampal/parahippocampal region. Although the network was bilateral, six out of eight activation clusters were found in the left hemisphere. The authors concluded that not only frontotemporal speech areas are involved in the mediation of AVH but also medial temporal areas associated with verbal memory.

In another meta-analysis, Kühn and Gallinat (2010) suggested a different neural substrate for state-versus-trait aspects of AVH. With "state" aspects, they meant the precise neural signature of hallucinations, as established by the comparison of periods of patient-reported hallucinations and absence of hallucinations in a within-subjects analysis (e.g., with the button-press method). With "trait" aspects, they referred to the comparison of brain activity among groups of patients with and without hallucinations, often during a cognitive task involving verbal material (we will discuss these studies in more detail in Sect. 20.3). Kühn and Gallinat (2010) included ten state and eight trait studies, and after meta-analytic integration of the results, they concluded that the state studies yielded activity in the bilateral inferior frontal gyri, the bilateral postcentral gyri, and the left parietal operculum. In contrast, the trait studies showed decreased activity in the left superior temporal gyrus, the left middle temporal gyrus, anterior cingulate cortex, and left premotor cortex. They suggested that the state of experiencing AVH is primarily related to speech-production regions such as Broca's area, whereas the trait that may render an individual prone to hallucinations is related to brain regions involved in auditory processing and speech perception.

With regard to visual hallucinations, studies (of which only a handful have been conducted) have consistently reported activity in extrastriate visual areas (Allen et al. 2008). For instance, ffytche et al. (1998) reported this in a study on visual hallucinations in patients with Charles Bonnet syndrome. Interestingly, they found that hallucinations of colors, faces, textures, and objects correlated with cerebral activity in ventral extrastriate visual cortex; that the content of the hallucinations reflected the functional specialization of the regions at hand (i.e., area V4 in the case of hallucinations in color and the fusiform face area in the case of hallucinated faces); that visual consciousness is a product of complex neuronal sequences influenced by top-down processing; and that those top-down processes may well take place in specialized areas of the brain rather than in distributed brain regions.

20.3 Perceptual-Interference Studies

Perceptual-interference studies investigate to which extent hallucinations and external sensory stimuli share common resources in brain areas involved in perception. Thus auditory or visual stimuli (depending on the sensory modality under study) are presented to patients who are actively hallucinating and to those who do not. If, for example, auditory hallucinations share a processing system with auditory sense perceptions, one would not expect an increase in activity upon external auditory stimulation in the auditory areas of patients actively experiencing AVH. In contrast, nonhallucinating patients can be expected to show an increase in the activity of auditory perceptual areas in response to external auditory stimulation. David et al. (1996) and Woodruff et al. (1997) confirmed this hypothesis in studies in which speech stimuli were presented to patients diagnosed with schizophrenia who experienced AVH. They found evidence of reduced responsivity of temporal cortex, notably the right middle temporal gyrus, to external speech during hallucinations, as compared to hallucination-free episodes. The authors conclude that "the auditory hallucinatory state is associated with reduced activity in temporal cortical regions that overlap with those that normally process external speech, possibly because of competition for common neurophysiological resources" (Woodruff et al. 1997, p. 1676). In another study in which patients with AVH were scanned while they were listening to external speech, Copolov et al. (2003) reported limbic regions as being more active in hallucinators. This observed pattern of activity may be interpreted as consistent with models of auditory hallucinations as misremembered episodic memories of speech. Plaze et al. (2006) replicated this finding in 15 patients diagnosed with schizophrenia who experienced hallucinations on a daily basis and who were investigated with the aid of fMRI while they were listening to spoken sentences. The severity of their hallucinations correlated negatively with activity in the left superior temporal gyrus in the speech-minus-silence condition. This suggests that auditory hallucinations would seem to compete with normal speech for processing resources in temporal cortex. A similar effect has been reported for visual hallucinations. Howard et al. (1995) reported reduced activity in visual cortex upon the presentation of visual stimuli concurrent with the experience of visual hallucinations in a patient diagnosed with schizophrenia. In their study with patients diagnosed with Charles Bonnet syndrome, ffytche et al. (1998) also found a decreased response in occipital cortex upon external visual stimulation during hallucinatory episodes. When external stimulation was presented during hallucination-free episodes in the same subjects, a normal increase in visual activity was detected.

20.4 Cognitive-Activation Studies of Processes Associated with Hallucinations

A number of studies have investigated brain activity during certain cognitive processes, such as language tasks, that may differ among people with a disposition toward hallucinations as compared to people without. Those studies aim at underlying mechanisms that may contribute to the mediation of hallucinations. Most of them have been directed at speech processing and verbal imagery in patients diagnosed with schizophrenia who were experiencing AVH. For example, McGuire et al. (1996) studied the neural correlates of inner speech and verbal imagery in patients diagnosed with schizophrenia, some of who did, and some of who did not experience AVH. The inner-speech task required volunteers to imagine speaking particular sentences. In the verbal-imagery task, they were asked to imagine sentences spoken in another person's voice, which, according to the authors, entails the monitoring of inner speech. During the verbal-imagery task, the hallucinators showed reduced activity in the left middle temporal gyrus and the rostral supplementary motor area, regions that were strongly activated in nonhallucinating, healthy volunteers and nonhallucinating patients. The authors concluded that a predisposition toward verbal hallucinations in psychosis is associated with a failure to activate areas implicated in the normal monitoring of inner speech. In an analogous study using fMRI, Shergill et al. (2000) investigated the functional anatomy of auditory verbal imagery in patients with AVH. They scanned patients diagnosed with schizophrenia who had a history of prominent AVH – as well as a healthy control group – while generating inner speech or imagining external speech. The patients showed no increased activity while they were generating inner speech, but concomitant with verbal imagery they showed a relatively attenuated response in the posterior cerebellar cortex, the hippocampi, the lenticular nuclei, the right thalamus, the temporal cortex, and the left nucleus accumbens. The authors concluded that this pattern of activity correlated with the monitoring of inner speech. These results were consistent with previous findings but suggested that a more distributed network of cerebellar and subcortical areas may be involved in the comparator function than previously assumed. In a parametric study of inner-speech generation, this group again examined brain areas implicated in the processing of inner speech in patients experiencing AVH (Shergill et al. 2003). The participants were trained to vary the rate of words during their inner-speech task. When the rate increased, the patients diagnosed with schizophrenia showed a relatively attenuated response in right temporal, parietal, parahippocampal, and cerebellar cortex. These findings were again interpreted as evidence of defective self-monitoring of inner speech in patients experiencing hallucinations. The healthy volunteers in that study showed activity in brain regions involved in speech generation (i.e., left inferior frontal cortex) and perception (temporal cortex) during the generation and monitoring of inner speech (Shergill et al. 2000, 2001). However, verbal self-monitoring seems to be particularly associated with temporal, parahippocampal, and cerebellar activity (Frith and Done 1988). Consistent with the self-monitoring hypothesis, patients prone to hallucinations showed relatively attenuated activity in these regions as compared to the control participants.

The neural correlates of explicit source/self-monitoring have also been addressed in a more direct way in healthy individuals and patients with and without hallucinations. McGuire et al. (1996) implemented a verbal self-monitoring task in a PET study with healthy volunteers. In the first condition, the volunteers were shown written words and asked to read them out loud. In the second condition, they were asked to read the words out loud while they were hearing the investigator saying the same word (alien feedback). In one half of the trials, the alien feedback was distorted by elevating the pitch. Distortion of the volunteers' speech while reading out loud led to bilateral activity of lateral temporal cortex. A similar pattern of activity was evident in the alien-feedback condition. These data suggest that self-generated and externally generated speech are processed in similar regions of temporal cortex. A subsequent fMRI study applying the same task to a healthy control group confirmed these results (Fu et al. 2006). In the latter study, the use of an event-related design allowed correct and misattributed source-judgment trials to be analyzed separately. Thus Fu et al. (2006) established that correct source attributions for self-generated speech were associated with greater temporal activity than misattributions, which supports the self-monitoring theory that a mismatch between expected (signaled via a feed-forward signal or "corollary discharge" signal, see Chap. 21) and perceived auditory feedback leads to an increase in temporal activity.

Recently, however, Frith's theory of self-monitoring as an explanatory model for AVH and other passivity phenomena has been criticized (see reviews by Pacherie et al. 2006, and Allen et al. 2007). In short, the model proposes that the experienced passivity results from a lack of awareness of having initiated an action and that the sense of externality results from a lack of sensory self-attenuation. But the model does not explain why a particular external author is experienced (Pacherie et al. 2006). Frith proposed that the experience of externality may be due to a default belief system. An alternative account by Jeannerod and Frak (1999), and by Jeannerod and Pacherie (2004), proposes that the attribution of one's own actions to an external agent is due to abnormal activity in neural networks involved in representing the actions of the self and others. Functional imaging evidence of such a shared system in humans has shown a significant overlap in the neural circuits involved in action execution and action observation (see Grézes and Decety (2001) for a review).

In an fMRI study in which the participants made judgments (self/other) about the source of prerecorded speech, Allen et al. (2007) studied the neural correlates of source misattribution in patients with and without AVH and in healthy controls. The patients with AVH were more likely to misattribute their own speech to an external source than the nonhallucinating patients and controls. Moreover, compared to both control groups, the patients experiencing hallucinations showed altered activity in the superior temporal gyrus and the anterior cingulate when making misattribution errors. The authors therefore suggested that the misidentification of self-generated speech in patients experiencing AVH is due to abnormal activity in the anterior cingulate and temporal cortex and may well be related to an impairment in the explicit evaluation of auditory verbal stimuli. An important finding of this study is the confirmation of a role for the anterior cingulate gyrus in source-monitoring processes.

An important question is whether inner speech generated by healthy individuals also activates speech-perception areas (i.e., Wernicke's region in left temporoparietal cortex). We investigated that issue using a performance-based task in healthy volunteers (Aleman et al. 2005). The results showed that making metrical stress judgments of visually presented words co-occurs with activity in speech-perception areas in the left superior temporal sulcus. The volunteers were asked to imagine hearing someone else reading a word out loud. The speech-perception areas did not show any increased activity in a control condition in which they were asked to make semantic judgments of the same visually presented words. This study suggests that auditory verbal imagery relies in part on phonological processing involving not only speech-production processes but also receptive processes subserved by temporal regions. Using the same task, we investigated the correlates with the severity of hallucinations in 24 patients diagnosed with schizophrenia (Vercammen et al. 2011). The results indicated that louder AVH were associated with reduced task-related activity in the bilateral angular gyri, the anterior cingulate gyrus, the left inferior frontal gyrus, the left insula, and the left temporal cortex. This might well be due to a competition for shared neural resources. On the other hand, the perceived reality of AVH was found to be associated with reduced language lateralization. Therefore, we concluded that activity in the inner-speech-processing network may contribute to the perceived loudness of AVH. However, right-hemisphere language areas may well be responsible for their more complex experiential characteristics, such as their apparent source and their perceived reality.

AVH also occur in nonpsychotic individuals in the absence of psychiatric or neurological disorder and/or substance abuse (see also Chap. 28). As such, the examination of verbal-imagery processes and hallucinatory processes in these individuals from the general population could shed light on the underlying mechanisms of hallucinations in relation to "normal" auditory-verbal processes, and on the extent to which AVH in psychotic patients differ from these subclinical experiences. Recently, Linden et al. (2011) carried out a functional imaging study of AVH and auditory imagery in seven healthy voice hearers. Using the button-press method, they found activity in the human-voice area in the superior temporal sulcus during both hallucinatory and imagery episodes. Other brain areas supporting both hallucinations and imagery included the frontotemporal language areas in the left hemisphere and their contralateral homologues, as well as the supplementary motor area (SMA). As hallucinations tend to be distinguished from imagery by the percipient's supposed lack of voluntary control, which is in turn thought to be represented by the prefrontal cortex, Linden et al. investigated whether that difference would be reflected in the relative timing of prefrontal and sensory areas. Activity of the SMA indeed preceded that of auditory areas during imagery, whereas during hallucinatory episodes, the two processes occurred simultaneously. Therefore, they suggested that voluntary control might be represented by the relative timing of prefrontal and sensory activation, whereas the sense of reality of the experiences may be a product of the activity in the voice area. Notably, this study did not report on the dorsolateral prefrontal cortex, which is also an important area for the willful extraction of perceptual information from memory (Kosslyn 1994). Indeed, it is remarkable that none of the functional neuroimaging studies of active hallucinations report on the involvement of the dorsolateral prefrontal cortex, whereas the mental-imagery studies generally do.

Although there appear to be some consistent differences between verbal imagery and hallucinations, new evidence suggests that subclinical hallucinations, as observed in healthy individuals, and the psychotic experiences of patients diagnosed with schizophrenia, are not easily differentiated on the basis of brain activity patterns. Diederen et al. (2011) conducted an fMRI study in 21 nonpsychotic subjects with AVH and 21 matched psychotic patients, who were asked to indicate the onset and offset of hallucinatory episodes with the aid of button presses. Interestingly, there were no significant differences in AVH-related brain activity between the groups. Common areas of activity for the psychotic and nonpsychotic groups during the experience of AVH comprised the bilateral inferior frontal gyri, the insula, the superior temporal gyri, the supramarginal and postcentral gyri, the left precentral gyrus, the inferior parietal lobule, the superior temporal pole, and the right cerebellum. These findings implicate the involvement of the same cortical network in the experience of AVH in both groups. It has been suggested that some experiential characteristics may still differentiate subclinical from clinical hallucinations. One such characteristic may be affective loading.

Escartí et al. (2010) focused on the emotional aspects of AVH in patients diagnosed with schizophrenia, by presenting emotionally laden words to patients with and without hallucinations, and to a group of healthy control subjects. In patients experiencing hallucinations, the parahippocampal gyrus and the amygdala were more strongly involved in the task-related network than in patients without hallucinations and in control subjects. This is consistent with the suggestion made by Aleman and Kahn (2005) that increased amygdala activation may well contribute to the mediation of positive symptoms in psychosis. Further research examining AVH in clinical and subclinical populations is warranted to clarify whether limbic contributions to AVH-associated activity can distinguish psychotic features from subclinical hallucinatory experiences.

20.5 Functional Connectivity and Hallucinations

Functional connectivity can be defined as a cross-correlation over time between spatially remote brain regions (Friston et al. 1993). Another concept is *effective* connectivity, which indicates the contributory influence of each region on a different one (Bullmore et al. 2000, Friston et al. 1996). In a study of functional connectivity by Lawrie et al. (2002), eight patients diagnosed with schizophrenia and ten control volunteers were studied with fMRI while they thought of the missing last word in 128 visually presented sentences. Although there were no differences in the regional brain responses among the two groups, correlation coefficients between left temporal cortex and left dorsolateral prefrontal cortex were significantly lower in the patients diagnosed with schizophrenia and were negatively correlated with the severity of AVH. Thus frontotemporal functional connectivity may well be reduced in patients diagnosed with schizophrenia and may also be associated with auditory hallucinations.

As regards effective connectivity, Mechelli et al. (2007) studied patients with and without AVH, as well as healthy volunteers. Using the fMRI data previously reported by Allen et al. (2007), they tested the hypothesis that source misattributions are associated with poor functional integration within the network of regions that mediate the evaluation of speech. In the healthy volunteers and patients without AVH, the connectivity between left superior temporal and anterior cingulate cortex was found to be significantly greater for alien speech than for self-generated speech. In contrast, a reverse trend was found in patients experiencing AVH. The authors concluded that in the latter group, the tendency to misattribute one's own speech to an external source is associated with an impaired effective connectivity between left superior temporal and anterior cingulate cortex. Although this finding is based on external rather than internal speech, a similar mechanism may well underlie the default appraisal of inner speech in AVH.

Five other studies have investigated functional connectivity in relation with hallucinations. Gavrilescu et al. (2010) restricted their analysis to cross-hemisphere resting functional connectivity, linking primary and secondary auditory cortices. They reported reductions in the hallucinating patients when compared with non-hallucinating patients and healthy control subjects. Vercammen et al. (2010) assessed functional connectivity during the resting state relative to a bilateral seed region located at the temporoparietal junction to compare patients with active AVHs and healthy control subjects. The patients showed a reduced connectivity of temporoparietal cortex and the right inferior frontal gyrus. Within the patient group, the severity of AVHs was correlated with the degree of reduction in the connectivity of temporoparietal cortex and anterior cingulate cortex. A study by Raij et al. (2009) found that subjective ratings of the perceived reality of AVHs were positively correlated with enhanced coupling between the left inferior frontal gyrus and bilateral auditory cortices, as well as the posterior temporal lobes. Finally, Hoffman et al. (2011) found greater connectivity between Wernicke's and Broca's areas (Wernicke's being the seed region) for hallucinating patients, as compared to nonhallucinating patients, but not compared to healthy control subjects during the resting state. They also found some evidence for greater connectivity summed along a loop linking Wernicke's and Broca's seed regions and the putamen for hallucinating patients as compared to nonhallucinating patients and healthy control subjects. They suggested that higher levels of functional coordination intrinsic to a corticostriatal loop may be an underlying factor in the mediation of AVHs. Their findings are complicated, however, by the fact that various relevant connectivity parameters in the hallucinating patients did not differ from those in the healthy controls and that for others they did not differ from those in nonhallucinating patients.

To summarize, although studies do not converge in all details, all published studies provide evidence for disrupted connectivity between temporal, prefrontal, and anterior cingulate cortex. Thus, alterations in the functional integration of language networks and attentional networks would seem to be crucial to the neural substrates of perceiving sounds that "are not there." These findings are relatively unsurprising, given the fact that schizophrenia has been described as a disorder of "disconnection" (Friston and Frith 1995), characterized by profound disruptions of larger-scale prefrontotemporal interactions. And yet further work is clearly required to parse out the exact nature of the disturbance. More research is needed to establish the nature of abnormalities in connectivity which may manifest as reduced versus increased connectivity depending on the networks studied and the task characteristics (or resting state) involved.

20.6 Conclusion

Functional neuroimaging techniques such as PET and fMRI have been employed successfully over the past two decades to reveal brain areas involved in the mediation of hallucinations. Numerous studies converge on the involvement of bilateral secondary sensory areas. Auditory verbal hallucinations in patients diagnosed with schizophrenia (and in those with schizophrenia spectrum disorders) have been studied most extensively. These studies have shown a central role for speech-production and speech-perception areas not only in the left hemisphere (see Fig. 20.2) but also their homotopes in the right hemisphere. The parahippocampal gyrus is also involved

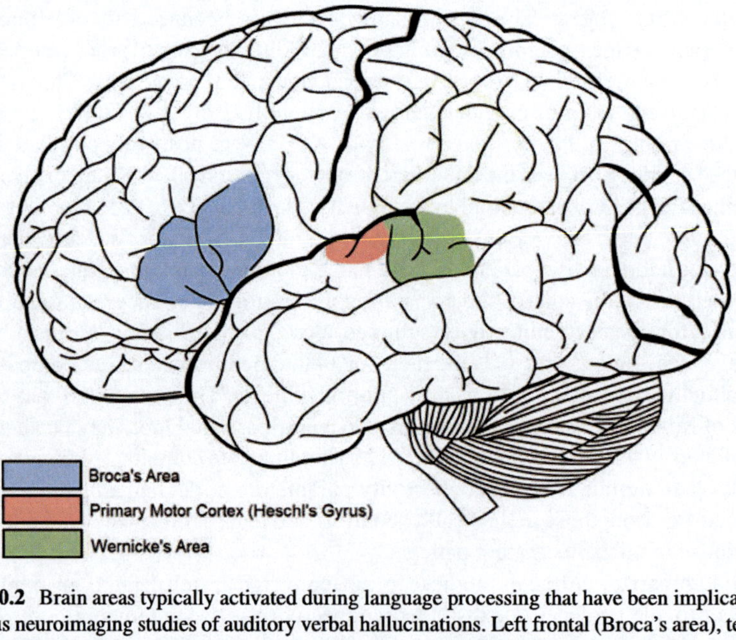

Fig. 20.2 Brain areas typically activated during language processing that have been implicated in various neuroimaging studies of auditory verbal hallucinations. Left frontal (Broca's area), temporal/parietal (Wernicke's area), and primary auditory cortex (Heschl's gyrus) (Brain map courtesy of the Free Software Foundation, 2010. Reproduced with permission)

in such hallucinations, presumably even in their initiation. Other relevant areas are the anterior cingulate, the insula, the cerebellum, and the thalamus. In conclusion, auditory verbal hallucinations would seem to depend primarily on distributed brain networks involved in perceptual attention and memory (which may well reflect top-down processing), in addition to various modality-specific sensory areas. Especially the role of monitoring systems and of the emotional connotations of hallucinations deserve further elucidation.

References

Aleman, A., Formisano, E., Koppenhagen, H., Hagoort, P., de Haan, E.H., Kahn, R.S. (2005). The functional neuroanatomy of metrical stress evaluation of perceived and imagined spoken words. *Cerebral Cortex*, 15, 221–228.

Aleman, A., Kahn, R.S. (2005). Strange feelings: Do amygdala abnormalities dysregulate the emotional brain in schizophrenia? *Progress in Neurobiology*, 77, 283–298.

Aleman, A., Larøi, F. (2008). *Hallucinations: The science of idiosyncratic perception*. Washington, DC: American Psychological Association.

Allen, P., Larøi, F., McGuire, P.K., Aleman, A. (2008). The hallucinating brain: A review of structural and functional neuroimaging studies of hallucinations. *Neuroscience and Biobehavioral Reviews*, 32, 175–191.

Behrendt, R.-P. (1998). Underconstrained perception: A theoretical approach to the nature and function of verbal hallucinations. *Comprehensive Psychiatry*, 39, 236–248.

Bullmore, E., Horwitz, B., Honey, G., Brammer, M., Williams, S., Sharma, T. (2000). How good is good enough in path analysis of fMRI data? *NeuroImage*, 11, 289–301.

Cherry, S.R., Phelps, M.E. (2002). Imaging brain function with positron emission tomography. In Toga A.W. & Mazziotta J.C. (Eds), Brain mapping: the methods. Amsterdam: Academic Press.

Copolov, D.L., Seal, M.L., Maruff, P., Ulusoy, R., Wong, M.T.H., Tochon-Danguy, H.J., Egan, G.F. (2003). Cortical activation associated with the experience of auditory hallucinations and perception of human speech in schizophrenia: A PET correlation study. *Psychiatry Research. Neuroimaging*, 122, 139–152.

David, A.S., Woodruff, P.W.R., Howard, R, Mellers, J.D., Brammer, M., Bullmore, E., Wright, I., Andrew, C., Williams, S.C. (1996). Auditory hallucinations inhibit exogenous activation of auditory association cortex. *NeuroReport*, 7, 932–936.

Diederen, K.M., Daalman, K., De Weijer, A.D., Neggers, S.F., Van Gastel, W., Blom, J.D., Kahn, R.S., Sommer, I.E. (2011). Auditory hallucinations elicit similar brain activation in psychotic and nonpsychotic Individuals. *Schizophrenia Bulletin* (Epub ahead of print).

Diederen, K.M., Neggers, S.F., Daalman, K., Blom, J.D., Goekoop, R., Kahn, R.S., Sommer, I.E. (2010). Deactivation of the parahippocampal gyrus preceding auditory hallucinations in schizophrenia. *American Journal of Psychiatry*, 167, 427–435.

Dierks, T., Linden, D.E., Jandl, M., Formisano, E., Goebel, R., Lanfermann, H., Singer, W. (1999). Activation of Heschl's gyrus during auditory hallucinations. *Neuron*, 22, 615–621.

Escartí, M.J., delaIglesia-Vayá, M., Martí-Bonmatí, L., Robles, M., Carbonell, J., Lull, J.J., García-Martí, G., Manjón, J.V., Aguilar, E.J., Aleman, A., Sanjuán, J. (2010). Increased amygdala and parahippocampal gyrus activation in schizophrenic patients with auditory hallucinations: An fMRI study using independent component analysis. *Schizophrenia Research*, 117, 31–41.

ffytche, D.H., Howard, R.J., Brammer, M.J., David, A., Woodruff, P., Williams, S. (1998). The anatomy of conscious vision: An fMRI study of visual hallucinations. *Nature. Neuroscience*, 1, 738–742.

Friston, K.J., Frith, C.D. (1995). Schizophrenia: A disconnection syndrome? *Clinical Neuroscience*, 3, 89–97.

Friston, K.J., Frith, C.D., Liddle, P.F., Frackowiak, R.S. (1993). Functional connectivity: the principal-component analysis of large (PET) data sets. *Journal of Cerebral Blood Flow Metabolism*, 13, 5–14.

Friston, K.J., Frith, C.D., Fletcher, P., Liddle, P.F., Frackowiak, R.S. (1996). Functional topography: multidimensional scaling and functional connectivity in the brain. *Cerebral Cortex*, 6, 156–164.

Frith, C.D., Done, D.J. (1988). Towards a neuropsychology of schizophrenia. *British Journal of Psychiatry*, 153, 437–443.

Fu, C.H., Vythelingum, G.N., Brammer, M.J., Williams, S.C., Amaro, E., Jr., Andrew, C.M., Yágüez, L., Van Haren, N.E., Matsumoto, K., McGuire, P.K. (2006). An fMRI study of verbal self-monitoring: Neural correlates of auditory verbal feedback. *Cerebral Cortex*, 16, 969–977.

Gavrilescu, M., Rossell, S., Stuart, G.W., Shea, T.L., Innes-Brown, H., Henshall, K., McKay, C., Sergejew, A.A., Copolov, D., Egan, G.F. (2010). Reduced connectivity of the auditory cortex in patients with auditory hallucinations: A resting state functional magnetic resonance imaging study. *Psychological Medicine*, 40, 1149–1158.

Grèzes, J., Decety, J. (2001). Functional anatomy of execution, mental simulation, observation, and verb generation of actions: A meta-analysis. *Human Brain Mapping*, 12, 1–19.

Hoffman, R.E., Anderson, A.W., Varanko, M., Gore, J.C., Hampson, M. (2008). Time course of regional brain activation associated with onset of auditory/verbal hallucinations. *British Journal of Psychiatry*, 193, 424–425.

Hoffman, R.E., Fernandez, T., Pittman, B., Hampson, M. (2011). Elevated functional connectivity along a corticostriatal loop and the mechanism of auditory/verbal hallucinations in patients with schizophrenia. *Biological Psychiatry*, 69, 407–414.

Howard, R., Williams, S., Bullmore, E., Brammer, M., Mellers, J., Woodruff, P., David, A. (1995). Cortical response to exogenous visual stimulation during visual hallucinations. *Lancet*, 345, 70.

Huettel, S.A, Song, A.W., McCarthy, G. (2004). *Functional magnetic resonance imaging*. Sunderland, MA: Sinauer Associates Publishers.

Jardri, R., Pouchet, A., Pins, D., Thomas, P. (2011). Cortical activations during auditory verbal hallucinations in schizophrenia: A coordinate-based meta-analysis. *American Journal of Psychiatry*, 168, 73–81.

Jeannerod, M., Frak, V. (1999). Mental imaging of motor activity in humans. *Current Opinion in Neurobiology*, 9, 735–739.

Jeannerod, M., Pacherie, E. (2004). Agency, simulation and self-identification. *Mind and Language*, 19, 113–146.

Kosslyn, S.M. (1994). *Image and brain.* Cambridge, MA: The MIT Press.

Kühn S, Gallinat J. (2010). Quantitative meta-analysis on state and trait aspects of auditory verbal hallucinations in schizophrenia. *Schizophrenia Bulletin*, doi: 10.1093/schbul/sbq152.

Lawrie, S.M., Buechel, C., Whalley, H.C., Frith, C.D., Friston, K.J., Johnstone, E.C. (2002). Reduced frontotemporal functional connectivity in schizophrenia associated with auditory hallucinations. *Biological Psychiatry*, 51, 1008–1011.

Lennox, B.R., Park, S.B., Medley, I., Morris, P.G., Jones, P.B. (2000). The functional anatomy of auditory hallucinations in schizophrenia. *Psychiatry Research*, 100, 13–20.

Linden, D.E., Thornton, K., Kuswanto, C.N., Johnston, S.J., Van de Ven, V., Jackson, M.C. (2011). The brain's voices: Comparing nonclinical auditory hallucinations and imagery. *Cerebral Cortex*, 21, 330–337.

McGuire, P.K., Shah, G.M., Murray, R.M. (1993). Increased blood flow in Broca's area during auditory hallucinations in schizophrenia. *Lancet*, 342, 703–706.

McGuire, P.K., Silbersweig, D.A., Wright, I., Murray, R.M., Frackowiak, R.S., Frith, C.D. (1996). The neural correlates of inner speech and auditory verbal imagery in schizophrenia: Relationship to auditory verbal hallucinations. *British Journal of Psychiatry*, 169, 148–159.

Mechelli, A., Allen, P., Amaro, E., Jr., Fu, C.H., Williams, S.C., Brammer, M.J., Johns, L.C., McGuire, P.K. (2007). Misattribution of speech and impaired connectivity in patients with auditory verbal hallucinations. *Human Brain Mapping*, 28, 1213–1222.

Plaze, M., Bartrés-Faz, D., Martinot, J.L., Januel, D., Bellivier, F., De Beaurepaire, R., Chanraud, S., Andoh, J., Lefaucheur, J.P., Artiges, E., Pallier, C., Paillère-Martinot, M.L. (2006). Left superior temporal gyrus activation during sentence perception negatively correlates with auditory hallucination severity in schizophrenia patients. *Schizophrenia Research*, 87, 109–115.

Pacherie, E., Green, M., Bayne, T. (2006). Phenomenology and delusions: Who put the 'alien' in alien control? *Consciousness and Cognition*, 15, 566–577.

Raij, T.T., Valkonen-Korhonen, M., Holi, M., Therman, S., Lehtonen, J., Hari, R., (2009). Reality of auditory verbal hallucinations, *Brain*, 132, 2994–3001.

Shergill, S.S., Brammer, M.J., Amaro, E., Williams, S.C., Murray, R.M., McGuire, P.K. (2004). Temporal course of auditory hallucinations. *British Journal of Psychiatry*, 185, 516–517.

Shergill, S.S., Brammer, M.J., Fukuda, R., Williams, S.C., Murray, R.M., McGuire, P.K. (2003). Engagement of brain areas implicated in processing inner speech in people with auditory hallucinations. *British Journal of Psychiatry*, 182, 525–531.

Shergill, S.S., Bullmore, E.T., Brammer, M.J., Williams, S.C., Murray, R.M., McGuire, P.K. (2001). A functional study of auditory verbal imagery. *Psychological Medicine*, 31, 241–253.

Shergill, S.S., Bullmore, E., Simmons, A., Murray, R., McGuire, P. (2000). Functional anatomy of auditory verbal imagery in schizophrenic patients with auditory hallucinations. *American Journal of Psychiatry*, 157, 1691–1693.

Shergill, S.S., Cameron, L.A., Brammer, M.J., Williams, S.C., Murray, R.M., McGuire, P.K. (2001). Modality specific correlates of auditory and somatic hallucinations. *Journal of Neurology, Neurosurgery, and Psychiatry*, 71, 688–690.

Silbersweig, D.A., Stern, E., Frith, C., Cahill, C., Holmes, A., Grootoonk, S., Seaward, J., McKenna, P., Chua, S.E., Schnorr, L., Jones, T., Frackowiak, S.J. (1995). A functional neuroanatomy of hallucinations in schizophrenia. *Nature*, 378, 176–179.

Sommer, I.E.C., Diederen, K.M.J., Blom, J.-D., Willems, A., Kushan, L., Slotema, K., Boks, M.P.M., Daalman, K., Hoek, H.W., Neggers, S.F.W., Kahn, R.S. (2008). Auditory verbal hallucinations predominantly activate the *right* inferior frontal area. *Brain*, 131, 3169–3177.

Suzuki, M., Yuasa, S., Minabe, Y., Murata M, Kurachi M. (1993). Left superior temporal blood flow increases in schizophrenic and schizophreniform patients with auditory hallucination: A longitudinal case study using 123I-IMP SPECT. *European Archives of Psychiatry and Clinical Neurosciences*, 242, 257–261.

Van de Ven, V.G., Formisano, E., Roder, C.H., Prvulovic, D., Bittner, R.A., Dietz, M.G., Hubl, D., Dierks, T., Federspiel, A., Esposito, F., Di Salle, F., Jansma, B., Goebel, R., Linden, D.E. (2005). The spatiotemporal pattern of auditory cortical responses during verbal hallucinations. *NeuroImage*, 27, 644–655.

Vercammen, A., Knegtering, H., Den Boer, J.A., Liemburg, E.J., Aleman, A. (2010). Auditory hallucinations in schizophrenia are associated with reduced functional connectivity of the temporo-parietal area. *Biological Psychiatry*, 67, 912–918.

Vercammen, A., Knegtering, H., Bruggeman, R., Aleman, A. (2011). Subjective loudness and reality of auditory verbal hallucinations and activation of the inner speech processing network. *Schizophrenia Bulletin*, 37, 1009–1016.

Woodruff, P.W., Wright, I.C., Bullmore, E.T., Brammer, M., Howard, R.J., Williams, S.C., Shapleske, J., Rossell, S., David, A.S., McGuire, P.K., Murray, R.M. (1997). Auditory hallucinations and the temporal cortical response to speech in schizophrenia: A functional magnetic resonance imaging study. *American Journal of Psychiatry*, 154, 1676–1682.

Chapter 21
Neurophysiological Research: EEG and MEG

Remko van Lutterveld and Judith M. Ford

21.1 Introduction

Imagine you are driving your car through a quiet suburb. Suddenly a soccer ball bounces onto the street with a child chasing it. Visual information about the unfolding scene travels from your eyes to the visual processing centers in the brain, and after involvement of many other brain regions, the motor cortex sends a signal to your foot to slam the brakes. The electrical activity of the brain processes involved in your rapid response takes place in a flash. Electroencephalography (EEG) and magnetoencephalography (MEG) are the only noninvasive neuroimaging techniques that allow tracking of such fast-changing brain activity (see Figs. 21.1 and 21.2).

EEG measures the electrical signals produced by groups of neurons in the brain, and MEG measures the concurrent magnetic signals elicited by these electrical signals. Both EEG and MEG are able to track brain activity on a millisecond timescale, with the same temporal resolution as the neural activity itself. For source localization, however, MEG may be a more suitable technique than EEG. Electrical signals related to neuronal activity are smeared out by the skull, hampering accurate EEG source localization, while magnetic signals measured by MEG are not substantially affected by the skull.

The focus of this chapter will be on auditory hallucinations because they are reported more often than visual, gustatory, or somatic hallucinations, and are a cardinal symptom of psychosis. Over time, several approaches evolved to study auditory

R. van Lutterveld, M.Sc. (✉)
Department of Psychiatry, University Medical Center, Utrecht, The Netherlands and Rudolf Magnus Institute of Neuroscience, Utrecht, The Netherlands
e-mail: R.vanlutterveld@umcutrecht.nl

J.M. Ford, Ph.D.
Brain Imaging and EEG Laboratory, VA Medical Center, Department of Psychiatry, University of California San Francisco, San Francisco, CA, USA
e-mail: Judith.Ford@ucsf.edu

J.D. Blom and I.E.C. Sommer (eds.), *Hallucinations: Research and Practice*,
DOI 10.1007/978-1-4614-0959-5_21, © Springer Science+Business Media, LLC 2012

Fig. 21.1 Research subject wearing an electroencephalography (EEG)-recording cap

Fig. 21.2 Magnetoencephalography (MEG) recordings. Every line represents data from an MEG sensor. Electroencephalography (EEG) recordings look similar

hallucinations using neurophysiological methods. The most intuitive strategy is to use symptom capture, in which patients indicate the presence of hallucinations. Brain activity during hallucinatory episodes is then compared to hallucination-free episodes. A second approach is to combine symptom capture with event-related potentials (ERPs) to assess the processing of auditory information during the active "state" of an auditory hallucination.[1] A third approach associates ERPs with the tendency or the "trait" to hallucinate. In the latter approach, the severity of hallucinations is correlated with an ERP index of auditory processing. In a fourth approach, repetitive transcranial magnetic stimulation (rTMS, a method that applies magnetic pulses to the brain in order to activate or deactivate brain regions, see also Chap. 25) is used to study EEG measures in the context of hallucinations. A fifth approach is to study basic neurophysiological mechanisms that may underlie the tendency to hallucinate. Each of these approaches will be described in detail.

21.2 Symptom-Capture Studies

21.2.1 Early Symptom-Capture Studies

Before the era of antipsychotic medications, depth-electrocorticography (ECoG) studies were sometimes conducted in conjunction with neurosurgery for relief of severe psychotic symptoms. In one such ECoG study, Sem-Jacobsen et al. (1955) reported that "a close relationship between the patient's acute episodes of psychotic behavior and the electric activity was found." As they continued, "The findings in this study draw attention to the presence of focal spike discharges in some chronically psychotic patients during episodes of disturbance or hallucinations or both, and to the presence of changes in the activity of the temporal lobe and probably the frontal lobe during hallucinations."

Thirteen years later, Marjerrison et al. (1968) used scalp-recorded EEG for the first time to capture the electrophysiological signal associated with auditory hallucinations. They reported that newly admitted or readmitted patients diagnosed with acute schizophrenia, who experienced hallucinations during the experiment, had lower variation in EEG-related brain activity than similar patients who were not hallucinating during the experiment. In the next two decades, EEG studies investigating hallucinations were scarce. In the 1970s, Whitton et al. (1978) recorded the spectral power preceding an auditory hallucination in six unmedicated patients. This was compared to EEG power preceding a response in healthy controls performing tests of creativity. They reported that EEG power was predominant in the delta and theta bands in the 4-s interval prior to reports of hallucinations and creative responses, and suggested that the intrusiveness of the hallucinatory experience may be similar to the sudden internal experience of solving a creative task.

[1]ERPs are measured with EEG. Its MEG counterparts are called event-related fields (ERFs).

In a landmark telemetry study, Stevens et al. (1979) equipped patients diagnosed with schizophrenia with EEG electrodes, and the EEG signal was sent through radio waves to a base station. With this system, the patients were able to walk freely about the ward or dayroom. Hallucinatory behavior (e.g., muttering) was coded by a trained observer, enabling the comparison of hallucination episodes with nonhallucination episodes ("symptom capture"). In this study, Stevens and her team published EEG recordings of a hallucinating patient, reporting power increases during hallucinations in all frequency bands and scalp derivations with the exception of alpha in the left temporal region. In a follow-up study using the same paradigm, Stevens and Livermore (1982) reported that hallucinations correlated with the presence of ramp spectra in the EEG, i.e., spectra characterized by a smooth decline in power from lowest to highest frequencies. According to the authors, such spectra have previously been found in conjunction with subcortical spike activity of epilepsy, suggesting hallucinations were present subsequent to some abnormal subcortical discharge.

On a methodological note, Serafetinides et al. (1986) investigated the influence of verbal versus button-press methods to indicate auditory hallucinations on oscillations in the EEG. The method used to determine the presence of hallucinations had a marked effect on the EEG results. Verbal reporting was associated with a bilateral increase of high-frequency activity, while nonverbal reporting was associated with an asymmetry in power between the left and right hemisphere. After this report was published, no study made use of verbal reporting of hallucinations anymore.

21.2.2 Contemporary Symptom-Capture Studies

With the advent of better analysis algorithms and greater computing power, EEG and MEG data can be decomposed into precise information in the time-frequency domain, while also providing better spatial resolution than the older clinical EEG methods. However, modern EEG and magnetoencephalography (MEG) symptom-capture studies investigating hallucinations are scarce. To date, only one EEG study and three MEG studies have been published. In the EEG study, Sritharan et al. (2005) reported an increase in alpha band power in the left superior temporal cortex during auditory hallucinations in seven patients diagnosed with schizophrenia. Moreover, an increase in synchronization between the left and right superior temporal cortices was found during auditory hallucinations, suggesting an increase in functional coupling between these brain regions during hallucinations.

Ishii et al. (2000) were the first to investigate auditory hallucinations using MEG in a symptom-capture design. In a case study, they reported an increase in theta-band activity in the left superior temporal cortex during hallucinations. In another case study, the same structure was implicated, albeit in the beta band (Ropohl et al. 2004). Reulbach et al. (2007) studied five patients with nonverbal auditory hallucinations (e.g., noise, music) and three patients with command hallucinations. Hallucinations in the former group were associated with an increase in beta-band activity in the left superior temporal cortex, while hallucinations in the latter group were associated with the same activation pattern extending into the left dorsolateral prefrontal cortex. According to the authors, these findings suggest that the lack of frontal-lobe involvement in nonverbal auditory hallucinations could be interpreted as a sign of diminished cortical involvement compared to the complex mechanisms involved in the generation of voices.

21.3 Combined ERP/ERF-Symptom-Capture Studies

Another approach to study auditory hallucinations is to combine symptom capture with EEG-based event-related potentials (ERPs). ERPs are evoked by a stimulus, and usually occur within 1 s after stimulus presentation. With the combined symptom-capture-ERP method, ERPs are studied during hallucinatory episodes and compared to ERPs during nonhallucinatory episodes. An ERP often used in this approach is the N100 component. The N100 is generated in the auditory cortex (Hari et al. 1984) and is considered to be a standard metric of auditory-cortex activation. As such, the N100 provides the opportunity to compare auditory-cortex activity during the hallucinatory state with activity during the nonhallucinatory state. Tiihonen et al. (1992) measured the N100 amplitude and latency to tones presented to two patients suffering from intense transitory auditory hallucinations. In both patients, the N100 was delayed during the experience of auditory hallucinations compared to when the patients were not hallucinating. In one of these patients, the N100 amplitude was also lower during hallucinations. In a larger study, Hubl et al. (2007) investigated the N100 amplitude in seven patients with a psychotic disorder with acute auditory hallucinations and found smaller amplitudes during hallucinations. Moreover, the largest differences in N100-source strength between episodes with and without hallucinations were located in the left superior temporal cortex. The authors concluded that these findings indicate competition between auditory stimuli and auditory hallucinations for physiological resources located in the primary auditory cortex and that abnormal activation of this brain region could be a component of auditory hallucinations.

Line et al. (1998) took advantage of the rapid timescale of EEG to study the timeframe surrounding auditory hallucinations. They presented eight patients diagnosed with schizophrenia with flickering visual stimuli, leading to the generation of electrical activity in the brain at the same frequency of the flashing stimulus. In the second before the onset of an auditory hallucination, patients showed a large and significant decrease in the latency of brain responses in the right temporoparietal area, suggesting involvement of this area in the genesis of hallucinations.

In a recent EEG study, transiently stable neuronal states were investigated (Kindler et al. 2011). The authors found that a so-called microstate associated with error monitoring was shorter during hallucinatory episodes compared to nonhallucinatory episodes. The authors speculated that the early termination of this microstate facilitated the misattribution of self-generated inner speech to external sources during hallucinations.

21.4 Associations Between Hallucinatory Trait and EEG/MEG Measures

Yet another strategy to study hallucinations is to investigate the association between EEG and MEG measures and the *tendency* to hallucinate. Lee et al. (2006) used quantitative EEG and source imaging to investigate 25 patients diagnosed with schizophrenia experiencing treatment-refractory auditory hallucinations and 23 who were hallucination-free for at least 2 years. Resting-state EEG in the hallucinating patients showed significantly increased beta-band activity in the left inferior parietal lobule and the left medial frontal gyrus compared to nonhallucinating patients. Moreover, gamma and beta frequencies were significantly correlated in hallucinating patients but not in nonhallucinating patients. The authors suggested that the strong correlation between gamma- and beta-frequency oscillations may indicate that the brains of hallucinating patients act as if they were experiencing real auditory stimulation, as previous studies have shown strong correlations between gamma- and beta-frequency oscillations in normal populations in response to auditory stimuli (Haenschel et al. 2000).

Various authors have used ERPs to study associations with auditory hallucinations. Still, the relationship between ERPs and clinical symptoms of psychosis remains controversial. Havermans et al. (1999) studied the P3b evoked potential, which is considered a standard measure of effortful attention, and reported reductions in P3b amplitude in chronic hallucinating patients compared to nonhallucinating patients. Turetsky et al. (1998) found a strong association between a frontal P3b subcomponent and the severity of auditory hallucinations. However, other studies failed to find any associations between P3b amplitude and positive symptoms (Eikmeier et al. 1992; Liu et al. 2004). As most patients diagnosed with schizophrenia who experience auditory hallucinations also experience other symptoms such as delusions, disorganization, and negative symptoms, the diverse P3b findings may be related to this diversity in symptomatology.

To circumvent this problem, Van Lutterveld et al. (2010) investigated P3b amplitude in nonpsychotic individuals experiencing auditory verbal hallucinations as an isolated symptom. Because a reduced P3b amplitude has consistently been demonstrated in patients diagnosed with schizophrenia, and hallucinating nonpsychotic individuals and patients share a single isolated symptom, the authors expected that the P3b amplitude would be reduced in these subjects compared to controls. Contrary to their hypothesis, however, they found an increase in P3b amplitude, which was interpreted as refuting a pivotal role of decreased effortful attention in the pathophysiology of auditory verbal hallucinations.

Finally, one study investigated the P3a event-related potential to speech sounds in hallucinating and nonhallucinating patients diagnosed with schizophrenia. Unlike the P3b, the P3a is not associated with effortful attention, but with involuntary shifts to auditory changes and the processing of novelties. Fisher et al. (2010) found that hallucinating patients had smaller P3a amplitudes than nonhallucinating patients, and that for the hallucinating patients, the P3a amplitude was negatively correlated with auditory hallucination trait scores. The authors suggested that auditory verbal hallucinations are associated with an impaired processing of external speech sounds, perhaps due to competition between external and internal auditory verbal stimuli (i.e., hallucinations).

Other studies have investigated mismatch negativity (MMN) and hallucinations. Mismatch negativity is an event-related potential related to automatic auditory change detection. However, as with the P3b findings, the results of these studies are inconsistent. Some of them reported an association between MMN amplitude and auditory hallucinations (Fisher et al. 2008a, b; Youn et al. 2003), whereas others did not (Kasai et al. 2002; Schall et al. 1999). These diverse findings may be at least partly explained by the different methodologies used. For instance, Schall et al. (1999) presented visual and auditory stimuli simultaneously while others did not.

Recently, interest has been growing in auditory steady-state evoked potentials elicited by click trains. With this paradigm, a steady stream of clicks is presented (hence the adjective "steady-state"), and the brain's responses are measured during the presentation epoch (Uhlhaas and Singer 2010). Spencer et al. (2009) presented click trains pulsing at 40 Hz to patients and healthy controls. They found that patients with higher gamma-band activity (~40 Hz) in the left primary auditory cortex had a greater propensity for experiencing auditory hallucinations. Moreover, this activity was influenced by delta-wave activity. The authors raise the possibility that aberrant oscillatory synchronization in the temporal cortex might interact with dysfunctional corollary discharge mechanisms (i.e., a malfunctioning in neural signals originating in frontal speech areas that indicate to sensory areas that a forthcoming thought is self-generated), leading to the experience of auditory hallucinations. The reported correlations in this study were based on lifetime hallucination ratings, and the medicated patients were not actively hallucinating at the time of the study. Still, these findings extended earlier results of the same laboratory, in which a correlation between gamma-band activity and hallucination severity of first-episode psychosis patients was found (Spencer et al. 2008).

21.5 Electrophysiology and Repetitive Transcranial Magnetic Stimulation

In the last decade, repetitive transcranial magnetic stimulation (rTMS) has emerged as a potential treatment option for auditory hallucinations. With rTMS, electromagnetic induction is used to noninvasively increase or decrease local brain activity (see also Chap. 25). Two studies have investigated the effects of rTMS on the EEG in the context of auditory hallucinatory activit (see also Chapt. 25). Jandl et al. (2006) reported that a subgroup of patients benefited from rTMS administered over the left superior temporal cortex, as revealed by a decrease in auditory hallucination severity, while no changes in whole-head EEG were reported. Horacek et al. (2007) applied rTMS to the left temporoparietal cortex for 10 days and reported a significant decrease in hallucination severity. TMS treatment caused a decrease in activity in the beta-1 and beta-3 bands in the left temporal lobe, whereas an increase was found for the beta-2 band in the right temporal cortex and the inferior parietal lobule, indicating transcallosal signal transmission. A possible explanation for the divergent findings of the two studies is that the data-analysis procedures differed significantly. For example, a source-localization procedure was used in the latter study, whereas in the former study, the EEG was assessed on sensor level.

21.6 Studies of a Basic Neural Mechanism That May Underlie Auditory Hallucinations

Feinberg (1978) suggested that malfunctioning of the corollary discharge mechanism might underlie the experience of auditory hallucinations. Corollary discharge is a basic feed-forward system involved in suppressing the sensory consequences of self-generated actions (Sperry 1950; Von Holst 1950). It has been documented across the animal kingdom (Crapse and Sommer 2008), and its action allows all species to suppress sensations that result from their own actions and to tag them as coming from oneself. For example, someone else can tickle you, but you cannot tickle yourself, as the corollary discharge predicts the forthcoming sensations, preventing the sense of surprise, and suppressing the intensity of the sensation.

Such feed-forward systems have been well described in the visual and somatosensory systems but also serve the auditory system across species from crickets (Poulet and Hedwig 2002) to songbirds (McCasland and Konishi 1981) to primates (Eliades and Wang 2003) and humans (Ford et al. 2007b; Paus et al. 1996). Because the corollary discharge mechanism operates on a rapid timescale, this theory has been investigated most extensively using neurophysiological recordings. In humans, EEG (Ford et al. 2010) and MEG (Curio et al. 2000; Houde et al. 2002) have been used for studies of the auditory system, but only EEG-based methods have been used in studies among patients diagnosed with schizophrenia.

While this mechanism explains the suppression and tagging of sensations resulting from overt motor acts, Feinberg (1978) suggested that thinking may conserve and utilize the computational and integrative mechanisms that evolved for the purpose of dealing with physical movement. In a well-functioning corollary discharge system, a signal is sent from frontal areas involved in thought generation to temporal speech reception areas, tagging the perception as self-generated. When this mechanism is malfunctioning, a person may experience an auditory hallucination through misperceiving his or her own thoughts as being externally generated.

Several lines of research support the hypothesis of corollary discharge dysfunction in psychosis. The first line explored whether this system is deviant in patients diagnosed with schizophrenia versus healthy controls. In these studies, control subjects and patients first uttered syllables and then listened passively to a recording of their own speech played back. EEGs were recorded during both talking and listening conditions, and the amplitude of the N100 component of the ERP to speech onset was used as a measure of auditory-cortical responsiveness. Consistent with the action of the corollary discharge system, the N100 amplitude was smaller during talking than listening in healthy controls. Interestingly, there was significantly less N100 suppression in the patients under study, suggesting aberrations in the corollary discharge system (Ford et al. 2001a, 2007a, b). In another N100 study, the effects of thinking on auditory-cortical responsiveness were investigated. It was shown that thinking affected the N100 amplitude in healthy controls but not in patients diagnosed with schizophrenia (Ford et al. 2001b).

In a second line of research, functional connectivity, as measured by coherence between frontal and temporal lobes in the gamma band, was found to be higher during talking than listening in healthy controls. This pattern was disrupted when the uttered syllables were pitch-shifted while the subjects were talking, resulting in a non-self experience of the spoken sounds. In patients diagnosed with schizophrenia, distortion of the auditory feedback did not result in alteration of gamma-band frontotemporal coherence, again suggesting a malfunctioning corollary discharge system (Ford and Mathalon 2005). In another coherence study, it was found that theta-band frontotemporal coherence was higher for talking than for listening in controls but not in patients diagnosed with schizophrenia. This effect was carried by the hallucinating patients, as the nonhallucinators tended to show the pattern seen in the healthy controls. The authors suggested that a failure in the frontal-temporal network during overt speech may also occur during covert speech, leading to a misattribution of self-generated thoughts to external sources (Ford et al. 2002). Given that the N100 recorded from auditory cortex is suppressed during talking, the net result of coherent communication between frontal and temporal lobes is to suppress auditory sensation.

The corollary discharge theory can also be investigated by examining the small time frame before the onset of speech. In one such study, prespeech neural synchrony was reported to be related to subsequent suppression of the N100 amplitude in healthy controls, but not in patients. Moreover, time-frequency analyses showed greater prespeech synchrony in healthy controls than in patients, especially those with severe auditory hallucinations. The authors interpreted these findings as suggesting that EEG synchrony preceding speech reflects the action of the corollary discharge system, which dampens auditory responsiveness to self-generated speech and is deficient in patients who hallucinate (Ford et al. 2007b).

Another line of research explored the influence of pitch-shifting auditory stimuli on auditory-cortex activation. In this paradigm, hallucinating and nonhallucinating patients diagnosed with schizophrenia as well as controls were asked to utter meaningless sounds. Simultaneously, they were presented with auditory feedback of their own sounds, pitch-shifted feedback of their own sounds, feedback of someone else's voice, or pitch-shifted feedback of an alien voice. It was found that the N100 amplitude to the unaltered self-voice was dampened relative to the altered self-voice and the alien auditory feedback. This pattern was not seen in hallucinating patients, and the degree of the imprecision correlated with the severity of hallucinations (Heinks-Maldonado et al. 2007).

Finally, in a recent study, subjects were asked to initiate auditory stimuli by button presses. It was found that the N100 suppression was normalized in patients after adding a delay of 50 ms in the presentation of the stimulus, suggesting a temporal delay in corollary discharge (Whitford et al. 2011). Moreover, this normalization correlated with white-matter integrity of the arcuate fasciculus, a fiber bundle connecting speech/motor initiation areas in the frontal lobe with the auditory cortex in the temporoparietal lobe. These data suggest that structural deficits of the arcuate fasciculus may lead to temporally delayed corollary discharges and that abnormalities in this fiber tract may be involved in the pathophysiology of auditory hallucinations. A recent study by De Weijer et al. (2011) supported this suggestion, demonstrating degraded fiber integrity of this bundle in hallucinating patients diagnosed with schizophrenia.

21.7 Electrophysiology and Auditory Hallucinations: What Does the Empirical Evidence Tell Us?

EEG and MEG studies of auditory hallucinations have yielded heterogeneous results regarding the involvement of various frequency bands. The theta, alpha, beta, and gamma bands have all been reported to be related to the experience of hallucinations. However, the results regarding location are more consistent. The area most consistently implied is the left temporal cortex. Symptom-capture studies, combined ERP-symptom-capture studies, and an rTMS/EEG study have consistently implicated this brain region. More specifically, most studies implicated the left superior temporal gyrus, consistent with the report that hallucinated voices sound loud and real. These results are in line with structural and functional magnetic resonance imaging (sMRI and fMRI) studies, in which this brain region is also frequently reported (Allen et al. 2008; Barta et al. 1990; Diederen et al. 2010; Dierks et al. 1999). As the left superior temporal cortex is implicated in speech perception, an aberrant corollary discharge mechanism may result in the experience of auditory hallucinations, leading to temporal-lobe abnormalities picked up by EEG and MEG studies. This idea is supported by studies that found alterations in the fiber bundle connecting speech/motor initiation areas in the frontal lobe with the auditory cortex in the temporoparietal lobe in hallucinating subjects.

Acknowledgment Parts of this chapter have been published in Van Lutterveld et al. (2011).

References

Allen, P., Larøi, F., McGuire, P.K., Aleman, A. (2008). The hallucinating brain: a review of structural and functional neuroimaging studies of hallucinations. *Neuroscience and Biobehavioral Reviews*, 32, 175–191.

Barta, P.E., Pearlson, G.D., Powers, R.E., Richards, S.S., Tune, L.E. (1990). Auditory hallucinations and smaller superior temporal gyral volume in schizophrenia. *American Journal of Psychiatry*, 147, 1457–1462.

Crapse, T.B., Sommer, M.A. (2008). Corollary discharge across the animal kingdom. *Nature Reviews. Neuroscience*, 9, 587–600.

Curio, G., Neuloh, G., Numminen, J., Jousmaki, V., Hari, R. (2000). Speaking modifies voice-evoked activity in the human auditory cortex. *Human Brain Mapping*, 9, 183–191.

Diederen, K.M., Neggers, S.F., Daalman, K., Blom, J.D., Goekoop, R., Kahn, R.S., Sommer, I.E. (2010). Deactivation of the parahippocampal gyrus preceding auditory hallucinations in schizophrenia. *American Journal of Psychiatry*, 167, 427–435.

Dierks, T., Linden, D.E., Jandl, M., Formisano, E., Goebel, R., Lanfermann, H., Singer, W. (1999). Activation of Heschl's gyrus during auditory hallucinations. *Neuron*, 22, 615–621.

De Weijer, A.D., Mandl, R.C.W., Diederen, K.M.J., Neggers, S.F.W., Kahn, R.S., Hulshoff Pol, H.E., Sommer, I.E.C. (2011). Microstructural alterations of the arcuate fasciculus in schizophrenia patients with frequent auditory verbal hallucinations. *Schizophrenia Research*, 130, 68–77.

Eikmeier, G., Lodemann, E., Zerbin, D., Gastpar, M. (1992). P300, clinical symptoms, and neuropsychological parameters in acute and remitted schizophrenia: a preliminary report. *Biological Psychiatry*, 31, 1065–1069.

Eliades, S.J., Wang, X. (2003). Sensory-motor interaction in the primate auditory cortex during self-initiated vocalizations. *Journal of Neurophysiology*, 89, 2194–2207.

Feinberg, I. (1978). Efference copy and corollary discharge: implications for thinking and its disorders. *Schizophrenia Bulletin*, 4, 636–640.

Fisher, D.J., Labelle, A., Knott, V.J. (2008a). Auditory hallucinations and the mismatch negativity: processing speech and non-speech sounds in schizophrenia. *International Journal of Psychophysiology*, 70, 3–15.

Fisher, D.J., Labelle, A., Knott, V.J. (2008b). The right profile: mismatch negativity in schizophrenia with and without auditory hallucinations as measured by a multi-feature paradigm. *Clinical Neurophysiology*, 119, 909–921.

Fisher, D.J., Labelle, A., Knott, V.J. (2010). Auditory hallucinations and the P3a: attention-switching to speech in schizophrenia. *Biological Psychology*, 85, 417–423.

Ford, J.M., Gray, M., Faustman, W.O., Roach, B.J., Mathalon, D.H. (2007a). Dissecting corollary discharge dysfunction in schizophrenia. *Psychophysiology*, 44, 522–529.

Ford, J.M., Mathalon, D.H. (2005). Corollary discharge dysfunction in schizophrenia: can it explain auditory hallucinations? *International Journal of Psychophysiology*, 58, 179–189.

Ford, J.M., Mathalon, D.H., Heinks, T., Kalba, S., Faustman, W.O., Roth, W.T. (2001a). Neurophysiological evidence of corollary discharge dysfunction in schizophrenia. *American Journal of Psychiatry*,158, 2069–2071.

Ford, J.M., Mathalon, D.H., Kalba, S., Whitfield, S., Faustman, W.O., Roth, W.T. (2001b). Cortical responsiveness during inner speech in schizophrenia: an event-related potential study. *American Journal of Psychiatry*, 158, 1914–1916.

Ford, J.M., Mathalon, D.H., Whitfield, S., Faustman, W.O., Roth, W.T. (2002). Reduced communication between frontal and temporal lobes during talking in schizophrenia. *Biological Psychiatry*, 51, 485–492.

Ford, J.M., Roach, B.J., Faustman, W.O., Mathalon, D.H. (2007b). Synch before you speak: auditory hallucinations in schizophrenia. *American Journal of Psychiatry*, 164, 458–466.

Ford, J.M., Roach, B.J., Mathalon, D.H. (2010). Assessing corollary discharge in humans using noninvasive neurophysiological methods. *Nature Protocols*, 5, 1160–1168.

Haenschel, C., Baldeweg, T., Croft, R.J., Whittington, M., Gruzelier, J. (2000). Gamma and beta frequency oscillations in response to novel auditory stimuli: A comparison of human electroencephalogram (EEG) data with in vitro models. *Proceedings of the National Academy of Sciences USA*, 97, 7645–7650.

Hari, R., Hämäläinen, M., Ilmoniemi, R., Kaukoranta, E., Reinikainen, K., Salminen, J., Alho, A., Näätänen, R., Sams, M. (1984). Responses of the primary auditory cortex to pitch changes in a sequence of tone pips: neuromagnetic recordings in man. *Neuroscience Letters*, 50, 127–132.

Havermans, R., Honig, A., Vuurman, E.F.P.M., Krabbendam, L., Wilmink, J., Lamers, Th., Verheecke, C.J., Jolles, J., Romme, M.A.J., Van Praag, H.M. (1999). A controlled study of temporal lobe structure volumes and P300 responses in schizophrenic patients with persistent auditory hallucinations. *Schizophrenia Research*, 38, 151–158.

Heinks-Maldonado, T.H., Mathalon, D.H., Houde, J.F., Gray, M., Faustman, W.O., Ford, J.M. (2007). Relationship of imprecise corollary discharge in schizophrenia to auditory hallucinations. *Archives of General Psychiatry*, 64, 286–296.

Horacek, J., Brunovsky, M., Novak, T., Skrdlantova, L., Klirova, M., Bubenikova-Valesova, V., Krajka, V., Tislerova, B., Kopecek, M., Spaniel, F., Mohr, P., Höschl, C. (2007). Effect of low-frequency rTMS on electromagnetic tomography (LORETA) and regional brain metabolism (PET) in schizophrenia patients with auditory hallucinations. *Neuropsychobiology*, 55, 132–142.

Houde, J.F., Nagarajan, S.S., Sekihara, K., Merzenich, M.M. (2002). Modulation of the auditory cortex during speech: an MEG study. *Journal of Cognitive Neuroscience*, 14, 1125–1138.

Hubl, D., Koenig, T., Strik, W.K., Garcia, L.M., Dierks, T. (2007). Competition for neuronal resources: how hallucinations make themselves heard. *British Journal of Psychiatry*, 190, 57–62.

Ishii, R., Shinosaki, K., Ikejiri, Y., Ukai, S., Yamashita, K., Iwase, M., Mizuno-Matsumoto, Y., Inouye, T., Yoshimine, T., Hirabuki, N., Robinson, S.E., Takeda, M. (2000). Theta rhythm increases in left superior temporal cortex during auditory hallucinations in schizophrenia: a case report. *Neuroreport*, 11, 3283–3287.

Jandl, M., Steyer, J., Weber, M., Linden, D.E., Rothmeier, J., Maurer, K., Kaschka, W.P. (2006). Treating auditory hallucinations by transcranial magnetic stimulation: a randomized controlled cross-over trial. *Neuropsychobiology*, 53, 63–69.

Kasai, K., Nakagome, K., Itoh, K., Koshida, I., Hata, A., Iwanami, A., Fukuda, M., Kato, N. (2002). Impaired cortical network for preattentive detection of change in speech sounds in schizophrenia: a high-resolution event-related potential study. *American Journal of Psychiatry*, 159, 546–553.

Kindler, J., Hubl, D., Strik, W.K., Dierks, T., Koenig, T. (2011). Resting-state EEG in schizophrenia: Auditory verbal hallucinations are related to shortening of specific microstates. *Clinical Neurophysiology*, 122, 1179–1182.

Lee, S.H., Wynn, J.K., Green, M.F., Kim, H., Lee, K.-J., Nam, M., Park, J.-K., Chung, Y.-C. (2006). Quantitative EEG and low resolution electromagnetic tomography (LORETA) imaging of patients with persistent auditory hallucinations. *Schizophrenia Research*, 83, 111–119.

Line, P., Silberstein, R.B., Wright, J.J., Copolov, D.L. (1998). Steady state visually evoked potential correlates of auditory hallucinations in schizophrenia. *NeuroImage*, 8, 370–376.

Liu, Z., Tam, W.C., Xue, Z., Yao, S., Wu, D. (2004). Positive and negative symptom profile schizophrenia and abnormalities in the P300 component of the event-related potential: a longitudinal controlled study. *Psychiatry Research*, 132, 131–139.

Marjerrison, G., Krause, A.E., Keogh, R.P. (1968). Variability of the EEG in schizophrenia: quantitative analysis with a modulus voltage integrator. *Electroencephalography and Clinical Neurophysiology,* 24, 35–41.

McCasland, J.S., Konishi, M. (1981). Interaction between auditory and motor activities in an avian song control nucleus. *Proceedings of the National Academy of Sciences USA*, 78, 7815–7819.

Paus, T., Marrett, S., Worsley, K., Evans, A. (1996). Imaging motor-to-sensory discharges in the human brain: an experimental tool for the assessment of functional connectivity. *NeuroImage*, 4, 78–86.

Poulet, J.F., Hedwig, B. (2002). A corollary discharge maintains auditory sensitivity during sound production. *Nature*, 418, 872–876.

Reulbach, U., Bleich, S., Maihofner, C., Kornhuber, J., Sperling, W. (2007). Specific and unspecific auditory hallucinations in patients with schizophrenia: a magnetoencephalographic study. *Neuropsychobiology*, 55, 89–95.

Ropohl, A., Sperling, W., Elstner, S., Tomandl, B., Reulbach, U., Kaltenhäuser, M., Kornhuber, J., Maihöfner, C. (2004). Cortical activity associated with auditory hallucinations. *Neuroreport*, 15, 523–526.

Schall, U., Catts, S.V., Karayanidis, F., Ward, P.B. (1999). Auditory event-related potential indices of fronto-temporal information processing in schizophrenia syndromes: valid outcome prediction of clozapine therapy in a three-year follow-up. *International Journal of Neuropsychopharmacology*, 2, 83–93.

Sem-Jacobsen, C.W., Petersen, M.C., Lazarte, J.A., Dodge, H.W. Jr., Holman, C.B. (1955). Intracerebral electrographic recordings from psychotic patients during hallucinations and agitation. *American Journal of Psychiatry*, 112, 278–288.

Serafetinides, E.A., Coger, R.W., Martin, J. (1986). Different methods of observation affect EEG measures associated with auditory hallucinations. *Psychiatry Research*, 17, 73–74.

Spencer, K.M., Niznikiewicz, M.A., Nestor, P.G., Shenton, M.E., McCarley, R.W. (2009). Left auditory cortex gamma synchronization and auditory hallucination symptoms in schizophrenia. *BMC Neuroscience*, 10, 85.

Spencer, K.M., Salisbury, D.F., Shenton, M.E., McCarley, R.W. (2008). Gamma-band auditory steady-state responses are impaired in first episode psychosis. *Biological Psychiatry*, 64, 369–375.

Sperry, R.W. (1950). Neural basis of the spontaneous optokinetic response produced by visual inversion. *Journal of Comparative and Physiological Psychology*, 43, 482–489.

Sritharan, A., Line, P., Sergejew, A., Silberstein, R., Egan, G., Copolov, D. (2005). EEG coherence measures during auditory hallucinations in schizophrenia. *Psychiatry Research,* 136, 189–200.

Stevens, J.R., Bigelow, L., Denney, D., Lipkin, J., Livermore, A.H. Jr., Rauscher, F., Wyatt, R.J. (1979). Telemetered EEG-EOG during psychotic behaviors of schizophrenia. *Archives of General Psychiatry*, 36, 251–262.

Stevens, J.R., Livermore, A. (1982). Telemetered EEG in schizophrenia: spectral analysis during abnormal behaviour episodes. *Journal of Neurology, Neurosurgery, and Psychiatry*, 45, 385–395.

Tiihonen, J., Hari, R., Naukkarinen, H., Rimon, R., Jousmaki, V., Kajola, M. (1992). Modified activity of the human auditory cortex during auditory hallucinations. *American Journal of Psychiatry*, 149, 255–257.

Turetsky, B., Colbath, E.A., Gur, R.E. (1998). P300 subcomponent abnormalities in schizophrenia: II. Longitudinal stability and relationship to symptom change. *Biological Psychiatry*, 43, 31–39.

Uhlhaas, P.J., Singer, W. (2010). Abnormal neural oscillations and synchrony in schizophrenia. *Nature Reviews. Neuroscience*, 11, 100–113.

Van Lutterveld, R., Oranje, B., Kemner, C., Abramovic, L., Willems, A.E., Boks, M.P., Glenthøj. B.Y., Kahn, R.S. Sommer, I.E. (2010). Increased psychophysiological parameters of attention in non-psychotic individuals with auditory verbal hallucinations. *Schizophrenia Research*, 121, 153–159.

Van Lutterveld, R., Sommer, I.E., Ford, J.M. (2011). The neurophysiology of auditory hallucinations - A historic and contemporary review. *Frontiers in Psychiatry*, 2, 1–7.

Von Holst, E.M.H. (1950). Das Reafferenzprinzip (Wechselwirkungen zwischen Zentralnervensystem und Peripherie). *Naturwissenschaften*, 37, 464–476.

Whitford, T.J., Mathalon, D.H., Shenton, M.E., Roach, B.J., Bammer, R., Adcock, R.A., Bouix, S., Kubicki, M., De Siebenthal, J., Rausch, A.C., Schneiderman, J.S., Ford, J.M. (2011). Electrophysiological and diffusion tensor imaging evidence of delayed corollary discharges in patients with schizophrenia. *Psychological Medicine*, 41, 959–969.

Whitton, J.L., Moldofsky, H., Lue, F. (1978). EEG frequency patterns associated with hallucinations in schizophrenia and "creativity" in normals. *Biological Psychiatry*, 13, 123–133.

Youn, T., Park, H.J., Kim, J.J., Kim, M.S., Kwon, J.S. (2003). Altered hemispheric asymmetry and positive symptoms in schizophrenia: equivalent current dipole of auditory mismatch negativity. *Schizophrenia Research*, 59, 253–260.

Chapter 22
Psychoactive Substances

Vicka Corey, John H. Halpern, and Torsten Passie

22.1 Introduction

This chapter discusses some substances that are particularly likely to induce hallucinations. Many such drugs are available in nature and have long histories of use; more have emerged from laboratories either accidentally or as the result of purposeful research. Many more substances than these *can* elicit hallucinations, given the natural corporeal variations among individuals. For example, antidepressants such as paroxetine (Shimizu et al. 2010) or citalopram (Capaldi and Carr 2010) occasionally induce auditory hallucinations, which tend to resolve quickly after the drug is discontinued. Even the lack of an ordinary biological component – such as water in dehydration – can cause similar effects.

This chapter focuses on a few substances often taken for the purpose of hallucination. Some effects that are typically sought by hallucinogen users are alterations of sensory perception, interpretation, and context. Changes in visual perception are perhaps the canonical hallucinogen-induced effect. For example, a normal aspect of human sight called "persistence of vision" enables us to perceive subsequent still images as a seamless motion picture, or a moving dot of light as a complete picture in a laser show. Some hallucinogens extend this ability so that a larger set of images is retained in perception – the line behind the laser dot becoming longer – an effect commonly known as "trailers." Spots or afterimages seen upon closing one's eyes

V. Corey, Ph.D. • J.H. Halpern, M.D. (✉)
Harvard Medical School, Boston, MA, USA

The Laboratory for Integrative Psychiatry, Division of Alcohol and Drug Abuse,
McLean Hospital, Belmont, MA, USA
e-mail: vicka@mclean.harvard.edu; john_halpern@hms.harvard.edu

T. Passie, M.D., M.A.
Department for Psychiatry, Social Psychiatry, and Psychotherapy,
Hannover Medical School, Hannover, Germany
e-mail: dr.passie@qmx.de

J.D. Blom and I.E.C. Sommer (eds.), *Hallucinations: Research and Practice*,
DOI 10.1007/978-1-4614-0959-5_22,

(Second rêve. — Une promenade dans le ciel.)

Fig. 22.1 *The Second Dream: A Stroll In The Sky*. Wood engraving by P. Soyer, after J.J. Grandville. Originally published in *Le Magasin Pittoresque*, 1847 may become elaborated into geometrical structures, architecture, faces, animals, or plants. Closed-eye visual effects are usually easier to discern and more elaborate than those perceived with eyes open, doubtless because of the lack of processing competition from ordinary sight. Alterations can also be found in the salience of signals, such as an increase in acuity or in the pleasure derived from visual stimulation – hence the common hallucinogen joke, "Wow, have you ever *really looked* at your hand?"

Auditory effects are less common than visual responses to hallucinogens, although there are substances that particularly evoke them (see Sect. 22.2.3.2). Other senses such as smell and touch are even less prone to their effects.

Most substance-induced alterations consist of "partial hallucinations" in which an actual sensory stimulus is involved; the hallucinator tends to be aware of the difference between the hallucinogen-affected experiences of that stimulus and of a usual sensorium. Sometimes a complete loss of contextual meaning is evoked, in which the hallucinator experiences an essentially different environment (i.e., a scenic or panoramic hallucination), and either loses awareness of his sensory surroundings or becomes unable to recognize them as such. In such cases, the hallucinator may appear to act upon such perceptions (e.g., by speaking to those who are not there) or may be apparently unconscious (in a state resembling dreaming sleep).

22.2 What Drugs Cause Hallucinations?

As noted earlier, many substances or chemical imbalances in the body *can* induce hallucinations. These substances are often ingested with hallucinogenic intent.

22.2.1 Historical Hallucinogens and Their Cultural Contexts

Many hallucinogens known today have long cultural histories, each with its own specific traditions and goals.

22.2.1.1 Mescaline

Mescaline occurs in substantive quantities in several types of American cacti, including the San Pedro cactus (*Echinopsis pachanoi*) in Peru, Argentina, Bolivia, Ecuador, and Chile. Its record of use dates back at least 2,000 years to the Moche culture. It is also present in *Lophophora williamsii*, the Peyote cactus native to southern Texas and northern Mexico. The Native American Church (NAC, see Fig. 22.2), established in the USA in the 1880s and incorporated in 1918, venerates peyote as its holy sacrament for all-night prayer services. In its religious context, the Peyote Road is a term for right living, emphasizing values such as love for others, devotion to family, hard work, and abstinence from alcohol (De Smet and Bruhn 2003).

Fig. 22.2 Native American Church (NAC) setting (Photograph by John Halpern)

The alkaloid mescaline is 3,4,5-trimethoxy-β-phenethylamine. It is known to cause nausea soon after onset, but the psychoactive effects are much longer lasting, up to 12 h. It increases the salience of music and intensifies colors and textures (see Fig. 22.3). Alterations of interpretation can also occur, such as seeing cars as having faces and personalities. Often users (even outside of a religious environment) report feeling a deepened sense of the divine in ordinary people and objects, a sense of the sacred within themselves and toward the world. Emotional effects increase with dose, although it is difficult to identify those as "hallucinatory" rather than genuine responses to the experiences.

Mescaline's neurological effects include blocking the release of acetylcholine and affecting the cell membrane's level of potassium ion conductance, as demonstrated in rat cortex and frog neuromuscular conjunctions (Ghansah et al. 1993). In live cats, behavioral changes in response to mescaline are blocked by pretreatment

Fig. 22.3 Mescaline-inspired artwork (Photograph by John Halpern)

with serotonergic or dopaminergic agonists, implying that those systems are involved in the perceptions and actions of those (and probably other) mammals (Trulson et al. 1983).

22.2.1.2 Psilocin and Amanitoxins

Mushrooms of many genera contain hallucinogens. By far the most common and well-known is psilocin (4-hydroxyl-dimethyltryptamine). Psilocin is a 5HT1A and 5HT2A/2C partial agonist, working in the brain's serotonergic systems. It can also be produced within the body as a dephosphorylated metabolite of the prodrug psilocybin (O-phosphoryl-4-hydroxy-N,N-dimethyltryptamine).

Psilocin and psilocybin are present in the fruit bodies of many mushroom species (Guzmán et al. 2000). Often these mushrooms are small, brown, and bitter-tasting. Common examples in the United States' Pacific Northwest are *Psilocybe cyanescens*, *P. azurescens*, and *P. stuntzii*; *Psilocybe caerulipes* in the East; and *Psilocybe cubensis* along the Gulf Coast (also frequently cultivated worldwide). In Japan, *Gymnopilus spectabilis* is known as *o warai take* or "the big laughter mushroom," while mere *warai take* is *Panaeolus papilonaceus* (Sanford 1972). At least some of the

Fig. 22.4 Two *Amanita muscaria* (Public domain image picture, retrieved May 4, 2011, from http://www.public-domain-image.com/flora-plants-public-domain-images-pictures/fungi-mushrooms-public-domain-images-pictures/two-amanita-muscaria.jpg.html)

distribution of these mushrooms may be due to human migration. In Europe, however, such mushrooms may have never entered, let alone altered, human consciousness.

The hallucinogenic properties of Chinese and Japanese psilocin-containing mushrooms are well recorded, dating back at least to a report by Chang Hua in the Chin Dynasty (approx. 245–400 AD), stating that "[mushrooms] growing on the Feng tree, when ingested, cause people to laugh unceasingly." Remedies such as strong tea with alum were prescribed, although since a typical course of psilocin hallucination is less than 6 h long, it is difficult to tell whether these treatments were effective or merely incidental. In Japan, an illustration of *Panaeolus papilonaceus* adorned the cover of the 1918 *Journal of Japanese Botany*. However, these fungi never became integrated into either traditional Chinese medicine or any other known traditions of use. An old folktale describes Buddhist nuns dancing and singing after eating some mushrooms they found, who are then joined by woodcutters whose experiences are the same. But the story only goes so far as that anecdote. A scattering of other stories repeat similar incidents throughout the centuries. The closest they seem to have come to having a place in those societies was very recently, after psilocin became a worldwide phenomenon. "Magic mushrooms" were available in Japan by vending machine and in "head shops" until illegalization in 2002. But even preprohibition, there appear to be no records of any outcomes more exotic than inappropriate public behavior.

The other common mycological hallucinogen comes from showy species of *Amanita* – large, bright-capped mushrooms decorated with white spots (e.g., *A. muscaria*, *A. pantherina*, and *A. gemmata*, see Fig. 22.4). These contain the active compounds muscimol and ibotenic acid (metabolized to ibotenate), constituting substrate analogues for gamma-aminobutyric acid (GABA) and N-methyl-D-aspartic acid (NMDA). The mycologist David Arora, in his classic work *Mushrooms Demystified*, describes their effects "on the central nervous system [as] confusion, mild euphoria, loss of muscular coordination, profuse sweating, chills, visual distortions, a feeling of greater strength, and sometimes hallucinations, delusions or convulsions. (An inordinate number of 'trippers' mistake themselves for Christ.)" (Arora 1990).

The human history involving the amanitas is likely vast, but laden with much mystery. They have been proposed as the basis for everything from the magical Soma of the Hindu Rig Veda texts to the Vikings' fearlessness in battle and also proposed as the original Tree of Knowledge. However, people who consume psychoactive amanitas often cannot recollect their experiences, perhaps as a result of the effects on the NMDA systems now known to regulate memory (Fei and Tsien 2009).

22.2.1.3 DMT and MAOIs

DMT – N,N-dimethyltryptamine – is widely found in nature, including in vines, barks, roots, a sea fan, and mammals. It binds nonselectively to at least eight subtypes of the serotonin receptor and is an agonist for at least three of them. It also shows affinity for the dopamine D1 and several adrenergic receptor subtypes, as well as for imidazoline-1 and trace-amine-associated receptors. Moreover, it is the only known endogenous sigma-1 opioid ligand (Fontanilla et al. 2009). It is a powerful and often context-altering hallucinogen. In traditional South American shamanism, it is sometimes intranasally insufflated. DMT is not orally psychoactive by itself, as monoamine oxidase (MAO) in the gut lining breaks it down. Intravenous injection, smoking, or inhalation of DMT can elude the body's MAO metabolism, but usually not for very long. Such experiences tend to span 30 min or less, while the effects experienced upon intramuscular injection last about 1 h.

DMT becomes psychoactive after oral ingestion when combined with a monoamine oxidase inhibitor (MAOI). Thus it is sometimes (though not always) an ingredient of a psychoactive beverage known as *ayahuasca*, along with reversible MAOI beta-carbolines from the vine *Banisteriopsis caapi*. The hallucinations produced by ayahuasca tend to last about 3 h. Typical effects include time dilation, visual and auditory hallucinations, and experiences that users cannot adequately render afterward, such as an "alternate reality" populated by "machine elves" or "beings of light" with whom they can interact (McKenna 1992).

Although DMT is illegal in much of the world, American members of the two largest Brazilian ayahuasca religions have won legal battles in Oregon and New Mexico protecting their right to practice their faiths freely and without government harassment. At present, the Santo Daime, the União do Vegetal, and the Barquinha are three of the best-known ayahuasca churches. Influenced by Christianity as well as South American indigenous shamanism and other religious traditions, their spiritual obligations include honoring nature in a personified way, such as in the following line from a Santo Daime hymn: "I venerate my sweet mama of the sky/ On the earth and in the astral." It is possible that such interpretations of the spiritual have a common neurological origin with the "machine elves" noted above.

Fig. 22.5 *From The Jungles Of West Africa – Tabernanthe Iboga*. Hand-printed with dark chocolate brown ink on archival coffee hand-stained paper (Copyright 2010 Dave Hunter. Reproduced with permission)

22.2.1.4 Ibogaine

The hallucinogen ibogaine, an indole alkaloid, is found in the religious sacrament of practitioners of Bwiti, a religion practiced across greater West Africa. Some Bwiti scholars believe the iboga plant (*Tabernanthe iboga*, see Fig. 22.5) to be the Biblical Tree of Knowledge.

At first onset, ibogaine offers visual hallucinations, and then, in later hours, an apparently unique introspective state allowing visions of the "true self" and a profound understanding of the consequences of one's actions. Here is an example of a user's account:

> I didn't see much. I travelled a red path and came to a village of one house with one door and one window. Two white men were sitting at either end of a table. They were writing. That was all. I returned then. But I was dissatisfied so I took a big dose of eboga again and this time I saw my mother and she was surrounded by many people. She died when I was young and I didn't recognize her. But men surrounding her said it was my mother. She came and stood at my right. Another woman came with a child and stood at my left. I reached for the child but she held it away from me. Then I became sick and had to pass out to the edge of the forest to throw up. As I came back I saw a host of small babies laughing and playing together in the air. That was all I saw.
>
> (Fernandez 1982)

During the 1880s, German soldiers in Cameroon did not fail to notice the stimulant effects of ibogaine: "Its exciting effect on the nervous system makes its use highly valued on long tiring marches, on lengthy canoe trips, and on difficult nightwatches." (District Officers from Kamerun 1888). From 1939 through 1970, it was marketed in France as *Lambarene* for fatigue, infections, and postdisease recovery. Incidentally, its hallucinatory effects were not discussed in this context.

During the 1960s, the Chilean psychiatrist Claudio Naranjo began to investigate ibogaine as an adjunct to psychotherapy, noting its utility in its particular tendency to evoke old memories associated with later-in-life bad choices, including addiction, and the opportunity to then decide to live differently (Naranjo 1974). Howard Lotsof unexpectedly achieved lasting abstinence from heroin dependence after an ibogaine experience himself in 1962 and subsequently obtained several US Patents for the "iboga cure" of addictions (Lotsof 1985, 1992, see also Fig. 22.6).

Ibogaine is not, as of this writing, an FDA-approved medication in the USA. This is unfortunate, as a potential goal and effect of ibogaine treatment is to end addictions, for which there are few useful techniques. Apparently, the tendency to see one's past in a detached way allows individuals to gain perspective on their needs and other ways to meet them. In this case, it is the drug's typical effects on thought that propel the cure. Other people's interpretations are secondary, although considerable support may be required to complete the process. There is also evidence that ibogaine and its long-lasting metabolites help suppress the physical symptoms of opiate withdrawal, making for a smoother transition to complete abstinence (Glick and Maisonneuve 1998). Supervised ibogaine treatment for addiction is presently available just south of the USA in Mexico and just north in British Columbia, as well as in many other parts of the world. Its legal status is variable and in some flux – at latest report ranging from the USA's Schedule I to no formal regulation in Canada, to government-licensed at certain treatment sites in the West Indies.

22.2.2 *Psilocybes and the Saint Children: A Transition Between Traditional and Modern Hallucinogen Cultures*

American hallucinogenic mushrooms have been part of Native American heritage since antiquity, though their origins and histories have been lost and cultural imperialism

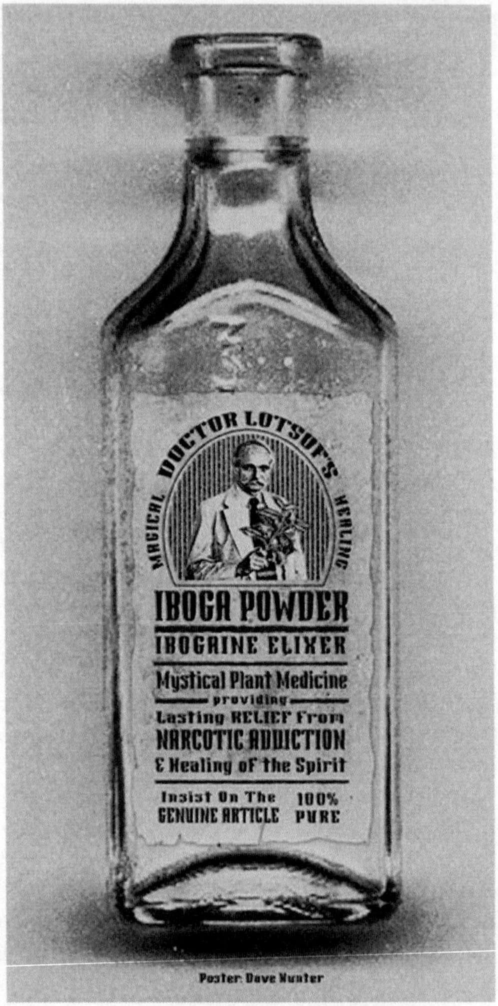

Fig. 22.6 Poster recognizing Howard Lotsof's discovery of the antiaddictive effects of ibogaine (Copyright 2004 Dave Hunter. Reproduced with permission)

destroyed a tremendous amount of both practice and oral records. However, traditions were also preserved, under some secrecy, in relatively peaceful and isolated groups such as the Mazatec-speaking villages in Oaxaca, Mexico. In their groundbreaking publication *Mushrooms, Russia, and History*, Valentina and R. Gordon Wasson ("RGW") discuss a smattering of historical references for psychoactive mushroom use in Oaxaca, mostly from Christian clerics from around the sixteenth century (Wasson and Wasson 1957).

After they had heard about a living mushroom tradition from a Mazatec-speaking Bible translator, in 1953, the Wassons traveled to Mazatec communities and, eventually, were permitted to participate in a psilocin-focused ritual called a *velada*. The *curanderas* who conducted these were reverent and proud, considering the mushrooms a magical medicine. In 1955, the Wassons went to Mesoamerica to enter deep into these practices and were introduced to María Sabina, said to be a *curandera* "of the first class." They wrote of RGW's first personal ingestion of the mushrooms:

> There was no inclination to sleep. At all times we [RGW and another American named Allan] were alert both to our subjective hallucinations and to the goings-on around us in the dark. RGW took imperfect notes intermittently and kept track of the hours. But he and Allan were both alive to the fact that they were not themselves. Though RGW had resolved to fight off any effects of the mushrooms and remain the detached observer, the mushrooms took full and sweeping possession of him. There is no better way to describe the sensation than to say that it was as though his very soul had been scooped out of his body and translated to a point floating in space, leaving behind the husk of clay, his body. 'Landslide', the designation of the Mazatecs for the mushroom we were using, had seemed to him a clumsy name before; now its awesome truth imposed itself. Our bodies lay there while our souls soared. We both felt nauseated; RGW twice made his way to the other room to vomit, and Allan three times. One or two others, not identified in the darkness, did likewise. But these episodes seemed of no moment. For we were both seeing visions, similar but not identical visions, and we were comparing notes in whispered interchanges. At first we saw geometric patterns, angular not circular, in richest colors, such as might adorn textiles or carpets. Then the patterns grew into architectural structures, with colonnades and architraves, patios of regal splendor, the stone-work all in brilliant colors, gold and onyx and ebony, all most harmoniously and ingeniously contrived, in richest magnificence extending beyond the reach of sight, in vistas measureless to man. For some reason these architectural visions seemed oriental, though at every stage RGW pointed out to himself that they could not be identified with any specific oriental country. They were neither Japanese nor Chinese nor Indian nor Moslem. They seemed to belong rather to the imaginary architecture described by the visionaries of the Bible. In the aesthetics of this discovered world attic simplicity had no place: everything was resplendently rich.
>
> (Wasson and Wasson 1957)

Later, in discussing a second velada, during which the visions were quite different, they wrote,

> For the world our visions were and must remain 'hallucinations'. But for us they were not false or shadowy suggestions of real things, figments of an unhinged imagination. What we were seeing was, we knew, the only reality, of which the counterparts of every day are mere imperfect adumbrations. At the time we ourselves were alive to the novelty of this our discovery, and astonished by it.
>
> (Wasson and Wasson 1957)

These hallucinations must all be described as "partial," drawing primarily upon the visual senses, and on the interpretive levels of cognition, where they invited the participants to compare their visions with Elizabethan aesthetics, Platonic ideals, and the imagery of the Bible. At no time, however, were they lost to their actual surroundings; RGW describes touching the wall of the house to reorient himself, which always worked, though it also tended to occasion nausea.

In 1958, Albert Hofmann succeeded in synthesizing psilocybin. He gave María Sabina some capsules of his product, and she attested that they were as efficacious as the fungal forms (Schultes and Hofmann 1973). But later, after word of these mushrooms had spread throughout the world, she said,

> From the moment the foreigners arrived, the 'saint children' [i.e., mushrooms] lost their purity. They lost their force; the foreigners spoiled them. From now on they won't be any good. There is no remedy for it.
>
> (Estrada 1981)

RGW agreed with her. In 1980, he wrote, "Since the white man came looking for the mushrooms, they have lost their magic" (Wasson 1980).

22.2.3 *Modern Hallucinogens: "Psychotomimetics" and Current Research Techniques*

During the twentieth century, great advances were made in medical chemistry, including the discovery of an extensive variety of novel psychoactive substances (Shulgin and Shulgin 1991, 1997). At first it was hoped that hallucinogens (especially lysergic acid diethylamide or LSD) would provide insight into the mechanisms of psychosis and hallucinations in general (Hoffer and Osmond 1967), although it soon turned out that these processes are far too complex to be so simply modeled (Geyer and Vollenweider 2008).

Significant work was done with the hallucinogens psilocin and psilocybin (Vollenweider et al. 1997a; Gouzoulis-Mayfrank et al. 1999), DMT (Riba et al. 2006; Gouzoulis-Mayfrank et al. 2005), and ketamine (Vollenweider et al. 1997b; Krystal et al. 1994). Most of these substances activate limbic and paralimbic structures, heightening arousal and leading, in turn, to intensified and/or additional endogenous stimuli which may be experienced as hallucinations. From the 1990s through 2008, it was commonly considered that changes in cerebral informational interplay are responsible for these experiences (Vollenweider and Geyer 2001), but more recently it was found that some substances (e.g., LSD, psilocin, and DMT) exert hallucinatory and other effects by directly stimulating serotonin receptors on cortical pyramidal cells (Geyer and Vollenweider 2008). Still, most modern hallucinogens have not received the kind of scientific attention they might warrant.

22.2.3.1 LSD

LSD (lysergic acid diethylamide, see Fig. 22.7) was first synthesized in 1938 by Albert Hofmann in an investigation of ergot- and squill-based bioactive chemicals, particularly in the hopes of finding an analeptic. On April 16, 1943, Dr. Hofmann, led by his intuition (Hofmann 1969), resynthesized the substance for further investigation and accidentally absorbed a small amount through his fingertips. To quote from his description, he felt

> a remarkable restlessness, combined with a slight dizziness. At home I lay down and sank into a not unpleasant intoxicated-like condition, characterized by an extremely stimulated imagination. In a dreamlike state, with eyes closed (I found the daylight to be unpleasantly glaring), I perceived an uninterrupted stream of fantastic pictures, extraordinary shapes with intense, kaleidoscopic play of colors. After some two hours this condition faded away.
>
> (Hofmann 1980)

"Fantasy pictures," "intense colors," and "dreamlike state" remain canonical for "hallucinogens" to this day.

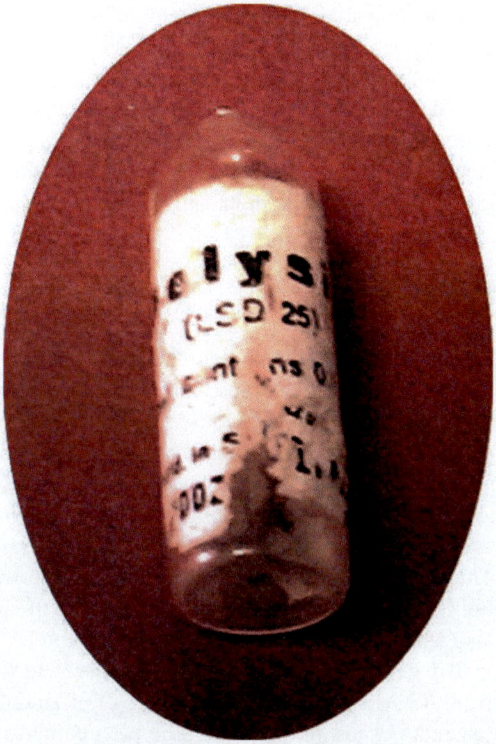

Fig. 22.7 An LSD ampule (Photograph by John Halpern)

On April 19, 1943, Hofmann purposely ingested 250 micrograms of LSD, which he expected to be too little for any bioactivity. As the effects began, he asked a lab assistant to escort him home. Because of wartime restrictions on motor vehicles, they rode bicycles. On the way, he began to experience intense, disturbing sensations and thoughts, for example, believing his next-door neighbor was a witch. He was examined by a physician, who assured him that his only apparent issue was pupil dilation, and then Hofmann was able to relax and enjoy the "fascinating images [….] rearranging and hybridizing themselves in constant flux." He came away convinced that LSD had a future in psychiatry because it was so intense and introspective. He could not imagine any recreational use.

As Hofmann described, hallucinations with LSD are predominantly visual. All other sensory spheres can be altered by LSD's effects, but contribute relatively slightly to its range. In contrast, psychotic individuals experience mainly auditory hallucinations (e.g., voices). True or "full" hallucinations are very rare with LSD; it typically alters sights from the physical environment and leaves the individual aware that the effects are drug-induced. Within the lower dose range, users may see trailers and other elaborations of normal visual processes, or primitive forms that look like webs, lattices, tunnels, or geometrical designs. There may be many more complex hallucinatory phenomena during more intense experiences, sometimes unfolding into mental fantasies or memories of life events. Most of LSD's hallucinogenic phenomena are described by knowledgeable users as rewarding and enjoyable, sometimes giving insights into their own psychological make-up and functioning.

LSD has a high affinity for serotonin receptors (mainly 5-HT2a and 5-HT1c) and a broad range of other receptors. Its complex pharmacodynamics are still not completely understood. Research has shown that the alterations of the serotonin system may play a major role in its hallucinatory effects (Passie et al. 2008).

22.2.3.2 DiPT

A member of the tryptamine chemical family, diisopropyltryptamine (DiPT) is a fascinating substance because, unlike most hallucinogens, its effects are predominantly auditory. It is also possibly less sensitive than other hallucinogens to the mindset of the user, the setting in which it is ingested, and other psychological considerations, perhaps because the auditory system has become less salient to the human organism as we have evolved into a vision-based species. In general, auditory pitch is perceived as lower than normal, and harmonious sounds lose their resonance with one another. This dissonance is even perceived by people with perfect pitch, which has some implications about where in the processing stream DiPT's effects occur. Voices are also altered and disharmonious with one another (Shulgin and Shulgin 1997).

DiPT has few other known effects; it would seem to call for further investigation from those interested in the neurology of sound, music, and verbal language processing. For example, it would be fascinating to know the effects of this substance on perceptions of tonal languages such as Chinese, Huichol, or Dogon; would it alter the words perceived as being spoken?

22.2.3.3 Ketamine

Ketamine (2-(2-chlorophenyl)-2-(methylamino)-cyclohexanone) is a noncompetitive NMDA receptor antagonist first synthesized in 1962. Its tendency to cause dissociation from the body and its actual state (in particular for people in pain) makes it a valuable type of medication in anesthesia. Because of these psychoactive properties, it is termed a "dissociative anesthetic."

Because it disrupts the normal stream of input from the peripheral to the central nervous system, ketamine can vastly distort the perception of the body. Experience of time and space as well as thinking are typically gravely altered. At higher doses, users may sometimes feel entirely disembodied, existing as a "point of consciousness" somewhere in the universe. They may experience the world and themselves as one entity, particularly with eyes closed to break another bond between themselves and their ordinary self-monitoring processes (see Chap. 7). Synesthesias may occur, particularly in the form of sound sensations with visual experiences, but auditory and open-eye visual hallucinations are rare. Users are generally aware of the influence of the drug, even though their reality-testing abilities may be limited (Kelly 1999).

As ketamine is metabolized, the individual slowly reorient, perhaps passing through moments where they feel as if they are surrounded by nonhuman consciousnesses. These unusual experiences resolve as the drug's course completes, although as memories they may remain compelling.

Dissociative phenomena might be expected to be distressing, but as one study of it as a postoperative pain medication stated, "A close supportive relationship with the surgeon and operating room personnel is probably as important as any pharmacologic manipulation in avoiding psychological mishap with low-dose ketamine." (Cunningham and McKinney 1983). Indeed, some mental distance from one's painful body in illness, injury, or surgery may come as a psychological mercy.

Ketamine is regulated in the USA at Schedule III, recognizing it as having a current medical use and stating that it has a relatively low abuse potential. In Canada, however, it is rated a Schedule I narcotic. Meanwhile, the World Health Organization considers ketamine a core element of its Essential Medicines List, as a general anesthetic.

22.3 What Happens After Drug-Induced Hallucinations?

Drug-induced hallucinations generally have a clear and typical duration, usually quite brief, after which the person's perceptions return to their baseline. Such time-limited effects are utterly unlike most psychiatric diseases, which tend to be more progressive and open-ended.

22.3.1 Effects of Experience

Hallucinatory experiences run the gamut from entirely unmemorable, to frightening, to triggering extensive life changes. All that can be added is that certain types of hallucinogens are prone to causing or emphasizing certain types of experience. However, no matter what hopes or fears or drugs one may bring to an experience, "few battle plans survive contact with the enemy," and this is as true among psychoactive substances as in other realms of human experience and endeavor.

22.3.1.1 The Good Friday Experiment

In 1962, Walter Pahnke, then an MD and also a PhD student in Harvard University's Religion and Society Program, gave psilocybin to 10 of 20 white Protestant male divinity students during Good Friday services, asking questions about the mysticism of their experience to determine the differential effects of the chemical. As detailed in Dr. Pahnke's resultant doctoral dissertation (Pahnke 1963), the Good Friday experiment found many elements of "deep mystical" experience reported by those who had received psilocybin, as opposed to those who had ingested a placebo.

In 1991, Rick Doblin published a follow-up study for which he tracked down 16 of the original participants. From his conclusions:

> For the psilocybin group, the long-term follow-up yielded moderately increased scores in the categories of internal and external unity, sacredness, objectivity and reality, and paradoxicality, while all other categories remained virtually the same as the six-month data. Several decades seem to have strengthened the experimental groups' characterization of their original Good Friday experience as having had genuinely mystical elements. For the controls, the only score that changed substantially was that of alleged ineffability, which decreased.
>
> A relatively high degree of persisting positive changes were reported by the experimental group while virtually no persisting positive changes were reported by the control group. In the open-ended portion of the long-term follow-up questionnaire, experimental subjects wrote that the experience helped them to resolve career decisions, recognize the arbitrariness of ego boundaries, increase their depth of faith, increase their appreciation of eternal life, deepen their sense of the meaning of Christ, and heighten their sense of joy and beauty. No positive persisting changes were reported by the control group in the open-ended section of the follow-up questionnaire.
>
> There was a very low incidence of persisting negative changes in attitudes or behavior in either group at either the six-month follow-up or the long-term follow-up. However, the one psilocybin subject reported to have had the most difficult time during the experiment was the one who declined this author's request to be interviewed in person or fill out a questionnaire, placing in question the generalizability of this finding for the long-term.
>
> (Doblin 1991)

In 2006, Griffiths and colleagues published a similar experiment using a more modern design, including methylphenidate for the placebo and 8-h private sessions that were comfortable and supportive, but not overtly religious (Griffiths et al. 2006). They found that psilocybin "produced a range of acute perceptual changes, subjective experiences, and labile moods including anxiety. Psilocybin also increased measures of mystical experience. At 2 months [post psilocybin], the volunteers rated the psilocybin experience as having substantial personal meaning and spiritual significance and attributed to the experience sustained positive changes in attitudes and behavior consistent with changes rated by community observers. [....] When administered under supportive conditions, psilocybin occasioned experiences similar to spontaneously occurring mystical experiences." This supports the notion that religious or mystical feelings may be based in physiological levels of perception and cognition.

22.3.1.2 Schizophrenic Break, Drug-Induced Psychosis, and HPPD

One particularly persistent piece of disinformation promoted by the War on Drugs is that hallucinogens make you mentally ill. This is not the case. Schizophrenia, for example, is remarkably consistent in its epidemiology at affecting 1% of the population all over the world (Patel et al. 2008).

However, it remains possible to have a hallucinogen-related experience so awful that it results in post-traumatic stress disorder, particularly if say one is victimized, arrested, or hospitalized during the process. The same is true of any traumatic experience, such as a car accident or rape, and in none of these cases is it appropriate to blame the victims. A difficult hallucinogenic experience may be helped with supportive care, potentially including medicine, though "talking through" is typically sufficient, as Bwiti and NAC clergy can attest.

Investigation is currently underway to discover if the so-called flashback syndrome, also known as hallucinogen persisting perception disorder (HPPD), is in fact a distinct clinical entity, or merely an interpretive change in ordinary peculiarities and individual differences in vision. In general, HPPD phenomena resolve spontaneously after the last exposure to a hallucinogen (over a time course of weeks to perhaps as long as a year), although there are also individuals reporting essentially permanent changes to visual perception. A number of HPPD patients describe milder forms of visual disturbance prior to hallucinogen exposure, suggesting that some individuals are more premorbidly susceptible to these rare adverse effects than others and that they may be related to changes in perceived salience as well as basic visual perception processing.

22.3.2 *Effects of Society*

Set, what persons bring of themselves to an experience, and setting, the environment in which the experience occurs, are critical to what happens in hallucinatory experiences (Zinberg 1986). This truism applies to other human undertakings as well, and in medicine in particular. The placebo effect depends heavily on the patient's expectations, whether for pain relief, the healing of a wart, or bronchodilation, or even unwholesome outcomes such as nausea and hypertension. This is unsurprising; we are a social species, and clues as to the appropriate metabolic state can be strongly affected by information from a trusted source – a physician, a priest, a parent. Brain structures such as the amygdala and neurochemicals such as endorphins can be regulated by social information as well as individual expectations and perceptions (Amanzio et al. 2001). Interestingly, those suffering from Alzheimer's disease lose the ability to respond to placebos, likely because their prefrontal cortices have lost the ability to form and maintain expectations (Benedetti et al. 2006).

Such varied sociobiological contexts may explain why a single substance such as psilocin can become an idle folktale in China and Japan, revered in Mesoamerica, and linked to criminality in the USA.

22.3.2.1 Religious Contexts

One common cultural response to hallucinogens is to treat them with reverence and surround them with ritual. Psilocin (as a psychoactive metabolite of administered psilocybin) increases mystical experiences relative to methylphenidate as an active placebo in drug-naïve adults, suggesting that this response is not arbitrary. Religions can also provide social support and control for experiences, such as the Bwiti iboga initiations, which can be physically and emotionally challenging, if also potentially rewarding. The Native American Church, with more than half a million adherents in the USA and Canada, attests to the scalability of this approach.

22.3.2.2 Illegal Contexts

In the majority of modern societies, the consumption of hallucinogens is illegal and socially stigmatized. Even in those subcultures in which it is not considered somehow immoral or dangerous, there is an outstanding threat that the hand of the law will descend upon a person who chooses to self-induce a hallucinatory state, depriving them of freedom, diminishing their social standing, and marking them as criminal.

Given this, it is nearly impossible to discuss a contemporary hallucinogenic experience without a certain amount of fear, except for those few which are protected by laws relating to freedom of religion, which necessarily fall into a class of their own. These fears must be recognized as part of the set and setting of modern hallucinogen consumption, but not as inherent to the substances or the experiences themselves.

While some cultures surrounded hallucinogen consumption with rules about appropriate usage, and others have essentially dismissed them as occasional aberrations in the course of daily life, modern Western culture is unique in its extent and breadth of condemnation. Disinformation distributed through both official and popular channels has been claimed as acceptable despite its untruth, simply because it might discourage potential consumption of a hallucinogen. The fear of hallucinations is quite powerful, as it can prevent general agreement on basic, empirically investigable scientific facts.

These factors must be considered in any analysis of psychoactive substances undertaken in the here and now. These stigmata may cause research subjects to lie, or researchers desirous of funding to spin their results to suit the dominant paradigm, whether by minimizing the hallucinatory potential of a drug in development or exaggerating the dangers of hallucinogen use. Needless to say, these tendencies can only hamper scientific understanding of hallucinogenic substances and their properties and endanger the health, safety, and freedom of society's members who might desire to experience hallucinations.

It can only be hoped that science's endless curiosity will continue to press the societal restraints upon these matters, and knowledge continue to accrue. The scientific process has tremendous resilience; those of us who would practice it need only to retain our faith and do our jobs. Further pursuit of such research may reveal that drug-induced hallucinatory phenomena offer a systematized route to observe not just how the brain processes hallucinations, but better clarify how our awareness of reality is consistently distorted.

References

Amanzio, M., Pollo, A., Maggi, G., Benedetti, F. (2001). Response variability to analgesics: a role for non-specific activation of endogenous opioids. *Pain*, 90, 205–215.

Arora, D. (1990). *Mushrooms demystified. Second edition*. Berkeley, CA: Ten Speed Press.

Benedetti, F., Arduino, C., Costa, S., Vighetti, S., Tarenzi, L., Rainero, I., Asteggiano, G. (2006). Loss of expectation-related mechanisms in Alzheimer's disease makes analgesic therapies less effective. *Pain*, 121, 133–144.

Capaldi, V.F. 2nd, Carr, R.B. (2010). Citalopram-induced hallucinations and delusions in a young adult. *General Hospital Psychiatry*, 32, 648.

Cunningham, B.L., McKinney, P. (1983). Patient acceptance of dissociative anesthetics. *Plastic and Reconstructive Surgery*, 72, 22–26.

De Smet, P.A.G.M., Bruhn, J.G. (2003). Ceremonial peyote use and its antiquity in the Southern United States. *HerbalGram*, 53, 30–33.

District Officers from Kamerun (1888). *Mitteilungen aus den Deutschen Schutzgebieten, Volume 1*, p. 49.

Doblin, R. (1991). Pahnke's "Good Friday Experiment": A long-term follow-up and methodological critique. *Journal of Transpersonal Psychology*, 23, Vol. 1.

Estrada, Á. (1981). *María Sabina: her life and chants.* Santa Barbara, CA: Ross-Erikson.

Fei, L., Tsien, J. (2009). Memory and the NMDA Receptors. *New England Journal of Medicine*, 361, 302–303.

Fernandez, J.W. (1982). *Bwiti: an ethnography of the religious imagination in Africa.* Princeton, NJ: Princeton University Press.

Fontanilla, D., Johannessen, M., Hajipour, A., Cozzi, N., Jackson, M., Ruoho, A. (2009). The hallucinogen N,N-Dimethyltryptamine (DMT) is an endogenous sigma-1 receptor regulator. *Science*, 323, 934–935.

Geyer, M.A., Vollenweider, F.X. (2008). Serotonin research: contributions to understanding psychoses. *Trends in Pharmacological Sciences*, 29, 445–453.

Ghansah, E., Kopsombut, P., Malleque, M.A., Brossi, A. (1993). Effects of mescaline and some of its analogs on cholinergic neuromuscular transmission. *Neuropharmacology*, 32, 169–174.

Glick S.D., Maisonneuve I.S. (1998). Mechanisms of antiaddictive actions of ibogaine. *Annals of the New York Academy of Sciences*, 844, 214–26.

Gouzoulis-Mayfrank, E., Heekeren, K., Neukirch, A., Stoll, M., Stock, C., Obradovic, M., Kovar, K.A. (2005). Psychological effects of (S)-ketamine and N,N-dimethyltryptamine (DMT): a double-blind, cross-over study in healthy volunteers. *Pharmacopsychiatry*, 38, 301–311.

Gouzoulis-Mayfrank, E., Schreckenberger, M., Sabri, O., Arning, C., Thelen, B., Spitzer, M., Kovar, K.A., Hermle, L., Büll, U., Sass, H. (1999). Neurometabolic effects of psilocybin, 3,4-methylenedioxyethylamphetamine (MDE) and d-methamphetamine in healthy volunteers A double-blind, placebo-controlled PET study with [18 F]FDG. *Neuropsychopharmacology*, 20, 565–581.

Griffiths, R., Richards, W., McCann, U. (2006). Psilocybin can occasion mystical-type experiences having substantial and sustained personal meaning and spiritual significance. *Psychopharmacology*, 187, 268–283.

Guzmán, G., Allen, J.W., Gartz, J. (2000). A worldwide geographical distribution of the neurotropic fungi, an analysis and discussion. *Annali di Museo Civico di Roverreto*, 14, 189–280.

Hoffer, A., Osmond, H. (1967). *The hallucinogens*. New York, NY: Academic Press.

Hofmann, A. (1969). LSD ganz persönlich. Translated from the original German by J. Ott (1996). In *Newsletter of the Multidisciplinary Association for Psychedelic Studies*, Vol. 6, Nr. 3.

Hofmann, A. (1980). *LSD: My problem child.* Translated by J. Ott. New York, NY: McGraw-Hill.

Kelly, K. (1999). *The little book of ketamine*. Berkeley, CA: Ronin Publishing.

Krystal, J.H., Karper, L.P., Seibyl, J.P., Freeman, G.K., Delaney, R., Bremner, J.D., Heninger, G.R., Bowers, M.B. Jr., Charney, D.S. (1994). Subanesthetic effects of the noncompetitive NMDA antagonist, ketamine, in humans. Psychotomimetic, perceptual, cognitive, and neuroendocrine responses. *Archives of General Psychiatry*, 51, 199–214.

Lotsof, H. (1985). US Patent 4,499,096, Rapid Method for Interrupting the Narcotic Addiction Syndrome.

Lotsof, H. (1992). US Patent 5,152,994, Rapid Method for Interrupting or Attenuating Poly-Drug Dependency Syndromes.

McKenna, T. (1992). *The archaic revival: Speculations on psychedelic mushrooms, the Amazon, virtual reality, UFOs, evolution, shamanism, the rebirth of the goddess, and the end of history.* New York, NY: HarperCollins.

Naranjo, C.C. (1974). *The healing journey*. New York, NY: Pantheon Books.

Pahnke, W.N. (1963). *Drugs and mysticism: An analysis of the relationship between psychedelic drugs and the mystical consciousness.* Thesis presented to the Committee on Higher Degrees in History and Philosophy of Religion. Cambridge, MA: Harvard University.

Patel, J., Pinals, D.A., Breier, A. (2008). Schizophrenia and other psychoses. In *Psychiatry, third edition.* Edited by Tasman, A., Kay, J., and Lieberman, J.A. Chichester: Wiley.

Passie, T., Halpern, J.H., Stichtenoth, D.O., Emrich, H.M., Hintzen, A. (2008). The pharmacology of lysergic acid diethylamide: A review. *CNS Neuroscience & Therapeutics*, 14, 295–314.

Riba, J., Romero, S., Grasa, E., Mena, E., Carrió, I., Barbanoj, M.J. (2006). Increased frontal and paralimbic activation following ayahuasca, the pan-Amazonian inebriant. *Psychopharmacology*, 186, 93–98.

Sanford, J.H. (1972). Japan's 'laughing mushrooms'. *Economic Botany*, 26, 174–181.

Schultes, R.E., Hofmann, A. (1973). *The botany and chemistry of hallucinogenic plants.* Springfield, IL: Charles C. Thomas.

Shimizu, S., Watanabe, K., Jibiki, I. (2010). Auditory hallucinations associated with clinical dosage of paroxetine: a case report. *Psychiatry and Clinical Neurosciences*, 64, 666–667.

Shulgin, A., Shulgin, A. (1991). *Phenethylamines I have known and loved.* Berkeley, CA: Transform Press.

Shulgin, A., Shulgin, A. (1997). *Tryptamines I have known and loved.* Berkeley, CA: Transform Press.

Trulson, M.E., Crisp, T., Henderson, L.J. (1983). Mescaline elicits behavioral effects in cats by an action at both serotonin and dopamine receptors. *European Journal of Pharmacology*, 96, 151–154.

Vollenweider, F.X., Geyer, M.A. (2001). A systems model of altered consciousness: integrating natural and drug-induced psychoses. *Brain Research Bulletin*, 56, 495–507.

Vollenweider, F.X., Leenders, K.L., Scharfetter, C., Maguire, P., Stadelmann, O., Angst, J. (1997a). Positron emission tomography and fluorodeoxyglucose studies of metabolic hyperfrontality and psychopathology in the psilocybin model of psychosis. *Neuropsychopharmacology*, 16, 357–372.

Vollenweider, F.X., Leenders, K.L., Scharfetter, C., Antonini, A., Maguire, P., Missimer, J., Angst J. (1997b). Metabolic hyperfrontality and psychopathology in the ketamine model of psychosis using positron emission tomography (PET) and [18 F]fluorodeoxyglucose (FDG). *European Neuropsychopharmacology*, 7, 9–24.

Wasson, G.W. (1980). *The wondrous mushroom: Mycolatry in Mesoamerica.* New York, NY: McGraw-Hill.

Wasson, V.P., Wasson, G.W. (1957). *Mushrooms, Russia, and history.* New York, NY: Pantheon Books.

Zinberg, N. (1986). *Drug, set, and setting: The basis for controlled intoxicant use*. New Haven, CT: Yale University Press.

Chapter 23
Examining the Continuum Model of Auditory Hallucinations: A Review of Cognitive Mechanisms

Johanna C. Badcock and Kenneth Hugdahl

23.1 Introduction

Faced with mounting evidence that auditory hallucinations occur both in health and in psychosis (Daalman et al. 2011; Sommer et al. 2010; Stip and Letourneau 2009; Van Os et al. 2009), the continuum model of psychotic symptoms has become the "accepted dogma" (David 2010). We would like to begin, however, by making a distinction between the term "auditory hallucinations," which we consider a symptom of psychosis, and "hearing voices," an experience that also occurs outside of the psychosis context (see Chap. 28). We would moreover like to make a distinction between the experience of hearing real voices (in the physical sense) and nonreal "voices" in the absence of an acoustic signal. Thus, we have used the terms auditory hallucinations, "hearing voices," and hearing voices to denote similar phenomenological experiences in patients diagnosed with schizophrenia and in individuals in the general population – in the absence and presence of a speech signal, respectively.

Despite the dominant influence of this model, careful phenomenological comparison suggests both similarities and differences between nonpsychotic and psychotic "voice hearers." For example, Daalman's study found that while the perceived location of the "voice" was the same in both groups, patients with psychosis experienced a more diminished sense of control (Daalman et al. 2011).

J.C. Badcock, Ph.D. (✉)
School of Psychiatry and Clinical Neurosciences,
University of Western Australia, Crawley, Australia

Centre for Clinical Research in Neuropsychiatry, Graylands Hospital, Claremont, Australia
e-mail: Jo.Badcock@health.wa.gov.au

K. Hugdahl, Ph.D.
Department of Biological and Medical Psychology, University of Bergen, Bergen, Norway

Division of Psychiatry, Haukeland University Hospital, Bergen, Norway
e-mail: Hugdahl@psybp.uib.no

J.D. Blom and I.E.C. Sommer (eds.), *Hallucinations: Research and Practice*,
DOI 10.1007/978-1-4614-0959-5_23, © Springer Science+Business Media, LLC 2012

The emotional characteristics of the experience also distinguished the two groups, with a more negative valence for "voices" experienced by individuals with psychosis. An intriguing (but also somewhat surprising) finding was that 78% of healthy "voice hearers" attributed the location of their "voices" to an external source, while only 58% of patients did so. Such differences begin to raise some doubts about the dimensional nature of these experiences; nonetheless, the influence of this model now extends to clinical care. For example, it is now recognized that the majority of hallucinations in the general population are transitory and do not necessitate a need for care, but those that persist are associated with an increased risk of developing psychosis (De Loore et al. 2011; Dominguez et al. 2011) – implying a continuum of disability. In practice, therefore, it makes sense that symptoms such as auditory hallucinations are used (e.g., in the clinical staging model) to ascertain an individual's risk for developing a psychotic illness (McGorry et al. 2007). Importantly, the continuum model of psychosis has also encouraged the view that auditory hallucinations and "hearing voices" in psychotic and nonpsychotic groups, respectively, rely on similar underlying cognitive and neural mechanisms. As a consequence, an increasing number of studies are being conducted on healthy "voice hearers," on the assumption that such studies will uncover the basic etiological mechanisms underlying all experiences of "hearing voices" – including symptoms of hallucinations in those diagnosed with schizophrenia – while confounding effects arising from medication, hospitalization, and illness duration are avoided.

Alternatively, phenomenological differences in psychotic and nonpsychotic groups suggest that some significant differences in the respective cognitive and neural mechanisms causing "voices" might also be expected. We decided, therefore, that a critical appraisal of this aspect of the continuum model would be timely and may have important clinical, etiological, and empirical implications. Consequently, we examined a range of studies from a variety of theoretical perspectives, including bottom-up, top-down, and memory-based models, as well those more specifically focused on the role of language, seeking comparable evidence from individuals in both groups.

23.2 Language Lateralization

Auditory hallucinations experienced by individuals diagnosed with schizophrenia usually entail "voices" talking to, or about, the patient. Not surprisingly, therefore, a large body of literature has examined the role of speech and language lateralization in psychosis (Sommer and Kahn 2009). There is a large body of literature showing that in healthy individuals the left hemisphere is dominant for language in general, with evidence from behavioral (Hellige 1990), lesion (Lezak 1994), and functional neuroimaging (Petersen et al. 1988) studies. There is a corresponding large literature showing a similar left-sided lateralization for speech perception, with evidence also from behavioral and neuroimaging studies (see Tervaniemi and Hugdahl (2003) and Cowell (2010) for reviews). Considering that the experience of "hearing voices" has, by definition, a profound perceptual characteristic, a straightforward prediction would be that a similar lateralization would be apparent for nonreal and real voice hearing. Studies of the lateralization of language and psychosis show, however, a complex picture, with both left- and right-hemisphere dominant activation. In a pioneering study Flor-Henry (1969) found that psychosis-like symptoms in patients diagnosed with temporal lobe epilepsy resembled those diagnosed with schizophrenia more when the focus of the epilepsy was in the left temporal lobe rather than in the right temporal lobe. Later studies have, however, not always replicated these original findings. Language processes seem to engage both the left (Hugdahl et al. 2008a, b) and right (Sommer et al. 2008, 2010; Sommer and Diederen 2009) hemispheres, or their disconnection (Ceccherini-Nelli et al. 2007) in patients diagnosed with schizophrenia, making it difficult to track down a primary pathology residing in one hemisphere or the other. However, in a recent review of both structural and functional brain-imaging data, Crow (2010) provides compelling evidence that a deficit of lateralization in patients diagnosed with schizophrenia in general seems to be associated with phonological rather than semantic/syntactic aspects of language. This observation would then nicely fit to a view of deficit in speech perception and phonology in patients diagnosed with schizophrenia as underlying the occurrence of auditory hallucinations. Such a view would, however, not be consistent with a continuum model of auditory hallucinations and "hearing voices" since there are no reports in the literature of a unique deficit in the lateralization of speech perception and phonology in "voice hearers" in the general population. A somewhat similar conclusion can be drawn from the study by Diederen et al. (2010), who compared brain activity in patients with auditory hallucinations with a group of nonclinical "voice hearers" and nonclinical non–"voice hearers." The results showed a reduced lateralization for language in the patient group, compared to the two nonclinical groups, which replicated previous findings of a deficit in left-hemisphere functioning in patients diagnosed with schizophrenia. More interesting, for the current discussion, is that there were no significant differences in language lateralization between the two nonclinical groups, thus supporting a noncontinuum view of auditory hallucinations in patients and "hearing voices" in the general population. The task used in the Diederen et al. (2010) study was, however, a verbal-fluency task requiring the subjects to covertly and overtly generate words that began with a certain letter shown on a screen in front of them. A word-production task will by its very nature also load on semantics in addition to phonology, which would engage different language processes and brain regions than a speech-perception task with a focus on phonology. Thus, it is still unsettled whether a patient group experiencing auditory hallucinations and a group of nonclinical "voice hearers" would differ in lateralization for a pure speech-perception, and thus pure phonetic task, for example, a dichotic listening task with presentations of simple consonant-vowel syllables that are devoid of semantic meaning (cf. Tervaniemi and Hugdahl 2003).

23.3 Bottom-Up Activity

The experience of "hearing voices" involves more than just speech. Voices, whether real or hallucinated, carry important information about the identity, emotion, and location of the speaker, which is processed in partially separable auditory pathways (Badcock 2010; Belin et al. 2011). Many patients report, for example, that they can identify the voice of another person, their gender or their accent, and phenomenological comparisons confirm that similar attributions to real and familiar people are made in healthy and psychotic "voice hearers" (Daalman et al. 2011). Consequently, it has been proposed that auditory hallucinations and "hearing voices" in patients and healthy individuals, respectively, may be viewed as the product of internal (bottom-up) activity in parallel voice-perception pathways which intrude into ongoing mental events (Badcock 2010; Hugdahl 2009).

In studies on cognition, such intrusive activity associated with auditory hallucinations has been examined in the guise of unwanted thoughts, images, and memories. Morrison and Baker (2000) were the first to examine the frequency of intrusive thoughts in patients diagnosed with schizophrenia, using the Distressing Thoughts Questionnaire (Clark and de Silva 1985). Hallucinating patients reported more frequent intrusions than nonhallucinating psychiatric or healthy controls, though subsequent studies – based on different questionnaires – have not consistently replicated this relationship (Linney and Peters 2007). Morrison (2005) has argued that auditory hallucinations can be conceptualized as a variation of normal intrusive thoughts, but that psychotic and nonpsychotic experiences differ as a result of the interpretation (or catastrophic misinterpretation) of intrusions that leads to distress and disability in individuals with psychosis. However, a recent meta-analysis has found little support for the role of metacognitive beliefs in hallucinatory experiences in either clinical or nonclinical samples (Varese and Bentall 2011). This is despite the fact that statistical modeling in healthy individuals prone to "hearing voices" has shown a strong and significant association with intrusive cognitions (Jones and Fernyhough 2009), suggesting that clinical and nonclinical "voice hearers" may indeed have a shared tendency to more frequent intrusive thoughts.

Auditory hallucinations have also been conceptualized as intrusive auditory imagery since both involve internal representations in the absence of external input. Initial reports suggested that individuals with clinical or nonclinical hallucinations experience more vivid auditory imagery. However, more recent evidence shows that mental-imagery vividness is independent of the presence of, or predisposition to, auditory hallucinations (Oertel et al. 2009). Following a critical analysis of the available data, Aleman and Larøi (2008) concluded that: "There is no convincing evidence of abnormalities in mental imagery ability in people who experience hallucinations. This does not imply that activation of mental images may not be central to hallucinations" (p. 91). In support of this conclusion, new evidence now shows that the neural networks engaged during auditory hallucinations show considerable overlap with those engaged during auditory imagery – both in psychotic and nonpsychotic individuals (Allen et al. 2008; Linden et al. 2011). Of note, this includes similar activity in the human-voice area in the superior temporal sulcus and adjacent areas, which may, in part, explain why hallucinated "voices" in both groups are typically perceived as real. Importantly, increased activation in voice-perception networks during auditory hallucinations would also be expected to add internal "noise" to ongoing cognitive processes, and might, therefore, be expected to result in impaired recognition of externally produced voices. Consistent with this prediction, the ability to distinguish familiar and unfamiliar voices was found to be impaired in patients diagnosed with schizophrenia who experienced auditory hallucinations, compared to nonhallucinating patients and controls (Zhang et al. 2008). However, similar evidence has yet to be reported in healthy "voice hearers," leaving open the possibility that both similarities and differences in voice identification may be present in clinical and nonclinical groups.

Another possibility is that auditory hallucinations involve intrusive memories (Hemsley 1993; Waters et al. 2006). A recent quantitative review of functional-imaging data has shown that auditory hallucinations in psychosis are associated with increased activation in the medial temporal lobes (Jardri et al. 2011), the region most often linked to verbal episodic memory. Memory-based models have sometimes been criticized (Jones 2008) on the basis that they can only account for intrusive memories of traumatic events, which form only a small proportion of auditory hallucinations, and are rather associated with anxiety disorders such as posttraumatic stress disorder (PTSD). However, such criticisms may be unfounded since the data clearly show that intrusions in both clinical and nonclinical "voice hearers" also arise in emotionally neutral conditions of free recall (Brébion et al. 2009, 2010). Furthermore, Waters et al. (2006) have shown that patients diagnosed with schizophrenia who experience auditory hallucinations often fail to form an integrated representation of an event in memory. Together, the findings suggest that intrusive recollections may comprise either individual features (words, voice identity) or complete episodes (memories of abuse) from memory and could potentially account for the diverse phenomenology across the continuum of hallucinated "voices."

23.4 Source Memory

Bottom-up models emphasize the fact that hallucinated "voices" involve multiple components (words, voice identity, emotion, and location) that must eventually be combined. One of the defining features of episodic memory is that events (e.g., words) are encoded along with details of the context in which they occurred (Ranganath 2010) – such as spatial location or voice pitch. This combined set of information allows the source of a memory to be correctly recalled (e.g., who said what, where the speakers were) and depends on distinct brain circuits in the medial temporal lobe (Ranganath 2010). For example, the parahippocampal and hippocampal cortices – in the auditory "where" pathway – play a critical role in encoding and binding spatial information in context. Spatial-source memory has received considerably less investigation in psychosis relative to the large number of studies assessing self/other source discrimination (reality monitoring) despite the fact that a recent meta-analytic review concluded there was no evidence for a differential deficit in reality monitoring in patients diagnosed with schizophrenia (Achim and Weiss 2008). Furthermore, phenomenologically, spatial-source memory has been identified as a key dimension of auditory hallucinations (Stephane et al. 2003), while structural brain imaging has shown that the perceived location of "voices" is associated with anatomical changes in the auditory "where" pathway (right temporoparietal junction) (Plaze et al. 2011). In contrast, other studies suggest that spatial-source memory is not impaired in healthy hallucination-prone individuals (Badcock et al. 2008; Chhabra et al. 2011; McKague et al. submitted). For example, Chhabra et al. showed intact integration of external voice and location information in memory in healthy adults predisposed to hallucinations, while McKague et al. manipulated the perceived internal/external location of auditory stimuli, and found no association between the accuracy or bias in performance and hallucination-proneness. Thus potentially important dissimilarities in spatial-source memory may exist in psychotic auditory hallucinations and "hearing voices," respectively, which clearly warrant further investigation.

23.5 Top-Down Control

It seems inevitable that intrusive activity associated with auditory hallucinations will require some form of top-down or "executive" control, and indeed this has been the focus of many cognitive accounts (Badcock 2010; Hugdahl 2009). Three key separable components of executive control have been identified, i.e., shifting between response sets, updating working memory, and inhibiting prepotent responses or response tendencies (Miyake et al. 2000). The notion that dysfunction of volitional inhibition may be a critical mechanism underlying auditory hallucinations has gained support from a variety of recent studies involving both patients diagnosed with schizophrenia (Badcock et al. 2005; Soriano et al. 2009; Waters et al. 2003) and healthy "voice hearers" (Paulik et al. 2007, 2008). For example, Soriano et al. used a directed forgetting task to measure the ability to suppress recently acquired information. The task involves the presentation of two lists of words for later recall. The first list is followed by an instruction to "forget" the items just learned, while the second list is accompanied by an instruction to "remember." At recall, the participants must remember words from both lists. Typically, fewer list-1 items are recalled in the forget condition than in the remember condition, indicating that suppression has indeed occurred. The results of the study by Soriano et al. (2009) showed that patients with auditory hallucinations failed to produce a reliable directed forgetting effect compared to patients without hallucinations, indicating a significantly impaired ability to intentionally suppress irrelevant information in memory. The degree of inhibitory failure was shown to be significantly correlated with the frequency of auditory hallucinations (but not with other symptoms), a finding that had previously been reported by Waters et al. (2003) based on different measures of inhibitory function. It seems likely, therefore, that at least some forms of inhibitory control are *specifically* associated with hearing "voices" both in clinical and nonclinical populations. In addition, it seems likely that difficulties in executive control reflect underlying abnormalities in prefrontal cortex (Pomarol-Clotet et al. 2010; Hugdahl 2009) since this region is usually associated with inhibitory processing (Koechlin et al. 2003).

Finally, it must also be noted that although these cognitive studies point to similar mechanisms of top-down control in healthy and psychotic "voice hearers," these groups differed in their subjective sense of control (Daalman et al. 2011). Hence, the mechanisms underlying objective and subjective control both in health and in psychosis may well be far more complex than is currently understood.

23.6 Discussion and Conclusions

Although the continuum model of psychotic symptoms has been quite influential and has accumulated evidence for similar etiological mechanisms in clinical and nonclinical populations, there is a risk of accepting the continuum view unchallenged because alternative causal models or experimental designs – and more effective clinical interventions – may not be developed, and inconsistent data may be inadvertently overlooked.

Against this background, our search revealed evidence of only partial overlap in the cognitive mechanisms associated with hallucinated "voices" in psychotic and healthy individuals (see Fig. 23.1), including elevated rates of intrusive cognitions and poor executive control. These results suggest that, regardless of the presence or absence of psychosis, a similar neural network is involved in generating auditory hallucinations and "hearing voices" in psychotic and nonpsychotic populations, respectively, including both left temporal (speech-perception and voice-selective areas) and prefrontal (cognitive control and executive areas) cortices (Allen et al. 2008; Jardri et al. 2011). Consistent with this conclusion, Diederen et al. (2011) recently identified several common areas of activity – including bilateral inferior frontal cortex, the superior temporal gyri, the inferior parietal lobe, and the insula – during hallucinations in psychotic and nonpsychotic individuals. Although these results appear to implicate the same cortical network in both groups, the authors caution that similar patterns of brain activity could be triggered by different causal mechanisms that merge in a final common pathway. In this context the results of the current review may be particularly valuable since they suggest subtle but significant differences in language lateralization and cognitive processing in psychotic and nonpsychotic hallucinators and "voice hearers." For example, the observed difficulties binding external sources of information (e.g., voice and location) are typically linked to abnormal activity in the hippocampus and the parahippocampal cortex (Ranganath 2010), pointing to a particular role of the medial-temporal-lobe memory system in the hallucinations of psychosis (Badcock 2010; Waters et al. 2006). Several independent lines of research converge on a similar conclusion, including evidence

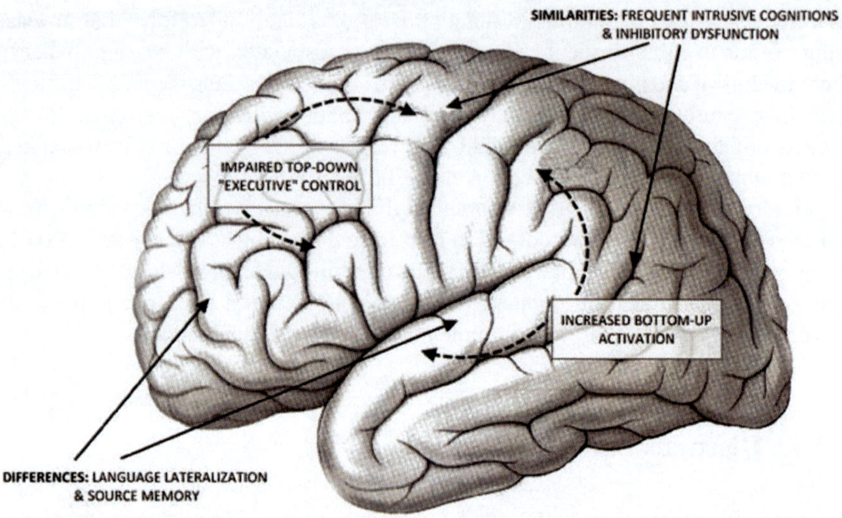

Fig. 23.1 Schematic illustration of similarities and differences in cognitive and neural processing in psychotic and healthy "voice hearers." Both groups share intrusive activations in separable auditory pathways and deficits in frontally mediated executive control. Reduced lateralization of word production and impaired spatial-source memory are only associated with auditory hallucinations in psychosis

of progressive deterioration in memory function (Frommann et al. 2011) and subtle signs of hippocampal pathology (Wood et al. 2010) associated with the onset of psychotic symptoms. Furthermore, significant deactivation of the parahippocampal gyrus occurs immediately prior to the onset of hallucinatory episodes (Diederen et al. 2010; Hoffman et al. 2008). Thus abnormal memory function may provide a critical "tipping point" or trigger in the genesis of auditory hallucinations.

On the basis of our findings, it is tempting to conclude that the similarities in voice features (e.g., attributions to real or familiar people) in psychotic and healthy individuals arise from dysfunctional activity in the same cognitive and neural substrates (e.g., intrusive representations from the human-voice area). We would like to take a more conservative standpoint, however, and suggest that a better model of the etiology of auditory hallucinations and "voice hearing" should combine evidence from both discontinuities and continuities models. However, this assumption must be empirically tested in future studies by examining psychotic and nonpsychotic subgroups that are closely matched phenomenologically. An important issue, when comparing clinical and nonclinical groups, with regard to auditory hallucinations and "hearing voices" is the extent to which the nonclinical subjects represent a "pure" nonclinical population. It is customary, therefore, to exclude individuals who have received a psychiatric diagnosis and/or medication for their "voices" (Sommer et al. 2010). This does not necessarily exclude individuals who consult a physician, but are not given a diagnosis. Thus, perhaps, a better criterion would be to exclude all subjects who have ever “contacted a physician or psychologist” because of worries about their “voices.” The argument is that it is not possible to get a diagnosis or have antipsychotic medication prescribed without “contacting” a physician (at least not legally), while the opposite is possible, i.e., contacting a physician without getting a diagnosis.

Our findings also highlight that research designs based solely on healthy “voice hearers” may be neither optimal nor sufficient for understanding hallucinations in clinical groups since they are likely to miss critical cognitive mechanisms, such as those linked to abnormal activity in the hippocampus and parahippocampal cortex in the medial temporal lobes. Conversely, many cognitive studies of psychosis compare patients with and without auditory hallucinations and conclude that only those cognitive processes *specific* to hallucinations are *relevant* to their etiology. Such conclusions may underestimate the importance of cognitive processes that underpin clusters of symptoms, including but not limited to auditory hallucinations. This point may be particularly salient when excluding other positive symptoms in the analysis since auditory hallucinations show high correlations phenomenologically (e.g., with delusions and conceptual disorganization), which should be expected to correlate with other cognitive domains and with brain activity in the same areas as for auditory hallucinations. Finally, wider recognition of the differences, as well as similarities, of “voice hearing” and auditory hallucinations in healthy and psychotic individuals, respectively, should encourage clinicians to conduct more detailed assessments of phenomenology and cognition in patients presenting with “voices” and develop more targeted (i.e., individualized) pharmacological and/or psychosocial interventions as necessary.

References

Achim, A.M., Weiss, A.P. (2008). No evidence for a differential deficit of reality monitoring in schizophrenia: A meta-analysis of the associative memory literature. *Cognitive Neuropsychiatry*, 13, 369–384.

Aleman, A., Larøi, F. (2008). *Cognitive-perceptual processes: Bottom-up and top-down.* In *Hallucinations: The science of idiosyncratic perception.* Washington, DC: American Psychological Association, pp. 91–108.

Allen, P., Larøi, F., Aleman, A., McGuire, P.K. (2008). The hallucinating brain: A review of structural and functional neuroimaging studies of hallucinations. *Neuroscience and Biobehavioral Reviews*, 32, 175–191.

Badcock, J.C., Chhabra, S., Maybery, M.T., Paulik, G. (2008). Context binding and hallucination predisposition. *Personality and Individual Differences*, 45, 822–827.

Badcock, J.C., Waters, F.A.V., Maybery, M.T., Michie. P.T. (2005). Auditory hallucinations: Failure to inhibit irrelevant memories. *Cognitive Neuropsychiatry*, 10, 125–136.

Badcock, J.C. (2010). The cognitive neuropsychology of auditory hallucinations: A parallel auditory pathways framework. *Schizophrenia Bulletin*, 36, 576–584.

Belin, P., Bestelmeyer, P.E.G., Latinus, M., Watson, R. (2011). Understanding voice perception. *British Journal of Psychology*, 102, 711–725.

Brébion, G., David, A.S., Bressan, R.A., Ohlsen, R.I., Pilowski, L.S. (2009). Hallucinations and two types of free-recall intrusion in schizophrenia. *Psychological Medicine*, 39, 917–926.

Brébion, G., Larøi, F., Van der Linden, M. (2010). Associations of hallucination proneness with free-recall intrusions and response bias in a non-clinical sample. *Journal of Clinical and Experimental Neuropsychology*, 32, 847–854.

Ceccherini-Nelli, A., Turpin-Crowther, K., Crow, T.J. (2007). Schneider's first rank symptoms and continuous performance disturbance as indices of dysconnectivity of left- and right hemispheric components of language in schizophrenia. *Schizophrenia Research*, 90, 203–213.

Chhabra, S., Badcock, J.C., Maybery, M.T. (2011). Context binding and hallucination predisposition: Evidence of intact intentional and automatic integration of external features. *Personality and Individual Differences*, 50, 834–839.

Clark, D.A., de Silva, P. (1985). The nature of depressive and anxious intrusive thoughts: Distinct or uniform phenomena? *Behaviour Research and Therapy*, 23, 383–393.

Cowell, P.E. (2010). *Auditory laterality: Recent findings in speech perception*. In *The two halves of the brain.* Edited by K. Hugdahl and R. Westerhausen. Cambridge, MA: MIT Press, pp. 349–379.

Crow, T.J. (2010). The origin of psychosis and the evolution of language: Cerebral asymmetry, sex chromosomes, and speciation. In *The two halves of the brain.* Edited by K. Hugdahl and R. Westerhausen. Cambridge, MA: MIT Press, pp. 621–661.

Daalman, K., Boks, M., Diederen, K., De Weijer, A., Blom, J.D., Kahn, R.S., Sommer I.E. (2011). The same or different? A phenomenological comparison of auditory verbal hallucinations in healthy and psychotic individuals. *Journal of Clinical Psychiatry*, 72, 320–325.

David, A.S. (2010). Why we need more debate on whether psychotic symptoms lie on a continuum with normality. *Psychological Medicine*, 40, 1935–1942.

De Loore, E., Gunther, N., Drukker, M., Feron, F., Sabbe, B., Deboutte, D., Van Os, J., Myin-Germys, I. (2011). Persistence and outcome of auditory hallucinations in adolescence: A longitudinal general population study of 1800 individuals. *Schizophrenia Research*, 127, 252–256.

Diederen, K., Daalman, K., De Weijer, A., Neggers, S.F.W., Van Gastel, W., Blom, J.D., Kahn, R.S., Sommer I.E. (2011). Auditory hallucinations elicit similar brain activation in psychotic and nonpsychotic individuals. *Schizophrenia Bulletin* (Epub ahead of print Apr 28).

Diederen, K.M., De Weijer, A.D., Daalman, K., Blom, J.D., Neggers, S.F., Kahn, R.S., Sommer, I.E. (2010). Decreased language lateralization is characteristic of psychosis, not auditory hallucinations. *Brain*, 133, 3734–3744.

Dominguez, M.D., Wichers, M., Lieb, R., Wittchen, H.U., Van Os, J. (2011). Evidence that onset of clinical psychosis is an outcome of progressively more persistent subclinical psychotic experiences: An 8 year cohort study. *Schizophrenia Bulletin*, 37, 84–93.

Flor-Henry, P. (1969). Psychosis and temporal lobe epilepsy: A controlled investigation. *Epilepsia*, 10, 263–395.

Frommann, I., Pukrup, R., Brinkmeyer, J., Bechdolf, A., Ruhrmann, S., Berning, J., Decker, P., Riedel, M., Möller, H.-J., Wölwer, W., Gaebel, W., Klosterkötter, J., Maier, W., Wagner, M. (2011). Neuropsychological profiles in different at-risk states of psychosis: Executive control impairment in the early- and additional memory dysfunction in the late-prodromal state. *Schizophrenia Bulletin*, 37, 861–873.

Hellige, J. (1990). Hemispheric asymmetry. *Annual Review of Psychology*, 41, 55–80.

Hemsley, D. (1993). A simple (or simplistic?) cognitive model for schizophrenia. *Behaviour Research and Therapy*, 31, 631–646.

Hoffman, R.E., Anderson, A., Varanko, M., Gore, J., Hampson, M. (2008). Time course of regional brain activation associated with onset of auditory/verbal hallucinations. *British Journal of Psychiatry*, 193, 424–425.

Hugdahl, K. (2009). "Hearing voices": Auditory hallucinations as failure of top-down control of bottom-up perceptual processes. *Scandinavian Journal of Psychology*, 50, 553–560.

Hugdahl, K., Løberg, E.M., Jørgensen, H.A., Lundervold, A., Lund, A., Green, M.F., Rund, B. (2008a). Left hemisphere lateralisation of auditory hallucinations in schizophrenia: A dichotic listening study. *Cognitive Neuropsychiatry*, 13, 166–179.

Hugdahl, K., Løberg, E.M., Specht, K., Steen, V.M., Wageningen, H., Jørgensen, H.A. (2008b). Auditory hallucinations in schizophrenia: The role of cognitive, brain structural and genetic disturbances in the left temporal lobe. *Frontiers in Human Neuroscience*, 1, 1–10.

Jardri, R., Pouchet, A., Pins, D., Thomas, P. (2011). Cortical activations during auditory verbal hallucinations in schizophrenia: A coordinate-based meta-analysis. *American Journal of Psychiatry*, 168, 73–81.

Jones, S.R. (2008). Do we need multiple models of auditory verbal hallucinations? Examining the phenomenological fit of cognitive and neurological models. *Schizophrenia Bulletin*, 36, 566–575.

Jones, S.R., Fernyhough, C. (2009). Rumination, reflection, intrusive thoughts, and hallucination-proneness: Towards a new model. *Behaviour Research and Therapy*, 47, 54–59.

Koechlin, E., Ody, C., Kouneiher, F. (2003). The architecture of cognitive control in the human prefrontal cortex. *Science*, 302, 1181–1185.

Lezak, M. (1994). *Neuropsychological assessment. Fourth edition.* Oxford: Oxford University Press.

Linden, D.E.J., Thornton, K., Kuswanto, C.N., Johnston, S., Van de Ven, V., Jackson, M.C. (2011). The brain's voices: Comparing nonclinical auditory hallucinations and imagery. *Cerebral Cortex*, 21, 330–337.

Linney, Y.M., Peters, E.R. (2007). The psychological processes underlying symptoms of thought interference in psychosis. *Behaviour Research and Therapy*, 45, 2726–2741.

McGorry, P.D., Purcell, R., Hickie, I.B., Yung, A.R., Pantelis, C., Jackson, H.J. (2007). Clinical staging: A heuristic model for psychiatry and youth mental health. *Medical Journal of Australia*, 187, S40–S42.

McKague, M., McAnally, K.I., Puccio, F., Bendall, S., Jackson, H.J. Hearing voices: Spatial source monitoring of words does not predict proneness to auditory-verbal hallucinations (submitted).

Morrison, A.P. (2005). *Psychosis and the phenomenon of unwanted intrusive thoughts.* In *Intrusive thoughts in clinical disorders: Theory, research and treatment.* Edited by D.A. Clark. New York, NY: Guilford Press, pp.175–198.

Morrison, A.P., Baker, C.A. (2000). Intrusive thoughts and auditory hallucinations: A comparative study of intrusions in psychosis. *Behaviour Research and Therapy*, 38, 1097–1106.

Miyake, A., Friedman, N.P., Emerson, M.J., Witzki, A.H., Howerter, A., Wager, T.D. (2000). The unity and diversity of executive functions and their contributions to complex "frontal lobe" tasks: A latent variable analysis. *Cognitive Psychology*, 41, 49–100.

Oertel, V., Rotarska-Jagiela, A., Van de Ven, V., Haenschel, C., Grube, M., Stangier, U., Maurer, K., Linden, D.E.J. (2009). Mental imagery vividness as a trait marker across the schizophrenia spectrum. *Psychiatry Research*, 167, 1–11.

Paulik, G., Badcock, J.C., Maybery, M. (2007). Poor intentional inhibition in individuals predisposed to hallucinations. *Cognitive Neuropsychiatry*, 12, 457–470.

Paulik, G., Badcock, J.C., Maybery, M. (2008). Dissociating the components of inhibitory control involved in predisposition to hallucinations. *Cognitive Neuropsychiatry*, 13, 33–46.

Petersen, S.E., Fox, P.T., Posner, M.I., Mintun, M., Raichle, M.E. (1988). Positron emission tomographic studies of the cortical anatomy of single-word processing. *Nature*, 331, 585–589.

Plaze, M., Paillère-Martinot, M., Penttilä, J., Januel, D., De Beaurepaire, R., Andoh, J., Bellivier, F., Galinowski, A., Gallarda, T, Artiges, E., Olie, J.-P., Mangin, J.-F., Martinot, J.-L., Cachia, A. (2011). "Where do auditory hallucinations come from?" - A brain morphometry study of schizophrenia patients with inner or outer space hallucinations. *Schizophrenia Bulletin*, 37, 212–221.

Pomarol-Clotet, E., Canales-Rodriguez, E.R,. Salvador, R., Sarro, S., Gomar, J.J., Vila, F., Ortiz-Gil, J., Iturria-Medina, Y., Capdevila, A., McKenna, P.J. (2010). Medial prefrontal cortex pathology in schizophrenia as revealed by convergent findings from multimodal imaging. *Molecular Psychiatry*, 15, 823–830.

Ranganath, C. (2010). Binding items and contexts: The cognitive neuroscience of episodic memory. *Current Directions in Psychological Science*, 19, 131–137.

Sommer, I.E.C., Daalman, K., Rietkerk, T., Diederen, K.M., Bakker, S., Wijkstra, J., Boks, M.P.M. (2010). Healthy individuals with auditory hallucinations: Who are they? Psychiatric assessments of a selected sample of 103 subjects. *Schizophrenia Bulletin*, 36, 633–641.

Sommer, I.E., Diederen, K.M. (2009). Language production in the non-dominant hemisphere as a potential source of auditory verbal hallucinations. *Brain*, 132, e124.

Sommer, I.E.C., Diederen, K.M.J., Blom, J.-D., Willems, A., Kushan, L., Slotema, K., Boks, M.P.M., Daalman, K., Hoek, H.W., Neggers, S.F.W., Kahn, R.S. (2008). Auditory verbal hallucinations predominantly activate the *right* inferior frontal area. *Brain*, 131, 3169–3177.

Sommer, I.E.C., Kahn, R.S. (2009). *Language lateralization and psychosis*. Cambridge: Cambridge University Press.

Soriano, M.F., Jiménez, J.F., Román, P., Bajo, T. (2009). Inhibitory processes in memory are impaired in schizophrenia: Evidence from retrieval induced forgetting. *British Journal of Psychology*, 100, 661–673.

Stephane, M., Thuras, P., Nasrallah, H., Georgopoulos, A.P. (2003). The internal structure of the phenomenology of auditory verbal hallucinations. *Schizophrenia Research*, 61, 185–193.

Stip, E., Letourneau, G. (2009). Psychotic symptoms as a continuum between normality and pathology. *Canadian Journal of Psychiatry*, 54, 140–151.

Tervaniemi, M., Hugdahl, K. (2003). Lateralization of auditory-cortex functions. *Brain Research Reviews*, 43, 231–246.

Van Os, J., Linscott, R.J., Myin-Germeys, I., Delespaul, P., Krabbendam, L. (2009). A systematic review and meta-analysis of the psychosis continuum: Evidence for a psychosis proneness persistence-impairment model of psychotic disorder. *Psychological Medicine*, 39, 179–195.

Varese, F., Bentall, R.P. (2011). The metacognitive beliefs account of hallucinatory experiences: A literature review and meta-analysis. *Clinical Psychology Review*, 31, 850–864.

Waters, F.A.V., Badcock, J.C., Maybery, M., Michie, P.T. (2003). Inhibition in schizophrenia: Association with auditory hallucinations. *Schizophrenia Research*, 62, 275–280.

Waters, F.A.V., Badcock, J.C., Michie, P.T., Maybery, M. (2006). Auditory hallucinations in schizophrenia: Intrusive thoughts and forgotten memories. *Cognitive Neuropsychiatry*, 11, 65–83.

Wood, S.J., Kennedy, D., Phillips, L.J., Seal, M.L., Yucel, M., Nelson, B., Yung, A.R., Jackson, G., McGorry, P.D., Velakoulis, D., Pantelis, C. (2010). Hippocampal pathology in individuals at ultra-high risk for psychosis: A multi-modal magnetic resonance study. *NeuroImage*, 52, 62–68.

Zhang, Z.-J., Hao, G.F., Shi, J.B., Mou, X.D., Yao, Z-J., Chen, N. (2008). Investigation of the neural substrates of voice recognition in Chinese schizophrenic patients with auditory verbal hallucinations: An event-related functional MRI study. *Acta Psychiatrica Scandinavica*, 118, 272–280.

Part IV
Treatment

Chapter 24
Classical Somatic Treatments: Pharmacotherapy and ECT

Iris E.C. Sommer and Jan Dirk Blom

24.1 Introduction

The treatment of hallucinations rests basically on psychoeducation, psychosocial interventions, psychotherapy, medication, and a number of additional somatic therapies. The present chapter will focus on medication and electroconvulsive treatment (ECT), whereas other types of treatment will be discussed elsewhere in this book. We will offer recommendations for the pharmacological and electroconvulsive treatment of hallucinations in psychotic disorders, Parkinson's disease, dementia, delirium, epilepsy, and sensory impairment, although the main focus will be on psychosis.

24.2 Assessing the Need for Treatment

Hallucinations occur in the context of many different disorders and syndromes. Therefore, the choice for a specific type of treatment does not only depend on the type of hallucination and its consequences for daily functioning, but also on the underlying disorder. If hallucinations occur in the context of epilepsy, for example, treatment should be focused on seizure control. In delirium, treatment should primarily be

I.E.C. Sommer, M.D., Ph.D. (✉)
Department of Psychiatry, University Medical Center Utrecht,
Rudolf Magnus Institute of Neuroscience, Utrecht, The Netherlands
e-mail: i.sommer@umcutrecht.nl

J.D. Blom, M.D., Ph.D.
Parnassia Bavo Academy, Parnassia Bavo Group and University of Groningen, Groningen,
Kiwistraat 43, 2552 DH, The Hague, The Netherlands
e-mail: jd.blom@parnassia.nl

J.D. Blom and I.E.C. Sommer (eds.), *Hallucinations: Research and Practice*,
DOI 10.1007/978-1-4614-0959-5_24, © Springer Science+Business Media, LLC 2012

directed at improving somatic health. In patients who develop hallucinations in the course of progressive vision or hearing loss, it can be helpful to apply interventions aimed at restoring the loss of function at hand, such as a cataract operation or the providing of hearing aids (Tuerlings et al. 2009; Cope and Baguley 2009). Some individuals who hallucinate only sporadically may be merely concerned that their experiences are a sign of mental disease, without being troubled by the hallucinations themselves. For others, the burden of their hallucinations may not outweigh the side effects of treatment. As a consequence, pharmacological treatment may not be necessary in all cases at hand. In psychotic patients, however, comorbid delusions may be an important reason to advocate antipsychotic treatment, even when the patient himself is reluctant to consider this. Meanwhile, one should be transparent about possible side effects. All in all, the potential benefits of treatment should be weighed carefully against acute as well as long-term side effects, especially in cases of mild and/or transient types of hallucination.

24.3 Pharmacological Treatment of Hallucinations in Schizophrenia Spectrum Disorders

The only type of medication known for its potential to effectively reduce the frequency and severity of hallucinations in schizophrenia spectrum disorders is antipsychotic medication. An important benefit of this type of medication is that it can also diminish concurrent delusions. So far, no clinical trials have been published that compare the efficacy of various antipsychotic drugs for the sole and specific indication of hallucinations. Therefore, we used the data from the European First Episode Schizophrenia Trial (EUFEST) to assess the potential of five antipsychotic agents to reduce the severity of hallucinations. The EUFEST study (Kahn et al. 2008) assessed 498 patients with a first psychotic episode, who were randomized to receive either haloperidol, olanzapine, amisulpride, quetiapine, or ziprasidone in an open-label design. The reduction of the total symptoms was virtually the same in all groups, lying around 60% after 12 months of treatment, although some major differences were observed in the discontinuation rate (Kahn et al. 2008). We reanalyzed those data with item P3 (Severity of Hallucinations) of the Positive and Negative Syndrome Scale (PANSS, Kay et al. 1987) as the primary outcome measure.

In the PANSS interview, the severity of hallucinations is scored on a six-point Likert scale (i.e., None, Questionable, Mild, Moderate, Marked, Severe). All subjects with a score of 3 or higher at baseline were included ($N=362$; 73% of the total sample). The number of subjects decreased over time, mostly due to treatment discontinuation. Latent growth curve (LGC) analyses (Muthén and Curran 1997) were performed to assess the change in hallucination item scores while including age (standardized), gender, and country as covariates. Even though 54% of the patients discontinued treatment within 12 months, unbiased parameter estimates were obtained under the assumption of missing at random. Out of these data, we extracted

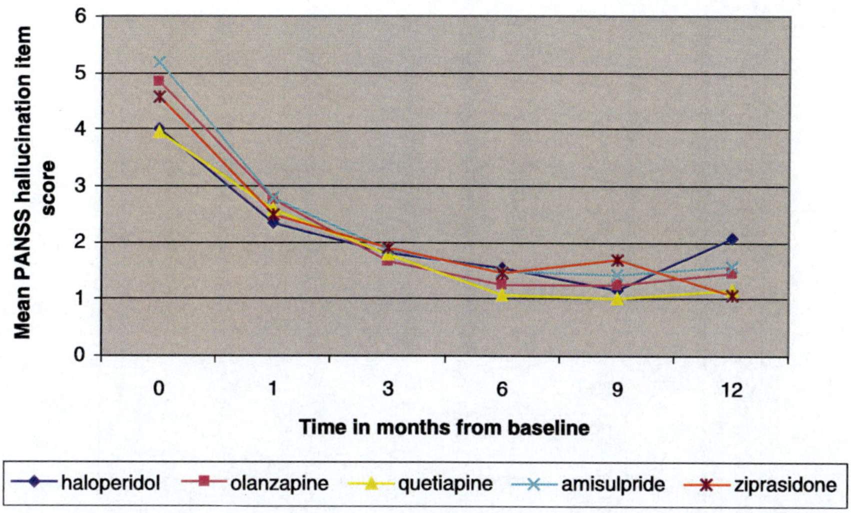

Fig. 24.1 Mean decrease in hallucination severity (item P3 of the PANSS) in first-episode patients with a nonaffective psychotic disorder after 1, 3, 6, 9, and 12 months on antipsychotic medication

a reduction of the severity of hallucinations from 4.4 on the PANSS item for hallucinations at baseline, indicating marked to severe hallucinations, to a mean value of 2.5, indicating mild to moderate hallucinations after 4 weeks. The severity of hallucinations continued to decline with prolonged treatment to mean values of around 1.5, reflecting the presence of questionable or mild hallucinations after 6 months of treatment (see Fig. 24.1 and Table 24.1). Likewise, the percentage of subjects with at least moderate levels of hallucinations decreased strongly over time from 100% at baseline to 8% after 12 months. These findings indicate that hallucinations in patients with a first episode of psychotic disorder respond fairly well to treatment, showing a sharp reduction in symptom severity occurring in the first month. After 1 year of pharmacotherapy, only 8% of the patients who continued their medication went on to experience hallucinations of moderate severity.

This result should be encouraging for patients suffering from hallucinations, and it might help them to decide in favor of pharmacotherapy. In parallel to the findings of the overall analysis of the EUFEST data, no differences in hallucination reduction between the five treatment groups were observed in those individuals who completed their treatment. Although these findings cannot be extrapolated to other antipsychotic drugs, they suggest that the most commonly prescribed antipsychotics are equally effective against hallucinations in patients with a first psychotic episode. However, the drop-out rates differed considerably among the five groups, with relatively high rates for the haloperidol group and relatively low rates for the amisulpride and olanzapine groups. This indicates that tolerability should be an important factor in the selection of an antipsychotic drug.

Table 24.1 Summary of augmentation strategies and their (mean) standardized differences (Taken from Sommer et al. 2011)

Augmentation strategy	Studies (N)	Subjects (N)	Hedges' g, 95% CI, and I^2 for PANSS/BPRS total	Hedges' g, 95% CI, and I^2 for positive subscores	Hedges' g, 95% CI, *p* and I^2 for negative subscores
Antiepileptics	7	189			
Lamotrigine	5	143	**0.53, 0.03–1.04, 60%**	0.38, −0.02–0.78, 39%	0.41, −0.13–0.94, 64%
Minus outlier	4	92	0.27, −0.10–0.65, 0%	0.15, −0.22–0.52, 0%	0.12, −0.25–0.49, 0%
Topiramate	3	89	0.75, −0.05–1.56, 69%	**0.63, 0.03–1.23, 47%**	0.66, −0.17–1.5, 71%
Minus outlier	2	57	0.38, −0.13–0.89, 0%	0.39, −0.24–1.01, 25%	
Antidepressants	4	129			
Citalopram	1	61	**0.81, 0.30–1.33**	0.28, −0.22–0.79	**0.81, 0.30–1.33**
Fluoxetine	1	33	No data	0.12, −0.55–0.79	0.19, −0.48–0.86
Mirtazapine	2	35	2.91, −2.69–8.51, 96%	0.04, −0.59–0.67, 0%	1.20, −0.25–2.66, 76%
Antipsychotics	10	548			
Amisulpride	1	20	0.13, −0.48–0.74	0.11, −0.50–0.72	0.21, −0.40–0.82
Aripiprazole	2	268	0.12, −0.12–0.36, 0%	0.22, −0.02–0.46, 0%	0.37, −0.19–0.93, 74%
Haloperidol	1	6	−0.15, −1.51–1.21	0.26, −1.11–1.62	−0.31, −1.68–1.06
Risperidone	5	226	0.18, −0.21–0.57, 53%	0.09, −0.24–0.74, 56%	0.22, −0.14–0.57, 43%
Sulpiride	1	28	**0.83, 0.07–1.59**	**0.77, 0.02–1.52**	**0.76, 0.01–1.51**
Glutamatergics	7	137			
CX516	1	18	**1.35, 0.32–2.38**	0.20, −0.74–1.14	**1.43, 0.38–2.46**
D-cycloserine	1	11	No data	No data	−0.76, −1.59–0.08
D-serine	1	20	No data	0.40, −0.45–1.24	0.33, −0.52–1.17
Glycine	3	68	−0.16, −0.62–0.30, 0%	−0.36, −1.19–0.46, 67%	−0.14, −0.60–0.32, 0%
Sarcosine	1	20	−0.21, −1.06–0.63	−0.07, −0.91–0.77	−0.07, −0.91–0.77

Significant effects are indicated in bold

As regards their side effects, antipsychotics can be divided roughly into those predominantly inducing weight increase and sedation (i.e., quetiapine, olanzapine, and clozapine) and those frequently associated with dystonia, parkinsonism, and akathisia (i.e., all other antipsychotic drugs). The guidelines issued for treatment in schizophrenia by the Patient Outcomes Research Team (PORT) do not recommend olanzapine or clozapine as drugs of first choice because of the severe weight gain they may induce (Buchanan et al. 2010). Sedation should never be a treatment goal in and of itself, although it may sometimes be welcomed as a side effect in agitated or aggressive patients (Lambert et al. 2008). Severe side effects, such as acute dystonia and epileptic seizures, tend to occur chiefly in association with steep dose increases and relatively high dosages (Ciranni et al. 2009; Hedges et al. 2003). As a consequence, dose increases should always be monitored carefully, especially when the clinical situation calls for a dramatic upward change (Chengappa 2004).

If remission is not obtained with the aid of the drug of first choice, a relatively quick switch is warranted (Buckley and Correll 2008). The exact moment at which this switch can be made is still under discussion (Derks et al. 2010), but contrary to the traditional view, there is cumulating evidence that antipsychotic drugs require only little time (i.e., on the order of hours rather than days or weeks) to manifest their potential (Agid et al. 2008). If true, this would imply that a switch can be considered after a relatively short period of time (i.e., after 2 or 3 weeks). A second antipsychotic is usually chosen from among a group of drugs with a different receptor profile, although any direct evidence to support this strategy is scarce (Buchanan et al. 2010). Expert guidelines issued in 2003 recommended risperidone as a "preferred second-choice drug" (Kane et al. 2003), but the 2008 guidelines did not include this recommendation any longer. For those patients who even fail to respond to a second antipsychotic agent, clozapine is considered the drug of choice. The landmark trial by Kane et al. (1998) demonstrated a superior efficacy of clozapine for this subgroup of medication-resistant patients in comparison with any other antipsychotic agents, a finding that has since been replicated consistently (Chakos et al. 2001; Lewis et al. 2006; McEvoy 2006). In order to optimize clozapine therapy, various studies have evaluated the relationship between blood levels and therapeutic response. Blood levels above 350–450 μg/mL are associated with superior treatment results (reviewed by Schulte 2003), not only for intractable hallucinations but also for intractable negative symptoms, disorganized behavior, and thought disorder (Chakos et al. 2001; Lewis et al. 2006; McEvoy 2006). However, despite those unique qualities, clozapine has failed to gain the status of a drug of first or second choice. This is due to its rare, but potentially severe side effects. One of these side effects is leukocytopenia, or even agranulocytosis, i.e., the cessation of the production of white blood cells, which can lead to treatment-resistant infections and even death (Atkin et al. 1996). For this reason, blood is sampled weekly in patients starting on clozapine therapy to count the white blood cells and sampled at larger intervals during the complete course of treatment. In case of a severe decrease in leukocyte numbers, the treatment can be interrupted, which usually leads to a complete recovery of the white cell count (Esposito et al. 2005). As most patients using clozapine have psychotic symptoms that are resistant to other drugs, it is advisable to restart clozapine treatment after the leukocytopenia has subsided. Dunk et al. (2006) restarted clozapine after neutropenia in 53 patients and reported that 55% of them did not have another blood dyscrasia. For the remaining group, addition of lithium (which has the potential to induce leukocytosis) may be an option. Alternatively, granulocyte colony–stimulating factor can be prescribed as comedication (Whiskey and Taylor 2007). Another severe side effect of clozapine is heart disease, presenting in the form of pericarditis, myocarditis, or even cardiomyopathy. If clozapine-induced heart disease occurs, it usually does so within the first 15 days of treatment (Kamphuis et al. 2010). These effects are not always reversible after the cessation of clozapine therapy and can occasionally be fatal (Layland et al. 2009). Because of them, and because of clozapine's reputation as a "last resort," clinicians tend to show a certain hesitancy to prescribe it. However, careful monitoring can effectively reduce the risk of side effects, and here again, the cons should be carefully weighed against the pros (Agid et al. 2007).

24.3.1 Maintenance Treatment

When successful, antipsychotic medication should be prescribed in an unaltered dose for a duration of at least 1 year (Buchanan et al. 2010). However, this does not imply that the treatment should be discontinued as soon as the year is over. As the propensity to hallucinate would seem to depend in large part on our genetic makeup, the vulnerability for developing symptoms such as these can be expected to remain in force. In times of distress, either mentally or physically, the risk of a new psychotic episode may be high in unmedicated patients. Therefore, as long as the side effects are tolerable, it is preferable not to discontinue the medication that has led to the initial improvement, not even after a prolonged psychosis-free episode.

To prevent relapses, either of two strategies can be followed: continuous maintenance treatment with antipsychotic medication or intermittent treatment, to be started as soon as any signs of potential relapse are detected. In a double-blind randomized study, maintenance treatment was found to be more effective than targeted intermittent treatment in preventing relapses, even in stable first-episode patients after their first year of maintenance treatment (Gaebel et al. 2011). As regards continuous maintenance treatment, there has been considerable discussion regarding the optimal dose to be prescribed. In an elegant study, Wang et al. (2010) randomized 404 patients diagnosed with schizophrenia-in-remission to three conditions: (1) initial optimal therapeutic dose continued throughout the study, (2) initial optimal therapeutic dose continued for 4 weeks and then reduced to 50%, and (3) initial optimal therapeutic dose continued for 6 months and then reduced to 50%. After 1 year, the relapse rates were 9.4% for group 1, 30.5% for group 2, and 19.5% for group 3. These findings indicate that a dose reduction of 50% may increase the risk for psychosis two- or threefold. Whether a dose reduction of *less* than 50% is as effective as continuation of the initial dose remains as yet unclear (Uchida et al. 2011). All in all, current evidence suggests that maintenance treatment with the initial optimal dose is the safest way to go.

24.3.2 Depot Medication

As psychotic relapses are most frequently associated with nonadherence to antipsychotic treatment (Morken et al. 2008), long-acting injectables (or depots) constitute a valuable alternative for oral medication. Studies comparing short-acting oral and long-acting injectable antipsychotics found the latter to be superior in terms of relapse prevention and improvement of social functioning (Emsley et al. 2008). This superiority is largely due to improved adherence. If patients forget or refuse to take their oral medication, it may take their caretakers weeks or even months to notice an exacerbation, whereas nonadherence to injectables is instantly noticeable. In clinical practice, however, the prescription of depot medication has declined since the introduction of second-generation antipsychotics. Although 40–60% of all patients suffering from psychosis are partially or totally nonadherent to their antipsychotic medication, less than 30% are now treated with long-acting injectables (Patel et al. 2009). Possible reasons for this decline include prejudices against injectables, as well as the erroneous assumption that one's own patient group shows better adherence than those treated by others (Patel et al. 2009). However, there is actually little reason to refrain from prescribing depot medication for patients with psychosis (Heres et al. 2010). On the contrary, its advantages deserve to be explained to all patients in need of maintenance treatment.

24.3.3 Poor Response to Antipsychotic Medication

Although clozapine is considered the most efficient antipsychotic agent for refractory patients, as many as 40–70% of them achieve only a poor or partial response, even with adequate blood levels of clozapine (Kane et al. 1998). Treatment of these patients has remained a persistent public health problem, as they tend to suffer severely from their symptoms and often have a significantly reduced quality of life (McGlashan et al. 1988). For these ultra-resistant patients, several treatment strategies are available, including psychotherapy (see Chap. 26), pharmacological augmentation, repetitive transcranial magnetic stimulation (see Chap. 25), and electroconvulsive therapy. In clinical practice, clozapine is often augmented with lithium, sodium valproate, benzodiazepines, selective serotonin reuptake inhibitors (SSRIs), risperidone, haloperidol, or aripiprazole. A recent meta-analysis quantitatively summarized all randomized controlled trials (RCTs) involving the pharmacological augmentation of clozapine (Sommer et al. 2011). That review included 29 RCTs reporting on 15 different augmentation strategies prescribed to 1,066 patients in total. Improvement of total symptom severity – in comparison with placebo – was found for lamotrigine, sulpiride, citalopram, and the glutamatergic agonist CX516. However, the superiority of lamotrigine turned out to depend on the inclusion of a single outlier, whereas the claim to superiority of sulpiride, citalopram, and CX516 was based on single RCTs. Significantly better efficacy on positive symptom severity was found for topiramate and sulpiride, although here it must be noted that the results for topiramate became nonsignificant after outlier exclusion, while those for sulpiride were based on a single RCT. Citalopram, sulpiride, and CX516 showed better efficacy for negative symptoms than placebo, all based on single studies. We must conclude, therefore, that pharmacological augmentation strategies in clozapine therapy are not (yet) supported by much convincing evidence from the literature.

24.4 Electroconvulsive Treatment for Hallucinations in Schizophrenia

In clinical practice, another augmentation strategy for the treatment of intractable hallucinations occurring in the context of psychosis is electroconvulsive treatment (ECT). Introduced as a treatment method with highly promising results during the 1930s, and subsequently discarded during the 1970s, it continues to be the most stigmatized therapeutic in psychiatry, although for a limited number of indications (notably catatonia and psychotic depression) it can be extremely helpful and potentially lifesaving (Payne and Prudic 2009).

During ECT, an electrical current is passed briefly through the brain via electrodes attached to the scalp, so as to induce a generalized seizure. The individual receiving treatment is under general anesthesia and given muscle relaxants to prevent body spasms. The ECT electrodes can either be placed on both sides of the head (bilateral placement) or on one side of the head alone (unilateral placement). Unilateral placement is usually over the nondominant half of the brain, with the aim of reducing any cognitive side effects. However, bilateral electrode placement tends to yield a faster improvement (Kellner et al. 2010) and may be preferable in urgent situations such as severe catatonia. The amount of current required to induce a seizure (called the seizure threshold) can vary largely among individuals and may increase during the course of treatment (Van Waarde et al. 2009). Cognitive impairments, especially memory problems, can occur immediately after the administration of ECT as well as afterward. However, pretreatment functioning levels tend to be reached within the first months following treatment (Semkovska and McLoughlin 2010).

Although ECT has been used in clinical practice since the 1930s, there is still no generally accepted hypothesis explaining its mechanism of action. In rat models, ECT (contrary to antidepressants, for example) can induce mossy fiber sprouting (Lamont et al. 2001), and there is growing evidence that it impacts brain-derived neurotropic factors capable of inducing neuroproliferation (Grønli et al. 2009). It is most frequently used as a treatment method for severe, medication-resistant depression, and it is also used for the treatment of mania and catatonia. There is no consistency whether persistent psychosis in patients diagnosed with schizophrenia should also be considered a valid indication for ECT. Recently, the National Institute of Clinical Excellence (NICE) concluded that "the current state of the evidence does not allow the general use of ECT in the management of schizophrenia to be recommended" (Young et al. 2010). Thus, despite 80 years of practice, only little systematic evidence for the efficacy of ECT in psychosis is available. However, various naturalistic studies have assessed the outcome of patients with medication-resistant schizophrenia receiving a combination of ECT and antipsychotic medication. König and Glatter-Götz (1990) reported "stable remission" in 9 out of 12 patients receiving ECT. Likewise, Kupchik et al. (2000) reported on 36 patients treated with ECT and clozapine and found that 67% showed "satisfactory clinical recovery." Finally, Hustig and Onilov (2009) described a naturalistic follow-up of 27 patients receiving ECT and antipsychotic medication and found that 63% improved at least two points on the Clinical Global Impression Scale (CGI, Guy 1976). After 1 year, 37% of the initial sample had consolidated this improvement. Understandably, few studies have assessed the effects of ECT in a double-blind, sham-controlled design. In 2005, Tayran and Adams published a systematic meta-analysis of double-blind randomized studies comparing ECT and antipsychotic medication to sham and medication. They included 10 RCTs with a total of 392 patients. The relative risk for clinical improvement was 0.78 in favor of real ECT.

It should be noted that none of the above-mentioned studies provided any details on the reaction of hallucinations to ECT. As a consequence, the reported clinical improvement in all those studies is not necessarily attributable to a reduction in the frequency or severity of hallucinations. In fact, we were unable to retrieve a single study demonstrating a specific relief of hallucinations in medication-resistant psychosis thanks to ECT. As a consequence, we must conclude that ECT as an augmentation to antipsychotic medication is capable of improving the clinical status of some patients with medication-resistant psychosis, although its effects on hallucinations per se are as yet unclear and might well be low.

24.5 Treatment of Hallucinations in Parkinson's Disease and Related Disorders

Hallucinations and other psychotic symptoms are quite common in patients with Parkinson's disease (PD), with reported lifetime prevalence rates of up to 80% (Forsaa et al. 2010). In Lewy body dementia, a condition closely associated with PD, these numbers are even higher, especially for visual hallucinations. Cross-sectional studies show that visual hallucinations occur in approximately one-third of PD patients, whereas up to three-quarters of all PD patients develop them during a 20-year period (Fénelon and Alves 2010). Auditory hallucinations are present in up to 20% (Fénelon and Alves 2010). Prospective longitudinal cohort studies suggest that hallucinations tend to persist and worsen in individual patients and that their prevalence increases over time (Fénelon and Alves 2010). Those hallucinations can have substantial psychosocial effects and historically constitute the main reason for the placement of patients in nursing homes (Diederich et al. 2009).

Our understanding of the pathophysiology of psychosis in PD and Lewy body dementia has expanded dramatically over the past 15 years, from an initial interpretation of symptoms as dopaminergic drug adverse effects to the current view of a complex interplay of extrinsic and disease-related factors. These include central dopaminergic overactivity and an imbalance of dopaminergic and cholinergic neurotransmission, dysfunction of the visual pathways (including specific PD-associated retinopathy and functional alterations of the extrastriate visual pathways), alterations of brainstem sleep-wake and dream regulation, and impaired attentional focus (Diederich et al. 2009). The most important extrinsic factor, however, is still the antiparkinson medication. While hallucinations can be triggered by amantadine and anticholinergics, they are more commonly experienced after changes in dopaminergic medication. Within the latter category, dopamine agonists have a greater potential to induce hallucinations than L-dopa (Poewe 2008).

The treatment of hallucinations in PD involves patient-initiated coping strategies, a reduction of antiparkinson medication, augmentation with atypical neuroleptics, and, potentially, augmentation with cholinesterase inhibitors. When the reduction of anti-PD medication to the lowest tolerated dose does not improve hallucinosis, further interventions may be warranted. Various atypical antipsychotic agents (i.e., clozapine, olanzapine, quetiapine) are used to decrease the severity and frequency of hallucinations in PD and Lewy body dementia. While the use of clozapine requires monitoring of the leukocyte count (see above), olanzapine tends to lead to an aggravation of the motor symptoms (Zahodne and Fernandez 2008). Studies of ziprasidone and aripiprazole use are limited to open-label trials and case reports and highly variable in outcome; while either drug may be effective in some patients, both are associated with various adverse effects (Zahodne and Fernandez 2008). While two randomized controlled trials could not demonstrate the efficacy of quetiapine, it is a common first-line treatment method for psychosis in the context of PD because of its tolerability, ease of use, and demonstrated utility in numerous open-label reports (Zahodne and Fernandez 2008). A small, double-blind RCT with a mean dose of 58 mg quetiapine showed a significantly larger reduction of the severity of hallucinations (Fernandez et al. 2009). Eng and Welty (2010) conducted a review of the literature, thus including 13 studies on antipsychotic treatment for PD patients, all involving clozapine and quetiapine. They concluded that patients with PD might well benefit from long-term clozapine therapy, whereas the results of the quetiapine studies were conflicting. However, when quetiapine and clozapine were compared head-to-head, no statistically significant differences in effectiveness were found.

The group of cholinesterase inhibitors currently represents the most promising pharmacological alternative to antipsychotics. Various open-label studies and one double-blind, placebo-controlled trial among 188 hallucinating PD patients are in support of the efficacy of rivastigmine (Burn et al. 2006). The cholinesterase inhibitor tacrine, however, has hardly been tested because of its hepatic toxicity, and controlled trials with donepezil have not yielded any significant reduction of psychotic symptoms due, perhaps, to methodological limitations (Burn et al. 2006). Thus, while the use of cholinesterase inhibitors, especially rivastigmine, appears to be a promising treatment method for hallucinations in PD, evidence-based studies support only the use of a single atypical antipsychotic drug, namely clozapine (Eng and Welty 2010).

24.6 Treatment of Hallucinations in Dementia

In Alzheimer's disease (AD), the occurrence of psychosis in 30–50% of the cases has serious consequences for both patients and caregivers (Spalletta et al. 2010), especially since the optimal type of treatment is still elusive. Interventions that optimize environmental and interpersonal factors can be helpful and should be attempted in all cases, although their overall effectiveness and applicability are not entirely clear. Cholinesterase inhibitors such as donepezil may have a beneficial effect on hallucinations while showing a relatively mild side effect profile (Wynn and Cummings 2004). In a similar vein, memantine has been shown to be more effective than placebo treatment without causing any disturbing side effects (Wilcock et al. 2008).

The Clinical Antipsychotic Trials of Intervention Effectiveness-Alzheimer's Disease (CATIE-AD) study included 421 AD outpatients with psychosis and agitated and/or aggressive behavior. The patients were randomized to obtain masked, flexible-dose treatment with olanzapine, quetiapine, risperidone, or placebo for up to 36 weeks. As regards the effects of those drugs upon the psychotic symptoms, risperidone appeared to be superior to the other two and placebo (Sultzer et al. 2008). Although antipsychotic medication can have a positive effect on hallucinations in dementia, several reports issue warnings against the excess risk of morbidity and even death associated with its use in older patients (Kalapatapu and Schimming 2009). As a consequence, it is strongly advised not to consider antipsychotic drugs as the first choice for treatment of psychotic symptoms in dementia. Extrapyramidal symptoms and arrhythmias due to QTc prolongation are well-known complications of the use of conventional antipsychotic agents, while cerebrovascular events appear to occur more frequently in association with atypical as well as conventional antipsychotics in comparison with placebo treatment (Kalapatapu and Schimming 2009). Nevertheless, a trial of these agents may be indicated when the severity of symptoms is extreme or when the symptoms fail to respond to other types of medication or to nonpharmacological interventions.

24.7 Treatment of Hallucinations in Delirium

Delirium is an acute neuropsychiatric syndrome, by definition due to organic disease, which is characterized by psychotic symptoms such as hallucinations and delusions in the presence of decreased attention, fluctuating consciousness, and other cognitive dysfunctions. It is very common in patients admitted to intensive care units, with a reported cross-sectional incidence of 32% (Salluh et al. 2010) and a marked association with poor prognosis and increased mortality (Kuehn 2010). The only causal treatment of delirium is the improvement of somatic health. Sometimes this can be accomplished by relatively simple means, for example, by restoring the volume of the blood plasma in dehydration or by treating a urinary tract infection. However, delirium frequently affects severely ill patients suffering from multiple somatic conditions such as cardiac failure complicated by asthma cardiale and diabetes, a combination which can be very hard to treat.

In such cases, the symptomatic treatment of hallucinations and other symptoms of delirium should commence with measures aimed at improving the patient's circadian rhythm and orientation. Symptomatic pharmacological treatment should preferably consist of haloperidol or olanzapine, as recommended by the latest NICE guidelines (Young et al. 2010). This type of treatment should be started at the lowest clinically effective dose and titrated cautiously. Although benzodiazepines are widely applied for the treatment of delirium, they are recommended only for delirium tremens (i.e., alcohol abstinence delirium). Two recent randomized controlled trials yielded good results for quetiapine in comparison with placebo (Devlin et al. 2010; Tahir et al. 2010), but so far no head-to-head comparisons with haloperidol or olanzapine have been published. Cholinesterase inhibitors are not recommendable, as demonstrated by an RCT with rivastigmine in delirious patients admitted to an intensive care unit. That trial was terminated at an early stage because of a significantly higher mortality and an increased duration of delirium in comparison with the control group (Van Eijk et al. 2010).

24.8 Treatment of Hallucinations in Epilepsy

The reported cross-sectional incidence of hallucinations and other psychotic symptoms in epilepsy is 3.3%, and in temporal lobe epilepsy, as high as 14% (Torta and Keller 1999). Those symptoms can occur shortly before, during, or after an epileptic seizure, and even independently of any motor seizures. Ictal hallucinosis is considered relatively rare. Postictal hallucinosis comprises some 25% of the hallucinatory episodes in epileptic patients. As post- and interictal psychotic episodes resemble those in patients diagnosed with schizophrenia, they are also designated as "schizophrenia-like psychoses of epilepsy."

The treatment of ictal as well as post- and interictal hallucinations should primarily consist of minimizing any medication capable of mediating these symptoms. Various antiepileptic drugs, such as phenobarbital, zonisamide, levetiracetam, and gabapentin, are known for their potential to induce hallucinations (Alper et al. 2002). In such cases, dose reduction or a switch to another antiepileptic drug may lead to a relatively quick cessation of the hallucinations. When antiepileptic drugs cannot be reduced or traded, or when such interventions are unsuccessful, antipsychotic medication is the next therapeutic step. Clozapine and chlorpromazine should be avoided, if possible, because of their epileptogenic properties. Molindone is the antipsychotic with the lowest epileptogenicity, making it very useful in epileptic patients (Alper et al. 2002). However, antipsychotics such as quetiapine, risperidone, and haloperidol also tend to be tolerated well. In all cases, the initial dose should be lower than in patients with hallucinations due to psychosis (Tadokoro et al. 2007).

24.9 Treatment of Hallucinations in Sensory Impairment

Visually impaired patients may experience complex visual hallucinations, a condition known as the Charles Bonnet syndrome (see Chap. 6). Likewise, individuals with progressive hearing loss may develop auditory hallucinations consisting of music (see Chap. 11), voices (see Chaps. 9 and 10), or other sounds. It is believed that such hallucinations are actually release phenomena due to a deafferentation of the visual or auditory association areas of the cerebral cortex, a process capable of yielding so-called phantom percepts (Menon et al. 2003). Cognitive defects and social isolation may act as additional risk factors. Release hallucinations generally affect the elderly, women more frequently than men. Patients who comprehend their unrealistic nature tend to be affected less severely by them, although they may still be distressed by the fear of imminent insanity. Reassurance and an explanation that the visions or auditory percepts do not imply any kind of mental illness may have a powerful therapeutic effect (Menon et al. 2003). Further therapeutic measures are not always necessary because release hallucinations may cease either spontaneously or upon the termination of social isolation. If warranted and possible, the treatment of first choice is the restoration of sight or hearing, for example, by carrying out a cataract operation, cleaning the meatus externus, or applying hearing aids (Tuerlings et al. 2009). In addition, one may consider the optimization of visual or auditory stimuli. When interventions such as these are unsuccessful, pharmacological treatment may be considered, although the pros do not always outweigh the cons of side effects. Both antipsychotic and antiepileptic drugs have been reported to be effective in case reports and open-label case series. There are currently no randomized trials on the efficacy of those types of medication in patients with release hallucinations. If pharmacological treatment is considered necessary, quetiapine may be a good choice (David and Fernandez 2000) as it is usually tolerated well in elderly populations (Rossom et al. 2010). However, as the dopamine receptor density tends to diminish in old age, initial doses should be very low and they should be increased only gradually. Pharmacological treatment in release hallucinations should preferably be carried out during a limited period of time (i.e., 2 or 3 months) and then tapered off when they are not effective in order to avoid unnecessary and potentially harmful side effects.

References

Agid, O., Kapur, S., Warrington, L., Loebel, A., Siu, C. (2008). Early onset of antipsychotic response in the treatment of acutely agitated patients with psychotic disorders. *Schizophrenia Research*, 102, 241–248.

Agid, O., Remington, G., Kapur, S., Arenovich, T., Zipursky, R.B. (2007). Early use of clozapine for poorly responding first-episode psychosis. *Journal of Clinical Psychopharmacology*, 27, 369–373.

Alper, K.R., Barry, J.J., Balabanov, A.J. (2002). Treatment of psychosis, aggression, and irritability in patients with epilepsy. *Epilepsy and Behavior*, 3, 13–18.

Atkin, K., Kendall, F., Gould, D., Freeman, H., Liberman, J., O'Sullivan, D. (1996). Neutropenia and agranulocytosis in patients receiving clozapine in the UK and Ireland. *British Journal of Psychiatry*, 169, 483–488.

Buchanan, R.W., Kreyenbuhl, J., Kelly, D.L., Noel, J.M., Boggs, D.L., Fischer, B.A., Himelhoch, S., Fang, B., Peterson, E., Aquino, P.R., Keller, W., Schizophrenia Patient Outcomes Research Team (PORT) (2010). The 2009 schizophrenia PORT psychopharmacological treatment recommendations and summary statements. *Schizophrenia Bulletin*, 36, 71–93.

Buckley, P.F., Correll, C.U. (2008). Strategies for dosing and switching antipsychotics for optimal clinical management. *Journal of Clinical Psychiatry*, 69(Suppl 1), 4–17.

Burn, D., Emre, M., McKeith, I., De Deyn, P.P., Aarsland, D., Hsu, C., Lane, R. (2006). Effects of rivastigmine in patients with and without visual hallucinations in dementia associated with Parkinson's disease. *Movement Disorders*, 21, 1899–1907.

Chakos, M., Lieberman, J., Hoffman, E., Bradford, D., Sheitman, B. (2001). Effectiveness of second-generation antipsychotics in patients with treatment-resistant schizophrenia: a review and meta-analysis of randomized trials. *American Journal of Psychiatry*, 158, 518–526.

Chengappa, K.N. (2004). Dosage and administration issues of antipsychotic agents (and adjunctive medicines) in the acute stabilization of psychoses. *CNS Spectrums*, 9(Suppl 9), 6–10.

Ciranni, M.A., Kearney, T.E., Olson, K.R. (2009). Comparing acute toxicity of first- and second-generation antipsychotic drugs: a 10-year, retrospective cohort study. *Journal of Clinical Psychiatry*, 70, 122–129.

Cope, T.E., Baguley, D.M. (2009). Is musical hallucination an otological phenomenon? A review of the literature. *Clinical Otolaryngology*, 34, 423–430.

David, R.R., Fernandez, H.H. (2000). Quetiapine for hypnagogic musical release hallucinations. *Journal of Geriatric Psychiatry and Neurology*, 13, 210–211.

Derks, E.M., Fleischhacker, W.W., Boter, H., Peuskens, J., Kahn, R.S., EUFEST Study Group (2010). Antipsychotic drug treatment in first-episode psychosis: should patients be switched to a different antipsychotic drug after 2, 4, or 6 weeks of nonresponse? *Journal of Clinical Psychopharmacology*, 30, 176–180.

Devlin, J.W., Roberts, R.J., Fong, J.J., Skrobik, Y., Riker, R.R., Hill, N.S., Robbins, T., Garpestad, E. (2010). Efficacy and safety of quetiapine in critically ill patients with delirium: a prospective, multicenter, randomized, double-blind, placebo-controlled pilot study. *Critical Care Medicine*, 38, 419–427.

Diederich, N.J., Fénelon, G., Stebbins, G., Goetz, C.G. (2009). Hallucinations in Parkinson disease. *Nature Reviews. Neurology*, 5, 331–342.

Dunk, L.R., Annan, L.J., Andrews, C.D. (2006). Rechallenge with clozapine following leucopenia or neutropenia during previous therapy. *British Journal of Psychiatry*, 188, 255–263.

Emsley, R., Oosthuizen, P., Koen, L., Niehaus, D.J., Medori, R., Rabinowitz, J. (2008). Oral versus injectable antipsychotic treatment in early psychosis: post hoc comparison of two studies. *Clinical Therapeutics*, 30, 2378–2386.

Eng, M.L., Welty, T.E. (2010). Management of hallucinations and psychosis in Parkinson's disease. *American Journal of Geriatric Pharmacotherapy*, 8, 316–330.

Esposito, D., Rouillon, F., Limosin, F. (2005). Continuing clozapine treatment despite neutropenia. *European Journal of Clinical Pharmacology*, 60, 759–764.

Fénelon, G., Alves, G. (2010). Epidemiology of psychosis in Parkinson's disease. *Journal of the Neurological Sciences*, 15, 12–17.

Fernandez, H.H., Okun, M.S., Rodriguez, R.L., Malaty, I.A., Romrell, J., Sun, A., Wu, S.S., Pillarisetty, S., Nyathappa, A., Eisenschenk, S. (2009). Quetiapine improves visual hallucinations in Parkinson disease but not through normalization of sleep architecture: results from a double-blind clinical-polysomnography study. *International Journal of Neuroscience*, 119, 2196–2205.

Forsaa, E.B., Larsen, J.P., Wentzel-Larsen, T., Goetz, C.G., Stebbins, G.T., Aarsland, D., Alves, G. (2010). A 12-year population-based study of psychosis in Parkinson disease. *Archives of Neurology*, 67, 996–1001.

Gaebel, W., Riesbeck, M., Wölwer, W., Klimke, A., Eickhoff, M., von Wilmsdorff, M., Lemke, M., Heuser, I., Maier, W., Huff, W., Schmitt, A., Sauer, H., Riedel, M., Klingberg, S., Köpcke, W., Ohmann, C., Möller, H.J, for the German Study Group on First-Episode Schizophrenia (2011). Relapse prevention in first-episode schizophrenia-maintenance vs intermittent drug treatment with prodrome-based early intervention: results of a randomized controlled trial within the German research network on schizophrenia. *Journal of Clinical Psychiatry*, 72, 205–218.

Grønli, O., Stensland, G.Ø., Wynn, R., Olstad, R. (2009). Neurotrophic factors in serum following ECT: a pilot study. *World Journal of Biological Psychiatry*, 10, 295–301.

Guy, W., ed. (1976). *ECDEU assessment manual for psychopharmacology*. Rockville, MD: Department of Health, Education, and Welfare.

Hedges, D., Jeppson, K., Whitehead, P. (2003). Antipsychotic medication and seizures: a review. *Drugs of Today*, 39, 551–557.

Heres, S., Reichhart, T., Hamann, J., Mendel, R., Leucht, S., Kissling, W. (2010). Psychiatrists' attitude to antipsychotic depot treatment in patients with first-episode schizophrenia. *European Psychiatry*, 26, 297–301.

Hustig, H., Onilov, R. (2009). ECT rekindles pharmacological response in schizophrenia. *European Psychiatry*, 24, 521–525.

Kahn, R.S., Fleischhacker, W.W., Boter, H., Davidson, M., Vergouwe, Y., Keet, I.P., Gheorghe, M.D., Rybakowski, J.K., Galderisi, S., Libiger, J., Hummer, M., Dollfus, S., López-Ibor, J.J., Hranov, L.G., Gaebel, W., Peuskens, J., Lindefors, N., Riecher-Rössler, A., Grobbee, D.E., EUFEST Study Group (2008). Effectiveness of antipsychotic drugs in first-episode schizophrenia and schizophreniform disorder: an open randomised clinical trial. *Lancet*, 29, 1085–1097

Kalapatapu, R.K., Schimming, C. (2009). Update on neuropsychiatric symptoms of dementia: antipsychotic use. *Geriatrics*, 64, 10–18.

Kamphuis, H., Arends, J., Timmerman, L., Van Marle, J., Kappert. J. (2010). Myocarditis en cardiomyopathie; ernstige complicaties van clozapinetherapie. *Tijdschrift voor Psychiatrie*, 52, 223–233.

Kane, J., Honigfeld, G., Singer, J., Meltzer, H. (1988). Clozapine for the treatment-resistant schizophrenic. A double-blind comparison with chlorpromazine. *Archives of General Psychiatry*, 5, 789–796.

Kane, J.M., Leucht, S., Carpenter, D., Docherty, J.P., Expert Consensus Panel for Optimizing Pharmacologic Treatment of Psychotic Disorders (2003). The expert consensus guideline series. Optimizing pharmacologic treatment of psychotic disorders. Introduction: methods, commentary, and summary. *Journal of Clinical Psychiatry*, 64(Suppl 12), 5–19.

Kay, S.R., Fiszbein, A., Opler, L.A. (1987). The positive and negative syndrome scale (PANSS) for schizophrenia. *Schizophrenia Bulletin*, 13, 261–276.

Kellner, C.H., Knapp, R., Husain, M.M., Rasmussen, K., Sampson, S., Cullum, M., McClintock, S.M., Tobias, K.G., Martino, C., Mueller, M., Bailine, S.H., Fink, M., Petrides, G. (2010). Bifrontal, bitemporal and right unilateral electrode placement in ECT: randomised trial. *British Journal of Psychiatry*, 196, 226–234.

König, P., Glatter-Götz, U. (1990). Combined electroconvulsive and neuroleptic therapy in schizophrenia refractory to neuroleptics. *Schizophrenia Research*, 3, 351–354.

Kuehn, B.M. (2010). Delirium often not recognized or treated despite serious long-term consequences. *Journal of the American Medical Association*, 28, 389–390, 395.

Kupchik, M., Spivak, B., Mester, R., Reznik, I., Gonen, N., Weizman, A., Kotler, M. (2000). Combined electroconvulsive-clozapine therapy. *Clinical Neuropharmacology*, 23, 14–16.

Lambert, M., Huber, C.G., Naber, D., Schacht, A., Wagner, T.T., Hundemer, H.P., Karow, A., Schimmelmann, B.G. (2008). Treatment of severe agitation with olanzapine in 166 patients with schizophrenia, schizoaffective, or bipolar I disorder. *Pharmacopsychiatry*, 41, 182–189.

Lamont, S.R., Paulls, A., Stewart, C.A. (2001). Repeated electroconvulsive stimulation, but not antidepressant drugs, induces mossy fibre sprouting in the rat hippocampus. *Brain Research*, 2, 53–58.

Layland, J.J., Liew, D., Prior, D.L. (2009). Clozapine-induced cardiotoxicity: a clinical update. *Medical Journal of Australia*, 16, 190–192.

Lewis, S.W., Davies, L., Jones, P.B., Barnes, T.R., Murray, R.M., Kerwin, R., Taylor, D., Hayhurst, K.P., Markwick, A., Lloyd, H., Dunn, G. (2006). Randomised controlled trials of conventional antipsychotic versus new atypical drugs, and new atypical drugs versus clozapine, in people with schizophrenia responding poorly to, or intolerant of, current drug treatment. *Health Technology Assessment*, 10, iii–iv, ix–xi, 1–165.

McEvoy, J.P. (2006). Risks versus benefits of different types of long-acting injectable antipsychotics. *Journal of Clinical Psychiatry*, 67(Suppl 5), 15–18.

McGlashan, T.H. (1988). A selective review of recent North American long-term followup studies of schizophrenia. *Schizophrenia Bulletin*, 14, 515–542.

Menon, G.J., Rahman, I., Menon, S.J., Dutton, G.N. (2003). Complex visual hallucinations in the visually impaired: the Charles Bonnet Syndrome. *Survey of Ophthalmology*, 48, 58–72.

Morken, G., Widen, J.H., Grawe, R.W. (2008). Non-adherence to antipsychotic medication, relapse and rehospitalisation in recent-onset schizophrenia. *BMC Psychiatry*, 30, 32.

Muthén, B.O., Curran, P.J. (1997). General longitudinal modeling of individual differences in experimental designs: a latent variable framework for analysis and power estimation. *Psychological Methods*, 2, 371–402.

Overall, J.E., Gorham, D.R. (1962). The Brief Psychiatric Rating Scale. *Psychological Reports*, 10, 799–812.

Patel, M.X., Taylor, M., David, A.S. (2009). Antipsychotic long-acting injections: mind the gap. *British Journal of Psychiatry*, 195, 1–4.

Payne, N.A., Prudic, J. (2009). Electroconvulsive therapy: Part I. A perspective on the evolution and current practice of ECT. *Journal of Psychiatric Practice*, 15, 346–368.

Poewe, W. (2008). When a Parkinson's disease patient starts to hallucinate. *Practical Neurology*, 8, 238–241.

Rossom, R.C., Rector, T.S., Lederle, F.A., Dysken, M.W. (2010). Are all commonly prescribed antipsychotics associated with greater mortality in elderly male veterans with dementia? *Journal of the American Geriatrics Society*, 58, 1027–1034.

Salluh, J.I., Soares, M., Teles, J.M., Ceraso, D., Raimondi, N., Nava, V.S., Blasquez, P., Ugarte, S., Ibanez-Guzman, C., Centeno, J.V., Laca, M., Grecco, G., Jimenez, E., Arias-Rivera, S., Duenas, C., Rocha, M.G., Decca (Delirium Epidemiology In Critical Care) Study Group (2010). Delirium Epidemiology in Critical Care (DECCA): an international study. *Critical Care*, 14, R210.

Schulte, P. (2003). What is an adequate trial with clozapine? Therapeutic drug monitoring and time to response in treatment-refractory schizophrenia. *Clinical Pharmacokinetics*, 42, 607–618.

Semkovska, M., McLoughlin, D.M. (2010). Objective cognitive performance associated with electroconvulsive therapy for depression: a systematic review and meta-analysis. *Biological Psychiatry*, 68, 568–577.

Sommer, I.E., Begemann, M.J., Temmerman, A., Leucht S. (2011). Pharmacological augmentation strategies for schizophrenia patients with insufficient response to clozapine: a quantitative literature review. *Schizophrenia Bulletin*, doi: 10.1093/schbul/sbr004.

Spalletta G., Musicco M., Padovani A., Rozzini L., Perri R., Fadda L., Canonico V., Trequattrini A., Pettenati C., Caltagirone C., Palmer K. (2010). Neuropsychiatric symptoms and syndromes in a large cohort of newly diagnosed, untreated patients with Alzheimer disease. *American Journal of Geriatric Psychiatry*, 18, 1026–1035.

Sultzer, D.L., Davis, S.M., Tariot, P.N., Dagerman, K.S., Lebowitz, B.D., Lyketsos, C.G., Rosenheck, R.A., Hsiao, J.K., Lieberman, J.A., Schneider, L.S., CATIE-AD Study Group. (2008). Clinical symptom responses to atypical antipsychotic medications in Alzheimer's disease: phase 1 outcomes from the CATIE-AD effectiveness trial. *American Journal of Psychiatry*, 165, 844–854.

Tadokoro, Y., Oshima, T., Kanemoto, K. (2007). Interictal psychoses in comparison with schizophrenia – a prospective study. *Epilepsia*, 48, 2345–2351.

Tahir, T.A., Eeles, E., Karapareddy, V., Muthuvelu, P., Chapple, S., Phillips, B., Adyemo, T., Farewell, D., Bisson, J.I. (2010). A randomized controlled trial of quetiapine versus placebo in the treatment of delirium. *Journal of Psychosomatic Research*, 69, 485–490.

Torta, R., Keller, R. (1999). Behavioral, psychotic, and anxiety disorders in epilepsy: etiology, clinical features, and therapeutic implications. *Epilepsia*, 40(Suppl 10), S2–20.

Tuerlings, J.H.A.M., Wijnen, H., Boerman, R., Verwey, B. (2009). Muziek als ongenode gast: de auditieve variant van syndroom van Charles Bonnet. *Tijdschrift voor Psychiatrie*, 51, 853–858.

Uchida, H., Suzuki, T., Takeuchi, H., Arenovich, T., Mamo, D.C. (2011). Low dose vs standard dose of antipsychotics for relapse prevention in schizophrenia: meta-analysis. *Schizophrenia Bulletin*, 37, 788–799.

Van Waarde, J.A., Verwey, B., Van der Mast, R.C. (2009). Meta-analysis of initial seizure thresholds in electroconvulsive therapy. *European Archives of Psychiatry and Clinical Neurosciences*, 259, 467–474.

Van Eijk, M.M., Roes, K.C., Honing, M.L., Kuiper, M.A., Karakus, A., Van der Jagt, M., Spronk, P.E., Van Gool, W.A., Van der Mast, R.C., Kesecioglu, J., Slooter, A.J. (2010). Effect of rivastigmine as an adjunct to usual care with haloperidol on duration of delirium and mortality in critically ill patients: a multicentre, double-blind, placebo-controlled randomised trial. *Lancet*, 376, 1829–1837.

Wang, C.Y., Xiang, Y.T., Cai, Z.J., Weng, Y.Z., Bo, Q.J., Zhao, J.P., Liu, T.Q., Wang, G.H., Weng, S.M., Zhang, H.Y., Chen, D.F., Tang, W.K., Ungvari, G.S., Risperidone Maintenance Treatment in Schizophrenia (RMTS) investigators (2010). Risperidone maintenance treatment in schizophrenia: a randomized, controlled trial. *American Journal of Psychiatry*, 167, 676–685.

Weiden, P.J. (2007). Understanding and addressing adherence issues in schizophrenia: from theory to practice. *Journal of Clinical Psychiatry*, 68(Suppl 14), 14–19.

Whiskey, E., Taylor, D. (2007). Restarting clozapine after neutropenia: evaluating the possibilities and practicalities. *CNS Drugs*, 21, 25–35.

Wilcock, G.K., Black, S.E., Hendrix, S.B., Zavitz, K.H., Swabb, E.A., Laughlin, M.A., Tarenflurbil Phase II Study investigators (2008). Efficacy and safety of tarenflurbil in mild to moderate Alzheimer's disease: a randomised phase II trial. Lancet. *Neurology*, 7, 483–493.

Wynn, Z.J., Cummings, J.L. (2004). Cholinesterase inhibitor therapies and neuropsychiatric manifestations of Alzheimer's disease. *Dementia and Geriatric Cognitive Disorders*, 17, 100–108.

Young, J., Murthy, L., Westby, M., Akunne, A., O'Mahony, R., Guideline Development Group (2010). Diagnosis, prevention, and management of delirium: summary of NICE guidance. *British Medical Journal*, 341, c3704.

Zahodne, L.B., Fernandez, H.H. (2008). Pathophysiology and treatment of psychosis in Parkinson's disease: a review. *Drugs and Aging*, 25, 665–682.

Chapter 25
Experimental Somatic Treatments: Transcranial Magnetic Stimulation in the Treatment of Auditory Verbal Hallucinations – A Meta-Analysis and Review

Christina W. Slotema and Z. Jeff Daskalakis

25.1 Introduction

Transcranial magnetic stimulation (TMS) is a technique in which a strong pulse of electrical current is sent through a coil (see Figs. 25.1 and 25.2). When the coil is placed over a person's skull, this induces a magnetic field pulse in a small brain area, depolarizing local neurons up to a depth of 2 cm. Barker and colleagues developed the first modern TMS device (Barker et al. 1985, 1990). TMS can be used as a brain mapping tool, as a tool to measure cortical excitability, as a probe of neuronal networks, and as a modulator of brain function. It is thought that it can induce longer-lasting effects as a result of long-term potentiation or depression at the neuronal level (Siebner and Rothwell 2003). TMS is non-invasive, has few side effects, and is a relatively safe technique. In the past, epileptic seizures have occurred during repetitive TMS, as it has been applied at high frequency, during a longer time, or at a high threshold. However, since Wassermann developed specific safety guidelines in 1998, seizures have become extremely rare (Wassermann 1998). Side effects such

C.W. Slotema, M.D., Ph.D. (✉)
Parnassia Bavo Group, The Hague, The Netherlands
e-mail: c.slotema@psyq.nl

Z.J. Daskalakis, M.D., Ph.D., F.R.C.P.
Department of Psychiatry, University of Toronto, Toronto, ON, Canada

Brain Stimulation and Research Program, Centre for Addiction and Mental Health, Toronto, ON, Canada

Schizophrenia Program, Centre for Addiction and Mental Health, Toronto, ON, Canada
e-mail: jeff_daskalakis@camh.net

J.D. Blom and I.E.C. Sommer (eds.), *Hallucinations: Research and Practice*,
DOI 10.1007/978-1-4614-0959-5_25, © Springer Science+Business Media, LLC 2012

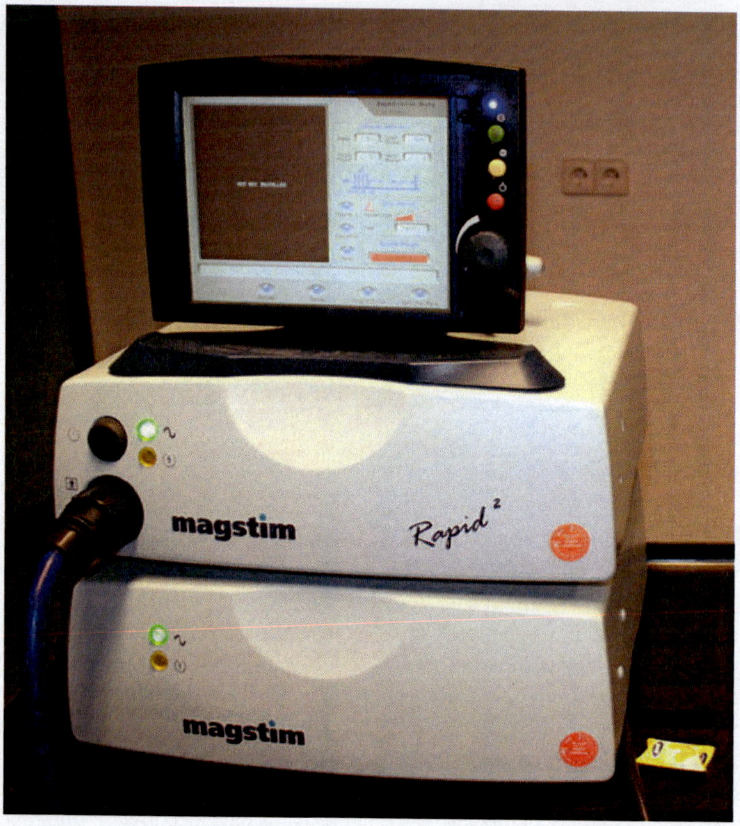

Fig. 25.1 Transcranial magnetic stimulation device

as headache, local discomfort due to direct stimulation of the facial musculature, and transient changes in the auditory threshold have been described, and in order to prevent the latter, earplugs are recommended during TMS treatment.

George et al. (1995) were the first to investigate the effects of rTMS for depression. This was followed by several other studies, and resulted in approval of rTMS by the Food and Drug Administration in October 2008 as a treatment option for depression.

In 1999, Hoffman et al. (1999) started to explore rTMS for the treatment of auditory verbal hallucinations (AVH). They directed the coil at the left temporoparietal cortex (see Fig. 25.3) overlying Brodmann's area 40 (Homan et al. 1987), which is critical to speech perception (Benson et al. 2001). Thus, they were able to ameliorate medication-resistant AVH. Since then, more studies on this subject have

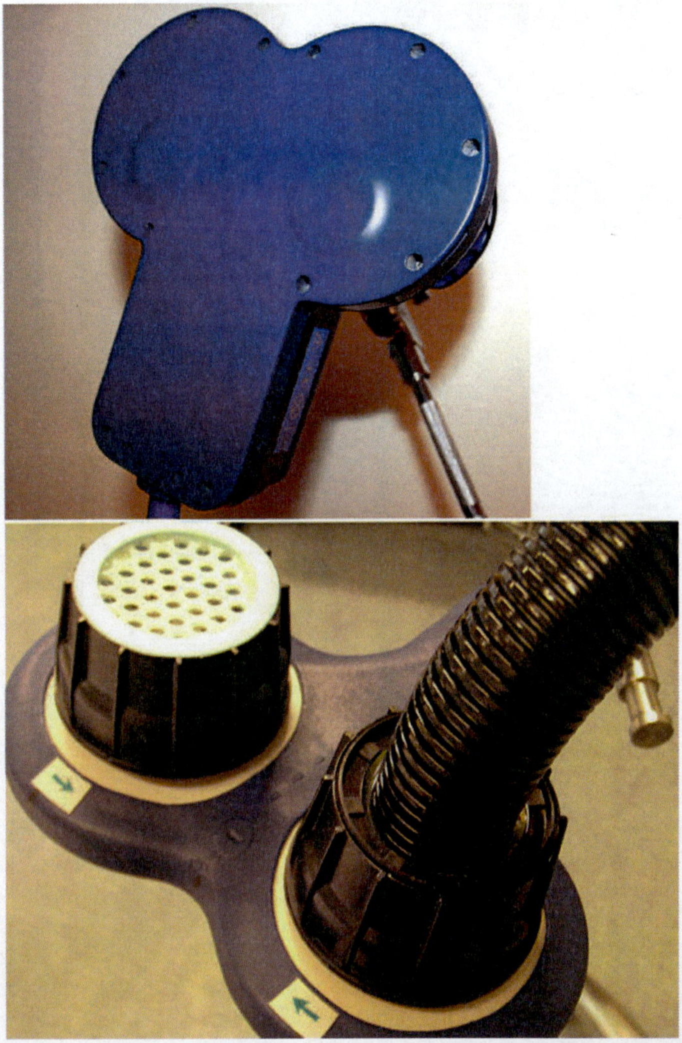

Fig. 25.2 Two types of figure-of-eight coils of the transcranial magnetic stimulation device

been published in the literature. The results of these studies have been summarized in four meta-analyses, which all conclude that rTMS has a moderate to good effect on AVH, with effect sizes ranging from 0.51 to 1.04 (Aleman et al. 2007; Freitas et al. 2009; Slotema et al. 2010; Tranulis et al. 2008). As the number of publications is still increasing, this chapter provides an up-to-date review and a new meta-analysis.

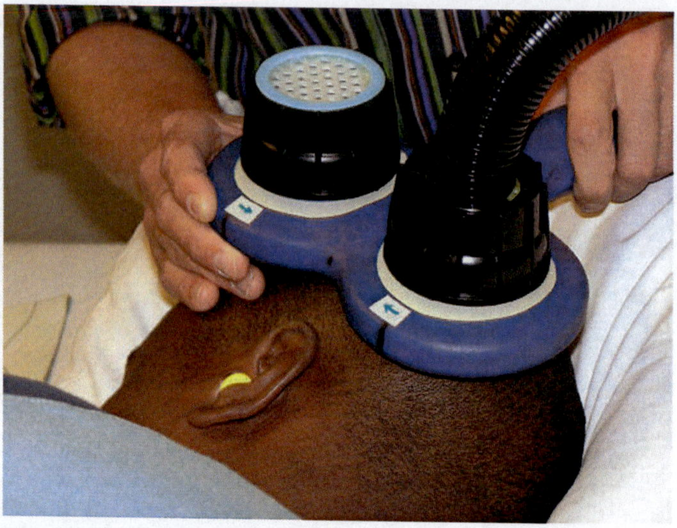

Fig. 25.3 Repetitive transcranial magnetic stimulation is applied to the left temporoparietal cortex

25.2 Methods

25.2.1 *Study Selection*

The methods for the present meta-analysis were adopted from Slotema et al. (2010). A literature search was performed in PubMed (1990 through January 2011), Ovid Medline (1990 through January 2011), Embase Psychiatry (1997 through January 2011), the Cochrane Central Register of Controlled Trials, the Cochrane Database of Systematic Reviews, the Database of Abstracts of Reviews of Effects, and PsycINFO (1990 through January 2011), using the search terms transcranial magnetic stimulation, TMS, repetitive TMS, auditory hallucination, and psychosis.

Criteria for inclusion were:

1. Treatment with rTMS.
2. Symptom severity of AVH was used as an outcome measure.
3. The study was performed in a parallel, double-blind, randomized controlled design using a placebo condition. (We chose for parallel designs only because patients usually do not remain completely blinded in crossover studies, which may influence the results.)
4. The data provided sufficiently exact data to compute Hedges's g (sample size, means, and standard deviations or exact t, F, or p values for rTMS main effect for change scores).

5. More than three patients were included per study.
6. Articles were written in English.

When various articles described overlapping samples, only the article with the largest sample size was included.

25.2.2 Data Extraction

The following data were acquired: number of treated patients, mean and standard deviation of the outcome measure at baseline and at the end of treatment (or exact F, t, or p value), study design, and treatment parameters (localization of treatment, frequency, intensity, number of stimuli per session, and number of treatment sessions). When a publication contained insufficient or incomplete data, the authors were contacted and invited to send additional data so that their study could be included in the meta-analysis. All meta-analyses were checked for cross-references.

25.2.3 Effect Size Calculation

Effect sizes were calculated for the mean differences (sham treatment versus rTMS) of the pretreatment-posttreatment change in rating scales. In a random effects model, the mean gain for each study was computed using Comprehensive Meta-Analysis Version 2.0. First, the individual effect sizes for each study were computed; after which, meta-analytic methods were applied to obtain a combined, weighted effect size (Hedges's *g*). The means of separate studies were weighed according to the sample sizes. A homogeneity statistic, I^2 (Higgins et al. 2003), was used to test whether the studies shared a common population effect size. An I^2 statistic (i.e. 30% or higher) indicated heterogeneity of the individual study effect sizes, which poses a limitation to a reliable interpretation of the results. If heterogeneity was high, a moderator analysis was performed wherever possible to investigate the potential influence of moderating factors. Sub-analyses were performed to investigate different treatment conditions, such as localization, frequency, number of stimuli, number of treatment sessions, and total number of stimuli. These parameters were correlated with Hedges's *g* using Pearson's correlations in SPSS18 (Statistical Package for Social Sciences version 18).

In studies comparing three treatment conditions, the two actual treatments were compared separately with the sham condition. We also computed a fail-safe number because effect sizes can be unreliable due to the omission of studies in which rTMS was not effective (Rosenthal 1979). This fail-safe number is an estimation of the number of missing studies that is needed to change the results of the meta-analysis to non-significant. Side effects and dropouts are presented according to rTMS localization.

25.3 Results

Ten studies were included, with a total number of 308 patients. One hundred and eighty-two of them received real rTMS, while 126 were included in the sham condition. Twenty studies did not fulfil the criteria for inclusion due to a crossover design ($n=5$), overlap with other studies ($n=4$), insufficient data ($n=3$), the use of rTMS as maintenance therapy ($n=3$), the absence of a sham condition ($n=3$), or the absence of 'severity of AVH' as an outcome measure ($n=2$).

Repetitive TMS was directed at the left temporoparietal cortex (i.e. T3P3 in terms of electroencephalogram electrode placements) in 137 patients, at the right temporoparietal cortex (T4P4) in 12 patients, at the left *and* right temporoparietal cortex in 14 patients, and at miscellaneous locations (using fMRI guidance) in 20 patients. Details of the rTMS paradigms are presented in Table 25.1. Most studies applied low-frequency rTMS at 1 Hz and at intensities below the motor threshold.

In three studies, a different localization was used. Lee et al. (2005) chose the right temporoparietal cortex (T4P4) as their focus of treatment. Only one study examined the effects of bilateral rTMS, directed at the left temporoparietal cortex (T3P3) during one half of each session, whereas during the second half, stimulation was switched to the right temporoparietal region (Vercammen et al. 2009). Recent functional magnetic resonance imaging (fMRI) studies indicate that the left temporoparietal cortex is not a general focus of activation during the experience of AVH (Sommer et al. 2008), but rather the right temporoparietal area. Based on these findings, a randomized controlled trial was performed in which rTMS was directed at the focus of maximal hallucinatory activity as assessed with the aid of individual fMRI scans (Slotema et al. 2011).

Table 25.1 rTMS parameters used in the treatment of auditory verbal hallucinations

Study	Location	Frequency (Hz)	MT (%)	Number of stimuli	Number of sessions
De Jesus et al. (2011)	T3P3	1	80	1,200 (day 1 480, day 2 960)	20
Slotema et al. (2011)	T3P3	1	90	1,200	15
Slotema et al. (2011)	fMRI-guided	1	90	1,200	15
Vercammen et al. (2009)	T3P3	1	90	1,200	12
Vercammen et al. (2009)	T3P3 and T4P4	1	90	1,200	12
Rosa et al. (2007)	T3P3	1	90	960	10
Brunelin et al. (2006)	T3P3	1	90	1,000	10
Chibbaro et al. (2005)	T3P3	1	90	900	4
Fitzgerald et al. (2006)	T3P3	1	90	900	10
Hoffman et al. (2005)	T3P3	1	90	900	10
Lee et al. (2005)	T3P3	1	100	1,600	10
Lee et al. (2005)	T4P4	1	100	1600	10
Saba et al. (2006)	T3P3	1	80	300	10

Hz Hertz, *MT* motor threshold, *T3P3* left temporoparietal cortex, *fMRI* functional magnetic resonance imaging, *T4P4* right temporoparietal cortex

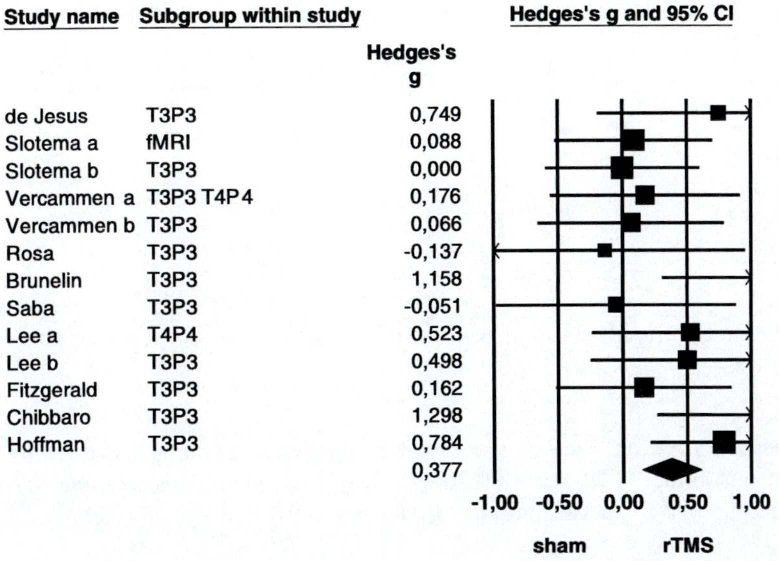

Fig. 25.4 Results of a meta-analysis of rTMS studies involving the treatment of auditory verbal hallucinations

Table 25.2 Reasons for dropout

	T3P3 (%)	fMRI-guided (%)	Sham (%)
Side effects	5 (3.6)	1 (5)	1 (0.7)
Worsening of psychosis	1 (0.7)	1 (5)	3 (2.2)
Other/unknown	1 (0.7)	0	2 (1.6)
Total	7/137 (5.1)	2/20 (10)	6/126 (4.8)

T3P3 left temporoparietal cortex, *fMRI* functional magnetic resonance imaging

The results of the present meta-analysis are presented in Fig. 25.4. Analysis showed that real rTMS is better than sham treatment, the mean weighted effect size being 0.38 (Hedges's g, $p=0.001$). Heterogeneity was moderately low ($I^2=17.9$, $p=0.26$). The fail-safe number was 372 studies.

Only the effects of rTMS directed at the left temporoparietal cortex turned out to be superior to those of sham treatment (Hedges's $g=0.431$, $p=0.005$), with a moderate heterogeneity ($I^2=31.8$, $p=0.15$). A correlation analysis of the effect sizes and the rTMS parameters 'motor threshold', 'number of stimuli', 'number of treatment sessions', and 'total number of stimuli' did not indicate that any of the paradigms used were superior to the others.

The dropouts and side effects are presented in Tables 25.2 and 25.3. It should be noted that these are listed per treatment condition, as a minority of the studies did not include any descriptions of the side effects and/or dropouts per treatment

Table 25.3 Side effects of rTMS treatment occurring in at least 1% of the participants per indication

	T3P3 (%)	fMRI-guided (%)	T3P3 and T4P4 (%)	Sham (%)
Headache	15 (10.9)	3 (15)	4 (33.3)	3 (2.1)
Dizziness	4 (3)	0	0	2 (1.6)
Facial muscular twitching	0	7 (35)	0	1 (0.8)
Scalp discomfort	0	1 (5)	0	0
Cervical pain	0	1 (5)	0	0
Tingling sensation in arm	0	0	1 (8.3)	0
Other	6 (4.4)	1 (5)	0	0
Total	25/137 (18.2)	13/20 (65)	5/14 (35.7)	6/126 (4.8)

T3P3 left temporoparietal cortex, *fMRI* functional magnetic resonance imaging, *T4P4* right temporoparietal cortex

condition. An equal percentage of patients dropped out of the real and sham TMS treatment groups (5.7% and 4.8%, respectively). Side effects were mentioned more frequently in the real TMS group (25.1% versus 4.8% in the sham condition).

25.4 Discussion

The aim of this chapter is to present an up-to-date meta-analysis of the results of rTMS in the treatment of AVH. A significant but moderate effect of and 0.43 respectively 0.38 was found for all included studies *and* for low-frequency rTMS directed at the left temporoparietal cortex alone. The magnitude of the effect sizes per study did not correlate with specific levels of the TMS treatment parameters such as frequency, intensity, or location.

These findings are in concordance with the results of four previous meta-analyses (i.e. Aleman et al. 2007; Freitas et al. 2009; Slotema et al. 2010; Tranulis et al. 2008). However, a note of caution may be in place here. When new treatment strategies are introduced, the initial reports tend to feature relatively small sample sizes and to provide favourable results, while small studies with negative findings do not tend to be published (Emerson et al. 2010). In the course of time, sample sizes tend to increase, and negative findings tend to become published as well. Such trends have led effect sizes to decrease per year of publication (Munafo and Flint 2010). As rTMS is a relatively young treatment method, future studies may show less favourable results. We indeed found a trend towards larger studies being published in recent years, yielding negative results. We therefore take into account that the initially reported positive effects may well disappear when more studies with larger patient samples will be published, but the present state of the evidence allows us to recommend low-frequency rTMS for AVH, especially when the relatively mild side-effect profile and the lack of other treatment options are taken into account.

Only few studies examined the effects of low-frequency rTMS targeted at other brain regions than the left temporoparietal cortex. A reduction in the severity of AVH after rTMS directed at the right temporoparietal cortex (Lee et al. 2005) could not be replicated by others (Jandl et al. 2006; Loo et al. 2009). Repetitive TMS directed at the left and right temporoparietal cortex (Vercammen et al. 2009) and rTMS directed at foci with maximal hallucinatory activity, as indicated by fMRI findings (Slotema et al. 2011), were not superior to sham treatment. Stimulation of the left temporoparietal cortex and the adjacent supramarginal gyrus, Broca's area, the left primary auditory cortex, and their contralateral homologues revealed only a greater improvement of AVH when the treatment was focused at the left temporoparietal cortex and the adjacent supramarginal gyrus (Hoffman et al. 2007). Furthermore, the effects of rTMS applied to either Broca's area or the left superior temporal gyrus were equal to those of sham treatment (Schönfeldt-Lecuona et al. 2004). It is possible that the facial musculature overlying the skull prevents rTMS from reaching Broca's area, as the electromagnetic pulse reaches a depth of no more than 2 cm (Hoffman et al. 2007). No firm conclusions can be drawn due to the small number of studies, but there is currently no evidence to suggest that locations other than the left temporoparietal cortex are suitable options for the treatment of AVH with rTMS.

In the majority of studies, rTMS was applied with a frequency of 1 Hz. But also, high-frequency rTMS has been studied as a treatment option for AVH (Montagne-Larmurier et al. 2009), yielding a strong clinical response when a frequency of 20 Hz was used. However, the patient population under study was small (11 participants), and no control condition was included, which precludes any firm conclusions regarding this type of treatment. A recent study from the Utrecht group assessed 20-Hz stimulation and 1-Hz stimulation in a double-blind, head-to-head comparison and found no differences between the two treatment arms (De Weijer et al. 2011). In a randomized controlled trial, the effects of low-frequency rTMS preceded by 5 min of 6-Hz rTMS was compared with low-frequency rTMS alone, and again no differences could be revealed between the two conditions (Slotema et al. submitted).

Two case reports described relief from chronic, intractable auditory hallucinations after bilateral and continuous theta-burst TMS (i.e. in a frequency of approximately 50 Hz), respectively (Eberle et al. 2010; Poulet et al. 2009). However, before any large sham-controlled RCTs become available, we cannot recommend high-frequency or theta-burst stimulations for the treatment of AVH.

The effects of rTMS for AVH using a motor threshold higher than 100% are unknown. However, so far, an increase in the number of stimuli and in the number of sessions has never resulted in a significantly stronger reduction of the severity of AVH (Slotema et al. 2011; Vercammen et al. 2009).

The majority of studies have been performed with the aid of a figure-of-eight coil (see Fig. 25.1). H-coils, on the other hand, are designed to maximize the electrical field in deep brain tissues by their ability to summate separate fields projected into the skull from several points around its periphery (Zangen et al. 2005). In an open-label study, eight patients were treated with deep-brain TMS using an H-coil, which resulted in a significant reduction of the severity of AVH (Rosenberg et al. 2011). However, this study also failed to employ a control group. As a consequence, its results need to be replicated in randomized, placebo-controlled, double-blind studies before any firm conclusions can be drawn from them.

25.5 Conclusion

The present meta-analysis of studies using rTMS for the treatment of auditory verbal hallucinations shows mildly favourable results. All of the included studies used low-frequency rTMS, and the majority of them directed the electromagnetic pulse at the left temporoparietal cortex. A significant yet moderate effect was found, which was significantly lower than the results of previous meta-analyses. This would seem to be primarily due to the publication of various new studies with larger samples and negative findings. We therefore take into account that this mildly favourable effect may well disappear when more studies with larger patient samples will get published. So far, no double-blind, randomized controlled trials have been performed applying other rTMS paradigms (using high-frequency rTMS, for example, or a higher motor threshold, or an H-coil) which may have the potential to yield an efficacy superior to that of low-frequency stimulation.

References

Aleman, A., Sommer, I.E., Kahn, R.S. (2007). Efficacy of slow repetitive transcranial magnetic stimulation in the treatment of resistant auditory hallucinations in schizophrenia: a meta-analysis. *Journal of Clinical Psychiatry*, 68, 416–421.

Barker, A.T., Freeston, I.L., Jarratt, J.A., Jalinous, R. (1990). *Magnetic stimulation of the human nervous system: An introduction and basic principles*. In: *Magnetic stimulation in clinical neurophysiology*. Edited by Chokroverty, S. Boston, MA: Butterworth's, pp. 55–72.

Barker, A.T., Jalinous, R., Freeston, I.L. (1985). Non-invasive magnetic stimulation of human motor cortex. *Lancet*, 11, 1106–1107.

Benson, R.R., Whalen, D.H., Richardson, M., Swainson, B., Clark, V.P., Lai, S., Liberman, A.M. (2001). Parametrically dissociating speech and nonspeech perception in the brain using fMRI. *Brain and Language*, 78, 364–396.

Brunelin, J., Poulet, E., Bediou, B., Kallel, L., Dalery, J., D'Amato, T., Saoud, M. (2006). Low frequency repetitive transcranial magnetic stimulation improves source monitoring deficit in hallucinating patients with schizophrenia. *Schizophrenia Research*, 81, 41–45.

Chibbaro, G., Daniele, M., Alagona, G., Di Pascuale, C., Cannavò, M., Rapisarda, V., Bella, R., Pennisi, G. (2005). Repetitive transcranial magnetic stimulation in schizophrenic patients reporting auditory hallucinations. *Neuroscience Letters*, 383, 54–57.

De Jesus, D.R., Gil, A., Barbosa, L., Lobato, M.I., Magalhães, P.V., Favalli, G.P., Marcolin, M.A., Daskalakis, Z.J., Belmonte-de-Abreu, P.D. (2011). A pilot double-blind sham-controlled trial of repetitive transcranial magnetic stimulation for patients with refractory schizophrenia treated with clozapine. *Psychiatry Research*, 188, 203–207.

De Weijer, A., Neggers, S.F., Diederen, K.M., Mandl, R.C., Kahn, R.S., Hulshoff Poll, H.E., Sommer, I.E. (2011). Aberrations in the arcuate fasciculus are associated with auditory verbal hallucinations in psychotic and in non-psychotic individuals. *Human Brain Mapping*, doi: 10.1002/hbm.21463.

Eberle, M.-C., Wildgruber, D., Wasserka, B., Fallgatter, A.J., Plewnia, C. (2010). Relief from chronic intractable auditory hallucinations after long-term bilateral theta burst stimulation. *American Journal of Psychiatry*, 167, 1410.

Emerson, G.B., Warme, W.J., Wolf, F.M., Heckman, J.D., Brand, R.A., Leopold, S.S. (2010). Testing for the presence of positive-outcome bias in peer review. A randomized controlled trial. *Archives of Internal Medicine*, 170, 1934–1939.

Fitzgerald, P.B., Fountain, S., Daskalakis, Z.J. (2006). A comprehensive review of the effects of rTMS on motor cortical excitability and inhibition. *Clinical Neurophysiology*, 117, 2584–2596.

Freitas, C., Fregni, F., Pascual-Leone, A. (2009). Meta-analysis of the effects of repetitive transcranial magnetic stimulation (rTMS) on negative and positive symptoms in schizophrenia. *Schizophrenia Research*, 108, 11–24.

George, M.S., Wassermann, E.M., Williams, W.A., Callahan, A., Ketter, T.A., Basser, P., Hallett, M., Post, R.M. (1995). Daily repetitive transcranial magnetic stimulation (rTMS) improves mood in depression. *Neuroreport*, 6, 1853–1856.

Higgins, J.P.T., Thompson, S.G., Deeks, J.J., Altman, D.G. (2003). Measuring inconsistency in meta-analyses. *British Medical Journal*, 327, 557–560.

Hoffman, R.E., Boutros, N.N., Berman, R.M., Roessler, E., Belger, A., Krystal, J.H., Charney, D.S. (1999). Transcranial magnetic stimulation of left temporoparietal cortex in three patients reporting hallucinated "voices". *Biological Psychiatry*, 46, 130–132.

Hoffman, R.E., Gueorguieva, R., Hawkins, K.A., Varanko, M., Boutros, N.N., Wu, Y., Carroll, K., Krystal, J.H. (2005). Temporoparietal transcranial magnetic stimulation for auditory hallucinations: safety, efficacy and moderators in a fifty patient sample. *Biological Psychiatry*, 58, 97–104.

Hoffman, R.E., Hampson, M., Wu, K., Anderson, A.W., Gore, J.C., Buchanan, R.J. Constable, R.T., Hawkins, K.A., Neayka, S. (2007). Probing the pathophysiology of auditory/verbal hallucinations by combining functional magnetic resonance imaging and transcranial magnetic stimulation. *Cerebral Cortex*, 17, 2733–2743.

Homan, R.W., Herman, J., Purdy, P. (1987). Cerebral location of international 10–20 system electrode placement. *Electroencephalography and Clinical Neurophysiology*, 66, 376–382.

Jandl, M., Steyer, J., Weber, M., Linden, D.E., Rothmeier, J., Maurer, K., Kaschka, W.P. (2006). Treating auditory hallucinations by transcranial magnetic stimulation: a randomized controlled cross-over trial. *Neuropsychobiology*, 53, 63–69.

Lee, S.-H., Kim, W., Chung, Y.-C., Jung, K.-H., Bahk, W.-M., Yun, T.-Y., Kim, K.-S., George, M.S., Chae, J.-H. (2005). A double blind study showing that two weeks of daily repetitive TMS over the left or right temporoparietal cortex reduces symptoms in patients with schizophrenia who are having treatment-refractory auditory hallucinations. *Neuroscience Letters*, 376, 177–181.

Loo, C.K., Sainsbury, K., Mitchell, P., Hadzi-Pavlovic, D., Sachdev, P.S. (2009). A sham-controlled trial of left and right temporal rTMS for the treatment of auditory hallucinations. *Psychological Medicine*, 40, 541–546.

Montagne-Larmurier, A., Etard, O., Razafimandimby, A., Morello, R., Dollfus, S. (2009). Two-day treatment of auditory hallucinations by high frequency rTMS guided by cerebral imaging: a 6 month follow-up pilot study. *Schizophrenia Research*, 113, 77–83.

Munafo, M.R., Flint J. (2010). How reliable are scientific studies? *The British Journal of Psychiatry*, 197, 257–258.

Poulet, E., Brunelin, J., Ben Makhlouf, W., D'Amato, T., Saoud, M. (2009). A case report of cTBS for the treatment of auditory hallucinations in a patient with schizophrenia. *Brain Stimulation*, 2, 118–119.

Rosa, M.O., Gattaz, W.F., Rosa, M.A., Rumi, D.O., Tavares, H., Myckowski, M., Sartorelli, M.C., Rigonatti, S.P., Elkis, H., Cabral, S.B., Teixeira, M.J., Marcolin, M.A. (2007). Effects of repetitive transcranial magnetic stimulation on auditory hallucinations refractory to clozapine. *Journal of Clinical Psychiatry*, 68, 1528–1532.

Rosenberg, O., Roth, Y., Kotler, M., Zangen, A., Dannon, P. (2011). Deep transcranial magnetic stimulation for the treatment of auditory hallucinations: a preliminary open-label study. *Annals of General Psychiatry*, 10, 3.

Rosenthal, R. (1979). The file drawer problem and tolerance for null results. *Psychological Bulletin*, 86, 638–641.

Saba, G., Verdon, C.M., Kalalou, K., Rocamora, J.F., Dumortier, G., Benadhira, R., Stamatiadis, L. Vicaut, E., Lipski, H., Januel, D. (2006). Transcranial magnetic stimulation in the treatment of schizophrenic symptoms: a double blind sham controlled study. *Journal of Psychiatric Research*, 40, 147–152.

Schönfeldt-Lecuona, C., Grön, G., Walter, H., Büchler, N., Wunderlich, A., Spitzer, M., Herwig, U. (2004). Stereotaxic rTMS for the treatment of auditory hallucinations in schizophrenia. *Neuroreport*, 15, 1669–1673.

Siebner, H.R., Rothwell, J. (2003). Transcranial magnetic stimulation: new insights into representational cortical plasticity. *Experimental Brain Research*, 148, 1–16.

Slotema, C.W., Blom, J.D., De Weijer, A.D., Diederen, K.M., Goekoop, R., Looijestijn, J., Daalman, K., Rijkaart, A.-M., Kahn, R.S., Hoek, H.W., Sommer, I.E.C. (2011). Can low-frequency repetitive transcranial magnetic stimulation really relieve medication-resistant auditory verbal hallucinations? Negative results from a large randomized controlled trial. *Biological Psychiatry*, 69, 450–456.

Slotema, C.W., Blom, J.D., De Weijer, A.D., Kahn, R.S., Hoek, H.W., Sommer, I.E.C. Priming does not enhance the efficacy of one-hertz repetitive transcranial magnetic stimulation for the treatment of auditory verbal hallucinations: results of a randomized controlled study (submitted).

Slotema, C.W., Blom, J.D., Hoek, H.W., Sommer, I.E.C. (2010). Should we expand the toolbox of psychiatric treatment methods to include repetitive transcranial magnetic stimulation? A meta-analysis of the efficacy of rTMS for psychiatric disorders. *Journal of Clinical Psychiatry*, 71, 873–884.

Sommer, I.E.C., Diederen, K.M.J., Blom, J.-D., Willems, A., Kushan, L., Slotema, K., Boks, M.P.M., Daalman, K., Hoek, H.W., Neggers, S.F.W., Kahn, R.S. (2008). Auditory verbal hallucinations predominantly activate the *right* inferior frontal area. *Brain*, 131, 3169–3177.

Tranulis, C., Sepehry, A.A., Galinowski, A., Stip, E. (2008). Should we treat auditory hallucinations with repetitive transcranial magnetic stimulation? A meta-analysis. *Canadian Journal of Psychiatry*, 53, 577–586.

Vercammen, A., Knegtering, H., Bruggeman, R., Westenbroek, H.M., Jenner, J.A., Slooff, C.J., Wunderink, L., Aleman, A. (2009). Effects of bilateral repetitive transcranial magnetic stimulation on treatment resistant auditory-verbal hallucinations in schizophrenia: a randomized controlled trial. *Schizophrenia Research*, 114, 172–179.

Wassermann, E.M. (1998). Risk and safety of repetitive transcranial magnetic stimulation: report and suggested guidelines from the International Workshop on the Safety of Repetitive Transcranial Magnetic Stimulation, June 5–7, 1996. *Electroencephalography and Clinical Neurophysiology*, 108, 1–16.

Zangen, A., Roth, Y., Voller, B., Hallett, M. (2005). Transcranial magnetic stimulation of deep brain regions: evidence for efficacy of the H-coil. *Clinical Neurophysiology*, 116, 775–779.

Chapter 26
Cognitive-Behavioral Therapy

Mark van der Gaag

26.1 Appraisal Is the Foundation of the Cognitive Model of Hallucinations

Medication is the treatment of choice for many types of hallucination, but a lack of results – due to noncompliance and the persistence of residual symptoms, for example – has led researchers to develop various ancillary treatment forms (Pantelis and Barnes 1996). Two decades ago, psychological research focused primarily on coping strategies. But although many patients do apply self-developed coping strategies, they tend to lose confidence in their efficacy over time, while patients with a long history of voice hearing do not always appear to select the most effective coping strategies (Carter et al. 1996; Farhall et al. 2007).

Cognitive-behavioral therapy (CBT) advocates a wholly different approach. In the context of the traditional medical model, which also infused the coping model, the primary variable to be treated is the symptom of disease (e.g., "hallucination") rather than the patient's depression, anxiety, or dysfunctional behavior, which are all considered *secondary* to the hallucinations at hand and only amenable to treatment when the hallucinations themselves are being treated. The cognitive model differs from more conservative variants of the medical model in that its primary goal is not the eradication of the primary symptoms per se but the reduction of distress and dysfunctional behavior. After all, symptoms only become "real" symptoms (as defined in the Diagnostic and Statistical Manual of Mental Disorders (DSM) and other psychiatric classifications) when they cause distress and/or dysfunctional behavior. However, within the context of the cognitive model, it is not so much the symptom itself that is held responsible for causing any distress or dysfunctional behavior but the patient's appraisal of that symptom (Birchwood and Trower 2006b).

M. van der Gaag, Ph.D. (✉)
University and EMGO Institute, Amsterdam, The Netherlands

Psychosis Research, Parnassia Bavo Group, The Hague, The Netherlands
e-mail: m.vander.gaag@vu.nl

J.D. Blom and I.E.C. Sommer (eds.), *Hallucinations: Research and Practice*,
DOI 10.1007/978-1-4614-0959-5_26,

Auditory hallucinations, for example, do not always occur in the context of disease (see Chap. 28). Actually, most people who hear voices lead perfectly normal lives and do not tend to seek any kind of treatment for them. Only when voices are appraised in a catastrophical way they are bound to cause depression, anxiety, social isolation, and so on (Honig et al. 1998; Lawrence et al. 2010). Those who are indifferent to their hallucinations, and live their lives the way they did before, will seldom become psychiatric patients. Those who do become patients are those who appraise their voices in a malevolent way and endow the latter with so much power that they develop a passive and submissive attitude (Mawson et al. 2010). For that reason, CBT for auditory hallucinations focuses on the ways in which voices are being appraised, and aims at a reduction of distress and dysfunctional behavior, even if the hallucinations themselves continue to be experienced.

26.2 Psychological Models for Hallucinosis

In recent years, various psychological models for hallucinations have been developed. Some of the earlier models were based on evidence – obtained from a number of highly intriguing studies – that the mediation of verbal auditory hallucinations may depend on subvocal speech. That finding set the stage for the subsequent neuroimaging of speech-related versus hallucination-related activity in the brain, the investigation of psychological processes underlying speech and hallucinations, and the investigation of the content, appraisal, and metacognitive aspects of auditory hallucinations. All those research avenues yielded valuable new data, but an integration into a unitary model has not yet been achieved, and to date many of the phenomena under study have remained unexplained. All in all, the psychological model of hallucinosis is still tentative and "under construction." And yet it boasts sufficient clinically relevant data to justify the application of cognitive-behavioral interventions, the efficacy and effectiveness of which are small to medium-sized, but stable (NICE 2009).

26.2.1 The Origin of Verbal Auditory Hallucinations

26.2.1.1 Inner Speech and Self-Monitoring Failure

A study of "tension in the throat" showed that automatic speech (or "inner speech") may well underlie the mediation of verbal auditory hallucinations (Gould 1948). Gould used an electromyograph with leads on the lower lip and chin to examine 100 patients diagnosed with schizophrenia. In that study, a hallucinating subgroup of 48 patients showed an increased tension in the muscles of the lip and chin in 83% of the cases, whereas the nonhallucinating group showed such tension in only 10% of the cases. The author therefore concluded that verbal auditory hallucinations are associated with an involuntary intensification of the psychomotor mechanisms of speech. Another study found similar changes in activity in the tongue and the chin, as well as changes in the patients' breathing amplitude, but no changes in the tension of their arm musculature (McGuigan 1966).

In the years following this pioneering work, many small intervention studies were carried out. Most of them were multiple-baseline studies comparing multiple-treatment conditions with multiple returns to baseline within series of hallucinating patients (Corrigan and Storzbach 1993). Those studies yielded some evidence for the effectiveness of systematic desensitization, operant conditioning (using reward and punishment), the voluntary use of the vocal cords (through humming and gargling), social conversation (which appeared to be incompatible with ongoing hallucinatory experiences), and the manipulation of auditory input by means of earplugs or loud music (see also Chaps. 11 and 28). Favorable short-term effects were recorded for all those interventions, but randomized controlled trials (RCTs) of sufficient power were never performed.

The concept of inner speech continues to play an important role in the cognitive-behavioral model of hallucinations, even though it fails to explain the occurrence of voices in the second or third person or the simultaneous occurrence of multiple voices. To account for the latter phenomena, the model now proposes that hallucinations may arise when the self-monitoring of inner speech is failing. The hearing of external voices and the awareness of one's own verbal thoughts both depend on activity in Wernicke's speech perception area. Wernicke's area "knows" the difference between the two because a corollary discharge feed-forward mechanism "tells" it that thoughts are on their way (Ford and Mathalon 2005). When Broca's speech production area "translates" a thought into a train of words and then sends them off to Wernicke's area, a collateral message is sent to that monitor to allow for its identification as "inner speech." When that monitor fails to inform Wernicke's area, the latter will "conclude" that the words must have an external origin (see also Chap. 21).

26.2.1.2 Intrusions from Traumatic Memories

And yet not all types of auditory hallucination appear to be associated with inner speech. Some individuals experience voices that are clearly related to earlier traumatic experiences. Such voices tend to constitute reexperienced verbal messages, conveyed priorly by people involved in the traumatic situation (Jones 2010). They can perhaps be conceptualized best as intrusions of traumatic memories (or "reperceptive hallucinations"). In contrast, voices attributed to inner speech tend to be involved with the planning and execution of one's present actions. They usually tell the percipients what to do next or threaten him with aversive consequences when he refuses to obey (Jones 2010).

26.2.2 Psychological Processes

26.2.2.1 Source Monitoring

Heilbrun found that 12 psychiatric patients with a history of hallucinations turned out to be insufficiently capable of identifying the words, meaning, and grammatical style of their own spoken sentences following a 1-week lapse of time, as compared to eight nonhallucinating psychiatric patients. Incidentally, the two groups did not differ with regard to their ability to remember verbal material, the stability of their opinions, or the level of their communication skills (Heilbrun 1980). A meta-analysis of 23 studies carried out over the past 30 years confirmed that individuals diagnosed with schizophrenia who are prone to hallucinations tend to attribute self-generated thoughts and utterances to other people (Waters and Badcock 2010). That cognitive bias tends to increase when their attention is directed to the self, and/or when strong emotions are aroused. It has been suggested that this source-monitoring bias may also be a source of verbal auditory hallucinations. However, a drawback of this model is that it does not explain why voices are actually *heard*. It would seem to explain experiences of thought insertion or "indirect Gedankenlautwerden" rather than the occurrence of hallucinations. The bias itself appears to be endophenotypical, as it has also been found during prodromal states and episodes of remission, as well as in siblings of patients diagnosed with schizophrenia (Brunelin et al. 2007).

26.2.2.2 Appraisal

Because verbal auditory hallucinations may differ from the percipient's "inner voice" and the utterances often convey specific themes that are coherent over time (Nayani and David 1996), voice hearers easily appraise them as coming from a different person or agent. When the voices are also appraised as omnipotent or omniscient and the percipient becomes convinced that they have malevolent intentions, anxiety is bound to set in. Once anxious, voice hearers often start to deal with the voices by making promises to them and by obeying their orders and commands (Birchwood and Chadwick 1997). The association between such appraisals and the severity of verbal auditory hallucinations has been repeatedly confirmed (Mawson et al. 2010). And yet changing those appraisals does not always lead to a reduction of the severity of the hallucinations at hand. In order to attain that goal, interpersonal aspects and self-esteem need to be addressed as well (Mawson et al. 2010).

26.2.2.3 Interpersonal Aspects

The relational aspects of hallucinations were noticed many years ago. In hospitalized patients, for example, who may experience extreme hallucinations, therapists found it hard to relate to the patient at all (Erickson and Gustafson 1968). The average therapist is available for no more than an hour per week, whereas in such cases, the voices may be present 24×7. Many patients do not feel free to discuss their voices because they appear to be listening – and speaking – all the time. That relational inequality was examined for the first time by Gilbert and Birchwood (2001), who demonstrated that the presence of powerful voices tends to correlate positively with subordinate behavior (including flight-or-fight reactions) and depression. They also found that the concomitant tendency to consider oneself as low-ranking in a hierarchy does not only characterize the relationship with the voices, but all of the patients' social relationships (Birchwood et al. 2000; Gilbert et al. 2001). Moreover, their research showed that this perceived low-ranking position is not due to depression, distress, or the power with which the voices are endowed. Nor do the voices themselves lead to depression, or the appraisal of power to the occurrence of voices. Instead, the authors found that the distorted perception of one's own social rank and power leads up to the appraisal that the voices are powerful, and also to distress and depression (Birchwood et al. 2004).

Self-criticizing thoughts occurring in the context of depression appear to have a role similar to that of voices in psychosis: they serve to punish the patient. In either case, the patients' own attempts to stop them tend to be futile. Gilbert (2009) developed a therapy based on the two-chair technique borrowed from Gestalt therapy, and combined it with evolutionary thoughts on the "soothing system" of social mammals. In young mammals, the "threat system" needs to be inactivated by a caring parent to allow for a normal development. Only when sufficiently caressed and comforted will the young mammal feel capable of exploring the world. By repeatedly experiencing safety and soothing, that feeling can be internalized, and the adult thus treated is able to sooth himself to overcome any inappropriate anxieties and feelings of threat. In patients who experience hallucinations, the threat system tends to be permanently active due to the presence of threatening voices. Gilbert's method consists of introducing the patient to thinking about what a caring and soothing voice might say. Thus, the patient may gradually learn to rely on the trustworthy, soothing "voices" and to escape the terror caused by the threatening ones (Gilbert 2009).

Two other interpersonal aspects are shame and stigmatization. Shame for being psychotic, and attempts to hide from others that one may in fact be different (or that one is using medication, etc.) tend to lead to a heightened self-awareness and social phobia anticipating the possible rejection by others (Birchwood et al. 2007).

26.2.2.4 Self-esteem and Depression

Low self-esteem plays a pivotal role in the continuation of hallucinatory experiences. Patients with low self-esteem also tend to be more severely depressed and to experience verbal auditory hallucinations with a higher degree of severity and associated distress (Smith et al. 2006). Beliefs about the omnipotence and malevolent intentions of voices, as well as low self-esteem, contribute independently to feelings of depression in patients experiencing persistent verbal auditory hallucinations (Fannon et al. 2009). Fannon et al. also conclude that low self-esteem is crucial to

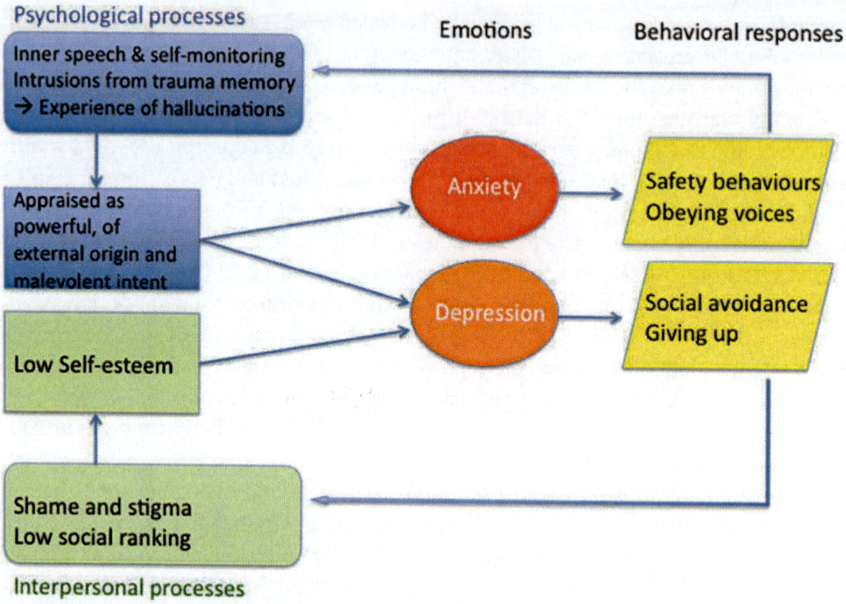

Fig. 26.1 The psychological model of verbal auditory hallucinations

our understanding of any affective disturbances that may occur in voice hearers and that therapeutic interventions therefore need to address the way the voices *and* the self are being appraised.

26.2.3 Conclusions Based on the Psychological Model

Although the psychological model of auditory hallucinations (see Fig. 26.1) is far from complete and leaves many aspects of their mediation unexplained, it has certainly aided to increase our insights into these phenomena. First of all, we have learned that the propensity to hear voices is distributed continuously in the general population (Van Os et al. 2010). Secondly, we now have two basic explanatory models at our disposal (the inner speech model and the self-monitoring model), whereas a third model (i.e., the reperception model) suits those cases where intrusive traumatic memories are reexperienced in the form of hallucinations. In the third place, we now know that hallucinations tend to lead to distress when they are being appraised as powerful, as having an external source, and as having malevolent intentions, whereas low self-esteem contributes independently to the severity of verbal auditory hallucinations and of depression. In the fourth place, we now know that anxiety tends to lead to submissiveness to the voices, as well as to a tendency to obey and appease them, while depression often entails feelings of defeat and surrender. Fifth, we have learned that feelings of shame and stigmatization may entail severe anxiety, a social phobia-like self-awareness, and attempts to avoid exposure to other people and their judgments. Sixth and finally, we have learned that those behavioral responses tend to lead to a consolidation of the appraisals and to the inability to experience any emotionally corrective experiences.

26.3 Cognitive-Behavioral Therapy

CBT targets the ways in which events are being appraised. It puts the appraisals in perspective through a collaborative approach in which other possible explanations for the events in question are considered together with the patient and in which any concomitant feelings and behavioral reactions are extrapolated when they are considered valid. To find out which type of appraisal may suit an event best, the patient is challenged to carry out behavioral experiments and to experience the adequacy or inadequacy of different approaches. The cognitive aspect of CBT involves the guided discovery of alternative explanations and alternative reactions. The behavioral part involves the testing of alternative ways to deal with particular situations and attempts to change one's feelings about them. The discovery that catastrophes fail to occur, e.g., that resisting the voices' commands does not lead to any kind of punishment, is a powerful way to diminish distress.

26.3.1 Education and Normalization

In addition, CBT for hallucinations has various normalizing and educational aspects. Many patients already feel better when they hear that each and every year, about 2% of the population start to experience hallucinations, and that at any given moment some 4% of the population are experiencing them (Van Os et al. 2010). It is also highly educational to tell patients that most of those people are not bothered very much by their voices, that they continue to live the way they did before, and that people with severely threatening voices can learn to ignore them and to continue with the things that make their lives worth living and meaningful.

When patients express the wish to learn something about their voices and the way they are mediated, it may be helpful to tell them about the sensitization of the dopamine system which gives rise to intrusions and hallucinations, but also makes them jump to conclusions, and renders them overly confident about those conclusions. One may also consider explaining that voices have a certain tendency to make people anxious or angry, i.e., that they "thrive" or "feed" on anxious and angry feelings. Patients may well have noticed that strong emotions make the voices louder and more persistent. Therefore, they tend to acknowledge that it is good advice to try to stay calm whatever the circumstances and to imagine that a "game" is being played where the one who shows any strong emotions will lose.

26.3.2 Challenging the Way Voices Are Appraised

Appraisals that tend to aggravate the severity of hallucinatory experiences are those that exaggerate their power, characterize them as omnipotent and omniscient, place their source in the external world, endow them with malevolent intentions, and make believe that they can actually do harm.

26.3.2.1 Power

Threatening voices and command hallucinations can pose a danger to the patient and his environment. Many patients are bent on resisting any dangerous and aggressive commands they may hear and on trying to appease their voices without letting themselves be lured into dangerous or aggressive behavior. But not all patients are able to resist their voices. The risk that they will comply with command hallucinations can be reduced by changing the patients' beliefs about the voices' power (Birchwood and Trower 2006a; Trower et al. 2004). Trower and colleagues tested the effectiveness of cognitive therapy in command hallucinations by randomizing 38 patients who had recently complied with their voices' commands and had suffered the consequences. The control condition was treatment-as-usual, and the patients were followed up after 6 and 12 months. In the cognitive therapy group, the authors found large and significant reductions in compliant behavior, with an effect size of 1.1. Improvements were also recorded in the patient group (but not in the control group) as regards the power attributed to the voices, the perceived need to comply, and the levels of concomitant distress and depression. No changes in the frequency, loudness, and content of the voices were recorded. The differences were still significant after 12 months of follow-up.

Beliefs about omniscience and omnipotence can be challenged with the aid of behavioral experiments. "Are those omniscient voices able to predict the headlines of tomorrow's newspaper?" "Do they possess knowledge that you do not already possess yourself?" "Is it possible to verify their knowledge?" Answers to questions such as these can be very revealing. The same method is applicable to beliefs about the voices' alleged omnipotence. Sometimes a shortcut can be taken by telling the patient that you yourself, as a therapist, have dealt with dozens of threatening voices, and that they are all mouth but no trousers. An alternative, and perhaps more elegant, way is to encourage the patient to look for examples of what happened when he refused to do what the voices commanded him to. He may then become aware that he has in fact experienced many occasions in which his refusal to obey failed to have any consequences.

Another elegant way is to draw the voices. Being unable to act, they do not have any arms or legs. Being only capable of seeing what the patient sees, they probably have no eyes of their own. Being reluctant to listen to the patient's pleas, they may well have no ears either. One ends up with the cartoon of a smiley, having nothing but a mouth that is solely capable of repeating the same old stuff over and over again.

26.3.2.2 External Hallucinations Versus Intrusions

A problem with external voices is that they do not consist of any actual sounds. If so, anyone would be able to hear them, which is obviously not the case. When the patient believes that he has a natural radio receiver inside his head – which is of course highly improbable – then one might suggest that it should be possible to disrupt the radio waves received by his brain. A cap made of aluminum foil (also known as a cage of Faraday) is sometimes used by physicists to create an anechoic space. Such a device prevents electrical fields to enter from the outside and vice versa, thus reducing the power of high-frequency radiation with 120 dB. Elaborating on that line of thought, myriad behavioral experiments can be thought of to test the patient's hypotheses and to challenge his beliefs about the alleged radio receiver.

26.3.2.3 Identity, Goals, and Meaning

Voices are there for a reason. What do they want to accomplish? Do they want to punish the patient? And if so, why? To destroy the patient? And do they have any reason to want this?

26.3.3 Challenging the Content of Voices

Voices can say many different things. Some of them constantly humiliate the patient by telling him that he is a loser, that nobody cares for him, that he is incompetent, or that he would be better off dead. Is all that true? Did the patient never gain anything? Has there never been anyone who cared for him? By exploring the voices' contents, many of the classical CBT questions can be put forth: Is it true what the voices say? And if it is true, is that something to worry about? And if it is something to worry about, is there a way to cope with it and preserve one's quality of life?

26.3.4 Changing the Patient's Attitude Toward His Voices and His Safety Behaviors

When it is possible to raise some doubt, it is time to encourage various behavioral changes. The patient might consider to limit the time spent with his voices, to pick up the routines of daily life, and to regain important social roles that lend meaning and fulfillment to his life. Also, safety behaviors must be broken down in order to accomplish the experience that any anticipated disastrous consequences fail to happen.

26.3.5 The Effects of CBT for Hallucinations

The effect sizes of CBT for voices are varying. The first report of CBT in psychosis was published some 60 years ago (Beck 1952). Structural research into the efficacy of CBT for this condition started about two decades ago in the UK, and in 2003, the results of meta-analyses entailed a recommendation for CBT in the psychosis guidelines of the National Institute of Clinical Excellence and in those of many other countries. In meta-analyses of trials with general CBT for psychotic symptoms in general, the effect sizes vary from small to medium-sized (NICE 2009), whereas trials with CBT for command hallucinations and other specific targets tend to show larger effect sizes (Trower et al. 2004). As the number of RCTs in those meta-analyses is large and the findings are robust, CBT for hallucinations is now accepted as an evidence-based treatment for hallucinations as well as for psychosis in general.

26.4 New Developments

CBT has always focused on the contents of thoughts and on thinking styles. Recently, the focus has shifted to issues such as interpersonal relations, loss and trauma, self-esteem, acceptance, metacognitive processes, and cognitive biases. There is a growing trend to work with schemas, emotion regulation, interpersonal relationships, and a more accepting relationship with one's thoughts and one's self.

26.4.1 Competitive Memory Training

Competitive memory training (COMET) is based on the notion that a therapy is successful when it succeeds to change the hierarchy of relevant neural networks and the order in which they are activated (Brewin 2006). In depression, it is quite easy to activate depressing thoughts and memories (i.e., mood-induced memories), to ruminate on them, and to induce feelings of defeat and entrapment. A successful treatment reinforces the neural networks that are incompatible with so-called negative networks. When instead "positive" networks are activated over and over again, they may well move up in the hierarchy of networks and come to surpass the negative ones. COMET teaches to reexperience personal memories that are incompatible with the dominant voices' messages. For instance, when the voices go on about a person's alleged incompetence, then the advice is to reexperience a memory of earlier competent behavior (e.g., scoring a winning goal) and to practice reexperiencing that memory five times a day. During the second stage of treatment, the positive and negative networks are activated in tandem. While reenjoying the winning goal, the patient is requested to "turn off" the audience's imagined cheering and to imagine the voices telling him again that he is a loser while he continues to enjoy the feelings of joy and pride associated with the winning goal. Most patients are able to accomplish that within a single session, and many of them report to then be able to tolerate the voices without being overcome by depressive feelings. During the third stage of treatment, patients are taught to distance themselves from the voices by "zooming out" – by turning the voices' volume down or by "watching" them from the last row in a theater with an attitude and posture that expresses boredom. COMET is a transdiagnostic technique that has proved to be effective for such varying conditions as eating disorders, panic disorder, and personality disorders. Applied to hallucinations, COMET primarily reduces depression. That effect is accomplished by the reduction of power attribution, the improvement of self-esteem, the acceptance of the voices, as well as a less submissive attitude (Van der Gaag et al. 2010).

26.4.2 Metacognitive Training

Metacognitive training (MCT) is based on scientific evidence regarding cognitive biases in psychosis and designed to teach patients to be aware of those biases and to prudently work around them (Moritz and Woodward 2007). The biases involved are working toward conclusions, attribution biases, problems with theory of mind, and biases against disconfirmatory evidence. The training consists of two blocks of eight sessions each, in which different issues are addressed in parallel. A small – and unfortunately underpowered – study found small to medium-sized effect sizes that were statistically nonsignificant but favored MCT above staying on a waiting list (Aghotor et al. 2010).

26.4.3 Mindfulness Training

Detached mindfulness has multiple components, requiring the activation of metacognitive knowledge, metacognitive monitoring and control, the suspension of conceptual processing, attentional flexibility, and a decentered relationship with one's own thoughts (Wells 2005). Those cognitive skills can be of aid to put hallucinations in perspective. The field is only just developing, and there are a small number of case studies as well as a single – underpowered – controlled feasibility study that found nonsignificant changes favoring the mindfulness condition over staying on a waiting list (Chadwick et al. 2009).

26.4.4 Compassionate Mind Training

Compassionate mind training (CMT) targets shame and self-criticism, as well as the ensuing submissiveness and negative affect. With the aid of the two-chair technique, a criticizing voice (or "inner bully") is interviewed, and the criticisms are compared with the patient's personal needs and distresses. The patient is then encouraged to reply to himself with warmth and compassion rather than with criticism. Case series reports are promising (Mayhew and Gilbert 2008), and an RCT is underway.

26.5 Concluding Remarks

Cognitive-behavioral therapy has a well-established track record that proves its efficacy, effectiveness, and cost-effectiveness for verbal auditory hallucinations. The effect sizes are small to medium-sized and in need of further improvement. The work that is currently going on is aimed at the further development of the cognitive model of hallucinations, which will hopefully allow for more targeted interventions to be developed and tested. Recently, a number of techniques have been developed that do not primarily focus on the content of thoughts and thinking styles but rather on emotional processing and the acceptance of persistent symptoms by finding a decentered perspective that gives room for recommitting to valuable personal goals and social roles in the community.

References

Aghotor J., Pfueller U., Moritz S., Weisbrod M., Roesch-Ely, D. (2010). Metacognitive training for patients with schizophrenia (MCT): Feasibility and preliminary evidence for its efficacy. *Journal of Behavior Therapy and Experimental Psychiatry*, 41, 207–211.

Beck, A.T. (1952). Successful outpatient psychotherapy of a chronic schizophrenic with a delusion based on borrowed guilt. *Psychiatry*, 15, 305–312.

Birchwood, M., Chadwick P. (1997). The omnipotence of voices: Testing the validity of a cognitive model. *Psychological Medicine*, 27, 1345–1353.

Birchwood, M., Gilbert, P., Gilbert, J., Trower, P., Meaden, A., Hay, J., Murray, E., Miles, J.N.V. (2004). Interpersonal and role-related schema influence the relationship with the dominant 'voice' in schizophrenia: A comparison of three models. *Psychological Medicine*, 34, 1571–1580.

Birchwood, M., Meaden, A., Trower, P., Gilbert, P., Plaistow, J. (2000). The power and omnipotence of voices: subordination and entrapment by voices and significant others. *Psychological Medicine*, 30, 337–344.

Birchwood, M., Trower, P. (2006a). Cognitive therapy for command hallucinations: Not a quasi-neuroleptic. *Journal of Contemporary Psychotherapy*, 36, 1–7.

Birchwood, M., Trower, P. (2006b). The future of cognitive-behavioural therapy for psychosis: Not a quasi-neuroleptic. *British Journal of Psychiatry*, 188, 107–108.

Birchwood, M., Trower, P., Brunet, K., Gilbert, P., Iqbal, Z., Jackson, C. (2007). Social anxiety and the shame of psychosis: A study in first episode psychosis. *Behaviour Research and Therapy*, 45, 1025–1037.

Brewin, C.R. (2006). Understanding cognitive behaviour therapy: A retrieval competition account. *Behaviour Research and Therapy*, 44, 765–784.

Brunelin, J., D'Amato, T., Brun, P., Bediou, B., Kallel, L., Senn, M., Poulet, E., Saoud, M. (2007). Impaired verbal source monitoring in schizophrenia: An intermediate trait vulnerability marker? *Schizophrenia Research*, 89, 287–292.

Carter, D.M., Mackinnon, A., Copolov, D.L. (1996). Patients' strategies for coping with auditory hallucinations. *Journal of Nervous and Mental Disease*, 184, 159–164.

Chadwick, P., Hughes, S., Russell, D., Russell, I., Dagnan, D. (2009). Mindfulness groups for distressing voices and paranoia: A replication and randomized feasibility trial. *Behavioural and Cognitive Psychotherapy*, 37, 403–412.

Corrigan, P.W., Storzbach, D.M. (1993). Behavioral interventions for alleviating psychotic symptoms. *Psychiatric Services*, 44, 341–347.

Erickson, G.D., Gustafson, G.J. (1968). Controlling auditory hallucinations. *Psychiatric Services*, 19, 327–329.

Fannon, D., Hayward, P., Thompson, N., Green, N., Surguladze, S., Wykes, T. (2009). The self or the voice? Relative contributions of self-esteem and voice appraisal in persistent auditory hallucinations. *Schizophrenia Research*, 112, 174–180.

Farhall, J., Greenwood, K.M., Jackson, H.J. (2007). Coping with hallucinated voices in schizophrenia: A review of self-initiated strategies and therapeutic interventions. *Clinical Psychology Review*, 27, 476–493.

Ford, J.M., Mathalon, D.H. (2005). Corollary discharge dysfunction in schizophrenia: Can it explain auditory hallucinations? *International Journal of Psychophysiology*, 58, 179–189.

Gilbert, P. (2009). Introducing compassion-focused therapy. *Advances in Psychiatric Treatment*, 15, 199.

Gilbert, P., Birchwood, M., Gilbert, J., Trower, P., Hay, J., Murray, B. et al. (2001). An exploration of evolved mental mechanisms for dominant and subordinate behaviour in relation to auditory hallucinations in schizophrenia and critical thoughts in depression. *Psychological Medicine*, 31, 1117–1127.

Gould, L.N. (1948). Verbal hallucinations and activity of vocal musculature; an electromyographic study. *American Journal of Psychiatry*, 105, 367–372.

Heilbrun, A.B.J. (1980). Impaired recognition of self-expressed thought in patients with auditory hallucinations. *Journal of Abnormal Psychology*, 89, 728–736.

Honig, A., Romme, M.A., Ensink, B.J., Escher, S.D., Pennings, M.H., DeVries, M.W. (1998). Auditory hallucinations: A comparison between patients and nonpatients. *Journal of Nervous and Mental Disease*, 186, 646–651.

Jones, S.R. (2010). Do we need multiple models of auditory verbal hallucinations? Examining the phenomenological fit of cognitive and neurological models. *Schizophrenia Bulletin*, 36, 566–575.

Lawrence, C., Jones, J., Cooper, M. (2010). Hearing voices in a non-psychiatric population. *Behavioural and Cognitive Psychotherapy*, 38, 363–373.

Mawson, A., Cohen, K., Berry K. (2010). Reviewing evidence for the cognitive model of auditory hallucinations: The relationship between cognitive voice appraisals and distress during psychosis. *Clinical Psychology Review*, 30, 248–258.

Mayhew, S.L., Gilbert, P. (2008). Compassionate mind training with people who hear malevolent voices: A case series report. *Clinical Psychology and Psychotherapy*, 15, 113–138.

McGuigan, F.J. (1966). Covert oral behavior and auditory hallucinations. *Psychophysiology*, 3, 73–80.

Moritz, S., Woodward, T.S. (2007). Metacognitive training in schizophrenia: From basic research to knowledge translation and intervention. *Current Opinion in Psychiatry*, 20, 619–625.

Nayani, T.H., David, A.S. (1996). The auditory hallucination: A phenomenological survey. *Psychological Medicine*, 26, 177–189.

NICE (2009). *Schizophrenia: Core interventions in the treatment and management of schizophrenia in the primary and secondary care: Update.* Edited by the National Institute for Clinical Excellence (NICE). http://www.nice.org.uk/CG82, retrieved May 30, 2011.

Pantelis, C., Barnes, T.R. (1996). Drug strategies and treatment-resistant schizophrenia. *Australian and New Zealand Journal of Psychiatry*, 30, 20–37.

Smith, B., Fowler, D.G., Freeman, D., Bebbington, P., Bashforth, H., Garety, P., Dunn, G., Kuipers, E. (2006). Emotion and psychosis: Links between depression, self-esteem, negative schematic beliefs and delusions and hallucinations. *Schizophrenia Research*, 86, 181–188.

Trower, P., Birchwood, M., Meaden, A., Byrne, S., Nelson, A., Ross, K. (2004). Cognitive therapy for command hallucinations: randomised controlled trial. *British Journal of Psychiatry*, 184, 312–320.

Van der Gaag, M., Van Oosterhout, B., Daalman, K., Sommer, I., Korrelboom, K. (2010). Competitive memory training (COMET) can change appraisals of voices. *Schizophrenia Research*, 2–3, 159–160.

Van Os, J., Kenis, G., Rutten, B.P. (2010). The environment and schizophrenia. *Nature*, 468, 203–212.

Waters, F.A., Badcock, J.C. (2010). First-rank symptoms in schizophrenia: Reexamining mechanisms of self-recognition. *Schizophrenia Bulletin*, 36(3), 510–517.

Wells, A. (2005). Detached mindfulness in cognitive therapy: a metacognitive analysis and ten techniques. *Journal of Rational-Emotive and Cognitive-Behavior Therapy*, 23, 337–355.

Chapter 27
Groundwork for the Treatment of Voices: Introducing the *Coping-With-Voices Protocol*

Willemijn A. van Gastel and Kirstin Daalman

27.1 Introduction

Kenny is 35 years old. He has been hearing voices since he enrolled in university. He has endured three psychotic episodes, which prevented him from successfully finishing his education. For 3 years, he has been working as a library assistant, but last autumn he was fired. Due to a lack of concentration, he tends to make many mistakes, and as part of his negative syndrome, he frequently oversleeps. He suffers from derogatory voices and has a delusional system in which he imagines himself to be a target of various secret organizations. His physician advised him to seek specialized treatment for those symptoms and referred him to our Voices Clinic. While his medication was being optimized, he also started with the Coping-With-Voices Protocol.

This chapter introduces the Coping-With-Voices Protocol as a first aid for patients who experience distressing voices. Each step in the program is designed to yield a quick reduction of distress through simple and practical interventions. The Coping-With-Voices Protocol aims for immediate effectiveness, right from the very first session, and requires no extensive prior training of the healthcare practitioner or the patient. The method was developed at the Voices Clinic of the University Medical Center Utrecht. It can be applied by family physicians as well as professionals working in psychiatric hospitals or any other type of mental health service and proceeds from various existing approaches such as psycho-education, stimulating daily activities, and (the shaping of) adaptive coping mechanisms (for an overview, see Farhall 2010). It shares some of the elements characteristic of cognitive-behavioral therapy (CBT) and Hallucinations-focused Integrative Therapy (HIT) (see Jenner 2010 for an overview). Its principal aim is the exploration – in close collaboration with the patient – of those occasions on which the voices are most disturbing and of ways in

W. A. van Gastel, M.Sc. (✉) • K. Daalman, M.Sc.
Psychiatry Division, Rudolf Magnus Institute of Neuroscience,
University Medical Center Utrecht, Utrecht, The Netherlands
e-mail: W.vanGastel@umcutrecht.nl; K.Daalman@umcutrecht.nl

J.D. Blom and I.E.C. Sommer (eds.), *Hallucinations: Research and Practice*,
DOI 10.1007/978-1-4614-0959-5_27,

which they can be dealt with. By assessing the efficacy of the coping mechanisms already applied, the most successful ones can be selected for further elaboration, while the unsuccessful or even harmful ones can be replaced by new and better ones. The first step of the Coping-With-Voices Protocol is to provide psycho-education for the hearing of voices, and – if applicable – for psychosis.

At our Voices Clinic, the Coping-With-Voices Protocol is offered as an initial treatment step to all voice hearers in order to lower the threshold for any further treatments that may be necessary and to establish a therapeutic alliance that is solid enough to allow for more demanding types of therapy. The treatment program offers the patient some basic knowledge about voices and psychosis, while the healthcare professionals involved are allowed to gain some insight into the unique and often complex world of the patient's voices and delusions. As the therapy also aims to optimize the patient's motivation for further treatment, quick results are of crucial importance. Other advantages of this treatment method are its low level of abstraction and its applicability to patients suffering from poor insight, concentration problems, or a lack of motivation.

While medication is usually prescribed to decrease the frequency and severity of voices, the Coping-With-Voices Protocol aims to help the patient cope with the impact that the voices may have on his daily life. As with CBT, the patient's active participation is mandatory. In contrast, however, the Coping-With-Voices Protocol does not require the patient to change his ideas about the voices, thus making it more easily applicable. In some cases, the therapy itself may suffice to bring the distress down to an acceptable level. For example, patients who only experience voices under exceptionally strenuous or stressful circumstances may learn to avoid those circumstances and to stop the voices effectively once they set in. More persistent types of hallucinations, however, require more extensive CBT or pharmacotherapy.

The Coping-With-Voices Protocol comprises psycho-education, mapping of the voices, coping techniques, stimulating healthy living, stimulating a positive outlook, and the prevention of relapse. In what follows, each of those issues will be explained in detail and illustrated with the aid of case vignettes.

27.2 Psycho-Education

Psycho-education basically involves the explanation of the results of scientific studies about auditory verbal hallucinations. Some of those studies report on the benefits of psycho-education for psychotic patients. Although an increase in insight through psycho-education appears to correlate with an increase of depression and a worsening of daily functioning (Trauer and Sacks 2000), the majority of therapists nonetheless consider it an effective treatment method. A review of studies reporting on psycho-education for patients diagnosed with schizophrenia by Lawrence et al. (2006) concludes that brief and preferentially groupwise treatment reduces positive as well as negative symptoms. A recent randomized controlled trial in an at-risk population suggests that types of psycho-education that aim at a normalization of symptoms aid to reduce the level of distress and to prevent any catastrophical delusional explanations (Rietdijk et al. 2010). Such a normalizing approach may well be the key element of the beneficial effects of psycho-education. In our clinic, we always emphasize that the hearing of voices does not necessarily imply losing control or developing a severe psychiatric disorder. Some examples of the topics we discuss are the following.

27.2.1 *Who Hears Voices?*

Voices are probably best known as a symptom of schizophrenia and other psychotic disorders. However, they are also experienced in the context of personality disorders (e.g., borderline personality disorder (see Chap. 10) and schizotypal personality disorder) and neurological disorders (e.g., the auditory variant of the Charles Bonnet syndrome, epilepsy, and tinnitus), as well as by otherwise healthy individuals. Approximately 15–20% of the general population hear voices at least once in their lives (Tien 1991). The majority of them are not bothered by those voices and therefore never seek any kind of help or treatment for them.

27.2.2 *Brain Activity*

With the aid of functional magnetic resonance imaging (fMRI), our research group demonstrated that the right cerebral hemisphere shows an increase in activity during verbal-auditory hallucinatory experiences, as opposed to the predominantly left-sided activity co-occurring with speech (Sommer et al. 2008; Diederen et al. 2010; see also Chap. 9). It may well be that, due to a lack of right-hemispheric inhibition, the homologues of Broca's and Wernicke's areas do not recognize the activity as coming from within the brain itself and thus misinterpret it as an externally generated voice (for an explanation of this default-corollary-discharge mechanism, see Chap. 21). Moreover, voices were found to be preceded by a change of activity in the parahippocampal gyrus, which would seem to imply that memories are involved in their mediation (Diederen et al. 2010).

27.2.3 *Treatment Options*

Broadly speaking, two kinds of treatment can be distinguished: the elimination or reduction of the voices themselves and the elimination or reduction of the ensuing distress. For the first purpose, antipsychotic medication can be prescribed, whereas for the second purpose, CBT or the Coping-With-Voices Protocol can be offered.

The hearing of voices is often accompanied by delusional ideas and formal thought disorders. Although such accompanying symptoms tend to be less recognized by patients, they may exert an even greater impact upon their social functioning. Therefore, we also provide information about those symptoms – illustrated with clear examples – to help patients to recognize them and find out how they apply to their own situation. Patients often wonder how it is possible that they hear voices while nobody in their vicinity is speaking. Knowing the answer about their endogenous origin can give them a sense of enormous relief. Those who foster idiosyncratic – and especially delusional – ideas about their origin may be convinced that they come from the neighbors or the police, from devils, criminals, and so on. In such cases, it may be difficult to discard the patients' own ideas. By providing them with our neurological explanation, an alternative and less threatening possibility is offered. The following case vignette describes how this may work out in actual practice.

For years, Kenny believed to be the only one who could overhear secret messages meant for the CIA. During psycho-education, he learned that the hearing of voices is a lot more common than he thought and that they often summon associations with secret agencies. He found himself trying to convince a fellow patient that she could not possibly hear her deceased uncle and was then told by her that she did not see why the CIA would be so interested in him. Although this did not immediately change his conviction, merely being provided with such an alternative explanation made him less certain about his own ideas.

27.3 The Mapping of Voices

To gain insight into the specific situations and moments of the day in which the voices are present, patients are asked to fill out a registration form (see Table 27.1) where they can indicate date and time and the activity or situation preceding or accompanying the voices. In addition, they are requested to rate the degree of ensuing distress (on a scale of 0–10). Patients often find this difficult and alarming because they now have to actively focus on those frightening experiences rather than ignoring them as much as possible. Moreover, chances are that they have had prior negative experiences when discussing the voices with others and therefore ceased to do so. Voices frequently prohibit the patient to talk about them. As a consequence, this treatment phase can be very difficult. However, it is also one of

Table 27.1 Mapping-of-voices registration form

Day and time	Location	Content of voice	Distress
Wednesday 5:30 p.m.	Supermarket	"They are looking for you, go home and hide."	I felt frightened, and went home
Thursday 1 p.m.	Library	"It's quiet, it's quiet, it's quiet, it's quiet."	Only a little bothered in the beginning, but able to ignore later on
Friday 9 p.m.	Birthday party of father	Various voices at once, making me feel insecure	Went upstairs for a while, voices calmed down only a little

the most important phases of the Coping-With-Voices Protocol because of its potential to identify strategies that can be used during the actual treatment phase that is to follow.

Kenny discovered that he was more bothered by the voices in crowded places and especially when he was sad and distressed. He noticed that they had less impact on him when he was at the library. They were also less prominent when he heard the sound of the traffic rushing by. The traffic noises appeared to drown out the voices, making them almost inaudible.

As voices tend to be more upsetting in the early morning and the late afternoon, at those moments coping techniques can be extra helpful.

27.4 Coping Techniques

Another important element of the Coping-With-Voices Protocol is the exploration of different techniques that can provide relief from the voices and associated distress. Those techniques can be physiological, cognitive, or behavioral in nature (Farhall et al. 2007). Although they are widely applicable, their effectiveness varies in individual cases. It is therefore helpful to assess each strategy individually with each patient. People who have already developed their own coping strategies – sometimes even unknowingly – tend to use only a limited number of them and can be quite oblivious to alternatives. By extending their repertoire, and thereby giving them a chance to try out different coping techniques, they have the opportunity to pick those that work best for them.

Patients participating in the Coping-With-Voices Protocol receive a list of coping techniques (see Table 27.2) and are asked which ones they have already adopted and which of those were successful. There may well be strategies on that list that they never tried before but may nonetheless prove helpful. Although the use of coping strategies has been studied extensively (e.g., Farhall 2010), it remains unclear to which degree they can contribute to the improvement of symptoms (Hayashi et al. 2001; Sayer et al. 2000; Nayani and David, 1996; Wiersma et al. 2001). Still, coping strategies continue to be an important aspect of most psychosis therapies.

The list we use in our Voices Clinic proceeds from information provided by our patients and from the literature (see the reference list at the bottom of this chapter).

After patients have experimented with all the coping strategies, they are requested to list their own "top five." This top five is then printed onto a small card for them to keep in their wallet as a reminder. Obviously, the strategies cannot be applied 24/7, but it may be comforting to have a few options at hand when the voices are most distressing.

By mapping his voices, Kenny learned that visiting the library and hearing the sound of traffic could bring him considerable relief. He now visits the library when the voices are particularly distressing. As sitting along the highway is not his favorite pastime, he recorded the sound of cars passing by and plays it on his headphones

Table 27.2 List of coping strategies

1.	Talking to others
2.	Sleeping
3.	Thinking about something else
4.	Reading (out loud)
5.	Using earplugs (right, left, or in both ears)
6.	Listening to music (with or without headphones)
7.	Listening to a certain type of music
8.	Singing
9.	Humming
10.	Whistling
11.	Listening to the radio
12.	Watching television
13.	Adjusting body position
14.	Relaxing
15.	Indulging in a hobby
16.	Holding water in the mouth or opening the mouth wide
17.	Chewing gum
18.	Physical activity (i.e., strolling, running, cycling)
19.	Talking back to the voices, negotiating with them
20.	Ignoring the voices
21.	Talking out loud (i.e., naming objects in the environment)
	Patients are always asked for their own unique coping strategies: "Are there any other things that you have tried, and that are not on the list?"
22.	…
23.	…
24.	…

when he needs to. Kenny also discovered that visiting the library worked out best in the morning, whereas in the evening the sound of traffic was more helpful. In addition, he found out that singing sometimes helped against the voices. This was new to him, and he was thrilled to use this strategy at home when he was alone.

27.5 Healthy Living

For patients who are bothered or distressed by their voices, it is important to have enough daytime activities and distractions to shift their focus of attention. Even for those with severe negative symptoms, minor changes in their daily routine may improve their mood and render them less vulnerable to their voices.

On good days, Kenny strolls around the city, making small talk with the people he meets. He has been taking the same route for years, so he has become acquainted with quite a few of them. On days such as these, he may well be hearing voices, but chatting helps him to ignore them. During the winter, however, he is liable to depression. One morning, when taking a stroll, a neighbor called out to him, "Awful weather, isn't it?" upon which the voices immediately claimed that the man had called him awful and that he was an awful person. The voices went on to comment on his dirty clothes and on the way he walked and suggested that he looked shabby. Kenny got very upset and tried to reach home as soon as possible. During the weeks that followed, he hardly left his home at all, which made him feel even worse. His therapist went over various aspects of healthy living with him.

27.5.1 Sleep

When Kenny became more psychotic he started to sleep during the daytime. As a consequence, he developed nighttime insomnia. With psycho-education on the beneficial effects of daylight on mood, and the influence of melatonin on sleep, he eventually became motivated to readjust his sleeping pattern.

27.5.2 Daily Activities

Kenny was asked how he spent his days and whether he enjoyed it. He had to admit that most of the time he was watching movies to avoid going out. Because of his mood, he had forgotten about the things that used to bring him joy, and so he had to think back: What did he use to like? He remembered enjoying soccer in high school and was advised to give that another try. With the help of his therapist, he contacted a nearby soccer club and now goes to practice once a week. The therapist also advised him to try out some new activities. As a result, he joined his cousin in a chess club.

27.5.3 Social Contacts

Having a social life is important for everyone. But it is particularly important for voice hearers because talking has the capacity to eliminate the hearing of voices by activating the same brain areas that are involved in their mediation. An actual conversation is therefore capable of stopping verbal-auditory hallucinatory activity in some patients. Moreover, talking with others helps us to stay in touch with reality and to correct delusional beliefs at an early stage. When listening to voices on one's own, chances are higher that one starts to believe them.

Kenny realized that he benefitted from his social contacts at the library and was pleased with the new ones at the soccer club and the chess club.

27.5.4 Physical Exercise and Eating Habits

Due to the release of endorphins, physical exercise is an effective way to improve one's mood. It also has a positive effect upon body weight, which can be a critical issue when using antipsychotic medication. The best way to lose weight, however, is to follow a healthy diet.

27.5.5 Cannabis

The use of cannabis and the risk of psychotic symptoms have consistently been found to be associated (for a review, see Moore et al. 2007). The use of cannabis is associated with higher symptom severity, an increase in the number of relapse episodes, and a worse outcome (for a review, see Linszen et al. 2004). Two of the psychoactive compounds of cannabis have been found capable of affecting psychotic symptoms, one being cannabidiol, the other, tetrahydrocannabinol (THC). Cannabidiol is thought to have some short-term beneficial effects upon anxiety and stress (Zuardi et al. 2006; Morgan and Curran 2008) but this advantage is outweighed by the increased risk for psychosis due to the other compound, THC (DiForti et al. 2009). In individuals who hear voices, quitting on cannabis can lead to a reduction of psychotic symptoms and to a considerable improvement of one's mental health in the long run.

Prior to his first psychotic episode, Kenny had never used any cannabis. However, during his stay in the hospital, the other patients had advised him to try it to alleviate his anxiety. Unfortunately, it took him some time to realize that the short-term benefits did not outweigh the long-term adverse effects. Psycho-education on this issue helped him to reduce his cannabis use.

27.6 Focusing on the Positive Aspects of Life

Patients who visit our Voices Clinic often suffer from a depressed mood due to the negative content of their voices. This can have a negative effect on their self-esteem. Unfortunately, it also works the other way around: having a low mood and self-esteem can make a person more vulnerable to the hearing of voices. To break this vicious cycle, patients can be stimulated to focus their attention on more positive aspects of life. This can be done, for example, with the aid of a diary, in which they write down three positive things about themselves that happened during the day.

Three positive things that Kenny wrote down are:

I did my groceries this morning.
My neighbor told me I looked happy.
I went for a drink with my best friend.

Table 27.3 Kenny's warning protocol

Name: Kenny
Which signals can predict a relapse in your case?
• *I can't sleep well, I wake up several times during the night, and I lie awake. I tend to stay at home and don't answer the phone. I get irritated quite easily, and become more paranoid.*
What are proper actions for you to undertake?
• *Talk about the voices with my father*
• *Keep taking my pills*
• *Play games on my computer*
What are the things that you shouldn't do?
• *Stop taking my pills*
• *Unplug my telephone and my doorbell*
• *Watch thrillers*
What should the people around you preferentially do?
• *Keep in touch with me, although not too much*
• *Ask whether I would like to talk about it, rather than just ask questions out of the blue*
• *Listen to what I have to say about my voices without immediately telling me that they are not real and that I am merely imagining them*
What shouldn't they do?
• *Call up on me various times a day*
• *Give me their advice or opinion all the time*
• *Urge me to come along to parties, shopping malls, or other crowded places*

27.7 Relapse Prevention

For those who are at risk for novel psychotic episodes, it is important to learn to recognize any symptoms and signals that may precede them and to take precautions to avert them. Patients in our clinic are therefore requested to write down which warning signals they (or their families or friends) are aware of and which actions should be undertaken once they occur. An example of the ensuing warning protocol can be found in Table 27.3.

27.8 Conclusion

In this chapter, we provided patients who hear voices an introduction to the Coping-With-Voices Protocol, a treatment method for patients who hear voices developed at the Voices Clinic of the University Medical Center Utrecht. The therapy's principal focus lies on the improvement of coping techniques, and each of its constituent parts aims to quickly and simply reduce the distress caused by the voices. The Coping-With-Voices Protocol can be applied by nurses, physicians, and psychologists, although it works best when the same professional can also offer further treatment (in the form of medication, cognitive-behavioral therapy, or both), thus benefitting from the working alliance established during this first treatment step.

References

Di Forti, M., Morgan, C., Dazzan, P., Pariante, C., Mondelli, V., Marques, T.R., Handley, R., Luzi, S., Russo, M., Paparelli, A., Butt, A., Stilo, S.A., Wiffen, B., Powell, J., Murray, R.M. (2009). High-potency cannabis and the risk of psychosis. *British Journal of Psychiatry,* 195, 488–491.

Diederen, K.M., Neggers, S.F., Daalman, K., Blom, J.D., Goekoop, R., Kahn, R.S., Sommer, I.E. (2010). Deactivation of the parahippocampal gyrus preceding auditory hallucinations in schizophrenia. *American Journal of Psychiatry*, 167, 427–435.

Farhall, J. (2010). Understanding and shaping adaptive coping with voices. In *Hallucinations. A guide to treatment and management.* Edited by F. Larøi and A. Aleman. Oxford: Oxford University Press, pp. 205–232.

Farhall, J., Greenwood, K.M., Jackson, H.J. (2007). Coping with hallucinated voices in schizophrenia: A review of self-initiated strategies and therapeutic interventions. *Clinical Psychology Review*, 27, 476–493.

Hayashi, N., Igarashi, Y., Suda, K., Nakagawa, S. (2001). Auditory hallucination coping techniques and their relationship to psychotic symptomatology. *Psychiatry and Clinical Neurosciences*, 61, 640–645.

Jenner, J. (2010). *Hallucinations-focused Integrative Therapy (HIT).* In *Hallucinations. A guide to treatment and management.* Edited by F. Larøi and A. Aleman. Oxford: Oxford University Press, pp. 163–182.

Lawrence, R., Bradshaw, T., Mairs, H. (2006). Group cognitive behavioural therapy for schizophrenia: a systematic review of the literature. *Journal of Psychiatric and Mental Health Nursing*, 13, 673–681.

Linszen, D., Peters, B., De Haan, L. (2004). *Cannabis abuse and the course of schizophrenia.* In *Marijuana and madness.* Edited by D. Castle and R. Murray. Cambridge: Cambridge University Press, pp. 119–126.

Moore, T.H., Zammit, S., Lingford-Hughes, A., Barnes, T.R., Jones, P.B., Burke, M., Lewis, G. (2007). Cannabis use and risk of psychotic or affective mental health outcomes: a systematic review. *Lancet*, 370, 319–328.

Morgan, C.J., Curran, H.V. (2008). Effects of cannabidiol on schizophrenia-like symptoms in people who use cannabis. *British Journal of Psychiatry*, 192, 306–307.

Nayani, T.H., David, A.S. (1996). The auditory hallucination: a phenomenological survey. *Psychological Medicine*, 26, 177–189.

Rietdijk, J., Dragt, S., Klaassen, R., Ising, H., Nieman, D., Wunderink, L., Delespaul, P., Cuijpers, P., Linszen, D., Van der Gaag, M. (2010). A single blind randomized controlled trial of cognitive behavioural therapy in a help-seeking population with an at risk mental state for psychosis: the Dutch Early Detection and Intervention evaluation (EDIE-NL) trial. *Trials*, 11, 30.

Sayer, J., Ritter, S., Gournay, K. (2000). Beliefs about voices and their effects on coping strategies. *Journal of Advanced Nursing*, 31, 1199–1205.

Sommer, I.E., Diederen, K.M., Blom, J.D., Willems, A., Kushan, L., Slotema, K., Boks, M.P., Daalman, K., Hoek, H.W., Neggers, S.F., Kahn, R.S. (2008). Auditory verbal hallucinations predominantly activate the *right* inferior frontal area. *Brain*, 131, 3169–3177.

Tien, A.Y. (1991). Distributions of hallucinations in the population. *Social Psychiatry and Psychiatric Epidemiology*, 26, 287–292.

Trauer, T., Sacks, T. (2000). The relationship between insight and medication adherence in severely mentally ill clients treated in the community. *Acta Psychiatrica Scandinavica*, 102, 211–216.

Wiersma, D., Jenner, J.A., Van de Willige, G., Spekman, M., Nienhuis, F.J. (2001). Cognitive behaviour therapy with coping training for persistent auditory hallucinations in schizophrenia: A naturalistic follow-up study of the durability of effects. *Acta Psychiatrica Scandinavica*, 103, 393–399.

Zuardi, A.W., Crippa, J.A., Hallak, J.E., Moreira, F.A., Guimaraes, F.S. (2006). Cannabidiol, a Cannabis sativa constituent, as an antipsychotic drug. *Brazilian Journal of Medical and Biological Research*, 39, 421–429.

Chapter 28
The Hearing Voices Movement

Sandra Escher and Marius Romme

28.1 Introduction

The Hearing Voices Movement is an international movement directed at creating opportunities for voice hearers to exchange experiences and knowledge about the hearing of voices. This is mostly done in groups of voice hearers where the participants can feel safe and respected and where their experiences are accepted rather than criticized. The participants also explore the personal backgrounds of their voices and learn coping strategies from each other.

The movement's initial development was sparked by Ms. Patsy Hage, who in 1987 had successfully convinced her psychiatrist (M.R.) that her voices were real, in the sense that she could really hear them. She wanted to learn to cope with those voices because she felt overwhelmed, powerless, and very afraid of them. Medication had not helped her sufficiently, and she had become more and more isolated because of all the things that the voices forbade her to do. She had also become suicidal, which was the reason for me (M.R.) to bring her into contact with another voice hearer, with the purpose of relieving her isolation and of being able to find out exactly how real the voices were to her. The two of them enjoyed talking about their experiences, which was a rather strange experience for me, sitting there and listening to all the things they had to say to each other. To my surprise, they comprehended each other perfectly well. We repeated this procedure a few times with other voice hearers, but at the time, neither Patsy nor any of the others knew how to cope with their voices or how to reduce the anxiety they provoked. This gave us

S. Escher, Ph.D. (✉)
Association Living with Voices, Gravenvoeren, Belgium board member of Intervoice, editor of de 'Klankspiegel' Resonance magasine member of Intar (international network towards recovery)
e-mail: sandra@hearing-voices.com

M. Romme, M.D., Ph.D.
Emeritus Prof. of Psychiatry, Maastricht University, Visiting Prof. Birmingham City University,
e-mail: www.hearing-voices.com

J.D. Blom and I.E.C. Sommer (eds.), *Hallucinations: Research and Practice*,
DOI 10.1007/978-1-4614-0959-5_28, © Springer Science+Business Media, LLC 2012

Fig. 28.1 Logo of the Dutch *Stichting Weerklank* ("Resonance Foundation")

(S.E. and M.R.) the idea to set out and search for people who had learned to successfully cope with experiences such as these. We found a Dutch national TV show willing to spend an episode on the voices experienced by Patsy, and they allowed me to invite people who were able to cope with their voices to comment. At the time, none of us were able to foresee the far-reaching consequences of this single broadcasting. Following the show, 700 people contacted the studio's special team of telephone counselors, with 500 of them saying that they themselves were familiar with the hearing of voices. We were overwhelmed by the sheer number of respondents and sent them a questionnaire with 30 questions, which were selected together with Patsy Hage, who went on to advise us and to explain the relevance of the remarks people had made during their telephone calls.

That same year, we organized the first Hearing Voices Congress in Utrecht, the Netherlands (see also Romme and Escher 1989). Prior to as well as during the congress, we learned from our personal contacts with voice hearers that they wished for their experiences to be accepted as actually perceived voices with a nonself quality. During the congress, we founded Resonance (*Stichting Weerklank* in Dutch), an organization for voice hearers and people who support them (Fig. 28.1).

The above mentioned questionnaire was eventually sent out to 450 voice hearers, 254 of whom replied, and 186 of whom answered sufficiently comprehensively to allow for a detailed analysis (the remaining 68 tended to send extensive letters). From the results of that analysis (see Romme et al. 1992), we learned that some 66% of the voice hearers under study could not cope with their voices, whereas 34% could.

In 1988, together with the chair of Resonance (Ans Streefland, a voice hearer who had never been a psychiatric patient, being able to cope well with her voices), we helped to establish the Hearing Voices Network in the UK. Soon thereafter, similar

Fig. 28.2 Logo of Intervoice

networks followed in Finland, Italy, Austria, Portugal, Sweden, Wales, Scotland, Switzerland, Germany, Japan, Norway, Spain, Denmark, Ireland, Brazil, and Greece. Ron Coleman from Scotland introduced the movement to Palestine, New Zealand, and Australia. The USA and Canada followed somewhat later. Currently, the UK boasts the highest number of self-help groups, around 180 in total, followed by Finland with 24 groups (which is quite large a number considering the relatively small population). Since 1996, the training and education of the voice hearers and professionals participating in those groups has been in the hands of Intervoice, an international organization which runs an informative website (www.intervoiceonline.org) and organizes annual meetings for its members (Fig. 28.2).

28.2 Basic Principles of the Hearing Voices Movement

The basic principles of the Hearing Voices Movement can be summarized as follows:

- The hearing of voices is not *in itself* a sign of mental illness.
- The hearing of voices is experienced by many people who lack any symptoms that would warrant a psychiatric diagnosis.
- The hearing of voices is often related to prior social-emotional problems in the percipient's life history.
- The hearing of voices can cause serious distress, although it is an experience that one can learn to cope with, and become able to change one's relationship with.

The first principle, which states that voices do not necessarily constitute a sign of mental illness, is based on our numerous meetings with voice hearers who had not become psychiatric patients and had never needed any kind of help for them. Those people were able to cope well with their voices, and to live their own lives, functioning well socially. It is also based on the second principle, about the occurrence of voices in the general population, which was later confirmed by various large-scale epidemiological surveys (Tien 1991; Eaton et al. 1991; Van Os et al. 2001). The third principle, which links the hearing of voices to prior life events, was based on a study of our own (i.e., Romme et al. 1992) in which we found that some 70% of the respondents to the above mentioned questionnaire related their voices to earlier traumatic experiences. That number was later confirmed by various additional studies (i.e., Romme and Escher 1993; Romme 1996; Escher 2005; Escher and Romme 2010; Romme et al. 2009). The 1993 study compared six voice hearers with a psychiatric diagnosis with six who had no diagnosis. They were all interviewed in depth about their experiences and life histories. In the study by Romme (1996), 33 voice hearers who had become patients were compared with 15 others who had not and who had been able to cope well with their voices. It indicated that the experiences of both groups were similar with regard to the phenomenological characteristics of the auditory hallucinations at hand, but that the patients tended to be afraid of their voices whereas the nonpatients did not. It also showed that both groups tended to relate their voices to earlier traumatic experiences. The Escher studies (2002, 2004, 2005), which describe a 3-year follow-up of 80 children who hear voices, show that no less than 80% of the young participants relate the hearing of voices to traumatic experiences. The fourth basic principle, about the distress that voices may cause and the possibilities to learn to cope with them, followed from our meetings with voice hearers who had succeeded to recover as well as from our own study of 50 recovered voice hearers (Romme et al. 2009).

28.3 Voices as a Problem of Living

As noted above, epidemiological studies indicate that the hearing of voices, in and of itself, is not a symptom of disease (Tien 1991; Eaton et al. 1991; Van Os et al. 2001). The phenomenon has been reported in 2–4% of the general population (with some studies yielding substantially higher estimates), while even more people (i.e., 8%) foster peculiar convictions, also known as delusions, without being "ill" in any way. Our own research has corroborated those findings (Romme and Escher 1989, 1993; Romme 1996). Over a third of the 350 voice hearers we interviewed with the Maastricht hearing voices interview (Escher et al. 2000) in the Netherlands had never been in contact with psychiatric services (Romme and Escher 1989; Romme et al. 1992; Romme 1996; Escher 2005). Most of them were able to cope well with their voices, and many of them even described them as life enhancing. Moreover, people who were able to cope well with their voices and those who were not showed marked differences in the way they related to their voices and the kinds of strategies they employed to manage them. Those who coped well tended to use active strategies such as setting limits and selective listening, but they also welcomed positive voices, especially those which gave them good advise (Romme 1989, 2000; Romme et al. 1992).

Meeting people such as these stimulated our interest in the hearing of voices itself. Hence, we designed a study (Romme 1996; Romme and Escher 2000) to compare voice hearers who had become patients with those who had not, and focused on three groups: one consisting of people diagnosed with schizophrenia ($n=18$), one with dissociative disorder ($n=15$), and one without any psychiatric diagnosis ($n=15$). The patients were all recruited from our Maastricht-based outpatient clinic, and the healthy voice hearers were volunteers we had met at congresses and other opportunities. All 48 participants were screened for pathology with the aid of the Composite International Diagnostic Interview (CIDI, Robins et al. 1988) and the Dissociative Experience Scale (DES, Bernstein and Putnam 1986). The results showed that the experiences among the three groups did not differ significantly. In all cases, the voices heard had the phenomenological characteristics of genuine auditory hallucinations. Nor were there any differences as regards the perceived location of the voices, i.e., inside or outside the head. The most significant difference we found was that the patients were all afraid of the voices they heard, whereas none of the healthy voice hearers were. We also found a reported relationship with traumatic experiences in 70% of the cases in the schizophrenia group, in a full 100% of the cases in the dissociative disorder group, and in only 53% of the nonpathological group. Due to a lack of statistical significance, however, they could not be used to differentiate between the three groups.

We also found that the long-term developmental processes of psychological adjustment, which typically precede the onset of the hearing of voices, tend to be governed by memories of "undigested" emotional events connected with key relationships (Romme et al. 2009). The types of trauma reported include sexual abuse, physical abuse, being bullied for longer periods of time, and emotional neglect (in the sense of being educated in a setting where the voice hearer was discouraged to express his emotions and/or criticized for longer periods of time). Our study among children identified similar traumatic experiences, but even more often a divorce, a bereavement phase, a love affair, or a pregnancy (Escher 2005).

28.4 Self-help Groups

Many people who hear voices, regardless of whether they are able to cope with them or not, feel an urgent need to gain a personal understanding of their experiences and to discuss them with others without being designated as "mad." The failure to do so in the presence of mental health professionals can be highly frustrating. Many service users report that their voices tend to be viewed quite exclusively as symptoms of a psychiatric disease and that they are discouraged to talk about them. This would still seem to be the case today, although we admit that we are aware of various notable exceptions. Instead, the Hearing Voices Movement advocates the following approach based on the recent work of Lucy Johnstone (2011):

- Voices can be considered part of a normal range of experiences.
- Voice hearers need to take responsibility for their own recovery.
- It is important for them to actively engage with the voices.
- It is important for them to rely on their own understanding of the voices.
- They should work with the unresolved trauma.
- If feasible, they should rely on self-help groups and workbooks.
- It is important to change people's relationship with their voices.

Fig. 28.3 Attendants of a Hearing-Voices Congress in Sweden, 2003. Photograph by Sandra Escher

Most importantly, however, we learned from our most recent study (Romme et al. 2009) that recovery is not a matter of the effectiveness of psychopharmacological agents but a personal process that is inextricably bound up with the nature of the experience at hand (see also Boyle 1990). As a corollary, we are convinced that the primary challenge for voice hearers is to meet people who are willing to accept that their voices constitute genuine experiences which are neither some strange product of their imagination nor a sign of mental disease or madness, and which can be viewed as reflecting something that has actually occurred to them in the past. The primary challenge for professionals is to accept that voices can be meaningful, in the sense that they may constitute a link with the voice hearer's prior life experiences. In our opinion, voices often represent an abuser, at least in those cases where the voice hearer has been the victim of sexual or physical abuse, while the age of the "person heard" may indicate the age at which the trauma was actually experienced. Our whole approach starts from the dictum that those vital links should not be eradicated with the aid of pharmaceuticals but explored with the aid of psychological therapy and self-help methods (see also Dillon and Longden 2011). The instruments we consider of crucial importance to that process are the groups of voice hearers which are now slowly spreading around the world (Fig. 28.3).

And yet most mental health professionals have not been trained to discuss their patients' voices with them. Instead, they are trained to treat the diseases supposedly underlying them, preferably with the aid of medication. A problem with that approach is that it fails to take into account the possibility that people are often perfectly capable to develop initiatives of their own and that those initiatives are equally capable to bring recovery within their reach. It also tends to alienate people from their experiences and to promote their social isolation. As Ron Coleman (2009) remarked, "They take my experience, mould it, and give it back unrecognisable."

The first self-help groups proved what we had been suspecting for quite some time, namely, that the traditional medical approach is far of the mark when it comes to the treatment of voices. The groups turned out to be extremely helpful in making people realize that they were not the only ones who were hearing voices, that it was possible to find acceptance for them, and that they were not in fact isolated. The participants were able to discuss their voices freely with each other, benefiting from each other's experiences, and exchanging and exploring numerous coping strategies. Their participation in the groups also stirred up their emotions, which were often so overwhelming that they could hardly be coped with. Soon, we realized that that seemed to be the core problem to be dealt with, the apparent inability to handle such powerful emotions. That was what the voices had been talking about all the time. Learning to cope with one's voices meant learning to cope with one's most difficult emotions. The fact that people wanted to learn to cope with their voices made them aware that that could only be done by reclaiming their own power. They felt that they had been rendered powerless by their voices – as well as by the mental health professionals who had tried to suppress their voices – and that they needed to regain their power if they ever wanted to recover.

28.5 The Need for a New Paradigm

As remarked by Ms. Jacqui Dillon, the coordinator of the Hearing Voices Network in the UK, "A starting point for me was creating a new paradigm for myself that honoured my resilience and capacity to heal." (Romme et al. 2009, p. 188). As it turned out, the need for such a new paradigm was shared by voice hearers and professionals alike. For voice hearers, that need springs naturally from their need to regain control over their lives. As regards professionals, the situation would appear to be slightly more complicated. In the first place, perhaps, because professionals tend to come into contact only with those voice hearers who have been rendered powerless by their voices. As a consequence, it may be difficult for them to believe that the hearing of voices deserves to be conceptualized and appreciated in a way that departs radically from the traditional medical model, and to grasp that prescribing medication may not be exclusively beneficial. Secondly, mental health professionals – not unlike other human beings – have a certain tendency to hold on to the models and assumptions that are familiar to them. Even when confronted with overwhelming proof, for example, of the lack of scientific validity of the schizophrenia concept (Blom 2003), they may be reluctant to abandon their familiar paradigms and choose to keep on working within the confines of the old ones.

Nevertheless, there are various advantages lying in store for those professionals who are brave enough to challenge their familiar assumptions about the hearing of voices. As summarized by Hoffman (2011), they – and we – should:

- Talk about "hearing voices," while the term "auditory hallucination" in therapy makes the thorny path into a good patient-therapist relationship even thornier, because the term hallucination does not make allowance for the fact that voice hearers actually do hear voices.
- Accept the reality of the experience. The development of a sound working relationship becomes easier with the staff member's simple explanation that he or she does not doubt that the voice hearer actually hears voices, and by pointing out that he or she knows quite a lot about people who hear voices.
- Accept the possibility of hearing voices as nonpathologic. In closing one's eyes to the fact that there are many people who have found ways to live with their voices without being "disturbed" or "ill," and who may not even want to lose this ability, professionals lose sight of a chance to observe how people can decide which voices they want to listen to and when, to mention some of the techniques mastered by those people.

28.6 The Need for Role Models

A final issue that we would like to discuss is the need for role models. From childhood onward, we all seek out role models that can help us make important decisions in life and shape our identity. In the book *Living With Voices* (Romme et al. 2009), one can find the stories of 50 voice hearers who managed to learn to cope with their voices and to recover from the patient roles many of them had played for up to 18 years. They are role models with whom many voice hearers can identify and which might inspire professionals in their daily practice. Voice hearers capable of viewing their own position from a certain distance, and of combining their individual experiences with general scientific data, may well be in a position to bridge the gap between mental health workers' daily practices and that which their professional education has taught them to believe. Some of the people we would like to recommend for that purpose are Peter Bullimore, Ron Coleman, Jacqui Dillon, Rufus May, and Rachel Weddingham in the UK; Olga Runciman and Johnny Sparvang in Denmark; Liz Bodil and Ami Rohnitz in Sweden; Flore Brummans en Frans de Graaf in the Netherlands; Debra Lampshire in New Zealand; and Marleen Janssen in Australia.

References

Bernstein, E.M., Putnam, F.W. (1986). Development, reliability, and validity of a dissociation scale. *Journal of Nervous and Mental Disease*, 174, 727–735.

Blom, J.D. (2003). *Deconstructing schizophrenia. An analysis of the epistemic and nonepistemic values that govern the biomedical schizophrenia concept.* Amsterdam: Boom.

Boyle, M. (1990). *Schizophrenia: A scientific delusion?* London: Routledge.

Coleman, R. (2009). *The need for organisational changes in mental health services to create conditions for recovery. Paper presented at the First World Hearing Voices Congress, Maastricht, September 17–18, 2009.*

Dillon, J., Longden, E. (2011). *Hearing voices groups: Creating safe spaces to share taboo experiences.* In *Psychosis as a personal crisis.* Edited by M. Romme and A. Escher. London: Routledge.

Eaton, W.W., Romanoski, A., Anthony, J.C. (1991). Screening for psychosis in the general population with a self-report Interview. *Journal of Mental and Nervous Disease*, 179, 689–693.

Escher, A. (2005). *Making sense of psychotic experiences. PhD Thesis, University of Maastricht.*

Escher, A., Hage, P., Romme, M. (2000). *The Maastricht Hearing Voices Interview.* In *Making sense of voices.* Edited by M. Romme and A. Escher. London: MIND Publications.

Escher, A., Morris, M., Buiks, A., Delespaul, Ph., Van Os, J., Romme, M. (2004). Determinants of outcome in the pathways through care for children hearing voices. *International Journal of Social Welfare*, 13, 208–222.

Escher, A., Romme, M. (2010). *Children hearing voices. What you need to know and what you can do.* Ross-on-Wye: PCCS Books.

Escher, A., Romme, M., Buiks, A., Delespaul, Ph., Van Os, J. (2002). Independent course of childhood auditory hallucinations: A sequential 3-year follow-up study. *British Journal of Psychiatry*, 181(suppl. 43), 10–18.

Hoffmann, M. (2011). *Changing attitudes in clinical settings. From auditory hallucinations to hearing voices.* In *Psychosis as a personal crisis.* Edited by M. Romme and A. Escher. London: Routledge.

Johnstone, L. (2011). People with problems not patients with illness. In *Psychosis as a personal crisis.* Edited by M. Romme and A. Escher. London: Routledge.

Robins, L.N., Wing, J., Wittchen, H.U., Helzer, J.E., Babor, T.F., Burke, J., Farmer, A., Jablenski, A., Pickens, R., Regier, D.A., Sartorius, N., Towle, L.H. (1988). The Composite International Diagnostic Interview: An epidemiologic instrument suitable for use in conjunction with different diagnostic systems and in different cultures. *Archives of General Psychiatry*, 45, 1069–1077.

Romme, M.A.J. (1996). *Understanding voices. Coping with auditory hallucinations and confusing realities.* Maastricht: Rijksuniversiteit Maastricht.

Romme, M., Escher, A. (1989). Hearing Voices. *Schizophrenia Bulletin*, 15, 209–216.

Romme, M., Escher, A. (1993). *Accepting voices.* London: MIND Publications.

Romme, M., Escher, A. (2000). *Making sense of voices.* London: MIND Publications.

Romme, M., Escher, A., Dillon, J., Corstens, D., Morris, M., eds. (2009). *Living with voices. 50 stories of recovery.* Ross-on-Wye: PCCS Books.

Romme, M.A., Honig, A, Noorthoorn, E.O., Escher, A.D. (1992). Coping with hearing voices: An emancipatory approach. *British Journal of Psychiatry*, 161, 99–103.

Tien, A.Y. (1991). Distributions of hallucination in the population. *Social Psychiatry and Psychiatric Epidemiology*, 26, 287–292.

Van Os, J., Hanssen, M., Bijl, R.V., Vollebergh, W. (2001). Prevalence of psychotic disorder and community level of psychotic symptoms: An urban-rural comparison. *Archives of General Psychiatry*, 58, 663–668.

Bibliography of Books on Hallucinations

Agadjanian, K. (1940). *Introduction à l'étude expérimentale du problème de l'hallucination.* Paris: J. Peyronnet et Cie.

Agadjanian, K. (1946). *Le mécanisme des troubles perceptive-associatifs en rapport avec l'origine de l'hallucination et du délire. Étude physio-pathologique.* Paris: J. Peyronnet et Cie.

Aleman, A. (2001). *Cognitive neuropsychiatry of hallucinations in schizophrenia. How the brain misleads itself.* Thesis University of Utrecht.

Aleman, A. (2011). *Hersenspinsels. Waarom we dingen zien, horen en denken die er niet zijn.* Amsterdam: Uitgeverij Atlas.

Aleman, A., Larøi, F. (2008). *Hallucinations. The science of idiosyncratic perception.* New York, NY: American Psychological Association.

Anders, R. (2003). *Wolkenlesen. Über hypnagoge Halluzinationen, automatisches Schreiben und andere Inspirationsquellen.* Niederstedt: Books on Demand.

Asaad, G. (1990). *Hallucinations in clinical psychiatry. A guide for mental health professionals.* New York, NY: Brunner/Mazel.

Aubanel, H. (1839). *Essai sur les hallucinations.* Thesis University of Paris.

Baker, P.K. (1996). *The voice inside. A practical guide to coping.* London: MIND Publications.

Barrett, W.F. (2005). *Visual hallucinations and phantasms of the living and dead.* Whitefish, MT: Kessinger Publishing. (Reprint of Chapter IX from Barrett, W.F. (1911). *Psychical research.* London: Williams and Norgate, pp. 111–132.)

Bauman, N.G. (2005). *Phantom voices, ethereal music & other spooky sounds.* Stewartstown, PA: GuidePost Publications.

Blackman, L. (2001). *Hearing voices, embodiment and experience.* London: Free Association Books.

Blom, J.D. (2010). *A dictionary of hallucinations.* New York, NY: Springer.

Bonnet, C. (1760). *Essai analytique sur les facultés de l'âme.* Copenhagen: Philibert.

Brierre de Boismont, A. (1862). *Des hallucinations ou histoire raisonnée des apparitions, des visions, des songes, de l'extase, des rêves, du magnétisme et du somnambulisme. Troisième édition.* Paris: Germer Baillère.

Brierre de Boismont, A. (1859). *On hallucinations. A history and explanation of apparitions, visions, dreams, ecstasy, magnetism, and somnambulism.* Translated by R.T. Hulme. London: Henry Renshaw.

Byrne, S., Birchwood, M., Trower, P.E., Meaden, A. (2006). *A casebook of cognitive behaviour therapy for command hallucinations: A social rank theory approach.* London: Routledge.

Canavero, S., Bonicalzi, V. (2011). *Central pain syndrome. Pathophysiology, diagnosis and management. Second edition.* Cambridge: Cambridge University Press.

Chadwick, P., Birchwood, M., Trower, P. (1996). *Cognitive therapy for delusions, voices and paranoia.* Chichester: John Wiley & Sons.

J.D. Blom and I.E.C. Sommer (eds.), *Hallucinations: Research and Practice*,
DOI 10.1007/978-1-4614-0959-5, © Springer Science+Business Media, LLC 2012

Clarke, E.H., Holmes, O.W. (1878). *Visions: A study of false sight (pseudopia).* Boston, MA: Houghton, Osgood & Company.

Clément, J.P., Michel, B.-F., Arnaud-Castiglioni, R., Derouesné, C., eds. (2002). *Hallucinations et maladies neuro-dégénératives. Monographies du Groupe de Recherche sur l'Alzheimer.* Marseille: SOLAL.

Cohen, S. (1964). *Drugs of hallucination.* London: Secker & Warburg.

Coleman, R., Smith, M. (1997). *Victim to victor: Working with voices.* Gloucester: Handsell Publications.

Connell, J.T. (1992). *The visions of the children: The apparitions of the Blessed Mother at Medjugorje.* New York, NY: St. Martin's Press.

Cramer, A. (1889). *Die Hallucinationen im Muskelsinn bei Geisteskranken und ihre klinische Bedeutung. Ein Beitrag zur Kenntniss der Paranoia.* Freiburg: Akademische Verlagsbuchhandlung von J.C.B. Mohr.

Critchley, E.M.R. (1987). *Hallucinations and their impact on art.* Cadley, Preston: Carnegie Press.

Critchley, M. (1953). *The parietal lobes.* London: Edward Arnold & Co.

Cytowic, R.E. (2002). *Synesthesia. A union of the senses. Second edition.* Cambridge, MA: MIT Press.

De Kroon, J. (2011). *De stem van de ander. Over (verbale) hallucinaties.* Apeldoorn: Uitgeverij Garant.

Dendy, W.C. (1847). *The philosophy of mystery.* New York, NY: Harper & Brothers.

Dheur, P. (1899). *Les hallucinations volontaires (l'état hallucinatoire). Suivi d'un chapitre sur les hallucinations. Notes manuscrites et inédites du Dr. J. Moreau (de Tours).* Paris: Société d'Éditions Scientifiques.

Diederen, K.M.J. (2011). *The hallucinating brain.* Thesis University of Utrecht.

Downs, J., ed. (2001). *Coping with voices and visions. A guide to helping people who experience hearing voices, seeing visions, tactile or other sensations.* Manchester: Hearing Voices Network.

Downs, J., ed. (2001). *Starting and supporting voices groups. A guide to setting up and running support groups for people who hear voices, see visions or experience tactile or other sensations.* Manchester: Hearing Voices Network.

Duits, Ch. (1994). *Vision et hallucination. L'expérience du peyotl en littérature.* Gordes: Question de/Albin Michel.

Durand, Ch. (1941). *L'Ėcho de la pensée.* Paris: G. Doin & Cie.

Ensink, B.J. (1992). *Confusing realities: A study on child sexual abuse and psychiatric symptoms.* Amsterdam: VU University Press.

Escher, A.D.M.A.C. (2005). *Making sense of psychotic experiences.* Thesis University of Maastricht.

Escher, A.D.M.A.C., Romme, M. (2010). *Children hearing voices. What you need to know and what you can do.* Ross-on-Wye: PCCS Books.

Esquirol, J.E.D. (1838). *Traité desmaladies mentales.* Paris: J.-B. Baillière.

Esquirol, J.E.D. (1965). *Mental maladies. A treatise on insanity. A facsimile of the English edition of 1845.* Translated by E.K. Hunt. New York, NY: Hafner Publishing Company.

Evans, H. (1984). *Visions, apparitions, alien visitors. A comparative study of the entity enigma.* Wellingborough: The Aquarian Press.

Ey, H. (1934). *Hallucinations et délire. Les formes hallucinatoires de l'automatisme verbal.* Paris: Félix Lacan.

Ey, H. (1973). *Traité des hallucinations. Tomes 1 et 2.* Paris: Masson et Cie., Éditeurs.

Faure, H. (1965). *Les objets dans la folie. Hallucinations et réalité perceptive.* Paris: Presses Universitaires de France.

Faure, H. (1966). *Les objets dans la folie. Les appartenances du délirant.* Paris: Presses Universitaires de France.

Ferriar, J. (1813). *An essay towards a theory of apparitions.* London: Cadell and Davies.

Fish, W. (2009). *Perception, hallucination, and illusion.* Oxford: Oxford University Press.

Gatiss, E., Dudley, R., McGregor, B., Freeston, M. (2005). *Positive experiences of voices.* Doctorate of Clinical Psychology, University of Newcastle Upon Tyne.

Giancoli, S. (2009). *Auditive und visuelle Halluzinationen.* Norderstedt: Books On Demand.

Gimenez, G. (2010). *Halluciner, persevoir l'impensé. Approche psychanalytique de l'hallucination psychotique*. Brussels: De Boeck.

Godderis, J. (2001). *Bestaan dingen alleen als men ze ziet? Historische, fenomenologisch-psychiatrische en metapsychologische reflecties inzake de waarneming, de verbeelding en het hallucineren.* Leuven: Garant.

Goldstein, K. (1912). *Die Halluzinationen, ihre Entstehung, ihre Ursachen und ihre Realität.* Wiesbaden: I.F. Bergmann.

Goodman, F.D., Henney, J.H., Pressel, E. (1982). *Trance, healing, and hallucination. Three field studies in religious experience*. Malabar, FL: Robert E. Krieger Publishing Company.

Green, C.E. (1968). *Lucid dreams.* Oxford: Institute of Psychophysical Research.

Grierson, M. (1991). *A report on the Manchester Hearing Voices Conference November 1990.* Manchester: Hearing Voices Network.

Gurney, E., Myers, F.W.H., Podmore, F. (1886). *Phantasms of the living. Volumes 1 and 2.* London: Trübner.

Haddock, G., Slade, P. (1996). *Empowering people who hear voices in cognitive behavioral interventions with psychotic disorders.* London: Routledge.

Harrison, J. (2001). *Synaesthesia. The strangest thing.* Oxford: Oxford University Press.

Hécaen, H., De Ajuriaguerra, J. (1952). *Méconnaissances et hallucinations corporelles. Intégration et désintégration de la somatognosie.* Paris: Masson et Cie.

Hibbert, S. (1824). *Sketches of the philosophy of apparitions; or, an attempt to trace such illusions to their physical causes.* Edinburgh: Oliver & Boyd, and G. & W.B. Whittaker.

Hoch, P.H., Zubin, J., eds. (1965). *Psychopathology of perception.* New York, NY: Grune & Stratton.

Holbrich, H.M., ed. (1987). *Halluzination und Wahn.* Berlin: Springer-Verlag.

Holmes, D. (1998). *Hearing voices: Hillary, angels, and O.J. to the voice-producing brain.* Rileyville, VA: Shenandoah Psychology Press.

Hoppe, J.I. (1871). *Hallucinationen und Illusionen. Begriffserklärung und Wesen. Ein Beitrag zur Erkenntnisstheorie. Auf Grund von Untersuchungen.* Basel: H. Amberger.

Hoppe, J.I. (1888). *Erklärung der Sinnestäuschungen. Hallucinationen und Illusionen aller fünf Sinne, bei Gesunden und bei Kranken. Vierte Edition.* Würzburg: Stuber.

Huxley, A. (2004). *The doors of perception and Heaven and Hell.* London: Vintage Books.

James, A. (2001). *Raising our voices. An account of the Hearing Voices Movement.* Manchester: Handsell Publishing.

Jardri, R., Thomas, P., Cachia, A., Pins, D., eds. (2012). *The neuroscience of hallucinations.* New York, NY: Springer.

Jaynes, J. (1976). *The origin of consciousness in the breakdown of the bicameral mind.* Boston, MA: Houghton Mifflin Company.

Jenner, J.A., ed. (2006). *Hallucinaties. Kenmerken, verklaringen, behandeling.* Assen: Van Gorcum.

Johnson, F.H. (1978). *The anatomy of hallucinations.* Chicago, IL: Nelson-Hall.

Join-Lambert, A. (2010). *Les expériences de mort imminente. Hallucinations ou lumières sur l'au-delà?* Namur: Fidélité.

Kandinsky, V. (1885). *Kritische und klinische Betrachtungen im Gebiete der Sinnestäuschungen. Erste und zweite Studie*. Berlin: Verlag von Friedländer und Sohn.

Karbowski, K., ed. (1982). *Halluzinationen bei Epilepsien und ihre Differentialdiagnose.* Bern: Verlag Hans Huber.

Kasten, E. (2008). *Die irreale Welt in unserem Kopf. Halluzinationen, Visionen, Träume.* Munich: Ernst Reinhardt Verlag.

Katona, C., Levy, R., eds. (1992). *Delusions and hallucinations in old age.* London: Gaskell.

Keup, W., ed. (1970). *Origin and mechanisms of hallucinations. Proceedings of the 14th annual meeting of the Eastern Psychiatric Research Association held in New York City, November 14–15, 1969.* New York, NY: Plenum Press.

Klemm, O. (1919). *Sinnestäuschungen.* Leipzig: Dürr'sche Buchhandlung.

Klüver, H. (1966). *Mescal and Mechanisms of hallucinations.* Chicago, IL: University of Chicago Press.

Lagache, D. (1934). *Les hallucinations verbales et la parole.* Paris: Félix Alcan.

Larøi, F., Aleman, A., eds. (2010). *Hallucinations. A guide to treatment and management.* Oxford: Oxford University Press.

Lelong, P. (1928). *Le problème des hallucinations.* Paris: Librairie J.-B. Baillière et Fils.

Lélut, L.F. (1846). *L'amulette de Pascal. Pour servir à l'histoire des hallucinations.* Paris: J.-B. Baillière et fils.

Leroy, E.B. (1926). *Les visions du demi-sommeil.* Paris: Félix Alcan.

Leubuscher, R. (1852). *ber die Entstehung der Sinnestäuschung. Ein Beitrag zur Anthropologie.* Berlin: Ferd. Dümmler's Verlagsbuchhandlung.

Leudar, I., Thomas, P. (2000). *Voices of reason, voices of insanity. Studies of verbal hallucinations.* London: Routledge.

Leuner, H. (1962). *Die experimentelle Psychose Ihre Psychopharmakologie, Phänomenologie und Dynamik in Beziehung zur Person.* Berlin: Springer-Verlag.

Leuner, H. (1963). *L'hallucinose optique et l'interprétation de son contenu.* Basel: Sandoz.

Lhermitte, J. (1951). *Les hallucinations. Clinique et physiopathologie.* Paris: G. Doin.

Lhermitte, J. (1959). *Les hallucinations dans leurs relations avec les lésions du lobe occipital.* Paris: Masson et Cie., Éditeurs.

Lungwitz, W. (1935). *Über extrakampine Halluzinationen, Mikropsie und Makropsie bei epileptiformen Anfällen.* Inaugural dissertation, University of Jena.

Maudsley, H. (1887). *Natural causes and supernatural seemings. Second edition.* London: Kegan, Paul, Trench & Co.

Maury, L.F.A. (1865). *Le sommeil et les rêves. Études psychologiques sur ces phénomènes et les divers états qui s'y rattachent. Troisième édition.* Paris: Librairie Académique Didier et Cie., Libraires-Éditeurs.

Mavromatis, A. (1987). *Hypnagogia. The unique state of consciousness between wakefulness and sleep.* London: Routledge.

McCreery, C. (2006). *Perception and hallucination. The case for continuity.* Oxford: Oxford Forum.

McKenna, T. (1993). *True hallucinations.* New York, NY: HarperCollins.

Meaden, A., Keen, N., Aston, R., Karen, B., Bucci, S. (2012). *Working with command hallucinations: An advanced practical companion.* London: Routledge.

Menninger-Lerchenthal, E. (1946). *Der eigene Doppelgänger.* Bern: Verlag Hans Huber.

Meyer, G.H. (1866). *Über Sinnestäuschungen.* Berlin: C.G. Lüderitz'sche Verlagsbuchhandlung A. Charisius.

Michea, C.F. (1851). *Du délire des sensations.* Paris: Labé.

Miller, F.P., Vandome, A.F., McBrewster, J., eds. (2010). *Auditory hallucination.* Beau Bassin: Alphascript Publishing.

Moreau de Tours, J.-J. (1899). *Quelques notes sur les hallucinations.* Paris: Société d'Éditions Scientifiques.

Mourgue, R. (1932). *Neurobiologie de l'hallucination. Essai sur une variété particulière de désintégration de la fonction.* Brussels: Maurice Lamertin.

Müller, J. (1826). *Ueber die phantastischen Gesichtserscheinungen.* Koblenz: Hölscher.

Myers, F.W.H. (1903). *Human personality and its survival of bodily death. Volumes I and II.* London: Longmans, Green, and Co.

Nelson, H.E. (2005). *Cognitive-behavioural therapy with delusions and hallucinations. A practice manual. Second edition.* London: Nelson Thornes.

Noë, A., ed. (2002). *Is the visual world a grand illusion?* Thorverton: Imprint Academic.

Őhman, R., Vogel, A., eds. (1984). *Temporal lobe epilepsy, psychotic and neurological manifestations, clinical tradition and new vistas. Proceedings of a symposium held in Copenhagen, 10–11 November 1982. Acta Psychiatrica Scandinavica supplementum no. 313, Vol. 69.* Copenhagen: Munksgaard.

Osis, K., Haraldsson, E. (1977). *At the hour of death.* New York, NY: Avon Books.

Parish, E. (1894). *Ueber die Trugwahrnehmung (Hallucination und Illusion).* Leipzig: Verlag von Ambrosius Abel.

Parish, E. (1897). *Hallucinations and illusions. A study of the fallacies of perception*. London: Walter Scott.

Parker, J.N., Parker, P.M., eds. *Hallucinations. A medical dictionary, bibliography, and annotated research guide to internet references*. San Diego, CA: ICON Group International.

Payne, M.S., ed. (2011). *Hallucinations. Types, stages and treatments*. New York, NY: Nova Science Publishers.

Pearson, R.S. (2005). *The experience of hallucinations in religious practice*. Seattle: Telical Books.

Penfield, W. (1975). *The mystery of the mind. A critical study of consciousness and the human brain*. Princeton, NJ: Princeton University Press.

Perceval, J. (1840). *A narrative of the treatment experienced by a gentleman, during a state of mental derangement; designed to explain the causes and the nature of insanity, and to explore the injudicious conduct pursued towards many unfortunate sufferers under that calamity*. London: Effingham Wilson.

Petit, G. (1913). *Sur une variété de pseudo-hallucinations. Les auto-aperceptions représentatives*. Bordeaux: Cadoret.

Pilon, L. (1959). *Contribution à la psychopathologie des hallucinations par l'étude de leur évolution sous chlorpromazine*. Paris: Dactylo-Sorbonne.

Pilowsky, D., Chambers, W., eds. (1986). *Hallucinations in children*. Washington, DC: American Psychiatric Press.

Piroux, A. (1861). *Fragments d'études sur les hallucinations*. Thesis University of Montpellier.

Plaut, F. (1913). *Über Halluzinosen der Syphilitiker*. Berlin: Verlag von Julius Springer.

Podmore, F. (1909). *Telepathic hallucinations: The new view of ghosts*. Halifax: London, Milner, & Co.

Poser, M. (1998). *Halluzinationen und Grenzerfahrungen im Alpinismus. Eine medizinisch-psychologische Untersuchung außergewöhnlicher menschlicher Erlebnisse*. Munich: Dt. Alpenverein.

Quercy, P. (1930). *L'hallucination. Tome 1. Philosophes. Mystiques*. Paris: Librairie Félix Alcan.

Quercy, P. (1930). *L'hallucination. Tome 2. Études cliniques*. Paris: Librairie Félix Alcan.

Ramachandran, V.S., Blakeslee, S. (1998). *Phantoms in the brain. Probing the mysteries of the human mind*. New York, NY: Quill.

Reed, G. (1972). *The psychology of anomalous experience. A cognitive approach*. London: Hutchinson & Co.

Ritti, A. (1873). *Théorie physiologique de l'hallucination*. Thesis University of Paris.

Roeckelein, J.E. (2004). *Imagery in psychology: a reference guide*. Westport, CT: Praeger.

Romme, M.A.J., Escher, A.D.M.A.C., eds. (1996). *Understanding voices. Coping with auditory hallucinations and confusing realities*. Manchester: Handsell Publications.

Romme, M.A.J., Escher, A.D.M.A.C. (1998). *Accepting voices. Second edition*. London: MIND Publications.

Romme, M.A.J., Escher, A.D.M.A.C. (2000). *Making sense of voices. A guide for mental health professionals working with voice-hearers*. London: MIND Publications.

Romme, M.A.J., Escher, A.D.M.A.C., Dillon, J., Corstens, D., Morris, M. (2009). *Living with voices. 50 stories of recovery*. Ross-on-Wye: PCCS Books.

Rudaleviciene, P. (2010). *Content of delusions and hallucinations in schizophrenia: Religious and socio-cultural background. Importance of personal religiosity*. Saarbrücken: Lambert Academic Publishing.

Rutten, F.J.Th. (1929). *Psychologie der waarneming. Een studie over gezichtsbedrog*. Thesis University of Utrecht.

Sacks, O. (1992). *Migraine. Revised and expanded*. New York, NY: Vintage Books.

Sacks, O. (2007). *Musicophilia. Tales of music and the brain*. New York, NY: Alfred A. Knopf.

Schorsch, G. (1934). *Zur Theorie der Halluzinationen*. Leipzig: Verlag von Johann Ambrosius Barth.

Seckel, A. (2005). *Super visions: geometric optical illusions*. New York, NY: Sterling Publishing Co.

Séglas, J. (1897). *Rapport sur la pathogénie et la physiologie pathologique de l'hallucination de l'ouïe*. Paris: Masson.

Séglas, J. (1901). *Sur les phénomènes dits hallucinations psychiques*. Paris: Félix Alcan.

Shulgin, A., Shulgin, A. (1991). *Pihkal. A chemical love story*. Berkeley, CA: Transform Press.

Sidgwick, H., Johnson, A., Myers, F.W.H., Podmore, F., Sidgwick, E. (1894). *Report on the Census of Hallucinations*. In *Proceedings of the Society for Psychical Research. Volume XXVI. Part X*. London: Kegan Paul, Trench, Trübner & Co, pp. 25–422.

Siegel, R.K. (1992). *Fire in the brain. Clinical tales of hallucination.* New York, NY: Dutton.

Siegel, R.K., West, L.J., eds. (1975). *Hallucinations. Behavior, experience, and theory.* New York, NY: John Wiley & Sons.

Simon, P.M. (1888). *Le monde des rêves. Le rêve, l'hallucination, le somnambulisme et l'hypnotisme, l'illusion, les paradis artificiels, le ragle, le cerveau et le rêve.* Paris: J.-B. Baillière et fils.

Slade, P.D., Bentall, R.P. (1988). *Sensory deception. A scientific analysis of hallucination.* London: Johns Hopkins University Press.

Slotema, C.W. (2011). *Treatment of patients with therapy-resistant auditory verbal hallucinations.* Thesis University of Utrecht.

Smith, D.B. (2007). *Muses, madmen, and prophets. Hearing voices and the borders of sanity.* London: Penguin Books.

Sollier, P. (1903). *Les phénomènes d'autoscopie*. Paris: Félix Alcan.

Sommer, I. (2011). *Stemmen horen.* Amsterdam: Uitgeverij Balans.

Spence, S.A., David, A.S., eds. (2004). *Voices in the brain. The cognitive neuropsychiatry of auditory verbal hallucinations.* London: Psychology Press/Taylor & Francis Group.

Spitzer, M. (1988). *Halluzinationen. Ein Beitrag zur allgemeinen und klinischen Psychopathologie.* Berlin: Springer-Verlag.

Spitzer, M. (1998). *Hallucinations.* New York, NY: John Wiley & Sons.

Starker, S. (1982). *Fantastic thought. All about dreams, daydreams, hallucinations, and hypnosis.* New York, NY: Prentice Hall.

Stephens, G.L., Graham, G. (2000). *When self-consciousness breaks: Alien voices and inserted thoughts.* Cambridge, MA: MIT Press.

Sully, J. (1881). *Illusions: A psychological study.* New York, NY: Humboldt Publishing Company.

Taylor, M. (1999). *Imaginary companions and the children who create them.* Oxford: Oxford University Press.

Tyrrell, G.N.M. (1953). *Apparitions. Revised edition.* London: The Society for Psychical Research.

Urbantschitsch, V. (1907). *Über subjective optische Anschauungsbilder.* Leipzig: Franz Deuticke.

Vercammen, A. (2009). *Cognitive and neural processes of auditory-verbal hallucinations in schizophrenia. Evidence from behavioral and neuroimaging experiments.* Thesis University of Groningen.

Vernon, J. (1963). *Inside the black room. Studies of sensory deprivation.* London: Souvenir Press.

Vernon, J.A., Sanders, B.T. (2001). *Tinnitus. Questions and answers.* Boston, MA: Allyn and Bacon.

Walter, W.G. (1960). *The neurophysiological aspects of hallucinations and illusory experience. The fourteenth Frederic W.H. Myers Memorial Lecture.* London: Society for Psychical Research.

Watkins, J. (2008). *Hearing voices. A common human experience.* Melbourne: Michelle Anderson Publishing.

Watkins, M. (2003). *Waking dreams. Third edition.* Putnam, CT: Spring Publications.

Weissman, J. (1993). *Of two minds. Poets who hear voices.* London: Wesleyan University Press.

West, L.J., ed. (1962). *Hallucinations.* New York, NY: Grune & Stratton.

Wiebe, P.H. (1997). *Visions of Jesus. Direct encounters from the New Testament to today.* Oxford: Oxford University Press.

Zenker, T.-K. (1993). *Halluzinationen bei alten Menschen.* Münster: Waxmann.

Zubek, J.P., ed. (1969). *Sensory deprivation: fifteen years of research.* New York, NY: Appleton-Century-Crofts.

Other Books by the Authors

Jan Dirk Blom
Deconstructing Schizophrenia (2004)
A Dictionary of Hallucinations (2010)

Iris Sommer
Language Lateralization And Psychosis (edited with René Kahn, 2009)
[*Hearing Voices*] (in Dutch, 2011)

Index

I

J

K

L

N

O

Q

R

T

Fig. 8 Resultant abnormal action frames from mixed background Dataset

SAIAZ video taken from the top angled single static camera in uniform illumination condition with simple background. Both Classroom violence and MBD dataset have a complex background with different illumination conditions when placed at different camera angles. Thus, the classification accuracy of *SVM* classifier on classroom violence data set and MBD set is not up to the level. Therefore, the ResNet*50* model was adopted to overcome those problems. The deep learning *ResNet50* model achieves better performance than *SVM* on given four types of datasets. But, in both cases, irrespectively, the result is pull down in MBD. The reason behind for the above case is that the training and testing frames are totally different. To test and improve the complexity of the system, earlier condition was entertained. To accomplish the above situation, the system is to be modified. The *ResNet50* shows good results on *SAIAZ and Classroom Violence* dataset. But, significantly decreases on mixed heterogeneous background. To handle these situations, the system will be modified in our future work.

References

1. Ke, S.-R., Thuc, H. L. U., Lee, Y.-J., Hwang, J.-N., Yoo, J.-H., & Choi, K.-H. (2013). A review on video-based human activity recognition. *Computers*, 88–131.https://doi.org/10.3390/computers2020088ComputersISSN2073-431Xwww.mdpi.com/journal/computer.
2. Popoola, O. P., & Wang, K. (2012). Video-based abnormal human behaviour recognition - a review. *IEEE Transactions on Systems, Man, and Cybernetics, Part C (Applications and Reviews), 42*(6), 865–878.
3. Hu, M.-K. (1962). Visual pattern recognition by moment invariants. *IRE Transactions on Information Theory, IEEE Explore*, 179–187.
4. Lecun, Y., Behgio, Y., & Hinton, G. (2015). Deep learning. *Nature, 521*(7553), 436-444.
5. Zheng, G. (2017). Fruit and vegetable classification system using image saliency and convolutional neural network. In *IEEE 3rd Information Technology and Mechatronics Engineering Conference* (pp. 613–617).
6. Zhang, J., Wu, C., & Wang, Y., et.al. (2018). Detection of abnormal behaviour in narrow scene with perspective distortion. *Journal of Machine Vision and Applications*, 1–12.

7. Wang, T., Chen, J., Honeine, P., & Snoussi, H. (2015). Abnormal event detection via multikernel learning for distributed camera networks. *International Journal of Distributed Sensor Networks, 2015*, 1–9. Article Id: 989450. https://doi.org/10.1155/2015/989450.
8. Tran, D., Yuan, J., & Forsyth, D. (2014). Video event detection: From subvolume localization to spatiotemporal path search. *IEEE Transaction on Pattern Analysis and Machine Int*elligence, *36*(2), 404–416.
9. Cheng, Y., & Fan, Q., Pankanti, S., & Choudhary, A. (2014). Temporal sequence modeling for video event detection. In *CVPR2014 Open Access version of Computer Vision Foundation.*
10. Szwed, P., & Komorkiewicz, M. (2013). Object tracking and video event recognition with fuzzy semantic petri nets. In *Proceedings of the 2013 Federated Conference on Computer Science and Information Systems* (pp. 167–174). IEEE. 978-1-4673-4471-5/$25.00 c.
11. Ravanbakhsh, M., Nabi, M., Sangineto, E., Marcenaro, L., Regazzoni, C., & Sebe, N. (2018). Abnormal event detection in videos using generative adversarial nets. In*Proceedings - International Conference on Image Processing, ICIP* (pp. 1577–1581).
12. Sabokrou, M., Fayyaz, M., Fathy, M., Moayed, Z., & Klette, R. (2018). Deep-anomaly: Fully convolutional neural network for fast anomaly detection in crowded scenes. *Journal of computer vision and Image Processing, 172*, 88–97.
13. Ravanbakhsh, M., Nabi, M., Mousavi, H., Enver, S., & Sebe, N. (2018, January). Plug-and-Play CNN for crowd motion analysis: An application in abnormal event detection. In *Proceedings - 2018 IEEE Winter Conference on Applications of Computer Vision,* WACV 2018 (pp. 1689–1698).
14. Mousavi, H., Nabi, M., Galoogahi, H. K., Perina, A., & Murino, V. (2015). Abnormality detection with improved histogram of oriented tracklets. In *International Conference on Image Analysis and Processing* (pp. 722–732). Springer.
15. Ravanbakhsh, M., Mousavi, H., Rastegari, M., Murino, V., Davis, L. S. (2015). Action recognition with image based CNN features. http://arxiv.org/abs/1512.03980.
16. Jia, Y., Shelhamer, E., Donahue, J., Karayev, S., Long, J., Girshick, R., Guadarrama, S., & Darrell, T. (2014). Caffe: Convolutional architecture for fast feature embedding. In *Proceedings of the ACM International Conference on Multimedia.*
17. Dubey, S., Boragule, A., & Jeon M. (2019). 3D ResNet with ranking loss function for abnormal activity detection in videos. In *2019 International Conference on Control, Automation and Information Sciences (ICCAIS)* (pp. 1–6). https://doi.org/10.1109/ICCAIS46528.2019.9074586.
18. Hu, Y. (2020). Design and implementation of abnormal behavior detection based on deep intelligent analysis algorithms in massive video surveillance. *Journal of Grid Computing, 18*, 227–237. https://doi.org/10.1007/s10723-020-09506-2.
19. Fu, Z., Robles-Kelly, A., & Zhou, J. (2010, December). Mixing linear SVMs for nonlinear classification. *IEEE Transactions on Neural Networks, 21*(12), 1963–1975.
20. Yu, H., & Kim, S. (2012, January). SVM tutorial: Classification, regression, and ranking. https://doi.org/10.1007/978-3-540-92910-9_15.
21. Ravichandran, M., & Krishnamoorthy, L. (2019, September 14). Detection and recognition of abnormal behaviour patterns in surveillance videos using SVM classifier. In *Proceedings of International Conference on Recent Trends in Computing, Communication & Networking Technologies (ICRTCCNT) 2019*. SSRN. https://ssrn.com/abstract=3453542, https://doi.org/10.2139/ssrn.3453542.
22. He, K., Zhang, X., Ren, S., & Sun, J. (2015, December 10). Deep residual learning for image recognition. arXiv:1512.03385v1 [cs.CV].
23. Zahangir Alom, Md., et al. (2018). The history began from AlexNet: A comprehensive survey on deep learning approaches. *Computer Vision and Pattern Recognition* (cs.CV). arXiv:1803.01164 [cs.CV].
24. Le Cun, Y., Denker, J. S., Henderson, D., Howard, R. E., Hubbard, W. E., & Jackel, L. D. (1989). Handwritten digit recognition with a back-propagation network. In *Proceedings of the Advances in Neural Information Processing Systems (NIPS)* (pp. 396–404).

RT-GATE: Concept of Micro Level Polarization in QCA

K. Bhagya Lakshmi, D. Ajitha, and K. N. V. S. Vijaya Lakshmi

Abstract Quantum-dot Cellular Automata is an evaluation paradigm in which transistors are not used and viable candidate for replacing the CMOS based technology. QCA is one of the boosting nanotechnology devices with the aim to replace the CMOS technology. QCA is implemented by utilizing the tunneling of the electrons with the given potential within the quantum cell. We made an attempt to suggest a multiplexer architecture in QCA using micro-level polarization. The proposed multiplexer design saves **16.67%** of effective area compared to the best designs reported till date. In this paper, new designs of universal gates are proposed. The proposed NOR and NAND gates requires less number of quantum cells which results in less effective area compared to the conventional majority gate based designs. By using these micro-level polarized gates, the multiplexer which is proposed in this paper is implemented. The proposed multiplexer is designed and simulated by QCA Designer.

Keywords Quantum-dot Cellular Automata (QCA) · RT-Gate (Rotate T-Gate) · Universal gates · Majority Voter Gate (MVG) · 2 × 1 Multiplexer · Operation cost (O-Cost)

1 Introduction

QCA is a proposed improvement on CMOS technology. *Moore's law* stated that the count of transistors present in an integrated circuit doubles about every eighteen months. But in the recent era, the size of the MOSFET is very minute. To satisfy the Moore's law, the size of the MOSFET must be very much shortened by reducing

K. Bhagya Lakshmi (✉)
Sasi Institute of Technology and Engineering, Tadepalligudem, India

D. Ajitha
Sreenidhi Institute of Science and Technology, Hyderabad, India
e-mail: ajithad@sreenidhi.edu.in

K. N. V. S. Vijaya Lakshmi
Sri Vasavi Engineering College, Tadepalligudem, India
e-mail: vijayalakshmi.kakaraparthi@srivasaviengg.ac.in

P. Nayak et al. (eds.), *IoT and Analytics for Sensor Networks*, Lecture Notes in Networks and Systems 244, https://doi.org/10.1007/978-981-16-2919-8_6

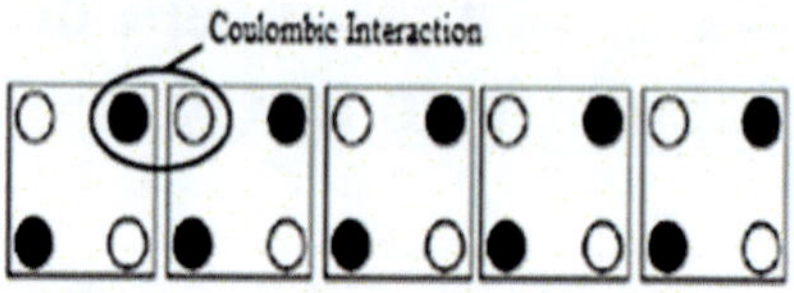

Fig. 1 Communication between the adjacent cells in a QCA

the length of its channel. By reducing the length of MOSFET's channel beyond certain limits, short channel effects come to existence. There comes the QCA technology which gives transistor less designs. The speed, density of devices and power consumption of QCA devices are at the forefront of the technologies of semiconductor like CMOS. In Fig. 1, the columbic repulsion rule indicates the interaction between the two neighbor cells in QCA. Based on the adjacent cell's electron repulsions, the position of electrons will be set in the quantum cells. This way, the message will be transferred from input side to output side. The different combinations of the quantum cells give rise to different devices.

In QCA technology, the wire is in the appearance of an array, a mixture of quantum cells. There are four corners of each quantum cell that contain four quantum dots. The electrons still occupy diagonal positions due to the Coulomb repulsion force. The polarization $P = +1$ is defined when one of the electrons is positioned at left corner lower side and the other electron is positioned at the right upper corner. In opposite case, it is defined as $P = -1$. These representations are shown in Fig. 2. Polarization of a quantum cell is given by "$Pi = ((P1 + P3) - (P2 + P4))/(P1 + P2 + P3 + P4)$" [1]. If two electrons reside in the opposite quantum dots of a quantum cell, then the polarization exits which defines logic 1 and logic 0. According to this, in Fig. 2, the first quantum cell polarization is $P = -1$ and $P = +1$ for the second quantum cell. In this paper, new designs of universal gates are proposed by changing the fixed polarization levels. The basic gates and the existing T-gate structure are discussed in the subsequent section. The proposed universal gates structure is discussed in Sect. 3 and the multiplexer implementation is discussed in Sect. 4 followed by the conclusion in Sect. 5.

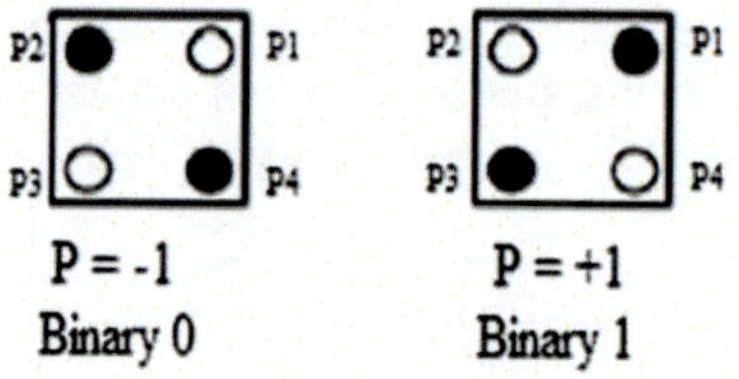

Fig. 2 QCA cell polarization states

2 QCA Basic Elements

The wire and Majority Voter Gate (MVG) are the basic elements [2] for logic implementation in QCA. Figure 3a is a QCA wire, an array of quantum cells. Figure 3b is QCA MVG whose logic is identical to Boolean expression $WX + XY + YW$ where W, X, Y are inputs. MVG implementation is shown in Fig. 3c. Cells A, B, and C are input cells, D is the output cell. The polarization of D is same as the majority of the input cells polarized value.

The basic gates AND & OR gates are implemented using majority voter gate shown in Fig. 4a. The majority voter gate has the drawback of having no provision for an inverter and has to base on displaced wire chain for inversion and it is shown in Fig. 4b.

2.1 *Universal Gates Based on Partial Polarization Technique*

Partial polarization is the action associated between two mutually orthogonal diagonal components [2]. The T-Gate can be implemented using 4 quantum cells [3]. However, to improve output polarization energy, one more quantum cell is to be added at the output. The maximum and minimum amplitude of output-simulated waveform is 9.52e-001e and –9.52e-001, respectively. Table1 gives the T-Gate implementation of NAND gate and NOR gate with 5 quantum cells. The existing T-gate structure was constructed with normal cells and 0.20 polarization value. Architecture for NAND/NOR is proposed using partial polarization of ±0.75 [4]. The NAND/NOR requires 5 quantum cells with 2 cells partially polarized [4]. However, it requires less number of cells a new gate structure is proposed to achieve better AUF factor

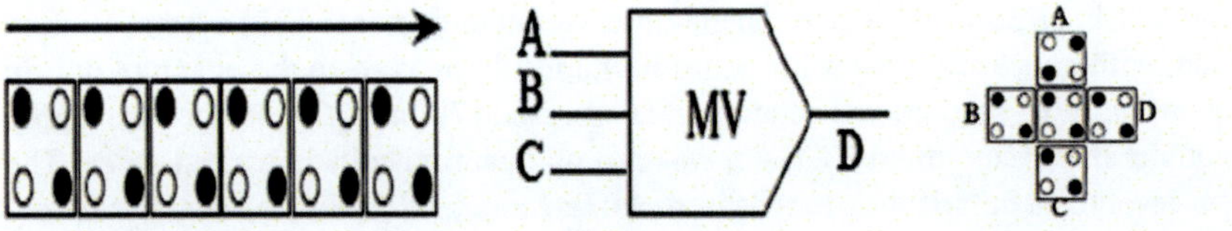

(a) Wire in QCA (b) MVG Symbol (c) Structure of MVG

Fig. 3 Basic QCA logic circuits

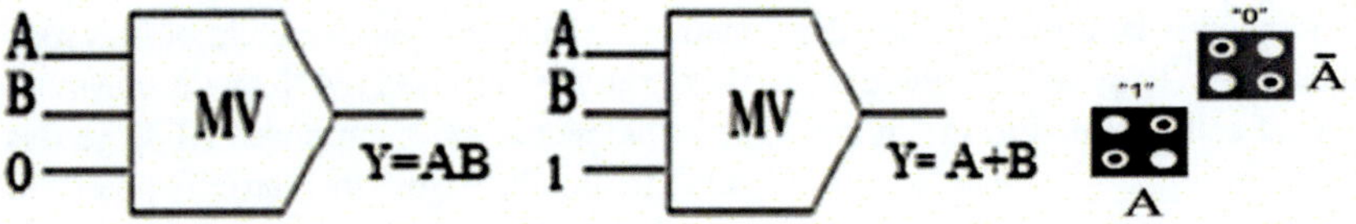

Fig. 4 **a** AND & OR gates. **b** NOT gate

Table 1 Universal gates using partially polarized T-gate

Logic gate	Symbol	QCA layout
RT⁺(A, B) = $\overline{A.B}$	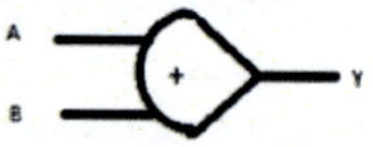	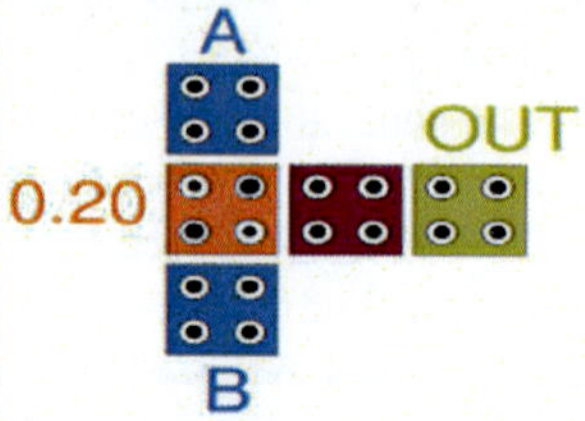
RT⁻(A, B) = $\overline{A+B}$		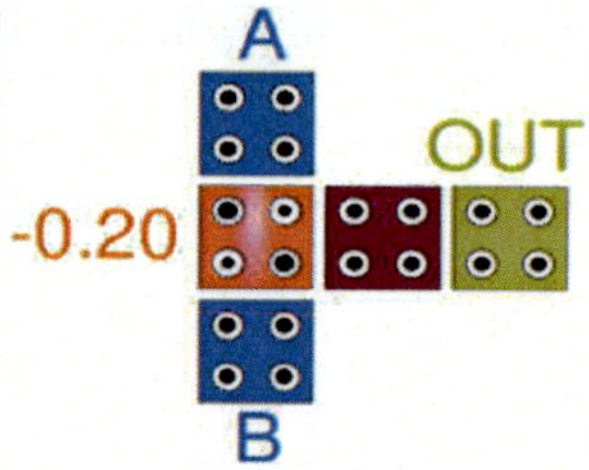

and discussed in Sect. 3. The clocking signal is a clipped signal sine wave that is a function of time is used as a default clock signal [5, 6].

3 Proposed RT-Gate

Quantum dots are part of a semiconductor device. The intrinsic electrostatics in the semiconductor device allows a "puddle" of electrons to form, near at semiconductor insulator boundary. The silicon metal-oxide-semiconductor (MOS) system is often used, with an additional level of gates in the insulator to form the quantum dot. In the presence of random cell rotation, it is assessed [7] that QCA devices are highly reliable. In the suggested RT-Gate, we used all quantum cells in rotated mode. The low-level fixed polarization ranges from ±0.001 to ±0.023. With the negative partial polarization NOR gate and by positive partial polarization NAND gate functionality can be achieved. The minimum waveform amplitude and maximum waveform output amplitude are –9.76e-001 and 9.76e-001, respectively. The proposed RT-Gate is shown as NAND Gate and NOR Gate, respectively, in Figs. 5 and 6, in which both are implemented with 5 quantum cells.

Fundamental logic gates are developed using the proposed structure of universal logic gates. To get AND gate functionality from the proposed NAND gate, one more rotated cell is added to the output. The implementation of proposed AND gate is shown in Fig. 7. Likewise, to get OR gate functionality from the partially polarized NOR gate, one more rotated cell is added at the output. The implementation of proposed OR gate is shown in Fig. 8.

Fig. 5 Implementation of RT gate as NAND gate

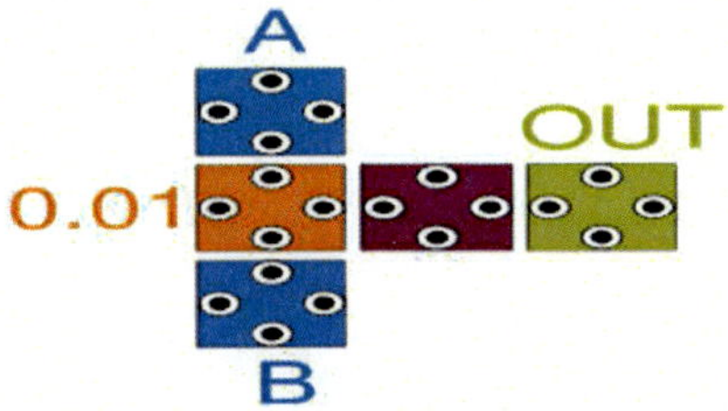

Fig. 6 Implementation of RT gate as NOR gate

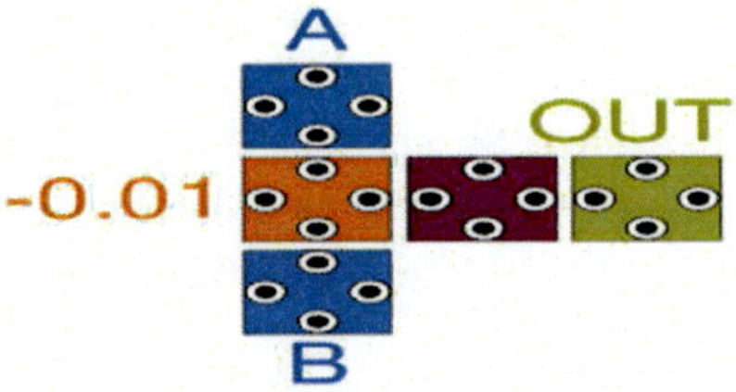

Fig. 7 Implementation of AND gate

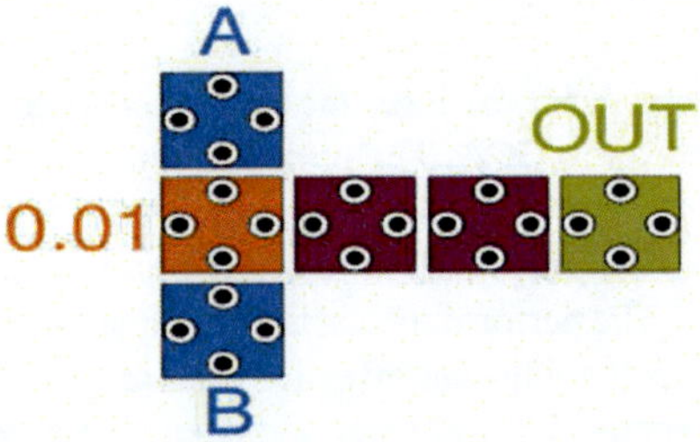

Fig. 8 Implementation of OR gate

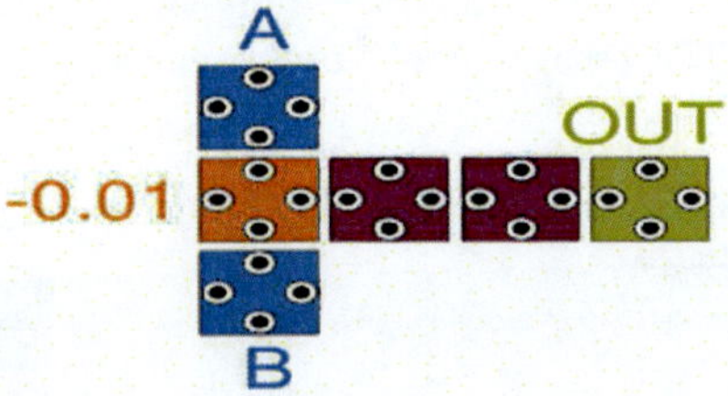

It is possible to enforce the NOT Gate structure by adding the same logic value to both of the inputs in the projected universal gate, and the same logic can also be extended to both NOR / NAND Logic. The effective area of the structure which is proposed is calculated to be 3,600 nm^2 against the T-gate [3] is 21769 nm^2 for effective implementation of the T-gate, the effective area required is 3,600 nm^2. The area utilization factor is 2.22. The area is calculated using the AUF formula method [8]. Using two clock zones, the output for the implemented gate is examined for minimum complexity. The power drop for proposed RT-Gate is 54.55 nwatts. The power drop of the proposed gate is calculated by using the algorithm that is used for computing the power drop in a coplanar wire [8]. As compared with the existing gate structures presented to design the serial adders and universal gate, the proposed gates will occupy less area with micro polarization [14, 15].

4 Multiplexer

Multiplexer is one of the key elements in digital circuit designs and widely used in the implementation of an FPGA circuits. The multiplexer design shown in Fig. 9 is proposed by Soudip Sinha Roy [3] using T-Gate requires 16 Quantum-dot cells and the maximum and minimum amplitude of the output waveform is simulated with 9.54e-001 and –9.54e-001, respectively.

4.1 Proposed Multiplexer

This paper explores the implementation of an area efficient design of a 2 × 1 multiplexer in QCA. The proposed multiplexer design and its simulated output are shown in Figs. 10 and 11 correspondingly. With low value polarizations and rotated cells, the proposed structure of 2:1 multiplexer is implemented.

The performance of the anticipated 2:1 multiplexer is compared with the preceding designs with the effective area and O-Cost. The analysis is shown in Table 2. The proposed multiplexer has been shown to be successful by using a smaller number of QCA cells. The existing multiplexer design [3] can be implemented by using 16 QCA. 2 to 1 multiplexer digital design [9] is implemented by using 34 QCA cells. QCA Designer tool is used to examine the design of QCA which is implemented [10].

Fig. 9 Existing Multiplexer design

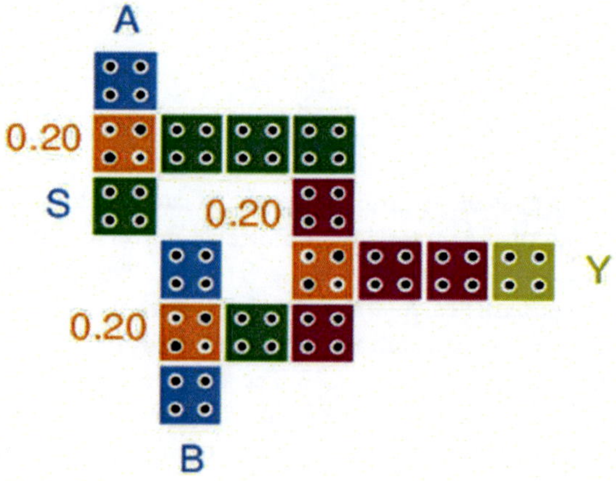

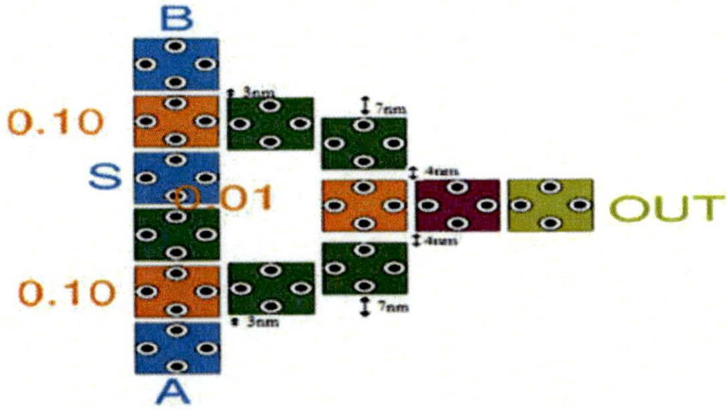

Fig. 10 Proposed multiplexer design

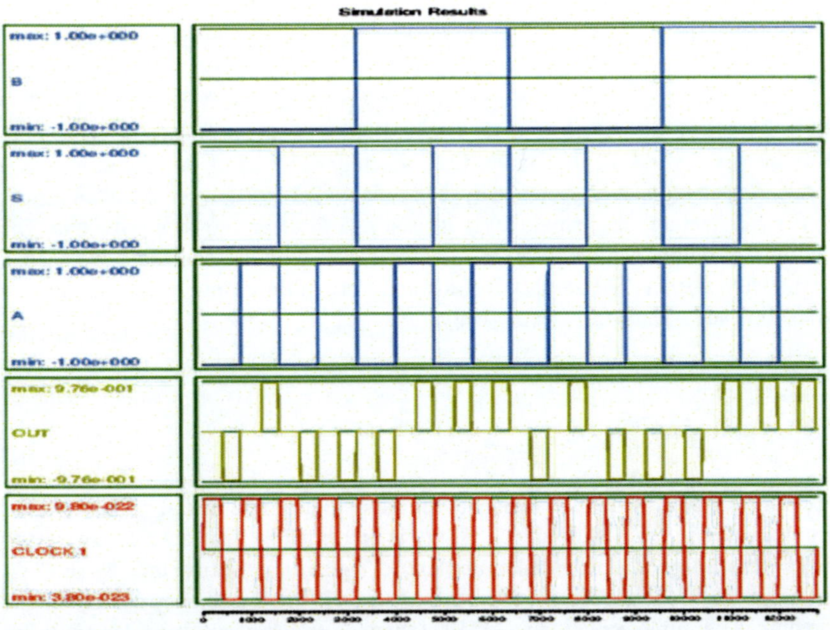

Fig. 11 Output waveform of proposed multiplexer

The constructed 2 to 1 multiplexer layout [11] is created by the QCA-LG method, which consists of 75 cells. The versatile architecture of the 2 to 1 multiplexer [12] comprises 41 quantum cells.

A novel design for 2:1 MUX which is implemented in [13] consists of 23 quantum cells, where as the proposed design requires only 13 cells. Length and breadth are the two dimensions of this layout. The normal spacing of 2 nmin between two quantum cells is the QCA meaning. 1 nm is the separation of the cells from the cell bed's terminal point. Because of the equal length and width of 18 nm resembles a perfect

Table 2 Analysis of 2:1 Multiplexer compared with the existing designs

S. No	No. of quantum cells for design implementation	Operation cost	Effective area in nm^2	Area improvement achieved in the projected RT-Gate
1	2 × 1 MUX [11]	75	121,764	90.14%
2	2 × 1 MUX [12]	41	45,924	73.86%
3	2 × 1 MUX [9]	34	40,764	70.56%
4	2 × 1 MUX [13]	23	23,360	55.03%
5	2 × 1 MUX [3]	16	21,769	44.87%
6	2 × 1 MUX [8]	14	14,400	16.67%
7	Proposed RT-mux	13	12,000	–

square, the cells are similar in terms of area. The region is, therefore, 18–18 n^2m^2. L= (1 + 18 + 2 + 18 + 2 + 18 + 2 + 18 + 2 + 18 + 18 + 2 + 18 + 18 + 1)= 100 nm is the estimated length of the proposed MUX. W= (1 + 18 + 2 + 18 + 2 + 18 + 2 + 18 + 2 + 18 + 2 + 18 + 2 + 18 + 2 + 18 + 18 + 1) = 120 nm is the width of the layout. Consequently the area of the circuit is 12000 $(nm)^2$. The effective area occupied by the proposed 2:1 multiplexer is **12,000 $(nm)^2$**. In the architecture, only single layer is used and there are no wire crossings. The multiplexer output can be observed in clock 1, i.e., only 2 clock zones are needed as shown in Fig. 11. The amplitude (Max and Min) of the output waveform here are 9.76e-001 and –9.76e-001, respectively.

5 Conclusion

Multiplexers are essential components for many logical and digital circuits. A 2 × 1 multiplexer in QCA with unique design is presented in this paper. In contrast to the present design, the proposed multiplexer design saves effective area by **16.67%**. By using this proposed design, the area and speed is optimized. Any combinational circuit can be implemented in an efficient manner with this proposed design. It can be considered that the anticipated design is more stable than the previous models, based on the implementation performance.

References

1. Lent, C. S., & Tougaw, P. D. (1997). A device architecture for computing with quantum dots. *Proceedings of the IEEE, 85*(4), 541–557.
2. Lahiri, M., Hochrainer, A., Lapkiewicz, R., Lemos, G. B., & Zeilinger, A. (2017). Partial polarization by quantum distinguishability. *Physical Review A, 95*(3), 033816.

3. Mukherjee, C., Roy, S. S., Panda, S., & Maji, B. (2016, May 24). T-gate concept of partial polarization in quantum dot cellular automata. *Proceedings of 20th International Symposium on VLSI Design and Test (VDAT)* (pp. 1–6). IEEE.
4. Ghosal, S., & Biswas, D. (2013). Study and defect characterization of a universal QCA gate. *International Journal of Computer Applications, 975*, 8887.
5. Lu, Y., Liu, M., & Lent, C. (2007). Molecular quantum-dot cellular automata: From molecular structure to circuit dynamics. *Journal of applied physics, 102*(3), 034311.
6. Blair, E. P., & Lent, C. S. (2003, September). Quantum-dot cellular automata: an architecture for molecular computing. In *Proceedings of International Conference on Simulation of Semiconductor Processes and Devices, 2003. SISPAD 2003* (pp. 14–18). IEEE.
7. Yang, X., Cai, L., Wang, S., Wang, Z., & Feng, C. (2012). Reliability and performance evaluation of QCA devices with rotation cell defect. *IEEE Transactions on Nanotechnology, 11*(5), 1009–1018.
8. Roy, S.S., 2017. pGate: An Introduction to A Novel Universal Gate and Power Drop Calculation of QCA circuits. *International Journal Of Engineering And Computer Science (IJECS)*, pp.20967–20972.
9. Askari, M., Taghizadeh, M., & Fardad, K. (2008, May 13). Digital design using quantum-dot cellular automata (a nanotechnology method). In *Proceedings of International Conference on Computer and Communication Engineering* (pp. 952–955). IEEE.
10. Walus, K., & Jullien, G. A. (2006). Design tools for an emerging SoC technology: Quantum-dot cellular automata. *Proceedings of the IEEE, 94*(6), 1225–1244.
11. Srivastava, S., & Bhanja, S. (2007). Hierarchical probabilistic macro modeling for QCA circuits. *IEEE Transactions on Computers, 56*(2), 174–190.
12. Teodósio, T., & Sousa, L. (2007, November 19). QCA-LG: A tool for the automatic layout generation of QCA combinational circuits. In *Proceedings of Norchip conference* (pp. 1–5). IEEE.
13. Kim, K., Wu, K., & Karri, R. (2006). The robust QCA adder designs using composable QCA building blocks. *IEEE Transactions on Computer-Aided Design of Integrated Circuits and Systems, 26*(1), 176–183.
14. Mostafaee, A., & Rezaei, A. (2017). Novel optimized designs for QCA serial adders. *International Journal of Information Technology and Computer Science*, 2(2), 38–46.
15. Bilal, B., Ahmed, S., & Kakkar, V. (2017). Optimal realization of universality of Peres gate using explicit interaction of cells in quantum dot cellular automata nanotechnology. *International Journal of Intelligent Systems and Applications, 11*(6), 75.

Comparative Performance Analysis of Tanh-Apodized Fiber Bragg Grating and Gaussian-Apodized Fiber Bragg Grating as Hybrid Dispersion Compensation Model

Baseerat Gul and Faroze Ahmad

Abstract Fiber optic systems are used for the prolonged reach transmission systems, but by increasing the bit rate which is the main requirement of the current time, dispersion gets arisen which results in intersymbol interference. Compensation of dispersion to improve the transmission capability of the fiber optic system provides a vast field for research. From the literature survey done, use of Dispersion compensation fiber has been found as the most reliable method for compensating the dispersion, but it becomes expensive as the length of Dispersion compensation fiber is increased for long distance transmission. The Fiber Bragg Grating is also used as a dispersion compensation module as reported in previous works but has been found inefficient method. However, the Performance of the Fiber Bragg Grating can be enhanced by adapting optimum Chirping technique and Apodization profile. From the previous reported works, Tanh-Apodized Fiber Bragg grating and Gaussian-Apodized Fiber Bragg grating are found to have optimum performance characteristics in terms of side lobe suppression and maximum reflectivity, which motivates us to analyze the respective Fiber Bragg Gratings for compensating the dispersion at various chirping techniques and variable grating lengths. In this work, Tanh-Apodized Fiber Bragg grating and Gaussian-Apodized Fiber Bragg grating are analyzed and simulated in various chirping techniques individually, as well as along with the Dispersion compensation fiber, in the hybrid model of dispersion compensation for a 100 km long optical fiber link at the data rate of 10Gbps. The simulation software used is optisystem. Also, the grating length has been varied and the different performance characteristics like Q-factor, BER, and Eye diagram are analyzed and compared. It has been observed that the Gaussian-Apodized quadratic-chirped Fiber Bragg Grating at the grating length of 26.6 mm along with the 11 km long Dispersion compensation fiber makes the cheaper dispersion compensation module with the finest performance.

Keywords Optical fiber communication · Dispersion · Dispersion compensation fiber · Fiber bragg grating

B. Gul (✉) · F. Ahmad
Department of Electronics and Communication Engineering, Islamic University of Science and Technology, Awantipora, J&K, India
e-mail: baseerat.gul@islamicuniversity.edu.in

P. Nayak et al. (eds.), *IoT and Analytics for Sensor Networks*, Lecture Notes in Networks and Systems 244, https://doi.org/10.1007/978-981-16-2919-8_7

1 Introduction

In the current era of Optical Fiber communication (OFC), the main issues with the increasing bit rate for the Conventional Single mode fiber (CSMF) are the Group velocity dispersion (GVD), Attenuation, and Non-linear effects [1]. However, the Attenuation problem has been solved by the introduction of Erbium doped fiber amplifier (EDFA) [2, 3] and the effect of non-linearity can be reduced by keeping the associated power-level below certain limits [2]. GVD, however, becomes negligible for bit rates below 2 Gbps, but for higher bit rates, it reduces the quality parameters of the signal and needs to be compensated. Abundant research work has been reported in the past regarding the compensation of dispersion to improve the performance characteristics of high speed optical communication systems. Dispersion compensation fiber (DCF) and Fiber Bragg grating (FBG) are most widely used for compensating the dispersion. DCF has shown superior performance in compensating the dispersion to a great extent but has been proven as an expensive method for compensating the dispersion and becomes costlier with its increasing length. FBG has shown inefficient performance as compared to DCF, but its performance can be improved by adapting proper Chirping technique and Apodization profile. From the work reported in paper [4], Gaussian-Apodized FBG and Tanh-Apodized FBG have shown optimum performance in terms of side lobe suppression and the maximum reflectivity. Paper [5] reports that the bandwidth of the compensated spectrum can be leveled up to optimal value by choosing the proper grating length and the chirp rate of the CFBG. In this paper, comparative study and analysis of five different dispersion compensation models have been made for 100 km long optical fiber communication systems at the data rate of 10 Gbps. In these dispersion compensation models, various chirping techniques of Tanh-Apodized FBG and Gaussian-Apodized FBG have been investigated individually and as a part of hybrid model along with the DCF to discover the dispersion compensation model which shows the finest performance both in terms of cost reduction and quality improvement. The lower order of the GVD, i.e., Chromatic dispersion is taken into consideration while ignoring the higher orders of GVD which have negligible effect at the data rate of 10Gbps and becomes prominent at the date rate of 40Gbps or above [2, 3, 6].

2 Theory

The Chromatic dispersion is caused due to the finite spectral width of the optical source and the non-linear variation in the refractive index of the core of CSMF with respect to the variation in wavelength parameter. Dispersion arisen in CSMF results in pulse broadening, which increases with the increase in transmission distance and the bit rate, hence making the high data rate long-haul optical communication difficult [1], which at present has the highest demand in the telecommunication field.

Although the chromatic dispersion is seen zero at Zero Dispersion wavelength (ZDW = 1300 nm), but the preferable wavelength parameter for the optical transmission is 1550 nm due to the least attenuation and other favorable parameters [2, 3]. The effect of dispersion at the 1550 nm wavelength window is the major constraint [3]. DCF and FBG are the extensively adapted methods for compensating the dispersion by connecting either one or both in cascade with the CSMF [3]. From the literature survey done, the use of DCF has been found the most reliable but expensive method for compensating the dispersion, and the use of FBG has been found an inefficient method [1–3, 7]. Performance of FBG as a dispersion compensation model, however, can be enhanced by adapting optimum Chirping techniques and Apodization profiles [1, 4, 8].

2.1 Dispersion Compensation Fiber

At ZDW (1300 nm), Group velocity (V_g) associated with the signal propagating through CSMF is seen maximum and decreases monotonically on the either side of wavelength change as shown in Fig. 1. If the operating wavelength chosen is 1550 nm, longer wavelength components in CSMF propagate with the slower speed and the shorter wavelength components propagate with the faster speed which results in dispersion. DCF has been designed in such a way by the variation in refractive index such that it allows longer wavelength components to propagate at the faster speed and vice versa, that is why it is said that the DCF is associated with negative dispersion [6]. DCF with negative dispersion when connected in cascade with the CSMF nullifies the dispersion arisen in CSMF. Negative dispersion in DCF can be increased by increasing the doping level of core with GeO_2, however, higher doping in turn increases the losses [6], hence the negative dispersion associated with the DCF cannot be increased beyond its optimal value which is approximately in the range of −70 to −80 ps/nm/km. For compensating higher dispersion produced by increasing length of CSMF, length of the DCF needs to be increased accordingly. However, increasing the length of DCF increases the cost associated with it.

Fig. 1 Group velocity verses wavelength

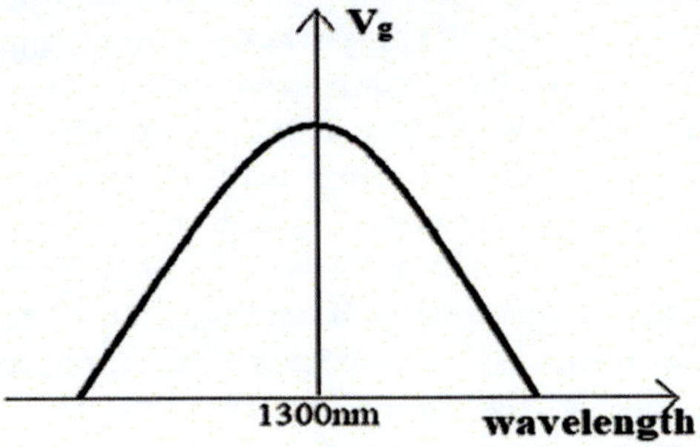

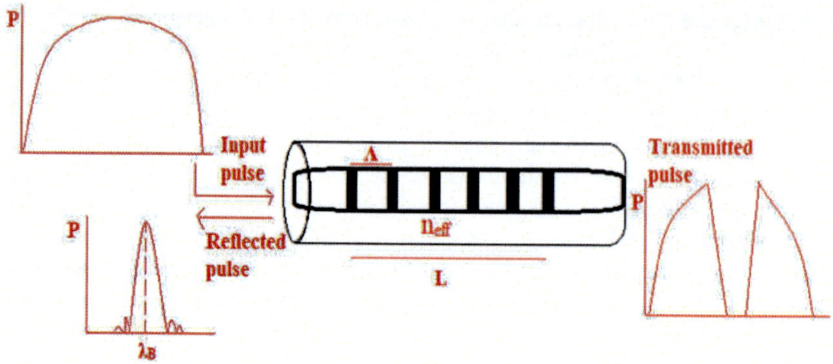

Fig. 2 Fiber Bragg Grating

2.2 Fiber Bragg Grating

FBG designed on a short segment of optical fiber, as shown in Fig. 2, is analogous to a stop band filter centered at wavelength called as Bragg's Wavelength (λ_B). It is a type of Bragg's reflector which works on the principle of Fresnel reflection [3]. FBG is fabricated by modulating the refractive index periodically and the respective period is named as Grating period (Λ). For a specific temperature and strain, Bragg's wavelength could be controlled by controlling the effective refractive index (n_{eff}) and the grating period [1, 9], as represented by the Eq. (1)

$$\lambda_B = 2\,\Lambda\, n_{\mathrm{eff}}. \tag{1}$$

If along with the Grating length (L) refractive index modulation is varied, bandwidth of the reflected spectrum broadens as each portion of the grating reflects different spectrum [1, 5, 7]. FBG of such kind is named as Chirped FBG (CFBG) and the technique adapted is known as Chirping. The various chirping techniques that are mostly used are linear chirping, quadratic chirping, square root chirping, and cubic root chirping. By controlling the chirp rate, peak reflectance can be controlled directly [5]. A CFBG provides longer delay to quick moving wavelength components and the shorter delay to fast moving wavelength components [1, 9, 10], hence finds a major application in dispersion compensation in long-haul optical communication. The reflection spectrum of the FBG exhibits side lobes which arises due to the rapid change in index modulation and can be removed by ramping up and down the strength of modulation smoothly [1], such an FBG is termed as Apodized FBG and the technique adapted is named as Apodization. To achieve a certain peak reflectance, overall grating length needs to be increased [4]. There is a trade-off between the side lobe suppression and the maximum peak reflectance for some restricted grating length [4], but both can be leveled to an optimum value by adapting optimum chirping and apodization techniques as shown in the current work.

3 Experimental Results

The reflected spectrum of the FBG is associated with the side lobes which can be suppressed by apodization. Also, the maximum reflectivity can be increased by increasing the grating length. For some constant grating length, side lobe suppression and peak reflectance shows trade-off and needs to be set out at the optimum value. From the work reported in paper [4], Gaussian-Apodized FBG and Tanh-Apodized FBG have shown best performance when we want to have both side lobe suppression, as well as the maximum reflectivity, at the optimum level. To use the FBG for dispersion compensation, bandwidth of the reflected spectrum needs to be increased which can be done by chirping. Paper [5] reports that the bandwidth of the reflected spectrum and the maximum reflectivity can be leveled up to optimum value by choosing the proper grating length and the chirp rate of the CFBG. Which chirping technique gives the best performance for the respective apodization techniques haven't been analyzed mostly? So our main aim in this work is to determine the chirping technique and the grating length of the FBG which gives the finest performance for compensating the dispersion in Gaussian-Apodized FBG and Tanh-Apodized FBG. In the current work, five dispersion compensation models have been implemented and simulated for a 100 km long optical fiber link at the data rate of 10Gbps as shown in Fig. 3 at the parameters described in Table 1. The simulation software used is OptiSystem, in which the performance characteristics of the designed links are observed and compared in terms of the Eye diagram. The Eye diagrams for the five simulated links are shown from Figs. 8 to 12, in which the red curve represents the Q-factor and alongside the various analytical parameters like Maximum Q-Factor, Minimum BER, Eye height, etc., are defined, so that the performance of the links can be analyzed and compared (Table 2).

In the link-1, DCF with the attenuation of 0.2 dB/km and dispersion of -70 ps/nm/km is used as a dispersion compensation module and is connected in cascade with the 100 km long CSMF. The link is analyzed at variable DCF lengths. It has been observed that the highest Q-factor of 36.02 comes into origin when the length of DCF chosen is 24.28 km. To compensate the losses introduced by the DCF itself, an extra EDFA of gain 4 dB is added to the link. The Eye diagram of the simulated link-1 is shown in Fig. 8.

In the link-2 and link-3, Tanh-Apodized FBG and Gaussian-Apodized FBG are used individually as dispersion compensation modules for compensating the dispersion of 100 km long optical fiber link. Both the links are observed for various chirping techniques to discover the optimum chirping technique for the respective FBG(s). Also, the grating lengths of the respective FBG(s) are varied. The analyzed response of the Q-factor at various chirping techniques verses the grating length for both the links are shown in Figs. 4 and 5. It can be seen that the link-2 gives the highest Q-factor of 23.71 for linear chirping at the grating length of 16 mm and the link-3 shows the highest Q-factor of 31.74 for quadratic chirping at the grating length of 41.8 mm. The Eye diagrams for the simulated respective links at the optimum performance characteristics are shown in Figs. 9 and 10.

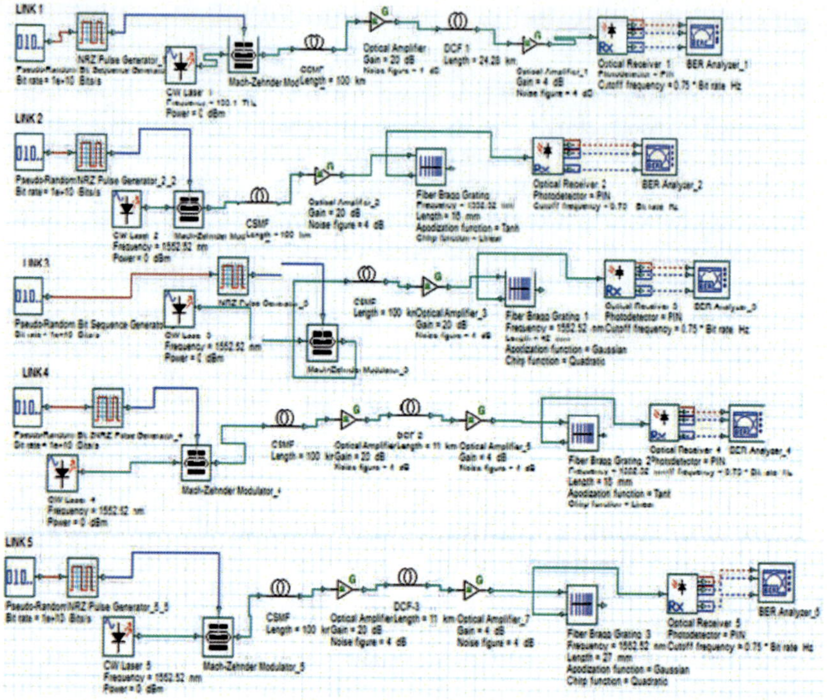

Fig. 3 Simulation set up of five dispersion compensation modules

In the link-4 and link-5, Tanh-Apodized-linear-chirped-FBG and Gaussian-Apodized-Quadratic-chirped-FBG along with the DCF are connected in cascade with the 100 km long CSMF as a hybrid dispersion compensation model. 10, 11, 12, and 13 km long DCF(s) are observed successively at various grating lengths of the respective FBGs as shown in Figs. 6 and 7. It has been observed that the Hybrid dispersion compensation model of 12 km long DCF and Tanh-Apodized linear-chirped FBG gives the highest Q-factor of 19.44 at the grating length of 16.2 mm for the link-4 and the Hybrid dispersion compensation model of 11 km long DCF and Gaussian-Apodized Quadratic-chirped FBG gives the highest Q-factor of 35.22 at the grating length of 26.6 mm for the link-5. The Eye diagrams for the simulated link-4 and link-5 at the optimum performance characteristics are shown in Figs. 11 and 12.

4 Conclusion

The dispersion compensation models used in the five simulated links and their performance are summarized in Table 2. In terms of performance characteristics, it can be

Table 1 Components, parameters and values

S.no	Components	Parameters and values
1	Data source	Bit rate 10Gbps
2	Pulse generator	NRZ
3	CW laser	Frequency 193.1THz Power 0 dBm Line width 10 MHz Noise bandwidth 0THz Noise threshold –100 dB Noise dynamic 3 dB
4	Mach-zender modulator (MZM)	Extinction ratio 30 dB
5	CSMF	Reference wavelength 1552 nm Length 100 km Attenuation 0.2 dB/km Dispersion 16.75 ps/nm/km
6	EDFA	Gain 20 dB Noise Fig. 4 dB
7	Optical receiver	Photodetector PIN Gain 3 Ionization ratio 0.9 Responsivity 1A/W Dark current 10nA Cutoff frequency 0.75* Bit rate

Table 2 Performance analysis of five simulated links

	Dispersion compensation module		
	DCF length	FBG used	Q-factor
Link-1	24.28 km	Not used	36.02
Link-2	0 km	Tanh-Apodized linear-chirped FBG with grating length of 16 mm	23.71
Link-3	0 km	Gaussian-Apodized Quadratic-chirped FBG with grating length of 41.8 mm	31.74
Link-4	12 km	Tanh-Apodized linear-chirped FBG with grating length of 16.2 mm	19.44
Link-5	11 km	Gaussian-Apodized Quadratic-chirped FBG with grating length of 26.6 mm	35.22

concluded that the Link-1, in which the 24.28 km long DCF alone is used as a dispersion compensation module and the Link-5, in which the 11 km long DCF along with Gaussian-Apodized Quadratic-chirped FBG at grating length of 26.6 mm is used as a dispersion compensation module gives the finest eye diagram and the highest Q-factor. The high cost of DCF is always an issue and it can be seen that the link-1 uses almost 2.2 times longer DCF as compared to the link-5. So in terms of cost effectiveness, it can be concluded that Link-5, in which the 11 km long DCF along

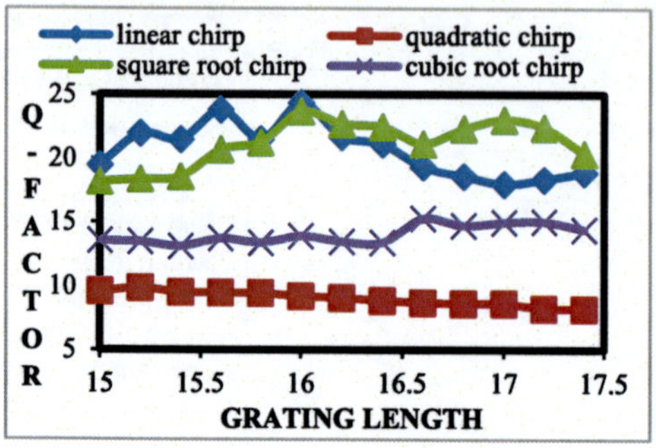

Fig. 4 Q-factor verses grating length for various chirping techniques in Tanh-Apodized FBG

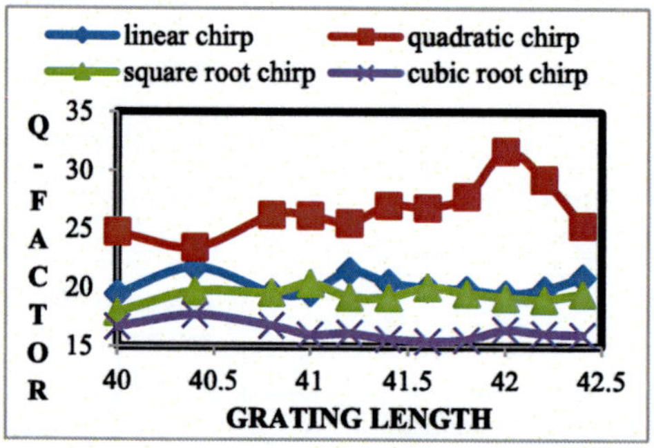

Fig. 5 Q-factor verses grating length for various chirping techniques in Gaussian-Apodized FBG

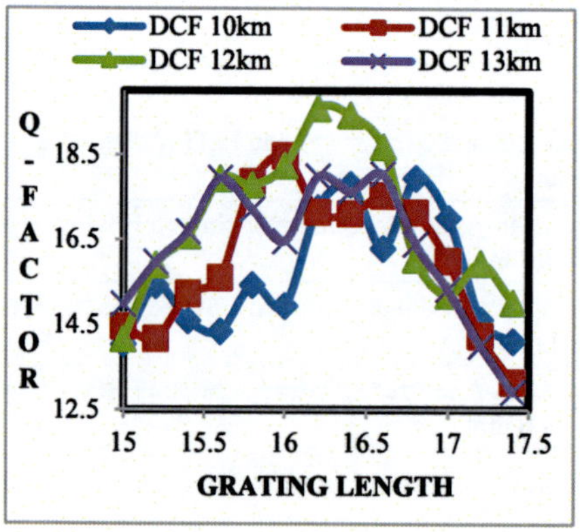

Fig. 6 Q-factor verses grating length for the hybrid model of DCF with Tanh-Apodized Linear-Chirped FBG

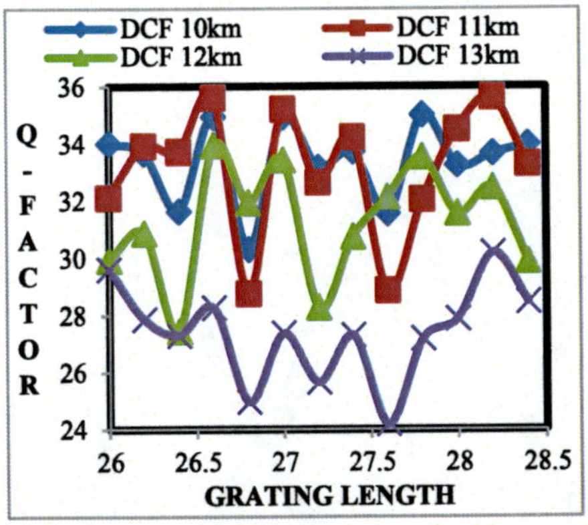

Fig. 7 Q-factor verses grating length for the hybrid model of DCF with Gaussian-Apodized Quadratic-chirped FBG

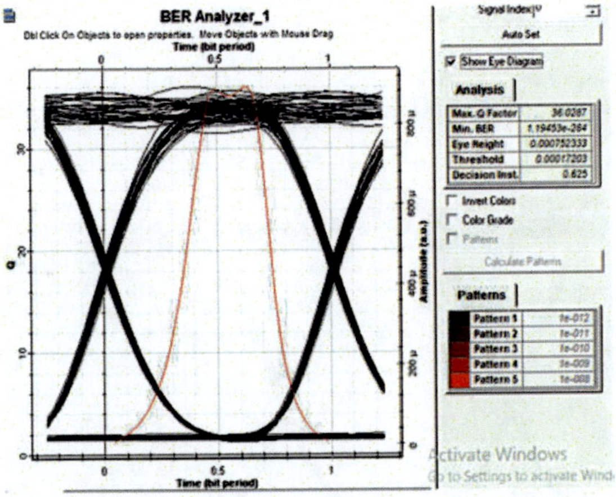

Fig. 8 Eye diagram of link-1

with Gaussian-Apodized Quadratic-chirped FBG at grating length of 26.6 mm is used as a dispersion compensation module makes the link cheaper as compared to Link 1. Hence, it can be concluded that the Gaussian-Apodized Quadratic-chirped FBG with grating length of 26.6 mm along with the 11 km long DCF makes the most efficient dispersion compensation model with the cost reduction of almost 54% as compared to using DCF alone for the 100 km long optical fiber link at the data rate of 10 Gbps Figs. 8, 9, 10, 11 and 12.

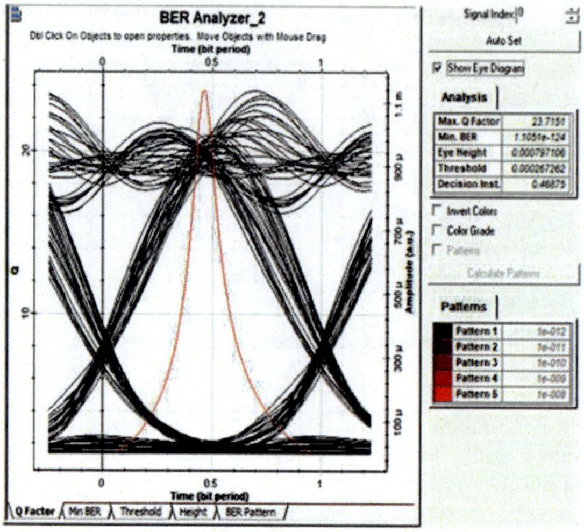

Fig. 9 Eye diagram of link-2

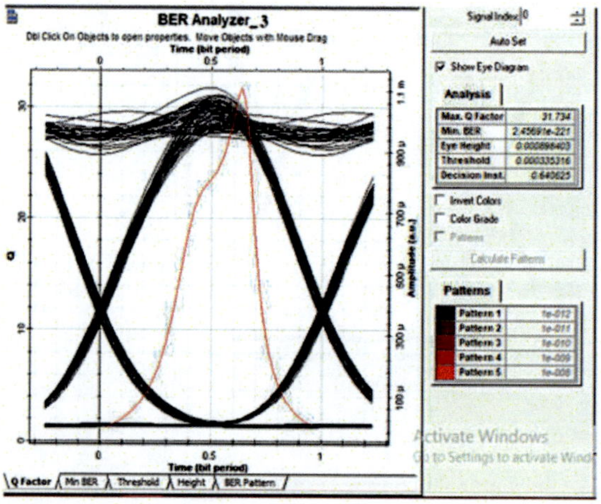

Fig. 10 Eye diagram of link-3

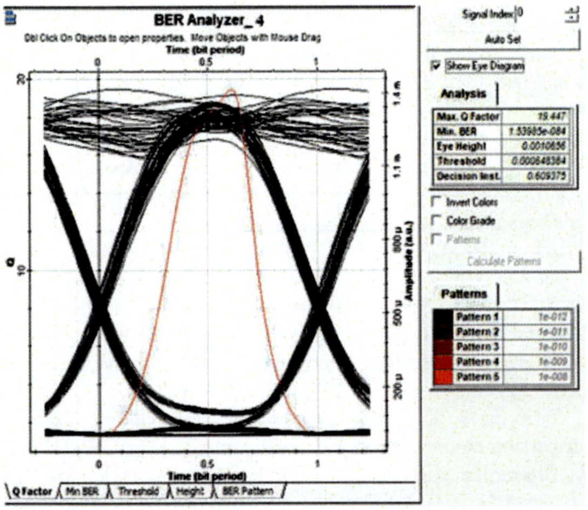

Fig. 11 Eye diagram of link-4

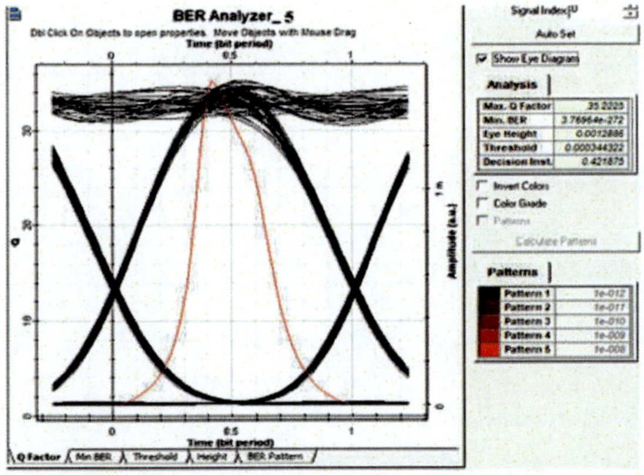

Fig. 12 Eye diagram of link-5

References

1. Mohammed, N. A., Solaiman, M., & Aly, M. H. (2014). Design and performance evaluation of a dispersion compensation unit using several chirping functions in a tanh apodized FBG and comparison with dispersion compensation fiber. *Applied Optics, 53*(9), 10.
2. Ahmed, R. K., & Mahmood H. A. (2017). Performance valuation of high data rate optical communication system utilizing FBG compensated Dispersion schemes under different Modulation techniques. *Diyala Journal of Engineering Sciences Iraq, 10*(02), 94–106.
3. Sharma, A., Singh, I., Bhattacharya, S., & Sharma, S. (2019). Performance comparison of DCF and FBG as dispersion compensation techniques at 100 Gbps over 120 km using SMF. In *Springer Nature Singapore Pte Ltd. 2019* V. Nath and J. K. Mandal (eds.), *Nanoelectronics, circuits and communication systems, Lecture notes in electrical engineering 511.*

4. Ashry, I., Elrashidi, A., Mahros, A., Alhaddad, M., & Elleithy, K. (2014). Investigating the performance of apodized Fiber bragg gratings for sensing applications. In *IEEE Proceedings of the Zone 1 Conference of the American Society for Engineering Education.*
5. Tosi, D. (2018). Review of chirped fiber bragg grating (CFBG) fiber-optic sensors and their applications. *Sensors, 18,* 2147.
6. Pal, B. P., & Pande, K. (2002). Optimization of a dual-core dispersion slope compensating fiber for DWDM transmission in the 1480–1610 nm band through G.652 single-mode fibers. *Elseiver, Optics Communications, 201,* 335–344.
7. Hussein, T. F., Rizk, M. R. M., Aly, M. H. (2019). A hybrid DCF/FBG scheme for dispersion compensation over a 300 km SMF. In *Optical and quantum electronics,* Springer Science + Business Media, LLC, part of Springer Nature.
8. Khan, S. S. A., & Islam, M. S. (2012). Determination of the best apodization function and grating length of linearly chirped fiber bragg grating for dispersion Compensation. *Journal of communications, 7*(11).
9. Meena, M. L., & Gupta, R. K. (2019). Design and comparative performance evaluation of chirped FBG dispersion compensation with DCF technique for DWDM optical transmission systems. *Optik—International Journal for Light and Electron Optics, 188,* 212–224.
10. Spolitis, S., & Ivanovs, G. (2011). Extending the reach of DWDM-PON access network using chromatic dispersion compensation. IEEE, pp. 29–33.

Performance Comparison of Adaptive Mobility Management Scheme with IEEE 802.11s to Handle Internet Traffic

Abhishek Majumder and Sudipta Roy

Abstract Wireless Mesh Network (WMN) is one type of mobile ad hoc network. So, if there is a link breakage, it can heal and reorganize the network by itself. Portals, Mesh Gates and Access Points, Stations (STAs) and external station (STA) are main components of WMN. Portals are used to connect the network to the internet. STAs are the nodes through which routing of packets take place. External STAs are the users of WMN. Mesh Gates and Access Points are the stations which route the packets and also act as Access Points for the external STAs. In case of WMN, IEEE 802.11 s is the standard. When the external STA moves in the WMN, it goes out of coverage of a Mesh Gate and Access Point and enters into the coverage of another one. Hybrid Wireless Mesh Protocol (HWMP), routing protocol of IEEE 802.11s, does not support external Station's (STA's) mobility. To integrate mobility, Adaptive mobility management technique had been proposed. The scheme considers session and mobility activities of the external STA by taking into account its session-to-mobility ratio (SMR). Based on SMR value, it makes the decision about transmission of route management packets. This paper presents numerical analysis of adaptive mobility management scheme and HWMP. Simulation of both techniques are carried out using NS-2. Performance comparison shows that adaptive mobility management scheme outperforms HWMP. Moreover, a comparison between the numerical analysis and simulation results have also been performed. It has been observed that results in case of HWMP are very close.

Keywords Portal · Mesh station · Mesh gateway · IEEE 802.11s · External station

---

A. Majumder (✉)
Department of Computer Science & Engineering, Tripura Universiy, Suryamaninagar, Tripura, India

S. Roy
Department of Computer Science & Engineering, Assam University, Silchar, Assam, India

P. Nayak et al. (eds.), *IoT and Analytics for Sensor Networks*, Lecture Notes in Networks and Systems 244, https://doi.org/10.1007/978-981-16-2919-8_8

1 Introduction

Portal, Mesh station (STA), external station (STA), Mesh Gate and Access Point form Wireless Mesh Network (WMN) [1–5]. Portals are the mesh nodes that connect the WMN to the internet. Mesh STAs receive and send packets. They also relay packets of other nodes. Mesh gates and access points (APMG) are the nodes which serve network connection to clients. External STAs are the client nodes to whom network services of WMN are provided. The internet packets are sent and received to and from the remote nodes.

In WMN, the users using external STA are free to move. Therefore, handoff process is performed. During this process, the online activities of external STA gets interrupted. So, it is necessary to have seamless mobility management. It is desirable to have minimum route management packet propagation for better performance. Reduced routing overhead will result in enhanced performance.

For IEEE 802.11s [6–10] the routing protocol is Hybrid Wireless Mesh Protocol (HWMP) [9, 11–13]. Uplink packets are transmitted by external STA towards APMG which are subsequently forwarded towards portal. APMG receives downstream internet packets from portal and forwards them towards external STA. Mobility is not supported in IEEE 802.11s [14]. Mobility [7, 14] management becomes an important challenge with increase in external STA's number and mobility. For mitigating the challenge, Adaptive mobility management technique [15] had been designed. In this work, adaptive mobility management technique and IEEE 802.11 s are numerically analyzed considering only internet traffic. Simulation results of both the schemes are also presented under varying data rate and node speed.

In this paper, Sect. 2 presents working of HWMP. In Sect. 3, working of adaptive mobility management scheme is presented. Section 4 discusses numerical analysis of both schemes. Section 5, discusses the simulation. In Sect. 6, comparison of results of simulation and numerical analysis is discussed. Conclusion has been presented in Sect. 7.

2 Hybrid Wireless Mesh Protocol

The scheme [9, 12, 13] utilizes the following message: path error (PERR), path request (PREQ), route announcement (RANN), path reply (PREP), proxy update confirmation (PXUC), and proxy update (PXU). Portal advertises its presence through broadcast of RANN. PREQ is broadcasted by source mesh node for finding route to destination. The destination in response sends back PREP message. For updating proxy information the mesh gate transmits PXU message. In response, the PXUC is sent back. PERR is broadcasted to propagate the information that some route is broken.

In proactive RANN mode following steps are carried out during handoff process:

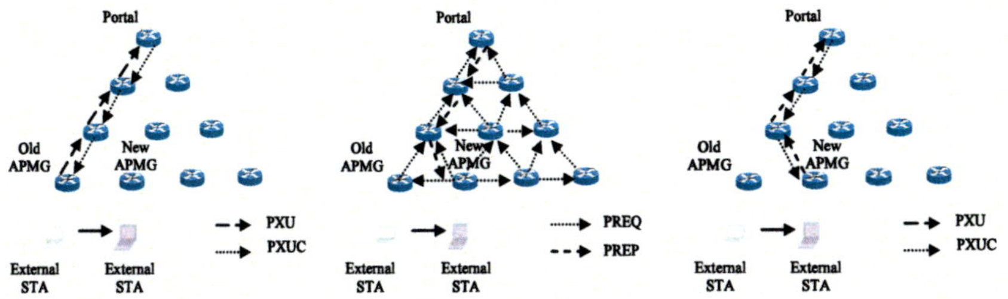

Fig. 1 **a** Previous APMG sending PXU to portal. **b** Process for setting up route when no route present towards destination. **c** Process for setting up route when route is present to the destination

i. Old APMG transmits PXU towards portal. Portal removes external STA's old proxy table entry and transmits back PXUC. Figure 1a shows the operation.
ii. When handoff occurs, external STA searches the route towards destination. If no route is found, PREQ is broadcasted. If portal receives PREQ, PREP is sent back. Figure 1b shows the process. If it has a route, PXU is transmitted to the portal. In response, PXUC is sent back towards source APMG. Figure 1c shows the process.

3 Adaptive Mobility Management Technique

Adaptive mobility management technique [15, 16] is presented in this section. It has been designed for handling internet traffic only. Routing and mobility management are its major components.

If handoff occurs, the external STA computes session-to-mobility ratio (SMR) [17] value. Then it sends SMR value with re-association message. When SMR value (SMR_{ESTA}) is received by APMG, it compares the value with threshold δth.

- If $SMR_{APMG} \leq \delta th$, new external STA's entry will be created in routing table of the new APMG. PREQ will be sent towards old APMG. When it receives PREQ, it updates external STA's routing table and next hop will be set to new APMG's address.
- If $SMR_{APMG} > \delta th$, in current APMG's proxy table external STA's information will be appended. PXU will be transmitted by new APMG towards portal. Portal updates its proxy table if it receives PXU message. So, forward chain gets reset.

Since RANN is periodically transmitted by the portal each APMG has route towards portal. Upstream packet is sent towards serving APMG by source external STA. It forwards packet to the portal. Downlink packet is received by portal. After that, it forwards received packet towards serving APMG of external STA. When packet is received, in proxy table of serving APMG, search is carried out for external STA's entry. If it is present, packet is transmitted towards external STA. Else, packet will be transmitted towards APMG which is next hop to reach external STA.

Fig. 2 Example scenario when $SMR_{APMG} \geq \delta th$

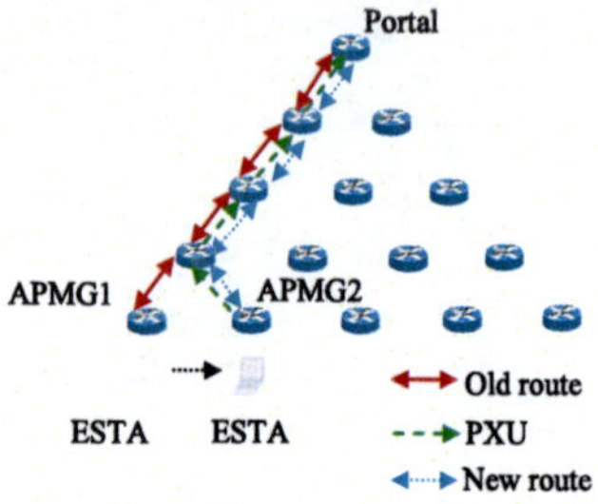

Fig. 3 Example scenario when $SMR_{APMG} < \delta th$

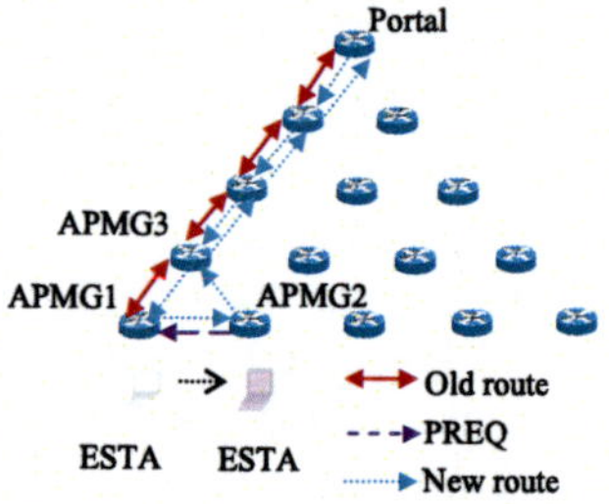

Figure 2 shows example of routing and mobility management. Let external STA (ESTA) moves from the coverage of APMG1 to APMG2. ESTA sends its current SMR value to APMG2 with re-association message. Let, external STA's SMR is greater than δth. So, APMG2 sends PXU message to portal. Portal updates proxy information of source eternal STA. Therefore, a path is established between source ESTA and the portal.

Figure 3 shows another scenario. This scenario is similar to the previous one. Only difference is that, here, the value of SMR is less than δth. When handoff takes place, APMG2 sends PREQ to APMG1. ESTA's entry in routing table of APMG1 will be updated and APMG2 is set as next hop. Upstream packets of internet are transmitted by APMG2 via APMG3. Portal sends downstream internet packets towards APMG1. It transmits the packet towards APMG2.

4 Numerical Analysis

Let, exponential distribution with rate λ_s [18] is followed by external STA's subnet residence time t_s. It is also considered that the inter session departure and arrival time follows exponential distribution having rates λ_d and λ_a, respectively [19]. Let t_a and t_b be the inter session arrival and departure time. Let the number of external STA is M. Between portal and arbitrary APMG, average hop count distance is considered as α. Let, γ is communication latency per hop. Value of threshold SMR is considered as δth. Let percentage of downstream internet packet is r_{inter}. In a session, average packet number per time unit is considered as λ_p [17].

4.1 Handoff Cost

In adaptive mobility management scheme, SMR is calculated by external STA during handoff. SMR is calculated as [17]

$$\mathrm{SMR}_{\mathrm{APMG}} = \frac{t_s}{t_a + t_d} \tag{1}$$

Therefore, probability of $\mathrm{SMR}_{\mathrm{APMG}}$ be less than δ_{th} can be calculated as [17],

$$\begin{aligned} P_{th} &= P(\mathrm{SMR}_{\mathrm{APMG}} < \delta\mathrm{th}) \\ &= \begin{cases} \dfrac{\lambda_a\lambda_d\lambda_s\delta\mathrm{th}}{\lambda_d - \lambda_a} \times \left[\dfrac{1}{\lambda_a(\delta\mathrm{th}\lambda_s + \lambda_a)} - \dfrac{1}{\lambda_d(\delta\mathrm{th}\lambda_s + \lambda_d)}\right] if \lambda_a \neq \lambda_d \\ 1 - \dfrac{\lambda_s^2}{(\lambda_s\delta\mathrm{th} + \lambda)^2} if \lambda_a = \lambda_d = \lambda \end{cases} \end{aligned} \tag{2}$$

In adaptive mobility management technique PXU is sent towards portal if external STA's SMR value becomes greater than or equal to δth. Otherwise, PREQ is sent towards old APMG by new APMG. Therefore, cost for this process is

$$\begin{aligned} C_{\mathrm{hSMRnewAPMG}} &= \gamma \text{if } \mathrm{SMR}_{\mathrm{APMG}} < \delta\mathrm{th} \\ &= \alpha \times \gamma \text{ if } \mathrm{SMR}_{\mathrm{APMG}} \geq \delta\mathrm{th} \end{aligned} \tag{3}$$

For this process number of messages transferred is

$$\begin{aligned} CM_{\mathrm{hSMRnewAPMG}} &= 1 \text{ if } \mathrm{SMR}_{\mathrm{APMG}} < \delta\mathrm{th} \\ &= \alpha \text{ if } \mathrm{SMR}_{\mathrm{APMG}} \geq \delta\mathrm{th} \end{aligned} \tag{4}$$

Therefore, per handoff cost is computed as

$$\begin{aligned} C_{\mathrm{hSMR}} &= \gamma \times P(\mathrm{SMR}_{\mathrm{APMG}} < \delta\mathrm{th}) + \alpha \times \gamma \times P(\mathrm{SMR}_{\mathrm{APMG}} \geq \delta\mathrm{th}) \\ &= \gamma \times P\mathrm{th} + \alpha \times \gamma \times (1 - P\mathrm{th}) \end{aligned} \tag{5}$$

Number of messages transmitted per handoff is

$$\begin{aligned} CM_{\mathrm{hSMR}} &= 1 \times P(\mathrm{SMR}_{\mathrm{APMG}} < \delta\mathrm{th}) + \alpha \times P(\mathrm{SMR}_{\mathrm{APMG}} \geq \delta\mathrm{th}) \\ &= P\mathrm{th} + \alpha \times (1 - P\mathrm{th}) \end{aligned} \tag{6}$$

In this work, internet traffic is considered only. So, in HWMP during handoff, PXU is sent to portal by old APMG. The portal responds by sending PXUC to old APMG. The cost for PXU and PXUC message is [20]

$$C_{\text{hHWMPoldAPMG}} = 2 \times \alpha \times \gamma \quad (7)$$

The number of message transferred during the procedure is

$$CM_{\text{hHWMPoldAPMG}} = 2\alpha \quad (8)$$

The new APMG transmits PXU toward portal when there is proxy information of destination or route to destination. PXUC is transmitted by portal in response. Otherwise, PREQ is broadcasted by APMG for finding route towards destination. Intermediate APMG rebroadcasts PREQ message. Let the probability of having proxy entry or route to the destination node in the new APMG is P_d. Therefore, cost incurred for transmission and reception of messages to and from the new APMG is

$$C_{\text{hHWMPnewAPMG}} = \{(1 - P_d) \times (M + \alpha) + P_d \times 2 \times \alpha\} \times \gamma \quad (9)$$

Number of messages transferred during this process is

$$C_{\text{hHWMPnewAPMG}} = (1 - P_d) \times (M + \alpha) + P_d \times 2 \times \alpha \quad (10)$$

So, per handoff cost can be calculated as

$$C_{\text{hHWMP}} = (C_{\text{hHWMPoldAPMG}} + C_{\text{hHWMPnewAPMG}}) \quad (11)$$

In case of HWMP, number of messages transferred during each handoff is

$$CM_{\text{hHWMP}} = (CM_{\text{hHWMPoldAPMG}} + CM_{\text{hHWMPnewAPMG}}) \quad (12)$$

4.2 *Packet Delivery Cost*

With respect of external STA's serving APMG, let its average displacement per APMG association be c_{move} [16]. For downlink internet, traffic packet delivery cost is [16]

$$C_{\text{pdSMR}} = \alpha \times \gamma + \gamma \times c_{\text{move}} \times Pth \times \left[\frac{1 - P_{\text{th}}^{\left(\frac{M-1}{c_{\text{move}}}\right)}}{1 - Pth} \right] \quad (13)$$

Uplink internet traffic is sent to portal without tunneling. So, the cost of packet delivery for uplink traffic is computed as [16]

$$C_{puSMR} = \alpha \times \gamma \tag{14}$$

So, per second packet delivery cost is

$$C_{pSMR} = \left\{ \frac{r_{inter}}{100} \times C_{pdSMR} + \left(1 - \frac{r_{inter}}{100}\right) \times C_{puSMR} \right\} \times (\lambda_a + \lambda_d) \times \lambda_p \tag{15}$$

In case of HWMP, downlink internet packet is directly transferred to external STA's current APMG. Similarly, uplink packets are transferred directly towards the portal by external STA's current APMG. So, in case of uplink and downlink traffic, the cost for packet delivery is computed as

$$C_{puHWMP} = C_{pdHWMP} = \alpha \times \gamma \tag{16}$$

So, cost per second for packet delivery is computed as

$$C_{pHWMP} = \alpha \times \gamma \times (\lambda_a + \lambda_d) \times \lambda_p \tag{17}$$

5 Comparison and Simulation

Simulation results have been presented in this section. NS-2 [21] is used for carrying out simulation. Random waypoint mobility model is used as mobility model of external STA. MAC layer uses IEEE 802.11. 10 times simulation has been repeated. Figure 4 shows the topology.

δth is set to 0.5. Packet size is considered as 258. Node 1 is the portal of the WMN shown in Fig. 4. CBR traffic is considered. External STA's pause time is 0 s. External

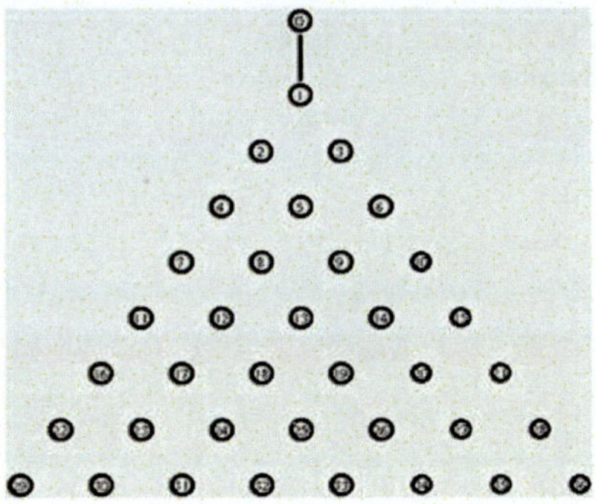

Fig. 4 Mesh STAs' topology used

STA's number in the topology is 36. 40 m has been set as two neighboring mesh STA's distance. Radio range of each of the mesh STA is 41.04 m. The simulation has been carried out for 100 s. During simulation, the maximum number of connections at a time is 10.

5.1 *Impact of Network Load*

External STA's maximum speed is considered as 10 m/s. Data rate is considered as 32, 64, 128, 256, 384, 512, 640, 768, 896, and 1024 kbps. External STA's number has been set as 10.

Figure 5 shows data rate's impact on throughput of IEEE 802.11s as well as adaptive mobility management technique. According to the figure, both scheme's throughput initially increases with rise in data rate, but reduces in later part. Through network, more data packets are sent, as data rate increases. Therefore, at initial part, as data rate increases, there is an increase in throughput also. If data rate is increased further, network's congestion also increases and throughput of network decreases. Since large quantity of PXUC, PXU, and PREQ messages are transmitted in IEEE 802.11s, congestion will be more. Therefore, as data rate increases throughput of IEEE 802.11s will increase at lower rate compared to adaptive mobility management scheme.

Data rate's impact upon normalized overhead for routing is shown in Fig. 6. As data rate increases, network gets congested. So, route management packets' retransmission will increase. In IEEE 802.11s, PPREQ, PXUC, and PPREQ packets are transmitted in large number. Therefore, retransmission of great quantity of route management packets also takes place. So, in IEEE 802.11s, since the data rate increases, normalized routing overhead will also be raised. Number of PXU and PREQ packets transmitted is limited in adaptive mobility management technique.

It can be seen from Fig. 7 that for IEEE 802.11s, as rate of data increases, the average cost of handoff also increases. In adaptive mobility management technique, route management packets' retransmission is limited. So, average handoff cost

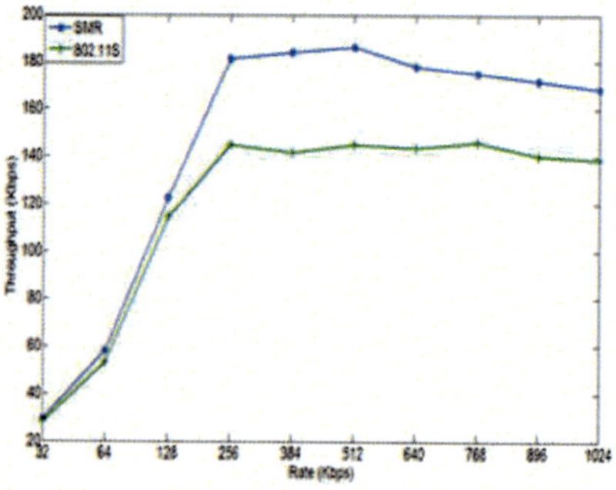

Fig. 5 Throughput versus data rate

Fig. 6 Normalized routing overhead versus data rate

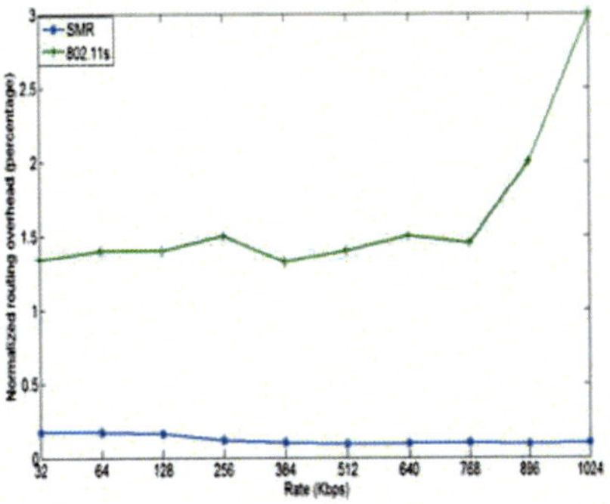

Fig. 7 Avg. handoff cost versus rate of data

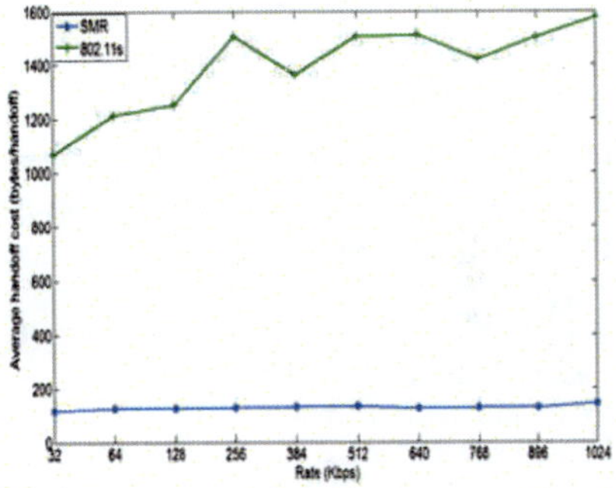

continues to be steady as data rate increases average. It can be seen from Fig. 8 that in IEEE 802.11s as rate of data rises, the average end-to-end delay increases. Reason behind this is the congestion caused by increasing rate of data and retransmission of great quantity of route management packets. In adaptive mobility management technique, route management packets are less propagated although the congestion increases as rate of data increases. So, average end-to-end delay also becomes higher as data rate increases. The increase is less in comparison with IEEE 802.11s.

Fig. 8 Avg. end-to-end delay versus rate of data

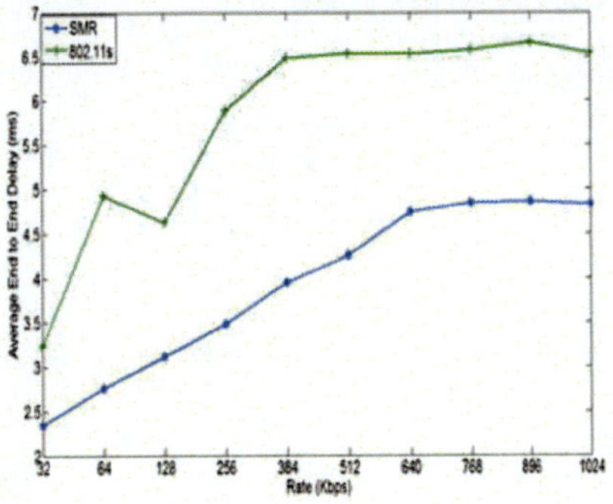

5.2 *Impact of External STA Speed*

The section discusses the impact of speed of external STA upon performance of adaptive mobility management technique and HWMP. 512 kbps is set as data rate. 10 has been set as external STA's number. External STA's maximum speed is set between 10 and 100 m/sec.

Figure 9 shows impact of node speed on normalized routing overhead. Handoff occurs more frequently if external STA's speed becomes high. Therefore, more PXUC, PXU, and PREQ messages are transmitted in IEEE 802.11s as external STA's speed increases. It leads to rise in normalized routing overhead. In the adaptive mobility management technique, there is very limited propagation of PXU and PREQ messages.

Figure 10 presents impact of node speed upon average cost for handoff. In IEEE 802.11s transmission of great quantity of management packets occur when handover takes place frequently due to increase in external STA's speed. This causes congestion and route management packets' retransmission in the network. Therefore, as external STAs' maximum speed increase, average cost for handoff also increases. In adaptive mobility management scheme, route management packet's propagation is very less.

Fig. 9 Normalized routing overhead versus external STA's maximum speed

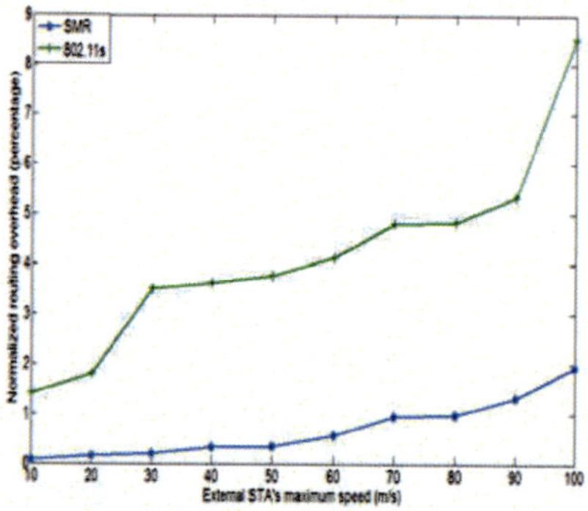

Fig. 10 Average handoff cost versus external STA's maximum speed

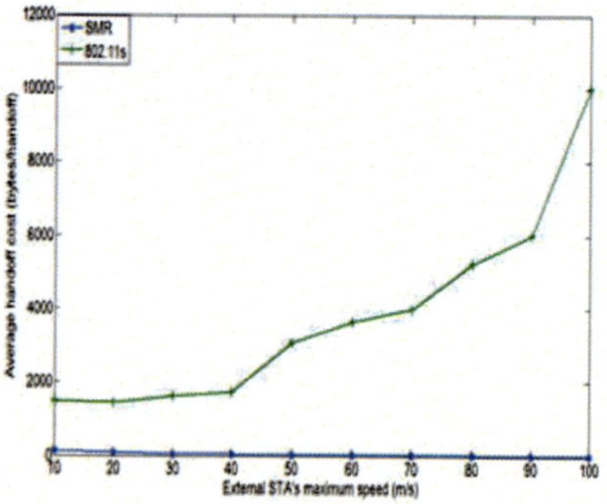

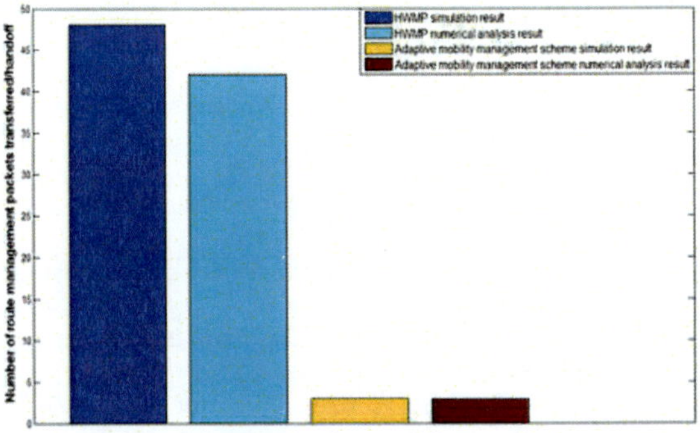

Fig. 11 Number of route management packets transmitted per handoff obtained from simulation and numerical analysis

This results in less congestion. So, with rise in external STA's maximum speed handoff cost remains stable.

6 Comparison of Numerical Results and Simulation

Figure 11 depicts comparison of numerical analysis and simulation results. Considering topology given in Fig. 4 the value of M and α are considered as 36 and 4, respectively. The value of λ_s, λ_a, λ_d, δth, and P_d have been considered as, 0.025882, 0.02, 0.02, 0.5, and 0.2, respectively. For minimizing the effect of congestion, data rate and maximum node speed have been set as 512 Kbps and 10 m/s, respectively. External STA's number is considered as 10. In SMR based scheme, results are very close. In HWMP, results are not that close. The main reason is packet dropping and congestion.

7 Conclusion and Future Work

Comparative study has been carried out in this paper between adaptive mobility management technique and IEEE 802.11s. Both the schemes are simulated using NS-2 and numerically analyzed. The schemes have been simulated in NS-2. Simulation results have been compared considering average handoff cost, normalized routing overhead, avg. end-to-end delay and throughput with respect to rate of data and external STA's speed. From the comparison, it was observed that the adaptive mobility management technique outperforms IEEE 802.11s. A comparison of results of numerical analysis and simulation was carried out. It has been found that in case of adaptive mobility management scheme, results obtained from simulation and numerical are very close. In HWMP, the results are not that close. In future comparison of adaptive mobility management technique will be carried out with other baselines considering more parameters.

References

1. Zachos, C. K., & Loo, J. (2016). Wireless mesh network: design, modeling, simulation, and analysis. In J. Loo, J. L. Mauri, J. H. Urtiz (eds.), *Mobile Ad Hoc networks: Current status and future trends*, CRC Press, pp. 449–466.
2. Akyildiz, I. F., Wang, X., & Wang, W. (2005). Wireless mesh networks: A survey. *Computer Networks, 47*(4), 445–487.
3. Akyildiz, F. (2005). A survey on wireless mesh networks. *IEEE Communication Magazine, 43*(9), S23–S30.
4. Zhang, Y., Luo, J., & Hu, H. (2006). *Wireless mesh networking architectures*. Protocols and standards: Auerbach Publications.
5. Benyamina, D., Hafid, A. S., & Gendreau, M. (2012). Wireless mesh networks design—A survey. *IEEE Communications Surveys and Tutorials, 14*(2), 299–310.
6. Hiertz, G. R., Denteneer, D., Max, S., Taori, R., Cardona, J., & Berlemann, L., et al. (2010). IEEE 802.11s: The WLAN mesh standard. *IEEE Wireless Communication, 17*(1), 104–111.
7. IEEE Std 802.11-2012. (2010). Information technology—Telecommunications and information exchange between systems—Local and metropolitan area networks—Specific requirements—Part 11: wireless LAN medium access control (MAC) and physical layer (PHY) specifications.
8. IEEE P802.11s/D1.07-2007. (2007). Information technology—Telecommunications and information exchange between systems—Local and metropolitan area networks—Specific requirements—Part 11: Wireless Medium Access Control (MAC) and Physical Layer (PHY) Specifications.
9. Camp, J. D., & Knightly, E. W. (2008). The IEEE 802.11s extended service set mesh networking standard. *IEEE Communications Magazine, 46*(8), 120–126.
10. Wang, X., & Lim, A. O. (2008). IEEE 802.11s wireless mesh networks: Framework and challenges. Ad Hoc *Networks, 6*(6), 970–984.
11. Sooriyaarachchi, S. J., Fernando, W. A. C., & Gamage, C. D. (2015). Evaluation of scalability of Hybrid Wireless Mesh Protocol in IEEE 802.11. In *Proceedings of the Fifteenth International Conference on Advances in ICT for Emerging Regions (ICTer)* (pp. 152–159), Colombo, Sri Lanka.
12. Bari, S. M. S., Anwar, F., & Masud, M. H. (2012). Performance study of hybrid wireless mesh protocol (HWMP) for IEEE 802.11 s WLAN mesh networks. In *Proceedings of the International Conference on Computer and Communication Engineering* (pp. 712–716), Kuala Lumpur, Malaysia.
13. Bahr, M. (2006). Proposed routing for IEEE 802.11s WLAN mesh networks. In *Proceedings of the 2nd Annual International Workshop on Wireless Internet* (p. 5), Boston, USA.
14. Sampaio, S., Soutp, P., & Vasques, F. (2015). A review of scalability and topological stability issues in IEEE 802.11s wireless mesh networks deployments. *International Journal of Communication Systems, 29*(4), 671–693.
15. Majumder, A., & Roy, S. (2020). Implementation of adaptive mobility management technique for Wireless Mesh Network to Handle Internet Packets. In *Proceedings of the 2nd International Conference on Communication, Devices and Computing* (pp 97–107), Haldia, India.
16. Majumder, A., Roy, S., & Dhar, K. K., (2013). Design and analysis of an adaptive mobility management scheme for handling internet traffic in wireless mesh network. In *Proceedings of the Annual International Conference on Emerging Research Areas* (pp. 1–6), Kanjirapally, India.
17. Majumder, A., & Roy, S. (2013). Design and analysis of a dynamic mobility management scheme for wireless mesh network. *Scientific World Journal, Article ID, 656259*, 1–16.
18. Pack, S., Jung, H., Kwon, T., & Choi, Y. (2005). SNC: a selective neighbor caching scheme for fast handoff in ieee 802.11 wireless networks. *ACM SIGMOBILE Mobile Computing and Communications Review, 9*(4), 39–49.

19. Kwon, S. J., Nam, S. Y., Hwang, H. Y., & Sung, D. K. (2004). Analysis of a mobility management scheme considering battery power conservation in IP-based mobile networks. *IEEE Transactions on Vehicular Technology, 53*(6), 1882–1890.
20. Majumder, A., & Roy, S. (2015). Implementation of forward pointer-based routing scheme for wireless mesh network. *Arabian Journal for Science and Engineering, 41*(3), 1109–1127.
21. Fall, K., & Varadhan, K. (2007). The network simulator (ns-2). http://www.isi.edu/nsnam/ns.

Automatic Attendance Management System Using Face Detection and Face Recognition

M. Varsha and S. Chitra Nair

Abstract Attendance plays a major role in every education system. Taking attendance of students manually can be a great burden for teachers. It may cause many problems like loss of time, repetition, incorrect markings, and difficulties in marking them. To avoid this, there is a need to design an automatic system that overcomes the issues with the traditional attendance system. There are many automatic methods available for this purpose like fingerprint systems, RFID systems, face recognition systems and iris recognition systems, etc. But among these Face Recognition proved to be more efficient. The main objective of this paper is to propose a model that captures images from videos, detect and recognize the faces, predict the recognized face, and then marks attendance. In this work, a basic step has been performed which uses fifteen classes from LFW Dataset and faces are detected, recognized, and then prediction is done on a randomly selected image from the used dataset. This system uses a combination of Multi Task Cascaded Neural Network (MTCNN) algorithm along with FaceNet that can be used to detect faces and extract facial features from images. SVM is used to predict the face of the person from the image. The proposed system obtains a classification accuracy of 99.177% for the training set and 100% on the test set. The accuracy, precision, recall, and F1-score are computed.

Keywords Face detection · Face recognition · MTCNN · FaceNet · SVM

1 Introduction

Students attendance plays a crucial role in every education system. When attendance is taken manually, it should be marked in a sheet which causes time loss, repetition, and incorrect marking. This may cause a great burden for teachers. So, there is a requirement for an automatic attendance management system. There are various automatic techniques available now like Face Recognition, Finger-print verification, RFID systems, Iris Recognition systems, Voice and Speech recognition systems [1, 2]. Among this, face recognition proved to be more efficient [2]. This is because in

M. Varsha (✉) · S. Chitra Nair
NSS College of Engineering, Palakkad, India

P. Nayak et al. (eds.), *IoT and Analytics for Sensor Networks*, Lecture Notes in Networks and Systems 244, https://doi.org/10.1007/978-981-16-2919-8_9

the case of the fingerprint verification system, the students have to form a queue to scan their thumb on the scanning device to mark their attendance [3], which causes time loss. In the case of RFID systems, the students will have to carry RFID cards, which is a weakness. There is a chance for the students to forget to take the RFID cards [4]. In the case of iris recognition systems, even though this method has very high accuracy, it is not practically possible. This is because there will be many pupils in the class and iris detection of each student is not feasible [3]. The major drawback of the voice/speech recognition system is that the speech features are sensitive to certain factors like background noise, and also the voice change that happens with age. These systems are not reliable and also may not work accurately if the students are suffering from throat infection [1].

Face detection is a technique of locating and localizing the faces in an image by creating a bounding box through the extent. A facial recognition is a technology that identifies and verifies a person from images or video frames. It is also specified as a Biometric Artificial Intelligence-based application. This is because, it will uniquely recognize a person by analyzing their facial features. There are many methods in which facial recognition works, but mainly they work by comparing the facial features with the known faces in a database to find a match.

In this work, an automatic system is developed using image processing techniques that can mark attendance automatically. An effective face recognition algorithm is proposed which can identify students efficiently. For image processing, an effective platform is used. The objective of this work is to propose a model that captures images from videos, detect and recognize the faces, predict the recognized face, and then marks attendance. This paper aims to build an automatic attendance management system that detects the faces of students from images using MTCNN, extracts the facial features using FaceNet model and classifies using SVM. After recognizing the faces, the attendance is marked. Initially, we worked on LFW dataset and the future work is to work on the image captured on real-time videos. The Sect. 2 of this paper presents related work, Sect. 3 presents motivation, Sect. 4 demonstrates the proposed system, Sect. 5 shows the Experimental results, and Sect. 6 shows the conclusion and future work.

2 Related Works

Over the last few decades, the research area is mostly focused on face detection and recognition. This section provides a detailed overview of the different methods used for implementing automatic attendance management systems.

2.1 Attendance System Using Image Processing Techniques

The idea behind this work [5], was to develop an automatic system which will handle the attendance of students using image processing techniques. Their aim was to develop a system that detects and recognizes the students faces using frames from videos and then recording their attendance by identifying them from their variant features. For this, they used Viola Jones algorithm and Fisher Face algorithm for detecting and recognizing faces, respectively. The proposed system achieved an accuracy of 45–50%. They were able to overcome the drawback of manual attendance systems. The efficiency of this system can be increased by improving the training process.

2.2 Attendance System Using Face Detection and Face Recognition

In this work [6], they used Haar filtered Adaboost for detecting faces and Principal Component Analysis (PCA) and Local Binary Pattern Histogram (LBPH) algorithms for identifying the detected faces. This system proposed an approach that provides better results than traditional attendance systems and other automatic attendance systems like biometric fingerprint and RFID systems. They obtained an accuracy in the range 75–90% during the experiments.

2.3 Attendance System Using Deep Learning Framework

In this work [7], they proposed a system using deep learning frameworks. They have used a state-of-the-art face detection model and a novel recognition architecture for detecting and recognizing faces respectively. CNN was used to develop the automated attendance system. They proposed a system that achieved an accuracy of 98.67% on LFW datasets and 100% on classroom datasets. They used a spatial transformer network to learn the alignment of faces which led to greater facial verification accuracy. They have mainly used the frontal face images of students.

2.4 Deep Learning Paradigm for Attendance Systems

In this work [8], they proposed a system using Convolutional Neural Networks. The captured frames were passed to the SRNet (Single Image Super-Resolution Network) for the image super resolution which results in SR images. The faces were detected from the SR image and cropped around the bounding boxes using MTCNN. Then

FaceNet was used for identifying the detected faces. They used RAISE and DIV2K for SRNet, VGGface2 for FaceNet and LFW and their own dataset for testing and validation. They produced an accuracy of 96.80% on LFW datasets.

2.5 *Attendance System Using FaceNet and SVM*

In this work [9], their objective was to get an improved accuracy for multi-face recognition. The proposed system uses FaceNet for feature extraction and SVM as classifier. They achieved an accuracy of 99.6% for multi-face recognition. The overall accuracy which they obtained by using CNN model was less than the accuracy which they obtained using FaceNet and SVM.

3 Motivation

In recent years, Image processing played a unique role in technological advancement as it deals with extracting useful information from a digital image. As the applications of these techniques are increasing day by day, a lot of research are being done in this field. Image capturing plays an important role in the educational field, robotics, the medical field, and in smart phones. A classroom consists of a huge number of students and marking their attendance manually wastes a lot of time. So, there is a need for an automatic system that can be built using current technology. Face recognition, which is an application derived from image processing, is one of the best methods used for human detection. It helps in detecting the face of each student. Face is multi-dimensional structure and it requires good computational analysis for recognition.

4 Proposed System

An automatic attendance management system using face detection and recognition are proposed in this paper. The proposed system aims to take the images of the students from classroom videos, to detect and identify their faces from images and upon successful recognition to mark their attendance automatically. Figure 1 illustrates the model representation of the proposed system. The system takes real-time classroom videos as input. These videos are captured and frames are created in the pre-processing phase. The faces of students are detected from the frames. After detection of a face, the system crops the face and performs certain image processing techniques. Then the features are extracted from the detected faces. Lastly, the features of the detected faces and the test faces are compared in the classifier and the faces are recognized. Then the prediction is generated.

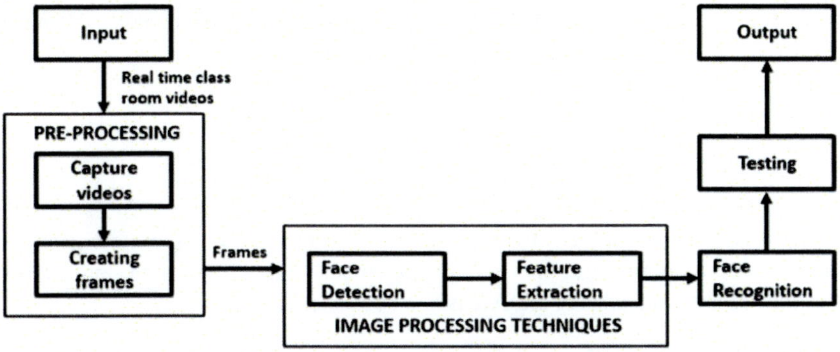

Fig. 1 Model representation

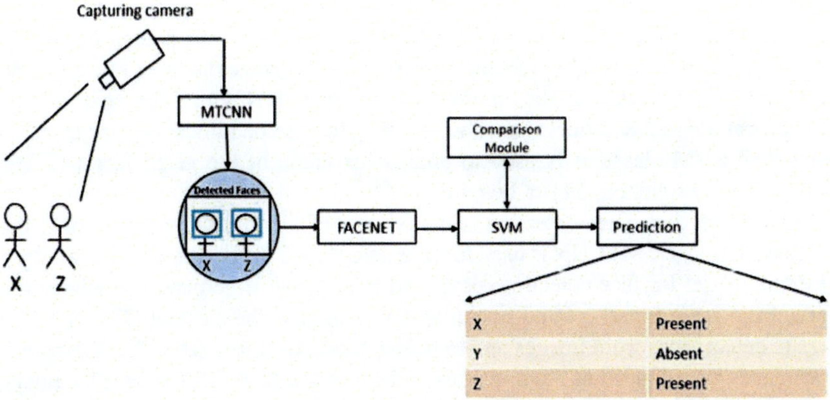

Fig. 2 System design

The proposed system requires a camera mount to a surface in the classroom at a point where it could capture the images of all the students in the classroom. The Fig. 2 illustrates the detailed system design of the proposed model. Here, MTCNN is used for face detection [8, 10]. FaceNet is used for creating face embeddings for a given image and SVM is used as a classifier for classification [9, 10]. The working of the proposed model is explained in brief below.

4.1 Input

The input to the model is real-time classroom videos of the students. Camera should be placed in the classroom in such a way that it captures all the students and their faces effectively. This camera needs to be interfaced to the computer system for further processing either through a wired or a wireless network.

4.2 Image Pre-Processing

In this phase, frames are created from videos. For this work, a sample video was taken from the internet and the frames were created. The size of the sample video was 0.34 s. The frames were created as one frame per 0.07 s.

4.3 Image Processing

In this phase, the faces are detected from the images and face embeddings are created from the detected faces. These features are used to recognize the faces of the people to mark their attendance. In the proposed system, MTCNN is used for creating Face Detector and FaceNet model is used for creating face embeddings for each detected face. And SVM classifier is used to predict the name of the given face [4]. The sub-steps are explained in brief below:

Face Detection. Face detection is the process of identifying and extracting the faces from images. MTCNN is used for face detection. This is a state-of-the-art deep learning model for face detection, described in [10, 11]. It helped in localizing the faces from images and created bounding boxes around their extent. The first step was to detect faces from images and to reduce the dataset to a set of faces. For this, the images were loaded as Numpy array and converted to RGB if in case the image is black and white. Then the MTCNN Face Detector was created. This detected the faces from the image and new dataset was obtained [4]. Figure 3 depicts the reduced dataset. Here, the faces are detected from each image of one class and the dataset is reduced to a series of faces only. The images are resized to (160 × 160 × 3) which is the input shape of FaceNet. All the images in the train and test sets are loaded and the faces are extracted.

Feature Extraction. In this section, face embeddings are created. Face embeddings are vectors that represent the features extracted from the face [4]. These embeddings are then compared with the vectors created for other faces. The FaceNet model will create the face embeddings for a face [12]. This model will return the face embeddings of the train and test datasets. Each face embedding is comprised of 128 vectors.

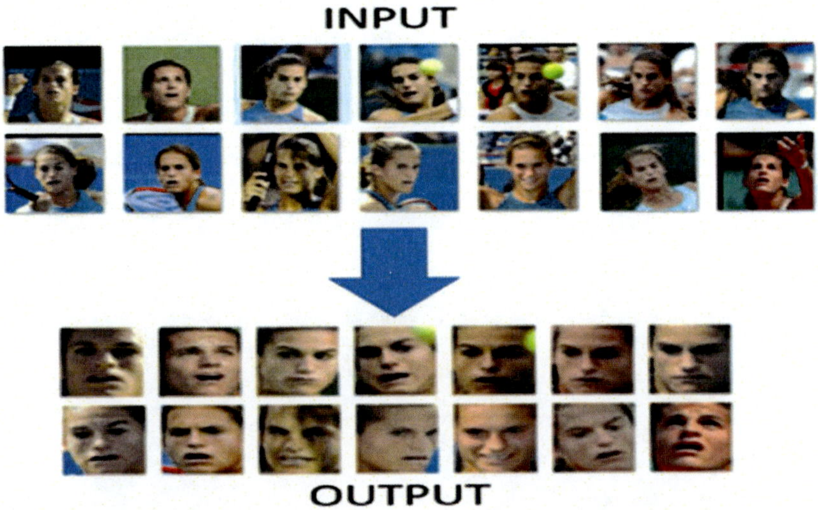

Fig. 3 Reduced dataset

4.4 Face Classification

The next step is to develop a model for classifying the face embeddings. First, the face embedding vectors are normalized. Here, SVM is used to work with face embedding. This is because SVM is efficient in separating the face embedding vectors [4]. Next, the model is evaluated and then the classification accuracy is calculated.

4.5 Testing

For testing, we selected a random image from the test dataset. The face embeddings are used as input to make predictions. The expected class name, predicted class name, and probability of the prediction is obtained.

5 Experimental Results

The dataset used here is LFW datasets [13]. For this work, 15 classes of image data were taken from the LFW datasets for the implementation of this system. The dataset was divided into two sets: Train and test sets. For training 80% of images were used and for testing 20% of the images were used. Table 1, describes the classification accuracy obtained with a different number of classes. This system produced a classification accuracy of 99.177% for training and 100% for testing

Table 1 Classification accuracy

LFW dataset	Train accuracy (%)	Test accuracy (%)
5 Classes	99.401	100
10 Classes	99.099	100
15 Classes	99.177	100

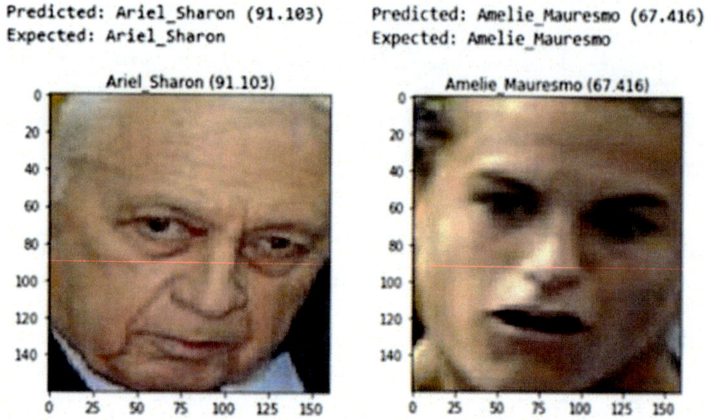

Fig. 4 Prediction done by SVM classifier [13]

with 15 classes. And in case of 10 classes, it provided an accuracy of 99.99% for training and 100% for testing.

FaceNet extracts high quality features from faces and creates facial embeddings. For prediction, a random image was selected from the test set. The system correctly recognized the faces from the images. The proposed system produced a good accuracy in recognizing the faces which were randomly taken from the test set [9]. The Fig. 4 depicts the face recognized from the random image taken from the test set. A plot of the selected face along with its expected name, predicted name, and probability is given. The accuracy, precision, recall, and F1-score of the system are also computed. The classification performance for the classes showing precision, recall and F1-score is shown in Table 2. This model provided an accuracy_score of 1.

6 Conclusion and Future Work

Automatic Attendance System has been proposed to reduce the drawbacks of traditional (manual) systems. The proposed system produced a good result with provided datasets. It saves time, especially during a lecture with a large strength of pupils. This attendance system illustrates the use of image processing techniques in a classroom. The system can be improved by implementing it in real time by capturing the images

Table 2 Classification performance

Classes	Precision	Recall	F1-score	Classes	Precision	Recall	F1-score
0	1.00	1.00	1.00	7	1.00	1.00	1.00
1	1.00	1.00	1.00	8	1.00	1.00	1.00
2	1.00	1.00	1.00	9	1.00	1.00	1.00
3	1.00	1.00	1.00	10	1.00	1.00	1.00
4	1.00	1.00	1.00	11	1.00	1.00	1.00
5	1.00	1.00	1.00	12	1.00	1.00	1.00
6	1.00	1.00	1.00	13	1.00	1.00	1.00

from real-time videos, detect faces from the created frames, recognize the faces, and then marks the attendance which is the future work.

Sentimental analysis and emotional analysis can also be implemented in the future. Sentimental analysis provides the feedback of a class which shows on which topics and at what time attracts more concentration of the pupils. Emotional analysis will help the faculties to acquire the feedback of their class and this helps them to change or improve their pedagogy. These developments can improve the applications of the work.

References

1. Jayant, N. K., & Borra, S. (2016). Attendance management system using hybrid face recognition techniques. In *2016 Conference on advances in signal processing (CASP)*, June 9, pp. 412–417.
2. Islam, M. S., Mahmud, A., Papeya, A. A., Onny, I. S., & Uddin, J. (2017). A combined feature extraction method for automated face recognition in classroom environment. In *International symposium on signal processing and intelligent recognition systems*, September, pp. 417–426.
3. Varadharajan, E., Dharani, R., Jeevitha, S., Kavinmathi, B., & Hemalatha, S. (2016). Automatic attendance management system using face detection. In *2016 Online international conference on green engineering and technologies (IC-GET)*, November 19, pp. 1–3.
4. Arsenovic, M., Sladojevic, S., Anderla, A., & Stefanovic, D. (2017). FaceTime—Deep learning based face recognition attendance system. In *2017 IEEE 15th International symposium on intelligent systems and informatics (SISY)*, September 14, pp. 000053–000058.
5. Hapani, S., Prabhu, N., Parakhiya, N., & Paghdal, M. (2018). Automated attendance system using image processing. In *2018 Fourth International Conference on Computing Communication Control and Automation (ICCUBEA)*, August 16, pp. 1–5.
6. Sanli, O., & Ilgen, B. (2018). Face detection and recognition for automatic attendance system. In *Proceedings of SAI intelligent systems conference*, September 6, pp. 237–245.
7. Sarkar, P. R., Mishra, D., & Subhramanyam, G. R. S. (2019). Automatic attendance system using deep learning framework. In *Machine intelligence and signal analysis*, pp. 335–346.
8. Gupta, R. K., Lakhlani, S., Khedawala, Z., Chudasama, V., Upla, K. P. (2018). A deep learning paradigm for automated face attendance. In *Workshop on computer vision applications*, December 18, pp. 39–50.
9. Nyein, T., Oo, A. N. (2019). University classroom attendance system using facenet and support vector machine. In *2019 International conference on advanced information technologies (ICAIT)*, November 6, pp. 171–176.

10. Brownlee, J. (2020). Machine learning algorithms in python, machine learning mastery. https://machinelearningmastery.com/machine-learning-with-python/. Accessed 8 June 2020.
11. Zhang, K., Zhang, Z., Li, Z., & Qiao, Y. (2016). Joint face detection and alignment using multitask cascaded convolutional networks. *IEEE signal processing letters, 23*(10), 1499–1503.
12. Schroff, F., Kalenichenko, D., & Philbin, J. (2015). Facenet: A unified embedding for face recognition and clustering. In *Proceedings of the IEEE conference on computer vision and pattern recognition;* pp. 815–823.
13. Huang, G. B., Mattar, M., Berg, T., Learned-Miller, E. (2008). Labeled faces in the wild: A database forstudying face recognition in unconstrained environments. In *Workshop on faces in 'Real-Life' Images: detection, alignment, and recognition,* October.

ESIT: An Enhanced Lightweight Algorithm for Secure Internet of Things

Manoja Kumar Nayak and Prasanta Kumar Swain

Abstract With the increasing use of sensors and intelligent devices, Internet of Things (IoT) becomes an important area of research to establish connectivity among connected devices. Traditional algorithms used for encryption are found to be highly complex and with higher number of rounds for encryption which is computationally expensive. In an IoT network, communicating nodes adapt fewer complex algorithms for maintaining security. Hence, lightweight security algorithms are used for this purpose. In this paper, we have proposed a method for encryption and decryption process called Enhanced Secure IoT (ESIT) of an image using shift (<< or > >) and bitwise binary modulo 2 $(+_2)$ operation for data transmission. It is a block cipher that accepts a 64-bit key which performs Left shift (<<) and $(+_2)$ function for encryption, and right shift (>>) and $(+_2)$ decryption. It is the normal bitwise left and right shift by removing the sequence of q-bits from the left and right side, respectively. The performance of the proposed method is established by correlation, entropy, and image histogram. Experimental evaluation on four available images clearly shows the advantages of the given approach.

Keywords IoT · Security · Encryption · Decryption

1 Introduction

The idea of IoT was coined by Kevin Ashton at Auto-ID center of MIT in 1999 [1]. In the twenty-first century, IoT becomes an important part in the lifestyle of the modern human-being to make their life smart and easy. It impacts on different fields of human life to make life smart, such as education, manufacturing automation, retail, financial technology, healthcare, governance, power utilities, agriculture, home, city, companies, and individuals. So, IoT has been a buzzword of the present days in communication technology due to intelligent computation. To provide services from Device to Devices (D to Ds), Devices to Device (Ds to D), Devices to Devices (Ds

M. K. Nayak · P. K. Swain (✉)
Department of Computer Application, Mharaja Sriram Chandra Bhanja Deo University, Baripada, India

P. Nayak et al. (eds.), *IoT and Analytics for Sensor Networks*, Lecture Notes in Networks and Systems 244, https://doi.org/10.1007/978-981-16-2919-8_10

to Ds) in an organization or to individual users, IoT employs millions of things like sensors, actuators, smart sensors, modern vehicles, bar-codes, smart cards, and RFID tags that are connected and providing benefits in various situations to exchange data among them in internet. These things will be responsible to gather and transfer related information to the gateway or cluster head or cloud potentially which are thousands of miles away [2]. The IoT allows "People Things" to be connected every time, every place with any things, and anyone using every services [3]. At present, smart military camp at eastern Ladhak is ready for Indian army people due to the technological innovation of IoT. As more and more people use the social media application and internet, there are number of challenges in IoT platform such as communication, management security, and monitoring. Among all these challenges, security is the top-most one. For sensitivity to IoT applications, it is desirable that devices are authenticated to ensure the data source. The sensible devices are known as resource constraint devices as they have low computation power, small size, less memory, low bandwidth, and limited battery life [4].

The use of tiny devices in IoT that are sensing heterogeneous access, data processing, services, and applications are prime parts in the platform. So IoT platforms are affected from information destruction level attacks such as interruption attack, eavesdropping attack, modification attack, fabrication, man-in-middle attack, and message Replay attack, and so on [5]. Due to above basis of attacks to information, the current state of research on key technology including encryption, secure communication, sensor data protection, and cryptographic procedure are the discussed challenges. To protect data from malicious event, it is a significant challenge for IoT technology. Nowadays, IoT is a smart technology, as peer-to-peer communication occurs efficiently due to the use of lightweight cryptographic algorithms that play a key role in achieving the security and privacy with constraint resources [6]. To reduce weakness of information transmission in IoT environment, the Enhanced lightweight cryptographic (ESIT) procedure is highly essential.

2 Related Works

A block cipher comprises of three sections. (1) Encryption process (2) Decryption process and (3) Key generation process. The large sub-keys are generated from the user secret key using the key generation algorithm. Through sub-keys, this algorithm allows all the secret key bits to impact encryption process in each round. Most secret key generation requires, once to serve both encryption and decryption process. Parameter of lightweight block ciphers which are found to be important are (1) block size (2) key length (3) operations (4) key scheduling (5) minimum number of rounds [7]. The lightweight ciphers are selected for resource constraint devices, and therefore, software and hardware implementation should be optimized. The major purpose of lightweight block cipher is to secure information confidentiality, integrity against security threat in IoT environment. By considering key parameters, such as size of block, key size, the architecture, and the number of rounds, we started an effort to include some of the light weight block ciphers as explained in the literature. A modern block cipher with a moderate code size requirement is proposed by Rijndael [8]. It is resistant against linear crypto analysis and differential attacks. Here, the key attack is possible when number of rounds is less than 9. With optimum parameters, it acts as an energy-efficient cipher. Authors propose an algorithm called Skipjack in [9], that is, the block cipher of 64-bit data operates on 80-bits key. Due to its shortest extended key and need of less memory and code, the space requirement is minimum. This uses energy efficiently than Rijndael, but it is found to be a more efficient cipher when speed is optimized. The fact is that it is not declared as secure by NSA. In 1999, Skipjack best attacked with lower than 32 rounds are known as Biham. RC5 is a cipher that supports both hardware and software implementation. It is represented as RC5-w/r/b where w denotes the word size, r denotes number of rounds, and b is the key length [10]. RC5 has poor energy efficiency due to Archille's heel of MSP430F149 used for multiplication and rotations. RC5 requires very small code memory. In 1998, it was attacked which breaks RC5-32/12/16 with just 244 chosen plain texts. Here, it is prescribed that the number of rounds must be greater than or equal to 18 for the security. Camellia is a Feistel structure cipher on 6 rounds that needs 13×28 plain text, with complexity of 2112 for cipher executions [11]. RC5 and RC6 are less energy-efficient than Camellia. This required large code size and more memory space. The attack to the Camellia is named as sequential attack. The resource constraint application for the lightweight algorithms like HEIGHT, TEA, KLEIN, and KATAN are executed on AVR Atmel. It is a tiny micro controller to assess memory efficiency, consumption of energy, and to evaluate level of confusion and diffusion for security investigation in [12]. Authors in [13] proposed an ultra-lightweight cipher named PRESENT which is a SP network having 31 rounds. The encryption algorithm DES is used based on 16 round Feistel cipher [14].

Due to key size of 56-bits, it is attacked by exhaustive attack that can break DES. So, DES is not secure today. 3-DES or Triple-DES is found to be more secure as it is of 48 rounds. In [15], the new block cipher SIT is a lightweight block cipher that uses 64-bits information with 64-bits key. To combine Shannon's confusion and di.

fusion principle, substitution-permutation (SP) network is used in AES block cipher. The Feistel structure is used by block ciphers Blowfish and DES taking the advantages of same encryption and decryption operations. SIT combines both Feistel Structure and SP network using principles of both to provide security. This procedure has two parts: Key generation and encryption. The use of key of 64-bits is segmented into 4-blocks and fed into SP network (Here F-function) organized in 4×4 matrices and generates new five unique keys.

3 Proposed Framework

We have proposed an enhanced version of lightweight algorithm for security in IoT data communication entitled Enhanced Secure IoT (ESIT) algorithm. Here, we have taken a plain text and 64-bit keys to generate a 64-bit block cipher and used in ESIT.

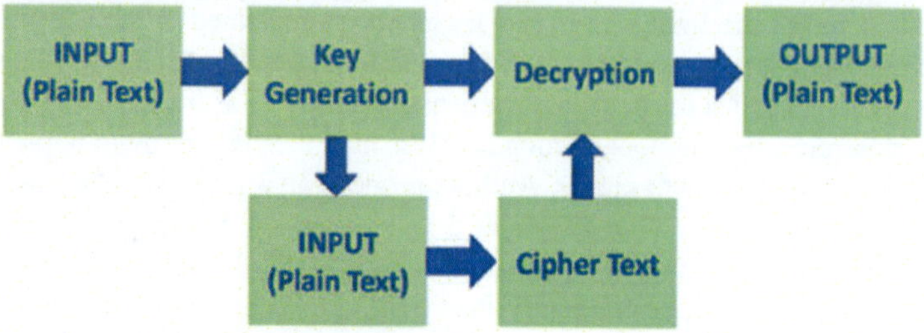

Fig. 1 Overview of the proposed ESIT framework

The proposed work follows the sequence such as: (1) Key Generation, (2) Encryption, (3) Decryption. Figure 1 depicts our proposed model. Each component of the process is described below.

3.1 Key Generation

Figure 2 shows the structure of the key generation. The detailed procedure of key generation is described as below:

In this case, a 64-bit key (K_c) is input to the key generation block and five unique keys are generated. Key generation block (as shown in Fig. 2b) uses Khazad block

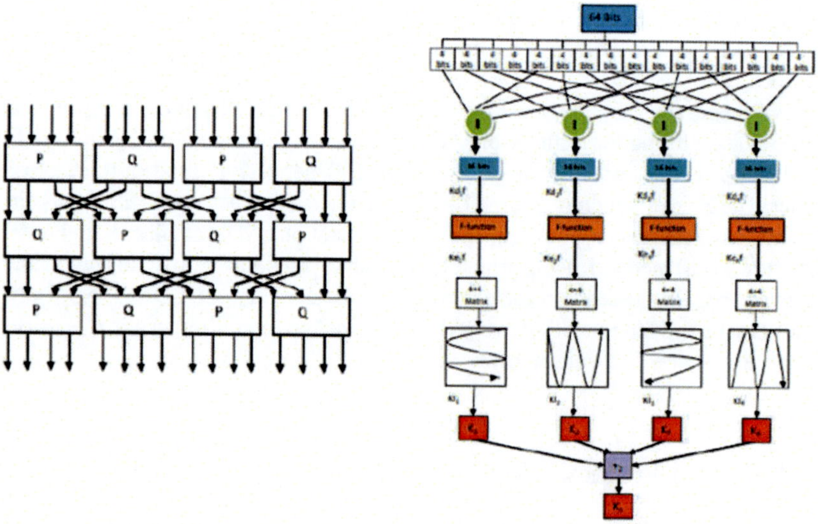

(a) F-Function

(b) Key Synthesis

Fig. 2 Flowchart of key synthesis

cipher [16]. To obtain multiple keys, 64-bits key (K_c) is divided into equal 4-bit units. The F-function (as shown in Fig. 2a) is implemented on 16-bits data. Here, all the 4 F-function blocks are used and can be represented as:

$$k_{(d_i)}f = \prod_{j=1}^{4} k_{c(j-1)+i} \tag{1}$$

where i deviates from 1 to 4, in case of initial four round as shown in Fig. 2a. In the next step, the 16-bits of $K_{di}f$ is used with the f-function to get K_{ei} and it is defined as:

$$k_{e_i}f = f\left(k_{d_i}f\right) \tag{2}$$

The F-function, the confusion, and diffusion are created and explained in [15] to transform linear and non-linear functions. Here, P and Q are represented from Khazads mini-box and presented in tables (a) and (b) (as explained in [16]). To prove F-function, a 4 × 4 matrix is used, called K_m as explained in [15].

The value of initial four round keys K_1, K_2, K_3, and K_4 is obtained by converting their matrices into 4 arrays of 16-bit. This is called as special key of round keys (Kr). Then $+_2$ operations are conducted among the K_1, K_2, K_3, and K_4, to get key K_5 and it is defined as:

$$K_5 = +_{2_{i=1}}^{4} K_i \tag{3}$$

3.2 *Encryption*

In the proposed method, the round keys that are generated from key generation process are applied for encryption. A plain text of size 64-bit (Pi) is segmented into four 16-bits text in the first round. Each of these can be noted as $P_{X0\text{-}15}$, $P_{X16\text{-}31}$, $P_{X32\text{-}47}$, $P_{X48\text{-}63}$. In each round, with the passing of bits to decrease data originality, the order of bits is swapped, and then the level of confusion increases in the originated cipher text. The respective round key K_i perform the bitwise XNOR to obtain key from the key generation process much before time. $P_{X0\text{-}15}$ and key is executed with K_i and $P_{X48\text{-}63}$ to obtain the respective results Ro_{11} and Ro_{14}. The result of XNOR is input to the F-function and output generated as Ef_{l1} and Ef_{r1} and shown in Fig. 2. Left shift function (<<) and bitwise binary modulo 2 ($+_2$) function is used between Ef_{l1} and $P_{X32\text{-}47}$ to get Ro_{12}, and Ef_{r1} and $P_{X16\text{-}31}$ to get Ro_{13} and is given by:

$$\boldsymbol{R}_{\boldsymbol{o}_{i,j}} = \begin{cases} P_{x_{i,j}} \odot K_i & for\ j = 1\ and\ 4 \\ \left(P_{x_{i,j+1}} \ll q\right) + 2Ef_{li} & for\ j = 2 \\ \left(P_{x_{i,j-1}} \ll q\right) + 2Ef_{ri} & for\ j = 3 \end{cases} \tag{4}$$

Similar process is imitated for the rest of the rounds by Eq. (4). The output of the first round added to generate Cipher Text (C_i) is given by (Fig. 3):

$$\boldsymbol{C_i} = \boldsymbol{R}_{51} + +\boldsymbol{R}_{52} + +\boldsymbol{R}_{53} + +\boldsymbol{R}_{54}$$

3.3 Decryption

In our proposed method, the symmetric key is used for encryption and decryption. So, the Decryption is the overturn process, i.e., on the move unintelligible cipher image back to plain image. It makes use of accurate keys (above five distinctive keys from the key generation part), where the diffusion function is primarily applied, afterward substitution operations, XNOR, right shift operation (>>) with Bitwise binary modulo2 ($+_2$), swapping, and left shifting operations in the same order.

4 Observations

Here, an experiment has been conducted with 4 images as input as in SIT [15]. The format of the pictures used are Portable Network Graphics (PNG) and each image with dimension of 256 X 256 pixels. The output is measured on the basis correlation, entropy, and histogram. All simulations are conducted on a PC with Intel core-i5-3.10 GHz and 8 GB of memory.

4.1 Evaluation Parameters

- **Correlation**: In our proposed algorithm, correlation coefficient id is determined for both the real and encrypted images. The output shows that for a plain image, the nearby pixels have better correlation. Hence, the encryption process of proposed (ESIT) algorithm is exciting. Table 1 shows the correlation and correlation coefficient with high delay.
- **Image Entropy**: Image entropy is used to express number of data coded by an encryption algorithm. After image encryption, it is measured with higher entropy for higher security of encryption method as shown in Table 1.

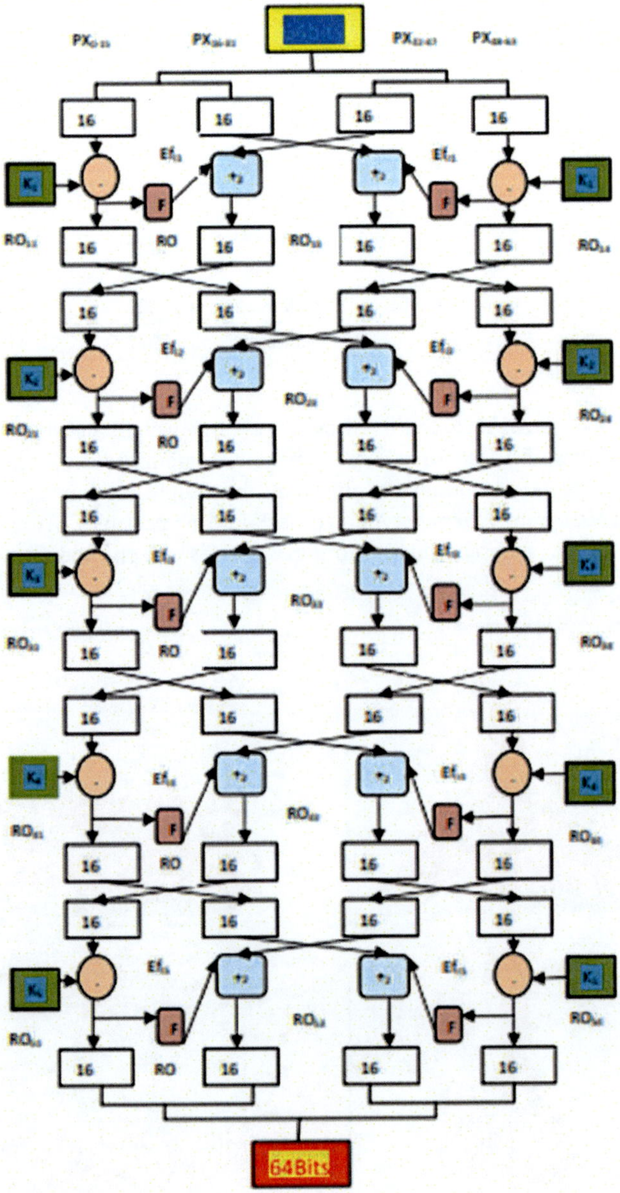

Fig. 3 Flowchart of encryption process

Table 1 Comparative study of Encrypted images with respect to correlation and entropy

Image	Measured Parameters	Usman et al. [15]	ESIT (ours)
Lena	Correlation	0.0012	0.0012
	entropy	7.9973	7.9979
Baboon	Correlation	0.0023	0.0031
	entropy	7.9972	7.9976
Cameraman	Correlation	0.0012	0.0012
	entropy	7.9973	7.9977
Panda	Correlation	0.0022	0.0031
	entropy	7.9971	7.9978

- **Image Histogram**: The security strength of a lightweight algorithm is detected through a powerful technique named image histogram. As shown in Fig. 4, the steep lines points to the number of pixels, and the straight line points to the intensity value for histogram. After encryption (attacks are not possible), pixels are equally speeded in the encrypted images (original images of Lena and Cameraman) in

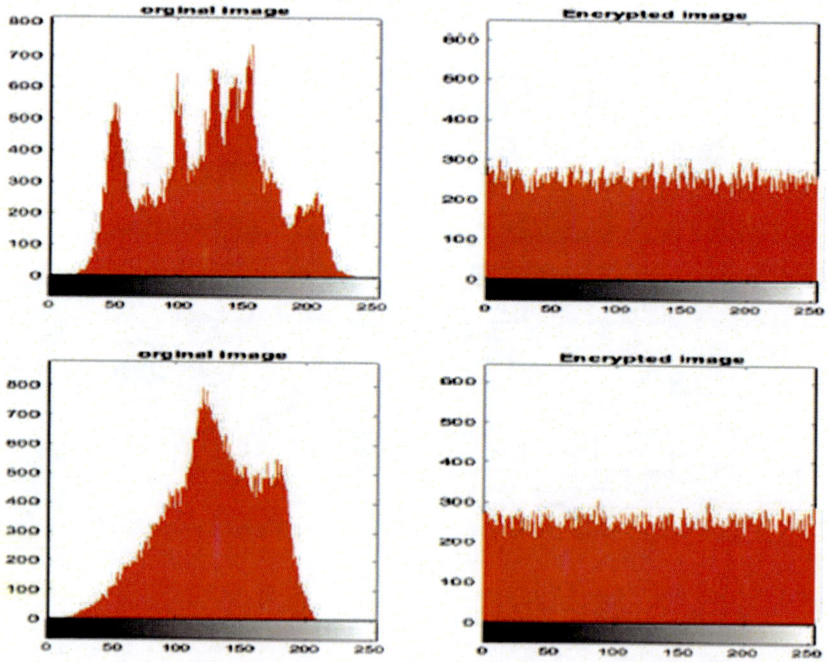

Fig. 4 Image histogram: X-axis represents intensity value and Y-axis represents number of pixels for image of Lena (first row) and Baboon (second row), First column (original Image) and second column (encrypted image)

left-hand side and corresponding encryption image in right end. Which indicates lightweight algorithms secured as the encryption is uniform.

5 Conclusion

This paper presents a lightweight cryptography technique, which is one of the important part to secure cipher for resource constraint applications in IoT system. With new developments, research proceeds with increasing implementation of IoT enabled systems; there is a significant demand of strong security for smart things, services, and applications. Therefore, our proposed lightweight block cipher technique, ESIT, will provide security and privacy in data communication in resource constraint applications of IoT platform.

References

1. Sundmaeker, H., Guillemin, P., Friess, P., & Woelfffle, S. (2010). Vision and challenges for realisin the internet of things. *Cluster of European Research Projectson the Internet of Things, European Commision, 3*(3), 34–36.
2. Singh, S., Sharma, P. K., Moon, S. Y., & Park, J. H. (2017). Advanced lightweight encryption algorithms for IoT devices: Survey, challenges and solutions. *Journal of Ambient Intelligence and Humanized Computing*, pp.1–18.
3. Sadeghi, A. R., Wachsmann, C., & Waidner, M. (2015). Security and privacy challenges in industrial internet of things. In *Proceedings of 52nd ACM/EDAC/IEEE design automation conference (DAC)*, pp. 1–6.
4. Rana, S., Hossain, S., Shoun, H. I., & Kashem, M. A. (2018). An effective lightweight cryptographic algorithm to secure resource-constrained devices. *Spectrum, 9*(11), 1–9.
5. Hossain, M., Hasan, R., & Skjellum, A. (2017). Securing the internet of things: A meta-study of challenges, approaches, and open problems. In *Proceedings of IEEE 37th international conference on distributed computing systems workshops (ICDCSW)*, pp. 220–225.
6. Dhanda, S. S., Singh, B., & Jindal, P. (2020). Lightweight cryptography: A solution to secure IoT. *Wireless Personal Communications, 112*(3), 1947–1980.
7. Cazorla, M., Marquet, K., & Minier, M. (2013) Survey and benchmark of lightweight block ciphers for wireless sensor networks. In *Proceedings international conference on security and cryptography (SECRYPT)*, pp. 1–6.
8. Daemen, J., & Rijmen, V. (1999). *AES proposal: Rijndael.*
9. Biham, E., Biryukov, A., Shamir, A. (1999). Cryptanalysis of skipjack reduced to 31 rounds using impossible differentials. In *Proceedings international conference on the theory and applications of cryptographic techniques* (pp. 12–23). Berlin, Heidelberg: Springer.
10. Biryukov, A., & Kushilevitz, E. (1998). Improved cryptanalysis of RC5. In *Proceedings of international conference on the theory and applications of cryptographic techniques* (pp. 85–99). Berlin, Heidelberg: Springer.
11. He, Y., & Qing, S. (2001). Square attack on reduced camellia cipher. In *Proceedings of international conference on information and communications security* (pp. 238–245). Berlin, Heidelberg: Springer.
12. Jha, V. K. (2011). Cryptanalysis of lightweight block ciphers. Aalto University School of Science Degree Programme of Computer Science and Engineering, Master's Thesis.

13. Bogdanov, A., Knudsen, L. R., Leander, G., Paar, C., Poschmann, A., Robshaw, M. J., Seurin, Y., & Vikkelsoe, C. (2007) PRESENT: An ultra-lightweight block cipher. In*Proceedings of international workshop on cryptographic hardware and embedded systems* (pp. 450–466). Berlin, Heidelberg: Springer.
14. Blaze, M., Diffie, W., Rivest, R. L., Schneier, B., & Shimomura, T. (1996). Minimal key lengths for symmetric ciphers to provide adequate commercial security. *A Report by an Ad Hoc Group of Cryptographers and Computer Scientists*. Information Assurance Technology Analysis Centre Falls Church VA.
15. Usman, M., Ahmed, I., Aslam, M. I., Khan, S., & Shah, U. A. (2017). SIT: A lightweight encryption algorithm for secure internet of things. arXiv:1704.08688.
16. Barreto, P. S. L. M., & Rijmen, V. (2000). The Khazad legacy-level block cipher. Primitive Submitted to NESSIE. 97, p.106.

A Novel Block Diagonalization Algorithm to Suppress Inter-user Interference in a Multi-user Mimo System

Harsha Gurdasani, A. G. Ananth, and N. Thangadurai

Abstract Diversity in MIMO applications tend to ameliorate the architecture of system for compensation with upgraded hardware and software requirements. Single-user MIMO systems allow one user to be serviced per transmission interval. This maximizes the throughput of a single user, but its disadvantage is that it does not take advantage of multi-user diversity. Multi-user MIMO systems (MU-MIMO) have become the main technique for meeting the requirements. In MU-MIMO, one of the biggest issues to deal with is eliminating co-channel interference. The block diagonalization (BD) is a linear precoding method used in broadcast channels of MU-MIMO that has been effective in removing multi-user interference (MUI), but is not computationally efficient. To counter this, we have developed a novel optimization algorithm (Bacterial Foraging Optimization), and implemented it with variation in the number of users and modulation orders. This paper exhibits the improvement in the diversity, and proves the efficiency of the proposed algorithm.

Keywords BFO · Block diagonalization · MIMO · MU · CSI

1 Introduction

Cellular systems have evolved a lot since the emergence of MIMO technology, but still limited by the amount of frequency bands allocated in the cell. If someone wants to communicate with several users within a cell, but not in the conventional way, i.e. using several channels. This time, communication is to be done with K users in the same band. This, in turn gives rise to co-channel interference within the same cell. With the innovation of MIMO technology, the cellular system has found a way to increase its data rates by applying MIMO technology, in addition to achieving high spectral efficiency within cellular networks, achieving great evolution of this system [1, 2]. MIMO technology consists of having more than one antenna

H. Gurdasani (✉) · N. Thangadurai
JAIN (Deemed-to-be University), Bangalore, India

A. G. Ananth
NMAMIT, Nitte, Mangalore, India

P. Nayak et al. (eds.), *IoT and Analytics for Sensor Networks*, Lecture Notes in Networks and Systems 244, https://doi.org/10.1007/978-981-16-2919-8_11

in the transmitter and the receiver [3]. MIMO increases the spectral efficiency of a wireless communication system by using the spatial domain [4]. In MU-MIMO, one of the biggest issues to deal with is eliminating co-channel interference [5]. This is because we will concentrate different users within the same frequency, increasing the interference between users [6]. If we manage to eliminate inter-user interference within the cell, then we are increasing the capacity of the system, the main objective of our algorithm to study. It is very important to have the Channel Status Information (CSI), since the development of our algorithm depends on this [7]. The solutions implemented so far, counteract the co-channel interference generated by different users, eliminating it almost entirely, but at the cost of additional signal processing algorithms on the transmitting and receiving sections [8–10].

Section 2 of the paper presents the proposed system model to suppress inter-user interference, Sect. 3, the results of the simulations, and finally, in Sect. 4, the conclusions.

2 Proposed Methodology

2.1 System Model

For our work, we will implement this model on multiple number of users. Also, with respect to users, it will be assumed that they are very far apart, so the user's channel will be independent with respect to the rest. We know that Channel Status Information (CSI) is very important in the transmitter, so it will be considered an ideal feedback channel between base station and user. Finally, the number of transmitting antennas would be equal to the product of the number of receiving antennas with the number of users, respectively.

Let us begin with an assumption of a multi-user communication system in which one base station supports several mobile stations.

Assuming that N_B and N_M are the base station and mobile station antennas, respectively.

As k independent user, $k.N_M$ antennas communicate with base station N_B antennas, where end-to-end communication for downlink is considered as $(k.N_M) \times N_B$ mimo system (Fig. 1).

'In a multiuser communication system, multiple antennas facilitate the base station to transmit multiple user data streams to be decoded by each user in downlink channel' [1].

By considering k independent users, where $X \in C^{N_B \times 1}$ is the transmit signal from the BS and $R_u \in C^{N_M \times 1}$ with received signal at the uth user, where,$u = 1, 2, 3, \ldots..k$ [3].

Let $Ch_u \in C^{N_M \times N_B}$ represent the channel gain between BS and the uth user. The signal received at the uth user position is expressed as:

$$R_u = Ch_u X + Z_u \tag{1}$$

$$u = 1, 2, 3, \ldots..k$$

where $Z_u \in C^{N_M \times 1}$ is the additive zero mean circular complex Gaussian random vector [.] for all user.

Where X is the set of transmitted signal $(Ch_1, \ldots Ch_k)$.

However, the main hindrance in data transmission in Broadcast Channel (BC) is that the signal detection on the receiver side is not linear, and therefore, interference cancellation becomes necessary. This paper utilizes Block Diagonalization precoding technique as the solution for the above stated problem.

2.2 *Broadcast Channel Transmission via Block Diagonalization*

Among the various solutions available to mitigate the problem of co-channel interference produced inside a cell by communicating several users in the same frequency band, we will deal with Block Diagonalization (BD) Technique. It uses a linear precoding matrix to transmit multiple frames of data to each user while removing inter-user interference. In the next section, we realize that multiplying the precoding matrix, with the channels of the interfering users, inter-user interference will be cancelled instantaneously, transmitting only the signal to the desired user. And, even if the interference from the other user is cancelled, inter-symbolic interference (ISI) will still be present in the receiver.

The channel inversion method is effective on its parts in clipping the interferences (any signal else the target signal). But it also introduces considerable noise enhancement in signals [11–14]. Block Diagonalization on other hand cancels only interferences of other user's signals at the stage of precoding. The inter-interference of signals from antenna, if occurred, could be tackled by various detection algorithm on-rolled in a MIMO network.

Let $N_{M,u}$ denote the number of antennas for the uth consumer. Where $u = 1, 2, 3, \ldots..k$.

For the uth signal $\tilde{x}_u \in C^{N_{M,u} \times 1}$, the received signal, $R_u \in C^{N_{M,u} \times 1}$ can be expressed as:

$$\begin{aligned} R_u &= Ch_u \sum_{k=1}^{K} p_k \tilde{x}_k + Z_u \\ &= Ch_u p_u \tilde{x}_u + \sum_{k=1, k \neq u}^{K} Ch_k p_k \tilde{x}_k + Z_u \end{aligned} \tag{2}$$

where $Ch_u \in C^{N_{M,u} \times N_B}$ is channel matrix between BS and uth user.

$w_u \in C^{N_a \times N_{M,u}}$ is the precoded matrix for the uth user and Z_u denotes the noise vector.

From Eq. (2), $\{Ch_u p_k\}_{u \neq k}$ increases interference to uth user unless,

$$Ch_u p_k = O_{N_M \times N_{m,u}}, \ \forall u \neq k \tag{3}$$

where $O_{N_M \times N_{M,u}}$ is a zero matrix.

To meet the total p

$$\widetilde{Ch}_u p_u = O_{(N_{M\,total} - N_{M,u}) \times N_{M,u}}$$

ower constraints, the precoder $p \in C^{N_B \times N_{M,u}}$ must be unitary, $u = 1, 2, 3, \ldots..k$.

From Eq. (3), the interference free received signal is,

$$R_u = Ch_u p_u \widetilde{x}_u + Z_u \tag{4}$$

$$u = 1, 2, 3, \ldots, k$$

For obtaining the value of $\widetilde{x}_u$, various signal detection algorithms now can be employed for estimation.

To obtain $[P_k]_{k=1}^{K}$, let us take channel matrix of all users except uth user.

$$\widetilde{Ch}_u = \left[(Ch_1)^{Ch} \ldots.(Ch_{u-1})^{Ch}(Ch_{u+1})^{Ch} \ldots..Ch_k\right]^{Ch} \tag{5}$$

where $N_{M,total} = \sum_{u=1}^{k} N_{M,u} = N_B$

$$\widetilde{Ch}_u p_u = O_{(N_{M\,total} - N_{M,u}) \times N_{M,u}} \tag{6}$$

$$u = 1, 2, 3, \ldots, k$$

Hence, precoding matrix $P_u \mathbb{C}^{N_B \times N_{M,u}}$ should exist in null space of $\widetilde{Ch}_u$ and precoders should satisfy the Eq. (6). For this, the singular value decomposition (SVD)$\widetilde{V}_u^{zero}$ of $\widetilde{Ch}_u$ is expressed in terms of non-zero and zero singular values and zero singular values [1].

$$\widetilde{Ch}_u = \widetilde{U}_u \widetilde{\wedge}_u \left[\widetilde{V}_u^{non\,zero} \widetilde{V}_u^{zero}\right]^{Ch} \tag{7}$$

Where $\tilde{V}_u^{non\,zero} \in C^{(N_{M,total}-N_{m,u})\times N_B}$ ***and*** $\tilde{V}_u^{zero} \in C^{N_{m,u}\times N_B}$ are comprising of right singular vectors that are denoted by non-zero singular values and zero singular values, respectively.

From Eq. (7), multiplying $\tilde{Ch}_u$ with $\tilde{V}_u^{zero}$, we get following term,

$$\tilde{Ch}_u \tilde{V}_u^{zero} = 0 \tag{8}$$

Multiplication of both the terms, i.e. channel gain and SVD results in zero. The zero received signals at destination end indicate the minimization of interference in signals. Thus, $P_u = \tilde{V}_u$ can be employed to precode the signal of uth user.

From Eq. (9), It can be seen that precoding matrix $P_u = \tilde{V}_u$ for the uth user. Where $\tilde{V}_u$ is composed of zeros and non-zero singular values. Size of $\tilde{V}_u$ depends on size of $\tilde{Ch}_u$. If $\tilde{V}_u$ is a large matrix than it consist greater number of non-zeros singular values which causes Eq. (3) to be:

$$\widetilde{Ch}_u P_K > 0 \quad \forall u \neq k \tag{9}$$

If $\tilde{V}_u$ is a smaller matrix than it consist less number of non-zeros singular values which causes Eq. (3) to be:

$$\widetilde{Ch}_u P_K < 0 \quad \forall u \neq k \tag{10}$$

Both the cases cause a significant co-channel interference for user u, since channel matrix is not completely block diagonalization. Thus, size of $\tilde{V}_u$ should be optimum for a better performance.

Since the size of $\tilde{V}_u$ depends on size of $\tilde{Ch}_u$, we can influence the magnitude of $\tilde{Ch}_u$ by configuring the number of receiving antennas N_{RX} for each user.

To find an optimal value of N_{RX}, an objective function can be drawn as follows:

$$\min f(\widetilde{Ch}_u) = \left|\widetilde{Ch}_u P_k\right| \tag{11}$$

where,

$$\tilde{Ch}_u = f(\dot{N_{RX}}) \tag{12}$$

The value of $\dot{N_{RX}}$ can be found as,

$$N_{RX} = G(N_{RX}) \tag{13}$$

Where G is an operator which is optimized by Bacterial Foraging Optimization.

In order to explain the food search strategy of bacteria (specifically Escherichia Coli, present in the intestines of humans), four processes are used:

(i) Chemotaxis
(ii) Swarming
(iii) Reproduction
(iv) Dispersion

Bacteria have a set of rigid flagella that allow movement. This movement used in the search for food can be classified in two ways or modes of movement: Swimming and the fall together called Chemotaxis. A movement in a certain direction is called Swimming or Running, while Chemotaxis is the movement of the set of Bacteria in different directions. During the life of the bacteria, both modes of movement alternate. Clockwise rotation of the flagella results in Chemotaxis, and counterclockwise rotational movement results in Swimming.

Successively, after many stages of Chemotaxis, the original set of bacteria is fit for reproduction. The reproduction process consists of dividing a Bacterium into two identical ones. This is reflected during the optimization process, replacing half of the bacteria with poorer fitness function, with the ones with better fitness function. During the evolution process, however, the population of bacteria remains stable.

The dispersal of a group of bacteria to a new location results in an alteration to the natural process of evolution [9]. This concept is used to replace bacteria close to the location of the food to avoid stagnation at local minimums.

Initialization of the total number of Bacteria (numBact) with (N + 1)/2 solutions (variables to optimize) that satisfy the restrictions of maximum and minimum values of the variables (var_{max} and var_{min}), their aptitudes are calculated, and the bacteria with the best aptitude as $var_{global}^{(0)}$.

Scatter-kill consists of going through and selecting the bacteria from the population to apply scatter-killing with a P_{el} probability. To do this, these bacteria are eliminated and new ones are dispersed in a random location, maintaining the population size.

The breeding process kills half the population of bacteria with the worst fitness, and doubles the half best adapted to maintain population size.

Swarming is applied to the abilities of the bacteria by the following equation [4].

$$J_i^{(j+1,k,l)} = J_i^{(j,k,l)} - \sum\nolimits_{numBact} d_{attract} x \left(e^{-w_{attract} x \sigma_i^{(j+1,k,l)}} \right) + \sum\nolimits_{numBact} d_{repellant} x \left(e^{-w_{repellant} x \sigma_i^{(j+1,k,l)}} \right) \quad (14)$$

where,

$$\sigma_i^{(j+1,k,l)} = \sigma_i^{(j,k,l)} + \sum_{numBact} \left(var_{global}^{(j,k,l)} - var_i^{(j,k,l)} \right)^2 \quad (15)$$

The chemotaxis movement that is applied to the variables (solutions) is as follows:

The variables of the next iteration are calculated imposing the established limits of maximum and minimum value (var_{max} and var_{min}) as:

$$var_i^{(j+1,k,l)} = var_i^{(j,k,l)} + \left(c^{var_i} - d_2\right) \times \frac{\Delta^{(var_i)}}{\Delta\Delta} \quad (16)$$

where,
$\Delta^{(var_i)} = (2.round(rand(.)) - 1) \bullet rand(\bullet);$

$$\Delta\Delta = \sqrt{\Delta^{(var_i)} \times \Delta'^{(var_i)}} \quad (17)$$

$$c^{var_i} = c_{max} - (c_{max} - c_{min}) \times \frac{j \times k}{max_{cycle}} \quad (18)$$

Swarming is applied obtaining $\Delta^{(var)}$ and $\Delta\Delta$ from Eq. (17), and the new variables are calculated with their restrictions:

$$var_i^{(j+1,k,l)} = var_i^{(j,k,l)} + \left(c^{var_i} + d_1\right) \times \frac{\Delta^{(var_i)}}{\Delta\Delta} \quad (19)$$

3 Simulation and Results

To verify the proposed Bacterial Foraging Algorithm in combination with Block Diagonalization Method, extensive and comprehensive simulations were carried out in MATLAB 2017a environment, and the following results were achieved.

Figure 2 exhibits the comparative performance of proposed approach using MMSE and ZF equalizers under Rayleigh Channel conditions.

The simulation result in Fig. 3 shows the performance comparison of SNR vs BER plots of different number of users (K) using the proposed approach, BFO optimized block diagonalization technique.

From the above graph, it is understood that due to the use of Block Diagonalization technique, the probability of occurrence of errors drops sharply for less number of users, but becomes almost constant as the no. of users increase. This can be a potential disadvantage.

The simulation result in Fig. 4 shows the performance comparison of SNR vs BER plots of different modulation orders (M) using the proposed approach.

The detailed analysis of the plots demonstrate that:

- To achieve a BER of 10^{-2}, the he SNR required is 4 dB for a modulation order M = 2, which increases to 9 dB to achieve a BER 10^{-3}.

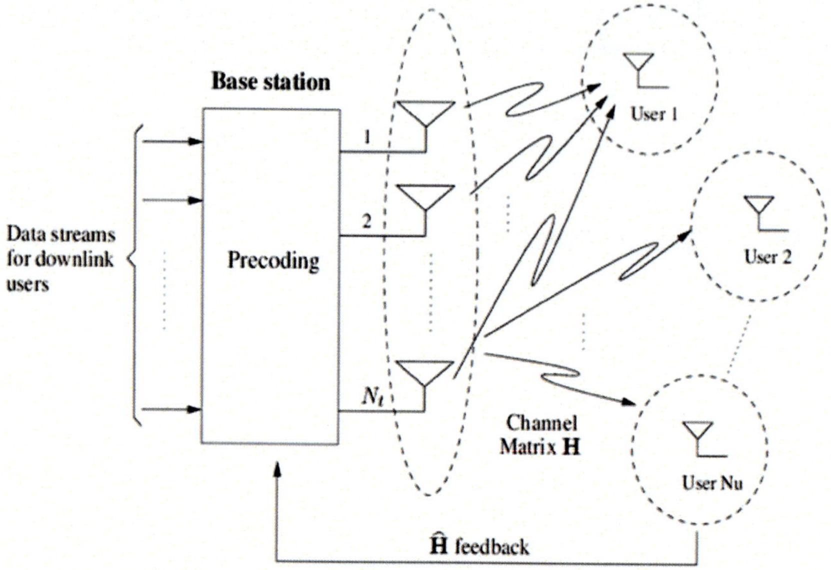

Fig. 1 Multi-user MIMO framework downlink channel model [1]

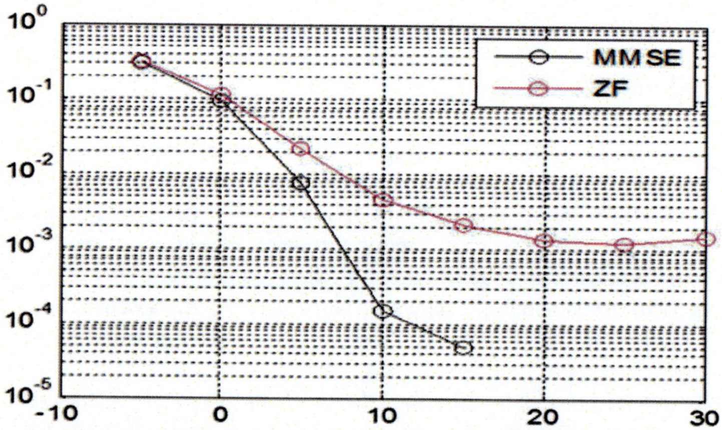

Fig. 2 MMSE and ZF equalizations under Rayleigh channel conditions

- Similarly, to achieve a BER of 10^{-2}, the SNR required is 8 dB for a modulation order M = 4, which increases to 11 dB to achieve a BER 10^{-3}.
- Also, to achieve a BER of 10^{-2}, the SNR required is 14 dB for a modulation order M = 8, which increases to 18 dB to achieve a BER 10^{-3}.
- Finally, to achieve a BER of 10^{-2}, the SNR required is 19 dB for a modulation order M = 16, which increases to 24 dB to achieve a BER 10^{-3}.

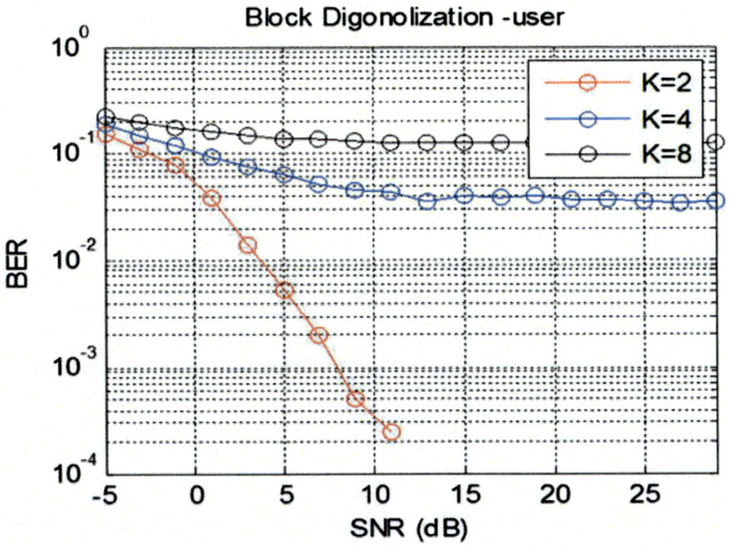

Fig. 3 Block diagonalization for various number of users

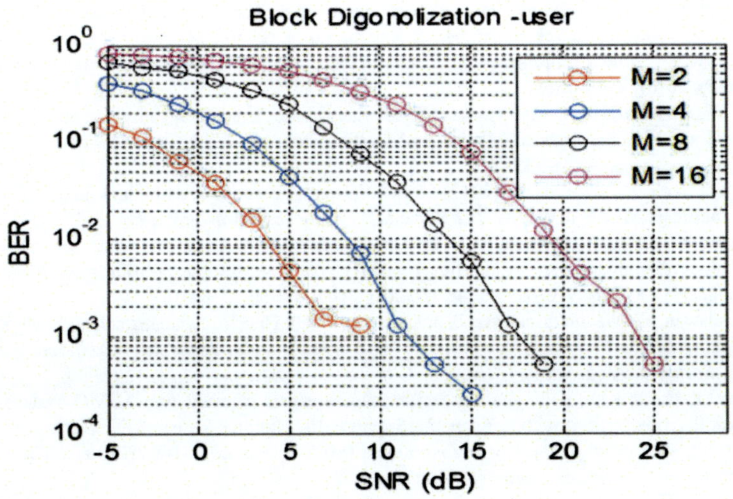

Fig. 4 BER curve with BFO optimized BD for different modulation order

4 Conclusion

From the above simulation results, the following can be inferred that the average value of BER is best achieved when the number of users is less, that is, the average value of BER is best achieved when the modulation order M = 2. Also, the SNR vs BER average value using MMSE equalizer is better than that of ZF equalizer.

References

1. Saxena, S., & Singh, R. P. (2018). Optimal block diagonalization precoding scheme for MU-MIMO downlink channel using particle swarm optimization. *International Journal of Engineering & Technology*
2. Ni, W., & Dong, X. (2016). Hybrid block diagonalization for massive multiuser MIMO systems. *IEEE Transactions on Communications, 64*(1), 201–211.
3. Murti, F. W., Astuti, R. P., & Nugroho, B. S. (2015). Enhancing performance of block diagonalization precoding in multi user MIMO (MU-MIMO) downlink. In *3rd International Conference on Information and Communication Technology (ICoICT), 2015* (pp. 102–106). IEEE.
4. Liang, L., Xu, W., & Dong, X. (2014). Low-complexity hybrid precoding in massive multiuser MIMO systems. *IEEE Wireless Communications Letters, 3*(6), 653–656.
5. Khan, M. H. A., Cho, K. M., Lee, M. H., & Chung, J. G. (2014). A simple block diagonal precoding for multi-user MIMO broadcast channels. *EURASIP Journal on Wireless Communications and Networking, 2014*(1), 1–8.
6. Larsson, E. G., Edfors, O., Tufvesson, F., & Marzetta, T. L. (2014). Massive MIMO for next generation wireless systems. *IEEE Communications Magazine, 52*(2), 186–195.
7. Saha, S. K., Kar, R., Mandal, D., Ghoshal, S. P. (2013). Bacteria foraging optimisation algorithm for optimal FIR filter design. *International Journal of BioInspired Computation.*
8. Caire, G., Jindal, N., Kobayashi, M., & Ravindran, N. (2010). Multiuser MIMO achievable rates with downlink training and channel state feedback. *IEEE Transactions on Information Theory, 56*(6), 2845–2866.
9. Li, W., & Latva-aho, M. (2011). An efficient channel block diagonalization method for generalized zero forcing assisted MIMO broadcasting systems. *IEEE Transactions on Wireless Communications, 10*(3), 739–744.
10. Park, J., Lee, B., & Shim, B. (2012). A MMSE vector precoding with block diagonalization for multiuser MIMO downlink. *IEEE transactions on communications, 60*(2), 569–577.
11. Gupta, A., Saini, G., Sachan, S. B. L. (2013). An efficient linear precoding scheme based on block diagonalization for multiuser MIMO downlink system. *International Journal of Engineering Research & Technology (IJERT), 2*(10). ISSN: 2278-0181
12. Kurniawan, E., Madhukumar, A. S., & Chin, F. (2009). Heuristic antenna selection algorithm for multiuser multi-antenna downlink. In *IEEE* 69th*Vehicular Technology Conference, 2009. VTC Spring 2009* (pp. 1–5). IEEE.
13. Kim, Y. H., & Kim, J. Y. (2011). Performance analysis of multi user MIMO system with block-diagonalization precoding scheme. Supported by National Research Foundation of Korea (NRF) grant funded by the Korea government (MEST) (No. 2011–0029329) and MKE (The Ministry of Knowledge Economy), Korea, 5, p. 6.
14. Khan, S., Chockalingam, A., Sundar Rajan, B. (2008). A low-complexity precoder for large multiuser MISO systems. In *Proceedings of the 67th IEEE vehicular technology conference, VTC Spring 2008, 11–14 May 2008, Singapore.*

Prediction of Chemical Contamination for Water Quality Assurance Using ML-Based Techniques

C. Kaleeswari and K. Kuppusamy

Abstract Big Data is used in a spacious, distinct set of information that is continuously evolving estimates. Excessively vast array of tools can be computed to reveal models, patterns, and relationships, exclusively linked to human attitudes and communication. Its applications are applied in a range of real-time applications such as agriculture, weather forecasting, healthcare, and water resource management. Each Big Data research-oriented process can provide the outcome of the prediction or forecasting process. Analyzing the quality of water is still so important to a human being, because human beings cannot live without water. Smart Cities have been growing in the recent period, but many rural development areas are not yet emerging and do not have a lot of facilities like a smart city, and these areas are affected by the chemical contaminations like Arsenic, Cadmium, and some heavy metals. In this research work, Prediction of Chemical contamination for Water Quality Assurance using ML-based Techniques is proposed. This article offers a brief overview of Machine Learning Algorithms such as Decision Tree, Support Vector Machine, Random Forest, Naïve bayes, K-Nearest Neighbor. These algorithms are used for classification and prediction using datasets to analyze which Algorithm provides the best Predicting Accuracy as seen in this paper.

Keywords Machine learning · Prediction · Contamination · Classification · Water · Quality assurance

1 Introduction

Predictive Analytics is a field of futuristic data used to prove things about anonymous future events. Predictive Analytics has used more methods from learning-based techniques like Machine Learning and Deep Learning, Statistics, Artificial Intelligence, Data Mining, and remote sensing to inspect prevailing analysis to predict the future [1–3, 5, 8, 9]. Generally, real data is used to frame an empirical model with a significant tendency. It also has received a broad range of concentration in recent

C. Kaleeswari (✉) · K. Kuppusamy
Department of Computational Logistics, Alagappa University, Tamilnadu, India

P. Nayak et al. (eds.), *IoT and Analytics for Sensor Networks*, Lecture Notes in Networks and Systems 244, https://doi.org/10.1007/978-981-16-2919-8_12

times due to current trends in technology support, especially in the field of Machine Learning and Big Data [5, 8, 9]. The Water Quality Assurance Framework is one of the applications of urban development. Nobody can define the term "smart city," but it can be used to make a normal city into a smart city with the right environmental setup using emerging technologies. Analyzing the water situation is still so important to a human being, because human beings are not living without water [10]. They use rivers, surface water, seawater, drinking water, ground water, reservoirs, ponds, and large aquatic areas. These services are impaired by some industrial exposures, such as chemical pollution, human trafficking, and lower chlorine levels. We know that seawater as well as its underlying water bodies are also salty. The Water Quality Assurance Framework is used to evaluate the use of water in agriculture and replicate it for drinking using the Systematic Approach. Rural Communities are not aware of the use of water and of the diseases induced by the use of these forms of water supplies polluted with chemicals such as Fluoride, Nitrate, Arsenic, Iron, Cadmium, Chromium, and Heavy Metals [2, 10–12]. In the study, the authors have established a smart city system for urban areas and several smart city initiatives are underway. Ability to predict the influence of water use in rural areas based on the quality of the water is more important. It is critical to determine whether the water is suitable for drinking or not, whether it is suitable for agricultural activities or not, and whether it can be recycled for routine use [6]. The study of the impacts faced by rural people can be used to recognize and raise awareness in the use of water.

In this research work, it is suggested that the Chemical Pollution Method be developed for the predictive analysis of water quality using Machine Learning. In the state-of-the-art, samples of water were collected from lakes and rivers only in smart cities. Instead, in this research work, the Water Sample Datasets obtained from the Water Board will be used as state-of-the-art water quality analysis in India. Urban areas have a lot of government-run services, but rural areas do not have an excessive development infrastructure resulting in certain health problems being posed and have little knowledge of them [9].

2 Related Works

Abba et al. [1], proposed the hybrid models GPELM and XGBELM. These two models were combined with various algorithms such as Extreme Gradient Boosting (XGB), Genetic Programming (GP), Extreme Learning Machine (ELM), Linear Regression (LR), and Stepwise Linear Regression (SLR). These hybrid models provide the highest accuracy for the minimum number of input variables. Libralato et al. [2], reported rare earth elements, nanomaterials, micro-and nano-plastics, PFAS-Perfluoroalkyl and polyfluoroalkyl substances, PPCPs-Pharmaceuticals and Personal Care Products, ionic liquids, artificial sweeteners, flame retardants, and illicit drugs contaminated with water. Most of these elements are discharged by sewage water, and the only solution to this problem is a wastewater treatment plant.

Miao et al. [3], have introduced the CWQI (Canadian Water Quality Index) which helps to assess the quality of water in a spacious and objective manner and to deter waste in water. This work also applied a remote sensing technique combined with CWQI using satellite data. Fatemi [4], noted the strategies and policies available for managing the quality of water in the context of environmental change. The author discusses the major impacts of water quality regions, such as soil erosion, floods, natural hazards, sediment loading, land use, and suggests some methods for problems such as phytoremediation, organic farming, wastewater treatment plants, etc. Chen et al. [5], said they were trying for the first time to achieve something that could be examined in contrast to the prediction of water quality. There are 7 traditional and 3 set models used in big data research (33,612 attentions) from the well-known rivers in China (2012–2018). Performance Evaluation of the F1-Score metrics, recall, precision and potential key exploration. On comparing the seven algorithms, Random Forest, Deep Cascade Forest, and Decision Tree provide slightly better forecasting performance. Zhang et al. [6], sought to improve the water quality index and the HHR (Human Health Risk) survey in the semi-arid regions of China with 51 samples collected. From the results, it has been shown that the collected water is significantly alkaline and provides a range of total dissolved solids. There are 94.12% of the samples that are suitable for drinking, and the major polluted parameters are NO2, NO3, F. The HHR reported that children are mainly affected by contaminated water compared to the male and female categories. Nenang et al. [7], used the DNA-based approach in their research and PCR–RFLP (Polymerase Chain Reaction–Restriction Fragment Length Polymorphism) as a biological indicator for regional weather. In the case of environmental pollution, the chemical parameters related to RFLP are determined by the PCA (Principal Component Analysis). Lerios and Villarica [8], reported the use of Machine Learning Algorithms such as Naïve bayes, Decision Tree, Random Forest, and Deep Learning Algorithm. Decision Tree Algorithm provides the best Accuracy and Precision compared with other Algorithms. Authors collected datasets using seven stations to evaluate seven quality attributes in a single reservoir named Laguna.

3 Study Site

The approach of the research work involves different stages. Dataset collection via Data.gov.in, Data.world.com, Kaggle, USDA Agricultural Research Service, and Water Board is one of the features of the system. Journal articles are collected and studied from various websites such as Springer, Elsevier, Research Gate, Wiley Publications, Bentham Science Publishers, Elsevier, Google Scholar, Taylor & Francis, and Emerald Publications from all around the world [4] (Fig. 1).

Fig. 1 Satellite image of the dataset description, India

4 Materials and Methods

Classification is used to forecast the result of a given data sampling when the variable of the outcome is in the pattern of categories. This model may look at the given data and to forecast labels like "Contaminated" or "Fresh." In this work, datasets are collected from various sources and applied for classification to predict the chemical contaminated accuracy using Machine Learning. For data integrity, unstructured format data is transformed to a structured format using pre-processing or filtering principles. Training data for pre-processing was used for the classification of Machine Learning Algorithms such as Decision Trees, Support Vector Machine, Naïve bayes, Random Forest, and K-Nearest Neighbor [5]. These ML-based algorithms are ideal for predictive analytics in the recent era and also have improved results and are comparable to other classification techniques. This scientific investigation has not been carried out recently, and a few alternative research papers have also been focused on this topic. The datasets are collected at first decision and preprocessed for the classification to be applied. The proportion of missing pre-processing values carried out by Anaconda 3 is shown in Fig. 2.

4.1 Machine Learning Algorithms

4.1.1 Naïve Bayes Classification

Naive Bayes is a classification algorithm developed on the basis of theorem of Bayes with the presumption of autonomy between forecasters [5, 8]. This classifier conjectures that the presence of a particular characteristic in a class is unrelated to the existence of all other characteristics. For example, a bird can be considered a peacock in

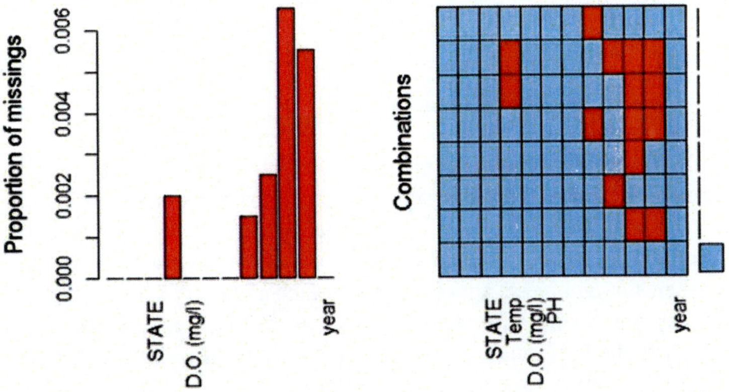

Fig. 2 Proportion of missing in preprocessing

the combined color of blue, green with long feathers and often spread its feathers. These features and properties clearly indicate the possibility that this is the bird and that is why it is known as Naïve. This model is very simple to develop and is especially useful for large volumes and a variety of datasets such as CSV files.

Figure 3 shows the result based on the Bayes theorem, which gives a path of scheming to manipulate Posterior Probability P_1 $(C_1|X_1)$ from P_1 (C_1), $P_1(X_1)$, and P_1 $(X_1|C_1)$. The following formula (1) is given below:

$$P_1(C_1|X_1) = \frac{P_1\ (X_1|C_1)P_1(C_1)}{P_1\ (X_1)} \tag{1}$$

$P_1(C_1|X_1)$ Rear Probability or Posterior Probability of class (C_1, target point) given forecaster (X_1, parameters).

$P_1(C_1)$ Previous or Prior Probability of a Class.

$P_1(X_1|C_1)$ Likelihood which is the Probability of Prediction of a given class.

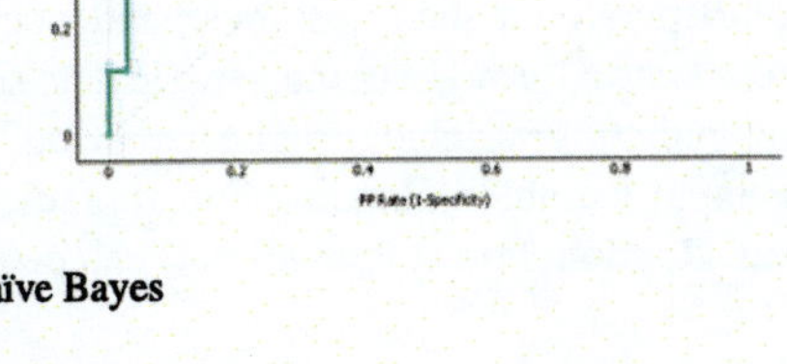

Fig. 3 ROC curve of Naïve Bayes

$P_1(X_1)$ Prior or Previous Probability of forecaster.

4.1.2 K-Nearest Neighbor Classification

KNN could be used for regression and classification time series prediction. But, it is also commonly applicable to classification problems of modern applications such as industry, social media applications for sentimental analysis, and also environmental monitoring. If KNN is used for classification, we can see the result as a lot of times when similar objects are connected to each and every data point [5, 8]. This technique captures a similarity plan called distance, range, and proximity or range. Figure 5 shows the ROC analysis based on the false positive rate and true positive rate following the below procedure.

1. Upload the dataset.
2. Boot K to selected number of instances called neighbors.
3. In each instance in the data

   3.1 Manipulate the range of data between the query sample and the present sample.
   3.2 Include the range and the ratio of the example to the samples collected.

4. Sort the values of the interval and range in ascending order.
5. Click on the first K entry in the sorted data order.
6. Take the labels from the selected k entries.
7. If Classification is required, provide the mode values of K.
8. If Regression is necessary, provide the mean value of K.

4.1.3 Random Forest Algorithm

Random Forest classification technique contains a set of smooth and simple prediction algorithm, and each one is effective in responding to a set of forecasting outcome values [5, 8]. The reaction of those belonging to a category as a class for classification issues, which collaborates or categorizes a set of individual forecasting outcomes with one of the divisions, resides in the variable. This model achieves the result like a large plant enclosed in bark and leaves. There are an inequitable number of simple trees, whatever the ultimate outcome can determine. The assembly technique of a simple tree defines the classification issues for the most prominent level. The use of tree collections must have a major evolution in the accuracy of forecasts. Each tree reacts by one relying upon this range of constraints selected individually (with a substitute) and the tree dispersion in the forest, independent of an array that is part of the massive reference scenario throughout the generic data set as shown in Fig. 4.

The optimal range for predicting variables is indicated as $\log 2\, M_1 + 1$, where "M_1" will be the range of the input variables. Random Forest predictions are reported as mediocre tree predictions. Random Forest Prediction is calculated using the formula (2).

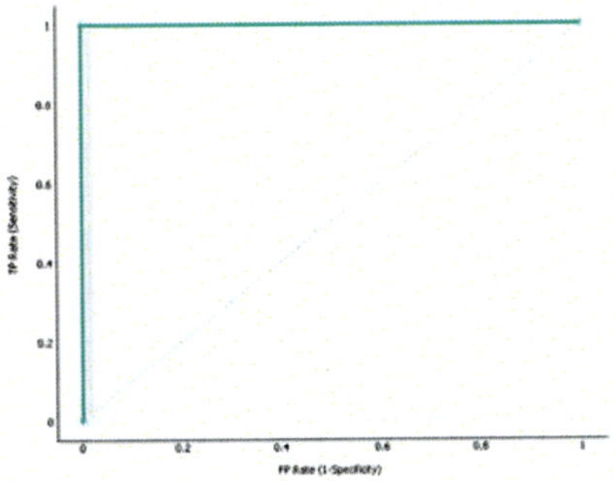

Fig. 4 ROC curve of random forest

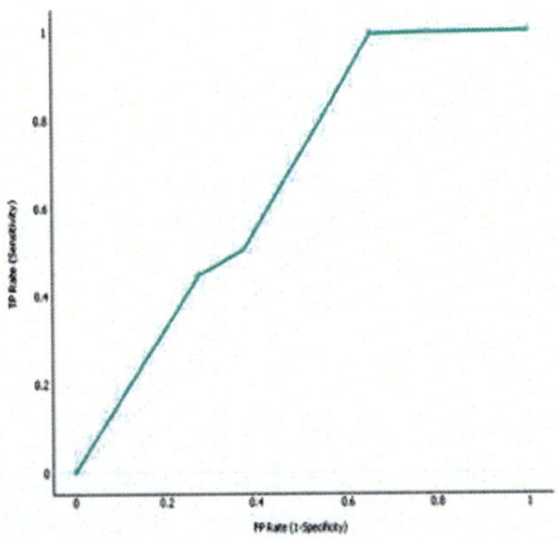

Fig. 5 ROC curve of K-nearest neighbor

RF Predict:

$$P_1 = \frac{1}{K_1} \sum_{K_1=1}^{K_1} K_1^{th}\,\text{TreeReaction} \tag{2}$$

where "K_1" has distinctive trees in the broad area like forest.

4.1.4 Decision Tree Algorithm

Decision Tree, as the name comes with it, appears like a tree structure operating on the key fundamentals of traditional legislation and laws. It is effective and has strong predictive analysis algorithms [5, 8, 10]. It has especially parameterized that of internal nodes, branches, and terminal nodes. Instead, every internal node has a

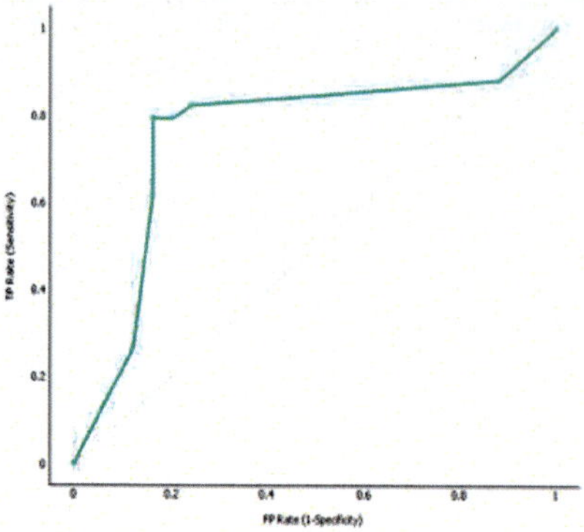

Fig. 6 ROC curve of decision tree

grip on a "test" on a function, branches have a test result, and so each leaf is a class tag as shown in Fig. 6. The main objective of using a DT would be to generate a training model that is being used to determine that hash value obtained by training basic decision rules implicit in the training data shown in Fig. 6.

4.1.5 Support Vector Machine

Support Vector Machine seems to have been a popular ML-based algorithm that conducts both regression and classification tasks [10]. This algorithm would be to generate an excellent outer boundary line or can detect n-dimensional space within classes to the eluded to as feature vector, which is why the method is described as SVM (Support Vector Machine). It is categorized into two forms, Linear SVM and Non-Linear SVM, and has applications such as Face Discovery Method, Image Categorization, Text Mining, and Classification. The ROC analysis of SVM is shown in Fig. 7.

5 Results and Discussion

5.1 Performance Evaluation

In this section, the performance metrics are analyzed for the estimation of the results shown by the individual algorithm in this research work, and comparative analysis of various machine learning approaches for predicting best accuracy and precision in the Python Framework is done. ROC (Receiver Operator Characteristics) curve plotting is given below based on the ML Techniques. False Positive Rate and True Positive Rate with threshold values are used for the calculation of the ROC. Proposed

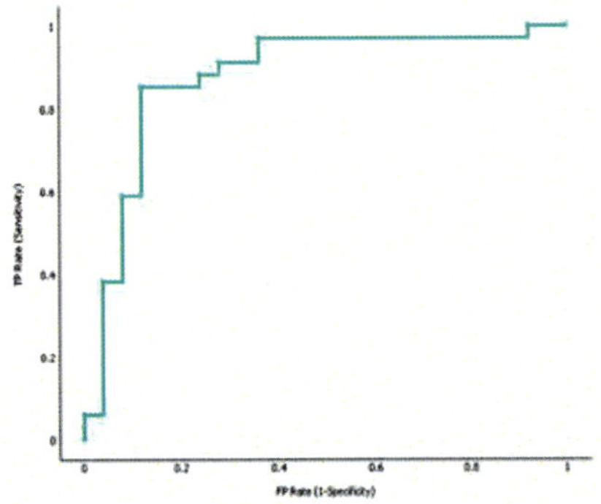

Fig. 7 ROC curve of support vector machine

work provides the perfect Prediction Results, depending on the Precision. Support Vector Machine, Naïve Bayes, Decision Tree, Random Forest, K-Nearest Neighbor are used in this work to estimate the accuracy of the classification [8, 9]. Random Forest has the highest Accuracy (98%) and Precision (98.04%) compared to other Algorithms shown in Table 2.

6 Conclusion

In this work, Machine Learning-based Water Quality Assurance framework has been formulated for measuring the chemical contamination and quality attributes in a specific station in state-wise in India using the datasets collected from various web sources. The collected datasets are preprocessed and classified by Machine Learning Algorithms [5, 8, 9]. This proposed methodology has been verified for its ability to measure Water Quality with high accuracy and precision. Chemical parameters of contamination levels estimated in this research, such as Arsenic, Fluoride, Cadmium, Chromium, Heavy Metals, etc., are shown in Table 1 [2, 10]. This method gives the good Prediction Results depending on the Precision. Support Vector Machine, Naïve Bayes, Decision Tree, Random Forest, K-Nearest Neighbor are applied in this work to predict the Classification Accuracy [8, 9]. Random Forest provides the best accuracy (98%) and precision (98.04%) compared with other algorithms as shown in Table 2.

Table 1 Dataset description

S. no	Attributes	Description
1	Locations	The location where samples of water are obtained
2	Date	Date for monitoring of water samples
3	State	The location of the region as in which state
4	pH	Potential Hydrogen Values ranging from 6.5 to 9.0
5	Temperature	14^0C to 36^0C
6	DO	Dissolved Oxygen mg/L (5)
7	EC	Property or power to conduct heat, electricity, or sound (EC above 3000 micro mhos/ cm)
8	(NO_3-H)	Nitrate (above 45 mg/l)
9	Salinity	Salty (above 900 mg/L)
10	F	Fluoride (above 1.5 mg/l)
11	As	Arsenic (above 0.01 mg/l)
12	Iron	Iron (above 1 mg/l)
13	Heavy Metals	Lead (above 0.01 mg/l)
14	Cr	Cadmium (above 0.003 mg/l)
15	Ch	Chromium (above 0.05 mg/l)

Table 2 Algorithm-wise classification accuracy

Algorithm	Accuracy (%)	Precision (%)
Naïve Bayes	72.58	73.26
Random forest	98	98.04
Decision tree	88.98	89
KNN	85.29	85.3
SVM	84.23	84.23

Acknowledgements This work was financially supported by Rashtriya Uchchatar Shiksha Abhiyan (RUSA, Phase 2.0) grant sanctioned vide letter No.F.24-51 / 2014-U, policy (TNMulti-Gen), Dept. of Edn. Govt. of India, Dt 09.10.2018.

References

1. Abba, S. I., Hadi, S. J., Sammen, S. S., Salih, S. Q., Abdulkadir, R. A., Pham, Q. B., Yaseen, Z. M. (2020). Evolutionary computational intelligence algorithm coupled with self-tuning predictive model for water quality index determination. *Journal of Hydrology* , *587* (2020) 124974. https://doi.org/10.1016/j.jhydrol.2020.124974.
2. Libralato, G., Freitas, R., Buttino, I., Arukwe, A., & Torre, C. D. (2020). Special issue on challenges in emerging environmental contaminants CEEC19. *Environmental Science and Pollution Research, 2020*(27), 30903–30906. https://doi.org/10.1007/s11356-020-09539-w.
3. Miao, S., Liu, C., Qian, B., & Miao, Q. (2020). Remote Sensing-based water quality assessment for urban rivers: A study in linyi development area. *Environmental Science and Pollution Research, 2020*(27), 34586–34595. https://doi.org/10.1007/s11356-018-4038-z.
4. Fatemi, A. (2020). Strategies and policies for water quality management of Gharasou River, Kermanshah, Iran: A review. *Environmental Earth Sciences (2020)* , *79*, 254. https://doi.org/10.1007/s12665-020-08997-2.
5. Chen, K., Chen, H., Zhou, C., Huang, Y., Qi, X., Shen, R., Liu, F., Zuo, M., Zou, X., Wang, J., Zhang, Y., Chen, Da., Chen, X., Deng, Y., & Ren, H. (2020). Comparative analysis of surface water quality prediction performance and identification of key water parameters using different machine learning models based on big data. *Water Research, 171*(2020), 115–454. https://doi.org/10.1016/j.watres.2019.115454.
6. Zhang, Q., Panpan, X., & Qian, H. (2020). Groundwater quality assessment using improved water quality index (WQI) and human health risk (hhr) evaluation in a semi-arid region of Northwest China. *Exposure and Health, 2020*(12), 487–500. https://doi.org/10.1007/s12403-020-00345-w.
7. Neneng, L., Nugroho, R. A., Komai, Y., Takayama, N., & Kawamura, K. (2019) Water quality measurements with a simple molecular analysis (PCR-RFLP) of the microbiome in a metropolitan river system in Japan. *Walailak Journal of Science and Technology 2020, 17*(3), 257–268.
8. Lerios, J. L., & Villarica, M. V. (2019). Pattern extraction of water quality prediction using machine learning algorithms of water reservoir. *International Journal of Mechanical Engineering and Robotics Research, 18*(6).
9. Koditala, N. K., & Pandey, P. S. (2018). Water quality monitoring system using IoT and machine learning. In *IEEE*.
10. Karthick, T., Gayatri, D., Kohli, T. S., & Pandey, S. (2018). Prediction of Water quality and smart water quality monitoring system in IoT environment. *International Journal of Pure and Applied Mathematics, 118*(20). ISSN: 1314–3395 (On-line Version).
11. Kumar, V., Sharma, A., Chawla, A., Bhardwaj, R., Kumar, A., & Thukral, A. K. (2016). Water Quality Assessment of River Beas, India, using Multivariate and Remote Sensing Techniques. *Environment Monit Assess*.
12. Gholizadeh, M. H., Melesse, A. M., & Reddi, L. (2016). A Comprehensive Review on Water Quality Parameters Estimation Using Remote Sensing Techniques. *Sensors, 16*, 1298. https://doi.org/10.3390/s16081298.

Design and Performance Analysis of Two-Port Circularly Polarized MIMO Antenna for UWB Applications

Madan Kumar Sharma, Aryan Sachdeva, Ayushi Ojashwi, and Mithilesh Kumar

Abstract It proposes and simulates a new Ultra-wide (UWB) Multiple Input Multiple Output (MIMO) circular polarization (CP) antenna. The developed antenna's configuration is uncomplicated and incorporates dual T-shaped feed ports at left and right edges to achieve both LHCP and RHCP at same frequency band. A wide circular slot with a hexagonal stub is created in the ground plane in order to achieve better bandwidth for the realization of UWB. The evolution measures of the antenna put forward are presented ahead in this research paper, and its parameters are designed to obtain the optimal return loss, isolation, polarization, and ARBW. Also, a parametric study is done on various types of materials chosen as substrate. The parametric study results, portrays that substrate with higher dielectric tends to give poor ARBW. Operating bandwidth (S11 < = –10db) of the planned antenna is obtained to be 6.7 GHz (3.4 to 10.1 GHz). An isolation of more than 15 dB is also achieved. 3 dB ARBW is obtained in the extent of 3.5–5.5 GHz. Furthermore, the circular polarization features of the antenna proposed are favorable. The surface current and radiation patterns obtained illustrates the circular polarization. Both RHCP and LHCP are obtained at the same frequency in this proposed antenna design. This proposed antenna is a solution to orientation problems of transmitter and receiver, complicated designing, multipath fading, and weather penetration. The simulated antenna is a low profile, compact sized antenna simulated for UWB circular polarization and next generation applications.

Keywords MIMO antennas · Ultra-wide band · Circular polarization · Axial ratio

1 Introduction

Nowadays, with the increase in the demand of the data, the high data transfer and transmission rate is of importance. In order to achieve this, ultra-wide bandwidth is

M. K. Sharma (✉) · A. Sachdeva · A. Ojashwi
Galgotias College of Engineering and Technology, Greater Noida, India

M. Kumar
Rajasthan Technical University, Kota, India

P. Nayak et al. (eds.), *IoT and Analytics for Sensor Networks*, Lecture Notes in Networks and Systems 244, https://doi.org/10.1007/978-981-16-2919-8_13

required. In modern usage, MIMO is widely used for its ability to convey and obtain more than one data signal concurrently over the identical channel by using multipath propagation.

At present, wireless networks are linearly polarized due to which they are less resistant to signal degradation and face multipath fading. In the communication system, the proper positioning of the antenna components, the transmitter, and receiver is of much importance and also a difficult task. Thus, by using a circularly polarized (CP) antenna, this problem can be solved, as till the time the transmitter and receiver is placed within the obtained range, it will observe equal amount of radiation in all directions. In comparison to linearly polarized antennas, the CP antenna has numerous advantages [1], such as higher performance, mobility, and good penetration strength. Apart from that, the multipath fading problem is also reduced using CP antenna. CP antennas are now more popular amongst other antennas for its better performance and benefits of the impact of depolarization in the modern communication system.

Different antennas had been studied which obtained broadband circular polarization by various techniques. Over the past years, antennas with different kinds of slots, such as Crescent moon shaped [2], C-shaped slot antenna [3], had been produced. A MIMO microstrip-fed antenna with both rectangular shapes cut on the ground on either side of the feed in [4] had been demonstrated for wideband circular polarization applications. A coplanar waveguide (CPW) feed antenna with three parallel coplanar ground planes was also reported in [5]. In [6], microstrip-fed monopole antenna has been claimed to CP mode excitation, an inverted L-shape parasitic strip has been used to achieve the CP. A square slot antenna with dual circular polarization was proposed in [7] using two asymmetric T-shaped feed lines. In [8], CPW-based monopole wideband antenna with an asymmetrical ground plane and a square-ring with an opening at the bottom edge was reported. A regular-hexagonal wide-slot antenna structure for broadband CP with CPW feed had been filed in [9], which also had slot protruded into an L-shaped monopole patch. By etching an L-slot, the dual-band function is obtained in the radiating patch of [10]. The characteristic mode theory is used in [11] to direct antenna design and evaluate performance. MIMO antenna comprising uneven antenna elements and two-sided shorted ground had been contemplated in [12]. Two MIMO antenna devices, one for 4G band and other for a possible 5G band, were analyzed in [13]. Authors in [14] have described the ways to reduce electrical coupling of the MIMO antenna. In [15], to prevent the current flow between the two ports of the same antenna, rectangular slots were engraved on either side of ground. The defected ground structure of a two-element orthogonally arranged MIMO had been presented in [16]. The MIMO efficiency analysis obtained by the use of orthogonal CP radiators in contrast to orthogonal linearly polarized radiators had been presented in [17]. The two salient performance parameters of the MIMO antenna, namely, isolation and bandwidth had been observed in [18]. On the ground plane, two ground stubs had been inscribed and that led to enhancement of isolation and increased impedance bandwidth in [19]. This concept helped us understand ways to attain better isolation for our built antenna.

Table 1 Comparative analysis of the proposed antenna

Ref. no	Antenna size (mm*mm)	Operating BW(GHz)	Axial Ratio	Isolation (dB)	Polarization
Midya et al. [1]	48 * 48	3–4.2	3–4.2	≤ −15	RHCP/LHCP
Saini and Dwari [7]	60 * 60	2–4.76	2–3.7	Below 15	RHCP/LHCP
Xu et al. [8]	90 * 100	0.88–3.08 3.64–5.58	0.98–2.68 3.69–5.5	–	RHCP/LHCP
Zhou et al. [9]	62 * 62	1.80–4.6	2.5–4.1	Below 15	RHCP/LHCP
Proposed antenna	48 * 48	3.4–10.1	3.5–5.5	Below 15	RHCP/LHCP

Although properties such as wide bandwidth and circular polarization had been observed and studied over in the above literature survey, they tend to face certain problems such as complex designs, complicated geometry, relatively narrow axial ratio bandwidth (ARBW), and larger size.

Table 1 shows the results of a comparison of the proposed antenna with quoted antenna in the literature. It can be clearly noticed from the table that the antenna suggested has comparatively better CP features, axial ratio bandwidth (ARBW), and an ultra-wide bandwidth with better size optimization.

Based on the paper [2], a novel UWB-CP antenna is proposed in the carried out work. A T-shaped feed is used in place of the rectangular feed to attain a better ARBW. Also, in proposed antenna, a hexagonal stub has been created which further helps in widening the antenna bandwidth. The CST microwave studio tool was used to build and configure the proposed antenna. This antenna is apt for next gen wireless applications and ultra-wide bandwidth circular polarization applications.

2 Antenna Design and Configuration

The proposed antenna conformation is outlined in Fig. 1. While designing the antenna, 1.6 mm depth of the substrate, Rogers RT5880LZ with loss tangent of 0.02, and relative permittivity of 2.0 is used. Measurements of the proposed antenna are as stated in Table 2. It is developed over a square substrate having the size of 48.0 × 48.0 mm^2.

At the lower edge, the antenna comprises a ground plane and two microstrip feed lines at the upper portion of the dielectric substrate. On the ground surface, a large circular shaped slot with two strip lines connecting the two edges is established to obtain ultra-wide bandwidth. A decagon has been added in the slot which acts as a hexagonal stub and initiates better result.

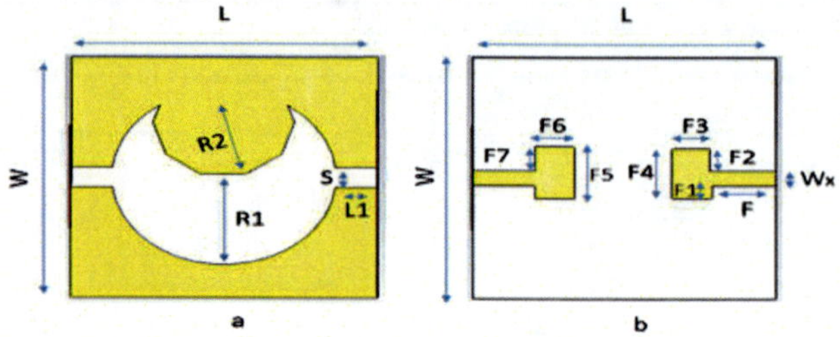

Fig. 1 Proposed antenna configuration

Table 2 Measurements of proposed antenna (in mm)

Parameter	Value	Parameter	Value	Parameter	Value
W	48	L1	6.61	F3	6
L	48	W_x	3	F4	10
R1	17.5	F	10.0	F5	10.5
R2	11	F1	2.5	F6	6.25
S	4	F2	4.5	F7	5

The hexagonal-shaped stub protruding from the ground surface helps in achieving CP. A 50 Ω feed line is used to excite the antenna. It is located from both the sides in the precinct of the hexagonal stub to broaden the ARBW of the single element. Also, the rectangular slots created in the ground thus decreases the feed-ground mutual coupling and helps in getting better parametric results.

2.1 *Evolution Process of the Proposed Antenna*

In Fig. 2, the four progression steps of the proposed antenna have been shown. Firstly, a two-feedline antenna with a circular slot is depicted as Ant1. It provides operating bands approximately from 4.38 to 9.5 GHz. Poor isolation above 15dB is obtained throughout in this case. Secondly, the decagon is added in the complete circular slot, thereby creating a stub as shown in Ant2. It has the return loss in the range of 3.64–7.9 GHz. However, ARBW below 3dB is not achieved in this antenna. Thirdly, in the ground plane of Ant2, two rectangular slots are integrated to decrease the mutual

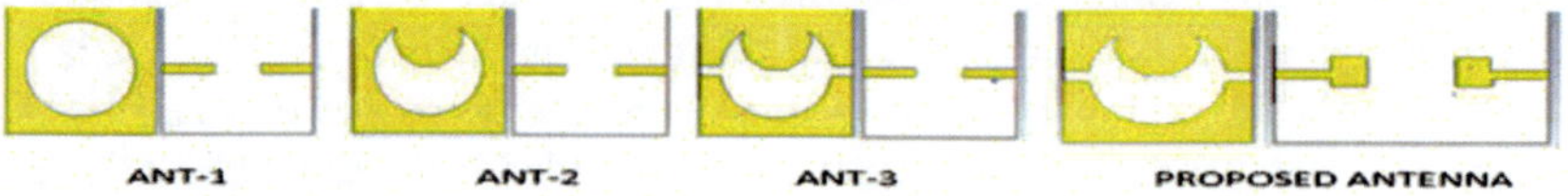

Fig. 2 Progression steps of the developed antenna

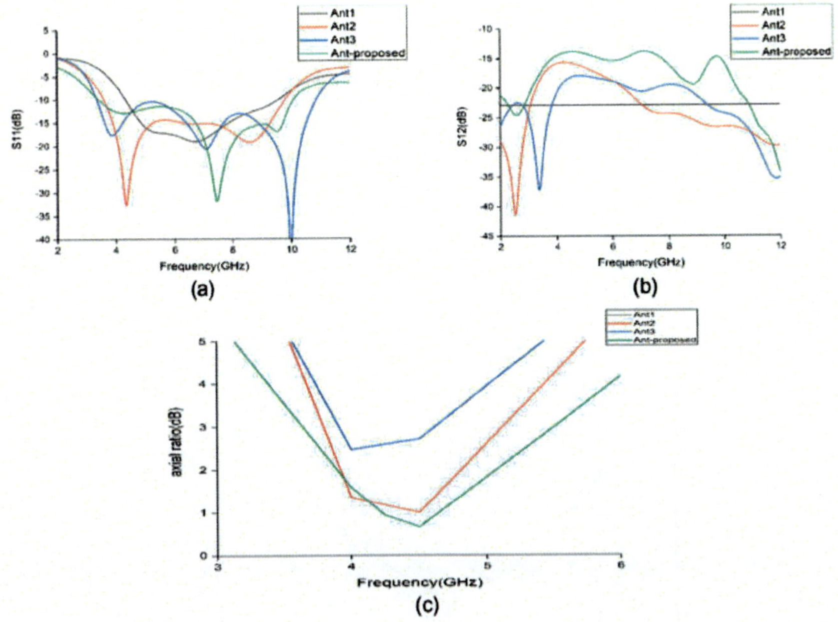

Fig. 3 Antenna performance parameters **a** Magnitude of S_{11}, **b** Magnitude of S_{12}, **c** Axial ratio

coupling between feed and ground. This is structured as shown in Ant3. Finally, to raise Axial ratio, the rectangular feed is designed as T-shaped feedline, parallelly facing each other.

The plots for s-parameter and ARBW with respect to every evolution change of antennas are shown in Fig. 3. Antenna 1 is named as Ant 1, and similarly Antenna 2 and 3 are named as Ant 2 and Ant 3, respectively.

2.2 *Parametric Study*

It is undoubtedly essential for us to research the performance of the developed antenna parameters. We shall therefore understand the variability of S_{11}, S_{21}, and AR in account of other materials with different value of permittivity.

2.2.1 Effect of the Substrate Permittivity

The S_{11}, S_{21}, and axial ratio variability with various other material type of the proposed antenna's substrate are depicted in Fig. 4. It had been seen that when

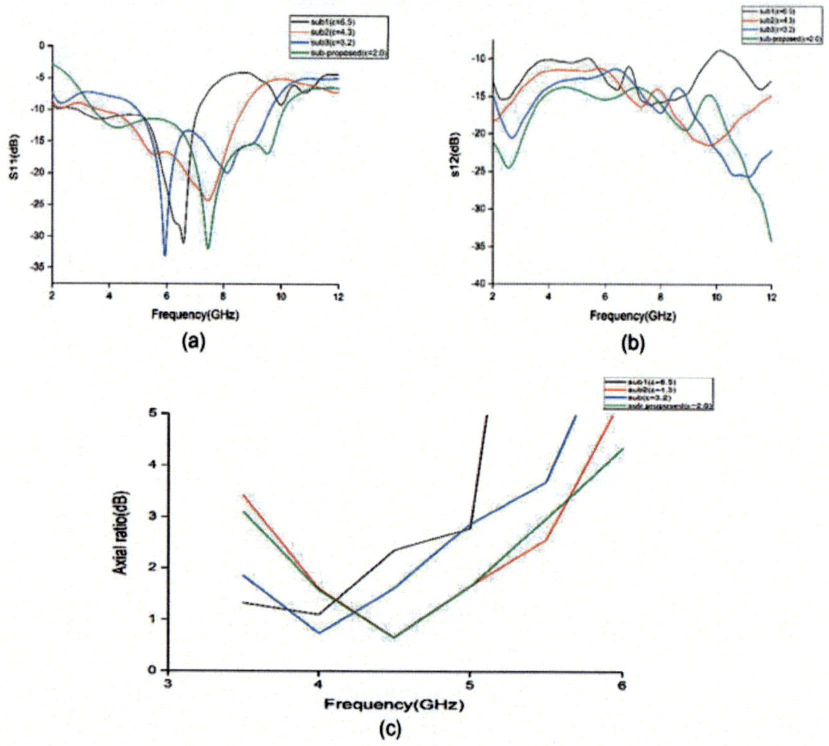

Fig. 4 Effect of substrate on **a** magnitude of S11 **b** magnitude of S12 **c** axial ratio

Rogers RO3006 with a permittivity of 6.5 had been used as substrate, the bandwidth gets limited to approximately 4.5 GHz. Also, it suffered with lesser ARBW. When the substrate had been changed to FR-4 (permittivity = 4.3), comparatively an increase in bandwidth had been observed. A better ARBW had been obtained but at the cost of poor isolation of the antenna. Rogers RO4830 substrate with a permittivity of 3.2 had been then selected as the substrate material. A better return loss, isolation as well as ARBW had been obtained in this case although the bandwidth achieved had been comparatively less than the proposed antenna. The best return loss had been observed in the proposed antenna where the substrate Rogers RT5880LZ had been used having a permittivity of 2.0. The ARBW, isolation, and bandwidth had been maintained in this case. Considering the results obtained, it had been made clear that the substrate with lower dielectric gives better return loss and isolation. In addition to this, substrate material has a big role to play in the ARBW. Material with higher dielectric value leads to poor ARBW in many cases.

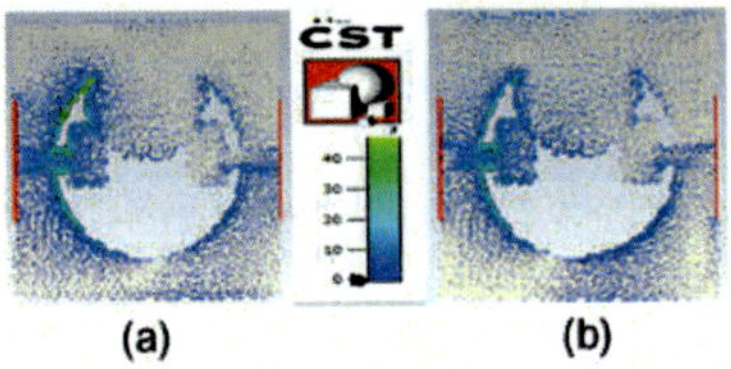

Fig. 5 Surface current at **a** 4.5 GHz **b** 5.5 GHz when port1 is excited

2.3 *Surface Current*

Figure 5 shows the simulated induced current distribution at the resonant frequency of 4.5 and 5.5 GHz. It should be remembered that at the same temporal moment, all the flows had been presented. When one port had been stimulated, the other port had been terminated through 50ohm matched load. Figure 6 clearly illustrates that when port1 is excited, the current density at port2 is lesser and vice versa. This antenna is thus said to have less mutual coupling. It was found by the effects of the surface current that antenna port which is not excited has low induced current. However, excited port has opposite rotation of the current for every step. This corresponded in achieving various modes of CP.

3 Results and Discussion

Figure 6 describes the obtained s-parameter of the antenna developed. The built antenna can be said to have an impedance bandwidth in the range of 3.4–10.1 GHz. In addition to this, the antenna provides adequate isolation (mutual coupling) less than -15dB within a range of 3.1–10.1 GHz. The axial ratio of the simulated antenna is put forward in Fig. 7. From axial ratio graph, it is clear that

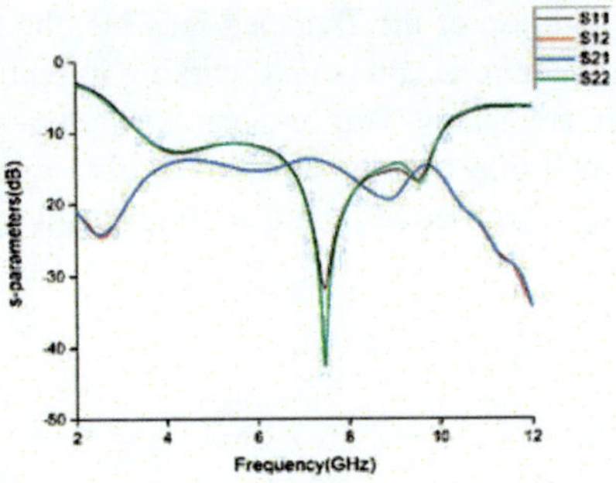

Fig. 6 Simulated S-parameters of the built antenna

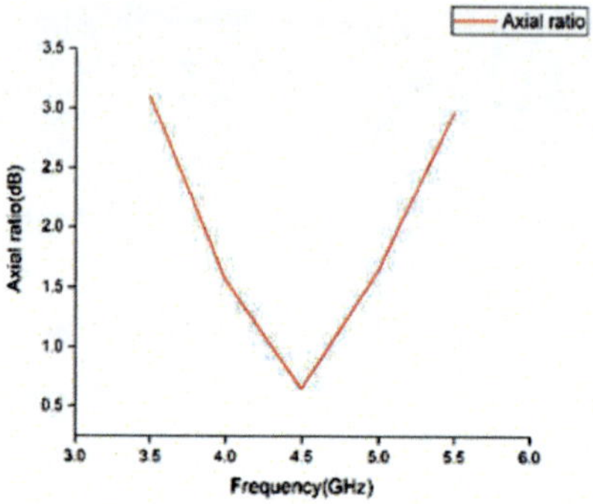

Fig. 7 Simulated AR of the built antenna

AR is below 3dB for the frequency spectrum 3.5–5.5 GHz, therefore it is offered CP characteristics for the wide band more than 2 GHz.

Radiation patterns at interested frequency of 3.5 GHz, 4.5 GHz, and 5.5 GHz are shown in Fig. 8. The results verify the correctness of the designed CP antenna with ultra-wide band. This antenna can exhibit LHCP and RHCP at the same frequency.

The gain and efficiency plot of the developed antenna is illustrated in Fig. 9 with frequency on the X-axis. The antenna is advantageous as it has peak gain of more than 4 dBi within the interested frequency range. The radiation efficiency of the antenna is also more than 70 %.

4 Conclusion

A new UWB-MIMO antenna with circular polarization is presented and implemented using CST successfully in this study. It has a very broad impedance bandwidth, a strong 3dB ARBW with good circular polarization, and port isolation performance. By using a circular slot with hexagonal stub, a wideband CP has been established. After the addition of the T-shaped feedline, the 3dB ARBW has been further expanded. The antenna is competent working in both modes of LHCP and RHCP at the very same frequency. This antenna is an ultra-wide bandwidth CP in nature, and therefore, will play significant role for ultra-wide bandwidth applications. Its expanded ARBW and good polarization makes it propitious for next gen applications.

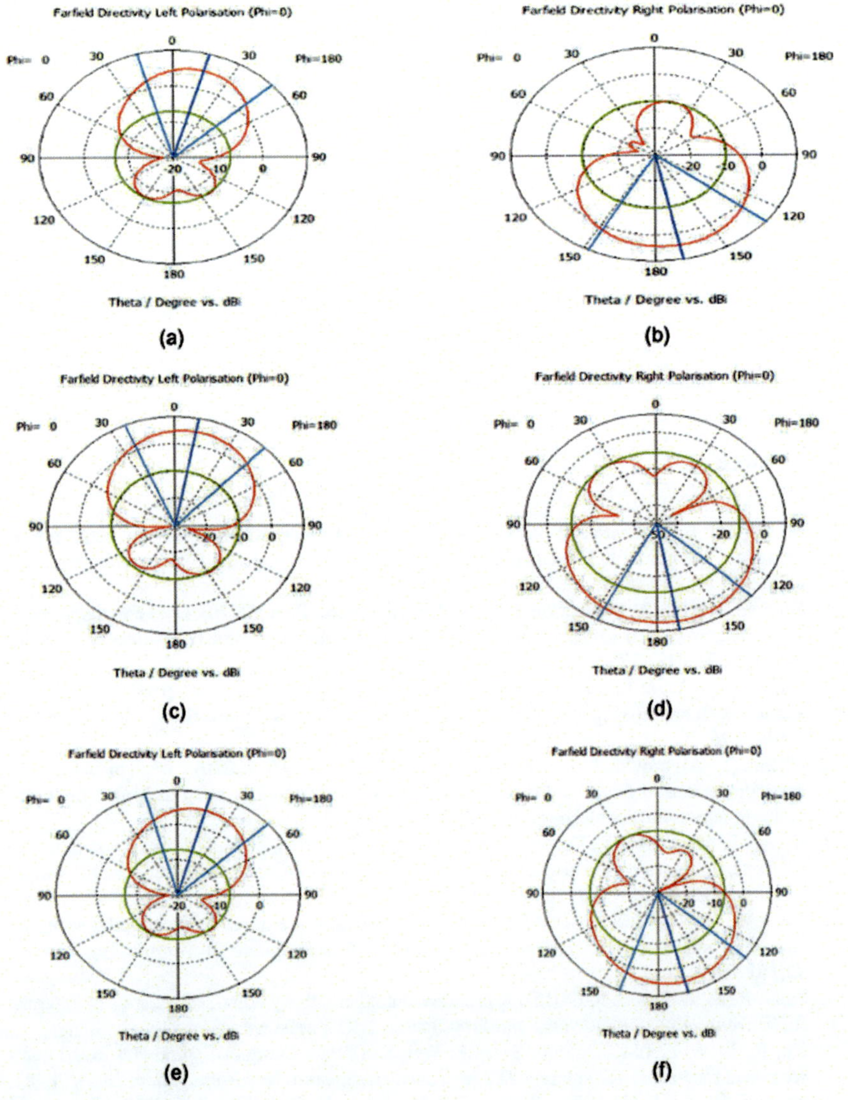

Fig. 8 Radiation patterns of the built antenna for LHCP and RHCP at the frequency **a** 3.5 GHz **b** 4.5 GHz **c** 5.5 GHz

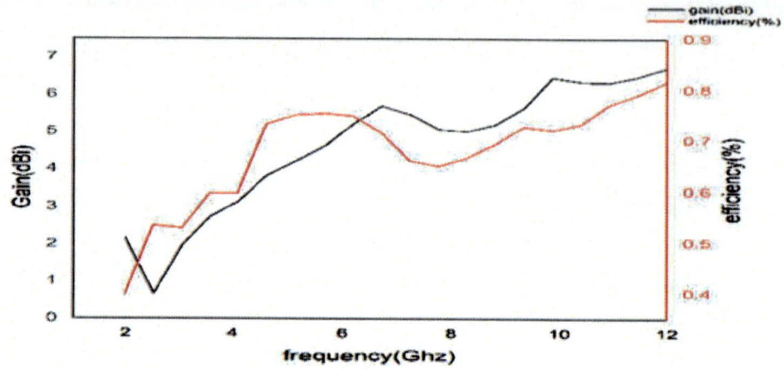

Fig. 9 Plot of gain and efficiency versus frequency

References

1. Midya, M., Bhattacharjee, S., & Mitra, M. (2019). Broadband circularly polarized planar monopole antenna with G-shaped parasitic strip. *IEEE Antennas and Wireless Propagation Letters, 18*(4), 581–585.
2. Ojaroudi Parchin, N., Basherlou, H. J., & Abd-Alhameed, R. A. (2020). Dual circularly polarized crescent-shaped slot antenna for 5G front-end systems. *Progress In Electromagnetics Research, 91*, 41–48.
3. Xu, R., Li, J. Y., & Liu, J. (2017). A design of broadband circularly polarized C-shaped slot antenna with sword-shaped radiator and its array for L/S-band applications. *IEEE Access, 6*, 5891–5896.
4. Ullah, U., Mabrouk, I. B., Koziel, S., & Al-Hasan, M. (2020). Implementation of spatial/polarization diversity for improved-performance circularly polarized multiple-input-multiple-output ultra-wideband antenna. *IEEE Access, 8*, 64112–64119.
5. Ullah, U., Al-Hasan, M., Koziel, S., & Mabrouk, I. B. (2020). Circular polarization diversity implementation for correlation reduction in wideband low-cost multiple-input-multiple-output antenna. *IEEE Access, 8*, 95585–95593.
6. Ullah, U., & Koziel, S. (2018). Design and optimization of a novel miniaturized low-profile circularly polarized wide-slot antenna. *Journal of Electromagnetic Waves and Applications, 32*(16), 2099–2109.
7. Saini, R. K., & Dwari, S. (2015). A broadband dual circularly polarized square slot antenna. *IEEE Transactions on Antennas and Propagation, 64*(1), 290–294.
8. Xu, R., Li, J. Y., Liu, J., Zhou, S. G., & Wei, K. (2018). Design of spiral slot-based dual-wideband dual-sense CP antenna. *IET Microwaves, Antennas and Propagation, 13*(1), 76–81.
9. Zhou, S. W., Li, P. H., Wang, Y., Feng, W. H., & Liu, Z. Q. (2011). A CPW-fed broadband circularly polarized regular-hexagonal slot antenna with L-shape monopole. *IEEE Antennas and Wireless Propagation Letters, 10*, 1182–1185.
10. Boukarkar, A., Lin, X. Q., Jiang, Y., Nie, L. Y., Mei, P., & Yu, Y. Q. (2017). A miniaturized extremely close-spaced four-element dual-band MIMO antenna system with polarization and pattern diversity. *IEEE Antennas and Wireless Propagation Letters, 17*(1), 134–137.
11. Li, H., Sun, S., Wang, B., & Wu, F. (2018). Design of compact single-layer textile MIMO antenna for wearable applications. *IEEE Transactions on Antennas and Propagation, 66*(6), 3136–3141.
12. Luo, X. S., Weng, Z. B., Zhang, W. J., & Yang, L. (2018). Compact planar multiband MIMO antenna based on composite right/left-handed transmission line for mobile phone applications. *Microwave and Optical Technology Letters, 60*(6), 1505–1511.

13. Sharawi, M. S., Ikram, M., & Shamim, A. (2017). A two concentric slot loop based connected array MIMO antenna system for 4G/5G terminals. *IEEE Transactions on Antennas and Propagation, 65*(12), 6679–6686.
14. Nadeem, I., & Choi, D. Y. (2018). Study on mutual coupling reduction technique for MIMO antennas. *IEEE Access, 7*, 563–586.
15. Chattha, H. T. (2019). 4-port 2-element MIMO antenna for 5G portable applications. *IEEE Access, 7*, 96516–96520.
16. Kumar J. P., & Karunakar G. (2018). Circular slotted ring shaped mimo antenna for ultra wideband applications. In *2018 IEEE International Conference on Computational Intelligence and Computing Research (ICCIC)*, vol. 13, pp. 1–5.
17. Dicandia, F. A., Genovesi, S., & Monorchio, A. (2017). Analysis of the performance enhancement of MIMO systems employing circular polarization. *IEEE Transactions on Antennas and Propagation, 65*(9), 4824–4835.
18. Sharma, M. K., Kumar, M., Saini, J. P., & Singh, S. P. (2020). Computationally optimized MIMO antenna with improved isolation and extended bandwidth for UWB applications. *Arabian Journal for Science and Engineering, 45*(3), 1333–1343.
19. Sharma, M. K., Kumar, M., & Saini, J. P. (2015). UWB-MIMO antenna with enhanced isolation for breast cancer detection. In *2nd International Conference on Computing for Sustainable Global Development (INDIACom)*, (pp. 787–790).

Intelligent Traffic Control System for Emergency Vehicles

Anuj Sachan and Neetesh Kumar

Abstract Traffic Control System has faced many issues, such as high waiting time, resulting in the emission of CO_2 (carbon) gas, which results in environment degradation and the loss of crucial time for the emergency vehicles, which increases impatience in drivers, which could be one of the primary causes of increase in accidents. There is also a considerable delay in vehicles' arrival time, particularly emergency vehicles that get stuck in the long queue. It is a matter of concern, as these vehicles are being categorized as priority vehicles, so their waiting time should be minimum. The current Traffic Control System does not use any specific algorithm to control the traffic for priority vehicles such as fire brigade, ambulance. These drawbacks can be resolved by replacing the existing Traffic Control System with a more intelligent Traffic Control System. In this work, we proposed an Intelligent Traffic Control System for Emergency Vehicles model (ITCSEV), which can identify the vehicles and separate them based on their types like "emergency," "passenger." It can change the duration and traffic signal accordingly. This model has been tested by us through a practical simulation using an open-source simulator, i.e., SUMO (Simulation of Urban MObility), on the map of India's Gwalior city, which demonstrates the efficiency and reliability of the model.

Keywords Traffic Control System · Emergency vehicles · Arrival time · Waiting time · SUMO

1 Introduction

The Traffic Control System is essential for organizing vehicles' movement and managing the chaotic situations of traffic congestion in cities' junctions. The continuous rise in vehicles' density has raised many concerns like air pollution, which causes environmental degradation, and noise pollution, which causes mental and

---

A. Sachan (✉) · N. Kumar
Indian Institute of Technology-Roorkee, Roorkee, Uttarakhand 247667, India

N. Kumar
e-mail: neetesh@cs.iitr.ac.in

P. Nayak et al. (eds.), *IoT and Analytics for Sensor Networks*, Lecture Notes in Networks and Systems 244, https://doi.org/10.1007/978-981-16-2919-8_14

physical stress on drivers. The current Traffic System divides the phase duration in an equal number of the time frame and then changes the Traffic Lights to Red/Green/Yellow accordingly. Often in case of malfunction of traffic signals, the light traffic intersections must be personally controlled by traffic police using hand signals. However, that is not the case every time. A particular lane does not always have congestion, so there is no need to fix any lane's time duration. It should be dynamic, real-time, and relative to the number of vehicles present at any amount of time in that lane.

Traffic Control System should be enhanced to establish a smart city in the future in the generation of advanced technology [1–7]. For this, the existing standard way of traffic control would not work [8, 9]. Thus, we have to shift to an Intelligent Traffic Control System which would sense the conditions and adjust relatively to the situation. ITCSEV (Intelligent Traffic Control System for Emergency Vehicles) model will be used to inform the traffic signals to change the time duration and color of the traffic signal accordingly when an emergency vehicle is approaching the junction. It will also result in less waiting time, reduced wastage of fuel, decreased noise and air pollution, and the cutback in road accidents. The rapid development in networking and data collection has permitted the execution of an Intelligent Traffic Control System that can have accurate records of the real-time traffic data daily of different types of vehicles, i.e., car, bus, police, fire brigade, etc. [4]. The proposed Intelligent Traffic Control System for Emergency Vehicle (ITCSEV) model detects the emergency vehicle from the vehicular traffic network through the induction loop detectors. It changes the traffic light accordingly to Red/Yellow/Green state. A functional simulation is being generated using an open-source simulator tool to check the feasibility of the proposed model, i.e., Simulation of Urban Mobility (SUMO) [10–12] on the map of Gwalior City, India. Also, the proposed ITCSEV model results are compared with the standard simulation, emergency simulation, and real-time traffic situation.

2 Related Work

Plenty of researches are being done to monitor the traffic lights in recent times efficiently and effectively. Previously, the researchers focused mainly on linear programming, fuzzy logic, deep reinforcement, etc. [10, 11]. However, all these techniques tried to handle the traffic scenario with a piece of finite and outdated knowledge, and none of them had any method to detect an emergency vehicle on the vehicular network [1, 2].

In this ITCSEV model, an algorithm detects an emergency vehicle through an induction loop detector that identifies the vehicle type. If the detected vehicle turns out to be an emergency vehicle [7], that lane's traffic light turns "Green" accordingly. This model also stores the vehicles' records like waiting time, CO_2 (carbon) emissions, average queue length, etc. Earlier, there was no way to have these records because of the unavailability of advanced cameras and sensors. Many issues in the vehicular network could not be solved due to the absence of this critical information.

There are mainly two drawbacks in the previous works. Firstly, most researches test the simulations in a simple traffic scenario that lacks reality, and secondly, there was no uniqueness given to any emergency vehicle. In contrast, in the ITCSEV model, the simulation is being tested on the map of Gwalior city of India and assigned siren to the emergency vehicle for proper identification [5, 6].

3 The Proposed Protocol

3.1 Problem Statement

This model focuses on prioritizing the vehicle based on vehicle types like the car, bus, emergency vehicle (ambulance, police, and fire brigade), and then dynamically decides to control the traffic suitably. The proposed ITCSEV model operates in two stages as shown in Fig. 1. In the first stage, the Traffic Control System collects various information of different vehicles from the vehicular network with the help of induction loop detectors, which work as a sensor. The second stage uses the captured data to decide whether to pass the particular vehicle by changing the traffic light.

The traffic light has three signals, i.e., Red/Yellow/Green, so a single traffic light cannot control the traffic, especially when the vehicles are coming from different directions at one junction. In such a case, an emergency vehicle cannot find its

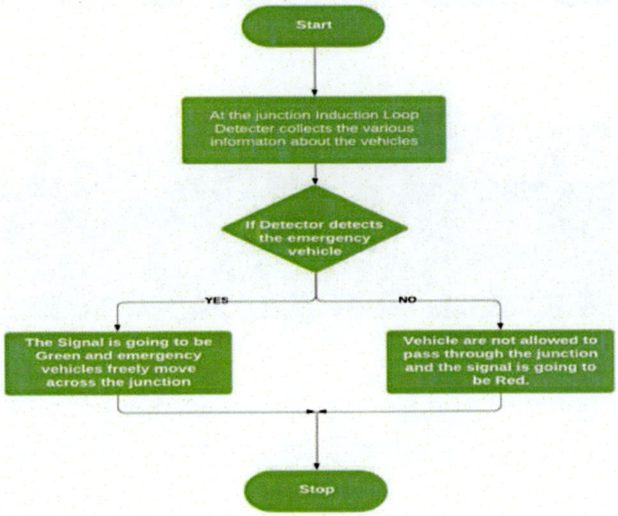

Fig. 1 Flowchart of recommended method

way out of that chaos. So, there should be a proper algorithm that can identify the emergency vehicle and let it pass from that congestion.

The time duration of stay at one specific traffic signal, i.e., Red, Green, Yellow, is known as a single phase. The number of potential signals mark the total number of phases at that junction. At that intersection, all phases begin to shift in a repetitive cyclic fashion. This technique aims to set the phase of the signal process dynamically. So, if the signal is green (G), the vehicle will travel freely through the junction, and if the traffic signal is Red (R) at any junction, then the vehicles are not permitted to drive through.

In Fig. 2, the traffic signal is Red as there are no emergency vehicles currently at the junction. The induction loop detector detects the vehicle and stores its information. Now in Fig. 3, an emergency vehicle "Ambulance" is detected by the induction loop. So, the traffic light at that junction will be changed to Green, and free passage will be given to that vehicle.

The Intelligent Traffic Control System for Emergency Vehicles (ITCSEV) model learns to identify an emergency vehicle and respond with a better decision, which allows the Traffic Control System to manage the traffic situations on the vehicular network and to save a significant amount of time.

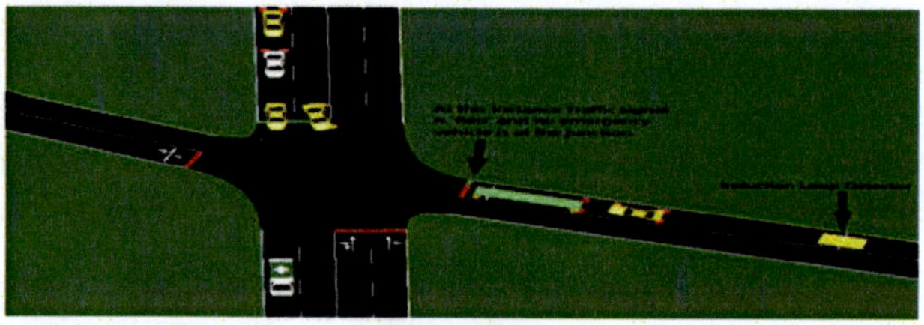

Fig. 2 Before ambulance is detected

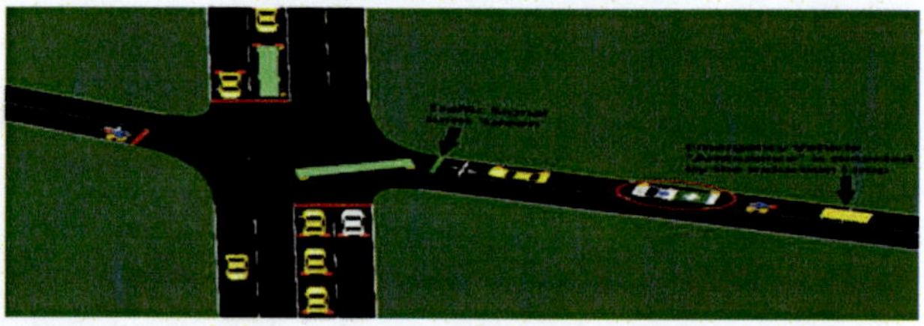

Fig. 3 After ambulance is detected

3.2 Intelligent Traffic Control Model

ITCSEV model is different from previous models, as in this model, there are induction loop detectors that keep the record of the traffic from the vehicular network. Numerous dictionaries of junction nodes are created in this model, which stores the information of the traffic light phases on different lanes like, for example, N1 = "R6": "GggggrrGrrr," "R1": "GrrrrRgGGGg," "R4": "GrrgGGgGrrr," so basically, here N1 is a dictionary which is used as a junction node, R1, R4, R6 are the keys which represent the lanes that intersect to form a junction, and "G," "g," "R," "r" are different phases of the traffic light.

Two parameters are used to measure the state of traffic at the intersection, i.e., the location and velocity of a vehicle in the vehicle network. The entire traffic intersection is broken into square grids of the same size. Two values (position, velocity) of the vehicle within the grid describe the state-space for each grid. The position is a binary value where zero (0) means that inside the grid there is no vehicle, and one (1) means the presence of the vehicle. The velocity dimension is a float value that shows the vehicle's current velocity in a meter/second. In Tables 1 and 2, the traffic conversion into state-space is shown.

The model is smart enough to pick a suitable action based on the traffic scenario for the smooth flow of the vehicles at the junction. The model scans the state of the traffic in this system and chooses one of the two acts:

0-does not change the phase of traffic signal, and

1-turn on the green lights sequentially for the next traffic light.

Firstly, the model calculates the load (weight) of the vehicles on the vehicular network using the function cal weight. In Algorithm 1, the conditional loop for takes in a vehicle id from a list of vehicles on the vehicular network list1. Then, the type of vehicle is initialized and compared with different vehicle types, i.e., police, ambulance, bus, etc., and weight is set accordingly to the respective vehicle types.

Table 1 Presence of the vehicle in the traffic grid

1	0	0	0	0	1	1
0	0	1	0	0	1	1
0	0	0	1	0	0	1
0	1	0	0	0	1	1

Table 2 Velocity of the vehicles in the traffic grid

0.9	0	0	0	0	0	0
0	0	0.6	0	0	0	0
0	0	0	0.5	0	0	0
0	0.5	0	0	0	0	0

Algorithm 1 Find weight

Input: list1 (list of vehicles)
Output: Load (weight) of different vehicles

```
Funtion cal_weight(list1)
    weight = 0;
    for vehicles in list1 do
        t = current vehicle type
        if t == police then
            weight = weight+3;
        end if
        if t == bus then
            weight = weight+5.5;
        end if
        if t == truck then
            weight = weight+6;
        end if
        if t == motorcycle then
            weight = weight+0.5;
        end if
        if t == ambu then
            weight = weight+5;
        end if
        if t == fire then
            weight = weight+5;
        else
            weight = weight+3;
        end if
    end for
    return weight
end
```

Algorithm 2 Find vehicle between detector and traffic signal

Input: list of detector id
Output: dict1 (contains detector ids)

```
Function detector()
    for detector id(i) in list of detectors do
        value = list of vehicle that were on the induction loop
        lane1 = id of lane the induction loop is on.
        if value == True then
            lane2 = id of lane the named vehicle was at the last step
            if lane1 = lane2 then
                add vehicle id of named vehicle to detector dictionary
                apply set() on detector dictionary
                convert detector dictionary to list (dict1)
            end if
        end if
    end for

    for detector id(i) in dict1 do
        lane1 = id of lane the induction loop is on
        for vehicle id(k) in dict1[i] do
            lane2 = id of lane the vehicle id(k) was at the last step
            if lane1 != lane2 then
                remove vehicle id(k) from dict1[i]
            end if
        end for
    end for
    return dict1
end
```

Now, the model finds the vehicle on each lane between the detector and traffic light with the help of induction loop detectors using function detector in Algorithm 2. The first conditional loop traverses the different detectors in the detector list. If the lane id of that lane on which the induction loop is on equivalent to the lane id of that lane on which the last vehicle was on then, the vehicle id of that vehicle is added to the dictionary of detectors and is converted to a list. The second conditional loop traverses the appended detectors list, and the lane is initialized with the id of the lane the induction loop is on. If the lane on which the induction loop is not equivalent to the lane of the last step vehicle, then the vehicle id is removed from the list of detectors.

Next, the model traverses the list of vehicles to find the type of detected vehicle through the detectors in Algorithm 3, and a counter is initialized to zero (0). If the type of vehicle is emergency like ambulance, then the counter is increased to one (1), and if the condition fails, the counter remains zero (0). Further in Algorithm 4, the waiting time and the total number of emergency vehicles are initialized to zero (0). The list of detectors is traversed, and the list of vehicles list1 is initialized with the value of the current detector id. Now that list of vehicles list1 is traversed and a variable is initialized for the type of vehicle that is detected, if that vehicle type matches an emergency vehicle then the total number of emergency vehicle is incremented by one (1), and the waiting time of that emergency vehicle is added to the total waiting time. Now, if a total number of emergency vehicles is greater than zero (0), the integer value of the average waiting time of the emergency vehicle is returned.

Algorithm 3 Find emergency vehicle

Input: list1 (list of vehicles)
Output: type of vehicle (emergency or other vehicle)

```
Function emergency_vehicle
    counter = 0
    for vehicle id in list1 do
        type = vehicle type of detected vehicle
        if type = vehicle (emergency) then
            counter = counter + 1
        else
            counter = 0
        end if
    end for
    return counter
end
```

Algorithm 4 Find waiting time of emergency vehicle

Input: dict1 (contains detector ids), list1 (list of vehicles)
Output: waiting time of emergency vehicle

```
Function avg_wt_vehicle
    waiting_time = 0
    total_e_vehicle = 0
    for detector id(i) in dict1 do
        list1 = detector id(i)
        for vehicle id in list1 do
            type = vehicle type of detected vehicle
            if type = emergency vehicle (ambulance) then
                total_e_vehicle = total_e_vehicle + 1
                waiting_time = waiting_time + (waiting time of detected vehicle)
            end if
        end for
    end for
    if total_e_vehicle > 0 then
        return integer value of (waiting time / total_e_vehicle)
    else
        return 0
    end if
end
```

4 Experimental Analysis

4.1 *Simulation Study*

The experimental study of the proposed ITSCEV through a practical simulation is discussed in this section. The simulation is prepared using the open-source simulator Simulation of Urban Mobility on the real map of Gwalior City in India (SUMO). The performance of the proposed ITSCEV model is evaluated in Normal Simulation, Emergency Simulation, and Real-time Simulation on the Waiting Time of the emergency vehicle. The simulated results demonstrate the feasibility of the model and how the waiting time of emergency vehicles on the vehicle network can be minimized.

4.2 *Evaluation Methodology and Parameters*

Using Python, the proposed ITSCEV model algorithms are implemented, and the simulation was developed in an open-source Urban MObility Simulation simulator tool (SUMO). In this model, the Fuzzy inference module is explicitly used to select an execution mode by observing the current status of vehicular traffic scenarios, and the model uses a free and open-source data flow software library, i.e., TensorFlow.

The simulation is carried out on the map of the Gwalior city of India by using a practical map that is converted into roads and traffic lights with the aid of SUMO conversion software, i.e., netconvert. The length of the vehicle depends on the type of vehicle present in the vehicles' network. With over 1000 vehicles in the entire vehicle network, the vehicles' arrival in the network follows randomly. SUMO also provides an interface to display the traffic flow in real time, i.e., TCI (Traffic Control Interface), and is also compatible with the Python API (Application Programming Interface), which is primarily used to retrieve vehicle's information from the traffic light control junction and to control the duration of the traffic light accordingly.

In addition, the combination of Python and SUMO is very powerful with the aid of the Python API (TCI), enabling the user to monitor the SUMO traffic flow in Python through written algorithms. The length of yellow signal in the Traffic Control System is set at 5 s, which is transition time from green to red for the signal to change. The goal of our model is to minimize the waiting time of the emergency vehicle by changing the traffic signal to green whenever emergency vehicle is detected via the induction loop detectors.

4.3 Experimental Results

This subsection presents the details of the experimental results of the ITSCEV model proposed along with the comparison of different parameters with the other model.

The figures show the comparative results of the ITSCEV model based on average waiting time, impatience level, speed, and time loss of emergency vehicles through a graphical representation. The X-axis represents the number of simulation steps, also known as episodes. The Y-axis indicates various parameters as (a) average waiting time in seconds (s), (b) impatience level, (c) speed (m/s), and (d) time loss (s) of the emergency vehicle, respectively. The model is being run on three different types of simulations, i.e., Standard simulation in which the phase duration of traffic lights at junctions is fixed, Emergency Simulation in which the emergency vehicle is given free passage on traffic lights, and Real-Time Simulation, which runs on a real-world traffic light scenario.

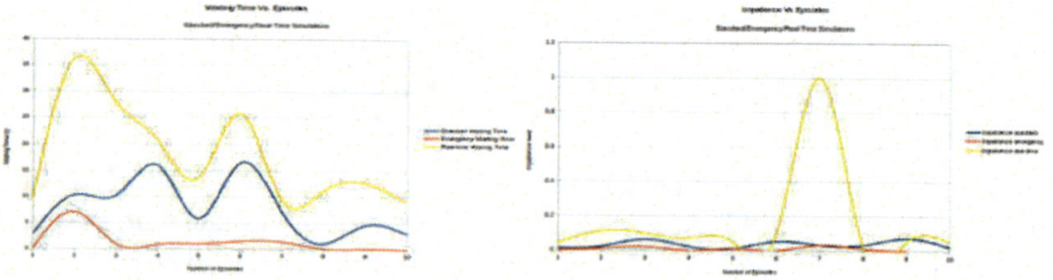

Fig. 4 Waiting time and impatience level

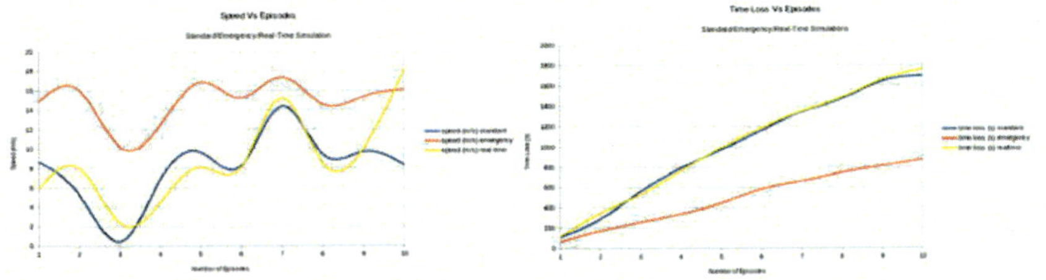

Fig. 5 Speed and time loss

In Fig. 4, the graph represents the average waiting time of the emergency vehicle at a junction. It is clearly visible that the emergency vehicle's waiting time in the ITSCEV model is meager compared to another scenario. In Fig. 4, the graph depicts drivers' impatience level on-road, which is very low in the emergency simulation compared to other scenarios. This will benefit the drivers as high impatience levels can cause stress and anxiety and can even lead to road accidents endangering people's lives.

In Fig. 5, the graph represents the speed of the emergency vehicle on the vehicular network. So, if an emergency vehicle has to reach its destination on time, it should advance at a certain speed. In this model, since the priority vehicles are given a free passage at the traffic junction, the vehicle's speed would be much higher than in other scenarios. In Fig. 5, the graph illustrates the time loss of emergency vehicles on the vehicular network in different scenarios. It can be seen that the time loss in the ITSCEV model is low compared to other simulations. This can perhaps result in saving many lives as the emergency vehicle can reach the destination on time.

It is evaluated through the series of experiments that the proposed model successfully reduces the average waiting time, the degree of impatience of drivers, and the time loss of emergency vehicles. It can also be used for high priority vehicles such as ambulances, police cars, and fire brigades in the future. Since the experiment consists of multiple vehicles, including emergency vehicles and high priority vehicles, the ITSCEV dramatically decreases the total waiting time.

5 Conclusion and Acknowledgement

The introduced model, i.e., Intelligent Traffic Control System for Emergency Vehicle (ITCSEV), explains the working of the induction loop detectors. This model will take data from different vehicles that will accordingly minimize the waiting time of emergency vehicles to a bare minimum. A practical simulation is generated using an open-source simulator tool, i.e., Simulation of Urban Mobility (SUMO) on the map of Gwalior city of India, to check the efficiency of the ITCSEV model. Normal simulation, emergency simulation, and real-time traffic simulation are compared with the results of the ITCSEV model.

Acknowledgements The authors would like to acknowledge Indian Institute of Technology-Roorkee, India for sponsoring the conference and the Council of Scientific and Industrial Research (CSIR), Govt. of India as this work is carried out under the project titled "Emergency Aware Intelligent Dynamic Traffic signal System for smart cities" with grant id 22(0817)/19/EMR-II).

References

1. Du, X., & Liang X. (2019). A deep reinforcement learning network for traffic lights' cycle control in vehicular network. IEEE Transactions on Vehicular Technology.
2. Azimirad, E., Pariz, N., & Sistani, M.B. (2010). A novel fuzzy model and control of single intersection at urban traffic network. IEEE Systems Journal, *4* 107–111.
3. Hammi, B., Khatoun, R., Zeadally, S., Fayad, A., & Khoukhi, L. (2017). IoT technologies for smart cities. IET Networks, *7*(1), 1–13.
4. Shah, S.H., & Nahrstedt, K. (2002) Predictive location-based QoS routing in mobile ad hoc networks. In *2002 IEEE International Conference on Communications. Conference Proceedings. ICC 2002 (Cat. No.02CH37333)*, vol. 2 (pp. 1022–1027). New York, NY, USA.
5. Kompella, V. P., Pasquale, J. C., & Polyzos, G. C. (1993). Multicast routing for multimedia communication. *IEEE/ACM Transactions on Networking, 1*(3), 286–292.
6. Roy, A., & Das, S. K. (2004). QM2RP: A QoS-based mobile multicast routing protocol using multi-objective genetic algorithm. *Wireless Networks, 10*, 271–286.
7. Kumar, V., & Kumar, S. (2016). Energy balanced position-based routing for lifetime maximization of wireless sensor networks. Ad Hoc Networks, *52*, 117–129, ISSN 1570-8705.
8. Haghighat, A.T., Faez, K., Dehghan, M., Mowlaei, A., & Ghahremani, Y. (2003). GA-based heuristic algorithms for QoS based multicast routing. Knowledge-Based Systems, *16*(5–6), 305–312, ISSN 0950-7051.
9. Meraihi, Y., Acheli, D., & Ramdane-Cherif, A. (2019). QoS multicast routing for wireless mesh network based on a modified binary bat algorithm. *Neural Computing and Applications, 31*, 3057–3073.
10. Fahad, M., Aadil, F., Rehman, Z.U., Khan, S., Shah, P.A., Muhammad, K., Lloret, J., Wang, H., Lee, J.W., & Mehmood, I. (2018). Grey wolf optimization based clustering algorithm for vehicular adhoc networks. Computers and Electrical Engineering, *70*, 853–870, ISSN 0045-7906.
11. Genders, W., & Razavi, S. (2016). Using a deep reinforcement learning agent for traffic signal control. arXiv preprint arXiv:1611.01142.
12. Kumar, N., & Rahman, S.S. (2019). Deep reinforcement learning with vehicle heterogeneity based traffic light control for intelligent transportation system. In*2019 IEEE International Conference on Industrial Internet (ICII)*, IEEE, pp. 28–33.

Digital Controller-Based Automated Drainage Water Monitoring and Controlling

T. Sairam Vamsi, Ch. Hari Krishna, P. Srinivasaraju, and G. Srinivasarao

Abstract The main purpose of this paper is to replace the manual work in sewage cleaning and monitoring with automation. In this era, there is a lack of utilization of technology for sewage disposal tasks. At present, manual effort has been used for drainage cleaning, which leads to loss of human life while cleaning the blockage in pipes. In order to overcome these types of problems, "Digital Controller based automated drainage water monitoring and controlling" will be helpful. This project includes PLC (Programmable Logic Controller), which plays a vital role in solving real-time problems. Using the resource in this paper, it is possible to monitor and control the sewage system continuously without using manual work.

Keywords PLC · Sewage disposal · Automation

1 Introduction

Drainage is the system by which excess water or other liquids are drained from a place. These are also in place to remove wastewater effectively, and this is referred to as a sewer system. One of the main functions of the drainage system is to collect and transport waste materials along with water [1]. Due to the solid waste, the flow of water will get blocked in drainages, these blockages lead to improper flow of drainage system. The drainage system must be cleaned time to time. In this present era, though there is high development in technology, cleaning of these drainages is done manually. The person who enters to clean the drainage is affected due to hazardous gases and faces serious health issues. Sometimes, there may be loss of

T. S. Vamsi (✉) · Ch. Hari Krishna · P. Srinivasaraju · G. Srinivasarao
SVECW, Bhimavaram, India

Ch. Hari Krishna
e-mail: harikrishnach@svecw.edu.in

P. Srinivasaraju
e-mail: viceprincipal@svecw.edu.in

G. Srinivasarao
e-mail: principal@svecw.edu.in

P. Nayak et al. (eds.), *IoT and Analytics for Sensor Networks*, Lecture Notes in Networks and Systems 244, https://doi.org/10.1007/978-981-16-2919-8_15

human life. So, there is a need to use automation in drainage systems to remove manual work.

There have been many methods used till now for cleaning of drainage systems but there is no completely automated method [2]. Here, in this paper, we are going to eliminate manual work entirely for total working of cleaning the drainage system. The method which is implemented in this paper helps in preventing the blockages in drainages automatically. This method is used to extract the solid waste materials from the wastewater. A series of mesh layers are employed at the flow of drainage water in order to collect the solid waste materials. The level of wastewater in the water tank is monitored using the level sensors. The collected solid waste will be dumped into a garbage bin with the help of mechanical motion based on the output from the sensors. There the waste materials will end up into tiny particles. The pumping of water into the mesh layers is controlled by the Level and Ultrasonic sensors [3].

2 Block Diagram

The block diagram for the proposed solution is shown in the Fig. 1.

1. The project utilizes two Level sensors, one Ultrasonic sensor, and one pH indicator.
2. One Level sensor is placed at the low level and one is at high level in the water tank to measure the level of the water [4].

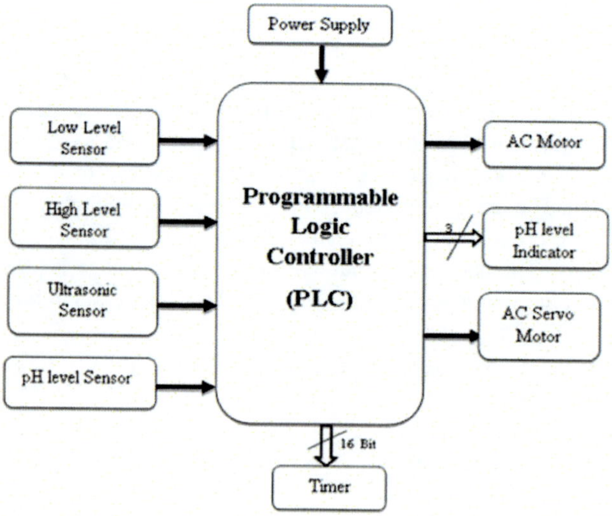

Fig. 1 Block diagram of the proposed solution

3. In the net collecting solid waste, an Ultrasonic sensor is placed in order to check the solid waste level.
4. The pH indicator is used to get the acidic and basic nature of the water, thereby LEDs are used to indicate the current range of pH [5].
5. By taking the input from the sensors through Programmable logic Controller (PLC), AC motor and AC servo motor will be swiveled [6].
6. Timer will be used to provide delay between the tasks.

3 Implementation of the Project Using PLC

The project "PLC Implementation of Control Logic for Monitoring and Controlling of Drainage Water System" is implemented in two stages [7].

1. Software Implementation.
2. Hardware Implementation.

The Software Implementation of the project is shown in Fig. 2. Initially, the values from the corresponding sensors were taken. If waste level sensor is in OFF state, then the slider will be in rest state, then depending on low-level sensor and high-level sensor conditions, motor will be either in OFF or ON state [8]. Otherwise, the waste level sensor will be in ON state, then simultaneously the slider is in ON state, and timer also starts working. After the timer stops, the waste level indicator will go to OFF state. The pH sensor will be continuously in working state irrespective

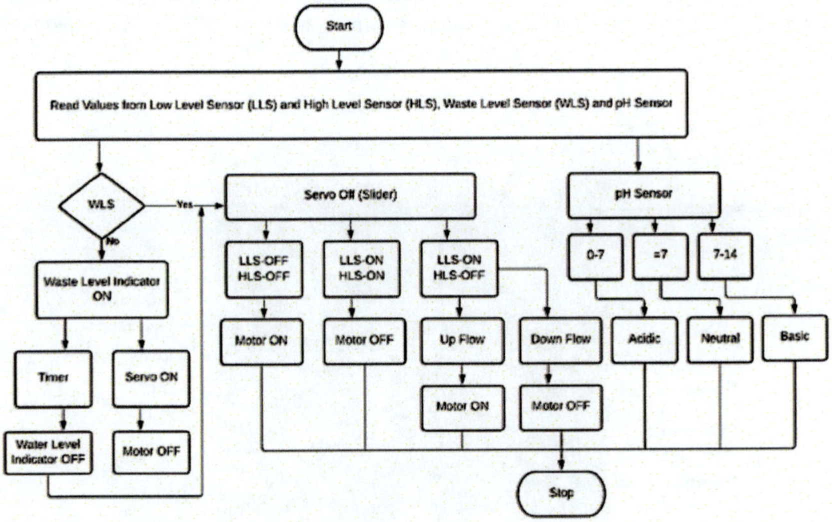

Fig. 2 Flow chart of software implementation

Table 1 Hardware components and its specifications

S. No	Name of the component	Specification
1	PLC	DVP-14SS11R2
2	8 channel relay cards	Single changeover
3	Terminal block	–
4	SMPS	230VAC–(19–26)V DC
5	Miniature circuit breaker (MCB)	5SL61027RC
6	Level sensor	–
7	Normally open and normally closed switches	–
8	DC motor	12 V
9	Inductive sensor	–
10	Buzzer	12 V DC
11	Indicator	Pilot Indicators

of the remaining sensors and always note corresponding pH values which will feed respective indicators (Table 1) [9].

The hardware implementation of the project is shown in the following Fig. 3.

In the Fig. 3, NC Switch, NO Switch, Inductive Sensor, Level Sensor are used as inputs, and Buzzer, DC Motor, Pivot Indicator are used as outputs. Here, NC Switch is used as low-level sensor to detect the wastewater level in the water tank, Level Sensor is used as High-level sensor to identify the water level at the top of the tank [10]. Inductive sensor acts as Mesh Level Sensor for identifying the level of waste in the mesh layer. In this prototype, DC Motor replaces the operation of AC Motor to pump the water. Based on the pH values of water, when the pH level is in between

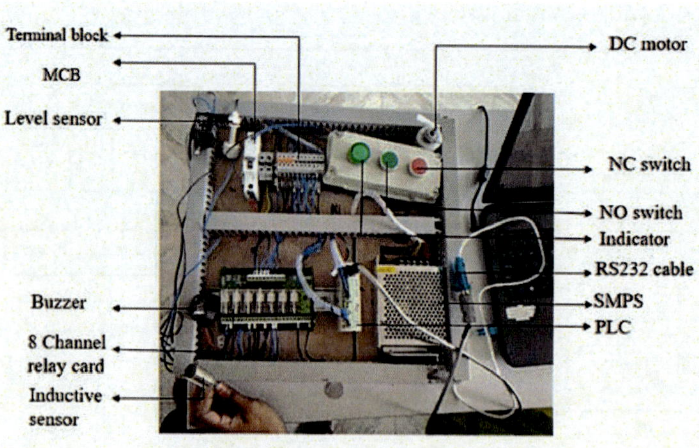

Fig. 3 Hardware implementation of the project

0 and 7, we will be able to hear a buzzing sound to indicate the nature of water as acidic. If the pH is 7, Pivot indicator will be ON to represent that the water is in neutral nature [11]. If the pH level is above 7 and below 14, both buzzer and pivot light will be in OFF condition to indicate the nature of water as Basic.

4 Results and Discussions

Initially, all the sensors are in their respective positions. Based on the inputs given by the sensors (Level sensors, pH sensor), the operation of ac motor and ac servo motor will be functioning. Table 2 shows sensor functionality with respect to servo motor with two different levels, Low and High.

4.1 Detailed Description of Stages

The following cases show the working stages of sensors and motor with respect to software implementation.

Case 1: If the Low-Level Sensor and High-Level Sensor are in OFF stage, then the motor will be in ON stage. Similarly, when LLS or HLS is OFF, then Motor will be in ON Stage. Figures 4 and 5 show the PLC outputs and inputs when the LLS and HLS are in different conditions. X0, X1 indicates LLS and MMS, Y0 indicates output, i.e., Motor.

Note: In figures, Color indicates OFF and Normal indicates ON (Input Side), Color Indicates ON and Normal indicates OFF (Output Side).

Table 2 Stages of sensors and motor functioning based on their conditions

AC Servo motor	Low-level sensor (LLS)	High-level sensor (HLS)	AC motor
ON	X	X	OFF
Upward flow			
OFF	OFF	OFF	ON
OFF	ON	OFF	ON
OFF	OFF	OFF	ON
OFF	ON	ON	OFF
Downward flow			
OFF	ON	OFF	OFF
OFF	OFF	OFF	ON

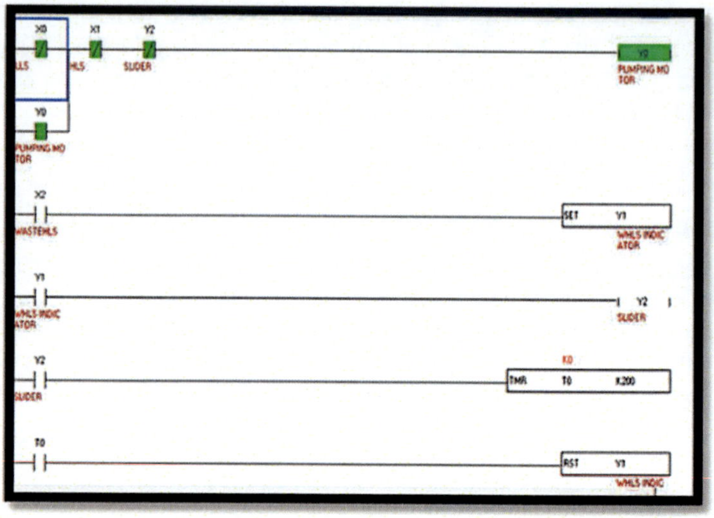

Fig. 4 PLC output when LLS-OFF, HLS-OFF

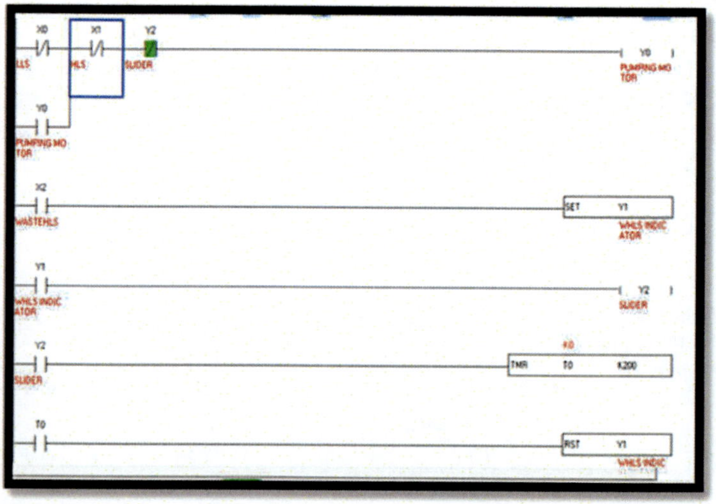

Fig. 5 PLC output when LLS-ON, HLS-ON

Case 2: Figure 6 shows the PLC output when servo motor is ON and Low-level Sensor, Servo Motor (Slider) are in ON Stage and High-level Sensor is in OFF Stage, then the Motor will be in OFF stage.

Case 3: If the Waste Level Sensor is in ON stage, then the Servo Motor (Slider) will be ON stage which drives the Motor into OFF stage automatically.

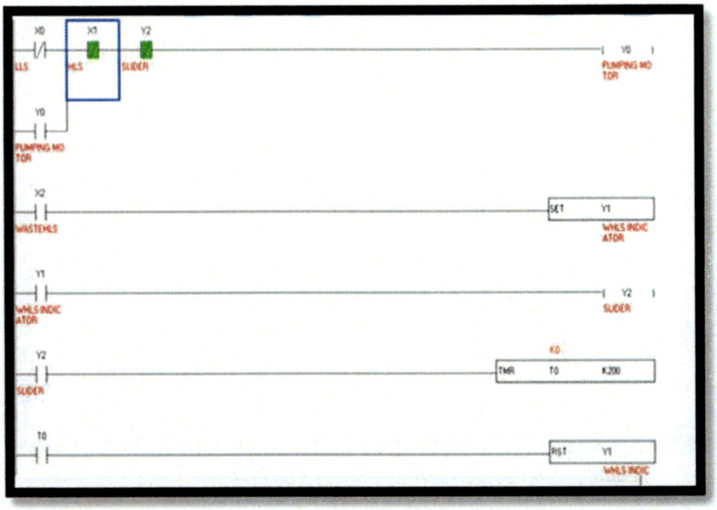

Fig. 6 PLC output when LLS-ON, HLS-OFF, servo motor-ON

Case 4: If the waste level sensor is in OFF stage and the Servo motor is in ON stage, then the Motor will be in OFF stage.

The above Figs. 7 and 8 show the PLC output when controlling with Waste level sensor which is indicated as X2.

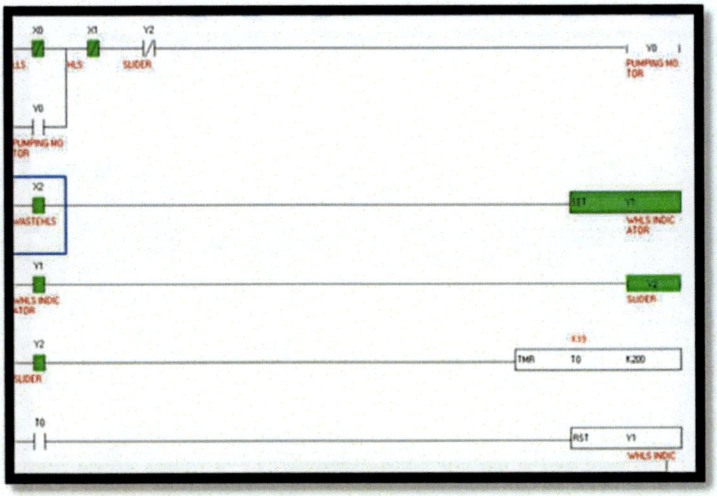

Fig. 7 PLC output when WLS-ON, servo motor-ON

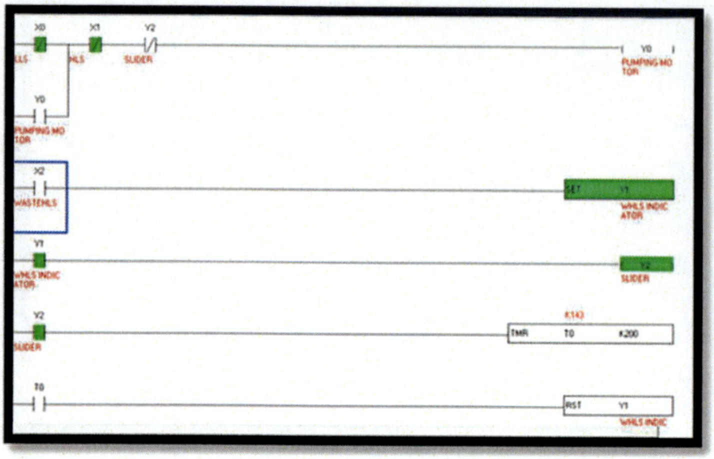

Fig. 8 PLC output when WLS-OFF, servo motor-ON

4.2 PH Level Indications

The following cases show the pH level indications of the water. Based on pH levels, there are three indications generated like Neutral, Acidic, and Basic, and we kept threshold as 7 to generate these outputs. Figures 9 and 10 indicate the Indicators enabling according to pH levels in the water.

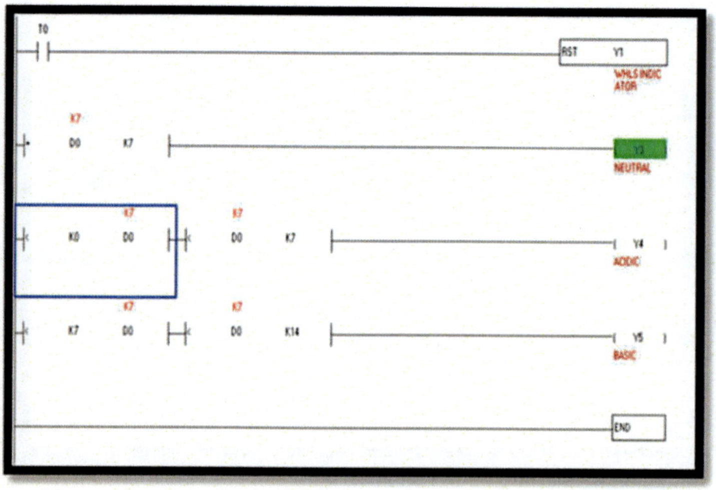

Fig. 9 pH neutral indication

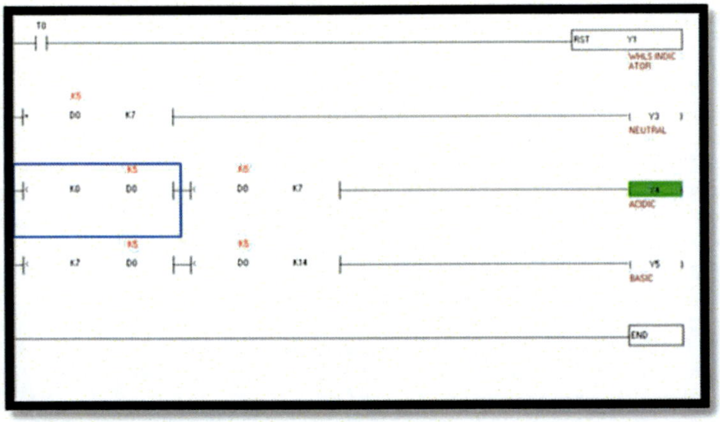

Fig. 10 pH acidic indication

Case 1: If the pH level of the water is equal to 7, then the Neutral Indicator will be ON.

Case 2: If the pH level of the water is below 7, then the Acidic Indicator will be ON, and If pH level of water is above 7, then the Basic Indicator will be ON.

5 Conclusions

The drainage system usually gets blocked because of the solid waste present in water. These get jammed in small holes in the path of water flow, which leads to back flow of water on to the roads. There are many cleaning techniques for the sewage disposal tasks. The project implemented in this paper is more reliable and fully automatic. This is mainly used in large workplaces like industries, hospitals, etc. It can separate the solid waste completely. A PLC logic for drainage system monitoring and controlling is designed in DELTA WPL Soft 2.49 in order to collect and separate the solid waste from drain water. The collected waste is made into tiny pieces at the garbage bin.

Acknowledgements This work is carried by Applied Robotic Control (ARC) Lab in Shri Vishnu Engineering College for Women (A).

References

1. Sathiyakala, R., et.al. (2016). Smart sewage cleaning system. International Journal of Innovative Research in Computer and Communication Engineering, *4*.
2. Xu, J., & Wang, W.B. Design of sewage treatment system based on arm controller. In *2016 28th Chinese Control and Decision Conference (CCDC)*.
3. Haswani, N.G. et al. (2018). Web-based realtime underground drainage or sewage monitoring system using wireless sensor networks. In *Fourth International Conference on Computing Communication Control and Automation*.
4. Deshpande, V. S., & Vibhute, A. S. (2014). Home automation using PLC and Amit S. Vibhute SCADA. *Multidisciplinary Journal of Research in Engineering and Technology, 1*(1), 111–118.
5. Nandgave, A., Deshbhratar, H., Khandare, S., & Heda, L. (2014). Industrial drives & automation using PLC. International Journal Of Engineering Research & Technology (IJERT), *3*(2).
6. Sairam Vamsi, T., & Srinivasa Raju, G.R.L.V.N. (2018). ARM-based industrial circuitous like robotic implementation using CAN bus for surveillance. In S. Satapathy, V. Bhateja, P. Chowdary, V. Chakravarthy, & J. Anguera (Eds.), *Proceedings of 2nd International Conference on Micro-Electronics, Electromagnetics and Telecommunications*. Lecture Notes in Electrical Engineering, vol 434. Springer, Singapore. https://doi.org/10.1007/978-981-10-4280-5_28.
7. Sairam Vamsi, T., Viswanadam, R., & Praveen kumar K. An industrial survivalence sensoric robot controlling through Arduino and GSM. International Journal of Contol Theory and Applications (IJCTE) Scopus Indexed with ISSN:0974-5572, 9(23), 87–96.
8. Du, Z.B. (2014). Study on treatment of wastewater system based on ARM. Advanced Materials Research, https://doi.org/10.4028/www.scientific.net/AMR.1006-1007.732.
9. Zhang, Q., Wang, X., & Fang, Z. Water treating control system based on predictive control theory of multi-model switching algorithm. Proceedings of the 5 World Congress on intelligent Control and Automation, *4*(5), 633–636.
10. Peng, X., Xiao, L., Mo, Z., & Liu, G. (2009). The variable frequency and speed regulation constant pressure water supply system based on PLC and fuzzy control. IEEE International Conference on Measuring Technology and Mechatronics Automation, 910–913.
11. Xinping, W. (2007). Application of PLC program control technology in the field of industrial water treatment. *Automatic Instrument, 28*, 168–171.

Cooperative Agent-Based Location Validation for Vehicular Clouds

Shailaja S. Mudengudi and Mahabaleshwar S. Kakkasageri

Abstract The evolution of Intelligent Technology Systems (ITS) has made possible the designing of vehicles with more sophisticated computing, communicating, and sensing capabilities. Vehicular Cloud (VC) magnifies these capabilities by cost-effectively sharing their resources. Location information about a node plays a prime role in many VC services which are based on the location of the user node. In this paper, we present a cooperative agent-based location verification framework for VC. Central Authority (CA), a trusted entity that coordinates the activities in VC, takes care of the location verification process with the help of special nodes known as verifier nodes. Verifier nodes are distributed in the network randomly and secretly. The presence of such nodes is known only to the CA, which makes it stronger in terms of security.

Keywords Vehicular Cloud · Cloud computing · Public key encryption · Software agent

1 Introduction

Vehicular Ad-hoc Network (VANET) is an area of mobile ad-hoc networks intending to provide safety for its users. The increase in the number of smart vehicles on road and evolution in VANET technology enables provision for applications, such as driver safety, enhanced traffic management, and infotainment. The most important of them is the safety application where resources with the vehicles are used effectively to increase the safety of the user in VANET. The vehicle nodes are converted to network devices with communicating and computing capabilities to help the user with safety messages like collision avoidance, obstacle detection, turn, and speed limit warnings. Along with safety, there is a huge demand for applications that ease the driving experience and provide entertainment [1, 2].

S. S. Mudengudi (✉)
Tontadarya College of Engineering, Gadag 582101, India

M. S. Kakkasageri
Basaveshwar Engineering College, Bagalkot 587102, India

P. Nayak et al. (eds.), *IoT and Analytics for Sensor Networks*, Lecture Notes in Networks and Systems 244, https://doi.org/10.1007/978-981-16-2919-8_16

The VANET infrastructure consists of Road Side Unit (RSU) installed at places like sides of the road, and On Board Unit (OBU) which are deployed in the vehicle nodes. OBU and RSU communication facilitate safety applications and other services. As safety related information contributes to the lives of humans participating in VANET, security is of utmost crucial importance [2]. A comprehensive and novel design where resources in VANET are effectively used in collaboration with the cloud is presented in [3], which is termed Autonomous Vehicular Clouds (AVC). Resources in AVC are dynamically pooled to form the cloud and provide services to the authorized user. The aim is the effective utilization of resources, with greater elasticity and from a commercial perspective [4]. The resources from the vehicles, RSU, and conventional cloud together from the Vehicular Cloud (VC). Fundamentals of VC, designing of VC architecture, operations, and services of VC are discussed in [5]. The author also further states, the VC is still under the primitive stage, and issues such as resource discovery, fault tolerance, or management of service-related content storage, and many others still need to be addressed. VC architecture presented in [6] is classified into on board unit, communication, and cloud service layers. The on board unit is installed on the vehicle node which consists of sensors, storage units, etc., with computing capabilities. The communication layer is responsible for the vehicle node to infrastructure and vehicle node to a vehicle node. The cloud computing layer is subdivided into resources, virtualization layer, services, application programming interface, and application layer. VC facilitates several services that may require or may be based on the location of a node. Several methods are present to get location information. It may be done by the node itself using technologies like GPS (Global Positioning System) or by neighboring nodes. VC services are economy-based on a greedy or malicious node that may manipulate its location information to enjoy more facilities than allotted or subscribed. Attacks such as position-spoofing intend to manipulate the position-related information of the node [7].

To the best of our knowledge, until now, there is no mechanism to prove or detect a vehicle nodes position in the VC network. We present an agent-based location verification service, which verifies the location of the vehicle node using verifier nodes deployed randomly in the network. A vehicle node may be denying about its position during a dispute after utilizing the cloud resources available at that location. In case of dispute, the following actions are being actuated:

- The Central authority which is the most trusted and secured entity sends a query regarding the location of the particular vehicle node at a required time.
- The location validation agency verifies the location of the node with the help of verifier nodes that are deployed randomly in the network.
- The knowledge about the presence of the verifier nodes is known only to CA.

This elevates the attacks which are node targeted and increases the security of VC. The VC architecture is shown in Fig. 1. VC consists of Vehicle Nodes (VN), Road Side Unit (RSU), Cloud Service Provider (CSP), conventional cloud, and Central Authority (CA). The vehicle nodes are capable of communicating with CSP, neighboring VN's, and RSUs. Some of the vehicle nodes or the RSU are chosen as verifier nodes by the CA. CA is the trusted entity that coordinated all the activities of VC.

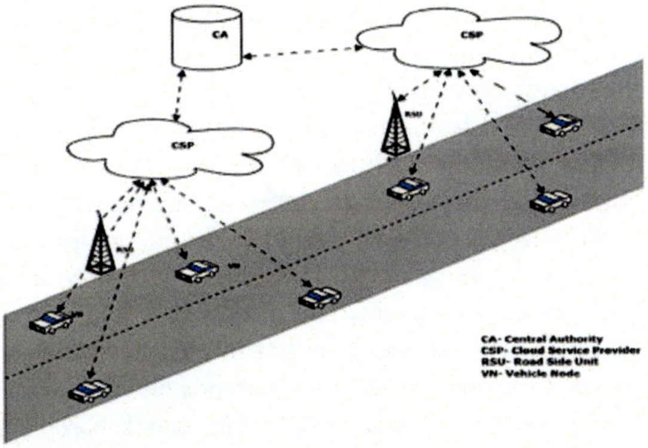

Fig. 1 Vehicular cloud network

It generates the public and private key pair for the verifier nodes and delivers it to respective nodes securely. The following assumptions are made:

- The verifier nodes are chosen by CA and known only to CA.
- The verifier nodes and CA have high storage and computational capabilities.

The paper is organized as follows. Section 2 presents related work, Sect. 3 presents the proposed cooperative agent-based location validation method, the conclusion is presented in Sect. 4.

2 Related Work

VANET is being attractive to elevate issues such as traffic congestion and accidents along with providing security with increased efficiency. Besides, it also faces attacks targeted towards the security and privacy of the user [8, 9]. Aiming to preserve the privacy of the user, identity-based group signatures (IBGS) method is proposed in [10]. The identity of the user is used in the generation of public keys, eliminating the certificate management issue. Intending to avoid malicious users accessing privileges or benefits using fake location claims, a block chain-based location proof generation with verification is presented in [11]. In case of dispute in location-based services, there would arise a scenario in which a node needs to prove its location at a specified instant. A prover node, which needs proof of its location, requests the neighboring nodes via short range communication methods such as Bluetooth. Nodes that are willing to respond to the prover node request authenticate the prover node and location proof singed by it is generated and broadcasted on to peer-to-peer. Location verification is employed as a security mechanism in [12] which uses the Global Navigation Satellite Systems (GNSS) for providing location verification securely with global coverage. The advantage of this system is it can be used by nodes that do not have a facility of navigation or GPS. A virtual force model-based location verification based on anchor promotion algorithm for WSN using incremental refinement is presented in [13]. The mechanism proposed re-locates the drifted nodes and elevates the malicious anchor-based attacks. A two-level verification of location information method is presented in Information Verification cum Security (LIVES) [14]. The first level of location verifications is done using tiles-based verification process where the position of a vehicle node is calculated in terms of latitude and longitude by itself and announced to its neighbors. Next, it is verified based on the signal strength received by the neighboring nodes using the shadowing propagation model. In the second layer, it uses a collective belief-based Transferable Belief list from the neighboring vehicles, for the two-fold verification.

A detailed study for data location verification in the cloud is presented in [15]. Different location verification methods along with vulnerabilities due to malicious CSP attacks are discussed. Further landmark-based approach like a landmark, types of landmarks, landmark distribution scale, landmark selection, Coordination of participant's training, location granularity is presented. In the position-based routing method, participating nodes need to know the destination node's location along with recognizing its geographic locations to route messages. Location service provides current geographic positions of nodes. In [16], a protocol for providing such location service securely using location information which is self-signed by the nodes is presented. But the protocol does not preserve the privacy of the node regarding location information. The location information over a geographic area that is visited by them frequently over a long period is observed and is defined as their Physical Social Location (PSL) which is further used in Mobile phone-based Physical Social Location (MPSL). In the location verification process, the nodes in the PSL act as witnesses to prove the presence of the node [17].

A detailed study of recent trends in cloud architecture is presented in [18]. It defines the cloud as a technology where a cloud user can access the resources from the pool of resources which may be application or storage or infrastructure. This can be achieved by cloud services such as software as a service, platform as a service, and infrastructure as a service which can be deployed in private or public or hybrid cloud structure. The location verification in the cloud is usually the verification of the location where the data is stored. In [19], geo-location-based approach, location verification is achieved using supervised classification algorithms. A literature survey in [20] presents the study of the entire spectrum of agent-based modeling and simulation tools. It also discusses the features, merits, and faults of the software. It also provides guidelines for selecting an appropriate agent-based model and its simulation tool. Public Key Infrastructure (PKI)-based authentication process is presented in [21] which find vast applications such as e-commerce, web services, etc.

3 Proposed Work

We present an agent-based location verification method which consists of a Location validation agency as shown in Fig. 2. The agent comprises of Location Validation

Fig. 2 Location validation agency

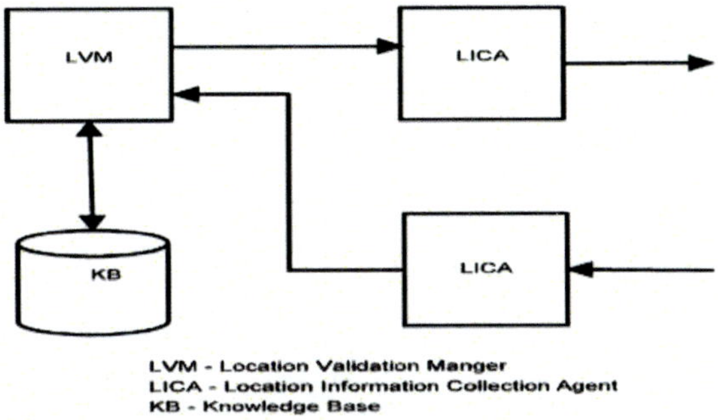

Manager (LVM), Location Information Collection Agent (LICA), Knowledge Base (KB).

Knowledge Base (KB): It consists of location-related information about vehicle nodes, which include entry and exit time into the geographical area, service taken or given to the cloud, trust rate, etc. The knowledge base is updated frequently by the information collected by LICA by the LVM.

Location Validation Manager (LVM): The Location Validation Agency is installed in the CA. The complete coordination between all the agents is executed by the LVM. When a query is received by the location validation agent, LVM triggers the LICA to collect location information from all the verifier nodes, which are distributed in the network. The knowledge about the verifier nodes is known only to the CA. The LICA visits these verifier nodes only instead of visiting all the nodes. This further reduces the cost of the verification process. The security aspect also increases because the knowledge about the verifier nodes is known only to CA. So attacks can be elevated which are node targeted.

Location Information Collection agent (LICA): The LICA collects information about the vehicle node location from the verifier nodes. The location details are encrypted using public key encryption.

The encrypted message is received by the CA. Table 1 shows the nomenclature. After decrypting the message, the location information is derived from the message. It consists of the ID of vehicle, location information, timestamp about arrival and exit of the vehicle node into the geographical area. The location of the vehicle is tracked from the location information collected by the TICA which is stored in KB.

Let $V_1, V_2, V_3 \ldots.. V_N$ be Verifier nodes. The location information of node 'M' is sent by node 'V' by signing the location information as shown in Eq. 1.

$$\text{LocM}_V = \text{SigV (V||PosM||TrustM||TVI||TVC||TMI||TMO)} \quad (1)$$

CA sends Query as shown in Eq. 2 to verifier node 'V' about Node 'M' location. This is encrypted using public key of 'V' as shown in Eq. 3. 'Type' tells the kind of message it is like, a location query message or a simple encrypted data, etc. 'seq' is a number used to keep track of all the query messages.

Table 1 Nomenclature used in proposed work

X	ID of node *V*
V_{PK}, V_{PR}	Public and Private key of node V
SigV(LOC)	Location-related information *LOC* signed by verifier node V using V_{PR}
CertV	Self-signed certificate of *A*'s public key
TVI	Timestamp representing the time *APR* was first generated
TVC	Timestamp representing the current time generated by *A*
PosV	*V*'s geographical location (coordinate) obtained from GPS
TMI	Timestamp representing the entry time vehicle node M into the geographical area of V
TMO	Timestamp representing the exit time vehicle node M into the geographical area of V
$Trust_M$	Trust factor of node M

$$\text{Loc request} = [\text{Type, seq, M}] \tag{2}$$

$$\text{Encrypted message} = \text{V}_{\text{PK}}[\text{Loc request}] \tag{3}$$

Only 'V' can decrypt Eq. 3 as the message is encrypted using public key of V. The Node 'V' decrypts the message using its private key. If Node 'V' has an encounter with Node 'M,' it responds as shown in Eq. 4, else it does not respond. This message is encrypted using public key of CA as shown in Eq. 5.

$$\text{Loc response} = [\text{Type, Seq, LocM}_{\text{V}}] \tag{4}$$

$$\text{Encrypted message} = \text{CA}_{\text{PK}}[\text{Loc response}] \tag{5}$$

The message received by CA can only be decrypted by CA as it has the private key. So the PKI elevates the malicious nodes manipulating the location information at any phase. So the public and private key pair is very important and is crucial. The public and private key pair is generated using RSA algorithm as shown in Table 2. The following steps show the generation of keys. The proposed work has the following advantages:

1. A separate pair of public and private keys can be used for the location verification process. As the number of verifier nodes is low, there is no significant increase in the storage requirement.

Table 2 RSA algorithm

Generation of keys: 1. Choose two large prime numbers 's' and 'r.' 2. Calculate n = sr. 3. Calculate $\phi(n)$ using $\phi(n) = (s-1)(r-1)$ *Here $\phi(n)$ is the totient function, also known as Euler's totient function.* It is the number of integers strictly smaller than 'n' that is relatively prime to 'n,' where 1 is counted as being relatively prime to all numbers 4. Choose a number 'e' which is relatively prime to $\phi(n)$ 5. To decrypt the message using $de \equiv 1(\text{mod } \phi(n))$ 'd' is the private key can be found using the extended Euclidean algorithm
Encryption of message: At the transmitter, the message 'L' is encrypted using the public key E = Le(mod n) Where E is an encrypted message and L is the location details of the node
Decryption of message: At the receiver, the secret information 'd' regarding the secret key is used for decryption $E^d \equiv (L^e)^d \equiv L^{ed} \equiv L^{N\phi(n)+1} \equiv L \text{ (mod n)}$

2. The CA has a list of all the verifier nodes location, public, and private key details in its database which is further divided, based on geographical area.
3. When there is a need to verify the location of a node, the CA checks the list of verifier nodes in the geographical location. Send query message including the ID of a vehicle about which information is needed. The query message is encrypted using the public key of the verifier node which ensures only the verifier node can read the query. As there is no need to request a public key certificate, there is a considerable reduction in the number of messages transferred between CA and the verifier node.
4. After decrypting the message, the verifier node checks its database and responds only if it has an encounter with that vehicle node in the given period. Due to this, the number of responses will be significantly reduced, and the computation time required to process the responses received by CA will again be greatly reduced. The response message is encrypted using the public key of CA.
5. Malicious nodes cannot manipulate the location information for the following reasons. The query message is visible to only the verifier node. The response message by the verifier node can be decrypted only by CA.

Even if any verifier node or any query or response message is compromised, it does not tamper the location verification process. Suppose, we have verifier nodes 'e,' 'f,' 'g,' 'h' which have an encounter with the vehicle node in a given duration of time and date. If verifier node 'f' or message from it gets compromised, the other three 'e,' 'g,' 'h' verifier nodes respond correctly. So we can trace the location of the vehicle easily.

4 Security Analysis

Privacy: The proposed framework ensures privacy of the verifier nodes identity. Even when the digital location evidence is accessed by a third party, it is unable to learn anything about either the query message or the identity of the verifier node responding to the query message. The completeness and importance of the location information is maintained. The involvement of hidden verifier nodes attains protection against misbehaving nodes trying to submit incorrect location data. The verifier nodes are highly trusted and authenticated nodes, the possibility of them being dishonest is rare.

Confidentiality: The entities involved in location verification are verifier nodes and CA. The messages cannot be decrypted by authorities with highest power such as RSUs. Thus, confidentiality against authorities which possess high power is achieved.

Computationally secure: RSA is computationally secure. It takes considerably very long period to break the RSA. A simple computer will take approximately 300 trillion years to break a RSA with 2048 bit encryption key. Meanwhile, the purpose of the encryption would have been attained.

Avalanche effect: It is defined as the change in number of bits of cipher text corresponding to the change in the number of bits in plain text. A slight change in the plain text must cause significant change in the cipher text. In Fig. 3, it can be observed that RSA produces good avalanche effect when compared to other standard encryption algorithms.

Comparison of different encryption algorithms is presented in Table 3. It can be observed that symmetric encryption method yield good results as compared to asymmetric encryption method. But in the proposed framework, it is crucial that the query messages are accessible only to the intended verifier nodes. In symmetric encryption, for some reason if the key is compromised, then the query message and the

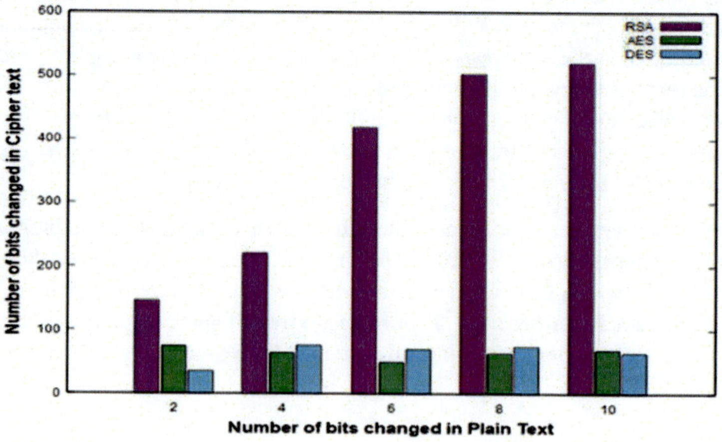

Fig. 3 Avalanche effect

Table 3 Comparison of different encryption methods

Parameters	Symmetric encryption			Asymmetric encryption	
	DES	AES	Blowfish	RSA	Diffie–Hellman
Key size	56 bits	128,192,256 bits	32 to 448 bits	>1024 bits	Key exchange management
Throughput	<AES	<Blowfish	High	Low	<RSA
Power consumption	Low	Low	Low	High	<RSA
Speed	High	High	High	High	Low
Secure against attack	Brute force attack	Chosen—Plain, known plaintext	Dictionary attack	Timings attacks	Eavesdropping
Ciphering and deciphering algorithm	Different	Different	Different	Same	Same
Security	As secret key is shared, risk of key compromise is high	As secret key is shared, risk of key compromise is high	As secret key is shared, risk of key compromise is high	More secure as private key is not shared	More secure as private key is not shared
Block size	64 bits	128 bits	64 bits	Minimum 512 bits	Not applicable

response message both will be compromised. This may lead to several attacks like masquerade, message modification flooding attack, etc. Thus asymmetric encryption method is chosen. RSA and Diffie Hellman are the two popular asymmetric encryption methods. But Diffie Hellman encryption method is used most widely for key exchange. RSA is computationally heavy and requires more power. It is also found that if not implemented correctly, they are easily prone to attacks. But the frequency of dispute arrival and number of location messages transferred are considerably low in number when compared to normal messages communicated. Thus, the extra overhead above factors can be sustained by the CA and verifier nodes.

5 Conclusion

Location-based services in VC face security issues such as location tampering, fake location claims to enjoy more benefits, denying about its location at a particular time and date. In this paper, we have presented a location verification method using special nodes called verifier nodes and trusted CA installed with location verification agent. The agent travels and collects information about nodes location and verifies the same. We use PKI to communicate between the CA and verifier nodes. As the presence of verifier nodes is unknown to normal nodes, security increases as manipulating the location information targeting the nodes are completely elevated.

References

1. Kopylova, Y., Farkas, C., & Wenyuan, Xu. (2011). Accurate accident reconstruction in VANET. In *IFIP Annual Conference on Data and Applications Security and Privacy* (vol. 26, pp. 271–279). Berlin: Springer.
2. Huang, B., Cheng, X., & Cheng, W. (2018). Enhancing negative messages broadcasting with meet-table and TTL in VANET. *Procedia Computer Science, 129*, 185–187.
3. VinothChakkaravarthy, G., Lavanya, R., & Alli, P. (2012). Communication efficient distributed decentralized key management framework for message authentication in vanet. In *International Conference on Advances in Communication, Network, and Computing* (vol. 3, pp. 405–408).
4. Soyturk, M., Muhammad, K. N., Avcil, M. N., Kantarci, B., & Matthews, J. (2016). From vehicular networks to vehicular clouds in smart cities. *SmartCities and Homes, 12*, 149–171.
5. Lee, E., Lee, E., Gerla, M., & Oh, S. Y. (2014). Vehicular cloud networking: Architecture and design principles. *IEEE Communications Magazine, 52*, 148–155.
6. Boukerche, A., & Robson, E. (2018). Vehicular cloud computing: Architectures, applications, and mobility. *Computer Networks, 135*, 171–189.
7. Xue, X., Lin, N., Ding, J., & Ji, Y. (2010) A trusted neighbor table based location verificationfor VANET Routing. In *IET 3rd International Conference on Wireless, Mobile and Multimedia Networks (ICWMNN 2010), Beijing* (pp. 1–5).
8. Ahmad, F., Kazim, M., Adnane, A., & Awad, A. (2015). Vehicular cloud networks: Architecture, applications and security issues. In *2015 IEEE/ACM 8th International Conference on Utility and Cloud Computing (UCC)* (vol. 5, pp. 571–576).
9. Wu, H.-T., & Homg, G.-J. (2016). Vehicular cloud network and information security mechanisms. *International Conference on Advanced Materials for Science and Engineering (ICAMSE), 7*, 196–199.
10. Qin, Bo., Qianhong, Wu., Domingo-Ferrer, J., & Zhang, L. (2011). Preserving security and privacy in large-scale VANETs. *International Conference on Information andCommunications Security, 8*, 121–135.
11. Nosouhi, M. R., Yu, S., Zhou, W., Grobler, M., & Keshtiar, H. (2020). Blockchain for secure location verification. *Journal of Parallel and Distributed Computing, 136*, 40–51.
12. Foresti, S., & Jajodia, S. (2010). Data and applications security and privacy XXIV. In *24th Annual IFIP WG 11.3 Working Conference, Rome, Italy, June 21–23, 2010, Proceedings* (vol. 6166, pp. 112–126).
13. Miao, C., et al. (2015). Collaborative localization and location verification in WSNs. *Sensors (Basel, Switzerland), 15*, 10631–49.
14. Sheet, D. K., Kaiwartya, O., Abdullah, A. H., & Hassan, A. N. (2015). Location information verification cum security using TBM in geocast routing. *Procedia Computer Science, 70*, 219–225.
15. Irain, M., Jorda, J., & Mammeri, Z. (2017). Landmark-based data location veryfication in the cloud: Review of approaches and challenges. *Journal of Cloud Computing, 1*, 31–34.
16. Lim, J., Kim, S., & Heekuck, Oh. (2007). A secure location service for ad hoc position-based routing using self-signed locations. *International Conference on Cryptologyand Network Security, 2*, 121–132.
17. Ni, X., Luo, J., Zhang, B., Teng, J., & Bai, X. (2012). Mpsl: A mobile phone-based physical-social location verification system. In *International Conference on Wireless Algorithms, Systems, and Applications* (vol. 5, pp. 488–499). Berlin: Springer.
18. Odun-Ayo, I., Ananya, M., Agono, F., & Goddy-Worlu, R. (2018). Cloud computing architecture: A critical analysis. In *2018 18th International Conference on Computational Science and Applications (ICCSA)* (vol. 9, pp. 1–7).

19. Ries, T., Fusenig, V., Vilbois, C., & Engel, T. (2011). Verification of data location in cloud networking. In *2011 Fourth IEEE International Conference on Utility and Cloud Computing* (pp. 439–444).
20. Abar, S., Theodoropoulos, G. K., Lemarinier, P., & O'Hare, G. M. (2017). Agent based modelling and simulation tools: A review of the state-of-art software. *Computer Science Review, 24*, 13–33.
21. Haidar, A. N., & Abdallah, A. E. (2009). Formal modelling of pki based authentication. *Electronic Notes in Theoretical Computer Science, 235*, 55–70.

An Analytical Approach for Traffic Grooming Problems Using Waiting Probability in WDM Networks

Priyanka Kaushal, Neeraj Mohan, Surbhi Gupta, and Seifedine Kadry

Abstract Traffic grooming is the optimization of resources in a network. An efficient traffic grooming provides better resource utilization, enhanced performance at a lower cost. Traffic grooming has become very significant for all types of computer networks. As WDM (wavelength division multiplexing) networks are providing very high speed for a huge amount of data transfer, so traffic grooming is even more important for these networks. There are so many parameters that may be considered for traffic grooming. The waiting probability is one of the key parameters for traffic grooming. Waiting probability is a measurement of time spent by a call to get the required resources for further communication. We have proposed an efficient traffic grooming technique. This technique is based on waiting probability calculations. Waiting probability is calculated, and then traffic grooming problems are addressed based on the waiting probability calculations. Some of the other key network performance parameters such as the number of servers required, ideal path length for a source–destination pair, number of free wavelengths required, etc., are also analyzed. It is a low complexity technique for handling traffic grooming problems efficiently in telecommunication and call center management.

Keywords Traffic grooming · Waiting probability · WDM networks · Network optimization

---

P. Kaushal
Chandigarh Engineering College, Landran, Ajitgarh, India

N. Mohan (✉)
IKGPTU, Kapurthala, India

S. Gupta
GRIRT, Hyderabad, India

S. Kadry
Beirut Arab University, Beirut, Lebanon
e-mail: s.kadry@bau.edu.lb

P. Nayak et al. (eds.), *IoT and Analytics for Sensor Networks*, Lecture Notes in Networks and Systems 244, https://doi.org/10.1007/978-981-16-2919-8_17

1 Introduction

Resource sharing and load distribution are the basic purposes of networking. WDM networks can easily share resources both within the network and with other networks. The capability to share resources efficiently makes the WDM networks a preferred choice. Earlier, frequency division multiplexing (FDM) and time-division multiplexing (TDM) were largely used for communication. But now the use of WDM and erbium-doped fiber amplifier (EDFA) has changed the entire networking world. WDM has provided the opportunity to use multiple light beams of different wavelengths in a single optical fiber. Whereas, EDFA can amplify different wavelength signals simultaneously [1, 2]. The WDM networks are very popular in modern era networking due to some factors such as exceptionally high bandwidth, extremely low loss rate, and very low bit error rate. A typical single fiber can provide speed in the range of Tera Bits per second. At the same time, a very low loss rate in the range of 0.2 dB/Km is offered by optical networks. The bit error rate is typically in the range of 109 to 1015. All optical network transmits data using lightpaths in the optical domain only [3].

Some applications need a large amount of data transfer in the range of Terabytes or Petabytes. At the same time, these applications are deadline-driven. The WDM networks along with advance reservation (AR) have provided a hopeful solution [4]. The major advantage of traffic grooming is maximum resource utilization of all the resources available in a network. If the resource utilization is at optimal level, then it will certainly enhance the overall network performance. Optical networks along with wavelength division multiplexing (WDM) technology has offered a platform for high-speed networks. An ultra-high-speed, huge capacity, and reliability are the key parameters that are expected from WDM networks. So, several traffic grooming techniques are applied to these networks so that performance may be optimized [5].

Traffic grooming is achieved through various multiplexing techniques, algorithms, methods, etc., in different domains of WDM networks. As the performance of a network depends on many parameters, it makes traffic grooming a challenging job. Further, it is categorized as static and dynamic. All of these categories have different attributes associated with them [6, 7].

2 Related Work

The authors have analyzed the blocking probability (BP) along with mean delay for a single-wavelength optical buffer. Poisson arrival process has been assumed for packets, whereas packet size follows the general distributions. A mathematical expression has been presented using the level crossing method [8]. An analytical model has been proposed for optical delay line buffer. It assumed that the packet spans are exponentially distributed random variables. It has been shown that the proposed model can reduce burst loss probability [9]. A large wireless mesh network was divided into numerous reduced groups. Each group used to be supported by an optical network unit of a passive optical network (PON). The proposed model was able to handle the heterogeneous node traffic load efficiently [10].

Basic Erlang C formula calculations have been used to determine the traffic route for call centers. The number of calls received per unit time and number of calls forwarded per unit time have given a performance parameter for a particular channel or agent. All these calculations have been done using basic Erlang C formula. Further, mean call handling time and the average number of the processed calls per unit time were also calculated [11]. A method was proposed to calculate the service waiting probability. It was based upon approximating the distribution function. Both service waiting probability and average delay time of packets were calculated [12]. A priority-based mode transformation procedure has been proposed to maximize network resource consumption. The proposed algorithm has improved the quality of service significantly. Different load state combinations were considered. Network resources were allocated depending upon the priority of requests [13, 14].

The Erlang C model has been used as it provided clear and accurate results. It was assumed that the skill level of agents in the contact center was not homogenous. Some of the key parameters such as average waiting time, queuing probability has been calculated for a given number of agents [15, 16]. A two-step method has been proposed to regulate the quantity of spectrum essential for the request of vast data allocations. The demands were generated with a target and blocking rate. Markov chain model was used to allocate variable bandwidth over a lightpath. The overall utilization of all the links was also calculated [17]. Approximate network blocking probabilities have been computed using the proposed mathematical model for heterogeneous all-optical WDM networks. A last-fit-first wavelength assignment scheme has also been proposed for dynamic traffic needs. The blocking probability has been minimized significantly [18].

A traffic grooming model has been introduced for dynamic multicast traffic grooming. This model has significantly reduced the blocking probability. A 0/1 knapsack-based multicast traffic grooming model was proposed [19, 20]. An algorithm based on spray and wait has been proposed to address routing problems. Relay node and token numbers were considered based on delivery probability, delivery time, and buffer status. The simulation results have shown significant improvement in the delivery and delay ratio. The planned algorithm has better efficiency for smaller buffer size [21].

The authors have proposed the utilization of Erlang B and Erlang C formulae for estimating and managing network traffic. It is described that the Erlang B formula does not contain the delay parameter. As the delays are omissible in ATM networks, so it makes Erlang B formula more suitable for ATM networks. The Erlang C formula is capable to estimate delay or waiting probability. So this formula is more suitable for IP networks. These formulae had played a vital role to monitor and ensure the Quality of Service in networks [22]. A model of the queueing system has been proposed with a non-full available server. The results of the proposed model were found quite accurate [23].

It may be summarized that researchers have mainly concentrated on blocking probability to address traffic grooming problems [8, 10, 18, 19, 23, 24]. Still, the waiting probability calculation based traffic grooming is addressed by few researchers [11, 12, 15, 25]. In the next section, the proposed technique is presented which is based upon the waiting probability calculation.

3 Proposed Technique

Several parameters such as blocking probability, server utilization, waiting probability, throughput, and response time determine the performance of a network. The optimal value of these parameters enhances network performance significantly. So, these parameters are known as performance measurement parameters. These parameters are defined as:

Blocking Probability: It is defined as ratio of the calls blocked to the total number of calls.

$$\text{Blocking Probability} = \frac{\text{Number of calls blocked}}{\text{Total number of call received}}$$

Server Utilization: The utilization in a network is usually associated with the utilization of the server. It is the fraction of time for which the server is active. If there are multiple servers, their utilization is calculated in a combined manner. It is calculated as:

$$\text{Server Utilization} = \frac{\text{Arrival Rate}}{\text{Service Rate}}$$

Waiting Probability: It is represented as the time spent by a request in a queue before getting resources. So the response time is the sum of waiting time in a queue and service time. It is calculated as:

$$\text{Waiting Probability} = \frac{\text{Number of calls in the waiting queue}}{\text{Total number of call received}}$$

Throughput: It is defined as the average number of handled requests per unit of time.

Response Time: It is the total time spent by a request in the system [26, 27].

We have proposed a model to calculate the waiting probability of a network. The waiting probability gives an estimation of the resources required to achieve a quality of service level. The call waiting probability (P_w) is described using Erlang C formula as [11]:

$$P_w = \frac{\frac{\mathrm{A^N N}}{\mathrm{N!\,(N-A)}}}{\sum_{i=0}^{N-1} \frac{A^i}{i!} + \frac{A^N N}{N!(N-A)}} \quad (1)$$

N is the total number of servers or agents and A is traffic load in Erlangs, and it may be defined as the ratio of the average number of calls per unit time (μ) to the average number of requests processed per unit of time (T), so

$$A = \frac{\mu}{T} \tag{2}$$

Further, the traffic load (A) of a network is dependent upon other parameters. So the traffic load (A) can be calculated as [24, 25, 28]:

$$A = \frac{\lambda}{l} \tag{3}$$

where λ is the number of free wavelengths and l is path length for a particular source–destination pair.

Now, substituting Eqs. (2) and (3) in Eq. (1), we get

$$P_w = \frac{\frac{(\frac{\mu}{T})^N * N}{N!\,(N-A)}}{\sum_{i=0}^{N-1}\frac{(\frac{\mu}{T})^i}{i!} + \frac{(\frac{\mu}{T})^N * N}{N!(N-A)}} \tag{4}$$

$$P_w = \frac{\frac{\left(\lambda/l\right)^N * N}{N!\,(N-A)}}{\sum_{i=0}^{N-1}\frac{\left(\lambda/l\right)^i}{i!} + \frac{\left(\lambda/l\right)^N * N}{N!(N-A)}} \tag{5}$$

Now, we have Eqs. (4) and (5) which are used to calculate waiting probability with respect to various network parameters. The results obtained may be further used for traffic grooming of the networks. These results are very useful for telecommunication problems and call center management.

4 Results and Discussions

The outcomes of the planned procedure have been verified on WDM networks. Certain parameters such as numbers of servers, the path length of the route, total network load, and the number of free wavelengths on the network are considered while calculating and analyzing the waiting probability of the network. The waiting probability and the blocking probability needs to be minimum for all networks as it maximizes the network performance.

First, we have fixed the number of servers as 20. The waiting probability is calculated when the path length is 3 and 4. The number of free wavelengths is increased from 25 to 55 with an interval of 5. It is observed that waiting probability is increased as the number of free wavelengths is increased (3). It is also observed that path length has an important role to play in the waiting probability calculations. If the path length is increased to 4 from 3, there is a significant fall in waiting probability. These results are shown in Table 1 and Fig. 1.

Network load is a significant network parameter. The performance of a network is largely affected by the network load. Usually, the overall performance of a network is decreased as the network load is increased because the network demands more resources to meet the enhanced network load requirements. We have calculated the waiting probability for network load from 30 to 55 Erlangs with a step of 5. The number of servers is fixed as 55 and 60. The waiting probability keeps on increasing as the network load is increased. Moreover, its value is decreased if the number of

Table 1 Effect of path length on waiting probability

No. of servers	No. of free wavelengths	Waiting probability (path length = 3)	Waiting probability (path length = 4)
20	25	0.000442	0.000096
20	30	0.003731	0.000115
20	35	0.018423	0.000801
20	40	0.062317	0.003732
20	45	0.160461	0.012869
20	50	0.338123	0.035259
20	55	0.614581	0.080719

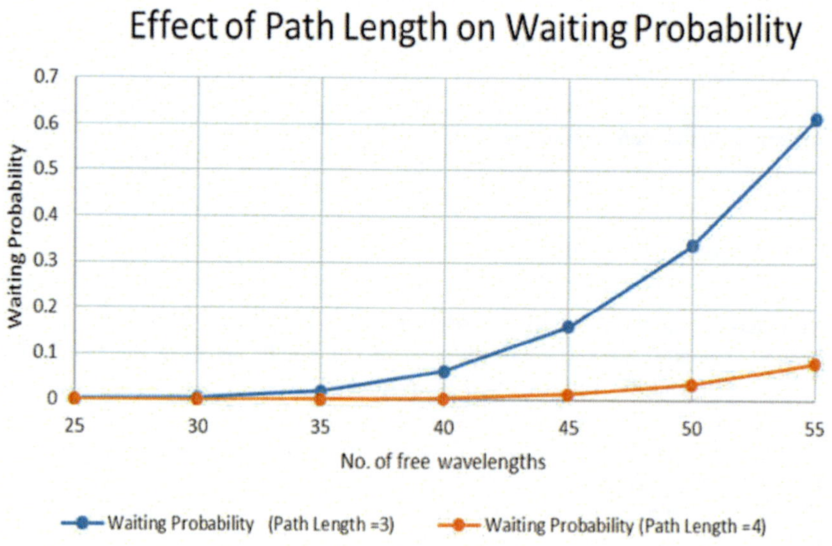

Fig. 1 Effect of path length on waiting probability

Table 2 Effect of network load on waiting probability

Network load (Erlangs)	Waiting probability (no. of servers = 55)	Waiting probability (no. of servers = 60)
30	0.000029	0.0000009
35	0.001146	0.000081
40	0.015896	0.002035
45	0.102372	0.021389
50	0.384547	0.117302

servers is increased. So, it is observed that the required number of servers may be increased to guarantee a minimum level of quality of service. These results are shown in Table 2 and Fig. 2.

The waiting probability is calculated for the different number of servers in the network. The decision about the number of servers is very crucial if the requests demand a minimum level of guaranteed quality of services. The waiting probability is calculated for the number of free wavelengths from 35 to 55 with a step of 5. The number of servers is fixed as 20 and 25. The path length is also fixed as 3. The waiting probability keeps on increasing as the number of free wavelengths is increased. Moreover, its higher value of the number of servers (25) provides better waiting probability. These results are described in Table 3 and Fig. 3.

The results of the proposed technique have been compared with existing techniques [12, 27]. In the existing techniques, the results are based upon approximation methods, whereas the proposed technique has produced calculated results with low complexity.

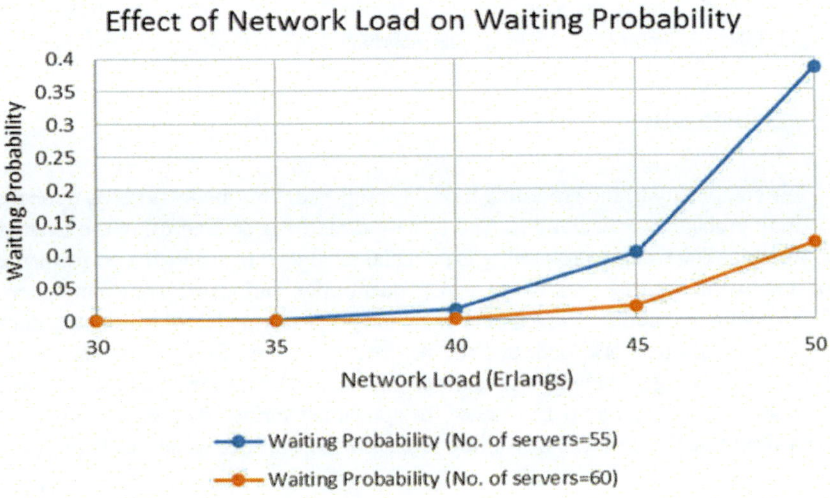

Fig. 2 Effect of network load on waiting probability

Table 3 Effect of number of servers on waiting probability

Path length	No. of free wavelength	Waiting probability (no. of servers = 20)	Waiting probability (no. of servers = 25)
3	35	0.018423	0.000489
3	40	0.062317	0.002973
3	45	0.160461	0.012434
3	50	0.338123	0.039083
3	55	0.614581	0.098504

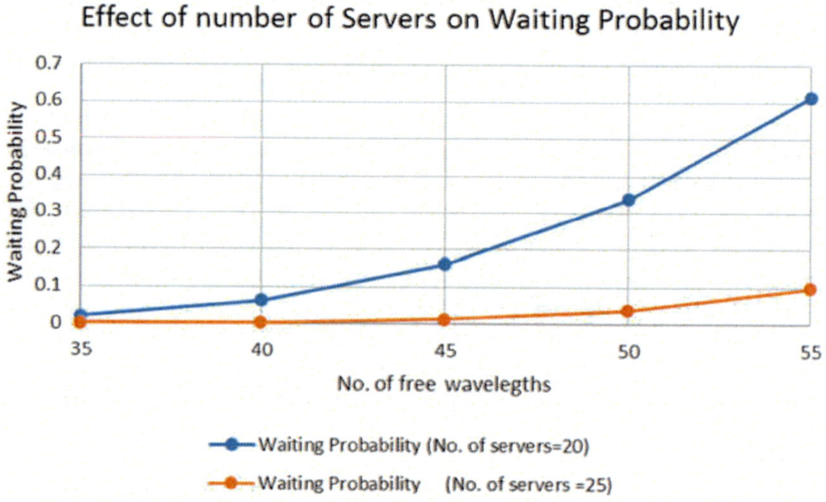

Fig. 3 Effect of number of servers on waiting probability

5 Conclusion

A low complexity traffic grooming model is proposed. This model is based upon the waiting probability calculations. The proposed model is derived using the Erlangs C formula. The waiting probability is a crucial parameter to access the performance of a network. Waiting probability is dependent upon various network parameters. The waiting probability is calculated using the proposed model for parameters such as path length, network load, and the number of servers. Each parameter has an impact on waiting probability calculations. The results have shown that the proposed model has better waiting performance for optical networks. The proposed model is providing calculated values of waiting probability rather than estimated values.

References

1. Mohan, N., Wason, A., & Sandhu, P. S. (2016). ACO based single link failure recovery in all-optical networks. *International Journal for Light and Electron Optics, Optik, 127*(20), 8469–8474.
2. Wason, A., & Kaler, R. S. (2010). Lightpath Rerouting algorithm to enhance blocking performance in all-optical WDM networks without wavelength conversion. *Optical Fiber Technology, 16*, 146–150.
3. Tripathi, T., & Sivarajan, K. N. (2000). Computing approximate probabilities in wavelength routed all-optical networks with limited range wavelength conversio. *IEEE Selected Areas in Communications, 18*(10), 2123–2129.
4. Kim, Y., Atchley, S., Vallée, G. R., Lee, S., & Shipman, G. M. (2017). Optimizing end-to-end bigdata transfers over terabits network infrastructure. *IEEE Transactions on Parallel and Distributed Systems, 28*(1), 188–201.
5. Barr, R. S., Kingsley, M. S., & Patterson, R. A. (2005). Grooming telecommunications networks: Optimization models and methods. In *Technical Report 05-emis-03*, 1–27.
6. Zhu, K., & Mukherjee, B. (2003). A review of traffic grooming in WDM optical networks: Architecture and challenges. *Optical Networks Magazine*, 55–64.
7. Modiano, E., & Lin, P. J. (2001). Traffic grooming in WDM networks. *IEEE Communications Magazine*, 124–129.
8. Tanga, S., & Tana, L. (2018). Single-wavelength optical buffers with general burst size distribution: Blocking probability and mean delay. *Optical Switching and Networking, 27*, 1–6.
9. To, H. L., Lee, S. H., & Hwang, W. J. (2015). A burst loss probability model with impatient customer feature for optical burst switching networks. *International Journal of Communication Systems, 28*(11), 1729–1740.
10. Chen, P. Y., & Reisslein, M. (2015). A simple analytical throughput-delay model for clustered FiWi networks. *Photon Network Communications, 29*, 78–95.
11. Chromy, E., Misuth, T., & Kavacky, M. (2011). Erlang C formula and its use in the call centers. *Information and Communication Technologies and Services, 9*(1), 7–13.
12. Lozhkovskyi, A. (2019). Calculation the service waiting probability with self-similar network traffic. *Journal of Engineering Science, 16*(2), 35–39.
13. Tong, L., Gan, C., Xie, W., & Wang, X. (2019). Priority-based mode transformation algorithm in a multisystem based virtual passive optical network. *Optical Switching and Networking, 31*, 162–167.
14. Xie, W., Gan, C., Xia, W., Zheng, W., & Yu, H. (2017). Mode transformation and united control mechanism supporting wavelength division multiplexing ethernet passive optical network and orthogonal frequency division multiplexing passive optical network in multi subsystem based virtual passive optical network. *IET Communications, 11*, 696–703.
15. Misuth, T., Chromy, E., & Baronak, I. (2010). Performance estimation of contact centre with variable skilled agents. In *Proceedings of the 33rd International Conference on Telecommunications and Signal Processing—TSP 2010* (pp. 391–396). Baden.
16. Misuth, T., Chromy, E., & Baronak, I. (2010). Method for fast estimation of contact centre parameters using Erlang C model. In *Proceedings of the Third International Conference on Communication Theory, Reliability, and Quality of Service, CTRQ 2010* (pp. 181–185). Glyfada: Greece.
17. Feng, D., Sun, W., & Hu, W. (2019). Dimensioning of store and transfer WDM networks with stratified ROADM node. *Optical Switching and Networking, 31*, 100–113.
18. Khan, A. N. (2017). An improved approximate network blocking probability model for all-optical WDM networks with heterogeneous link capacities. *Optical Fiber Technology, 38*, 7–16.
19. Pradhan, A. K., Keshri, S., Das, S., & De, T. (2016). A heuristic approach based on dynamic multicast traffic grooming in WDM mesh networks. *Journal of Optics*, 1–11.

20. Pradhan, A. K., Chatterjee, B. C., Oki, E., & De, T. (2018). Knapsack based multicast traffic grooming for optical networks. *Optical Switching and Networking, 27*, 40–49.
21. Derakhshanfard, N. (2020). Erlang based buffer management and routing in opportunistic networks. *Wireless Personal Communications, 110*, 2165–2177.
22. Chromy, E., Misuth, T., & Weber, A. (2012). Application of Erlang formulae in next generation networks. *International Journal of Computer Network and Information Security, 1*, 59–66.
23. Hanczewski, S., Stasiak, M., & Weissenberg, J. (2019). Non-full-available queueing model of an EON node. *Optical Switching and Networking, 33*, 131–142.
24. Mohan, N., Gupta, O. P., Wason, A., & Sandhu, P. S. (2015). Traffic grooming and blocking optimization in all-optical networks. *International Journal of Scientific & Engineering Research, 69*(1), 409–413.
25. Mohan, N., & Kaushal, P. (2019). Waiting probability based traffic grooming in optical networks. In *Proceedings of the International Conference on Sustainable Computing in Science, Technology & Management, February 26–28, 2019, Jaipur, India,* (pp. 777–781).
26. Chromy, E., & Baronak, I. (2018). Mathematical model of the contact center. *International Journal of Internet of Things and Web Services, 3*, 17–23.
27. Nag, K., & Helal, M. (2017). Evaluating Erlang C and Erlang A models for staff optimization: A case study in an airline call center. In *Proceedings of the IEEE Industrial Engineering and Engineering Management (IEEM), Singapore* (pp. 1–5).
28. Mallika, N. M. (2013). Link failure recovery in WDM networks. *International Journal of Computer Science and Electronics Engineering, 1*(5), 599–602.

Flow-Based Detection and Mitigation of Low-Rate DDOS Attack in SDN Environment Using Machine Learning Techniques

K. Muthamil Sudar and P. Deepalakshmi

Abstract Software Defined Networks (SDN) have become more efficient and popular by having effective controller. Centralized controller makes the decision to handle traffic in data plane by analyzing the entire network. In addition, handling network attack is equivalently complex for controllers. Low-rate Distributed Denial of Service (LR-DDoS) attack restricts legitimate users to access resources by sending unwanted and half-opened request towards the devices in data plane. Hence, it is vital to detect and mitigate LR-DDoS attack in its early stage because nature of attack is very similar to original request. The nature of this attack exhausts the network resources and leads to resource unavailability or delay while processing the legitimate requests. In this paper, we propose a flow-based detection and mitigation framework using machine learning models like Support Vector Machine (SVM), C4.5 Decision tree and Naïve Bayes as classifiers to detect LR-DDoS attack. From every traffic flow samples, we extract the essential features to detect attack. In mitigation phase, we handle the attack flow information and install the mitigation rules to avoid LR-DDoS attack from same source. Our experimental results show that SVM mechanism achieves better accuracy compared to C4.5 and Naïve Bayes techniques.

Keywords SDN · DDoS · Machine learning · SVM · C4.5 decision tree · Naïve Bayes

1 Introduction

Software-Defined Networking (SDN) is an emerging technology which overcomes some drawbacks faced in traditional network environment by providing centralized network controller and programming interface to adopt the dynamic changes in

K. M. Sudar (✉) · P. Deepalakshmi
Kalasalingam Academy of Research and Education, Krishnankoil, Tamilnadu, India
e-mail: k.muthamilsudar@klu.ac.in

P. Deepalakshmi
e-mail: deepa.kumar@klu.ac.in

P. Nayak et al. (eds.), *IoT and Analytics for Sensor Networks*, Lecture Notes in Networks and Systems 244, https://doi.org/10.1007/978-981-16-2919-8_18

network [1]. However, providing security is one of the complex tasks in SDN environment. SDN requires proper security mechanisms to protect controllers and other devices in data plane. Distributed Denial of Service (DDoS) attack [2] plays a major role in disrupting the services offered in SDN environment. DDoS attackers generate either high-rate or low-rate traffic towards the target in order to stop processing legitimate requests. Many researchers already addressed about detection of high-rate DDoS attacks in both traditional as well SDN environments. On the other side, only few researchers explored about the impact of low-rate DDoS detection method and proposed detection or mitigation methods. Hence, in this paper, we concentrate on detection of low-rate DDoS attacks in SDN environment by analyzing the essential features.

It is very hard to detect low rate attacks because the features of such attack's traffic are very similar to normal traffic. In recent days, machine learning techniques help to improve the efficiency and accuracy of attack detection system. In this work, we propose a low-rate DDoS detection and mitigation system using machine learning models such as Support Vector Machine (SVM), C4.5 Decision tree and Naïve Bayes with the help of essential features extracted from SDN traffic flows. Since we update every detected attack flow information in blocklist table, we could avoid the attack from same source in the beginning itself.

The following contributions are done in this paper:

(i) We introduce flow management module to extract the essential features like relative distribution of packet interval, number of packets, duration of the flow and relative distribution of match bytes.

(ii) We design and implement machine learning models like SVM, C4.5 and Naïve Bayes techniques to detect the LR-DDoS attacks.

(iii) We introduce blocklist table to gather the information about the attack and malicious traffic flows.

(iv) We discuss and compare the performance of our proposed work in terms of accuracy, precision and recall.

2 Background and Related Work

2.1 Software Defined Networking

SDN is a promising paradigm which is powerful, dynamic, directly programmable and supports high-bandwidth characteristics of today's applications. SDN architecture isolates control from data plane devices and facilitates the control using programmable interface. Abstract control from data plane devices helps network administrators to adjust traffic flows dynamically based on the needs of changes in network topology and functions. SDN architecture allows the administrators to manage and optimize network resources with the help of automated SDN logical programs. Unlike traditional network architecture, SDN also abridges the design and network operation because logics are provided by SDN controllers. Network Function Virtualization (NFV) [3] takes apart network functions from hardware devices in the form of virtualized functions. SDN architecture consists of three planes such as application, data and control plane. Southbound API (Application programming Interface) and northbound API are used to enable communication between these planes. Application plane contains all the applications and services running on the network. Control plane contains logically centralized controller which acts as core part of network and decides how to handle traffic based on the information from applications and devices in data plane. Data plane contains forwarding devices like switches and routers which forwards the traffic upon receiving routing information from control plane.

2.2 Low-Rate DDoS Attack

Many researchers and studies reported DDoS attacks as one of the prime threats created by cyber criminals to make massive destruction and have impact in all kinds of network platform. Now a days, the intensity of attack generate by attacker varies from high-level traffic to low-level traffic. DDoS attack with smart, low-level traffic is called low-rate DDoS attack. In low-rate DDoS attack, amount of attack traffic generate towards the target network accounts for 10–20% only. Identifying the behaviour of LR-DDoS attack is quite difficult. Even though the nature of attack is small, it will cause heavy destruction to target system or network. This type of low-rate attack is referred to shrew attack. The nature of shrew attack is extremely opposite to the DDoS flood attack. This attack traffic accounts 10–20% of legitimate network traffic and has strong concealment. L-DDoS attack shows traces of a periodic pulse shape which mainly concentrates on attack energy and leads to small average attack traffic. L-DDoS attack pulses generated from various attack sources with small average flows is associated together to form attack pulse for certain period of time. This leads to bottleneck in the target terminal and degrades the quality of service. L-DDoS attack traffic mainly concentrates on pulse period (i.e.) rectangular pulse in which several packets are sent towards the target. So, power of the attack mainly depends on duration and amplitude of the pulse period. Currently, L-DDoS attack mainly focuses on high speed and centralized services like cloud, big data. The reason is that low rate attack traffic can easily conceal into large traffic generated by cloud and big data environments. The effect of this attack results in huge traffic loss and economical loss to high speed computing environments. The mechanism used to detect general DDoS attack is not sufficient to detect L-DDoS attack because number of attack packets, and average attack rate is very low which looks like legitimate traffic. Figure 1 represents the L-DDoS attack in which L_{ab} indicates length of attack burst, R_a indicates rate of attack, B indicates beginning time of attack and F_a indicates frequency of attack. X and Y axis represents packet rate and time, respectively.

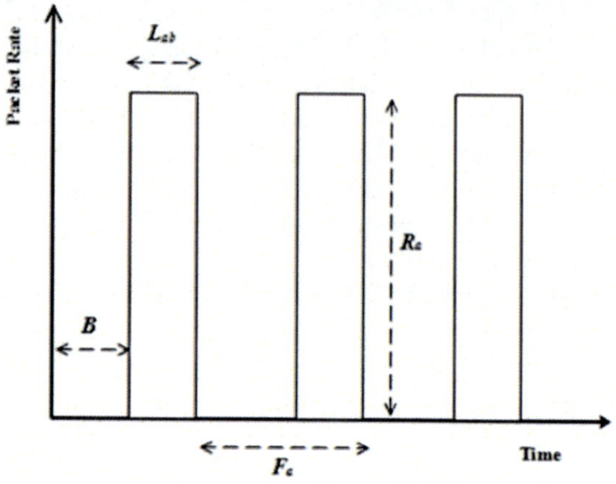

Fig. 1 L-DDoS attack representation

Attackers generally use sockstress attack, slowloris tool, and RUDY tool to generate low-rate attack [20]. Slowloris tool can generate partial HTTP headers after connecting to server and makes the connection open for long period of time. RUDY tool can generate HTTP POST requests and makes the server connection open for long time by sending the data in very slow manner. Sockstress attack mainly identifies the vulnerability in Transmission control protocol (TCP) handshake process and creates a numerous connection. Table 1 indicates the different types of LR-DDoS attack with their effect.

2.3 *Related Work*

Many researchers proposed different techniques to detect the L-DDoS attack by analyzing diverse parameters in both traditional and SDN environment. In [4], the authors attempted in SDN environment to identify L-DDoS attacks using machine learning models like Support Vector Machine (SVM), J48, Multi-layer Perceptron, REP tree, Random forest and Random tree. They claimed that Multi-Layer Perceptron model achieved an accuracy of 95% against Canadian Institute of Cyber security DoS Dataset. The authors of [5] proposed packet size measurement technique by analyzing the distribution difference in packet size to detect low rate DDoS attack. They demonstrated with real time datasets for various tolerance factors. They claimed that proposed work can detect L-DDoS attacks in short span of time with low false alarm rate.

In [6], the authors used entropy-based optimal technique to classify L-DDoS attacks in traditional environment. The authors of [7] used metric as information distance using generalized SDN control plane. They stated that proposed technique could achieve high accuracy compared to traditional entropy technique like Shannon entropy. In [8], the authors proposed correlation based detection technique to discover

L-DoS attack. By applying this technique they analyzed the traffic characteristics of attack and normal traffic by calculating linear relationship between the variables. They claimed that PRCD technique works better than relative entropy-based scheme.

The authors of [9] proposed statistical-based detection mechanism to detect L-DDoS attack in application server. They used t-statistic measure based on hypothesis test to identify L-DDoS flows. They claimed that the proposed distribution function technique detects the attack in efficient manner. In [10], the authors employed ML techniques to detect low-rate DDoS attack in SDN enabled IoT networks. They demonstrated with different machine learning techniques. They concluded that SVM achieve similar accuracy of 91% in both switch and controller with low false alarm rate.

From the literature work, we have identified that low-rate DDoS attack is one of the notable threat in all kinds of networking environment. Since the nature of attack is looking like original traffic instances, it is very difficult to detect and mitigate. It tries to leave many connections in open state as long as possible. The impact of this attack is very high in terms of bringing down the target but in slow manner. Few researchers have proposed statistical and ML techniques to detect the attack. Statistical-based technique like entropy, probability distribution, t-statistic measure achieved good accuracy but at the cost of high computation overload. Machine learning-based techniques like SVM, RF, KNN, Naïve Bayes, and J48 achieved better accuracy with low false alarm rate and at the same time with low computation overhead also.

3 Architectural Design

To detect and mitigate the L-DDoS attack, we proposed a framework which separates the process of detection and mitigation from network application. This reduces the processing overhead in SDN control plane. The framework consists of two independent phases such as detection phase and mitigation phase. Detection phase consists of flow management for extracting the essential flow information from every flow and ML models to identify the flow as attack or normal flow. Mitigation phase consists of malicious attack management and mitigation rules which will define the flow drop rules for the malicious flow.

3.1 Detection Phase

Low-rate DDoS attack mainly focuses on saturating the data plane devices like switches. Open Flow (OF) switch forwards data packets based on rules specified by controller. Every switch contains flow table listed with flow rules that are installed by the centralized controller after analysing the entire network. General OF switches use Ternary Content-Addressable Memory (TCAM) which can store 1500–3000 flow information [11]. So, data plane devices like switch become the target for attackers

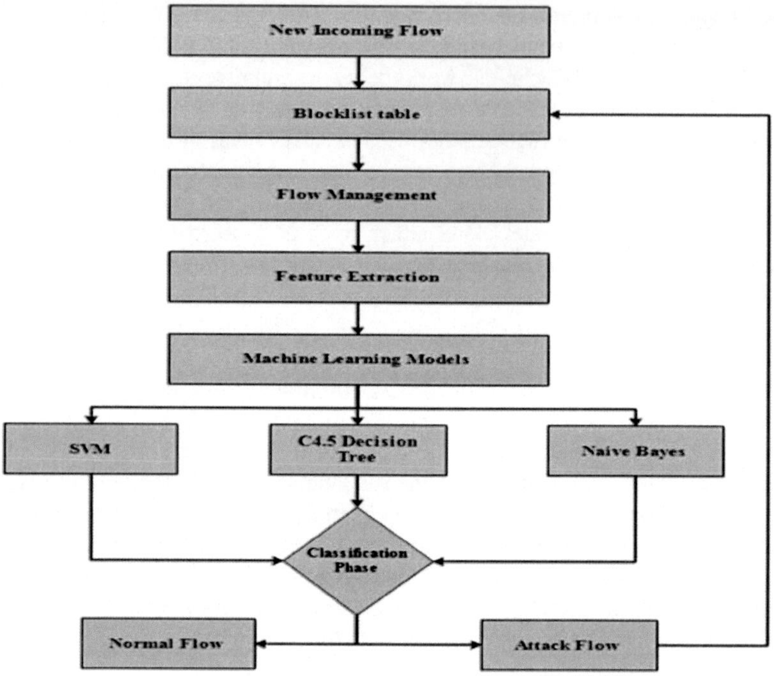

Fig. 2 Proposed architecture

to generate LR-DDoS attacks. This attack is also referred to low TCAM saturation attack. Figure 2 illustrates about architecture diagram of our proposed work.

The attackers generate low-rate attack in following conditions.

(i) The number of bots controlled by the attacker should be very less to generate the attack.
(ii) Every bot will send similar set of packets to target switch using real IP addresses. Upon receiving these packets, flow rules get installed in switch's flow table.
(iii) The rate of packets generating from bots are not too high. So, the attackers do not use spoofed IP address to generate the LR-DDoS attack.
(iv) Attackers used to generate the data packets from bots in less time interval compared to flow entry idle time out. The idle time out interval is the duration of entry available in the flow table in no-match state.

3.1.1 Flow Management

Flow management in the detection module will extract the essential features from flow samples. The four essential features used in this work are relative distribution of packet interval, number of packets, flow duration and relative distribution of match bytes. Duration of a flow indicates the time period of corresponding flow available in a flow table. The ultimate aim of LR-DDoS attack in SDN data plane is to exhaust the resources of flow table. So, attackers will send new data packets and makes the connection as open, which occupies flow table for long time. By calculating relative distribution match bytes, we can identify the difference between normal flow and attack flow. Because normal flow matches large data compared to attack flow. Relative distribution match bytes can be calculated using Eq. 1.

$$RDB = \frac{\sum_{i=1}^{N} (P_i - \mu)^2}{N} \tag{1}$$

Here, P_i indicates each packet size in bytes matched by flow rule; N indicates total number of packets matching with flow rules in a particular time and μ indicates average value of matched packet sizes again in a particular time. Original flow will follow the random and different packet interval based on the size of data packets. But, attack flow will follow periodic and smaller interval in order to act like original flow. So, packet interval also plays a vital role in detection of LR-DDoS attack. Relative distribution packet interval can be calculated using Eq. 2.

$$RDP = \frac{\sum_{i=1}^{N} |D_i - \lambda|}{N} \tag{2}$$

Since attack flows are very similar to legitimate flows, we have considered all flow samples to detect the attack by using the above mentioned features.

3.1.2 Machine Learning Models

In this module, we have implemented ML models like support vector machine (SVM), C4.5 Decision tree and Naïve Bayes to detect the attack by analyzing the features extracted from flow management module. SVM is one of the commonly used classification technique by placing the best hyper plane between two classes such as attack and normal [12]. C4.5 is a variant of decision tree technique which helps to classify the classes by constructing tree using if-then rules [13]. Naïve Bayes classifier is the simplest classification method based on the probability of each feature [14]. To implement these machine learning models, we used Scikit-learn 0.23 library provided by python. The output of flow management module is passed as input to machine learning models. Once the attack is detected, it will alert the mitigation phase for further processing of attack packets.

3.2 Mitigation Phase

3.2.1 Malicious Attack Management and Mitigation Rules

Machine learning models will return 0 for normal flow and 1 for attack flow after analyzing the traffic features. Once the flow is identified as malicious, assuming that it could be an attacker, detection module will forward the corresponding source IP address to blocklist table. Mitigation module will alert the switch to block the particular flow by creating blocking port rule immediately. Source IP address of every flow will be matched with the entries in blocklist table before forwarding for further process. The installed flow mitigation rules are set to get removed from flow table after seven days. This helps to analyze and update the flow in regular manner. Since flow management and analysis is already part of SDN, this module will not increase the complexity of SDN process.

3.3 Algorithm

Input: Incoming Traffic Flows
Output: Attack flow or Normal Flow
Begin
For each P_i
If P_i*already on BT* **then**
Drop the flow
Else
Forwards to Flow management module
Extract Features from the traffic flow
Send the selected features value to ML models
If *flow* $==$ *malicious* **then**
Add source ip address to BT
Update the mitigation rules
Else
Label as Normal flow
End
End
End

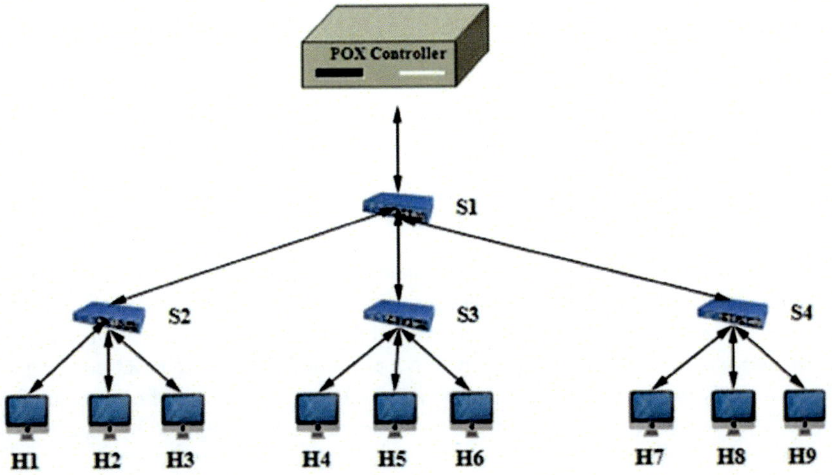

Fig. 3 Network topology

4 Experimental Setup

4.1 Environment Setup

We use Mininet 2.3.2 [15] tool, which is a commonly used tool to simulate the SDN environment by creating virtual network topologies. We configure hosts in data plane as both legitimate and attack devices. OpenFlow switch is another standard virtual component available in Mininet. We use POX—a python-based controller [16, 22] to simulate the functionalities of controller in control plane. The experimental network topology consists of nine host machines (H1–H9) and four switches (S1–S4). Figure 3 portrays the network topology of proposed work. All hosts connect to switch with the data rate of 10 Mbps. We use Scapy tool [17, 18] to generate the low rate attack by forging the fields of packets and rate of transmission. We have considered average rate of 1.6 to 2.2 Mbps to generate the low-rate DDoS attacks from different hosts.

4.2 Dataset

To train our machine learning models, we have used CIC DoS 2019 dataset (Canadian Institute of Cyber security) [19]. CIC dataset contains both benign and common DDoS attacks. CIC dataset contains LR-DDoS attacks compared to commonly used datasets like KDD-CUP'99 and NSL-KDD dataset. CIC dataset consists of eight different low rate attacks and normal instances. Totally, it contains 80 features to provide statistical information about the flows. Table II specifies the distribution of attack and normal flows in CIC dataset. We use Flowtbag method [4] to convert conventional traffic instances into flow instances with same set of features. This works well in SDN environment too.

4.3 Evaluation Metrics

The performance of proposed work is evaluated using standard metrics like accuracy, recall, precision and F-Score [21]. Accuracy (AC) determines rightly classified data (both attack and normal) over all the classified data. Precision (P) determines righty classified positive data over all the positive classified data. Recall (R) determines how correctly the model classifies true positives. F-score is a mean of both recall and precision. Accuracy, Recall, Precision and F-score can be computed by Eqs. 3, 4, 5 and 6 , respectively.

$$Accuracy = \frac{TP + TN}{TP + TN + FP + FN} \tag{3}$$

$$Precision = \frac{TP}{TP + FP} \tag{4}$$

$$Recall = \frac{TP}{TP + FN} \tag{5}$$

$$\text{F - Score} = 2 * \frac{P * R}{P + R} \tag{6}$$

Here, TP indicates correctly classified normal flows, TN indicates correctly classified attack flows, FP indicates wrongly classified normal flows, and FN indicates wrongly classified attack flows.

Table 2 depicts the performance of our proposed work. SVM achieves better accuracy to detect low rate DDoS attack compared to C4.5 and Naïve Bayes techniques. Figure 4 indicates the confusion matrix for C4.5, SVM and Naïve Bayes models. Figure 5 indicates the accuracy comparison of different machine learning models for our proposed work.

Table 1 LR-DDoS Attacks and Effects

LR-DDoS Types	Effect of Attack
Shrew DDoS	Reduce Quality
Fake Session	Delay in service
Session attack	Delay in service
HTTP Fragmentation	Delay in service e
Bandwidth Congestion	Reduce Quality

Table 2 CIC dataset traffic flow distribution

Type of Traffic	Number of Flows
DDoSSIM	2185
Goldeneye	661
H.U.L.K	2088
R.U.D.Y	2070
Slowbody2	4391
Slowheaders	3185
Slowloris	2912
Slowread	1464

Table 3 Performance evaluation

Algorithm	Accuracy	Precision	Recall	F-Score
SVM	0.93	0.93	0.99	0.95
C4.5 Decision Tree	0.91	0.91	0.98	0.94
Naïve Bayes	0.87	0.87	0.97	0.92

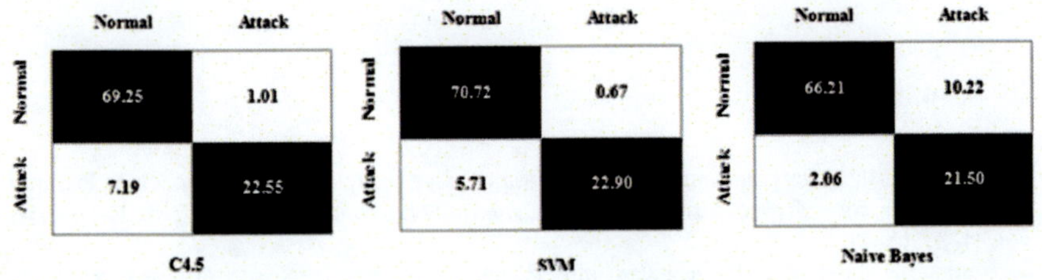

Fig. 4 Confusion matrix

Fig. 5 Accuracy comparison of our proposed work

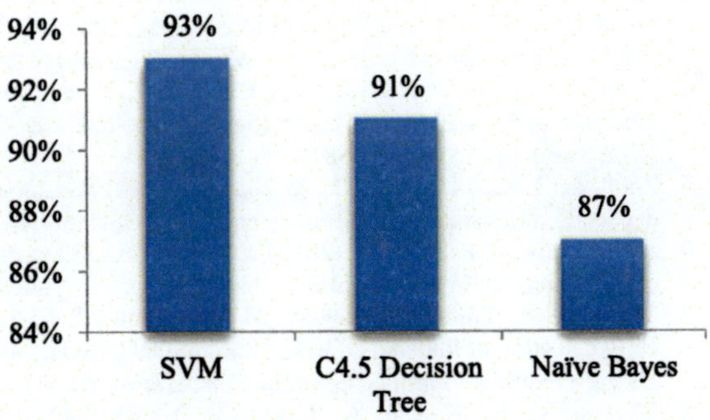

5 Conclusion

In this paper, we propose mechanism to detect and mitigate LR-DDoS attack in SDN environment using machine learning techniques. Since normal features like port number, source and destination address are not sufficient to detect attack, we have used four essential features viz. relative distribution of packet interval, number of packets, duration of the flow and relative distribution of match bytes from the SDN flow. In this paper, we used machine learning techniques like SVM, C4.5 Decision tree and Naïve Bayes as classifiers to detect attacks. We trained machine learning models with the help of CIC DoS dataset, which contains special types of low rate DDoS attacks. Our experimental results indicate that SVM aids to detect attack with better accuracy compared to C4.5 and Naïve Bayes. Once attack is detected, mitigation module adds particular attack flow details to blocklist table and alerts the controller to drop particular flow from flow table by entering mitigation rules. Our experimental results illustrate the efficacy of our proposed work. In future, we plan to implement multi-layer machine learning-based techniques to distinguish low rate and high rate DDoS attacks using SDN flow information.

References

1. Sudar, K. M., & Deepalakshmi, P. (2020). Comparative study on IDS using machine learning approaches for software defined networks. *International Journal of Intelligent Enterprise, 7*(1–3), 15–27.
2. Muthamil Sudar, K., & Deepalakshmi, P. (2020). A two level security mechanism to detect a DDoS flooding attack in software-defined networks using entropy-based and C4. 5 technique. Journal of High Speed Networks, (Preprint), 1–22.
3. Deepa, V., Sudar, K. M., & Deepalakshmi, P. (2018, December). Detection of DDoS attack on SDN control plane using Hybrid Machine Learning Techniques. In 2018 International Conference on Smart Systems and Inventive Technology (ICSSIT) (pp. 299–303). IEEE.
4. Pérez-Díaz, J. A., Valdovinos, I. A., Choo, K. K. R., & Zhu, D. (2020). A flexible SDN-based architecture for identifying and mitigating low-rate DDoS attacks using machine learning. *IEEE Access, 8*, 155859–155872.
5. Zhou, L., Liao, M., Yuan, C., & Zhang, H. (2017). Low-rate DDoS attack detection using expectation of packet size. Security and Communication Networks, 2017.
6. Jadhav, P. N., & Patil, B. M. (2013). Low-rate DDOS attack detection using optimal objective entropy method. International Journal of Computer Applications, 78(3).
7. Sahoo, K. S., Puthal, D., Tiwary, M., Rodrigues, J. J., Sahoo, B., & Dash, R. (2018). An early detection of low rate DDoS attack to SDN based data center networks using information distance metrics. *Future Generation Computer Systems, 89*, 685–697.
8. Bhuyan, M. H., Kalwar, A., Goswami, A., Bhattacharyya, D. K., &Kalita, J. K. (2015, April). Low-rate and high-rate distributed DoS attack detection using partial rank correlation. In 2015 Fifth International Conference on Communication Systems and Network Technologies (pp. 706–710). IEEE.
9. Bhushan, K., & Gupta, B. B. (2018). Hypothesis test for low-rate DDoS attack detection in cloud computing environment. *Procedia computer science, 132*, 947–955.
10. Cheng, H., Liu, J., Xu, T., Ren, B., Mao, J., & Zhang, W. (2020). Machine learning based low-rate DDoS attack detection for SDN enabled IoT networks. *International Journal of Sensor Networks, 34*(1), 56–69.

11. Deepa, V., Sudar, K. M., & Deepalakshmi, P. (2019, March). Design of Ensemble Learning Methods for DDoS Detection in SDN Environment. In 2019 International Conference on Vision Towards Emerging Trends in Communication and Networking (ViTECoN) (pp. 1–6). IEEE.
12. Sudar, K. M., Deepalakshmi, P., Nagaraj, P., &Muneeswaran, V. (2020, November). Analysis of Cyberattacks and its Detection Mechanisms. In 2020 Fifth International Conference on Research in Computational Intelligence and Communication Networks (ICRCICN) (pp. 12–16). IEEE.
13. Singh, S., & Gupta, P. (2014). Comparative study ID3, cart and C4. 5 decision tree algorithm: a survey. International Journal of Advanced Information Science and Technology (IJAIST), 27(27), 97–103.
14. Jiang, L., Wang, D., Cai, Z., & Yan, X. (2007, August). Survey of improving naive bayes for classification. In International Conference on Advanced Data Mining and Applications (pp. 134–145). Springer, Berlin, Heidelberg.
15. Mininet Team Mininet: An Instant Virtual Network on your Laptop (or other PC). Available online: http://mininet.org/ (accessed on 3 January 2021).
16. S. Wilson Prakash and P. Deepalakshmi, Flow-based Dynamic Load balancing algorithm for the Cloud networks using Software Defined Networks, International Journal of Cloud Computing, 8(4) (2019), 299–318.
17. Scapy-Packet Crafting Tool. Available online: https://scapy.net/ (accessed on 3 January 2021)
18. S. Wilson Prakash and P. Deepalakshmi, DServ-LB: Dynamic server load balancing algorithm, International Journal of Communication Systems, 32 (1) (2019), 1–11.
19. Leevy, J. L., & Khoshgoftaar, T. M. (2020). A survey and analysis of intrusion detection models based on CSE-CIC-IDS2018 Big Data. *Journal of Big Data, 7*(1), 1–19.
20. Zhijun, W., Wenjing, L., Liang, L., & Meng, Y. (2020). Low-Rate DoS Attacks, Detection, Defense, and Challenges: A Survey. *IEEE Access, 8*, 43920–43943.
21. Muthamil Sudar, K., & Deepalakshmi, P. An intelligent flow-based and signature-based IDS for SDNs using ensemble feature selection and a multi-layer machine learning-based classifier. Journal of Intelligent & Fuzzy Systems, (Preprint), 1–20.
22. Balakiruthiga, B., Deepalakshmi, P., Mohanty, S. N., Gupta, D., Kumar, P. P., & Shankar, K. (2020). Segment routing based energy aware routing for software defined data center. *Cognitive Systems Research, 64*, 146–163.

Design and Simulation of MEMS Based Capacitive Accelerometer

S. Veena, Newton Rai, Amogh Manjunath Rao Morey, H. L. Suresh, and Habibuddin Shaik

Abstract Accelerometer is an electromechanical device, which is used for physical measurement along the orthogonal coordinates. Micro Electro Mechanical Systems (MEMS) based capacitive accelerometers are embedded in many modern technological applications. This paper presents the comparison between two single axis MEMS based capacitive accelerometers, which have the natural frequencies of 7 and 2.2 kHz. This work includes design, simulation, analytical modelling, and finite element modelling of each MEMS comb type capacitive accelerometer with different operating frequencies. The accelerometer was designed using COMSOL Multiphysics and MATLAB simulator tool.

Keywords Accelerometer · MATLAB · COMSOL

1 Introduction

Accelerometer is an electromechanical device, which is used for physical measurement like vibration, acceleration or force of a moving solid. Most accelerometers are based on the principle of mechanical vibration. The basic structure of mems accelerometer contains the seismic mass supported by beams. The mass is frequently appended to a dashpot that gives the essential damping impacts [1, 2]. The spring and the dashpot are in turn connected to frame as shown in Fig. 1 [3]. An accelerometer, which is stored at rest on the surface of earth, will measure acceleration governed by gravity ($g \approx 9.81\ \text{ms}^2$) and in contrast, accelerometers, which are in free descend state, will measure to zero [3]. The mathematical analysis was executed with accordance of damping [5] in order to analyse the characterization of the device. The

S. Veena · N. Rai (✉)
Nitte Meenakshi Institute of Technology, Bengaluru, India

A. M. R. Morey · H. L. Suresh · H. Shaik
Sir. M Visvesvaraya Institute of Technology, Bengaluru, India
e-mail: hl_suresh@rediffmail.com

H. Shaik
e-mail: habibuddin.shaik@nmit.ac.in

P. Nayak et al. (eds.), *IoT and Analytics for Sensor Networks*, Lecture Notes in Networks and Systems 244, https://doi.org/10.1007/978-981-16-2919-8_19

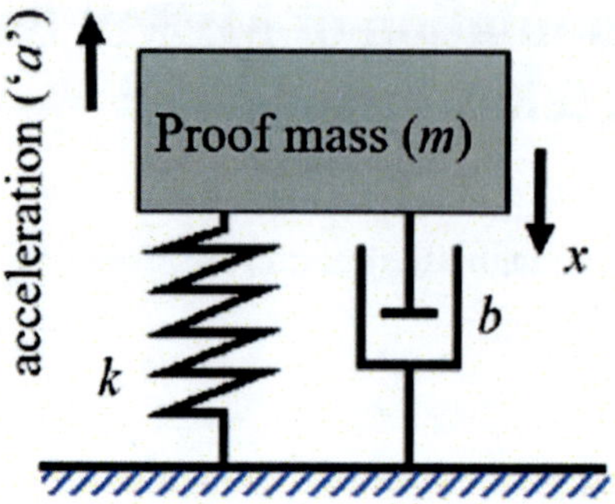

Fig. 1 Typical spring mass damper model of accelerometer (*Source* MDoF [3], p. 3)

capacitive based MEMS accelerometers measures capacitance fluctuation between a proof mass and an anchored conductive electrode disembodied by a small gap [6]. Here, the sensitivity of the accelerometer is a measure of displacement with respect to acceleration [8]. Although commercially different accelerometers are available, the main aim for selecting capacitive based MEMS accelerometers are high sensitivity with respect to resonant frequency [8] and reliability. This work presents a comparison between two novel single folded beam type capacitive MEMS Accelerometers. There are a wide scope of utilizations that require acceleration measurement such as automotive industry, biomedical applications [11], vibration analysis, navigation system, robotics, and this measurement is also achievable with different accelerometers fabricated through the CMOS micromachining process [9]. Hence, the need of exploration with respect to this field is rampant, as the configurations are getting more diverse. Air was elected as the di-electric medium between the combs [13]. This work was carried out, starting with the structural design and mathematical analysis along with the subsequent simulations to verify the same and an overview about the comparison between the two accelerometers.

2 Structure Design and Working Phenomenon

The structure design of proposed capacitive based MEMS accelerometer is as shown in Fig. 2 [1]. The proposed design of MEMS accelerometer has unique structure in comparison to former design, where two proof masses is suspended on either side by a single folded beam. The structure of mems comb accelerometer is symmetry in shape. Movable parts comprises two proof mass, single folded beam, movable fingers and Fixed parts comprises anchors, pad metal, fixed fingers attached to pad metal [9]. The two proof mass are symmetrically suspended to central anchor through single folded beam. The fixed fingers are connected to the left and the right anchors.

The two proof masses are responsible for vibrating in response to the incident inertial force. The portable fingers are attached on one side of the each proof mass,

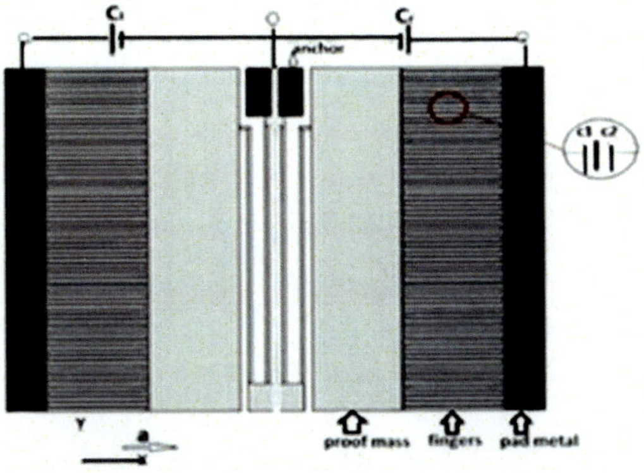

Fig. 2 Design of proposed MEMS comb accelerometer

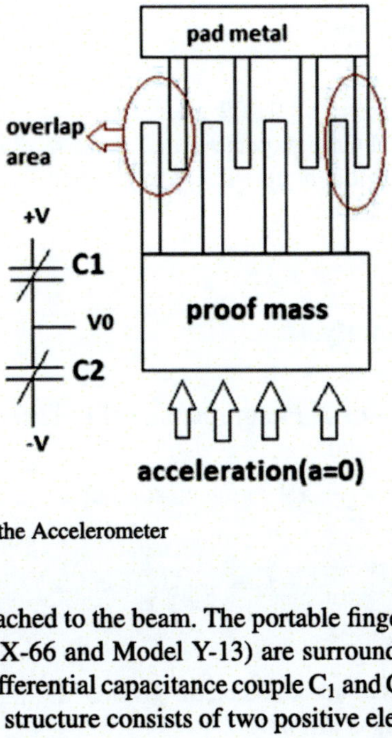

Fig. 3 Resting position of the Accelerometer

and the other side is attached to the beam. The portable fingers (Number of portable fingers [N_f] for Model X-66 and Model Y-13) are surrounded by anchored fingers which establishes the differential capacitance couple C_1 and C_2 with anchored fingers on either side [10]. The structure consists of two positive electrodes and one ground electrode [11].

Keeping this in check, the models were reconstructed in Comsol Mutliphysics 5.1a. Silicon material was chosen from the designated Comsol library function.

In order to draw comparisons between the two different frequency oriented accelerometers, the working phenomenon should be utilized. At steady state (when acceleration = 0), the proof mass does not experience any inertial force so the portable fingers are in rest. As there is no significant change in area between fixed and portable

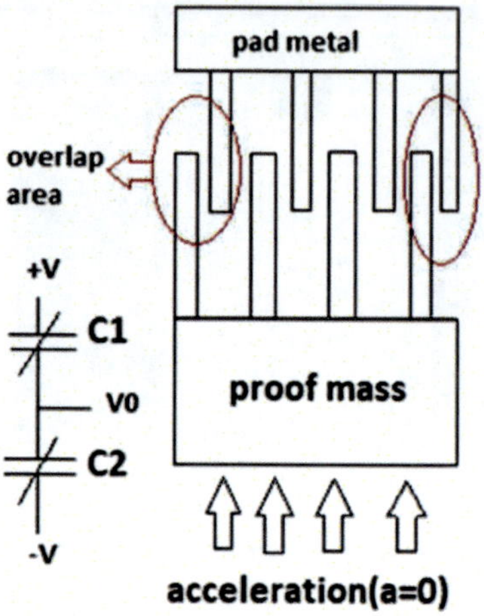

Fig. 4 Accelerometer in movement

fingers, the capacitance couple C1, C2 are equal. Under steady state an external driving voltage is applied through positive and negative electrode, which develops an electrostatic force and this force will try to displace the sensing electrodes and the corresponding capacitance.

3 Mathematical Analysis

3.1 To Find the Natural Frequency of the Accelerometers

The natural frequency of a general spring mass damper system [1] is

$$f_2 = \frac{1}{2\pi}\sqrt{\frac{k}{m}}, \tag{1}$$

where

the total sensing mass of accelerometer is given by [12], m = mass
K = stiffness of spring is given by

$$K = \frac{EtW^3}{4L^3} \tag{2}$$

where L, t, and W are the length,thickness, and width of the beam respectively, E = Young's modulus of elasticity(silicon) = 170* $10^9 N/m^2$.

3.2 To Find the Damping Ratio of the Accelerometers

1. **Model X Accelerometer**

Squeeze film damping: The pressure is developed due to the compression of air between moving fingers and fixed fingers, and this effect is called damping effect. It acts as a dragging force and opposes the movement of sensing fingers [13]. An analytical formula [6] for the air squeeze film damping is given as

$$b = N_f n_{eff} l_f \left(\frac{t}{d_o}\right)^3, \tag{3}$$

where

n_{eff} is the effective viscosity of air 18.5* 10^{-6} Ns/m^2, number of portable fingers (N_f) and l_f, t, d_o-Table [1],after substituting all values in Eq. 3 the damping coefficients is b = 3.815625* 10^{-5} Ns/m.

At Steady state:

$$S = \frac{1}{w_n^2}, \tag{4}$$

where, S = sensitivity.

At dynamic state: The dynamic behaviours of spring mass damper system is defined by parameters like angular frequency (w_n) and damping ratio (). The dynamic analysis of this system is done in Laplace domain of equation [2, 9]

when x → F

when x → a

$$\frac{X(s)}{F(s)} = \frac{1}{ms^2 + bs + k}, \tag{5}$$

$$\frac{X(s)}{a(s)} = \frac{1}{s^2 + \frac{b}{m}s + \frac{k}{m}}, \tag{6}$$

where

$$w_n = \sqrt{\frac{k}{m}}, \tag{7}$$

$$Damping\,ratio\,(\zeta) = \frac{b}{2mw_n}. \tag{8}$$

The value of damping ratio (ζ) directly controls the nature of the system response.

For 0 ≤ <1, the system is under damped system

For = 1, the system remains critically damped

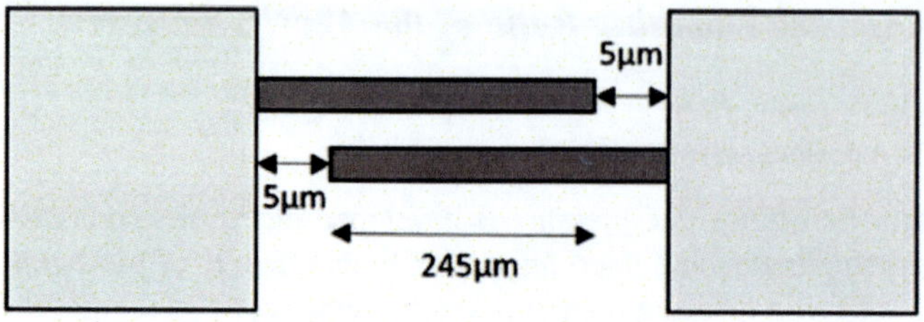

Fig. 5 Positioning of the combs in the accelerometer structure

For ≥ 1, the system remains overdamped
K, m = Table [2]
On substituting in above equations, we get damping ratio as **= 0.03208.**

2. **Model Y Accelerometer**

The length of the comb and the distance between the comb and the proof mass is shown in Fig. 5.

Here the value of damping coefficient is given by [13]

$$b = \frac{64\sigma p_a L w}{g\pi^6 w_n} \sum_{min} \frac{m^2 + \left(\frac{n}{B}\right)^2}{(mn)\left[\left(m^2 + \frac{n^2}{B^2}\right)^2 + \frac{\sigma^2}{\pi^4}\right]}\pi^6, \tag{9}$$

where

L = 245um, W = 10um, σ = compression ratio of air, Pa = Ambient pressure (1.01x 10^{-5} Pa),*and* W_n = natural frequency.

And natural frequency is given as

$$W_n = 2\pi \mathrm{f} \tag{10}$$

For f = 7 kHz, W_n = 14000π Hz.

And this compression ratio is given as

$$\sigma = \frac{12uw^2 w_n}{p_a g^2}. \tag{11}$$

Here,

u = coefficient of dynamic viscosity (1.85 x 10^{-5}), $\sigma = 0.19576$.

The variable B is given by

$$\mathrm{B} = L/W \tag{12}$$

B = 245 μm/10 μm, B = 24.5.

Table 1 Dimensions of the capacitive MEMS Accelerometers

Accelerometer parameters	Dimensions (μm) of Model X accelerometer	Dimensions(μm) of Model Y accelerometer
Right proof mass	225X1000X25	225X400X25
Left proof mass	225X1000X25	225X400X25
Gap between two fingers (d_0)	5	5
Pad metal size	100X1000X25	100X400X25
Thickness of device (t)	25	25
Beam length (L)	550	250
Beam width (W)	10	10
Length of finger (l_f)	245	245
Breadth of finger (b_f)	10	10

Substituting this in the main damping coefficient equation we get b = 7.304* 10^{-6} Ns/m; on substituting in the equations [7, 8] we get damping ratio as = **0.0549.** The damping ratio for both the systems is less than one (< 1), hence the proposed accelerometers are underdamped systems.The above calculations for both the accelerometers was carried out to execute the simulations in MATLAB which is discussed further (Sect. 4) in this paper.

3.3 *To Find the Static Capacitance of the Accelerometers*

The static capacitance is expressed as

$$C_0 = \frac{EN_f l_f t}{d_o} \tag{13}$$

where

C_0 is the static capacitance, Length of finger (l_f), Number of movable fingers (N_f), Thickness (t), distance between the plates (d_o)

Referring Table 1, the static capacitance obtained for Model X Accelerometer is $C_0 = 0.730455$pf and Model Y Accelerometer $C_0 = 0.28199$pf.

3.4 *To Find the Displacements Sensitivity of the Accelerometers*

$$F_{inertia} = M * a \tag{14}$$

$$F_{damping} = K * x \tag{15}$$

The displacement sensitivity of proof mass is given by

$$S_d = x = \frac{M}{K} * a, \tag{16}$$

where

M = mass, K = stiffness of the spring.

It is found that the displacement sensitivity of the proof mass of Model X Accelerometer is $S_d = 3.53625 * 10^{-8}\text{m/g}$ and of model Y Accelerometer is $S_d = 4.5096 * 10^{-9}\text{m/g}$.

4 Simulation and Results

4.1 Model X

1. **Frequency simulation**: Silicon (Single crystal) as the base material, as shown in Fig. 6, the simulated frequency obtained was 2.1084 kHz. The simulated frequency values when compared to the calculated values referred from Table 2 gives us an error difference of about 0.492 kHz for the model X accelerometer.

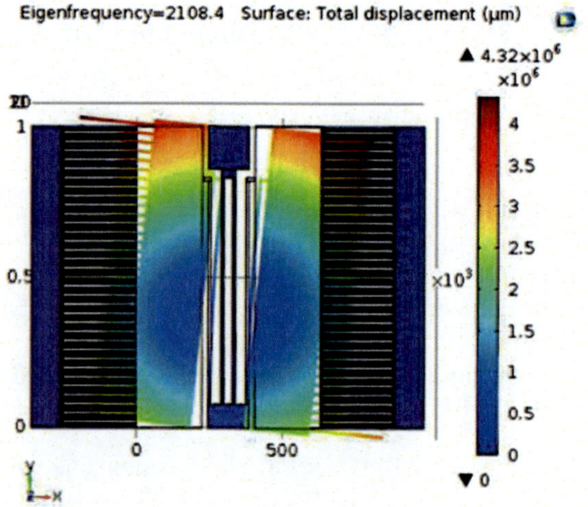

Fig. 6 Frequency simulation of Model X comb

Table 2 Natural frequencies of the accelerometers

Model X Accelerometer	Model Y Accelerometer
K = 10 N/m and m = 3.53625* 10^{-8} Kg	K = 31 N/m and m = 1.42651 * 10^{-8} Kg
f_n = 2.6 kHz	f_n f7 = 7.41 kHz

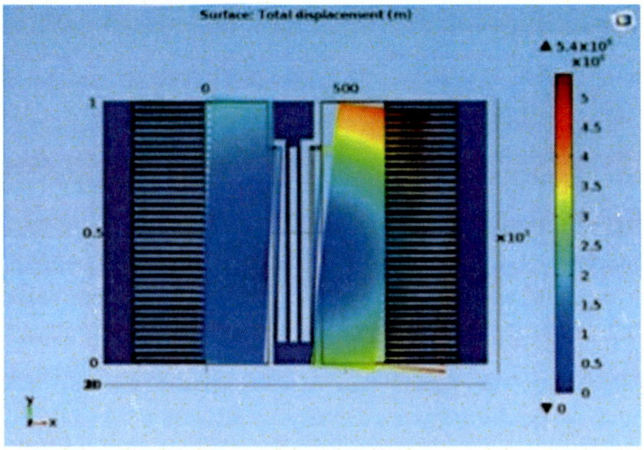

Fig. 7 Total Displacement for the Model X comb

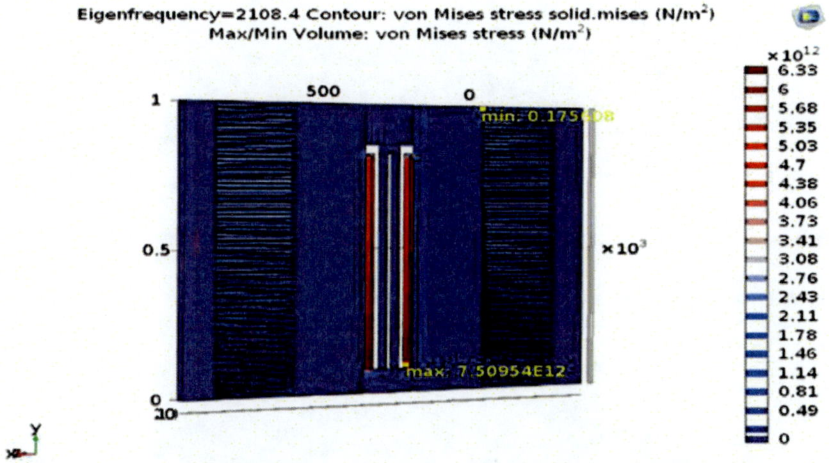

Fig. 8 Stress evaluation in Model X com

2. **Displacement simulation**: Silicon (Single crystal) as the base material, as shown in Fig. 7, the total displacement was observed to be 5.4x 10^{-8}.
3. **Stress simulation:** Here the yield point is up to 7.509X 10^{-12} N/m^2, beyond which failure could occur in the bottom region of Model X. The module can toggle with a stress range of 7.33X 10^{-12} N/m^2 given the min value is 0.175 10^{-12} N/m^2.

4.2 *Model Y*

1. **Frequency simulation**: Silicon (Single crystal) as the base material, as shown in Fig. 9, the simulated frequency obtained was 7.074 kHz.
2. **Displacement simulation**: Silicon (Single crystal) as the base material, as shown in Fig. 10, the total displacement was observed to be 6.03x 10^{-9} m.
3. **Stress simulation**: As shown in Fig. 11, here the yield point is up to 6.361X 10^{-12} N/m^2, beyond which failure could occur in the upper region of Model Y. The module can toggle with a stress range of 7.33X 10^{-12} N/m^2 given the min value is 0.175 10^{-12} N/m^2 around the corners.

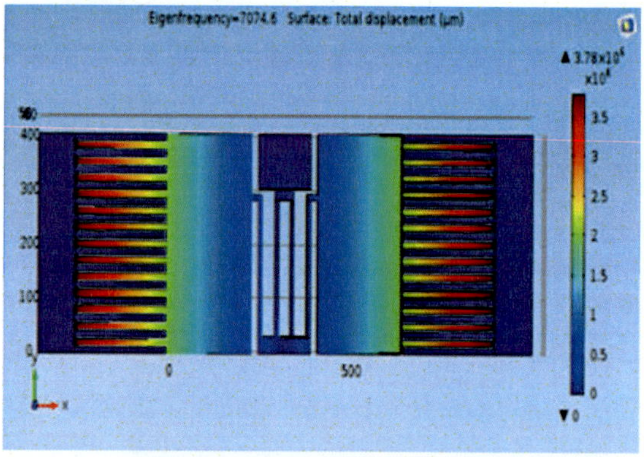

Fig. 9 Frequency simulation of Model Y accelerometer

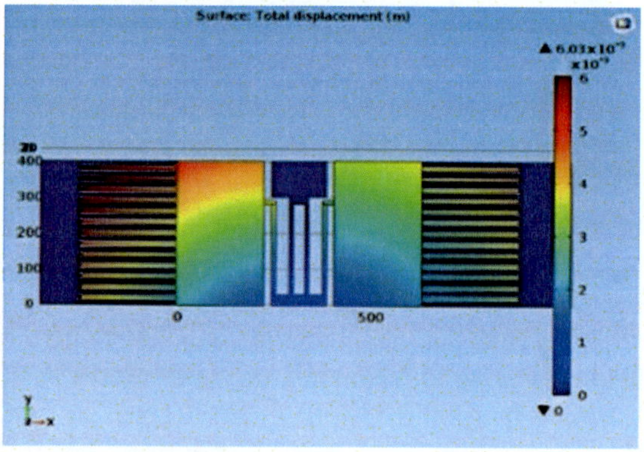

Fig. 10 Total Displacement of Model Y comb

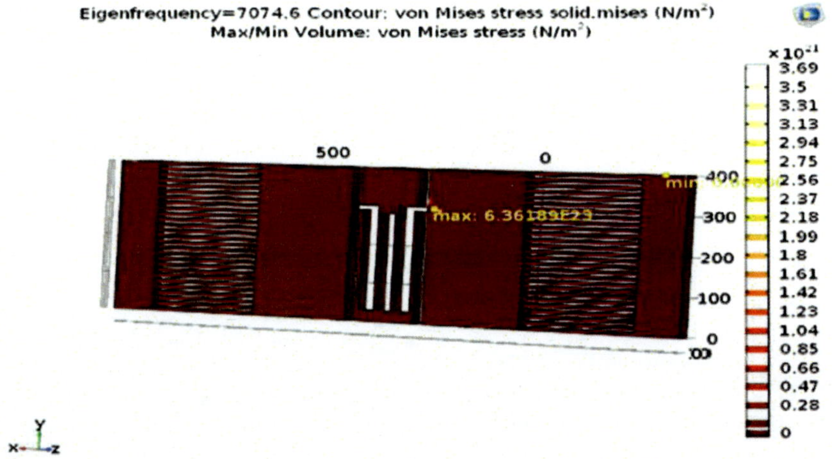

Fig. 11 Stress evaluation in Model Y comb

4.3 *Simulations to Check Variation of Mass and Dimensions of the Proof Mass Over Resonant Frequency*

1. Width of proof mass is reduced to half keeping the beam length unaltered: Given that both the accelerometers are underdamped systems, one of them (Model Y) was considered to the check variation of mass and dimensions of the proof mass over resonant frequency.

2. **Both Width of Proof Mass and Beam Length Reduced to Half**

As shown in Figs. 12, 13, 14, and 15, it was observed that the resonant frequency of accelerometer depends on its design parameter like beam length and sensing mass.

Fig. 12 Resonance frequency is 10.8 kHz

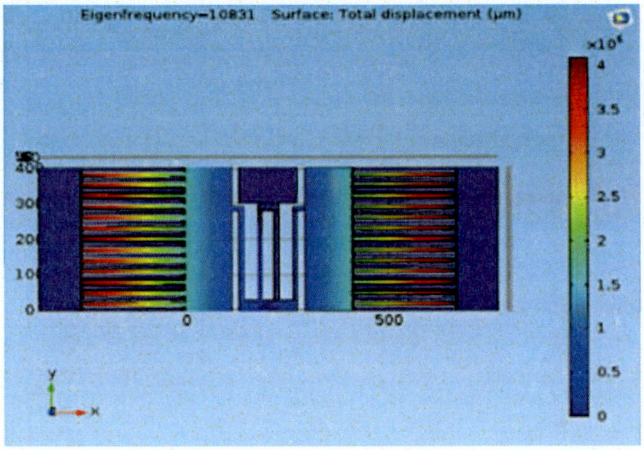

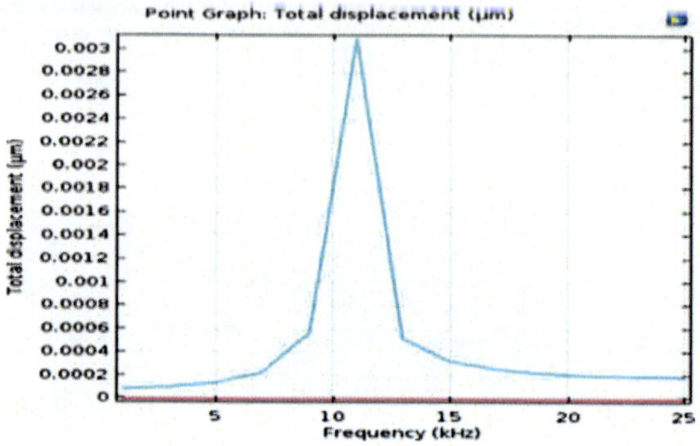

Fig. 13 Max displacement at resonance frequency 10.8 kHz

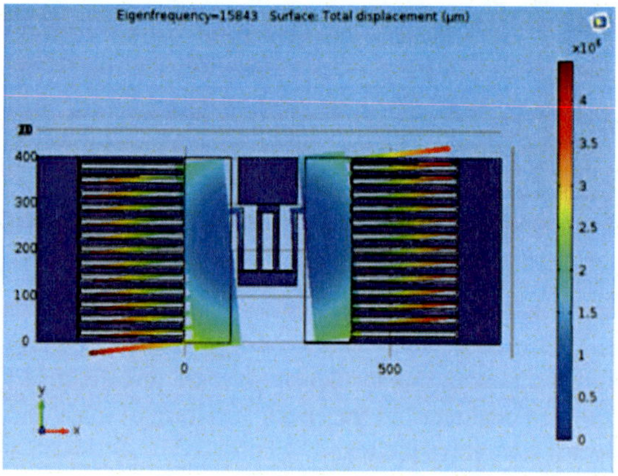

Fig. 14 Resonance frequency is 15.8 kHz

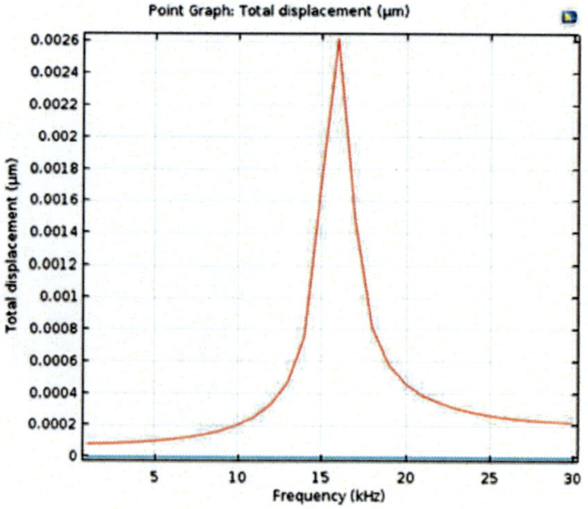

Fig. 15 Max displacement at resonance frequency

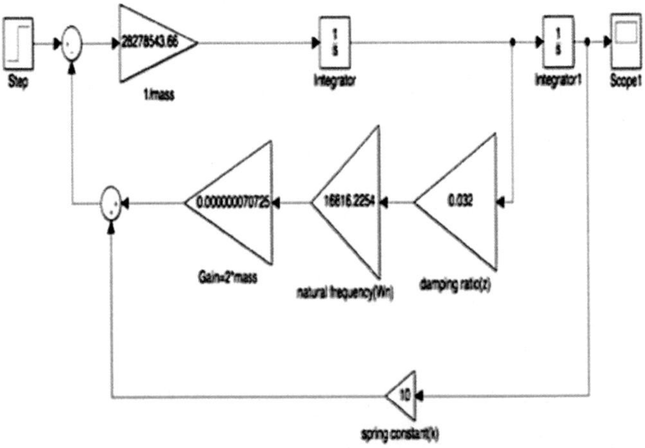

Fig. 16 Simulink model of a

Fig. 17 Step response for $\zeta = 0.5$(Model X)

4.4 *MATALAB Simulations*

1. Open loop model of 2 KHZ accelerometer:

In an open-loop system no additional control circuits are required, the electrical output signal is directly interfaced to analog front end circuit (AFE) for capacitance to voltage conversion and thereafter output voltage is used for measurement analysis [5]. The Eq. (7) can be rewritten into a standard form with z and w_n as

$$\frac{X(s)}{a(s)} = \frac{1}{s^2 + 2\zeta s w_n + w_n^2}. \tag{17}$$

The open loop Simulink model of two accelerometers are as shown in Figs. 16 and 18. The step responses of both the accelerometers as shown in Figs. 17 and 19 are analogous to that of an underdamped system, where increasing the damping ratio stabilized the Model X more.

2. Open loop model of 7 kHz accelerometer
   See Figs. 18 and 19.

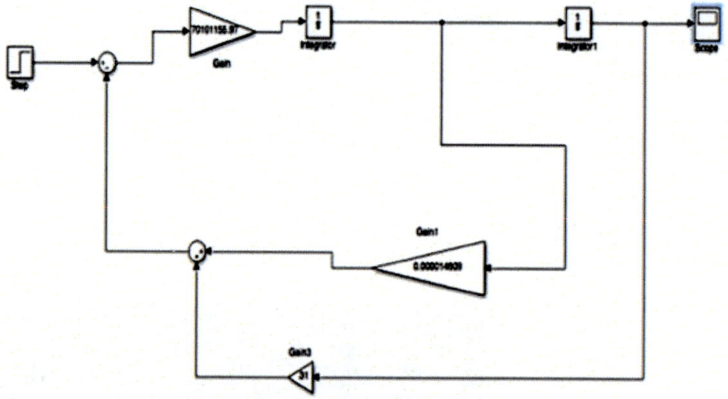

Fig. 18 Simulink model of an open-loop. 7 kKhz accelerometer

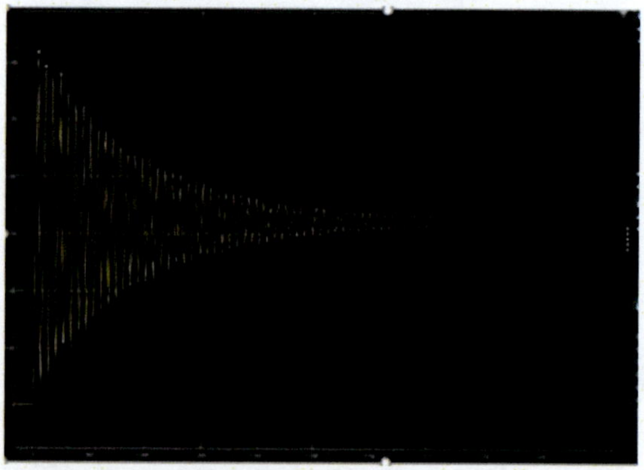

Fig. 19 Step response for $\zeta = 0.5$(Model Y)

5 Results and Discussions

In this work an attempt is made to employ the detailed design, simulation, mathematical modelling and sensitivity analysis of single axis mems based capacitive accelerometer of different operating frequencies (2 kHz, 7 kHz). Mathematical modelling of frequency, displacement and capacitance is done and found that it is almost closer to the values obtained from simulation. The frequency, displacement and capacitance values are gained by simulating the desired accelerometer in COMSOL Multi-physics software. Both the accelerometers were analysed to be underdamped systems. The error difference between the simulated and calculated value is on an average of 0.045 kHz. Considering the effect of the physical dimensions on sensitivity, it is concluded that though the total displacement remained more or less the same (0.03 + um for 10.8 kHz and 0.028um for 15.8 kHz), the resonant frequency changed in regard to multiple configurations of physical parameters. Moreover, the displacement was found to be more in regard to 2 kHz than 7 kHz; it could be concluded that lower the resonant frequency, greater will be the sensitivity of the accelerometer. In addition, this is cross-verified by comparing the displacement sensitivity factor which is relatively more with respect to the Model X accelerometer making it more susceptible to displacement. Model X has more stable capacitive range as its static capacitance is more, which allows it to be used to detect multiple voltage surges caused by various displacement factors. Moreover, taking the von mess stress into consideration, the Model Y accelerometer can withstand considerably lower stress, where the maximum stress hold point is quite opposite in Model X (lower with respect to the beam). Analytically, the damping ratio for 2 kHz accelerometer is calculated and found to be 0.03208 and concluded this device to be underdamped. It was also observed that by improving to 0.5, the system becomes more stable. The comparison between the simulated and calculated values of the two accelerometers is displayed in Tables 3 and 4.

Table 3 Comparison between simulated and calculated values

Model X		Model Y	
Simulated value	Calculated value	Simulated value	Calculated value
$f_n = 2.1$ kHz	$f_n = 2.6$ kHz	$f_n = 7.0$ kHz	f_n f7 = 7.4 kHz

Table 4 Comparison of the two accelerometers

Parameters	Model X	Model Y
Natural frequency	$f_n = 2.6$ kHz	f_n f7 = 7.41 kHz
Static capacitance	Co = 0.730455pf	Co = 0.28199pf
Displacement sensitivity	$S_d = 3.53625 * 10^{-8}$m/g	$S_d = 4.5096 *10^{-9}$m/g
Stress range	Min:0.0175 N/m^2 to max-7.509E12 N/m^2	Min-0.00 N/m^2 to max -6.361E23 N/m^2

6 Conclusion and Future Scope

The Model X accelerometer surpasses the Model Y in many parameters, making it a viable selection over the other; it can be concluded to extend this work towards improving the of the model X accelerometer. Although, great number of research papers was available on improvement of sensitivity of capacitive MEMS accelerometers, still there is a lot of scope to carry out the research on capacitive accelerometers to improve its sensitivity. Therefore, the future development of this work will be to enhance performance and sensitivity of capacitive accelerometers that includes change of finger width, shape of proof mass, applied load and by using a complex pair of folded beams structure and produce the more optimized structure.

Acknowledgement We take immense pleasure in expressing our gratitude to Dr.Veda Sandeep Nagaraja, Associate Professor, ECE department, NMIT, and all the staff members of EEE Dept, NMIT, and Bangalore for the assisting in our learning. We also thank the staff members of Nano materials and MEMS COE for the support extended. We further acknowledge the guidance provided by the management of NMIT, Bengaluru.

References

1. Padmanabhan Y (2017) Mems based capacitive accelerometer for navigation. Dissertation University of Windsor. https://doi.org/10.13140/rg.2.2.35625.49769.
2. Kei-Ming Kwong (2017) MEMS Accelerometer Specifications and Their Impact in Inertial Applications Dissertation, Master of Applied Science, Department of Computer Engineering, the University of Toronto
3. Zakriya M, Ibrahim (Abe) ME, Mahmoud R (2018) Monolithic Multi Degree of Freedom (MDoF) Capacitive based MEMS Accelerometers. J Micromachines (Basel). 10.3390/mi9110602
4. Senturia (2001) Microsystem Desig. ISBN 0-7923-7246-8 Publisher: Kluwer Academic Publishers
5. Akila Kannan BE (2005) Design and modelling of a mems-based accelerometer with pull in analysis by (Electrical Engineering), the Anna University
6. Albarbar A, Mekid S, Starr A. (2008) Suitability of MEMS Accelerometers for Condition Monitoring: An experimental study. Journal of Sensors. https://doi.org/10.3390/s8020784
7. Amini BV (2006) A mixed-signal low-noise sigma-delta interface IC for integrated sub-micro-gravity capacitive SOI accelerometers. Dissertation, School of Electrical Engineering, Georgia Institute of Technology
8. Wong WC, Azid IA, Majlis BY (2010) Theoretical analysis of stiffness constant and effective mass for a round-folded beam in MEMS Accelerometer. J Mech Eng 57(2011)6, 517–525 Paper received: 03.11.2009 https://doi.org/10.5545/sv-jme.2009.151.
9. Benmessaoud, M., & Nasreddine, M. M. (2013). Optimization of MEMS capacitive accelerometer. Microsyst Technology, *19,* 713–720. https://doi.org/10.1007/s00542-013-1741-z.
10. Gupta V, Mukherjee T (2012) Layout synthesis of CMOS MEMS accelerometers. Department of ECE, Carnegie Mellon University, Pittsburgh, PA-15213.
11. Nagaraja VS, Rudresha KJ, Pinjare, S. L. (2018). Design, Fabrication and Characterization of a Biologically Inspired MEMS Directional Microphone. IEEE SENSORS, New Delhi, 1–4

12. Sharma, K., Macwan, I. G., Zhang, L., Hmurcik, L., & Xiong, X. (2012). *Design optimization of a MEMS comb Accelerometer*. Department of electrical engineering: University of Bridgeport, Bridgeport.
13. Zhou W., Chen Y., Peng B., Yang H., Yu H., Liu H., and He X. (2014). Air Damping Analysis in Comb Microaccelerometer. Journal of Advances in Mechanical Eng
14. Kumar D. P. and Karat (2006). System Level Simulation of Servo Accelerometer in Simulink. Department Of Earthquake Engineering, I.I.T.Roorkee, India. Journal of Physical Sciences, Vol. 10, 2006, 143 – 150.

Transport Tracking Using RFID and GSM Based Technique

N. Subbulakshmi, R. Chandru, and R. Manimegalai

Abstract Nowadays, the crime has increased and the occurrence of the accidents are more in the cities. People are afraid about the current scenario. Moreover parents are worried on children when they travel to schools. In order to provide protection and safety to the children, there is a necessity for a technology based security and alert. In this paper, an idea to identify the status of the children when they are out to school so that they can identify the location of their children. Using GSM and RFID the tracking of transport is done. It has two units; one is bus unit for tracking the activity and alerting parents during the school travel. Another is school unit for tracking child inside the school. The results show that our approach will provide safety and security to school children.

Keywords RFID · GSM module · Sensors · GPS module · PIC controller · Tracking system

1 Introduction

At present accidents become predominant, and people are conscious about the safety and life. Parents are very much concerned about their children, and they admit their children in highly reputed schools which has hi-tech amenities. Recently, most of the schools have transport services in around that city peripheral. Though children go by school bus, parents are worried whether the child reaches the school safely or is in some danger situation, i.e., met with any accidents [1]. This paper identifies such accidents and alert the parents. This method conveys the alert messages which are used to send pick up and drop information to the parents. As discussed above, bus unit

N. Subbulakshmi (✉)
Francis Xavier Engineering College, Tirunelveli, Tamilnadu, India
e-mail: subbulakshminammalwar@francisxavier.ac.in

R. Chandru
Sri Ramakrishna Engineering College, Coimbatore, Tamilnadu, India

R. Manimegalai
PSG Institute of Technology and Applied Research, Coimbatore, Tamilnadu, India

P. Nayak et al. (eds.), *IoT and Analytics for Sensor Networks*, Lecture Notes in Networks and Systems 244, https://doi.org/10.1007/978-981-16-2919-8_20

consists of RFID (Radio-Frequency Identification) reader that uses electromagnetic fields to automatically track and identify the tags that is being attached in ID cards of the children so that once the card is read it means child has entered the bus. It has a local power source like battery.

2 Literature Review

Various types of sensors like temperature, gas sensor, and accelerated tilt sensors are also placed in the transport any accidents during driving. When there is such a danger, the alert message will reach the parents with help of GSM and GPS modules. GSM (Global System for Mobile Communications) is used for mobile communication which can operate over a wide range of network. GPS (Global Positioning System) is used to identify the geographical location of the child. This ensures that each movement of the child is being identified buy their parents. Next the school entity comprises of RFID reader and GSM module. Both the units are being processed by the PIC Controller 16F877A. RFID tag is worn by the child and when it is placed [2]. The sensors and the RFID tag are interfaced through the controller. There will be a LCD display for drivers to view where the child is getting down. If a child gets down somewhere else other than school the message of the location will be displayed to the driver. The GSM modem send message to authorized person according to the data received [3].

Parents will get message when child enters the bus, leaves the bus, when enters the school premises and leaves the school. The RFID reader reads the ode in all these cases. Each activity and location of the child is monitored by their parents through the GPS. If any accident on the way to school, that time also alert message will reach the parents. This system will be efficient and ensure security and safety of the child. The paper is proposed to reduce the risk of accidents and loss of life of school children without any reason. There were so many school bus accidents that occurred which has taken life of so many children. There were so many safety measures taken for school children to prevent accidents like speed control and checking devices to avoid over speeding of bus, limitation of children count in bus, etc. But there were no alert systems and tracking systems. The RFID based this system will track the activity of the child and will report if there is any danger [2]. The live location is also tracked for quicker recovery. It helps in intimating school management and parents regarding the danger in bus. And also when there is tilt in the bus driving, that is also alerted. SMS based kids' tracking and safety system is introduced in [4]. RFID GSM imparted school children security system is described in [5]. Efficient real time tracking is designed in [6]. A robust tactic-based real time framework for public transport synchronization is discussed in [7].

3 Design and Implementation

The block diagram of the proposed system is shown below as depicted in Fig. 1.

The main components used in the system is shown in Fig. 1. The main component is PIC 16F877A. It acts as a main control unit of the whole proposed system. All the other components like sensors and GPS modules are fed as input to the controller. The alert messages can be carried to parents along with GSM modem. The details of the components in the system is described below.

3.1 Pic 16f877a

The cost of this controller is low, and its handling is also easy. Because of these features it is used in the system. It processes the data and then sends output data to GSM modem and LCD display. Figure 2 shows PIC 16F877A.

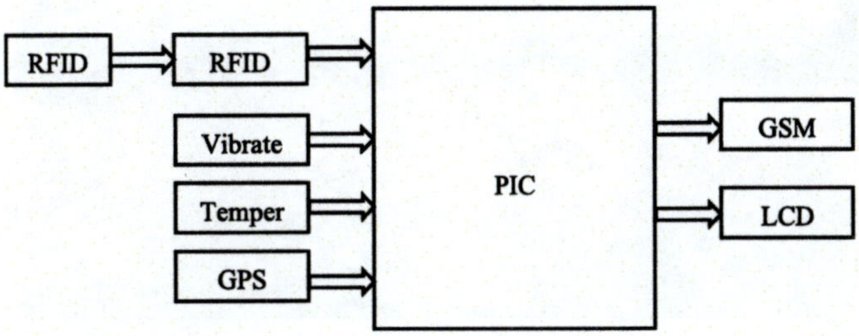

Fig. 1 Block diagram of the system

Fig. 2 PIC 16F877A

3.2 RFID Tag

The RFID tag is worn by child. RFID tag is active tag, and another is passive tag. The active tags have battery, and they provide longer read range than the latter and moreover it broadcast its own signals [3]. Whereas passive tag is opposite of active tags. The passive tag is shown in Fig. 3.

The process of RFID is shown in Fig. 4. The passive tag consists of a chip and an antenna. The antenna collects the signal from the reader and controls the chip. Then the signal will be adjusted based on unique information which is available the block. This signal is again retransmitted to the antenna. Passive tag will be appropriate for this system because it has short reading range which is necessary for this application. So that it can identify the person who comes near to tag reader.

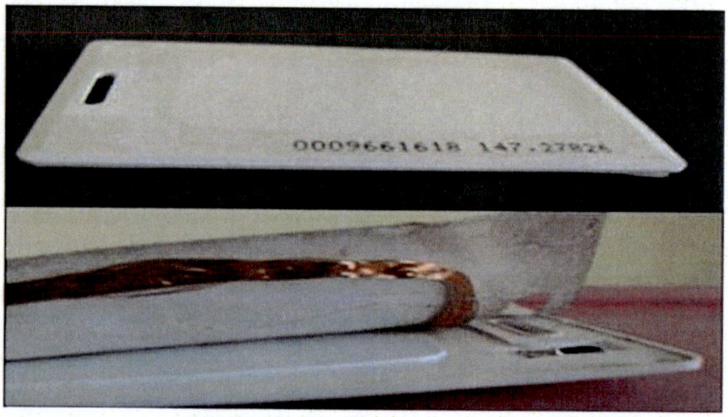

Fig. 3 RFID TAG

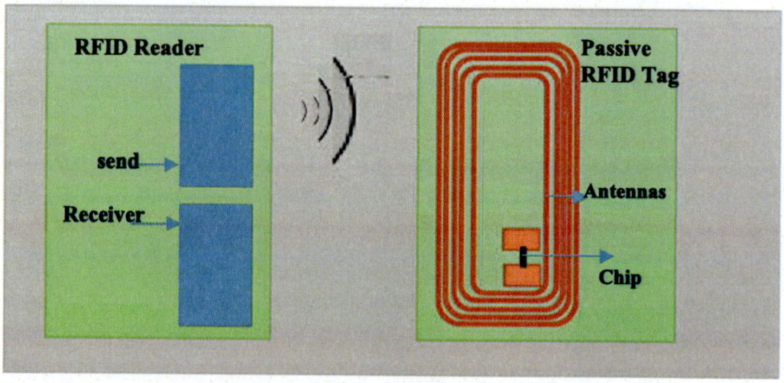

Fig. 4 Process of RFID

3.3 RFID Reader

The RFID reader is used to read the signal that is being sent from tag. It is of three types that is classified based on their frequency range. The types are low frequency, high frequency and ultrahigh frequency readers. The reader that is suitable for this application is low frequency reader. EM-18 is being used which has a frequency range of 125 KHz. It has serial output with range of distance 8–12 cm. Another feature is that it has an in built antenna which can be directly connected to the PC by RS232 protocol [4] and [3]. The reader also has on-chip antenna and can be powered up with help of a 5 V power supply, and moreover it is inexpensive [5].

3.4 Temperature Sensor

LM35 is a precision temperature sensor that is used for this proposed system. It monitors the temperature exactly when compared to thermistor. The operating voltage is from 4 to 30 V. Its accuracy is approximately about 0.5 Celsius. The range of the system is from –55 to 150 C. It has three pins in which 1st is power supply pin, 2nd pin is output that is being connected to PIC, and 3rd pin is ground pin.

3.5 Accelerated Tilt Sensor

Accelerometer is a device that measures the proper acceleration. The ADXL330 is a smallest triaxle accelerometer. It is thin, and it is low power device. It measures the dynamic and static acceleration. Dynamic acceleration results from vibration, shock or motion. Whereas static acceleration is gravity in tilt-sensing. This sensor is being interfaced with the PIC. The sensed data is processed by the PIC16F877A and the output signal, i.e., the alert message is being transmitted to parents using GSM.

3.6 Gas Sensor

Gas sensor is used to detect the presence of gases. It will detect toxic gases, flammable, combustible and oxygen depletion. It is widely used for safety purpose. MQ6 is being used here which has high sensitivity to gases like Butane, LPG and Propane. It can detect concentration of gases from 200 to 1000ppm. This is also interfaced with PIC and the sensed output information will be transmitted to parents through GSM.

3.7 GPS Module

GPS stands for Global Positioning System. It helps in tracking and identifying the location of the child. It has coordinates, and it is captured. The GSM modem sends the message to parents and school management. The location is accurately determined by it. GPS is a satellite navigation system that accurately identifies location and time [4]. The GPS receiver identifies the location of the students and also there are many other satellites and ground stations to support this.

3.8 GSM Module

GSM clearly explains about the 2G networks (cellular) which are used in mobile communication. It has a wide range of usage in around 219 countries and union territories. This technology was implemented using one of the narrow band techniques named as Time Division Multiple Access (TDMA). Its process is as follows. The data is digitized, and then it is compressed and then is passed through the transmitting channel. The frequency band is in the range 900 MHz or 1800 MHz. Since it is present globally it allows the users to use their same mobile phone all over the world but by changing the SIM (Subscriber Identity Module) card [4].

4 Hardware Implementation

The GPS module, GSM module, the various types of sensors and the RFID reader are connected to the PIC controller. The gas sensor and accelerated tilt sensor detect the gases entering into the bus and the tilting of the bus, respectively, and send the received signal to the PIC controller. Figure 5 represents the hardware implementation of the model.

The controller processes the received information from all those inputs and gives relevant output to the GSM module and the LCD display. When the RFID tag is placed near the reader it reads the code and it will interface it to PIC.

At the same time if there is any fire accidents, road accidents or any gas leakage into the bus, the system has different types of sensors to detect the accidents. As discussed previously all these are interfaced with controller. This controller issues a command to the GSM modem in order to deliver the received messages for respective persons [2]. The GPS plays a major role here. It along with the messages sends the location. At last the parents and the school management receives the messages from the GSM modem [4]. In paper [3] also this block can be implemented for the automatic design.

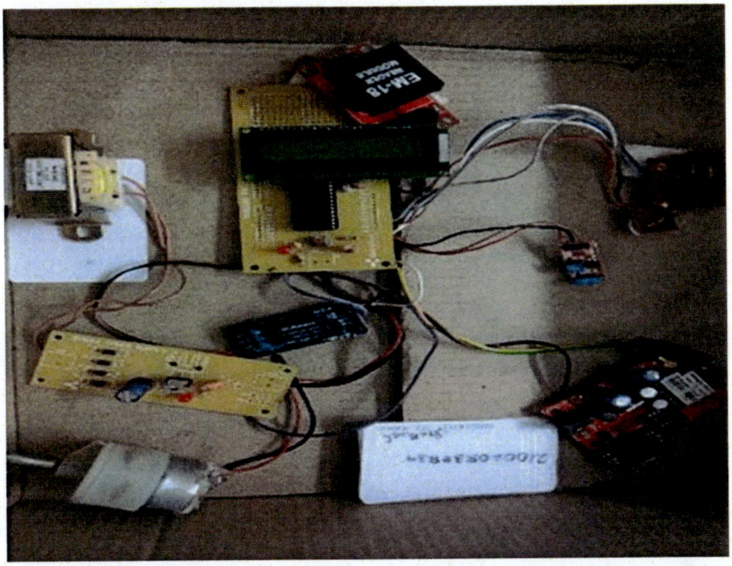

Fig. 5 Hardware implementation

5 Results and Discussions

Some of the results of the system is shown below. When the child safely boarded and de-boards the bus the following message is received by the parents [6]. The current situation of the child is being mentioned, and it is done through GSM modem. This controller sends some instruction to the GSM modem. According to this the alert message will be received by the school authorities and parents. References [8, 9] describes the survey of Filter bank algorithms. Low power techniques have been discussed in [10] and [11].

This shows the fire accident alert message. When the fire is being detected in the transport is identified by the sensor present, and it will send a message to the GSM modem. The temperature sensor that is used here will identify the changes in temperature will sense that and gives information to the controller. From the information of the controller, the information will reach the authorized persons through GSM modem. Similarly, the gas sensor detects the presence of gas in the transport and accelerated tilt sensor detect the change in acceleration in the traveling bus and sends the information in a similar way to the controller which in turn sends alert message to parents [7]. All three sensors work in a similar way in transmitting the messages. Many algorithms were discussed for the gearing aid applications in [9].

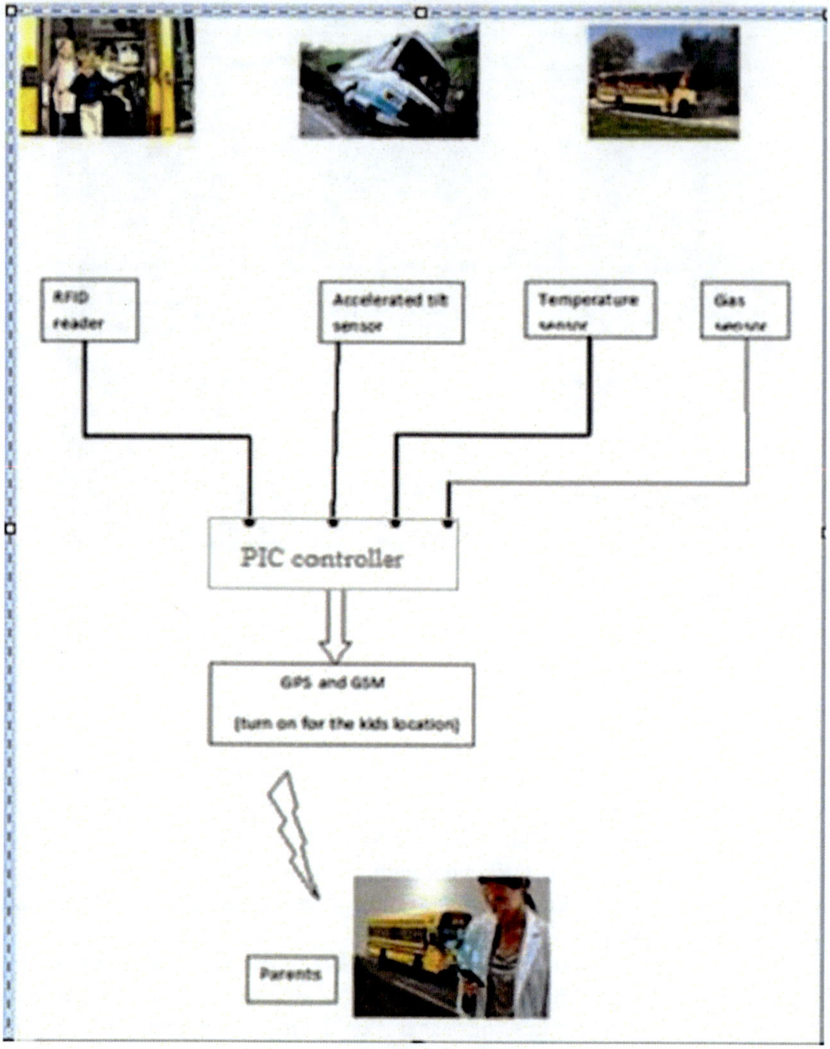

Fig. 6 PIC controller

6 Conclusion

This paper focuses on safety of the school children during their school travel. GSM module is responsible for issuing messages to the parents during boarding and de-boarding of the bus. Similarly, the gas sensor detects the presence of gas in the transport and accelerated tilt sensor detect the change in acceleration in the traveling bus and sends the information in a similar way to the controller which in turn sends alert message to parents. The types of sensors identify the accidents. It has two units

Fig. 7 Alert message of Boarding and Deboarding

Fig. 8 Accident message

one is Bus unit for tracking the activity and alerting parents during the school travel. Another is school unit for tracking child inside the school. The results show that our approach will provide safety and security to school children. The location of the accident or the status of the child is detected using GPS module. This will enhance the safety of the children.

References

1. Anusha, R., & China Appala Naidu, R. (2016). GPS and RFID Based School Children Tracking System. *International Journal of Advanced Research in Computer Engineering & Technology (JARCET)*, *5*(6).
2. Malliga, R., & Narmatha, T. (2016). RFID-based system for school children transportation safety enhancement. *International Journal of Advanced Research in Computer Science and Software Engineering*, *6*(4).
3. Subbulakshmi, N., & Manimegalai, R. (2020). Novel lifting filter bank for bird call analysis. In L. Ashok Kumar, L. S. Jayashree, & R. Manimegalai (Eds.), *Proceedings of international conference on artificial intelligence, smart grid and smart city applications* (pp. 871–880). Springer, Cham. https://doi.org/10.1007/978-3-030-24051-6_82.
4. Shyam, N., Kumar, N., Shashi, M., & Kumar, D. (2015). SMS based kids tracking and safety system by using RFID and SGSM. *International Journal of Innovative Science, Engineering & Technology*, *2*(5).
5. Vidyasagar, K., Balaji, G., & Narendra Reddy, K. (2015). RFID- GSM imparted school children security system. *Communications on Applied Electronics (CAE).*
6. James, J. G., & Nair, S. (2017). Efficient, real-time tracking of public transport, using LoRaWAN and RF transceivers. In *Proceedings of the IEEE region 10 conference TENCON* 2017–2017, Penang (pp. 2258–2261). https://doi.org/10.1109/TENCON.2017.8228237.
7. Nesheli, M. M., (Avi) Ceder, A., & Liu, T. (2015). A robust, tactic-based, real-time framework for public-transport transfer synchronization. *Part of special issue: Papers selected for Poster Sessions, The Transportation Research Procedia, 9*(2015), 246–268.
8. Elleuch, W. (2014). Mining road map from bug database of GPS data. In *14th international conference on hybrid intelligent systems* (pp. 193–198). IEEE Press, Kuwait.
9. Subbulakshmi, N., & Manimegalai, R. (2014). A survey of filter bank algorithms for biomedical applications. In *2014 international conference on computer communication and informatics, Coimbatore* (pp. 1–5). https://doi.org/10.1109/ICCCI.2014.6921778.
10. Rooban, S., Subbulakshmi, N., & Poorna Vamsi, Y. (2021). Low power circuit design for dynamic comparator. *International Journal of Performability Engineering*, *17*(5), 444.
11. Subbulakshmi, N., & Manimegalai, R. (2017). MACGDI: low power MAC based filter bank using GDI logic for hearing aid applications. *International Journal of Electronics and Electrical Engineering*, *5*(3), 235–239.

An Ultra-Wide Band Patch Antenna for Commercial Communication Applications

L. Diana Evangeline, G. Shine Let, and C. Benin Pratap

Abstract Modern wireless systems require a single antenna to perform multi-band functionality considering different applications. As per the FCC report, the designed antenna works in the ultra-wide band from 0.45 to 14.37 GHz with an impedance bandwidth of 187.8%. For the designed antenna, efficiency is greater than 50% and gain is greater than 1 dBi for the entire UWB band. By tuning the feed line width, the same antenna can operate in eight narrow bands covering various mobile communication bands, military applications, global positioning system, satellite communication, and some part of Ku band that is from 0.45 to 14.37 GHz. The designed antenna provides a reflection coefficient of less than −10 dB for all eight narrow bands and has good impedance matching. Moreover, the proposed structure can also support many mobile applications like LTE, Wi-Fi, Wi-Max, and 5G.

Keywords Ultra-wide band · Narrow band · Antenna · Communication · Microstrip patch

1 Introduction

Federal Communication Commission (FCC) in 2002 reported, frequency range from 3.1 to 10.6 GHz is called Ultra-wide band (UWB) and can be used for various commercial communication applications. Within the UWB band, ITU-T specifies various narrow bands for different applications [1]. Table 1 shows the different frequency ranges and their applications.

Many UWB antennas have been researched and developed for industry and academic purposes [2]. The system integrated with cognitive radio capability should have a UWB antenna and a narrowband antenna for proper communication. For efficient spectrum sensing purposes, a UWB antenna is preferred. To have data communication, a narrowband functionality antenna is used. Due to the rapid advancements in the standards of technology, it would be easier to have a single antenna covering

L. D. Evangeline · G. S. Let (✉) · C. B. Pratap
School of Engineering and Technology, Karunya Institute of Technology and Sciences, Coimbatore, India

P. Nayak et al. (eds.), *IoT and Analytics for Sensor Networks*, Lecture Notes in Networks and Systems 244, https://doi.org/10.1007/978-981-16-2919-8_21

Table 1 Frequency range and its applications

Bands	Frequency range (GHz)	Applications
L band	1–2	RADAR, GPS, GSM-1800 band
S band	2–4	LTE, 5G, Bluetooth, Wi-Fi, ISM band
C band	4–8	Satellite communications, WiMax, Wi-Fi, ISM band
X band	8–12	Military applications
Ku band	12–18	Broadcasting, satellite communication

multiple applications [3]. Researchers have carried out various wideband designs with strips, slots, metamaterial structures, and meander designs to encounter several operating bands. For high-speed communication, a coplanar waveguide (CPW) fed UWB antenna is proposed in the literature [4]. The author used an elliptical radiating structure with a partially bent ground plane. Wideband antenna structures are required for cognitive radio spectrum sensing applications. Various shapes such as circular, elliptical, rhombic, semi-circular crown-shaped radiating structures are designed for wideband spectrum sensing application [5]. Rectangular shape low profile dielectric resonator antenna is proposed by authors for spectrum sensing in cognitive radio applications [6]. A U-shaped tunable radiating stub antenna is designed for cognitive radio applications [7].

Half circular ring and half square ring 2-port monopole UWB antenna are in literature for automotive communications [8]. Electromagnetic bandgap structures near the feed point are used to make the antenna operate in the UWB band [9]. Square shape metamaterial structure is reported in the literature to make the antenna operate in UWB for wireless body area network (WBAN) application [10]. 4-port rectangular shape CPW fed monopole antenna is proposed for multiple-input multiple-output (MIMO) applications [11]. A defective ground structure UWB antenna is proposed by authors for 4G applications [12]. Generally, the slot is a desirable technique to make an antenna operate over a larger bandwidth. The slots can be of any interesting shape. A circular slot structure is used in radiating patch to operate in the UWB band [13–15].

A lotus-shaped printed monopole antenna is designed to operate in the ultra-wide band and is proposed in this article. By varying the feed line width, the same antenna can function in multiple narrow bands within the UWB band. The design considerations, various analyses carried out on the deigned lotus-shaped antenna are discussed in the resulting sections.

2 Design of Lotus-Shaped UWB Antenna

The deigned lotus-shaped UWB antenna structure is shown in Fig. 1. The radiating patch exploits a lotus-shaped geometry, and a partial ground plane is considered. Copper material of thickness 35 μm is used for radiating patch and ground plane.

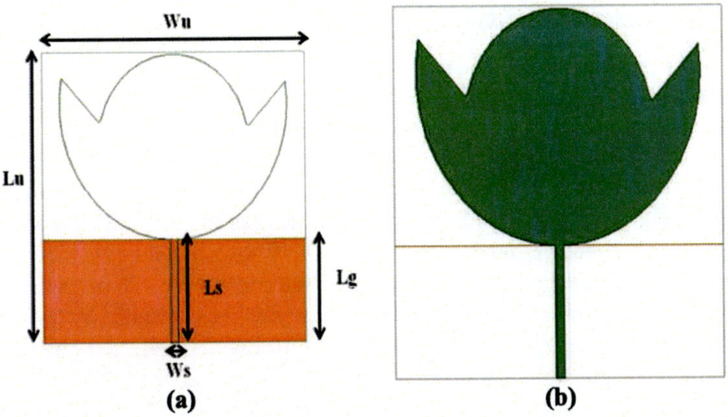

Fig. 1 Lotus-shaped UWB antenna structure **a** ground plane **b** radiating structure

Table 2 Design specifications

Parameter	Dimensions (mm)
Lu	140
Wu	150
Ls	50
Ws	4
Lg	50

Glass epoxy-based commonly available low profile flame retardant-4 (FR4) is used as a substrate. FR4 substrate has a thickness of 1.6 mm, dielectric constant 4.4, and dielectric loss tangent 0.02. Lu × Wu (mm^2) is the total area of the proposed lotus-shaped antenna. Table 2 provides the design specification considered for the antenna design. The trimetric view and the microstrip feed provided to the antenna are shown in Fig. 2.

Fig. 2 Lotus-shaped UWB antenna—feed point

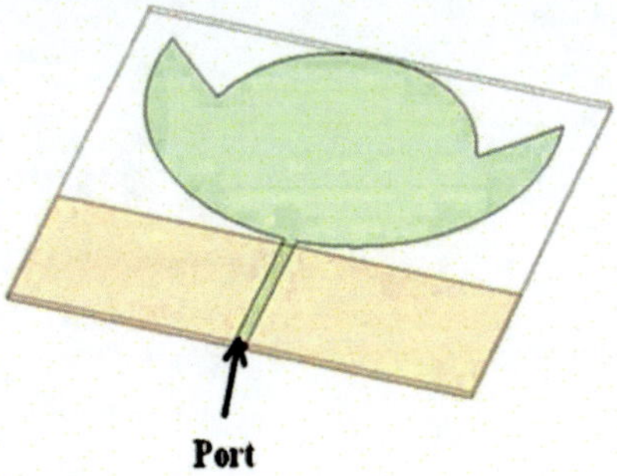

3 Results and Discussions

The lotus-shaped UWB antenna is simulated using ANSYS HFSS software. The optimum dimensions of the antenna are obtained by using parametric analysis. Using parametric analysis, the length and width of the antenna are varied within a particular range. The simulated performance of the antenna is verified at every step size which is within the given range. From the literature, the operating band of the antenna is decided when the voltage standing wave ratio (VSWR) is less than two or the reflection coefficient is less than −10 dB. The reflection coefficient −10 dB indicates that out of the total transmitted power, 90% power is effectively radiated by the antenna, and 10% power is reflected. Hence, if $S_{11} < -10$ dB, the reflected power by the antenna is less than 10% and transmitted power is greater than 90%. The reflection coefficient is based on the antenna terminal impedance (Z_t) and characteristic impedance (Z_0) and is shown in Eq. (1). The relation between VSWR and S_{11} is given by Eq. (2). If the above-said criterion is met for the frequency range between 3.1 and 10.6 GHz, the antenna is accurately operating at the UWB band.

$$S_{11} = \frac{Z_t - Z_0}{Z_t + Z_0} \tag{1}$$

$$VSWR = \frac{1 + S_{11}}{1 - S_{11}} \tag{2}$$

As the width of the feed line (Ws) is varied, there is a vital change in the operating frequency range. Figure 3 shows the reflection coefficient (S_{11}) characteristics of the lotus-shaped UWB antenna for the width Ws = 4 mm. From Fig. 3, the operating band of the designed antenna is from 0.45 to 14.32 GHz and the obtained impedance bandwidth is 187.8%. Figure 4 shows the VSWR characteristics plot.

Microstrip feed is provided to the designed antenna through a feedline of width (Ws) 4 mm and length (Ls) 50 mm. Figure 5 shows the current flow through the

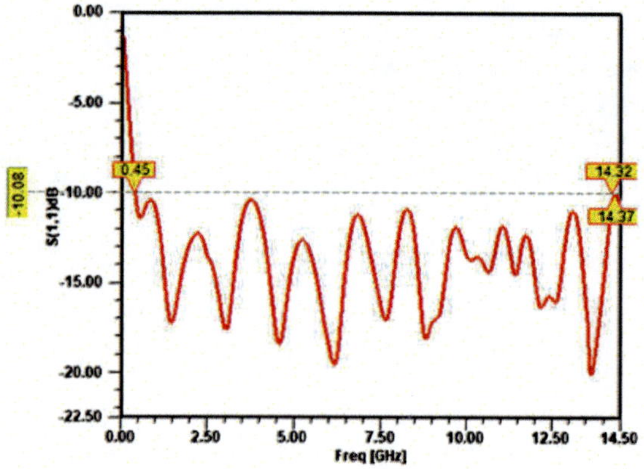

Fig. 3 Reflection coefficient characteristics for Ws = 4 mm

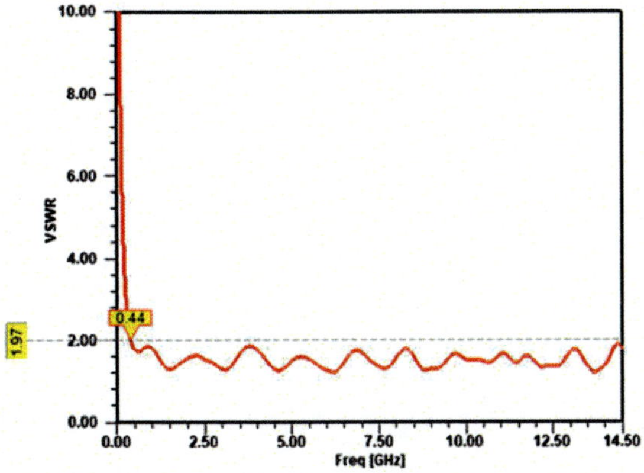

Fig. 4 VSWR characteristics for Ws = 4 mm

Fig. 5 Current distribution

radiating patch as the feed is provided. Figure 6 depicts the far-field radiation pattern of the lotus-shaped UWB antenna at $\phi = 90°$ and $\phi = 0°$. When the power is fed to the antenna more than 90% of the input power is effectively radiated and the direction of radiation is shown in Fig. 6.

The shape and area of the antenna determine the directivity of the antenna for the entire frequency of operation. Gain (G), efficiency (η) and directivity (D) are closely related to each other and are shown in Eq. (3). The figure of merit of the antenna is represented by the gain of the antenna. Figs. 7 and 8 shows the gain and efficiency of the antenna, respectively, for the operating UWB frequency range.

$$\eta = \frac{G}{D} \tag{3}$$

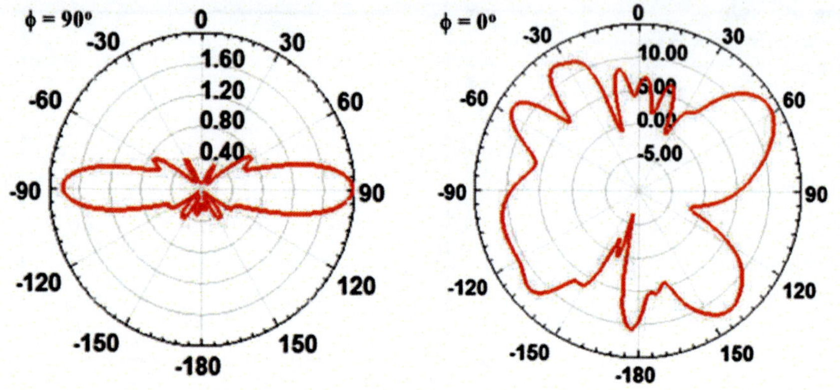

Fig. 6 Radiation pattern at $\phi = 90°$ and $\phi = 0°$

Fig. 7 Lotus-shaped UWB antenna gain versus frequency

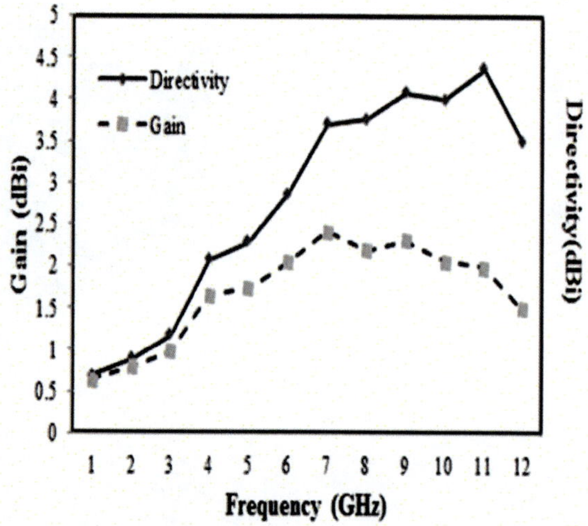

The same UWB antenna can be used to operate in multiple narrow bands if the feed width (Ws) is increased to 6 mm. In eight narrow bands, the antenna can operate with a good reflection coefficient. The narrow bands are working in different communication applications mentioned in Table 1. Figure 9 shows the reflection coefficient characteristics for Ws = 6 mm.

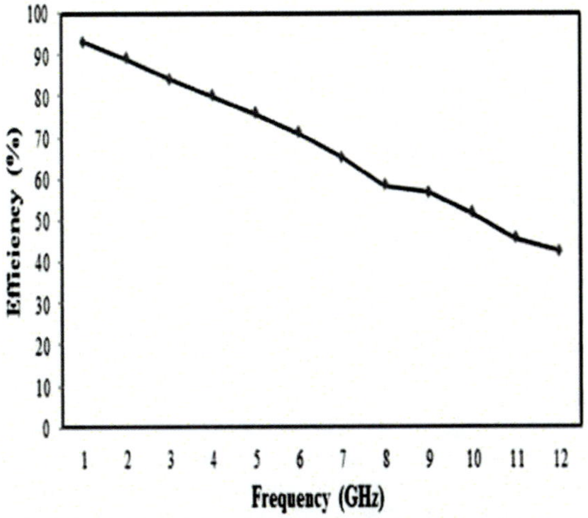

Fig. 8 Lotus-shaped UWB antenna efficiency versus frequency

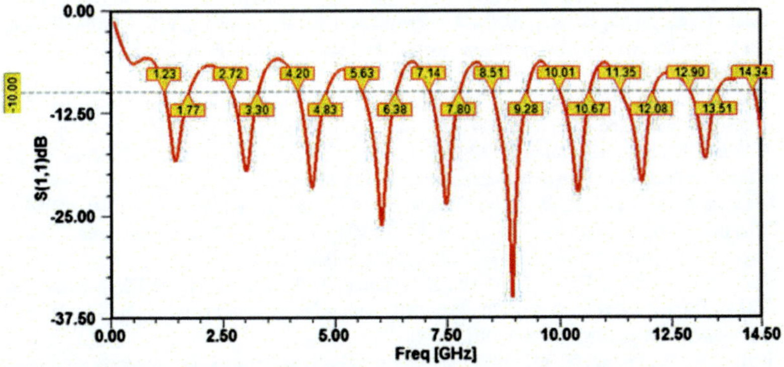

Fig. 9 S_{11} versus frequency for width Ws = 6 mm

4 Conclusion

In this paper, a simple lotus-shaped ultra-wide band antenna operating at the frequency range of 0.45–14.32 GHz is proposed. The antenna is imprinted on a low-cost, low-profile FR4 substrate. The performance of the Lotus-shaped UWB antenna is validated by showing the reflection coefficient characteristics, VSWR characteristics, current distribution, radiation pattern, gain, directivity, and efficiency. For the entire operating impedance bandwidth of 187.8%, S_{11} is less than −10 dB, efficiency is greater than 50%, and gain is greater than 1 dBi. The proposed antenna can be used for UWB commercial communication applications.

References

1. I.T.U. (ITU-R). (2014). Fixed service use and future trends (F.2323-0).
2. Li, Y., Li, W., Ye, Q., & Mittra, R. (2014). A survey of planar ultra-wideband antenna designs and their applications. *Forum for Electromagnetic Research Methods and Application Technologies (FERMAT), 1*, 1–16.
3. Porcino, D., & Hirt, W. (2003). Ultra-wideband radio technology: Potential and challenges ahead. *IEEE Communications Magazine, 41*(7), 66–74.
4. Ibrahim, A.A. (2019). Compact planer UWB antenna for high speed communications. In *2019 International Conference on Innovative Trends in Computer Engineering (ITCE)* (pp. 266–269).
5. Durbhakula, M.K.C., & Nalam, V.K.R. (2017). Wide band patch antenna structures for cognitive radio applications. In *2017 IEEE 7th International Advance Computing Conference (IACC)* (pp. 434–439).
6. Atallah, H.A., Abdel-Rahman, A.B., Yoshitomi, K., & Pokharel, R.K. (2014). Novel compact UWB monopole RDRA for cognitive radio spectrum sensing applications. In *2014 IEEE Fourth International Conference on Consumer Electronics Berlin (ICCE-Berlin)* (pp. 462–466).
7. Bharathi, A., & Pandharipande, V.M. (2016). Design and development of antenna with U-shaped tuning stub for UWB applications. In *2016 IEEE Annual India Conference (INDICON)* (pp. 1–4).
8. Alsath, M. G. N., & Kanagasabai, M. (2015). Compact UWB monopole antenna for automotive communications. *IEEE Transactions on Antennas and Propagation, 63*(9), 4204–4208.
9. Dalal, P., & Dhull, S. K. (2020). Upper WLAN band notched UWB monopole antenna using compact two via slot electromagnetic band gap structure. *Progress in Electromagnetics Research, 100*, 161–171.
10. Yalduz, H., Koç, B., Kuzu, L., & Turkmen, M. (2019). An ultra-wide band low-SAR flexible metasurface-enabled antenna for WBAN applications. *Applied Physics A, 125*(9), 609.
11. Naktong, W., & Ruengwaree, A. (2020). Four-port rectangular monopole antenna for UWB-MIMO applications. *Progress in Electromagnetics Research, 87*, 19–38.
12. Ibrahim, M.A., Swelam, W., Abd El-Azeem, M.H., & El Hennawy, H. (2019). Ultra-wide band microstrip antenna for 4G applications. In *IOP Conference Series: Materials Science and Engineering* (Vol. 610, No. 1, p. 012025). IOP Publishing.
13. Florencio, R., Boix, R. R., Carrasco, E., Encinar, J. A., Barba, M., & Pérez-Palomino, G. (2014). Broadband reflectarrays made of cells with three coplanar parallel dipoles. *Microwave and Optical Technology Letters, 56*(3), 748–753.
14. Jang, J. W., & Hwang, H. Y. (2009). An improved band-rejection UWB antenna with resonant patches and a slot. *IEEE Antennas and Wireless Propagation Letters, 8*, 299–302.
15. Guo, Z., Tian, H., Wang, X., Luo, Q., & Ji, Y. (2013). Bandwidth enhancement of monopole UWB antenna with new slots and EBG structures. *IEEE Antennas and Wireless Propagation Letters, 12*, 1550–1553.

8-Bit Carry Look Ahead Adder Using MGDI Technique

P. Ashok Babu, V. Siva Nagaraju, and Rajeev Ratna Vallabhuni

Abstract High-performance and low power consumption are major factors that describe the significance of a design in VLSI. At low and ultra-low power applications, power consumption and logic delays are a fundamental problem. Nowadays, higher performance designs are built on the concept of computation units like ALU, where adders and multipliers are the essential components. To optimize low-power and high-performance applications, adders and multipliers need to be engineered to meet specifications such as speed and power dissipation. Using Modified Gate Diffusion Technique, this paper proposes a carry look-ahead adder (CLA). Compared with other adder models, the carry look ahead adder has much less propagation delay. The proposed adder is implemented on 90 nm technologies using the cadence environment. This paper's main objective is to use Modified Gate Diffusion Input (MGDI) Technology to develop and implement a CLA.

Keywords ALU · EDP · MGDI · Carry look ahead adder · PDP

1 Introduction

Nowadays, in our day-to-day life, we are using many electronic devices of low power and high-performance VLSI circuits. Due to modernization, there have been improvements in designs, and the need to make various transistors in one single chip has increased. This also affected the increase in operating frequency. Even the size of the construction chip is gradually reduced. As a result of all these factors, it has caused an increase in energy dissipation [1–5].

P. A. Babu (✉) · V. S. Nagaraju
Institute of Aeronautical Engineering, Dundigal, Hyderabad 500043, India
e-mail: p.ashokbabu@iare.ac.in

V. S. Nagaraju
e-mail: v.sivanagaraju@iare.ac.in

R. R. Vallabhuni
Bayview Asset Management, LLC, Coral Gables, FL 33146, USA
e-mail: rajeevratna@ieee.org

P. Nayak et al. (eds.), *IoT and Analytics for Sensor Networks*, Lecture Notes in Networks and Systems 244, https://doi.org/10.1007/978-981-16-2919-8_22

Carry look-ahead adder is a fast and resilient among various adders with very less propagation delay and low power dissipation. The main objective of this design is to reduce the propagation delay. Other adders more than 1-bit are designed using serial connection of 1-bit full adders. When we compare it with the existing adder designs the output carry of 1-bit adder is given as input carry to the next adder and so on [6–9].

2 Literature Review

P. Balasubramanian and J. John have proposed a low-power design of 2-input *OR* gate and 2-input *XOR* gate in MGDI technology, which are to be used to design the various complex circuits like arithmetic and logical circuits. The focus of this paper is on power efficiency. They created the logic gates using the MGDI to obtain a low power than the existing gate diffusion input technology (GDI). They have worked on the 0.35 mm CMOS technology and compared the results with the static CMOS logic. They have calculated power consumption, propagation delay, transistor count and have done the area comparison and the PDP comparison for different schemes. Power efficiency is a significant problem in designing digital devices, and high power dissipation can lead to device failure. Hence to overcome those issues, the design of the systems is developed at the transistor level [10–15].

The source station of the PMOS is always connected towards the VDD in the MGDI cell, and the NMOS remains forever connected towards the ground to reduce the consumption of electricity. The 2-input or and *XOR* gate design is carried out in design architect on the Linux platform [3]. S. Sapana and A. Smitha (2015) have proposed low power NAND and NOR gates using 16 and 32 nm MGDI Technology. The designed logic gate performance is compared Conventional Metal Oxide Semiconductor (CMOS) technology. In this paper, the dynamic and static power remain calculated using the tanner EDA tool and showed that technology requires less power for electronic devices' applications. This paper also showed different design considerations for various logic gates and observed the static, dynamic power, and the area of the 16 and 32 nm Technology [16–21].

Ebrahim Abiri et al. (2014) have proposed Logic Gates' design using the 32 nm CNTFET technology in Modified Gate Diffusion Input (MGDI) Technology. In this paper, the design of the MGDI cell is carried out using a nano process. It has shown that the area of the networks has reduced to 80% for the pull up section and up to 50% for the pull down section when compared to the GDI technology. In this paper, the calculations of the power dissipation, delay, and PDP are carried out using the H-SPICE software at 0.9 V supply. It proved that these transistors have less power dissipation than that of the conventional Si-MOSFET, and the area of the networks also decreased drastically. The PDP of the multiplexer is very low. The designs of the GDI, conventional CMOS and the MGDI technology are compared. The power dissipation, propagation delay, and the PDP of the logic gates and the multiplexer are calculated and showed that the MGDI cell best solution of the VLSI designers [5, 22].

2.1 Working of CLA

CLA is fast and resilient among various adders with significantly less propagation delay and low power dissipation. The addition of two parallel binary numbers means that all levels and additive effects are obtained simultaneously. The signal must propagate through the gates as with any merging period until the output edges' correct output value is obtained. The total stream time, periods the amount of gate stages now the round, is proportional to the standard gate propagation delay. The most prolonged delay in additions is the time it takes to refine complete add-ons [23, 24]. As the total output bit depends on the number of input entries, it is only after the entries have been added to that category that the Si sum in any additional category in the addendum will be at its final provincial value [25].

Using the subscriptions, the input and output variables define a standard add-on group. Signs on the P_i and G_i after spreading through their gates are still up to their influential state principles. These two characteristics remain corporate towards altogether half-adders also are based exclusively on adding and adding parts to the auction. At the exit, the signal after the input carries the C_i carrying the C_{i+1} propagating finished the OR gate and the OR gate, producing two stages of the gate [26].

The carrier transmission period remains a significant feature of the additive since it decreases the rapidity when two digits remain added. Though the add-on-or, designed aimed at that matter, somewhat integration circuit-will still consume approximately value in its output workstations, unless signals remain assumed extended period towards transmit complete the output gateways, the results will not be nice. Because with consecutive addition, all other arithmetic operations are used, the time spent during the process is significant. Using a faster gate with reduced delay is the apparent option for decreasing the carrier delay time [27]. Duration of a transmitted stream in the same add-on can be shortened in several respects. The principle of moving on is the most commonly used technique. If we define two new binary variables, consider the complete adder's circuit [28].

$$P_i = A_i \oplus B_i$$
$$G_i = A_i B_i.$$

The production sum and carry can remain conveyed as, respectively,

$$S_i = P_i \oplus C_i$$
$$C_{i+1} = G_i + P_i C_i.$$

G_i remains called a generate carry, and when both A_i and B_i are 1 regardless of the input carry C_i, it generates a carry of 1.

P_i is known as a carry propagate because it decides whether to propagate a carry into stage $i+1$.

Boolean functions hold outputs for each step and replace the value of each C_i with the following equations:

$$\begin{aligned}
C_1 &= G_0 + P_0C_0 \\
C_2 &= G_1 + P_1C_1 = G_1 + P_1(G_0 + P_0C_0) = G_1 + P_1G_0 + P_1P_0C_0 \\
C_3 &= G_2 + P_2C_2 = G_2 + P_2G_1 + P_2P_1G_0 + P_2P_1P_0C_0 \\
C_4 &= G_3 + P_3C_3 = G_3 + P_3G_2 + P_3P_2G_1 + P_3P_2P_1G_0 + P_3P_2P_1P_0C_0 \\
C_5 &= G_4 + P_4C_4 = G_4 + P_4G_3 + P_4P_3G_2 + P_4P_3P_2G_1 + P_4P_3P_2P_1G_0 \\
&\quad + P_4P_3P_2P_1P_0C_0 \\
C_6 &= G_5 + P_5C_5 = G_5 + P_5G_4 + P_5P_4G_3 + P_5P_4P_3G_2 \\
&\quad + P_5P_4P_3P_2G_1 + P_5P_4P_3P_2P_1G_0 + P_5P_4P_3P_2P_1P_0C_0 \\
C_7 &= G_7 + P_7C_6 = G_7 + P_7G_6 + P_7P_5G_5 + P_7P_6P_5G_4 + P_7P_6P_5P_4G_3 \\
&\quad + P_7P_6P_5P_4P_3G_2 + P_7P_6P_5P_4P_3P_2G_1 + P_7P_6P_5P_4P_3P_2P_1G_0 \\
&\quad + P_7P_6P_5P_4P_3P_2P_1P_0C_0 \\
C_8 &= G_8 + P_8C_7 = G_8 + P_8G_7 + P_8P_7G_6 + P_8P_7P_6G_5 + P_8P_7P_6P_5G_4 \\
&\quad + P_8P_7P_6P_5P_4G_3 + P_8P_7P_6P_5P_4P_3G_2 + P_8P_7P_6P_5P_4P_3P_2G_1 \\
&\quad + P_8P_7P_6P_5P_4P_3P_2P_1G_0 + P_8P_7P_6P_5P_4P_3P_2P_1P_0C_0.
\end{aligned}$$

As the Boolean function is represented in the sum-of-product form for each output carrier, each purpose can be executed through one level of AND gates shadowed by an OR gate or a two-level NAND gate. The three Boolean occupations are implemented in the carry look ahead generator for C1, C2, and C3. The circuit can be introduced in less time as C3 does not have to wait on behalf of propagation by C2 and C1; C3 is also transmitted on the similar period as C1 also C2. At the cost of extra complexity, this increase in operating speed is achieved.

In the designing of 8-bit CLA, an 8-bit input bit stream is required A_0–A_7 and B_0–B_7 and a carry input signal (Cin). In this 8-Pis and 8-Gis are generated as per the equations. The schematic is implemented by the 8-bit Carry Look Ahead Adder. Block diagrams of 4-bit CLA and 8-bit CLA are represented as the basic architecture is common in CMOS, GDI and MGDI technique; common block diagrams are used to represent the three methods that are to avoid the redundancy.

3 Methodology

3.1 Mod-GDI Technology

The simple equation of the improved gate input method is given by

$$\mathrm{D} = \mathrm{P}\bar{G} + NG \tag{1}$$

Due to its special configuration, power dissipation is becoming a major concern in complex VLSI circuits that can be overcome using the MGDI technique. The low voltage terminal [SP] is still associated with the V_{DD} in MGDI, and the high voltage terminal [SN] is always associated towards the ground, as shown in Fig. 1. The three terminals left as they are and can be used as input to the transistor. Thus, as the number of inputs has increased, the number of transistors required to perform a particular operation are decreased drastically [8, 9]. Hence, using the MGDI method, the number of transistors required is reduced to a greater extent, which reduces the power dissipation and propagation delay. So, it is the best suitable solution for the VLSI designers in this rapid growth of technology.

The various operations that can be performed using MGDI cell are shown in the Table 1 given above. All these operations are obtained using the Eq. (1).

Fig. 1 Basic MGDI cell

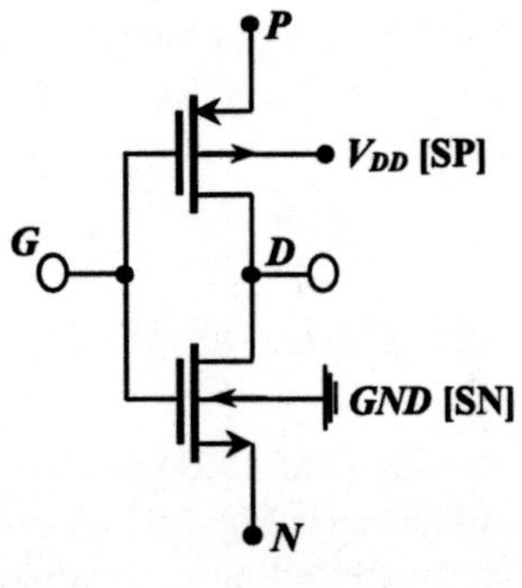

Table 1 Operations of MGDI cell

N	SN	P	SP	G	Result	Function
0	0	1	1	A	A′	Inverter
A	0	0	1	B	AB	AND
A	0	B	1	A	A + B	OR
B′	0	1	1	A	A′B′	NOR
A′	0	A	1	B	A′B + AB′	XOR
B	0	A	1	S	AS′ + BS	2X1 MUX

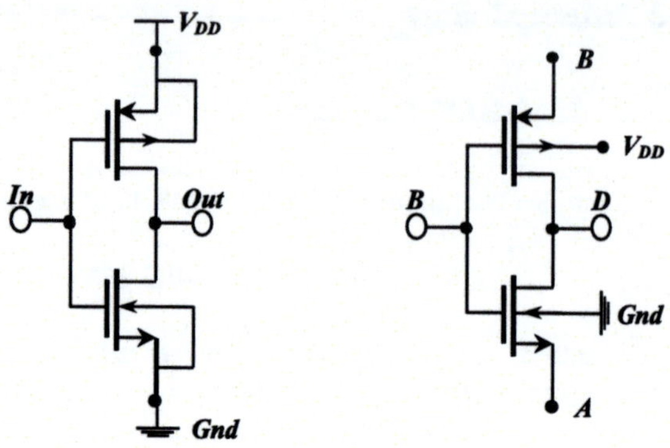

Fig. 2 **a** Schematic of MGDI using FinFET based Inverter. **b** Schematic of MGDI based AND gate

3.2 Design of 8-bit CLA

The design of inverter gate using improved gate diffusion input method using FinFET is depicted in Fig. 2a. It consists of two transistors, one pMOS and one nMOS, which are connected in series where the nMOS source terminal is associated towards the ground also the drain of the pMOS is associated to the V_{DD}. When the input of logic 0 is applied, then the nMOS acts as an off-state connecting the output to the V_{DD} through pMOS, providing the output of logic 1. When the input is of logic 1 is applied, then the pMOS acts as an off state connecting the output to the ground through nMOS, providing the output of logic 0. The implementation is done using cadence virtuoso tool at 18 nm technology for 1 V as supply voltage, as shown in Fig. 2.

3.3 Design of AND Gate in MGDI Technique

The design of AND gate using modified gate diffusion input technique using FinFET is depicted in Fig. 2b. It consists of two transistors, one pMOS and one nMOS, which are connected in series where the nMOS source terminal is connected to the input signal A and the drain of the pMOS is connected to the input signal B. The pMOS' SP terminal is associated with the V_{DD}, and the SN terminal of the nMOS is related to the ground. The common gate terminal is connected to the input signal B. If any one of the input is of logic 0 then the nMOS acts on connecting the output to ground hence providing the output as logic 0. When both the inputs are of logic 1, then the pMOS connects the output to the V_{DD} hence providing the output as logic one. The implementation is done using cadence virtuoso tool at 18 nm technology for 1 V as supply voltage.

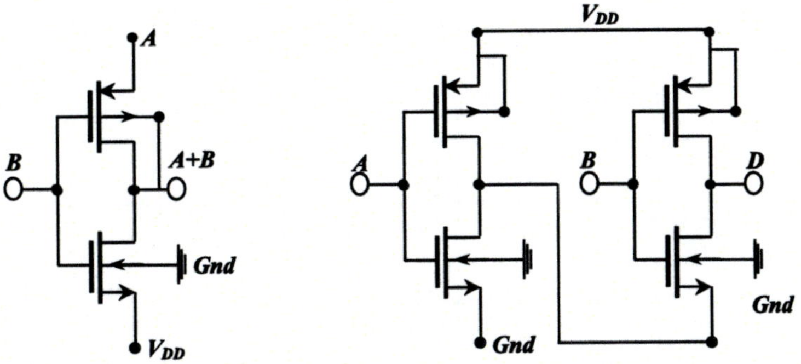

Fig. 3 **a** Schematic of MGDI based OR gate. **b** Schematic of MGDI based XOR gate

3.4 Design of OR Gate in MGDI Technique

The OR gate design using modified gate diffusion input technique using fin field-effect transistor is depicted in Fig. 3a. It consists of two transistors, one pMOS and one nMOS, linked in sequence, connecting the nMOS source terminal to V_{DD} and connecting the pMOS drain to the A input signal. The pMOS SP-terminal is associated with the output; the nMOS SN-terminal remains associated towards the ground. It is connected to the input signal B by the common gate terminal. The implementation is carried out as a supply voltage using cadence virtuoso tool at 18 nm technology for 1 V.

3.5 Design of XOR Gate in MGDI Technique

The inverter is used to get the invert of the input signal A that is provided to the nMOS source; also the drain station of the pMOS is associated to the input signal A. Figure 3b shows the XOR gate configuration. The common gate terminal is connected to the input signal B. The pMOS SP terminal is connected to the V_{DD}; also the nMOS SN terminal is associated with the deck. Hence based upon the fundamental equation of the MGDI Eq. (1), the output is obtained as below.

$$D = (A.B') + (A'.B) \tag{2}$$

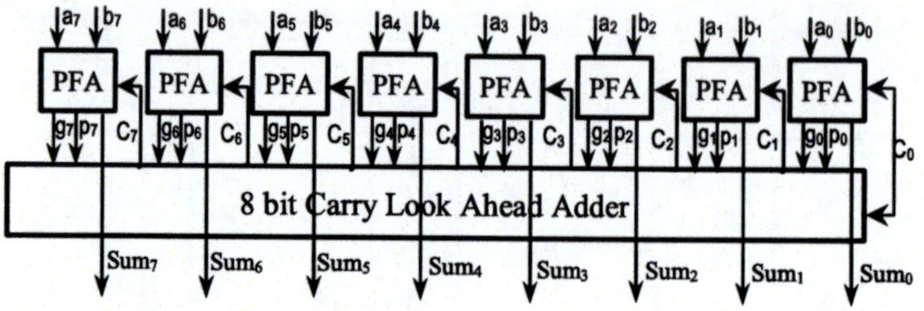

Fig. 4 Schematic of 8 bit carry look ahead adder

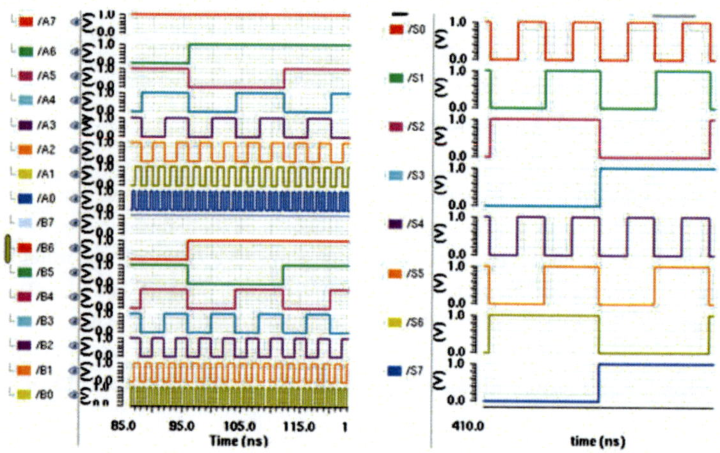

Fig. 5 **a** Inputs for 8 bit CLA. **b** Outputs of 8 bit CLA

3.6 *Design of 8-bit CLA in MGDI Technique*

Schematic of 8 bit CLA and internal circuit is presented in Fig. 4b. Inputs for 8 bit CLA are provided in Fig. 5a, b shows the outputs of 8 bit CLA.

4 Results and Discussions

This chapter tabulates the total power dissipation, propagation delay, and transistor count of each circuit implemented by the three techniques, i.e., CMOS, GDI, and MGDI technology. The calculations are performed directly using the cadence virtuoso method. Table 2 offers a comparison of the total power dissipation, propagation delay, and a number of circuit transistors using the FinFET with and without the Modified gate diffusion technology. We can observe from these tables that there is a drastic difference in the values of the circuits using different technologies; hence, we can say that MGDI is best solution for designing complex circuits with low power dissipation and propagation delay. The obtained results below prove that MOD-GDI is the efficient technique to implement the CLA with low power consumption.

Table 2 Comparison of 8 bit CLA using different technologies

Parameter	CMOS	GDI	MGDI
Power dissipation in nW	250×10^3	248.85	190.42
Propagation Delay in ns	155	95	72.35
Transistor count	294	246	152

Table 2 shows that the power dissipation of the circuits designed using MGDI technique is very low compared to that of the other technologies. This is mainly due to the reduction of the leakage power and the area of the circuits. Hence the power dissipation is significantly less in circuits using MGDI technique.

5 Conclusion

In this paper, the high-performance adder called Synchronous Carry Generate Adder/Carry look-ahead adder remains calculated by exhausting existing techniques (CMOS and GDI) and proposed MGDI technique. The functionalities of MGDI based designs have been verified. It is observed from Table 2 that MGDI based designs offer less power dissipation, high speed, and require less no. of transistors. Hence, it is concluded that MGDI based adders can be used in high-performance applications and can be used in the multipliers. The focus was on reducing the size of the transistor and speed, which later increased the problem of power intemperance in the circuits. It remains becoming the major problem as the device's lifetime and performance are getting significantly affected. The future work can be implemented using the Tunnel FET (TFET), which is used to decrease the power dissipation further. The implementation of the CLA can be done using the TFET.

References

1. Manchala, S., & Vallabhuni, V. (2020). A unique approach to provide security for women by using smart device. *European Journal of Molecular and Clinical Medicine, 7*(1), 3669–3683.
2. Siva, N.V., et al. (2020). Design and implementation of low power 32-bit comparator. In *ICICNIS 2020,* 10–11 Dec 2010 (pp. 1–8). Palai, India.
3. Rani, B.M.S, et al. (2020). *Retinal vascular disease detection from retinal fundus images using machine learning*. Australia patent 2020101450.
4. Venkateswarlu, S. C., et al. (2019). Implementation of area optimized low power multiplication and accumulation. *International Journal of Innovative Technology and Exploring Engineering, 9*(9), 2278–3075.
5. Vallabhuni, R.R., et al. (2020). High speed energy efficient multiplier using 20nm FinFET technology. In *Proceedings of the International Conference on IoT Based Control Networks and Intelligent Systems (ICICNIS 2020),* 10–11 Dec 2020 (pp. 1–8). Palai, India.
6. Rajeev, R.V., et al. (2020). Comparative validation of SRAM cells designed using 18nm FinFET for memory storing applications. In *Proceedings of the 2nd International Conference on IoT, Social, Mobile, Analytics and Cloud in Computational Vision and Bio-Engineering* (pp. 1–10).

7. Vallabhuni, V., & Avireni, S. (2019). A low power waveform generator using DCCII with grounded capacitor. *International Journal of Public Sector Performance Management, 5*, 134–145.
8. Vallabhuni, V., et al. (2019). System and method to improve performance of amplifiers using bias current. The Patent Office Journal No. 43/2019, India. International classification: C12Q1/6869. Application No. 201941042648 A.
9. Babu, P.A., et al. (2021). Realization of 8 × 4 barrel shifter with 4-bit binary to gray converter using FinFET for low power digital applications. *Journal of Physics: Conference Series, 1714*(1). IOP Publishing.
10. Rajeev, R.V., et al. (2020). An advanced computing architecture for binary to thermometer decoder using 18nm FinFET. In *2020 Third International Conference Smart Systems and Inventive Technology (ICSSIT)*, 20–22 Aug 2020 (pp. 510–515). Tirunelveli, India.
11. Vallabhuni, V., et al. (2019). A simple and enhanced low-light image enhancement process using effective illumination mapping approach. *Lecture Notes in Computational Vision and Biomechanics, Cham, Switzerland* (pp. 975–984).
12. Vallabhuni, V. (2017). *Second generation differential current conveyor (DCCII) and its applications*. Vignan's Foun. Sci., Tech. & Res. (Deemed to be University).
13. Rajeev, R.V., et al. (2020). Performance analysis: D-Latch modules designed using 18nm FinFET Technology. In *2020 International Conference on Smart Electronics and Communication (ICOSEC), September 10–12* (pp. 1171–1176). Tholurpatti, India.
14. Vallabhuni, V., & Avireni, S. (2017). A novel square wave generator using second generation differential current conveyor. *Arabian Journal for Science and Engineering, 42*(12), 4983–4990.
15. Vijay, V., et al. (2012). Performance evaluation of the CMOS full adders in TDK 90 nm technology. *International Journal of Systems, Algorithms and Applications, 2*(1), 711.
16. Vallabhuni, V., et al. (2020). High performance 2:1, 4:1 and 8:1 binary and ternary multiplexer realization using CNTFET technology. *Journal of Critical Reviews, 7*(6), 1159–1163.
17. Ratna, V.R., et al. (2021). High speed energy efficient multiplier using 20nm FinFET technology. In *ICICNIS 2020*. Retrieved from SSRN https://ssrn.com/abstract=3769235, https://doi.org/10.2139/ssrn.3769235
18. Rajeev, R.V., et al. (2021). Universal shift register designed at low supply voltages in 20nm FinFET using multiplexer. *Lecture Notes in Networks and Systems.*
19. Siva, N.V., et al. (2020). Design and implementation of low power 32-bit comparator. In *ICICNIS 2020*. Retrieved from SSRN https://ssrn.com/abstract=3769748
20. Saritha, P., et al. (2020). 4-Bit Vedic multiplier with 18nm FinFET technology. In *2020 International conference on Electronics and Sustainable Communication Systems (ICESC)* (pp. 1079–1084). Coimbatore, India.
21. Vijay, V., et al. (2012). A review of the 0.09 µm standard full adders. *International Journal of VLSI Design & Communication Systems, 3*(3), 119.
22. Shaker, P.C., et al. (2020). Realization and comparative analysis of thermometer code based 4-bit encoder using 18 nm FinFET technology for analog to digital converters. *Advanced Intelligent Systems and Computing (AISC).*
23. Mohammad, K., et al. (2020). Design of carry select adder based on a compact carry look ahead unit using 18 nm FinFET technology. *Journal of Critical Reviews, 7*(6), 1164–1171.
24. Rajeev, R.V., et al. (2020). Design of comparator using 18 nm FinFET technology for analog to digital converters. In *2020 7th International Conference on Smart Structures and Systems (ICSSS)* 23–24 July 2020 (pp. 318–323). Chennai, India.
25. Rajeev, R.V., et al. (2020). Smart cart shopping system with an RFID interface for human assistance. In *2020 3rd International Conference on Intelligent Sustainable Systems (ICISS)*, 4–5 Dec 2020 (pp. 165–169). Palladam, India.
26. Vallabhuni, V., et al. (2019). System for minimizing crosstalk effects of shells and designing multiwalled carbon nanotube models. The Patent Office Journal No. 43/2019, India. Int. classification: B82Y10/00. Application No. 201941042460 A.

27. Rajeev, R.V., et al. (2020). Comparative analysis of 8-bit manchester carry chain adder using FinFET at 18 nm technology. In *2020 3rd International Conference Intelligent Sustainability Systems (ICISS)*, 4–5 Dec 2020 (pp. 1579–1583). Palladam, India.
28. Rajeev R.V., et al. (2020). Transistor SRAM cell designed using 18 nm FinFET technology. In *2020 3rd International Conference Intelligent Sustainability System* (pp. 1584–1589). Palladam, India.

Improved Scientific Workflow Scheduling Algorithm with Distributed Heft Ranking and TBW Scheduling Method

Ramandeep Sandhu and Kamlesh Lakhwani

Abstract Scheduling is a process that manages the workflow tasks during execution on different resources. Virtual infrastructure is a dynamic mapping of system resources to applications in order to maximize its utilization. In today's technological world, cloud has taken a long stride on the success towards maximum throughput as well as highest qualitative services to its consumers. Yet, approaches for maximizing the utilization of cloud resources are at peak demand. Each cloud service provider focuses on maximum utilization with minimum consumption of cloud resources, although managing and providing computational resources to maximum number of users and to execute such huge applications is a challenging one. In this paper, a scheduling algorithm with name TBW (Tabu Bayesian Whale Optimization) has been proposed. Basically, the algorithm is used to target the improvement in scheduling of scientific workflows. The complete framework has firstly used a ranking algorithm named distributed HEFT ranking and then applied TBW algorithm on ranked tasks of input workflows. The work has been executed for five scientific workflows LIGO, MONTAGE, Epigenomics, SIPHT and Cybershake. TBW is using tabu method on workflow tasks for fast local search in cloud system, and Bayesian Optimization is used to find out best possible combinations of resources where tasks are mapped and then whale optimization maps the tasks on the resources in a smart way. In the whole process, total execution time and cost parameters are minimized under deadline constraints.

Keywords Cloud VMs · Task mapping · Tabu search · Bayesian optimization · Whale optimization

R. Sandhu (✉) · K. Lakhwani
Lovely Professional University, Phagwara, India

P. Nayak et al. (eds.), *IoT and Analytics for Sensor Networks*, Lecture Notes in Networks and Systems 244, https://doi.org/10.1007/978-981-16-2919-8_23

1 Introduction

As cloud is a new trend with myriad satisfactory features for consumer as well as for provider, the best is on demand service with pay-as-per-use. Workflow scheduling under deadline constraints is like a fast medium for easy and efficient management of cloud environment with big data [1]. Moreover, parameters like time, cost, response time and energy (TCRE) are highly constrained to implement a cloud in capable as well as proficient way. Cloud computing provides features like elasticity, most favorable resource consumption, cost lessening and better-quality flexibility. Cloud computing is used factually everywhere in the world like big data analysis, file storage, backup, development of projects, testing, etc. [2–4]. Effective resource utilization makes it a prior requirement in industrial world. Albeit, the world today is fast-moving towards success in each and every era, yet without cloud, the technological world is not near satisfaction. Almost each business is feeling smooth and flexible execution in cloud environment. On the other side of cloud users, cloud-based services are if less complex and with genuine expenditure then each attached user will become a pleased member of cloud. So, visibility is an important fact [5].

To attain it, convex optimization of cloud is highly supportive. It will automatically thrust high the delivery on time as well as best utilization of resources. Furthermore, task mapping across means movement of data across machines. It should be complete secure, with zero response time as well as time and cost in movement must be negligible. At each glance, consistency is major concern [6].

Disregards confronted by researchers in Scientific Applications

Researchers in their studies have given many openings that the cloud has brought in, such as superior consumption of resources, improved responsiveness thereby improving user practice, enabling a generation of collaborative scientific workflows and reducing the cost in challenges and opportunities in running scientific workflows on cloud. The challenges faced by the applications are architectural challenges, service challenges, challenge in compute intensive applications, challenge for data management, and service management challenge. Even, Qi Zhang et al. [7] have discussed a variety of cloud computing skills and commercial products in detail. The profitable products have been evaluated on parameters like cloud provider, computing classes, target application, computation, auto scaling and storage. The research challenges expressed in this work are managing energy and providing. Furthermore, as noticeable concern of high cloud efficiency, Christian Vecchiola et al. [8] have given a judgment of computing enlightenment such as Amazon EC2, Google App Engine, and Microsoft Azure. This evaluation is completed on the basis of parameters like type of tune, value added provider, if PaaS, ability to organize on third party IaaS, platform, Virtualization, consumption Model and interface for user access. Even, the aneka architecture, deployment model and application model are discussed in detail. Programming models like task model, map reduce model, thread model, parameter sweep model, and workflow have been compared on the basis of execution services, applications and execution unit. Apart from this, Suraj Pandey et al. [9] have worked on reducing the computing cost of the function by using particle swarm optimization (PSO) algorithm. Researchers in [10] have discussed dependency of cost on execution models in this work. In this work, the cost is calculated as a function of number of processors. The cost to execute montage workflow has been anticipated by running simulation using GridSim tool. The cost of running each of the data management models have been compared graphically. The Whale Optimization Algorithm (WOA) is a recently developed meta-heuristic swarm-based algorithm based on the bubble-net hunting manoeuvre technique of humpback whales to solve the complex problems of optimization [11].

2 Literature Review

Alkhanak et al. [12] proposed a cost optimization approach for scientific workflow scheduling in cloud computing. The proposed approach employs the four meta-heuristic algorithms which are based on the population. The approach helps in reducing cost and time of the service providers. The execution cost and time are reduced as compared to baseline approaches (Rimal et al. [13]). The result of the proposed work is compared with existing approaches and algorithms. The recreation result of the proposed approach shows more effective results than the existing approaches. Casas and Israel et al. [14] focused on Parallelization and proposed a scheduling approach called Balanced and file Reuse–Replication scheduling. This approach is used to schedule the scientific application workflows. It splits the workflows into sub-workflows which help in proper utilization via parallelization process. This approach provides the facility if data reuse and replication which helps in optimization of data and transfer it at run time. The optimization process is based on execution time and monetary cost of workflows. The work in [15–19] reduced time and cost and introduced a gravitational search algorithm for workflow scheduling in the cloud environment. The optimizations in workflow reduce the cost and make span. Two algorithms are hybridized GSA and HEFT for workflow scheduling. The performance evaluation is done on the basis of two metrics that are monetary cost ratio and schedule length ratio. The justification of result is also tested by ANOVA test, and it shows that the proposed approach does better. Sagnika et al. [20] proposed BAT algorithm for workflow scheduling in cloud computing which helps to handle the large size of data. The scheduling process decides that which task is executed first and which is last according to their requirement of the resources. It manages the resources according to the task size and execution time. The result of the proposed algorithm is compared with particle swarm optimization algorithm and Cat swarm optimization algorithm. The convergence of current study is better than the existing algorithms. Thiago Genez et al. [21] have worked with selection of CPU frequency configuration for resources carefully to reduce the total make span. It is combination of PSO and HEFT schedulers to make it better in case of time. A fitness function without any parameter value is imposed to measure performance of various particles.

Less is fitness value, better is solution. So, particles of swarm are moved towards less fitness value region the workflows with different sizes are accepted for simulation. Cybershake, SIPHT and LIGO are the workflows on which experiments are performed.

3 Proposed Ideology

In today's advance world of technology, cloud and its applications have provided various techniques. In the literature survey, many papers have been studied related to workflow scheduling in which cloud environment depends on the static configuration of virtual machine, which is not a real condition. Rather than optimizing and searching virtual machine utilization criteria, it is formulated that scanning and maximizing migration of tasks among VMs should be major concern. Also, distribution of tasks should not be random. Even, optimization initialization in a mannered way is more important which will enhance the resources utilization. Although, both consumers and providers of cloud are highly linked comfortably through the best way utilization of cloud resources yet the issue of underutilized virtual machines (called cloud resources) cannot be ignored. By considering it, few limitations have been mentioned below which are basic pillar for proposed study. Random distribution of tasks [22] has been done in existing works. Also, optimization of the task depends on single objective, which is sometimes conflict like time and cost [23]. Furthermore, only workflow dependency is main aspect on which deadlines are totally dependent. In existing approaches, optimization use local VM or global data center [14, 24–36].

3.1 Proposed Representation

The TBW approach has been proposed for mappings tasks to VMs in cloud. As depicted in above Fig. 1, task scheduling using Tabu Search, Bayesian Optimization and Whale Optimization algorithm has been done.

3.2 Proposed Flowchart and Phases of Proposed Methodology

Various phases of proposed methodology are listed below: Phase 1: Total five workflows are accepted as input named CYBERSHAKE, MONTAGE, LIGO, SIPHT and EPIGENOMICS.

Phase 2: Parsing is a step of analyzing input. In proposed cloud workflow framework, parsing will occur at initial stage. It will show tasks with dependencies.

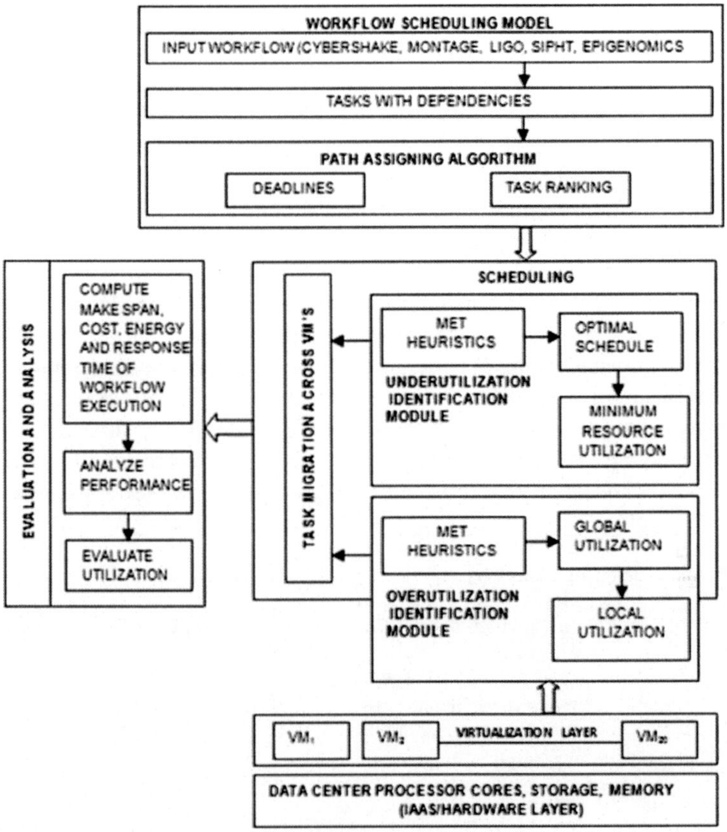

Fig. 1 Graphical representation of proposed methodology

Phase 3: A path assignment algorithm will be used for ranking various tasks of workflow. Here, deadlines are prior concern. Moreover, budget and heuristics for initializations are also important factors.

Phase 4: Scheduling is performed in two scenarios, and the outcome in both cases is mapping on Virtual Machines (VMs). It means tasks migration across VMs should be optimized so that under-utilized VMs can be scanned.

Phase 5: It is basically evaluation and analysis. After performing VM migration in optimized manner, Makespan, Cost, energy and response time of Workflow Execution will be calculated. So, these four parameters will help to analyze performance of proposed framework. In whole system, utilization of resources will be increased to much high level.

So, proposed framework improves the utilization of resources which reduce the cost and time of execution. Also, improves the virtual machine migration because of optimization schedule. Apart from this, it reduces the decision time of Virtual machine mapping. The proposed study has inclined the research towards energy-efficient scheduling for optimum resource utilization in cloud environment by designing and deploying an effective task ranking algorithm. By creating correlation between time and deadline of input tasks, it has been evaluated that results are better than HEFT method. In HEFT, TET (Total execution Time) parameter of all input tasks is also calculated on cloudsim [25].

As shown in Tables 1 and 2, algorithms for finding critical path have been used.

In the proposed ideology, three algorithms are used: Use of Tabu Search, Use of Bayesian Optimization and Use of Whale Optimization.

Then all three steps have been combined to make it TBW Scheduling Approach.

Tabu Search: it takes Input: ranked tasks mapped to VMs. Then apply tabu search algorithm to find neighbours of current VM. It works iteratively. If neighbour is less utilized than current VM then adds such neighbour in tabu list. Result after applying tabu search—we get a final list of VMs which are not efficiently utilized.

After applying tabu search method, next step targets towards try to shift tasks from these VMs on other VMs (efficiently utilized) but without making Queue. Here

Table 1 Initialization algorithm (W)

```
Begin
Initialize number of fog nodes and workflows
N←VM
W← Number of workflows
Parse workflows
   while (W)
     start
       Parse W_i
       Extract task
T_i←critical path(W_i )

       End
   X←∑_{i=1}^{n} T_i + W_i  .............................(1)
Return X
End
```

Table 2 Algorithm to find critical path (W_i)

```
Begin
Cp=Ø, task_i = task_exit, Critical path= Ø
While(task_i != task_entry) do
Calculate critical parent task_p of task_i
Criticalparent(taski) =
{taskp|max taskp {finishtime+totaltime}}.............(2)
    Cp=Cp U task_p
    task_i= task_p
End While
Return Cp
End
```

the meaning of queue is that increase in total execution time. So, use of Bayesian Optimization is helpful here. Basically Bayesian optimization works for choosing best combination for mapping tasks of not properly utilized VMs on other utilized VMs. It means, it targets tasks of the VMs which are not efficiently utilized and targets those VMs which are in the underutilized category. It provides all combinations where tasks can be shifted. The next part of the TBW algorithm is whale optimization. Whale works in an intelligent way. It takes input from Bayesian Optimization. Based on its objective functions, it chooses best combinations and shifts tasks on VMs efficiently without increasing TEC and TET. The overall results are better than GA-PSO for different scientific workflows.

4 Conclusion

In this paper, we have introduced our ranking algorithm named distributed HEFT ranking also presented a new approach for better scheduling in cloud system (named as TBW—Tabu Bayesian Whale Algorithm) included three optimization approaches (Tabu Optimization, Bayesian Optimization and Whale Optimization). As researchers are already using various techniques for cloud optimization and for minimizing the time and cost parameters, here, in this study we have provided the working methodology of TBW. Also study of various algorithms which have used the smart optimization techniques has been done. In further work, results with TBW advance approach will be calculated and also will be compared with existing optimization approaches like GA, PSO, GA-PSO and WOA.

References

1. Deelman, E., Singh, G., Livny, M., Berriman, B., & Good, J. (2008). The cost of doing science on the cloud: The montage example. In *SC'08: Proceedings of the 2008 ACM/IEEE Conference on Supercomputing, Austin, TX* (pp. 1–12).
2. Bölöni, L., & Damla, T. (2017). Value of information based scheduling of cloud computing resources. *Future Generation Computer Systems, 71*, 212–220.
3. Chirkin, M., Adam, B., Sergey, K., Marc, M., Mikhail, M., Alexander, V., & Denis, A. (2017). Execution time estimation for workflow scheduling. *Future Generation Computer Systems, 75*, 376–387.
4. Zhang, L., Kenli, L., Changyun, L., & Keqin, L. (2017). Bi-objective workflow scheduling of the energy consumption and reliability in heterogeneous computing systems. *Information Sciences, 379*, 241–256.
5. Vöckler, J., Juve, G., Deelman, E., Rynge, M., & Berriman, B. (2011). Experiences using cloud computing for a scientific workflow application. In *Proceedings of the 2nd International Workshop on Scientific Cloud Computing* (pp. 1–10). ACM.
6. Wang, Y., Jiajia, J., Yunni, X., Quanwang, W., Xin, L., & Qingsheng, Z. (2018). A multi-stage dynamic game-theoretic approach for multi-workflow scheduling on heterogeneous virtual machines from multiple infrastructure-as-a-service clouds. *International Conference on Services Computing* (pp. 137–152). Cham: Springer.

7. Zhang, C., & Raouf, B. (2018). Cloud computing: State-of-the-art and research challenges. *Journal of Internet Services and Applications, 1*(1), 7–18.
8. Vecchiola, C., Suraj, P., & Buyya, R. (2009). High-performance cloud computing: A view of scientific applications. In *2009 10th International Symposium on Pervasive Systems, Algorithms, and Networks (ISPAN)* (pp. 1–9). IEEE.
9. Pandey, S., Wu, L., Guru, S.M., & Buyya, R. (2010). A particle swarm optimization-based heuristic for scheduling workflow applications in cloud computing environments. In *2010 24th IEEE International Conference on Advanced Information Networking and Applications (AINA)* (pp. 1–10). IEEE.
10. Goyal, M., & Mehak, A. (2017). Optimize workflow scheduling using hybrid ant colony optimization (ACO) & particle swarm optimization (PSO) algorithm in cloud environment. *International Journal of Advance Research, Ideas and Innovations in Technology, 3*(2).
11. Alameer, Z., Elaziz, M., Ewees, A., Ye, H., & Jianhua, Z. (2019). Forecasting gold price fluctuations using improved multilayer perceptron neural network and whale optimization algorithm. *Resources Policy, 61*, 250–260.
12. Alkhanak, E., Lee, S., Rezaei, R., & Parizi, R. (2018). Cost optimization approaches for scientific workflow scheduling in cloud and grid computing: A review, classifications, and open issues. *Journal of Systems and Software, 113*, 1–26.
13. Israel, C., Javid, T., Rajiv, R., Lizhe, W., & Albert, Z. (2017). A balanced scheduler with data reuse and replication for scientific workflows in cloud computing systems. *Future Generation Computer Systems, 74*, 168–178.
14. Mao, M., & Humprey, M. (2011). Auto-scaling to minimize cost and meet application deadlines in cloud workflows. In *SC'11: Proceedings of 2011 International Conference for High Performance Computing, Networking, Storage and Analysis, Seatle, WA* (pp. 1–12).
15. Kapoor, N., & Kumar, Y. (2020). The efficient management of renewable energy resources for Vanet-cloud communication. In *Nature-Inspired Computing Applications is Advanced Communication Networks* (pp. 228–253).
16. Kumar, Y., & Kaul, S. (2019). Effective use of the machine learning approaches on different clouds. In *International Conference on Sustainable Computing in Science, Technology and Management* (pp. 892–897).
17. Kumar, Y., & Mahajan, M. (2019). Intelligent behavior of fog computing with IOT for healthcare system. *International Journal of Scientific and Technology Research, 8*, 674–679.
18. Liu, L., Zhang, M., Buyya, R., & Fan, Q. (2016). Deadline-constrained co evolutionary genetic algorithm for scientific workflow scheduling in cloud computing. *Concurrency and Computation Practice and Experience*, 1–9.
19. Masadari, M. (2017). Towards workflow scheduling in cloud computing: A comprehensive analysis. *Journal of Network and Computer Applications*, 1–27.
20. Sagnika, S., Saurabh, B., & Bhabani, S. (2018). Workflow scheduling in cloud computing environment using bat algorithm. In *Proceedings of First International Conference on Smart System, Innovations and Computing* (pp. 1–13). Springer, Singapore.
21. Vinothina, V., & Sridaran, R. (2018). An approach for workflow scheduling in cloud using ACO. In *Big data analytics* (pp. 525–531). Springer.
22. Genez, T., Pietri, I., Sakellariou, R., Bittencourt, F., & Madeira, E. (2015). A particle swarm optimization approach for workflow scheduling on cloud resources priced by CPU frequency. In *Data Science and Symptoms* (pp. 1–9).
23. Kumar, B., Mala, K., & Poonam, S. (2017). Discrete binary cat swarm optimization for scheduling workflow applications in cloud systems. In *2017 3rd International Conference on Computational Intelligence & Communication Technology (CICT)* (pp. 1–6).
24. Verma, A., & Sakshi, K. (2010). A hybrid multi-objective particle swarm optimization for scientific workflow scheduling. *Parallel Computing, 62*, 1–19.
25. Sandhu, R., & Lakhwani, K. (2020). Optimal cloud system enhancement using improved workflow task ranking system. *Test Engineering and Management, 83*, 10092–10101.
26. Choudhary, A., Gupta, I., Singh, V., & Jana, P.K. (2018). A GSA based hybrid algorithm for bi-objective workflow scheduling in cloud computing. *Future Generation Computer Systems*, 1–10.

27. Dillon, T., Elizabeth, C., & Chen, W. (2010). Cloud computing: Issues and challenge. In *24th IEEE International Conference on Advanced Information Networking and Applications.*
28. Garg, J., & Gurjit, B. (2017). Research paper on genetic based workflow scheduling algorithm in cloud computing. *International Journal of Advanced Research in Computer Science, 8*(5), 1–7.
29. Ghose, M., Verma, P., Karmakar, S., & Sahu, A. (2017). Energy efficient scheduling of scientific workflows in cloud environment. In *2017 IEEE 19th International Conference on High Performance Computing and Communications; IEEE 15th International Conference on Smart City; IEEE 3rd International Conference on Data Science and Systems (HPCC/SmartCity/DSS), Bangkok* (pp. 170–177).
30. Iosup, A., Ostermann, S., Yigitbasi, M., Prodan, R., Fahringer, T., & Epema, D. (2011). Performance analysis of cloud computing services for many-tasks scientific computing. *IEEE Transactions on Parallel and Distributed Systems, 22*(6), 931–945.
31. Jiang, J., Yaping, L., GuoqiXie, L., & Junfeng, Y. (2017). Time and energy optimization algorithms for the static scheduling of multiple workflows in heterogeneous computing system. *Journal of Grid Computing, 15*(4), 435–456.
32. Quang, H., Nguyen, S., & Nam, T. (2017). Energy-saving virtual machine scheduling in cloud computing with fixed interval constraints. In *Transactions on large-scale data-and knowledge-centered systems XXXI* (pp. 124–145). Springer.
33. Scott, C., Ewa, D., Dan, G., Gideon, J., Philip, M., Christopher, B., Karan, V., Kevin, M., Robert, G., Edward, F., David, O., & Thomas, J. (2010). Scaling up workflow-based applications. *Journal of Computer and System Sciences, 76*, 428–446.
34. Zhao, Y., et al. (2011). Opportunities and challenges in running scientific workflows on the cloud. In *International Conference on Cyber-Enabled Distributed Computing and Knowledge Discovery (CyberC)*. IEEE.
35. Ahmed, M., Houssein, E., Hassanien, A., Taha, A., & Hassanien, E. (2019). Maximizing lifetime of large-scale wireless sensor networks using multi-objective whale optimization algorithm. *Telecommunication Systems, 72*(2), 243–259.
36. Ben Oualid Medani, K., Sayah, S., & Bekrar A. (2017). Whale optimization algorithm based optimal reactive power dispatch: A case study of the Algerian power system. *Electric Power Systems Research, 163*, 696–705.

Selection of OLAP Materialized Cube by Using a Fruit Fly Optimization (FFO) Approach: A Multidimensional Data Model

Anjana Yadav and Anand Tripathi

Abstract The Online Analytical Processing (OLAP) based multidimensional examination hassles for several stockpiling magnificence over huge data. For as much to recognize queries answering time companionable by OLAP framework users and understanding entire business perceive mandatory, OLAP data is structured as a data cube (a multidimensional model). The OLAP queries are responded in speedy and steady time by utilizing the cube materialization for assessments takers. But, this also involves unendurable expenses, regarding to stockpile memory and period, and as a data depot, OLAP has an average dimension and dimensionality which is to be significant on query processing. Consequently, cube assortment has got to be finished motivating to diminish inquiry management expenses, maintaining as a restraint the materializing gap. Several techniques and heuristics like deviationist and insatiable algorithms have been utilized to offer an estimated result. In this work, a Fruit Fly Optimization (FFO) approach is implemented in a lattice structure to obtain an optimal materialized data cube for reducing the query processing expenses. The results illustrate that FFO generates better performance than Particle Swarm Optimization (PSO) in terms of frequency and number of dimensions.

Keywords Cube materialization · Data cube · Fruit fly optimization · OLAP · Multidimensional model

1 Introduction

Data warehouse [1] is a collection of huge information of assorted data about several organizations utilizing for assessments. These assessments are taken by complex queries applied to data warehouse [2] to reduce the answer time. The materialization in data cubes is well utilized for query processing in an efficient manner with lesser time consuming. Entire views of data cubes can be materialized to access the data quickly with minimizing the answering time of queries. The Online Analytical Processing (OLAP) [3] is used for query processing over data warehouse and

A. Yadav (✉) · A. Tripathi
P. K. University, Shivpuri, India

P. Nayak et al. (eds.), *IoT and Analytics for Sensor Networks*, Lecture Notes in Networks and Systems 244, https://doi.org/10.1007/978-981-16-2919-8_24

extracting the useful information for analysis work. It is also helpful for assessment support to provide the users overviews. OLAP is combined with Key Performance Indicators (KPIs) [4, 5] to provide a dashboard application over historical data. The production efficiency is explained with numerical examples and developed in the Java programming language.

OLAP is also utilized for energy cost analysis in commercial areas to reduce the expenditure with improving the power efficiency [6]. A multidimensional cube model is utilized for evaluating the power consumption at several stages of generalization. This time OLAP [7] is used with association rules to generate feasible solutions for huge data about buildings in commercial sectors. Unified Modeling Language (UML) [8] and Structured Query Language (SQL) are also introduced for query processing in OLAP. The web based software is implemented for analyzing the students' results based on the object oriented methodology. The analytical comparison of huge student data is easily performed by OLAP to minimize the stress and workload of schools. Lectures are easily delivered to students through this object oriented platform [9]. Another application of OLAP is a police intellect, a decision scheme to catch the criminals and take an efficient decision about crime [10].

The analytical process is also utilized for intellectual study over multidimensional data for business purposes [11] in Social Business Intelligence (SBI). The social, private and public data can be analyzed easily using OLAP semantic analysis to model the huge information of companies for business point of views [12, 13]. The health care data [14] is also formulated and structured by mining in OLAP. The Internet of Medical Things (IoMT) systems are well suited for providing health information of patients and also predict the treatment of diseases by using medical data analysis [15]. The extraction of solar radiation data from huge amount of information is performed by query processing in several geographical positions and structures. The sensor data of this system accesses by user to initiating the query on solar data and analyzing the data at several stages [16].

In the above analysis of works, materialized data cubes are generated from important data extraction from the huge amount of data. These data cubes are selected by using optimization approaches in large space to provide efficient query processing in an OLAP multidimensional data model. The PSO is one of the nature inspired optimization approach initiating on data cubes to generate optimal materialized data cubes. Here, we proposed another optimization technique FFO, which generates better results than PSO over OLAP multidimensional model in terms of frequency and number of dimensions to reduce the query processing expenditure in several constraint spaces.

2 The Fruit Fly Optimization (FFO) Approach for Selection of OLAP Materialized Cube (A Multidimensional Data Model)

2.1 FFO Approach

A bio-inspired Fruit Fly Optimization (FFO) approach is utilized for generating the global optimal solutions on the basis of foraging inspired by fruit flies. Various realistic explanations to optimization quandary are illustrated by foraging of fruit flies in FFO. The fruit fly gaits by plunging to the food, exploits its vigilant spirit to realize food and where it correlates flock and then it gaits by plunging into a route.

Start

Step 1. Put primary standards of position, fitness function, smell, generation and population randomly of entire fruit flies (Eq. (1)).

$$\begin{aligned} X_{\emptyset} &= X_{axis\_value} + random\_value \\ Y_{\emptyset} &= Y_{axis\_value} + random\_value \end{aligned} \tag{1}$$

Step 2. Compute the fitness standards on the basis of distance (***Dist***) and smell (***Sml***) for entire fruit flies. Hence, optimized solutions are obtained with fitness of individual and population (Eqs. (2), (3) and (4)).

$$Dist = \sqrt{X_{\Phi}^2 + Y_{\Phi}^2} \tag{2}$$

$$Sml = \frac{1}{Dist} \tag{3}$$

$$Fitness\_Function = Function(sml) \tag{4}$$

Step 3. Change the standards of best index and position for entire fruit flies (Eqs. (5) and (6)). Hence, update the fruit fly's position (Eqs. (7) and (8)).

$$\begin{aligned} BestIndex(t+1) = \mu \times BestIndex(t) + \eta_1 \times Rand \\ \times (Dist - X_{\Phi}(t)) + \eta_2 \times Rand \times (Sml - X_{\Phi}(t)), \end{aligned} \tag{5}$$

where μ = indolence coefficient.
η_1 and η_2 = Constant.

$$[BestSmell\ BestIndex] = minimum(Sml) \tag{6}$$

$$X_{axis\_value} = X(BestIndex) \tag{7}$$

$$X_{\Phi} = Round\left(X_{\Phi} \times \left(\begin{array}{l} EntireNodes - EntireSourceNodes \\ -EntireDestinationNodes \end{array}\right)\right) + EntireSourceNodes + 1. \tag{8}$$

Step 4. Establish optimized result, if not, go over 2.

If established the optimized result, found best node.

End

2.2 *Lattice Structure*

The lattice structure combines entire probable data cubes at dissimilar stages of concentricity by describing cubes on their reliance. Two cubes C_x and C_y are connected by a route for generating reliance association ($C_x \leq C_y$). This association illustrates that an answer to the OLAP query given by C_x, can also be provided by C_y. In lattice structure, 2^N data cubes are possible for N dimensions for an information association. Here, $2^3 = 8$ data cubes are obtained from three dimensional data "Sales" (Customer (C), Supplier (S), Part (P)) (Fig. 1).

The minimum concentricity represents by bottom cube (e.g., CSP) and maximum concentricity represented by the top cube. The structure illustrates that an answer

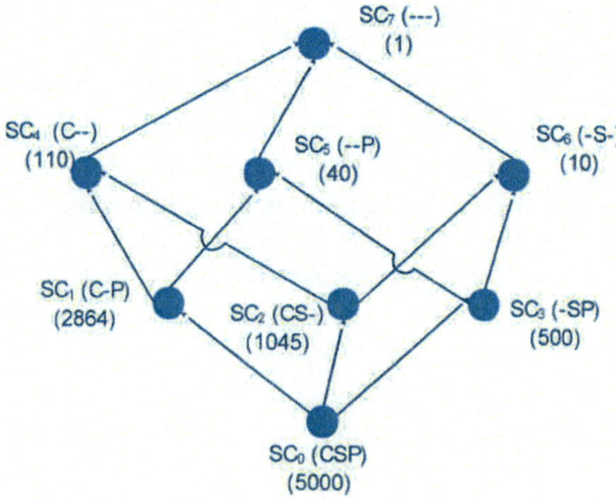

Fig. 1 Lattice structure

to the OLAP query given by child cube (*, S,*) can also be provided by any parent cubes (*, S, P), (C, S, *) or (C, S, P) by concluding data beside few dimensions.

2.3 Cube Selection Using FFO Approach

The objective of an OLAP model is to reduce expenditure of query and preservation with fulfilling a constriction like materialized space. The Entire Query Expenditure (EQE) for an OLAP query is evaluated by utilizing answering cost and frequency of query using Eq. (9).

$$EQE = \sum_{x=1}^{N} fq_x \times E(q_x, C_M) \tag{9}$$

Here,
fq_x = query q_x frequency.
$E(q_x, C_M)$ = query q_x answering expenditure (cost) with C_M (Materialized Cube) which is calculated by using Eq. (10).

$$E(q_x, C_M) = Minimum(|SC_x|, Lpre(q_x, C_M)) \tag{10}$$

Here,
$|SC_x|$ = Number of sub-cubes of query q_x.
$Lpre(q_x, C_M)$ = Least Predecessor (Lpre) conception of query q_x with C_M.
The Preservation Expenditure (PE) of a materialized cube $C_x \in C_M$ is evaluated in terms of least predecessor (Lpre) of C_x using Eq. (11).

$$PE(C_x, C_M) = |Lpre(C_x, C_M)| \tag{11}$$

After that Entire Preservation Expenditure (EPE) is evaluated by Eq. (12).

$$EPE = f_{PE} \sum_{C \in C_M} PE(C, C_M) \tag{12}$$

Here,

f_{PE} = frequency of inclusion in support association.
Hence, the Entire Cost Function (ECF) is calculated by combining the EQE (Entire Query Expenditure) and EPE (Entire Preservation Expenditure) using Eq. (13).

$$\text{Minimum} \quad ECF = \sum_{x=1}^{N} fq_x \times E(q_x, C_M) + f_{PE} \sum_{C \in C_M} PE(C, C_M) \tag{13}$$

The optimal value of ECF is obtained by applying FFO approach on this objective function ECF. The initial values of expenditure can be evaluated from pedestal association and the value of C_M is empty initially. After that optimal values are evaluated using FFO. The ECF function is utilized as variation function, as it discovers the solution fitness throughout the following of the several targets. The ECF function is minimized to obtain least expenditure of query with highest fitness.

3 Result and Analysis

The FFO and PSO approaches have implemented in MATLAB 2019a environment and analyzed in terms of frequency and number of dimensions. The results illustrate the best quality performance of FFO over PSO with minimum query processing expenditure. Here, we measured several belongings of space constrictions as 10, 20, 30, 40, 50 and 60% to analyze the FFO and PSO approaches.

We performed the implementation of FFO and PSO on the multiple dimensional data like three dimensions ((Customer (C), Supplier (S), Part (P)), and four dimensions (Customer (C), Supplier (S), Part (P), Time (T)) evaluate the results.

Figures 2 and 3 illustrate that the FFO generates better results for selecting optimal data cubes with minimum OLAP query processing expenditure as compared to PSO for all dimensional data.

We also evaluated the results of FFO and PSO in terms of identical and arbitrary frequencies (range 0–1).

Figures 4 and 5 illustrate that the FFO obtains more efficient performance for choosing optimal data cubes as compared to PSO in all frequencies with least OLAP query processing expenditure.

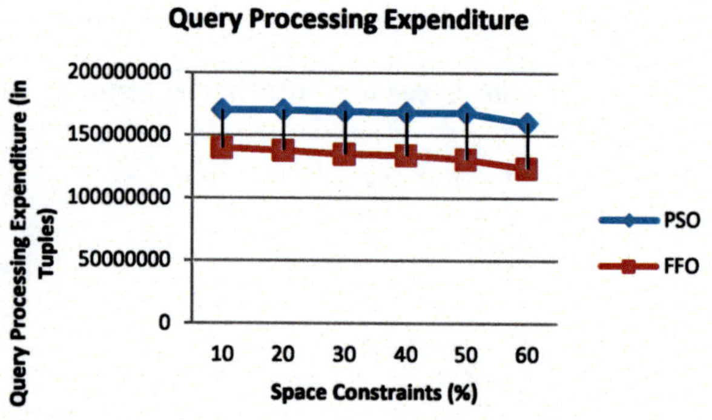

Fig. 2 Query processing expenditure for PSO and FFO approaches (3 Dimensions)

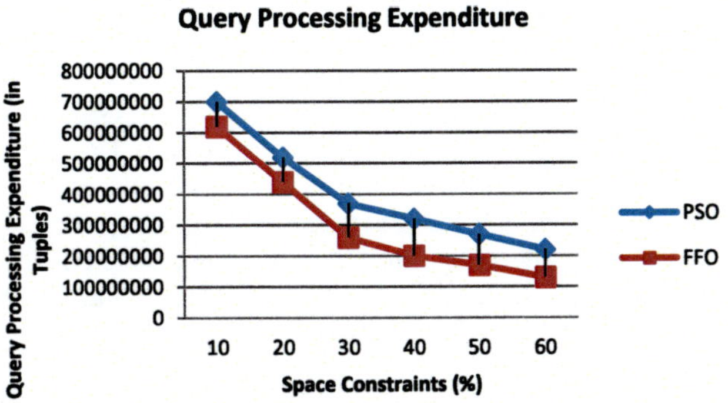

Fig. 3 Query processing expenditure for PSO and FFO approaches (4 Dimensions)

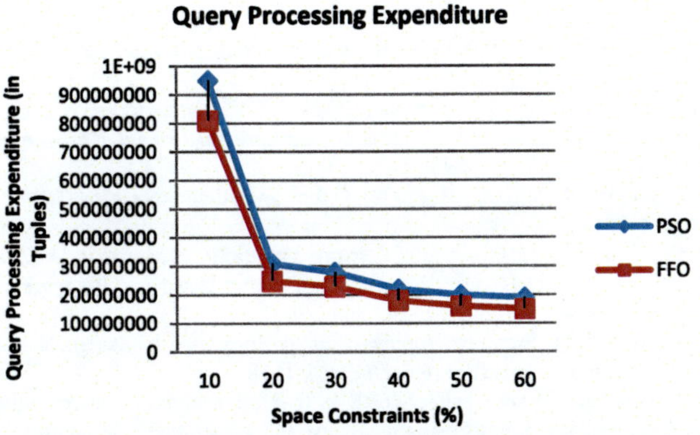

Fig. 4 Query processing expenditure for PSO and FFO approaches (Identical frequencies)

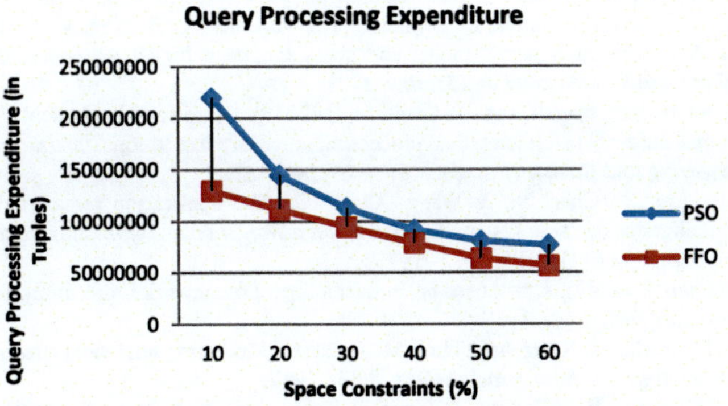

Fig. 5 Query processing expenditure for PSO and FFO approaches (Arbitrary frequencies)

4 Conclusion

In this work, FFO based optimal materialized cube selection is performed on lattice structure with multidimensional data over OLAP framework. The results are evaluated on multidimensional data in terms of frequency and number of dimensions. The analysis of performance of FFO illustrates the improved quality, efficiency of FFO to reduce the OLAP query processing expenditure as compared to PSO. In the future, several optimization approaches will be implemented for optimized cube selection by considering the time complexity as a factor.

References

1. Vaisman, A., & Zimanyi, E. (2019). Mobility data warehouses. *International Journal of Geo-Information, MDPI, 8*(170), 1–22.
2. Agapito, G., Zucco, C., & Cannataro, M. (2020). COVID-warehouse: A data warehouse of Italian COVID-19, pollution, and climate data. *International Journal of Environment Research and Public Health, MDPI, 17*(5596), 1–22.
3. Jukic, N., Jukic, B., & Malliaris, M. (2008). Online analytical processing (OLAP) for decision support (pp. 1–25).
4. Papacharalampopoulos, A., Giannoulis, C., Stavropoulos, P., & Mourtzis, D. (2020). A Digital twin for automated root-cause search of production alarms based on KPIs aggregated from IoT. *Applied Science, MDPI, 10*(2377), 1–16.
5. Stefanovic, N. (2014). Proactive supply chain performance management with predictive analytics. *The Scientific World Journal, Hindawi,* 1–18.
6. Noh, B., Son, J., Park, H., & Chang, S. (2017). In-depth analysis of energy efficiency related factors in commercial buildings using data cube and association rule mining. *Sustainability, MDPI, 9*(2119), 1–20.
7. Almeida, D. R. D., Baptista, C. D. S., Andrade, F. G. D., & Soares, A. (2020). A survey on big data for trajectory analytics. *International Journal of Geo-Information, MDPI, 9*(88), 1–24.
8. Ciferri, C., Ciferri, R., Gomez, L., Schneider, M., Vaisman, A., & Zimanyi, E. (2012). Cube algebra: A generic user-centric model and query language for OLAP cubes. *International Journal of Data Warehousing and Mining,* 1–23.
9. Emmanuel, N. E., Obiageli, J. A., & Osinachi, V. (2019). Design and implementation of multi-dimensional students result analytical processing of tertiary institutions. *International Journal of Engineering and Computer Science, 8*(8), 24814–24828.
10. Shen, L., Liu, S., Chen, S., & Wang, X. (2012). The application research of OLAP in police intelligence decision system. In *International Workshop on Information and Electronics Engineering (IWIEE)* (Vol. 29, pp. 1–6). Elsevier.
11. Venkatraman, S. (2017). A proposed business intelligent framework for recommender systems. *Informatics, MDPI, 4*(40), 1–12.
12. Cruz, I. L., Berlanga, R., & Aramburu, M. J. (2018). Modelling analytical streams for social business intelligence. *Informatics, MDPI, 5*(53), 1–17.
13. Fuertes, W., Reyes, F., Valladares, P., Tapia, F., Toulkeridis, T., & Perez, E. (2017). An integral model to provide reactive and proactive services in an academic CSIRT based on business intelligence. *Systems, MDPI, 5*(52), 1–20.
14. Qwaider, W. Q. (2012). Apply on-line analytical processing (OLAP) with data mining for clinical decision support. *International Journal of Managing Information Technology (IJMIT), 4*(1), 1–13.

15. Rubi, J. N. S., & Gondim, P. R. L. (2019). IoMT platform for pervasive healthcare data aggregation, processing, and sharing based on OneM2M and OpenEHR. *Sensors, MDPI, 19*(4283), 1–25.
16. Cervantes, J. L. S., Radzimski, M., Enriquez, C. A. R., Hernandez, G. A., Mazahua, L. R., Ramirez, C. S., & Gonzalez, A. R. (2016). SREQP: A solar radiation extraction and query platform for the production and consumption of linked data from weather stations sensors. *Journal of Sensors, Hindawi*, 1–19.

Fault Tolerant Multimedia Caching Strategy for Information-Centric Networking

Dharamendra Chouhan, Sachinkumar Hegde, N. N. Srinidhi, J. Shreyas, and S. M. Dilip Kumar

Abstract Extensive usage of mobile multimedia-based applications is increasing the user's demand on the internet. Edge computing makes it possible to bring storage of data and computation in close vicinity to mobile users. To improve the experience of the users, the proposed method uses edge Internet of Things equipment-assisted Fault Tolerant Multimedia Caching Strategy (FTMCS) for information-centric networking. To accomplish this, we propose a new technique composed of optimized task distribution and fault tolerant network services. The task distribution method efficiently distributes the downloading task among the edge nodes having required data. Fault tolerant service makes it possible for the network to withstand node failures that may occur during the file downloading, and by using this service, fast and reliable transfer of content happens seamlessly according to user preferences. The experimental results show that the FTMCS provides better network performance compared to the existing solution in terms of download speed and throughput.

Keywords Caching · Edge computing · Task distribution · Fault tolerant · ICN · Internet of things · Multimedia

1 Introduction

The demand for multimedia applications is continuously increasing on the internet, which increases the data volume on the internet [1]. Nevertheless, in conventional Transmission Control Protocol/Internet Protocol (TCP/IP), repeated transmission of data occurs during transmissions involving massive information. Recognizing a method that empowers various applications to become smart during usage is an essential topic focused on future networking administration.

In this era, Information-Centric Networking (ICN) is a type of internet architecture, which has gained significant attention. In ICN, the location of the content is not

D. Chouhan · S. Hegde · J. Shreyas (✉) · S. M. D. Kumar
University Visvesvaraya College of Engineering, Bangalore, India

N. N. Srinidhi
Sri Krishna Institute of Technology, Bangalore, India

P. Nayak et al. (eds.), *IoT and Analytics for Sensor Networks*, Lecture Notes in Networks and Systems 244, https://doi.org/10.1007/978-981-16-2919-8_25

considered by the user, but only the content. Based on the content, ICN will be able to route, position, and transmit the data. In general, ICN offers caching with seamless and built-in functionality to speed up the delivery of data. It also maximizes the network resource usage to avoid congestion [2] generated by the growing network traffic.

With the advancement in Information-Centric Networking with respect to mobility, the need for a technique to get superior user experience through caching has gained significant importance. Scholars have done considerable work to provide better transmission by modifying the cache entities. Nevertheless, it still remains challenging to obtain a good caching strategy for mobile users because of the movement of the users and limited device resources.

The development of edge computing gives a new perspective on these issues. Edge computing [3] is a modern computing technique that relies upon cloud and IoT services [4]. These services are required to be deployed at the edge of networks. This offers advantages like better response time, improved battery life, and minimized usage of bandwidth. Edge computing is effective in offering a possible improvement in network throughput and enhanced experience of the users. In addition, edge computing plays an important role in fostering new and popular solutions, like smart homes and smart cities, based on a smart environment [5]. Caching focuses on learning how to efficiently reduce the time needed for repetitive traffic to access content to achieve efficient content transmission. Caching decisions, therefore, play an important role in enhancing the user experience.

Caching mechanism is an innovation that exchanges time for space, and it concentrates on adequately diminishing the time needed to acquire information and accomplish the proficient transmission of information. In the proposed work, research is done on the issues related to the caching methodology for portable multimedia in ICN with edge computing. The outline of overall contributions of this paper can be summed up as follows:

- FTMCS for Edge IoT, a standard framework that enables task distribution is proposed. It can effectively distribute the downloading task among the edge nodes having required data.
- An optimal solution to deal with the failure of IoT nodes is proposed. It allows smooth relocation of existing networks, such as peer-to-peer, and thoroughly explores their potential.
- The performance of FTMCS is evaluated by conducting the experiments. The obtained results clearly show that the proposed FTMCS provides clients with higher download speed and throughput compared to the existing system.

The rest part of the paper is summarized as follows. Related research work and problem statement have been explained in Sects. 2 and 3, respectively. Sections 4 and 5 provide the system model and proposed method along with evaluation in Sect. 6. The conclusion part has been provided in Sect. 7.

2 Related Work

There has been much research works done on caching technology in ICN. Among them, Ahgren et al. [6] conducted a survey of Information-Centric networking. There are two main functions in ICN which must be done by resolution of name and routing when a particular NDO is requested. The first step is to search for a node that contains a copy of NDO and deliver the request to that node. The next step is to find a route back to the requestor from that node from which the NDO can be delivered. To offer better performing and more efficient transport services, ICN typically uses in-network storage. NDOs can be cached on the path, but by notifying them in a routing protocol or registering them in a name-resolution service, ICN can also make cached objects accessible for off-path requests. Research also reveals that there are no end-to-end linkages to the ICN approach. Xylomenos et al. [7] established a set of problems that motivates a reimagining of how the internet can function to meet the continuously evolving requirements. ICN is convenient for the infrastructure of the future Internet. ICN enables the proper and on-time transmission of information to the end-users. In this survey, the main targets are: (i) Identifying the important features of information-centric networking designs, (ii) To explain the main objectives of the information-centric networking in the form of tutorials, representing the differences and common factors between them with reference to those key features, (iii) Shortcomings of the information-centric networking proposals should be detected, and the unsolved research problems in this area of network research are identified. Wu et al. [8] proposed a Grey Relational Analysis (GRA)-based cross-layer caching strategy and a cross-layer cooperating caching technique for Content-Centric Networking. User priority is enforced in the form of the application layer, physical layer, and network layer metrics. Then, among all nodes along the distribution road, a caching probability of caching function is developed based on the grey relational analysis. GRA-based cooperating caching technique is developed based on the above-mentioned three-layered caching parameters. Grey relational analysis is a part of the grey method theory, which is best suited for resolving problems of complex connections between multiple variables and factors. This strategy does not provide scalability. Quan et al. [9] conducted a study that focuses on finding the possibility for social collaboration between neighboring highway vehicles to facilitate the use of resources for caching. In particular, the authors propose a cooperative caching mechanism based on the highway customized ICN, namely, ICoC, aimed at optimizing the quality of video streaming without excessive startup time and reducing the playback buffering ratio. In order to allow advanced QoE optimization, this study primarily characterizes information-centric delivery and highway traffic. Partner assist cooperation and courier-assist cooperation are being established in order to improve the efficiency of caching in highway VANETs. The performance of the caching mechanism proposed in this study is poor.

3 Problem Statement and Objectives

The problem considered in the proposed work is to formulate an advanced caching mechanism for information-centric networking by developing fault tolerance and task distribution features to achieve better performance. To accomplish the research goal, the following are the objectives set forth:

- To design an efficient caching strategy with task distribution to maximize the download speed.
- To develop a fault tolerance feature to the caching strategy to withstand the node failures in the network.
- To achieve reduced download time and increased throughput.

4 System Model

The architecture of the proposed system is shown in Fig. 1. The proposed system has three parts. (i) User storage, (ii) Edge node, and (iii) Server.

The user storage constitutes the private space of a client, which can store something from the cache system that they like. In addition, all the user instructions are transferred through the user configuration to the system.

The edge node is the second element. Where, each edge node, updates the cached multimedia content, can be composed of customized storage space and computation power.

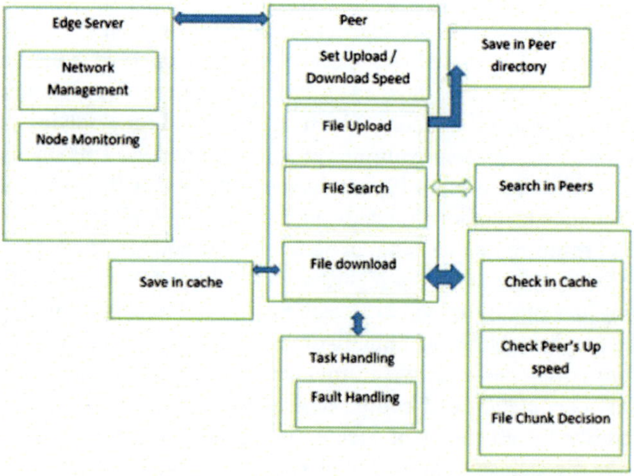

Fig. 1 System architecture

The server is the last part of the device that involves the computation of data, cache, and storage. It stores the information resources, and also archives all clients' interest content and mobile traces. It monitors the nodes and manages the network.

The multimedia data is cached in the peers to respond to user queries faster, and implemented fault tolerant ICN will save network bandwidth. The fault tolerance feature equips the network with the ability to withstand the unexpected node failures that may occur in the network. It essentially monitors the ongoing network operation and can assign the task to the next available peer in case of unexpected node failures. The proposed system also has a task distribution feature. A data request is divided among nodes, and parallel processing of data collection and caching mechanism makes the overall network efficient. The task is divided among the selected nodes equally, and all the tasks are processed simultaneously so that the data chunks from different nodes are collected and merged together in less span of time. The proposed system provides better response time, battery life, transmission capacity, throughput, and enhanced client experience.

5 Implementation

In this section, we discuss the details of the proposed algorithm. The size of the chunk file which is to be downloaded from available peers is calculated, and then data will be downloaded from them through parallel processing.

5.1 Calculation of Chunk Size

Algorithm 1 shows the way in which chunk size is calculated. While downloading the content, all the peers having the required cached data will be involved. The file will be downloaded in the form of chunks from each of the peers, and the size of the chunk that needs to be downloaded from each of the peers has to be calculated. This is done on the basis of the upload speed of the respective peer. Upload speed depends upon the configuration done by the user. Total upload speed can be calculated by using the formula (1).

$$Total_{Up\_Speed} = \sum_{i=0}^{n}(Up\_speed(node_i)) \tag{1}$$

The size of chunks from each peer can be calculated by using the formula (2).

$$Chunk_{size} = \frac{file_{size}}{Up_{speed(node)}} \tag{2}$$

A chunk of higher size will be downloaded from the peer with the highest upload speed. Likewise, chunk size will be calculated for all of the available peers based on their upload speed limit set by the user. The calculated chunk size information will be used later while downloading the file.

```
Algorithm 1: Chunk Size Calculation Algorithm
Input: num_peers, upload_speed, size_file
Output: sorted peers, chunk size

1. Assign arr ← sorted_selected_list
2. t←0
3. for i=0 to length (node) do
4. total_up_speed= up_speed (node_i)
5. end for
6. t←(size_file/ t_up_speed)
7.        St←0
8. End←0
9. for each peer
10.       Bytes←up_speed*t
11.       end←st+Bytes
12.       st←end+1
13. end for
14. save peers_name and chunk_size
15. Stop
```

5.2 Download File

Based on the previous chunk size calculation, the file will be downloaded as shown in Algorithm 2. The requester sends the request for a chunk from the assigned peer. When the assigned peer receives the chunk request, extracts the corresponding chunk from the file, and sends it back to the requester. The requester on receiving the chunk assigns the chunk id and then, it saves the chunk in the buffer. Later, it sorts all the received chunks based on the chunk id. After it receives a complete set of chunks, merges them in sorted order to get the complete file, and saves them in the application directory as well as in the cache memory.

The network is able to address the issues arising due to node failures. After assigning the task to the selected peers, if any network failure occurs with any of the peers, the requester assigns the task to get the chunk to the next available peer and follows the previously mentioned procedures. The requester continuously monitors the progress of the download of chunk and is able to detect the node failures. Thus, it ensures the integrity of data.

Algorithm 2: Download file algorithm with task distribution and fault tolerance
Input: filename, peer_name , ip , port, chunk_size
Output: downloaded file

```
1. Send chunk from source node to destination node
        a. Create object soc
        b. soc ← peer_ip, peer_port
        c. Create ObjectOutputStream object
        d. write chunk_details, chunk_id, chunk_size to object
2. Read file
        a. Load the file f from dir
        b. Read file f using BufferedReader
        c. for index = 0 to length (index) do
        d.     read f[index]
        e. end for
3. transfer chunk to requested peer
        a. soc ← peer_ip, peer_port
        b. Create ObjectOutputStream object
        c. write chunk_data to object
4. monitor peer transfer
5. if connection_lost
6.      Search(next_peer)
7.          if found
8.              Send_chunk_request(fname,chunk_id)
9.          end if
10. end if
11. Sort chunks using chunk id
12. Merge chunks
```

6 Performance Evaluation

Performance of the Proposed FTMCS is assessed by comparing it with the existing SCM [10]. The following performance metrics have been used to evaluate the performance: Download time with cache and throughput. The results obtained are depicted in the form of graphs.

6.1 Download Time

In Fig. 2, a comparison of the proposed system and the existing system is done using files of different sizes. Here, it can be observed that the time taken to download the same size file is less in the proposed system. In the proposed system, the download task will be distributed among all the peers having the required data, and while downloading, data will be downloaded from all those peers in the form of chunks. In the existing system, download can happen from a single peer at a time. This makes it possible for the proposed system to achieve a faster download speed.

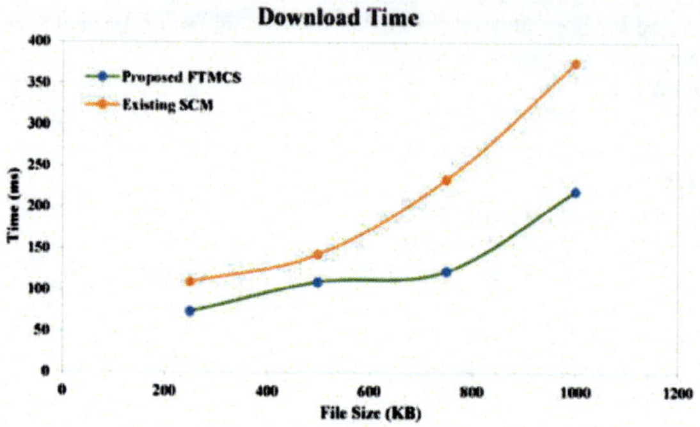

Fig. 2 Download time

6.2 Throughput

In Fig. 3, it can be observed that the throughput in the proposed system is improvised. The proposed system is tolerant to node failures in the network. If any node fails while transmitting the data, the proposed system will continue to download from the next available node. The existing system is not tolerant of node failure. This enables the proposed system to achieve increased throughput than the existing system.

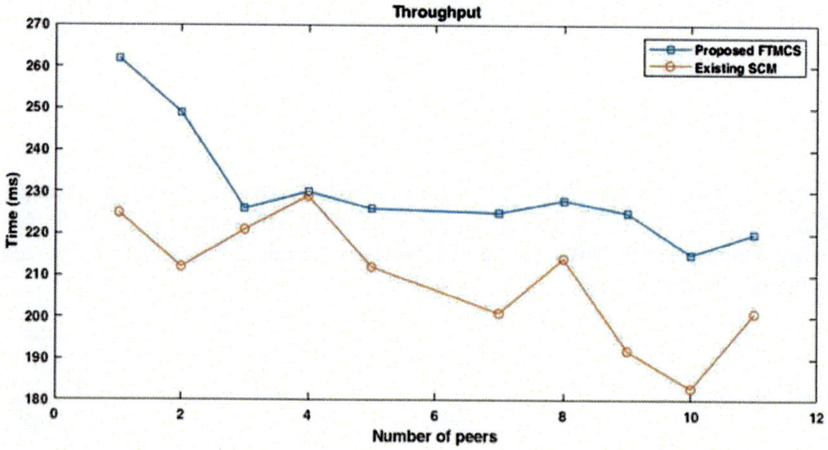

Fig. 3 Throughput

7 Conclusion

In this paper, an optimized solution for caching multimedia data in information-centric networking has been proposed. The caching strategy in the proposed work is inclusive of task distribution and fault tolerance. Task distribution essentially divides the task among the peers having the required file. Fault tolerance ensures the proper transmission of data even if there is a node failure in the network. A major issue is being addressed by the algorithm presented in this paper. Multimedia data is cached in the peers for a faster response to the queries. The proposed algorithm will increase performance dramatically, optimizes the time for data collection, and improves efficiency. A comparison between the existing system and the proposed system shows that the proposed system provides better download speed and throughput. In-network caching for ICN can be used as a further extension to the proposed method.

Acknowledgements This research work has been funded by SERB-DST Project File No: EEQ/2017/000681.

References

1. Shreyas, J., Jumnal, A., Kumar, S. M. D., & Venugopal, K. R. (2020). Application of computational intelligence techniques for internet of things: An extensive survey. *International Journal of Computational Intelligence Studies, 9*(3), 234–288.
2. Shreyas, J., Singh, H., Bhutani, J., Pandit, S., Srinidhi, N. N., & Kumar, S. M. D. (2019). Congestion aware algorithm using fuzzy logic to find an optimal routing path for IoT networks. In *International Conference on Computational Intelligence and Knowledge Economy (ICCIKE)*, December 11–12 (pp.141–145). Dubai, United Arab Emirates.
3. Shi, W., Cao, J., Zhang, Q., Li, Y., & Xu, L. (2016). Edge computing: Vision and challenges. *IEEE Internet of Things, 3*(5), 637–646.
4. Srinidhi, N. N., Sagar, C. S., Chethan, S. D., Shreyas, J., & Kumar, S. M. D. (2019). Machine learning based efficient multi-copy routing for oppiot networks. In *International Conference on Computational Intelligence, Security and Internet of Things*, December 13–14 (pp. 288–302). Agartala, India.
5. Srinidhi, N. N., Sunitha, G. P., Nagarjun, E., Shreyas, J., & Kumar, S. M. D. (2019). Lifetime maximization of IoT network by optimizing routing energy. In *IEEE International WIE Conference on Electrical and Computer Engineering (WIECON-ECE)*, November 15–16 (pp. 1–4). Bangalore, India.
6. Ahlgren, B., Dannewitz, C., Imbrenda, C., Kutscher, D., & Ohlman, B. (2012). A survey of information-centric networking. *IEEE Communications Magazine, 50*(7), 26–36.
7. Xylomenos, G., Ververidis, C. N., Siris, V. A., Fotiou, N., Tsilopoulos, C., Vasilakos, X., et al. (2014). A survey of information-centric networking research. *IEEE Communications Surveys & Tutorials, 16*(2), 1024–1049.
8. Wu, L., Zhang, T., Xu, X., Zeng, Z., & Liu, Y. (2015). Grey relational analysis based cross-layer caching for content centric networking. In *IEEE/CIC International Conference on Communications in China (ICCC)*, November 2–4 (pp. 1–6). Shenzhen, China.

9. Quan, W., Xu, C., Guan, J., Zhang, H., & Grieco, L. A. (2014). Social cooperation for information-centric multimedia streaming in highway VANETs. In *Proceeding of IEEE International Symposium on a World of Wireless, Mobile and Multimedia Networks*, June 19–19 (pp. 1–6). Sydney, Australia.
10. Tang, Y., Guo, K., Ma, J., Shen, Y., & Chi, T. (2019). A smart caching mechanism for mobile multimedia in information centric networking with edge computing. *Future Generation Computer Systems, 91*(2), 590–600.

Sizing of Wireless Networks with Sensors for Smart Houses with Coverage, Capacity, and Interference Restrictions

Jhonatan Fabricio Meza Cartagena, Deepa Jose, and J. S. Prasath

Abstract This research work proposes an effective solution for sizing in wireless networks supported by IEEE 802.15.4g, thus presenting the possibility of communicating networks with remote sensors in a completely transparent way for end devices, considering for this analysis restriction parameters such as the capacity, coverage, and interference knowing that the field of application of Wireless Networks is recently emerging, considerably gaining much popularity, which is increased as their features increase and new applications are discovered for them. This article has also included the application of a practical scenario, which demonstrates the communication of the devices remotely.

Keywords Wireless sensor networks · ZigBee · IEEE 802.15.4g · HAN domestic area networks · Sizing

1 Introduction

In this document, we have tried to provide an optimal solution to the sizing of wireless sensor networks for smart homes (Smart home), Fig. 1, based on the IEEE 802.15.4g, taking into account the characteristics and its restrictions allowing us to analyze the different parameters that may exist [1].

Considering the scenario based on the internal location of wireless sensors, we say that these are small devices that consume very little power, low data speed, and low mobility and are capable of measuring environmental parameters and communicating

J. F. M. Cartagena
Instituto Superior Tecnológico17 de Julio, Ibarra, Ecuador
e-mail: jmeza@ist17dejulio.edu.ec

D. Jose (✉)
Department of ECE, KCG College of Technology, Chennai, India

J. S. Prasath
Department of EIE, KCG College of Technology, Chennai, India
e-mail: prasath.ei@kcgcollege.com

P. Nayak et al. (eds.), *IoT and Analytics for Sensor Networks*, Lecture Notes in Networks and Systems 244, https://doi.org/10.1007/978-981-16-2919-8_26

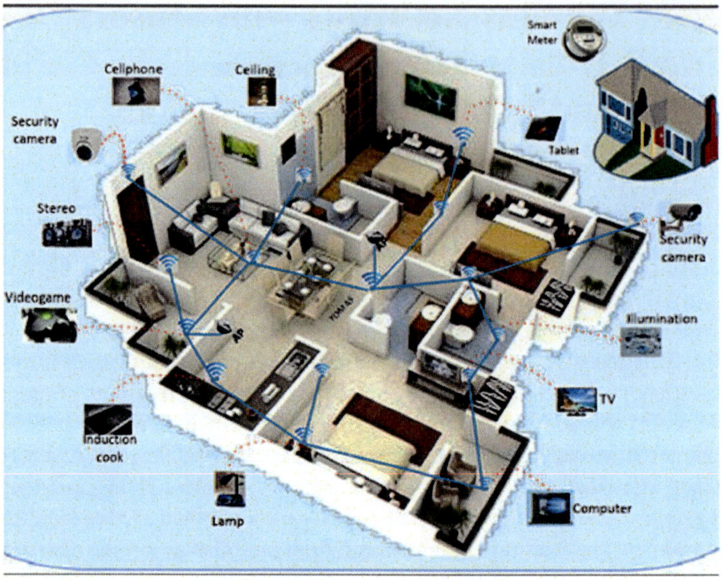

Fig. 1 Architecture of a WSN home network

with each [2]. Therefore, the environment of these sensors can be associated with their positions allowing the creation of many interesting applications.

On the other hand, within a group of simple protocols emerges the IEEE 802.15.4, where one of the most relevant characteristics of this standard are the flexibility of the network, low costs, and low energy consumption. It should be recognized that this standard has many home automation and industrial applications [3], which require a low data transmission rate, the key to its use being the reduction of expenses and information exchange with minimal effort [4, 5]. There is also talk of the introduction of 4G in communication technologies in the fourth generation.

On the other hand, if we mention the IEEE 802.15.4, we also have to mention that it is ZigBee [6]. It is a network technology that emerges as a low-cost wireless communication standard [7] that is capable of satisfying needs, such as the advantages of low cost, low power, and a broader coverage, making it very suitable for the development and use in a house.

In other words, smart grids today make up a very sophisticated energy system connecting a number of networks and devices that imply a tendency to be driven by their great ability to integrate wireless components, resembling the success of mobile wireless communication systems, considering intelligent systems capable of analyzing a situation, making calculated decisions, and responding to anomalies based on implemented models or algorithms.

It is necessary to know the methods that exist to extend the useful life of the devices of a WSN wireless sensor network [8], the processing of their data, routing, topologies, and the placement of the control devices in the determination of coverage.

Therefore, the problem presented below aims to provide a solution to the detection of a minimum energy consumption in order to improve the excess energy in a connected wireless sensor network (WSN), and designed for improving performance and coverage for indoor users (where most mobile users spend much of their time), as well as overall system capacity by downloading data [9].

On the other hand, considering our study, we have to talk about the wireless access points (Wireless Access Point) known or called AP for its acronym in English, since these will serve as intermediaries serving as transmitters and receivers of the signals sent in the radio of the network.

We also have to consider the interference that the devices of our wireless network system may receive, considering that this parameter can be caused by other services that also operate within our band, and it is imperative to know that a primary requirement of IEEE 802.15 applications 0.4 is to try to control and reduce interference caused by communications within a common channel. Therefore, it is very important in the design and efficient control within the topologies of a wireless network of network sensors. The rest of the work is organized as follows. In Sect. 2, the work related to the sizing of wireless networks is presented. Section 3 describes the formulation of the optimization problem to deploy a network infrastructure with wireless sensors. In Sect. 4, we propose the optimization of the problems posed in a sensor network based on the analysis of results. Finally, we conclude this document in Sect. 5.

2 Network Dimensioning Wireless

In this section, we briefly examine smart home networks, and based on their main contributions, we discuss the main task approaches based on capabilities described in Sect. 3. Furthermore, in this document, a series of methods have been proposed to find coverage, capacity, and interference connected to a wireless sensor network (WSN). The scenario may be optimal under certain criteria, therefore, the main objective is to cover a total detection area by placing a minimum of sensors, which when deployed tend to be efficiently connected so that each sensor can find a connection path and can arrive at a station considered as a base. In fact, in a WSN, where the devices are densely deployed, a subset of the devices can cope with coverage and connectivity issues. The rest of the devices can be switched to a sleep state to conserve power. With the use of device placement and routing techniques, it is difficult to find coverage, when using heterogeneous WSNs. Due to this a connectivity problem occurs. Therefore, we consider all the traffic generated by the applications that run smart grids with a heterogeneous nature in terms of minimum performance, reliability, etc. [10].

3 Problem Formulation

Wireless sensor networks constitute the fundamental basis of our study, thus determining the necessary parameters to be able to achieve a modeling focused on the optimal dimensioning of the WSN, considering restrictions such as coverage, capacity, and interference.

For the proposed problem, a set N of users located randomly in the region is considered. Assuming likewise M possible locations (candidate sites) of access points in the region. A possible location is a place where an AP could be placed, but not where one will necessarily be located. A coverage radius is highlighted depending on the configured power of the AP equipment, depending on the objective of covering a certain area.

a. Coverage Problem

In this way, for this case, we propose a metaheuristic approach considering an algorithm that allows us to cover a 2-dimensional area (length and width), and where it can be.

Identify a potential position for the placement of a sensor point, so that our target points are k-coverage and all sensors are our n-connectors.

The algorithm that we propose [11] considers technical and modeling criteria, necessary to carry out a sizing process where a design problem is formulated in a concrete way. Therefore, the type of objective function and restrictions must meet some characteristics or certain initial parameters such as:

- For each candidate site yes, the quantity is defined $Z_i \in \{0.1\}$. If the value is 1, it indicates that the candidate site i is an active site.
- For each sensor u_j, the quantity is defined $Y_j \in \{0, 1\}$. If the value is 1, it indicates that the user is covered by at least one candidate site.
- The relationship between the list of active sites and user coverage is as follows:

$$Y_j \leq \sum_{i=1}^{M} \propto_{i,j} \sum Z_i \tag{1}$$

The above equation is defined as a constraint. If there is no active candidate site covering the sensor, Y_j is necessarily equal to 0, that is, the sensor is not covered. To indicate that it covers at least 90% of the sensors, another restriction must be added, which counts the percentage of sensors covered.

$$N \sum_{j=1}^{1N} Y_j \geq 0.9 \tag{2}$$

Finally, the goal of the problem is to minimize the number of candidate sites that are active. To do this, we write another equation that is now called the objective function.

Therefore, we write what we call the optimization problem that we completely express as:

$$\min \sum_{i=1}^{M} Z_i \tag{3}$$

$$\text{s.t. } Y_j \leq \sum_{i=1}^{M} \alpha_{i,j} Z_i ; \forall j \in U \tag{4}$$

$$N \sum_{j=1}^{1N} Y_j \geq 0.9 \tag{5}$$

where, Eq. (3) seeks to minimize the number of candidate sites that are active. Equation (4) defines the coverage constraint. Equation (5) tells us that 90% of the sensors can be covered (Fig. 2).

b. Capacity Coverage Problem

The next step that we can consider with the same algorithm raised above is to include a capacity restriction in the existing network wireless local (WLAN) claiming benefit of AP devices placed on stage. In this parameter, our scenario raises the problem considering the capacity of the WLAN network, arbitrarily placing APs, determining a radius and a limited capacity to the sensors. We describe the optimization problem below.

$$\min \sum_{i=1}^{M} Z_i \tag{6}$$

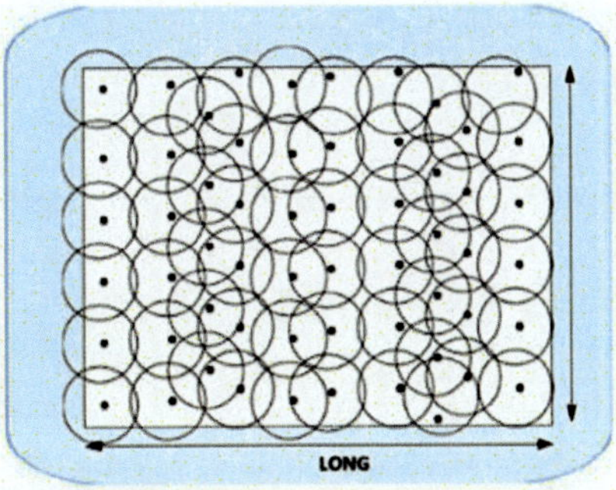

Fig. 2 Illustration of a full coverage

$$\text{s.t. } Y_j \leq \sum_{i=1}^{M} X_{i,j};\ \forall j \in U \tag{7}$$

$$\sum_{j=1}^{N} X_{i,j} \leq C\, Z_i;\ \forall i \in S \tag{8}$$

$$\sum_{j=1}^{N} Y_j \geq N \cdot Porc \tag{9}$$

$$X_{i,j} \leq \propto_{i,j}\ Z_i;\ \forall i \in S;\ \forall j \in C \tag{10}$$

Equation (6) corresponds to the objective function. Equation (7) is the one that is subject to the restrictions.

iii. Coverage, Capacity, and Interference Problem

This section further investigates the impact of interference on the performance of the desired connection. We base the interference reduction algorithm on the ZigBee home network architecture, where one or multiple network-related functionalities are implemented. In the architecture, we can assume that at least one WLAN AP is associated with the proposed scenario.

In this scenario, a network is proposed where nearby APs interfere with each other, this being the base model for others who were looking forassign channels.

In fact, an AP is considered to allow connectivity with sensors that are within a certain coverage radius [12, 13]. The capacity of an AP is limited and is given by the number of simultaneous sensors that can be attended, considering that the APs operate on the same channel, so there cannot be two APs within an interference distance, since they would produce interference with each other. The sensors can be located in any position within the region and seek to connect to an AP that is available and capable of serving it.

It is therefore necessary to consider the parameters previously raised in the problem of coverage with capacity, and we assign the following consideration:

Fig. 3 Scenario of the problem to perform the optimal sizing of networks with wireless sensors

$$Zi + Zk \leq 2 - \beta i, k; \forall i \in S; \forall k \in S \quad (11)$$

Universal data aggregation point optimization algorithm considering coverage, capacity, and interference

Step 1: R = Coverage Radius.

Step 2: P = We define the coverage probability between 0 and 10, we define a capacity for the second restriction, and we define a percentage of interference.

Step 3: Determine the number of candidate sites S = (S1, S2, ... Sn) in the i-nth position given by *(XSi, YSi)*.

Step 4: We define a set U = (U1, U2, ... Un) of sensors. The j-nth sensor position given by *(XUj, YUj)*.

Step 5: We establish the relationship between AP condition and sensor coverage.

Step 6: The problem performs AP minimization by means of the function.

Objective (Fig. 3).

4 Analysis of Results

Figure 4 shows a scenario with n meters by n meters, where we place WLAN APs to provide connectivity to the different sensors considering a coverage radius of R meters in any position within our scenario. In addition, the scenario includes 21 sensors and 45 candidate sites for the deployment of the wireless sensor network within a map made under the AutoCAD platform.

In fact, the main limitation of our proposed scenario is its scalability. The problems that due to their simplicity can generate millions of decisions, and according to the models, both real and approximate solutions can be found. To simulate the coverage optimization problem, the MATLAB programming language was used, and the parameters of the problem are initialized with previously determined values, while the degree of coverage in the destination field and dimension of the access point were entered by the sensor, starting from the previous problem with 21 sensors and 45 candidate sites in the scenario.

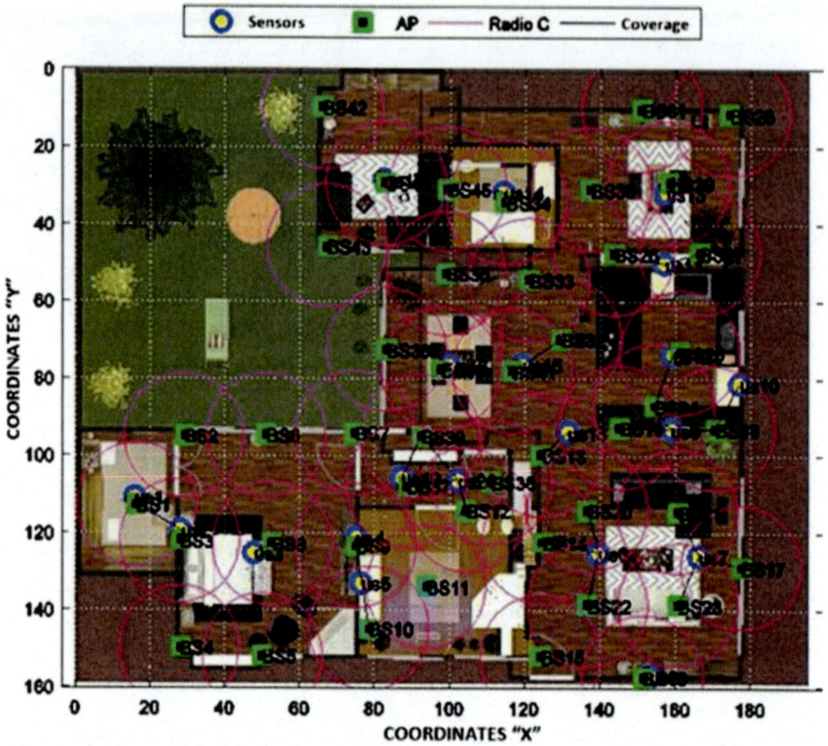

Fig. 4 Location of multiple APs to cover WSN

Figure 5 finds a minimum of AP active sites that cover at least a percentage of sensors at starting from the previous problem with 21 sensors and 45 candidate sites in the scenario.

Figure 6 locates an AP arbitrarily with each access point considering a certain radius and a capacity limited to this with the number of sensors.

Therefore, when analyzing the results obtained in the scenario in Fig. 6, we consider that to correct this problem and maintain the minimum speeds required, there are some tips and recommendations that they make, such as finding the best location for the access point, placing either your antenna or even change the latter for a better performance one. It is evident that if we choose the perfect place to place it, we will have more opportunities to improve (or optimize) the wireless network (Fig. 7).

Figure 8 shows us the last scenario where we place the same conditions as the previous ones with the addition of the interference restriction, which will allow us to understand what happens in this case.

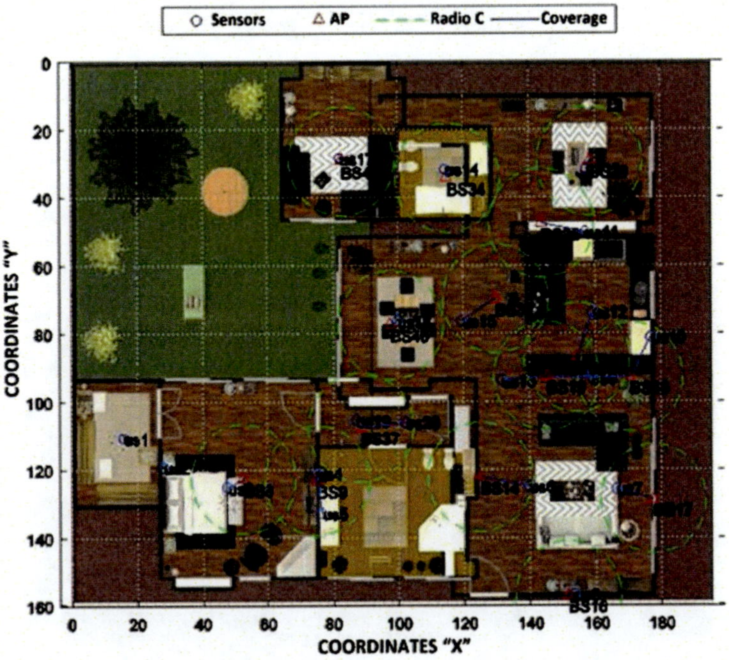

Fig. 5 Coverage of a minimum percentage of area

Figure 9 solves the interference problem that arises in our scenario, but clearly shows the coverage problem. This means that the APs that are within a distance cannot interfere with each other but limits the coverage.

As we can see that under the restriction of coverage, capacity, and interference, the solution is not as simple as putting another channel. The channels overlap each other, to avoid 100% interference, it should be more than 5 channels apart.

That is, if we take channels 6 and 11 as an example, they do not interfere, but 6 and 7 do. Some APs have an automatic channel selection function that tries to automatically select the least saturated channel, which in other words is meant to say that wireless connections are made over radio waves and there are several channels available. If your wireless network connection shows interference or disconnects regularly, then an attempt is made to change channels.

5 Conclusions

In this document, we propose an algorithm with heuristic techniques that seeks a WSN deployment with optimal coverage through an optimal initial distribution of the wireless network, which through computer simulations found a better solution to

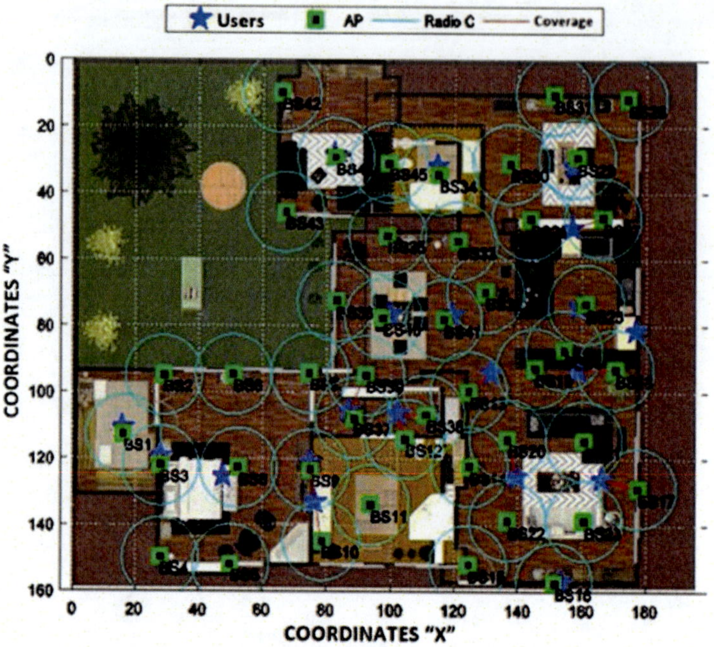

Fig. 6 Scenario graph not optimized

move our AP devices with maximum coverage and a minimum of energy consumption for communication. On the other hand, the analysis in the proposed scenario leads us to say that the scope of communication and the range of interference in IEEE 802.15.4g is based on channel models. The realistic intention is to include analysis with interference minimization techniques that are capable of relaxing the range of interference. However, the WSN application at various smart grid sites presented in this document consider the architecture of wireless sensors for integration with candidate sites to sustainably satisfy the smart grid. Analyzing the themes and scenarios carried out in relation to Wireless Sensor Networks and taking into consideration the restrictions previously raised, it is important to highlight that it is necessary to take into account that without the implementation of said technology in these environments, many of the homes, they may not have the benefits that these technologies have provided us.

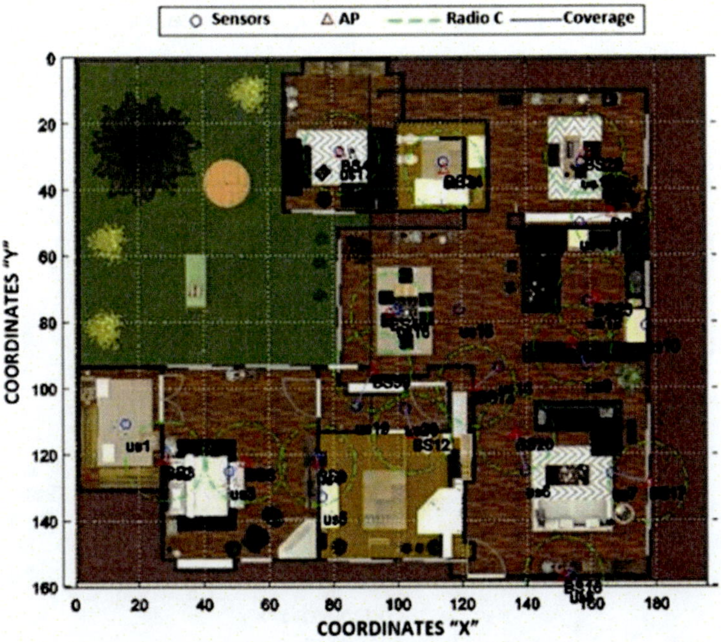

Fig. 7 Network capacity simulation with APs that connect each sensor

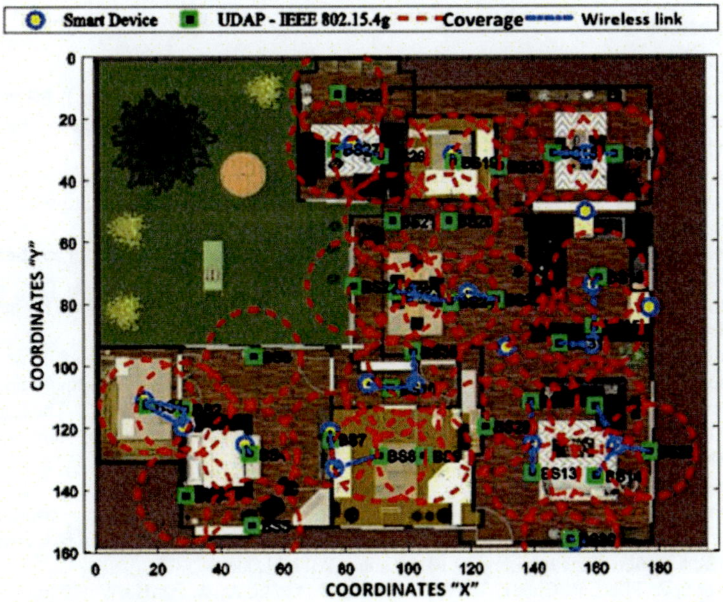

Fig. 8 Scenario graph not optimized

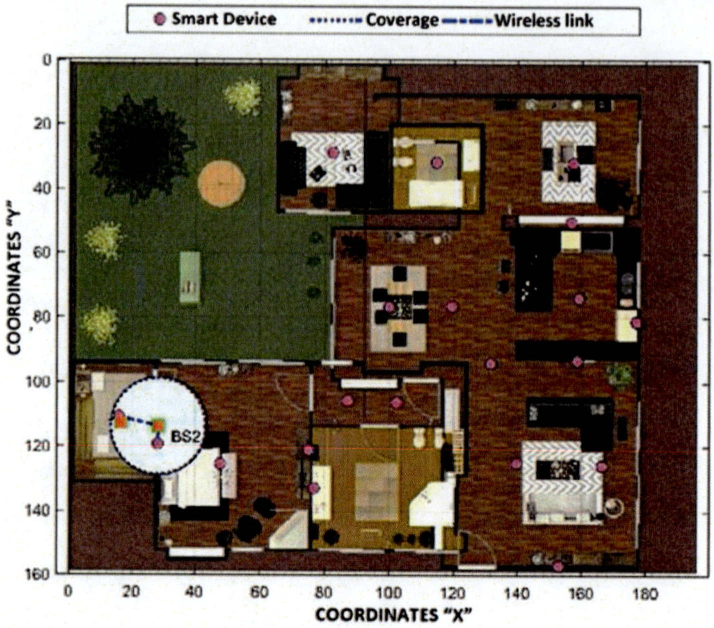

Fig. 9 Interference simulation with coverage limitation

References

1. Cuomo, F., Cipollone, E. & Abbagnale, A. (2009). Performance analysis of IEEE 802.15.4 wireless sensor networks: An insight into the topology formation process. *Computer Networks, 53*(18), 3057–3075. https://doi.org/10.1016/j.comnet.2009.07.016
2. Cheng, L., Wu, C. D., & Zhang, Y. Z. (2011). Indoor robot localization based on wireless sensor networks. *IEEE Transactions on Consumer Electronics, 57*(3), 1099–1104. http://ieeexplore.ieee.org/xpls/abs_all.jsp?arnumber=6018861
3. Erol-kantarci, M., & Mouftah, H. T. (2011). Wireless sensor networks for cost-efficient residential energy management in the smart grid. *2*(2), 314–325
4. Asad, O., Erol-kantarci, M., & Mouftah, H. (2011). Sensor network web services for demand-side energy management applications in the smart grid. 1176–1180.
5. Galinina, O., et al. (2014). Optimizing energy efficiency of a multi-radio mobile device in heterogeneous beyond-4G networks. *Performance Evaluation, 78*, 18–41. http://linkinghub.elsevier.com/retrieve/pii/S016653161400056X. Garroppo, R. G., et al. (2015). The impact of the access point power model on the energy-efficient management of infrastructured wireless LANs. *Computer Networks,* 1–13. http://linkinghub.elsevier.com/retrieve/pii/S13891286150 0451X
6. Cuomo, F., Abbagnale, A., & Cipollone, E. (2013). Ad hoc networks cross-layer network formation for energy-efficient IEEE 802.15.4. ZigBee wireless sensor networks. *Ad Hoc Networks, 11*(2), 672–686. https://doi.org/10.1016/j.adhoc.2011.11.006
7. Baronti, P., Pillai, P., Chook, V. W. C., Chessa, S., Gotta, A., & FunHu, Y. (2006). Wireless sensor networks: A survey on the state of the art and the 802.15.4 and ZigBee standards.
8. Akilandeswari, M., & Srikanth, U. (2012). Metaheuristic approach for maximizing the lifetime of heterogeneous wireless sensor networks.

9. Dudnikova, A., Panno, D., & Mastrosimone, A. (2015). Measurement-based coverage function for green femtocell networks. *Computer Networks, 83*, 45–58. http://www.sciencedirect.com/science/article/pii/S1389128615000730
10. Amin, R., & Martin, J. (2012). Smart grid communication using next generation heterogeneous wireless networks, 229–234.
11. Cheng, H., et al. (2016). Energy-efficient node scheduling algorithms for wireless sensor networks using Markov random field model. *Information Sciences, 329*, 461–477. http://linkinghub.elsevier.com/retrieve/pii/S0020025515006945
12. Murugan, R., & Mohan, M. R. (2012). Artificial bee colony optimization for the combined heat and power economic dispatch problem. *ARPN Journal of Engineering and Applied Sciences, 5*(7), 1353–1366.
13. Jose, D., Kumar, P. N., & Ramkumar, S. (2014). Reliability aware self-healing FFT system employing partial reconfiguration for reduced power consumption. In *Proceedings of the 2014 IEEE Students' Technology Symposium* (pp. 31–36).

Cloud-Based Parkinson Disease Prediction System Using Expanded Cat Swarm Optimization

Ramaprabha Jayaram and T. Senthil Kumar

Abstract Parkinson disease is identified as the second most severe neurodegenerative disorder that affects the nerve system of people. This disease could mainly affect the walking, speech, and vision of patients followed by body nervousness, handwriting, harsh voice quality, depression, and sleeping problems. The proposed research study focuses on early detection and diagnosis of disease from the accelerometer sensor-based data by evaluating the deviations present in patient's motor symptoms. A cloud-based Parkinson disease prediction system is developed for a clinical decision-making process that helps the doctor to diagnose the Parkinson-influenced patient from a remote place. Gait parameters of the patient were extracted along to provide input vectors to the classifier model for onboard Parkinson disease prediction and diagnosis. An effective expanded cat swarm optimization (ECSO)-based feature selection technique is explored to overcome the problem of dataset dimensionality. It could select the most appropriate features from the patient dataset according to a logically inspired evolutionary algorithm. Using this feature selection technique in the k-Nearest Neighbor (k-NN), classifier model could significantly improve the disease prediction accuracy, and also minimizes the disease prediction time against the existing classifier models.

Keywords Parkinson disease · Feature selection · Cloud-based prediction · Cat swarm optimization · k-nearest neighbor classifier

1 Introduction

Parkinson disease is treated as the most severe neurological disease caused due to increase in motor dysfunction. This disease can be recognized by exploiting handwriting dynamics, speech, motion, and gait signals. To support early identification

R. Jayaram (✉) · T. Senthil Kumar
SRM Institute of Science & Technology, Kattankulathur, Chengalpattu, India
e-mail: ramapraj@srmist.edu.in

T. Senthil Kumar
e-mail: senthilt2@srmist.edu.in

P. Nayak et al. (eds.), *IoT and Analytics for Sensor Networks*, Lecture Notes in Networks and Systems 244, https://doi.org/10.1007/978-981-16-2919-8_27

and diagnosis of Parkinson disease, there is a need for continuous monitoring and assessment of changes occurring in the patient's gait observation. Motor symptoms of this disease can be easily characterized by the occurrence of resting tremor, body stiffness, and bradykinesia (slowness of movement) [1]. Some of the non-motor symptoms can be characterized by the depression, fatigue, and sleep disorder. In addition, the Parkinson patient may also have some common problems like postural instability, coordination, and balance. Also, the patient having disease used to walk with leisurely gait speed, decreased step length, and height. In order to improve the healthcare system, accelerometer and gyroscope sensors were used to monitor the physical activity and fall prediction of remote patients by measuring the spatial and temporal parameters of gait prototype [2]. The key parameters such as gait velocity, cadence, time, distance, stride length, and step length were considered to be the major variables for analyzing the gait cycle.

Most researchers put emphasis on using the accelerometer sensor to measure the balancing factor and posture transitions. In some cases, gait evaluation was done through live video recording of patients to track the gait movements for human fall prediction and real-time activity monitoring. Recognizing such human activity through gait analysis could help the therapists in general context, but does not provide any kind of clinical assistance for them. There are different types of feature sets like time domain and frequency domain to encode the gait signals in the regression-based model [3]. To make better assistance to the clinicians during prediction and diagnosis of Parkinson disease, various machine learning classifiers such as Bayesian, decision tree, support vector machine, and artificial neural networks were used to identify the hidden patterns associated with the dataset. In order to avoid the over fitting problem and improve the accuracy and prediction time classification, various feature selection techniques such as information gain, principal component analysis, linear discriminant analysis, and correlation feature subset selection were deployed over the dataset [4]. Afterwards, various stages of Parkinson disease are identified by the classifier model for further diagnosis and rehabilitation monitoring purposes.

The key contributions of this research study includes: (1) a cloud-based Parkinson disease prediction system for real-time prediction and diagnosis of Parkinson disease; (2) a novel ECSO-based feature selection method for handling dimensionality reduction and over fitting problems; and (3) an efficient k-NN-based classifier model is proposed for predicting various stages of Parkinson disease severity. The remainder of this research study is organized as follows. The next section presents the related works pertaining to the Parkinson disease prediction system. Section 3 briefs the architecture of cloud-based Parkinson disease prediction system along with ECSO-based feature selection and k-NN based classifier model proposed for this research study. In Sect. 4, results and discussions of the proposed healthcare system is assessed in the context of prediction time and accuracy. The final section gives the conclusion and future enhancement of this research study.

2 Related Works

Nowadays, cloud-based remote patient monitoring and diagnosis have become more popular with various diseases. Especially, people suffering from neurological disorder need to be identified for regular monitoring and diagnosis according to severity stages [5]. In order to predict the intended target variable, a feature selection method is used to minimize the number of input variables by removing non-sensitive information from the data samples. The Principal Component Analysis (PCA) feature selection is used to find the ways of maximum variation. It performs well in case of having very less data samples for the available classes. But, sometimes it ignores the class labels also due to unsupervised methods [6]. In some cases, Linear Discriminant Analysis (LDA)-based feature selection method is preferred for the sake of dimensionality reduction. This linear transformation technique takes the supervised methods to find feature subspace by maximizing the class separation based on the normally distributed classes assumption [7].

The Naive Bayes' classifier has independent features among the classes, and expresses the event probability according to its prior knowledge. At the time of analysis, this classifier could make the classification based on the combined source of information such as likelihood and prior probability values. Next, a multiplicative function is used to combine these values to produce the posterior probability as a product [8]. The Support Vector Machine (SVM) classifier can be used to make classification on both linear and nonlinear data samples. It needs to identify the hyper plane that could differentiate the classes through maximum origin [9]. Further, the SVM is upgraded with particle swarm optimization and genetic algorithms to select the optimal features for searching among the appropriate continuous variables [10]. Therefore, the proposed research focused on developing optimization selection techniques along with k-NN classifier models to enhance the Parkinson disease prediction systems.

3 Cloud-Based Parkinson Disease Prediction System

An architecture of cloud-based Parkinson disease prediction system is proposed as presented in Fig. 1. It clearly shows that the remote patient can interact with the doctor using smart phones or video-enabled devices over the cloud platform. All the monitoring and diagnosis activities related to the healthcare system is controlled by the cloud manager through appropriate resource allocation policy. Usually, the medical data samples obtained from the remote patient using smart phones will be controlled by the multimedia content server to control the live streaming of data over the internet. Then, the Parkinson disease manager service could handle the relevant operations by running the feature extraction and classifier models deployed in the virtual machines. In this research study, an ECSO-based feature selection method is explored to reduce the dimensionality of medical data before going for disease

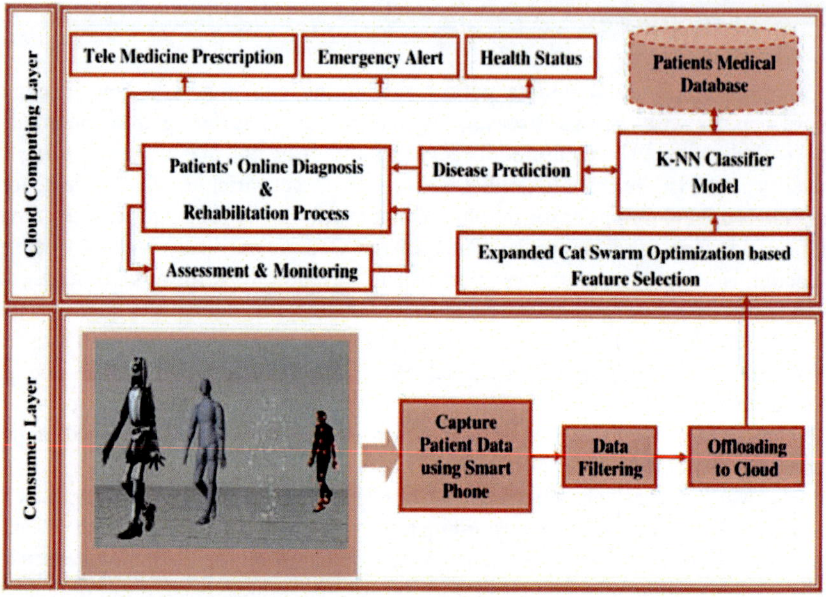

Fig. 1 Cloud-based Parkinson disease prediction system

prediction. Then, the proposed k-NN-based classifier model is used for predicting the severity of Parkinson disease. Accordingly, it will suggest the therapist to make appropriate diagnosis pertaining to severity levels.

As shown in the architectural process, the remote patient has to upload the medical data sample using a smartphone to know whether he/she is affected by the Parkinson disease or not. After receiving the data sample at cloud layer, the foremost thing is to remote the noise present in the signal, and then go for further feature selection and classification process. After prediction, discriminated medical data is stored in the electronic health record over the cloud platform as patient history. So that the healthcare professionals and therapist could periodically monitor and diagnose the patient from remote places. When the user tries to access the healthcare system using the internet, then the cloud manager will authenticate the user and initiate the privacy preserving access controls to establish the collaboration between the user and healthcare system. Afterwards, virtual machines will be launched with ECSO-based feature extraction and k-NN-based classifier model deployments to handle the medical session in the form of web services. After disease prediction, the multimedia content server will notify the severity and assessment values during diagnosis.

3.1 ECSO-Based Feature Selection

An N-dimensional integer vector represents the important features of lively captured gait data of patients. It can belong to integer space as given in Eq. (1).

$$S^N = \{(x_1, \ldots x_n, \ldots, x_N) \mid x_n \in I, x_n^{\min} \le x_n \le x_n^{\max}, 1 \le n \le N\} \quad (1)$$

Let I denote the set of integers, $x_n^{\min}$ and $x_n^{\max}$ represent the lower and upper limit ranges of integer variables. Here, the multi-objective optimization of this research problem can be mathematically expressed as shown in Eq. (2).

$$f(X) = \begin{cases} Min_X\ f_1(X) \\ Min_X\ f_2(X) \\ \quad \ldots \\ Min_X\ f_n(X) \end{cases} \quad (2)$$

where X be the vector is present over the integer space S^N. Here, the mth objective function can be represented as $f_m(X)$ for all $m \in \{1, 2, \ldots, n\}$ with N-dimensional search vectors X. The solution to the minimization problem is said to be the dominate vector if and only if satisfies Eqs. (3) and (4).

$$f_p(X_u) \le f_p(X_v),\ \forall p \in \{1, 2, \ldots, n\} \quad (3)$$

$$f_p(X_u) < f_p(X_v),\ \exists p \in \{1, 2, \ldots, n\} \quad (4)$$

Let X_u and X_v be the two candidate solution with its dominant vector $f(X_u)$ (represented by $f(X_u) < f(X_v)$). A solution X_u is said to Pareto optimal if it is not dominated by any other possible solution. It means there are no other possible solutions available in the space with respect to multiple objectives. Therefore, an ECSO is proposed with a multi-objective integer optimization context.

The proposed ECSO feature selection uses two sets of cats with seeking and tracing mode to find the best solution. A mixing ratio MR parameter is defined as the ratio of cats in tracing mode to the cats in seeking mode. Other parameters like MI denote the mixing iteration, φ denote the probability of asynchronous updating, N be the decision variables, and SS be the swarm size denoting the number of cats available in the pool. Let us consider the position and velocity of ith cats available in the N-dimensional space $N = \{1, 2, \ldots, n\}$ can be represented as $P_i = \{p_{i1}, p_{i2}, \ldots, p_{in}\}$ and $V_i = \{v_{i1}, v_{i2}, \ldots, v_{in}\}$. In tracing mode, the position of copied j^{th} cat is defined by $P_j = \{p_{j1}, p_{j2}, \ldots, p_{jn}\}$, and the best global position of cat swarm is defined by $P_g = \{p_{g1}, p_{g2}, \ldots, p_{gn}\}$. During the seeking mode of the cats, the parameters like seeking memory pool (SMP) denotes the copies of current cat, self-position consideration (SPC) denotes whether the cat is selected again or not, count dimension change (CDC) indicates varying dimension of cat position, and selected dimension range (SDR) represents the mutation ratio of desired dimension. Then, create $i = SMP - SPC$ copies of jth cat such as $p_{ln} : 1 \leq l \leq i, 1 \leq n \leq N$, where p_{ln} denotes the nth dimension of lth copy with mutation operation. All copies will be checked for bounds as defined in Eq. (5).

$$p_{ln} = \begin{cases} round(p_{ln}) & if\ round(p_{ln}) \in R_n \\ p_{in} & otherwise \end{cases} \tag{5}$$

where R_n denotes the set of integer value range $R\left[p_n^{\min}, p_n^{\max}\right]$ with condition $p_n^{\min} \leq p_n \leq p_n^{\max}$. Finally, choose any candidate randomly from the ith copies and place it in the appropriate position of jth cat.

At the tracing mode, the ith cat will start chasing the target according to the updating of velocity as defined in Eq. (6).

$$v'_{in} = v_{in} + c_1 * r_1 * \left(p_{gn} - p_{in}\right) \tag{6}$$

Let p_{in} and v_{in} be the position and velocity of the ith cat with nth dimension, c_1 be the constant, and r_1 be the uniformly distributed range of random numbers [0, 1]. Then, the position of ith cat will get updated as stated in Eq. (7).

$$p_{in} = \begin{cases} round\left(p_{in} + v'_{in}\right) & if\ round\left(p_{in} + v'_{in}\right) \in R_n \\ p_{in} & otherwise \end{cases} \tag{7}$$

According to the evaluation of Pareto front from the newly built ith cat fitness, further updating is done until the current solution is non-dominating.

3.2 Decision Prediction Classifier Model

The K-NN classifier model is proposed for classification of Parkinson disease. The patient data containing incomplete patients with p assessment. Assume the patient data d_i consisting of class information represented as $\left\{\omega_1, \omega_2, \ldots, \omega_p\right\}$. To estimate the d_i^g and ω_g class related to the training set T^g in which each patent fits in to ω_g. So, the incomplete patent of k-NN can be expressed as given in Eq. (8).

$$T_i^g = \left\{t_1^g, t_2^g, \ldots, t_k^g\right\} \subset T^g \tag{8}$$

Let the parameter t_k^g denote the kth neighbor of the pattern d_i belonging to training subset T^g. As a result, the total p assessment of the pattern d_i can be defined by d_i^g. To obtain the class estimation ω_g with a bigger distance will lead to a discounting factor with weight function as defined in Eq. (9).

$$\alpha_k^g = \frac{e^{-||d_i - t_k^g||}}{\sum_{k=1}^{K} e^{-||d_i - t_k^g||}} \tag{9}$$

where $||.||$ denotes the Euclidean distance. After obtaining the weight α_k^g and class ω_g of each neighbor by the k-NN classifier, the missing pattern of ω_g can be estimated as given in Eq. (10).

$$d_{io}^g = \sum_{k=1}^{K} \alpha_k^g . t_{ko}^g \tag{10}$$

where d_{io}^g denotes the oth missing attribute of d_i pertaining to complete attribute in t_k^g. Based on the complete estimation vector, the sub-classification result of the classifier model over p pieces could result in d_i as given in Eq. (11).

$$P_i^g = \Gamma(d_i^g | T) \tag{11}$$

Let Γ be the standard classifier model exploited to classify the d_i^g, P_i^g be the output classification obtained in the form of probability value of d_i^g.

To measure the performance of k-NN classifier models, evaluation metrics like sensitivity, specificity, and accuracy are measured as defined in Eqs. (12), (13), and (14), respectively.

$$C_{\text{Sensitivity}} = \frac{T^P}{T^P + F^N} \tag{12}$$

$$C_{\text{Specificity}} = \frac{T^N}{T^N + F^P} \tag{13}$$

$$C_{\text{Accuracy}} = \frac{T^P + T^N}{T^P + F^P + T^N + F^N} \tag{14}$$

where T^P, T^N, F^P, and F^N represents the number of true positive, true negative, false positive, and false negative obtained during classification. In future, research study can be enhanced with behavioral learning negotiation framework for choosing the best and cost-effective healthcare service by negotiating with multiple healthcare service providers [11, 12].

4 Experimental Evaluation

Experimental implementation of ECSO feature selection and k-NN classifier model is implemented using python coding. In order to verify the effectiveness of the proposed healthcare system, benchmark dataset containing extracted gait features are taken from the public repository [13]. The dataset consists of a 10 s accelerometer measurement of Parkinson disease patients taken by keeping their hands in the position of postural tremor. Also, their arms are outstretched with palms facing in upward direction. Afterwards, this dataset is supplied to k-NN classifier model by extracting the necessary features using ECSO approach for further training purpose. After training, the new patient's sample is given at run time that acts as a testing dataset. Then, the model will predict whether the new patients' samples are affected by Parkinson or not. The observed results of various classifier models with respect to prediction time and prediction accuracy are given in Table 1.

The obtained result shows that the proposed k-NN classifier model with ECSO feature selection method takes very less prediction time over the existing classifiers. Similarly, the proposed k-NN classifier model outperforms the existing classifier models with respect to prediction accuracy. This type of disease severity level prediction could bring some impact on diagnosing the Parkinson disease. Moreover, the proposed cloud-based Parkinson disease prediction system presented a prediction model with the aid of machine learning to make improvement over the human accuracy and scope of early detection. It can be more useful in developing rapid prediction models that can help a doctor to minimize the long process of diagnosis and also eradicate the human error. In future, a hybrid wrapper feature selection method can be used to improve the performance of proposed classifier model to implement in the real-time tele-monitoring healthcare applications. Therefore, this research study provides disease identification from remote places, and also diagnoses the disease with low cost investment by saving doctors risk and increasing patient appointments.

Table 1 Performance of classifier models

Classifier models	Feature selection	Prediction time (Seconds)	Prediction accuracy (%)
Proposed k-NN	PCA	69	84.56
	LDA	64	91.65
	ECSO	62	93.82
NB	PCA	85	74.34
	LDA	82	79.20
	ECSO	78	81.57
SVM	PCA	89	81.77
	LDA	82	83.20
	ECSO	76	90.98

As per observation, the performance of various classifiers were measured with respect to the disease prediction time as shown in Fig. 2. It is very much clear from the graph, the proposed k-NN classifier with ECSO feature selection method takes very less prediction time compared to existing NB and SVM classifier models with PCA and LDA feature selection methods. Since the proposed ECSO feature selection method arrives to optimally possible solutions available in the space while comparing to any other methods in terms of multiple objectives. Similarly, the performance measurements of various classifiers were made with respect to the disease prediction accuracy as shown in Fig. 3. More evidently, the performance graph shows that the proposed k-NN classifier with ECSO feature selection method obtains high prediction accuracy against the existing classifier models with PCA and LDA feature selection

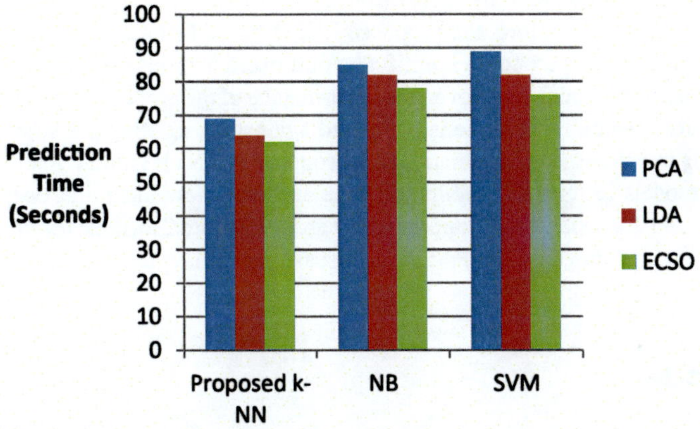

Fig. 2 Cloud-based Parkinson disease prediction system

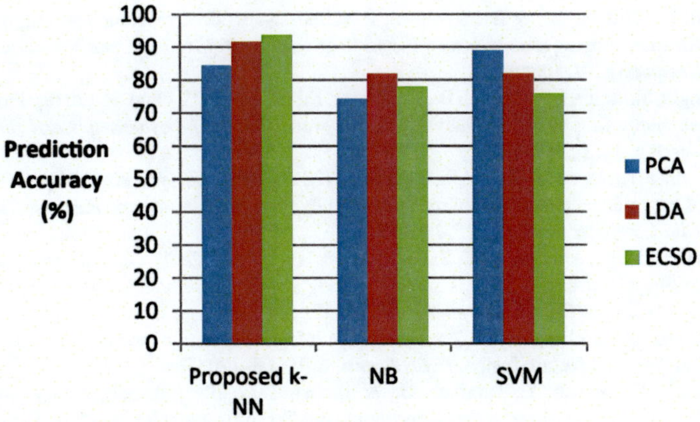

Fig. 3 Cloud-based Parkinson disease prediction system

methods. This dramatic improvement is achieved by the proposed classifier model due to appropriate management of medical data containing incomplete patents with an assessment.

5 Conclusion and Future Works

The proposed research study presents the cloud-based Parkinson disease prediction system to identify and diagnose the various stages of Parkinson disease. In order to minimize the misclassification of disease and to assist the neurologist, a k-NN classifier model is explored. This research problem needs to handle the multiclass classification and quantify the Parkinson disease severity stages over the higher dimensional data. Therefore, the proposed ECSO-based feature selection technique from the medical data set could identify the optimized features for directly applying in k-NN classifier model. Moreover, the performance of the proposed classifier model with feature selection technique is compared against the existing classifier models. According to experimental evaluation, the proposed k-NN classifier outperforms the existing models in terms of prediction time and accuracy. In future, deep feature selection and Bayesian-based approaches can be explored to rank the features of original data and diagnose the Parkinson disease problem.

References

1. Gondim, I. T. G. O., Souza, C. C. B., Rodrigues, M. A. B., Azevedo, I. M., Coriolano, M. G. W. S., & Linsa, O. G. (2020). Portable accelerometers for the evaluation of spatio-temporal gait parameters in people with Parkinson's disease: An integrative review. *Archives of Gerontology and Geriatrics, 90*, 104097.
2. Balaji, E., Brindha, D., & Balakrishnan, R. (2020). Supervised machine learning based gait classification system for early detection and stage classification of Parkinson's disease. *Applied Soft Computing, 94*, 106494.
3. Asuroglu, T., Acıcı, K., Erdaş, C. B., Toprak, M. K., Erdem, H., & Ogul, H. (2018). Parkinson's disease monitoring from gait analysis via foot-worn sensors. *Biocybernetics and Biomedical Engineering, 38*(3), 760–772.
4. Tejeswinee, K., Jacob, S. G., & Athilakshmi, R. (2017). Feature selection techniques for prediction of neuro-degenerative disorders: A case-study with Alzheimer's and Parkinson's disease. *Procedia Computer Science, 115*, 188–194.
5. AlMamun, K. A., Alhussein, M., Sailunaz, K., & Islam, M. S. (2017). Cloud based framework for Parkinson's disease diagnosis and monitoring system for remote healthcare applications. *Future Generation Computer Systems, 66*, 36–47.
6. Li, Y., Dai, W., & Zhang, W. (2020). Bearing fault feature selection method based on weighted multidimensional feature fusion. *IEEE Access, 8*, 19008–19025.
7. Tumar, I., Hassouneh, Y., Turabieh, H., & Thaher, T. (2020). Enhanced binary moth flame optimization as a feature selection algorithm to predict software fault prediction. *IEEE Access, 8*, 8041–8055.

8. Uddin, S., Khan, A., Hossain, M. E., & Moni, M. A. (2019). Comparing different supervised machine learning algorithms for disease prediction. *BMC Medical Informatics and Decision Making, 19*(281), 1–16.
9. Ozcift, A. (2012). SVM feature selection based rotation forest ensemble classifiers to improve computer-aided diagnosis of Parkinson disease. *Journal of Medical Systems, 36*, 2141–2147.
10. Ye, F. (2018). Evolving the SVM model based on a hybrid method using swarm optimization techniques in combination with a genetic algorithm for medical diagnosis. *Multimedia Tools and Applications, 77*, 3889–3918.
11. Rajavel, R., & Mala T. (2021). Agent-based automated dynamic SLA negotiation framework in the Cloud using the stochastic optimization approach. *Applied Soft Computing, 101*, 107040.
12. Rajavel, R., & Mala, T. (2016). Adaptive probabilistic behavioural learning system for the effective behavioural decision in cloud trading negotiation market. *Future Generation Computer Systems, 58*, 29–41.
13. Devpost. (2017). Parkinson's telemonitoring—monitor severity of patients' Parkinson's tremors through their mobile phones. *MedHacks*. https://devpost.com/software/parkinson-s-telemonitoring

Electric Vehicle Monitoring System Based on Internet of Things (IoT) Technologies

Yogesh Mahadik, Mohan Thakre, and Sachin Kamble

Abstract Electric vehicles (EVs) are emerging as a preferred way to reduce environmental concern's needs. Concern and energy insufficiency, and in the foreseeable term, this pattern is expected to grow. However, this inadequate charging infrastructure is now a major barrier to the adoption of EVs. Deployment of this infrastructure is expected to maximize the adoption of EVs to promote community access. Connectivity between charging substations (CS) is therefore mandatory. Entertainment real-time status of CSs can provide useful information, such as availability of charging provisions, reserves, and time to meet the SC. The purpose of this paper is to have a better solution related to EV charging mechanism leveraging the benefits of the Internet of Things (IoT) technologies. The IoT is a model that gives the present facilities a real-time worldwide communication view of the physical world employing the sensors and the transmitting networks. This article suggests a real-time server-based prediction of EV infrastructure.

Keywords Charging substations · Forecasting of system · EV charging · Internet of things

1 Introduction

Countries around the world have introduced in recent years regulatory criteria for solving transport problems, such as the increasing non-renewable fuel demand and metropolitan emissions. So, a lot of attention was paid to eco-friendly EVs due to availability issues of fossil fuel, climate, and pollution control. Many countries have laid down essential criteria for facilitation of EV extension and approval [1, 2]. For instance, several North American cities are in the evolving time to easily substitute hybrid plug-in conventional cars and EVs [3], then. EVs have great cost advantages and convenience. EVs have ease, in comparison to internal combustion engine (ICE)

Y. Mahadik (✉) · S. Kamble
Government Polytechnic Malvan, Malvan, MH, India

M. Thakre
K. K. Wagh Institute of Engineering Education and Research, Nashik, India

P. Nayak et al. (eds.), *IoT and Analytics for Sensor Networks*, Lecture Notes in Networks and Systems 244, https://doi.org/10.1007/978-981-16-2919-8_28

automobiles, and maintenance cost reduces for an EV [4]. The range of EVs also expands faster than in the past. The lightweight Mahindra Reva car with 16 kWh power will have 80 miles of driving range. In Model Tesla X (SUV) vehicles, the mileage with a power of 94 kWh is 400–450 km. Figure 1 indicates a block diagram of the system. Though the driving ranges of the EVs depend on the capability of the battery, the complexity of the powertrain plays a significant part. New recent study concentrates on improving driving battery control and management systems [5]. EVs spectrum needed a detailed review on the Li-ion batteries used. In [6], discussed with EVs are the fundamental principles of service, assembly, and the efficiency of several Li batteries. In [7], fast CSs can load battery from 20% to about 80% of the state of charge level (SOC) 30 min. Naomi et al. studied the best loader mode in [8]; this ensures quality and minimal time for charging electric Cars and Hybrid Vehicles (HEVs). SOC is one of the major players in delivering the present battery condition for safe charging and discharge that increases battery life in turn [9, 10]. Different methods of SOC estimation are discussed in [11]. In the Lobby Automotive business, intelligent communication networks (1) vehicle-to-infrastructure (V2I), (2) vehicle-to-driver (V2D), and (3) Vehicle to vehicle (V2V) are the types of networks. Due to major government expenditures, connectivity systems such as V2V and V2I were well studied. The key purpose of using these communication mechanisms is to minimize congestion of roads, preservation of road transport [12], and avoid collisions with cars [13, 14]. However, few works confirm the reliability of vehicle-to-device systems in an EV, and additional work is required in this field of study. The data correspondence deliberately serves the EV customer and the CS effective power control scheduling. The SOC calculation information to assign the slot is transmitted to the closest CS based on grid power demand. In [15, 16], the approximate SOC information is sent to three separate operators, to exchange the EV information, the supplier of charging facilities (CSP), delivery system retailer (RET), and Operator (DSO). EV user info is transmitted to an automatic charge controller infrastructure of connectivity such as the Internet or cellular networks [17]. After all, information has been obtained; the charge controller has an extra degree of flexibility to tax the EV. In [18], the article aggregator has a precondition to alert the aggregator besides the planned departure with the available SOC period when a car is attached in or out. The lightweight charging of EVs was carried out in [19]. Optimum online preparation was introduced in [20] using a cost-responsive fee method. Thereby, the EV plug-in pattern can be explored by the aggregator. Along with the advice of the CS, the algorithm suggested can alert the EV driver to the safe SOC limit, and a detailed calculation of the EV charging time can be given. For EV, the geographical positioning and path views of various available CSs are required. The disadvantages of the current EV Charging methods are – EV location, user phone number, information of the vehicle, and sufficient charging power can trigger different. EV load estimation is a crucial problem for efficient operation and energy business executives. EV charging has been advised based on an accurate range predictor system in real-time. Not in the static, the device cannot provide the driver with any details of CS, i.e., CS works or does not work, how many cars are waiting for charging, and what charging points are available. From the above, it was observed that an application is in real-time Battery status EV needed to collect transportation details and the CS status for the estimation of future CS.

This work suggests a server-based forecast in real-time Application: (i) the administration of scheduling to avoid waiting time; and (ii) have a CS recommendation in real-time.

Economic risks, lowered charges for night time EVs, these benefits relative to other approaches—this is what we are talking about. The PHP programming generates a synergetic application language of the operating system Linux UBUNTU 16.04 LTS and cloud processes and handles all related details of Google cloud platform structured query language (CSQL). The utility of this application is also verified by a low-cost LTC 4150, Wi-Fi ESP 8266 trial device, and Adriano.

2 System Description

The emphasis of this paper is on the IoT aspect of deciding the value of the SoC and data transfer to IoT networks. The data can be accessed inside the App. The user can also find the Locations of the nearest charging station using the smartphone. The user once knowing the state of his car battery, can quickly assess whether to continue with the supply of power to the grid or based on the tariff rates, sinks electricity from the grid. There will be various prices for selling electricity to the grid and power is taken from the grid. Bidirectional, the grid would have converters for power transfer [5, 6]. Some grids are available, which uses solar energy as a source as well [7]. Figure 1 indicates a block diagram of the system.

The significant and visible connectivity, sensors, and small devices are all IoT features. New enabling networking technologies, especially IoT networking is a kind of network that is not connected to major network suppliers [8]. Small and large networks are created by IoT. IoT uses sensors to collect the data from its machine modules over a wide area. It is the primary working machine, sensor or controller is used. More controllers, such as Arm Mbed and Arduino, achieve more in the area of IoT, such as phones or tablets. The output or outcomes are used to view the output that minimizes the effort to get the details.

2.1 EV Load Forecasting and SOC Measurement

Load prediction is an important process that can improve distribution network performance. The precision of load forecasting has an immense effect on the functioning of an organization. In general, the cost depends on the peak request. During a precise time, a particular CS may have additional power demand. It must be noted that the unit cost is expressed in kilowatt-hour rupees (Rs/kWh), where the rupee is India's official currency.

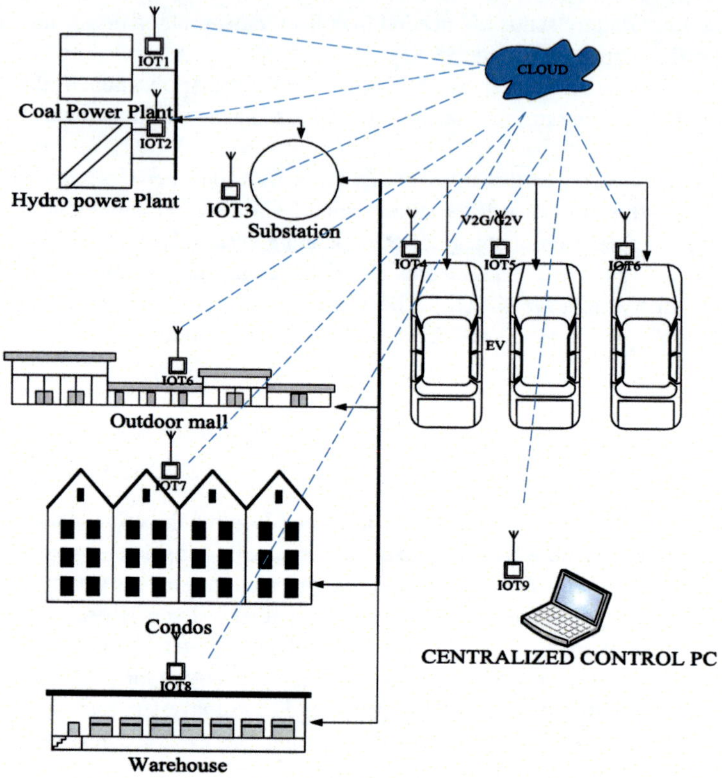

Fig. 1 Block diagram of IoT-based V2G and G2V system

The communication network proposed would work to minimize the variations. It updates the user about the accessibility of cost and slots for each CS. A consumer can therefore prefer a station where there is local supply. Of course, if intimated, the accessibility of slots will prevent congestion over a precise period impressively during peak hours at the CSs. Therefore, inadequate number of CSs can lead to load forecasting complications in the delivery of energy to EV consumers. The energy delivery crisis can arise from the debate above and affect the recipient of the EV. A stronger battery has been able to solve this problem. Therefore, the battery SOC is to be tracked. Evaluation of the SOC is a big concern for the use of the battery. The available SOC of a battery provides main control method parameters [21–23]. The output of the battery is calculated by the SOC so that the exact SOC calculation can improve life and also save energy by activating the device intelligent management strategy [24]. But the important thing is the battery problem that is driven by chemical energy, which cannot be directly measured. This contributes to the difficulty in the SOC calculation.

There are several methods for determining a battery's SOC Setting [10, 11]. These techniques can be categorized further as: (I) electrochemical, (II) adaptive, (III) Electrical. The electrochemical techniques are extremely precise but need access to the chemical composition of the battery. So, the technology is difficult to introduce. Techniques of adaptation need a circuit model equivalent and solution procedure battery. The precision of the adaptive techniques is determined by the equivalent model efficiency. Electrical techniques only need measurable parameters like this present load/to unload terminal voltage and internal resistance. Because of its simple application and low complexity, Coulomb counting (electrical) is one of the most common SOC estimation techniques. This approach is known as well as the new integration technique of ampere-hour (Ah). An algorithm is complex and easy to deploy. The current that is charged or is measured in a cell is known as the current integration technique. If the first SOC is known, correct SOC can then be obtained [25, 26]. The current readings are incorporated mathematically, SOC values can be used to calculate as normal the usable SOC is found by measuring the moving charge. When determining the battery cell, say, it has 60 Ah, 216,000 A seconds can be estimated. Coulomb count 0% SOC thus means battery counts 0 C and 100% SOC is a count of the 216,000, which is used for calculating the charge rate. The method of counting coulomb is simple and of low complexity. In theory, the procedure for coulomb counting is incorporation and charging quantification by detecting the battery's current input and output. It operates by integrating the current active over time to measure the cumulative amount of battery power input or output. This leads to a capacity measured usually in ampere-hours. The method is sound, as long as it is consistent with current measurements, it can be used on all the EV device batteries [27, 28]. The instrument tracks the current through an external resistor between the positive end of the battery and the load. On the Volt-Frequency converter, it transforms the present sensory stress through several interrupt pin output pulses. These pulses fit at a fixed charge quantity to or from the battery. The device has an Arduino connection. The exit of the digital format is available for Arduino codes. Surveillance of the SOC is important because the remaining is used to measure energy and the maximum possible driving range to achieve the best output (speed or distance).

3 Proposed IoT-Based EV Management

By the year 2030, the GSM Association has estimated that every EV will be connected, and this will earn living, earnings of over 19B euros. Several criteria and standards have been set in recent years. Now technologies have emerged for V2V and V2I communications. Various forms of communications already available for exchange are V2G contact to achieve a successful EV charging based on standards of the Society of Automotive Engineers (SAE), the International Association for Standards (ISO), and Proposals from the International Electrotechnical Commission (IEC) [29, 30]. Several standards have been defining different communication

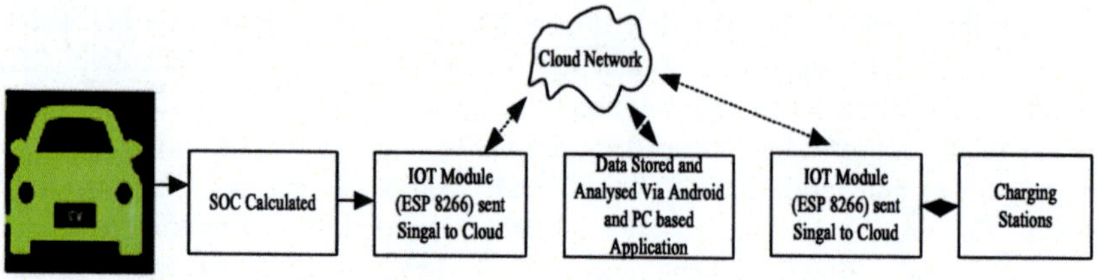

Fig. 2 EV management unit

processes. The basics are different norms, such as charging contact costs, protocols, connector charging, protection, and charging topology [31, 32]. The developed application has three divisions (web, android), (i) SOC estimation, (ii) CS selection, and (iii) database selection and the processing of data. The total planned flowchart of the scheme is portrayed in Fig. 2.

The Coulomb counter of the LTC is related to the Adriano and this will measure the battery's SOC. The information read by the laptop or PC will be stored in the server database using the IoT Wi-Fi module ESP8266. It will run via an Internet protocol at the same time. Depending on the request of the user, the web page will display the CSs positions with the aid of GPS. Based on availability, the user will have the choice of slots, charging costs, and distance, choosing a specific CS. Once the request is received by the CS from the EV, the charging slot will be assigned based on the requirements for the infrastructure for EV charging. This system is built to ensure the ideal EV charging, and it would also boost the grid by avoiding peak charges.

3.1 Test Method for Calculating SOC

The experiment is conducted with a battery of 9 V, 300 mAh Li-ion, Arduino Uno, ESP 8266, and a 9 V DC motor with a coulomb counter. The Li-ion battery is currently rated at 300 mAh, which means, for 60 min, it can supply 300 mA, and this is said to be the battery cell's initial charge. The load on the motor is 70 mAh, and with the charge available, it can run for 4.28 h. Now the price for the battery charge is given by:

$$\mathrm{Cc} = \mathrm{Rp} * \mathrm{E_c(t)} \tag{1}$$

where, $\mathrm{E_c(t)}$ is the cost of energy at a given hour.

If the motor is running for 1 h, so 70 mAh will be consumed.

The remaining SOC value, therefore, is 76.66% of the observed capacity, it is 0.63 kWh.

3.2 Algorithm of CS Selection

Relevant information will be given by the CS selection algorithm to the end user's EV. When engaging the charging points in the CSs with the EV, the points convey data to the server. Large networks known as network service providers (NSPs) help in sharing of knowledge between CSs and the server. The NSP is linked to exchanges in the metropolitan area (MAEs) via the access points of the network (NAP's). The initial NAP's were the internet links the dots. The references to both NAPs and MAEs are as points for internet sharing (IXs). Following the data, the domain name system will be passed through (DNS). The DNS records the names of the computers and their corresponding IPs on the internet, emails. Finally, to guarantee the transferred data stays private, it uses a stable socket layer (SSL). In the meantime, the EV position information is tracked via GPS. A significant parameter is the battery drain rate. To calculate the SOC that is the available form of Coulomb counting research. It is used to measure the rate of battery drain and the SOC available at routine intervals. With the open SOC, it is possible to calculate the remaining distance which the EV will cover. The results are stored in the localhost and the same will be communicated with the application that was created. The position of EVs and CSs are carried to the application using the Maps and the API. An API is a code that makes two distinct APIs computer applications for communicating with one another using a hypertext transfer protocol to access the API (HTTP). The website uses protected information to ensure the security of the information Protocol of hypertext transfer (HTTPS). Therefore, everything in the application has reviews from the CSs and the EVs. An Index of the Availability of the slot is given, and this will be compared with the remaining power of the battery to find the optimized CSs.

Below is the step-by-step method.

Step 1: Dynamically collecting the input from the CSs and EVs for a one-minute cycle.

Step 2: The percentage of the SOC is the key input from which the battery will calculate the drain rate (BDR).

Step 3: The approximate distance to empty, based on the BDR (DTE). As given in Eq, it will be determined.

Capacity of the battery (B_c) = X (kWh)

$$\text{Approximate Range } (E_r) = Y \text{ (km/kWh)} \tag{2}$$

$$(\text{DTE}) = \text{km } (X * Y * \text{SOC}) \tag{3}$$

Step 4: The overall distance measured (DTE) will be that with the available SOC and any EV up to CS, the EV will drive inside the limit will be drawn.

Step 5: Then in the CS, an index of availability (AI) will be retrieved for each station.

Step 6: Both the DTE and AI details are taken into consideration and optimized, as seen in the sample as shown in Fig. 3.

Fig. 3 Optimal point finding

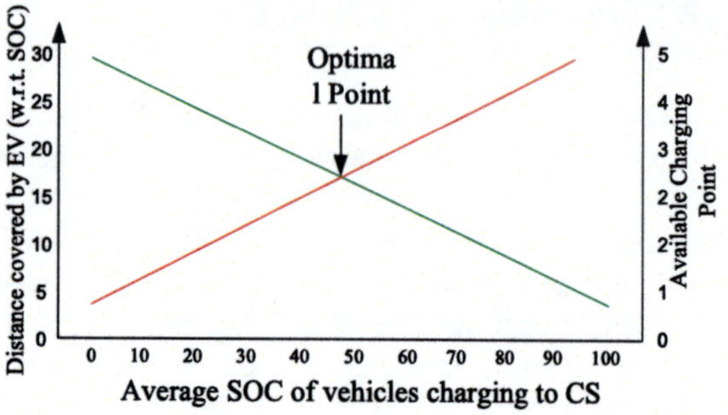

Step 7: The EV will fly up to a maximum distance of 27 km SOC Remaining. The battery would be empty at the average speed, 40 min, and the EV would not be able to move. In the meantime, the availability of charging slots also varies with the individual CS.

Step 8: The booking option and all related details of the booking option CS are shown for each CS, this process is repeated.

Step 9: The station will be provided with the lowest optimal point as the best choice, and in ascending order, the results will be sorted. Now as needed, the EV user can pick the CS. (Note: The calculation of the total time is given in the appendix for two cases in which a user chooses a CS based on the shortest distance required).

It should be noted that if an EV user hits the booked user, CS will use reserve points approved by the CS for EV earlier, no-booking, or early-arrival users. Such points of reserve support the CS with quick chargers in the event of an emergency and anomalies. Non-flexible booking drivers can pay, however, using the program rather than drivers. Time-based arrival in such instances uses the priority (ATP) algorithm. ATP enables the Once vehicles to arrive, they enter service, and therefore reduce the wait. It takes time and prevents the charger from being idle. Lineups for ATP algorithms continued until all the charging points are connected by all the vehicles. Engaged based on the vehicles arrival time. Once a vehicle is present, the algorithm, assigned to a charging point, updates the waiting points. For other cars, algorithm will repeat with the same procedure before it charges all the cars. Figure 4 reveals the Flowchart of the algorithm of ATP, the EV charging plan.

3.3 Applications

(i) Provide schedule management to reduce waiting time; and (ii) provide real-time management; CS adviced for less distance and shortened charging time. Moreover, the proposed scheme prevents interference by third parties and preserves the anonymity and complexity of EV users. Exchange of information between the customer and the CS. End-users can use the CS conveniently based on their requirements for this. This synergistic application is built in the programming language

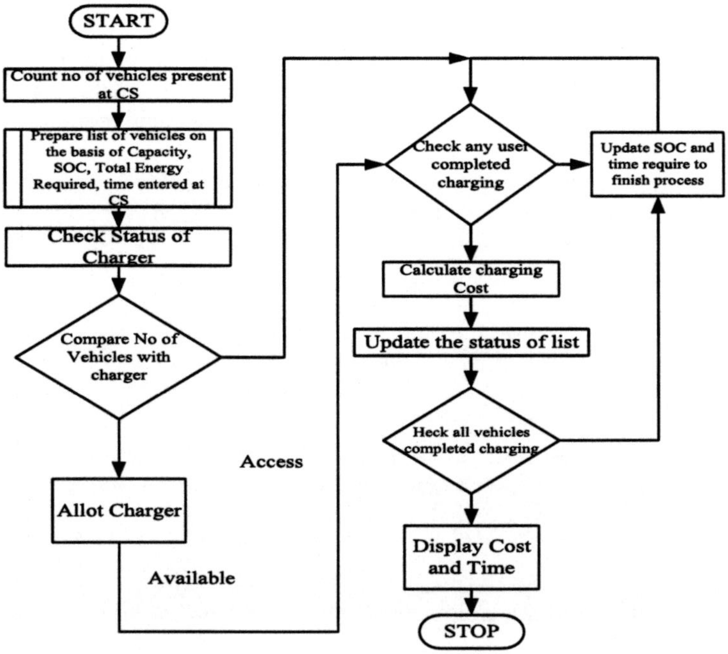

Fig. 4 ATP-based algorithm

of PHP, Linux UBUNTU 16.04 LTS operating system, and all related information shall be stored and handled with Cloud Structured Query Language (CSQL) from the Google cloud platform.

4 Conclusion

Increasingly, the IoT paradigm is integrated with real-world applications, particularly in the real world, in the field of electricity. In this article, a real-time solution is suggested that, the charge scheduling of EVs should be improved based on modern state-of-the-art with reliable communication between vehicles and CSs-Technologies-the-art. More protection is offered by the proposed scheme, it is possible to stop unnecessary rush in charge stations. The user will receive the data of all nearby CSs based on various variables, such as consumer comfort, electricity variations, the user's ability to pay, and prices This robust framework not only is useful for EV users, but also CSs to predict future loading, congestion avoidance, and energy consumption with reducing and/or balancing loads. To ensure the stability of the grid, V2V, and V2V in future work, the developed application can be integrated into V2G. The introduction of smarter grids has been slowed down. The mentality is that we need to provide some form of contact from each unit back to the operator of the grid. Furthermore, the efficient processing of data from both EVs and CSs takes time. To achieve a solution that is optimized, the appliances, therefore, by linking to a web page anywhere, it can be made intelligent for that client, and then a computer is programmed to respond to that client, also suitable for the grid operator to better balance the load and grid.

The complexity and computation pressure of this centralized approach would exponentially increase due to the rising number of EVs. Since the cellular network is inexpensive, it can cause over-congestion that will degrade the consistency of the relationship.

References

1. Thomas, C. E. S. (2009). Transportation options in a carbon-constrained world: Hybrids, plug-in hybrids, biofuels, fuel cell electric vehicles, and battery electric vehicles. *International Journal of Hydrogen Energy, 34*(23), 9279–9296.
2. Dickerman, L., & Harrison, J. (2010). A new car, a new grid. *IEEE Power Energy Magazine, 8*(2), 55–61.
3. Vrazic, M., Vuljaj, D., Pavasovic, A., & Paukovic, H. (2014). Study of a vehicle conversion from an internal combustion engine to electric drive. In *IEEE International Energy Conference* (pp. 1544–1548).
4. Dost, P. K-H., Spichartz, P., & Sourkounis, C. (2018). Charging behavior of users utilizing battery electric vehicles and extended-range electric vehicles within the scope of a field test. *IEEE Transactions on Industry Applications, 54*(1).
5. Hannan, M. A., Hoque, Md. M., Hussain, A., Yusof, Y., & Ker, P. J. (2018). State-of-the-art and energy management system of lithium-ion batteries in electric vehicle applications: Issues and recommendations. *IEEE Access, 2018*, 19362–19378.
6. Navid, R., Kashanizadeh, B., Vakilian, M., & Barband, S. A. (2013). Optimal placement of fast-charging station in a typical microgrid in Iran. In *International Conference on the European Energy Market* (pp. 1–7).
7. Mohamed, N., Aymen, F., Mouna, B. H., & Alassaad, S. (2017). Review on the autonomous charger for EV and HEV. In *International Conference on Green Energy Conversion Systems, 2017*, 1–6.
8. Ramadan, H. S., Becherif, M., & Claude, F. (2017). Energy management improvement of hybrid electric vehicles via combined GPS/rule-based methodology. *IEEE Transactions on Automation Science and Engineering, 14*(2), 586–597.
9. Gholizadeh, M., & Salmasi, F. R. (2014). Estimation of state of charge unknown nonlinearities and state of health of a lithium-ion battery based on a comprehensive unobservable model. *IEEE Transactions on Industrial Electronics, 61*(3), 1335–1344.
10. Tang, X., Liu, B., & Gao, F. (2017). State of charge estimation of LiFePO4 battery based on a gain-classifier observer. In *Energy Proceedings-105* (pp. 2071–2076). https://doi.org/10.1016/j.egypro.2017.03.585.
11. Di Lecce, V., & Amato, A. (2011). Route planning and user interface for an advanced intelligent transport system. *IET, 5*(3), 149–158.
12. Taleb, T., Benslimane, A., & Letaief, K. B. (2010). Toward an effective risk-conscious and collaborative vehicular collision avoidance system. *IEEE Transactions on Vehicular Technology, 59*(3), 1474–1486.
13. Barrios, C., Motai, Y., & Huston, D. (2015). Trajectory estimations using smartphones. *IEEE Transactions on Industrial Electronics, 62*(12), 7901–7910.

14. The International Electrotechnical Commission (IEC) (2001) Electric vehicle conductive charging system—Part 1: General requirement. In *IEC 61851-1 the International Electrotechnical Commission.*
15. Clairand, J., Rodríguez-García, J., & Álvarez, B. C. (2018). Smart charging for electric vehicle aggregators considering users' preferences. *IEEE Access, 6*, 54624–54635.
16. Rotering, N., & Ilic, M. (2011). Optimal charge control of plug-in hybrid electric vehicles in deregulated electricity markets. *IEEE Transactions on Power System, 26*(3), 1021–1029.
17. Han, S., Han, S., & Sezaki, K. (2010). Development of an optimal vehicle-to-grid aggregator for frequency regulation. *IEEE Transactions on Smart Grid, 1*(1), 65–72.
18. Sundstrom, O., & Binding, C. (2012). Flexible charging optimization for electric vehicles considering distribution grid constraints. *IEEE Transactions on Smart Grid, 3*(1), 26–37.
19. Shaobing, Y., & Yang, S. (2019). Price-responsive early charging control based on data mining for the electric vehicle online scheduling. *Electric Power Systems Research, 167*, 113–121.
20. Xydas, E. S, Marmaras, C. E., Cipcigan, L. M., Hassan, A. S., & Jenkins, N. (2013). Forecasting electric vehicle charging demand using support vector machines. In *Proceedings of the 48th International Universities' Power Engineering Conference.*
21. Quiros-Tortos, J., Ochoa, L., & Lees, B. (2015). A statistical analysis of EV charging behavior in the UK. In *Innovative Smart Grid Technologies Latin America (ISGT LATAM) 2015 IEEE PES* (pp. 445–449).
22. Cai, Z. H., Liu, G. F., & Luo, J. (2010). Research state of charge estimation tactics of nickel-hydrogen battery. In *Proceedings of International Symposium Intelligence Information Processing and Trusted Computing* (pp. 184–187).
23. He, H. W., Xiong, R., & Guo, H. Q. (2012). Online estimation of model parameters and state-of-charge of LiFePO4 batteries in electric vehicles. *Applied Energy, 89*(1), 413–420.
24. Hu, X., Zou, C., Zhang, C., & Li, Y. (2017). Technological developments in batteries: A survey of principal roles types and management needs. *IEEE Power Energy Magazine, 15*(5), 20–31.
25. Maurya, D. S., Jadhav, P. D., Joshi, R. S., & Bendkhale, R. R. (2020). A detailed comparative analysis of different multipulse and multilevel topologies for STATCOM. In *International Conference on Electronics and Sustainable Communication Systems (ICESC)* (pp. 1112–1117). Coimbatore, India. https://doi.org/10.1109/ICESC48915.2020.9155708.
26. Hadke, V. V., & Thakre, M. P. (2019). Integrated multilevel converter topology for speed control of SRM drive in plug in-hybrid electric vehicle. In *Proceedings of the 3rd IEEE International Conference on Trends in Electronics and Informatics (ICOEI 2019)* (pp. 1013–1018). SCAD Clg. of Engg. & Tech., Tirunelveli, Tamil Nadu, India CFP19J32-ART; ISBN: 978-1-5386-9439-8.
27. Mahadik, Y. V., & Vadirajacharya, K. (2017). Battery life enhancement using hybridization of battery and UC. In *International Conference on Circuit, Power and Computing Technologies (ICCPCT).* https://doi.org/10.1109/iccpct.2017.8074393.
28. Mahadik, Y., & Vadirajacharya, K. (2019). Battery life enhancement in a hybrid electrical energy storage system using a multi-source inverter. *World Electric Vehicle Journal, 10*(2), 17. https://doi.org/10.3390/wevj10020017
29. Vaibhav, S. G., & Mohan, P. T. (2019). A novel 4-level converter for switched reluctance motor drive in plug-In HEVs. In *IEEE International Conference on Intelligent Computing and Control Systems (ICICCS 2019)* (pp. 626–631). Vaigai College of Engineering, Madurai, Tamil Nadu, India. ISBN: 922-1-7281-2214-6.
30. Mahadik Y. V., Yeole D. S., & Chowdhary P. K. (2020). Fast charging systems for the rapid growth of advanced Electric Vehicles (EVs). In *International Conference on Power, Energy, Control and Transmission Systems (ICPECTS), Chennai, India* (pp. 1–6). https://doi.org/10.1109/ICPECTS49113.2020.9336979.
31. Chowdhary, P. K., & Thakre, M. P. (2020). MMC-based SRM drives for hybrid-EV with decentralized BESS in battery driving mode. In *International Conference on Power, Energy, Control and Transmission Systems (ICPECTS)* (pp. 1–6). Chennai, India. https://doi.org/10.1109/ICPECTS49113.2020.9337029.

32. Mane, J., & Hadke, V. (2020). Performance analysis of SRM based on asymmetrical bridge converter for plug-in hybrid electric vehicle. In *International Conference on Power, Energy, Control and Transmission Systems (ICPECTS)* (pp. 1–6). Chennai, India. https://doi.org/10.1109/ICPECTS49113.2020.9337059.

Dynamic Analysis and Projective Synchronization of a New 4D System

M. Lellis Thivagar, Ahmed S. Al-Obeidi, B. Tamilarasan, and Abdulsattar Abdullah Hamad

Abstract A new 4D dissipative hyperchaotic system with an unstable equilibrium point is introduced. The proposed system consists of ten terms including three quadratic nonlinearities which constructed through using a nonlinear state control algorithm in the known Lorenz system, exhibits self-excited attractors and chaotic system attractors, with two positive derivations of Lyapunov. The dynamical properties of this system are analyzed using theoretical and numerical simulations based on equilibrium points, stability, dissipative, Lyapunov exponents, and phase portrait. Besides, various coexisting attractors or multistability with different initial conditions under the same parameters are investigated. Furthermore, Projective Synchronization (PS) of an identical proposed system is realized by nonlinear control strategy and Lyapunov stability theory.

Keywords 4D hyperchaotic system · Self-excited attractors · Multistability · Projective synchronization

1 Introduction

The attractors of nonlinear dynamic systems have attracted a lot of attention from researchers due to their possible control and optimization applications [1, 2], electronic circuits [3–6], conforming dynamics [7], and encryption [8–11]. One of the important applications in secure communication is messages with simple chaotic systems are not always safe. To overcome this problem, a higher n-dimensional system ($n \geq 4$) with more than two positive Lyapunov exponents are required to increase randomness and unpredictability [12–17].

Since 1963, Lorenz system is the first physical of a 3D chaotic attractors, and other 3D chaotic systems were discovered such as the systems of Chen (1999) Lu

M. L. Thivagar · B. Tamilarasan · A. A. Hamad (✉)
Madurai Kamaraj University, Madurai, India

A. S. Al-Obeidi
Specialty of Mathematics, Gifted School of Nineveh, Nineveh, Iraq
e-mail: ahmedsedeeq@uomosul.edu.iq

P. Nayak et al. (eds.), *IoT and Analytics for Sensor Networks*, Lecture Notes in Networks and Systems 244, https://doi.org/10.1007/978-981-16-2919-8_29

(2002), Liu (2004), Pan (2009), all of these mentioned systems contain only one positive Lyapunov exponent (LE) [18–20]. Rössler system is considered the first 4D hyperchaotic system with two positive LE which was introduced in 1979. The 4D hyperchaotic systems are more complex and interesting compared with other chaotic systems because it contains two positive LE to take advantage of it in the secure communication. Several systems have been introduced with various types of the dimension: 4D [12–15], 5D [21–23], and 6D [24–27].

The exponent of these systems is calculated from Wolf Algorithm [27], and it plays an essential and effective role in classification of the attractors of these systems as periodic, quasi-periodic, chaotic, hyperchaotic, and chaotic with 2-torus. Every 4D system is possible to be chaotic or hyperchaotic attractors based on the sign of LEs. So, the sign of LEs (+, + , 0, −) refers to hyperchaotic, whereas the sign (+, 0, −, −) indicates to chaotic attractor. A hyperchaotic system with n-dimensional ($n \geq 4$) and n−2 positive Lyapunov exponent is considered the best system.

The above discussion motivated us to introduce a new 4D system. The following points summarize the results of the paper:

- A new 4D dissipative system with $n-2$ of positive Lyapunov exponent is constructed;
- The proposed system has self-excited attractors;
- Multistability of the new system is studied with various types of initial conditions; and
- Projective Synchronization (PS) of an identical proposed system is realized utilizing nonlinear control and Lyapunov stability theory.

2 Description of the System

Lorenz chaotic system is the first-ordinary differential equation which has three-dimensional depicted as:

$$\begin{cases} \dot{x}_1 = -ax_1 + ax_2 \\ \dot{x}_2 = cx_1 - x_1x_3 - x_2 \\ \dot{x}_3 = -bx_3 + x_1x_2 \end{cases} \tag{1}$$

In which $x_i, i = 1, 2, 3$ are variables, and a, b, c positive parameters, device (1) is chaotic at the parameters $a = 10, b = \frac{8}{3}, c = 28$. By adding nonlinear feedback controller strategy to Lorenz system, a new 4D system with (+, + , 0, −) sign of LEs is obtained and consists of ten terms including three nonlinearities which is depicted as:

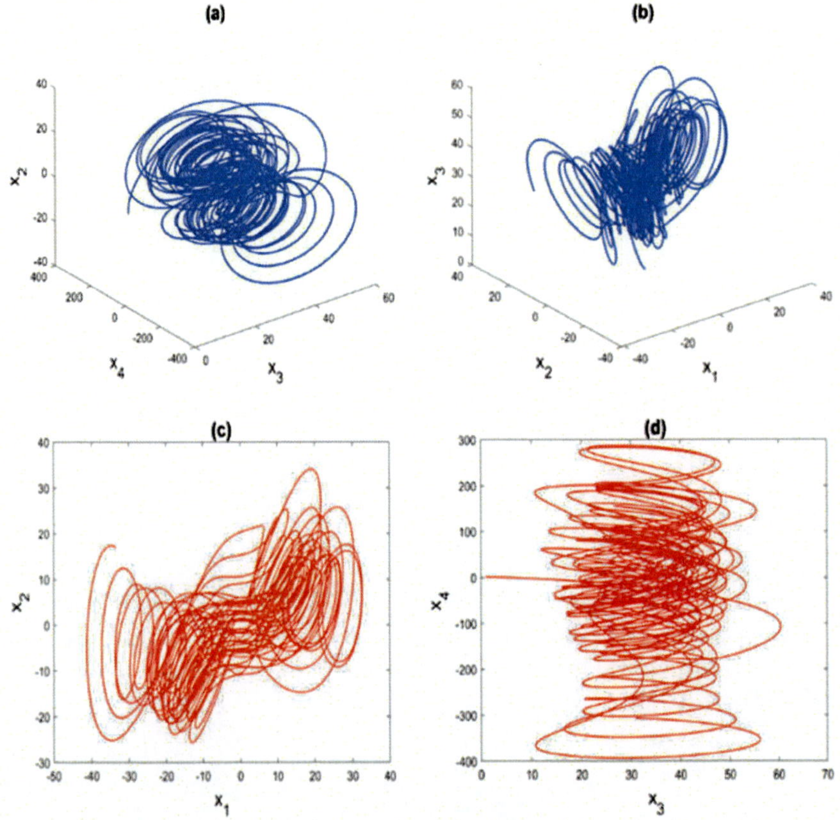

Fig. 1 Phase portrait of new system with a $= 10, b = \frac{8}{3}, c = 35, h = 2.39$

$$\begin{cases} \dot{x}_1 = -ax_1 + ax_2 + x_4 \\ \dot{x}_2 = cx_1 - x_1x_3 - x_2 \\ \dot{x}_3 = -bx_3 + x_1x_2 \\ \dot{x}_4 = hx_4 - x_1x_3 \end{cases} \tag{2}$$

In which $x_i, i = 1, 2, 3, 4$ are variables, and a, b, c are positive parameters, and h is the control parameter. Based on phase portrait, the attractors of system (2) are shown in Fig. 1.

2.1 Equilibrium Points, the Jacobian Matrix, and Divergence

The system (2) possesses three equilibrium points, only one is real as

$$\begin{cases} P_1(0,0,0,0) \\ P_2(-13.9721i, +6.7725i, 35.4847, -207.4866i) \\ P_3(+13.9721, -6.7725i, 35.4847, +207.4866i) \end{cases} \tag{3}$$

The Jacobian matrix, characteristic equation, and roots at P_1, and typical parameters are Eqs. (4), (5), and (6), respectively, as

$$J = \begin{bmatrix} -a & a & 0 & 1 \\ c-x_3 & -1 & -x_1 & 0 \\ x_2 & x_1 & -b & 0 \\ -x_3 & 0 & -x_1 & h \end{bmatrix} \tag{4}$$

$$(-b-\lambda)(\lambda^3 + (a+1+h)\lambda^2 + (a - h(a+1) - ca)\lambda + ah(c-1)) = 0 \tag{5}$$

$$\begin{cases} \lambda_1 = -8/3 \\ \lambda_2 = 2.39 \\ \lambda_3 = -24.7419 \\ \lambda_4 = 13.7419 \end{cases} \tag{6}$$

Since some roots are posiive, the equilibrium point P_1 is unstable, that means it is classified as self-excited attractors. The trace or divergence of system (2) is written as Eq. (7):

$$\nabla v = \sum_{i=1}^{4} \frac{\partial \dot{x}_i}{\partial x_i} = -a - 1 - b + h \tag{7}$$

So, system (2) is dissipative under the condition$-(a+1+b) > h$. The value of control parameter h plays an important role in divergence and $\nabla v = -11.2767$.

2.2 Lyapunov Exponents and Dimensions

The numerical simulation was implemented via Wolf Algorithm with MATLAB R2020a, under $a = 10, b = \frac{8}{3}, c = 35$ and $h = 2.39$, the initial value at $(0.1, 0.1, 0.1, 0.1)$ with step $= 0.5$ and tend $= 200$. The proposed system has (+, +, 0, −) sign of Lyapunov exponents and of these exponents are Eqs. (8) and (9), respectively, as

$$\begin{cases} LE_1 = 0.8246 \\ LE_2 = 0.2776 \\ LE_3 = 0.0000 \\ LE_4 = -12.376 \end{cases} \tag{8}$$

And

$$\sum_{i=1}^{4} LE_i = -11.2737 \tag{9}$$

It is clear that the sum of Lyapunov exponents is almost equal to divergence, i.e., Eq. (7), which indicates the validity of our results and the sign of the Lyapunov exponents as (+, +, 0, −), refer to the new system as hyperchaotic. The plot of the LEs is shown in Fig. 2. Table 1 gives more details about compute LEs with various values of control parameters. Dimensions of Lyapunov are found as:

$$D_{LE} = j + \frac{1}{|LE_{j+1}|}\sum_{i=1}^{j} LE_i = 3 + \frac{LE_1 + LE_2 + LE_3}{|LE_4|} = 3.0891 \tag{10}$$

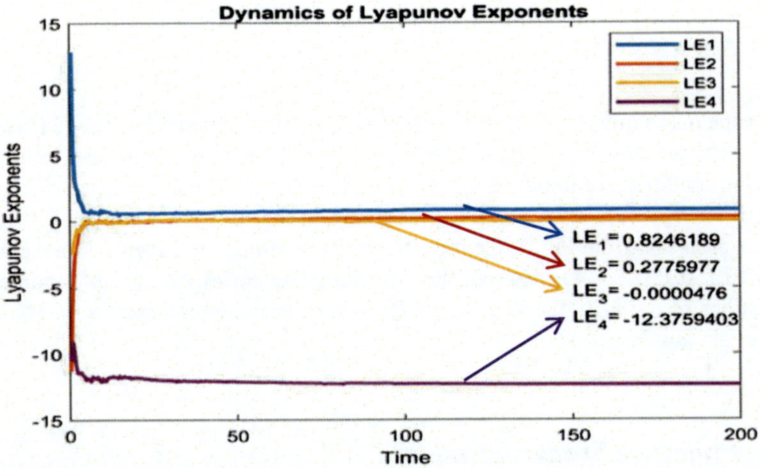

Fig. 2 Lyapunov exponents of 4D hyperchaotic system

Table 1 LE_S of system (2) various initial $(0.1, 0.1, 0.1, 0.1)^T$

No.	h	LE_1	LE_2	LE_3	LE_4	behavior	Sign of the LE_S
1	0.093	0.2378	0.0000	−0.9989	−12.8085	chaotic	(+, 0, −, −)
2	0.107	0.3056	0.0000	−1.0800	−12.7813	chaotic	(+, 0, −, −)
3	1.269	0.3550	0.0007	−0.0913	−12.6580	chaotic	(+, 0, −, −)
4	1.287	0.2477	0.1377	0.0002	−12.7618	hyperchaotic	(+, +, 0, −)
5	3.034	1.1918,	0.0880	−0.0003	−11.9103	hyperchaotic	(+, +, 0, −)
6	3.516	1.2018	0.0008	−0.3634	−10.9886	chaotic	(+, 0, −, −)
7	3.55	1.1030	−0.0002	−0.3919	−10.8268	chaotic	(+, 0, −, −)

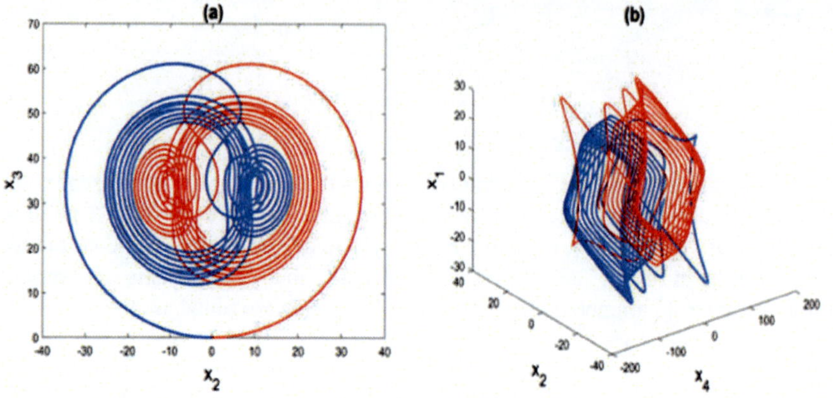

Fig. 3 Phase portraits of different coexisting for conditions of initial $(\pm 0.1, 0.1, 0.1, 0.1)$ (red, blue) in the: (a) $x_2 - x_3$ plane, (b) $x_1 - x_2 - x_4$ space

2.3 Multistability

A change in the initial conditions under the same set of parameters is known as.

coexistence attractors or multistability. System (2) exhibits coexistence attractors with the conditions of initial $(\pm 0.1, 0.1, 0.1, 0.1)$ (red, blue), and the parameters as $a = 10, b = \frac{8}{3}, c = 35$, and $h = 1$ as shown in Fig. 3.

It is clear that from Fig. 3, the red attractors refer to one symmertic initial conditions $(0.1, 0.1, 0.1, 0.1)$, whereas the blue attractors correspond to one symmertic conditions of initial $(-0.1, 0.1, 0.1, 0.1)$, under the same prameters $a = 10, b = \frac{8}{3}, c = 35$, and $h = 1$.

2.4 Projective Synchronization

The word synchronization means that one system follows another system, therefore, two systems are required. The first and second systems are known as the drive and response systems, respectively. Consider system (2) as follows:

$$\begin{bmatrix} \dot{x}_1 \\ \dot{x}_2 \\ \dot{x}_3 \\ \dot{x}_4 \end{bmatrix} = \underbrace{\begin{bmatrix} -a & a & 0 & 1 \\ c & -1 & 0 & 0 \\ 0 & 0 & -b & 0 \\ 0 & 0 & 0 & h \end{bmatrix}}_{A} \begin{bmatrix} x_1 \\ x_2 \\ x_3 \\ x_4 \end{bmatrix} + \underbrace{\begin{bmatrix} 0 & 0 & 0 \\ 1 & 0 & 0 \\ 0 & 1 & 0 \\ 0 & 0 & 1 \end{bmatrix}}_{B} \underbrace{\begin{bmatrix} -x_1 x_3 \\ x_1 x_2 \\ -x_1 x_3 \end{bmatrix}}_{C} \tag{11}$$

A is parameters matrix and $B.C$ is the nonlinear part, whereas the response system is:

$$\begin{bmatrix} \dot{y}_1 \\ \dot{y}_2 \\ \dot{y}_3 \\ \dot{y}_4 \end{bmatrix} = A \begin{bmatrix} y_1 \\ y_2 \\ y_3 \\ y_4 \end{bmatrix} + \left(B \begin{bmatrix} -y1y3 \\ y1y2 \\ -y1y3 \end{bmatrix} + \begin{bmatrix} u1 \\ u2 \\ u3 \\ u4 \end{bmatrix} \right) \tag{12}$$

and let $U = [u_1, u_2, u_3, u_4]^T$ is the controller that should be constructed. The error dynamics for synchronization between the 4D systems (11) and (12) are

$$e_i = y_i - Sx_i, \quad i = 1, 2, 3, 4,$$

S is scaling matrix $S1 = \begin{bmatrix} 2\,0\,0\,0 \\ 0\,3\,0\,0 \\ 0\,0\,4\,0 \\ 0\,0\,0\,5 \end{bmatrix}$ and satisfied that, $\lim_{t\to\infty} e_i = 0$.

Then we obtain the error dynamics as:

$$\begin{cases} \dot{e}_1 = ae_2 + ax_2 - ae_1 + e_4 + 3x_4 + u_1 \\ \dot{e}_2 = ce_1 - cx_1 - e_2 - y_1y_3 + 3x_1x_3 + u_2 \\ \dot{e}_3 = -be_3 + y_1y_2 - 4x_1x_2 + u_3 \\ \dot{e}_4 = he_4 - y_1y_3 + 5x_1x_3 + u_4 \end{cases} \tag{13}$$

Theorem 1 System (12) follow to system (11) (Projective Synchronization) with the following nonlinear control:

$$\begin{cases} u_1 = -ax_2 - 3x_4 - ce_2 \\ u_2 = -ae_1 + cx_1 + y_1y_3 - 3x_1x_3 \\ u_3 = -y_1y_2 + 4x_1x_2 \\ u_4 = -e_1 - 2he_4 + y_1y_3 - 5x_1x_3 \end{cases} \tag{14}$$

Proof Rewrite the error dynamics (13) with control (14) as:

$$\begin{cases} \dot{e}_1 = -ae_1 + (a - c)e_2 + e_4 \\ \dot{e}_2 = (c - a)e_1 - e_2 \\ \dot{e}_3 = -be_3 \\ \dot{e}_4 = -e_1 - he_4 \end{cases} \tag{15}$$

Construct a Lyapunov function as:

$$V(e) = e^T Pe = \frac{1}{2}e_1^2 + \frac{1}{2}e_2^2 + \frac{1}{2}e_3^2 + \frac{1}{2}e_4^2 \tag{16}$$

$$\dot{V} = e_1\dot{e}_1 + e_2\dot{e}_2 + e_3\dot{e}_3 + e_4\dot{e}_4$$

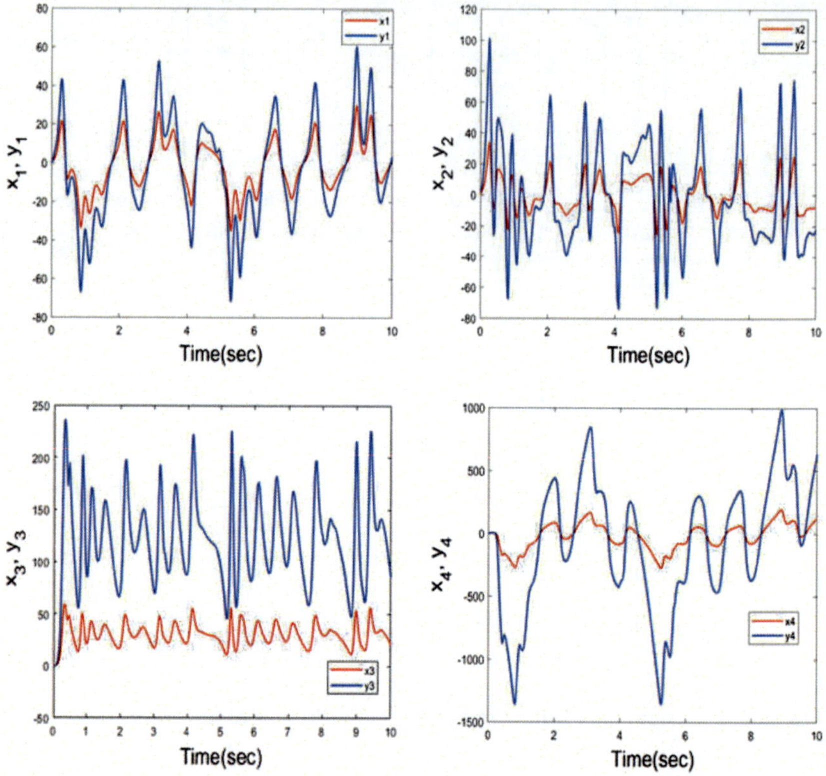

Fig. 4 Projective synchronization between systems (11) and (12) with control (14)

$$\begin{aligned}\dot{V} &= e_1(-ae_1 + (a-c)e_2 + e_4) + e_2((c-a)e_1 - e_2) + \\ &\quad e_3(-be_3) + e_4(-e_1 - he_4) \\ \dot{V} &= -ae_1^2 - e_2^2 - be_3^2 - he_4^2 = -e^T Qe \end{aligned} \tag{17}$$

where $Q = diag(a, 1, b, h)$, so $Q > 0$. Therefore, $\dot{V}(e_i) > 0$. The controller (14) is acceptable and the PS is realized as shown in Figs. 4 and 5.

3 Conclusions

A new 4D dissipative hyperchaotic system with only one unstable equilibrium point is constructed from the famous 3D Lorenz system by using a state feedback control strategy. The proposed system belongs to self-excited attractors and consists of ten terms including three quadratic nonlinearities. The dynamics behavior of the system

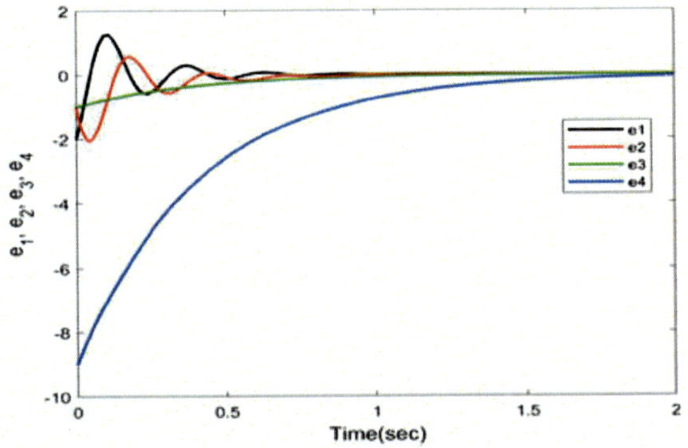

Fig. 5 The convergence of system (13) with controllers (14)

are analyzed theoretically and numerically, i.e., sensitivity to initial conditions, equilibrium points, stability analysis, Lyapunov exponents, Kaplan-Yorke dimension, multistability, and phase portrait. Moreover, a projective synchronization on an identical new system is carried out.

References

1. Abed, K. A., & Ahmad, A. A. (2020). The best parameters selection using Pso algorithm to solving for Ito system by new iterative technique. *Indonesian Journal of Electrical Engineering and Computer Science, 18*(3), 1638–1645.
2. Ahmad, A. A. (2020). By using a new iterative method to the generalized system Zakharov-Kuznetsov and estimate the best parameters via applied the Pso algorithm. *Indonesian Journal of Electrical Engineering and Computer Science, 19*(2), 1055–1061.
3. Zhang, S., Zeng, Y., Li, Z., Wang, M., Zhang, X., & Chang, D. (2018). A novel simple no-equilibrium chaotic system with complex hidden dynamics. *International Journal of Dynamics and Control, 6*(4), 1465–1476.
4. Vaidyanathan, S., Dolvis, L. G., Jacques, K., Lien, C. H., & Sambas, A. (2019). A new five-dimensional four-wing hyperchaotic system with hidden attractor, its electronic circuit realization and synchronisation via integral sliding mode control. *International Journal of Modelling, Identification and Control, 32*(1), 30–45.
5. Zhang, S., Wang, X., & Zeng, Z. (2020). A simple no-equilibrium chaotic system with only one signum function for generating multidirectional variable hidden attractors and its hardware implementation. *Chaos: An Interdisciplinary Journal of Nonlinear Science, 30*(5), 053129.
6. Thivagar, M. L., Ahmed, M. A., Ramesh, V., & Hamad, A. A. (2020). Impact of nonlinear electronic circuits and switch of chaotic dynamics". *Periodicals of Engineering and Natural Sciences, 7*(4), 2070–2091.
7. Al-Kateeb, Z. N., & Jader, M. F. (2020). Encryption and hiding text using DNA coding and hyperchaotic system. *Indonesian Journal of Electrical Engineering and Computer Science, 19*(2), 766–774.
8. Al-Kateeb, Z. N., & Mohammed, S. J. (2020). Encrypting an audio file based on integer wavelet transform and hand geometry. *TELKOMNIKA Telecommunication, Computing, Electronics and Control, 18*(4), 2012–2017.
9. Al-Kateeb, Z. N., & Mohammed, S. J. (2020). A novel approach for audio file encryption using hand geometry. *Multimedia Tools and Applications, 79*(27–28), 19615–19628.

10. Al-Kateeb Z. N., AL-Shamdeen M. J., & Al-Mukhtar F. S. (2020). Encryption and steganography a secret data using circle shapes in colored images. In *Proc. FISCAS, Karbala, Iraqi Journal of Physics: Conference Series* (vol. 1591, p. 012019).
11. Singh, J. P., & Roy, B. K. (2017). Coexistence of asymmetric hidden chaotic attractors in a new simple 4-D chaotic system with curve of equilibria. *Optik, 145*, 209–217.
12. Hamad, A. A., Al-Obeidi, A. S., Al-Taiy, E. H., Khalaf, O. I., & Le, D. N. (2020). Synchronization phenomena investigation of a new nonlinear dynamical system 4D by Gardano's and Lyapunov's methods. *66*, 3311–3327.
13. Al-Talib, Z. S., & AL-Azzawi, S. F. (2020). Projective synchronization for 4D hyperchaotic system based on adaptive nonlinear control strategy. *Indonesian Journal of Electrical Engineering and Computer Science, 19*(2), 715–722.
14. Al-Azzawi, S. F., Mohamed, M. A., Rubiani, H., Mamat, M., Titaley, J., & Langi, Y. A. R. (2020). Chaotic Lorenz system and it's suppressed. *Journal of Advanced Research in Dynamical and Control Systems, 12*(2), 548–555.
15. Wang, Z., Cang, S., Ochola, E. O., & Sun, Y. (2012). A hyperchaotic system without equilibrium. *Nonlinear Dynamics, 69*(1–2), 531–537.
16. Zhang, S., Zeng, Y., Li, Z., Wang, M., & Xiong, L. (2018). Generating one to four-wing hidden attractors in a novel 4D no-equilibrium chaotic system with extreme multistability. *Chaos: An Interdisciplinary Journal of Nonlinear Science, 28*(1), 013113.
17. Wei, Z. (2011). Dynamical behaviors of a chaotic system with no equilibria. *Physics Letters A, 376*(2), 102–108.
18. Thivagar, M. L., & Abdullah Hamad, A. (2020). A theoretical implementation for a proposed hyper-complex chaotic system. *Journal of Intelligent & Fuzzy Systems, 38*(3), 2585–2595.
19. Abed, K. A. (2020). Controlling of jerk chaotic system via linear feedback control strategies. *Indonesian Journal of Electrical Engineering and Computer Science, 20*(1), 370–378.
20. Singh, J. P., Rajagopal, K., & Roy, B. K. (2018). A new 5D hyperchaotic system with stable equilibrium point, transient chaotic behaviour and its fractional-order form. *Pramana, 91*(3), 33.
21. Ojoniyi, O. S., & Njah, A. N. (2016). A 5D hyperchaotic Sprott B system with coexisting hidden attractors. *Chaos, Solitons & Fractals, 87*, 172–181.
22. Hu, G. (2009). Generating hyperchaotic attractors with three positive Lyapunov exponents via state feedback control. *International Journal of Bifurcation and Chaos, 19*(02), 651–660.
23. Yang, W. M., Osman, & Chen, C. (2015). A new 6D hyperchaotic system with four positive Lyapunov exponents coined.*International Journal of Bifurcation and Chaos, 25*(4), 1550060.
24. Al-Azzawi, S. F., Thivagar, M. L., Al-Obeidi, A. S., & Hamad A. A. 2020. Hybrid synchronization for a novel class of 6D system with unstable equilibrium points. *Materials Today: Proceedings.*
25. Al-Obeidi, A. S., & Al-Azzawi, S. F. (2020). Hybrid synchronization of high-dimensional chaos with self-excited attractors. *Journal of Interdisciplinary Mathematics, 23*(8), 1569–1584.
26. Al-Azzawi, S. F., & Al-Obeidi, A. S. (2020). Chaos synchronization in a new 6D hyperchaotic system with self-excited attractors and seventeen terms. *Asian-European Journal of Mathematics*, 2150085.
27. Wolf, A., Swift, J. B., Swinney, H. L., & Vastano, J. A. (1985). Determining Lyapunov exponents from a time series. *Physica D: Nonlinear Phenomena, 16*(3), 285–317.

Improving the Protection of Wireless Sensor Network Using a Black Hole Optimization Algorithm (BHOA) on Best Feasible Node Capture Attack

Ankur Khare, Rajendra Gupta, and Piyush Kumar Shukla

Abstract Wireless Sensor Network (WSN) is an area of research that connects mutually huge subareas of communication, routing, security, and attacks. WSN is conceivably the most susceptible network to node capture attack due to its dynamic nature in huge area. A node capture attack is introduced by seizing few nodes through an intruder to capture entire WSN by extracting the useful information like keys, routing mechanism, and data from WSN. To improve the protection of WSN, we proposed a Black Hole Optimization Algorithm (BHOA) on best feasible node capture attack to discover the optimal nodes having superior possibility of attack. The BHOA is applied on a function Vertex Participation. The experiment is performed on MATLAB 2019a environment, and the results show the better quality, efficiency of BHOA against MA, OGA, MREA, GA, and FFOA based on traffic compromised ratio, power consumption cost, and attacking time.

Keywords Black hole · Compromised · Feasible · Seizing · Wireless Sensor Network (WSN)

1 Introduction

The hasty growth of WSN (Wireless Sensor Network) [1, 2] has been achieved in the last few years for enormous application fields like disaster monitoring, remedial use, and manufacturing sections. The vast deployment of WSN has been provided the greater chance to attacker for attacking on network vulnerabilities to devastate the total communication of WSN [3, 4]. The MA (Matrix Algorithm) is developed to increase the efficiency of node capture attack [5, 6] by generating node-path group with minimum execution time. Node-path group is evaluated on the basis of direct and indirect combination of vertices and paths, and after that the cost of confiscating vertices is obtained. Therefore, MA is used to obtain minimum cost of confiscating

A. Khare (✉) · R. Gupta
Rabindranath Tagore University, Raisen, India

P. K. Shukla
University Institute of Technology, RGPV, Bhopal, India

P. Nayak et al. (eds.), *IoT and Analytics for Sensor Networks*, Lecture Notes in Networks and Systems 244, https://doi.org/10.1007/978-981-16-2919-8_30

vertices for malicious use of WSN. The attacking effectiveness is not obtained by MA [7]. So, another Opti-Graph Attack (OGA) approach is applied to generate advanced efficiency of node capture attack with least cost. Energy expenditure is not considered in OGA [8], further MREA (Minimum Resource Expenditure node capture Attack) [9] is implemented with shortest path calculation through traveling sales person and Hamilton path approach. The performance of MREA is further enhanced in terms of least node capturing by utilizing an optimization approach GA (Genetic Algorithm) [10]. GA is applied on functions having more than one objective like vertex and key contribution and resource expenditure to obtain malicious vertices [11]. Moreover, the efficiency of attacking method of GA is increased by using another optimization approach FFOA (Fruit Fly Optimization Algorithm) in terms of compromised traffic and attacking rounds [12].

Therefore, we developed a Black Hole Optimization Algorithm (BHOA) to get better quality outcomes on node capture attack in WSN. BHOA utilizes vertex participation for evaluating the optimal value of each vertex for selection of malicious vertices over WSN.

2 BHOA for Node Capture Attack

Various vertices (sensor nodes) V_{sn} and links L_{sn} are merged to generate a wireless sensor network $W_{sn} = (V_{sn}, L_{sn})$ with single and multipath ways for message passing. A message is transmitted in the form of packets in which a packet is passed through a link $L_{e,f}$ between vertices V_e and V_f within an area of transmission T^a after encryption via common key $K_{e,f} = K_e \cap K_f$. Different keys $K_e \subset K_{sn}$ are used to perform encryption of different packets before passing via vertex $V_e \in V_{sn}$. The intruder model is a multifarious structure to initiate on WSN for attacking the vertices to take away the essential information like cryptographic keys, and encryption methods. A single key $K_{e,f} \in K^{cm}$ of a link $L_{e,f}$ is compromised to capture that link $L_{e,f} \in L_{sn}$. A single link $L_{e,f} \in L^{cm}$ of a path P_e is compromised to capture that path.

$P_e \in P_{sn}$. A single path $P_e \in P^{cm}$ of a route $R_{\text{source,dest}}$ is compromised to capture that route $R_{\text{source,dest}} \in R_{sn}$.

The key objective of intruder is to attack small number of vertices for controlling the whole WSN. Therefore, we implemented a BHOA to pick the optimal vertices based on maximum vertex participation for compromising the full WSN.

2.1 Vertex Participation

A vertex $e \in V_{sn}$ has a direct connection (link) (e, vp_{sn}) (symbolized as $(e \rightarrow vp_{sn})$) having a vertex of path $vp_{sn} \in P_{sn}$. A vertex e is seized to capture the vp_{sn}, which is called as direct connection from e to vp_{sn}, and represented by a function $P^{cm}(e, vp_{sn})$: $V_{sn} \times P_{sn} \rightarrow R_{sn}^{\{0,1\}}$ through Eq. 1.

$$P^{cm}(e, vp_{sn}) = \begin{cases} 1, e \rightarrow vp_{sn} \\ 0, \text{Otherwise} \end{cases} \tag{1}$$

A vertex $e \in V_{sn}$ has an indirect connection having a vertex of path $vp_{sn} \in P_{sn}$. A vertex e is seized to capture the vp, which is called as partial compromise (indirect connection) from e to vp gratifying some situation:

- vp is not captured.
- At least one key is provided to vertex e which is mutually used through a connection in the path vp. A connection (f, g) of vp. is shown as $1 \leq \left|K_{f,g} \cap K'^{cm}\right| < \left|K_{f,g}\right|$, n which K'^{cm} is compromised key set with vertex e by attacker. It is called as partial compromise (indirect connection) from e to vp

The direct and indirect vertex participation are calculated. Direct connections are utilized to obtain number of captured paths for generating Direct Vertex Participation $(DP_v(e))$, and indirect connections are utilized to obtain Indirect Vertex Contribution $(IP_v(e))$. $DP_v(e)$. is calculated through Eq. 2 by utilizing Eq. 1.

$$DP_v(e) = \sum_{e \in V_{sn}, vp_{sn} \in P_{sn}} P^{cm}(e, vp_{sn}) \tag{2}$$

$IP_v(e)$ is calculated for the vertex e and entire path connections through Eq. 3.

$$IP_v(e) = \sum_{e \in V_{sn}, vp_{sn} \in P_{sn}} IP^{cm}(e, vp_{sn}) \tag{3}$$

Here, $IP^{cm}(e, vp_{sn}) : V_{sn} \times P_{sn} \rightarrow R_{sn}^{\{0,+\infty\}}$ is calculated through Eq. 4.

$$IP^{cm}(e, vp_{sn}) = \frac{\left|K'_{vp_{sn}}\right|}{\left|K_p(vp_{sn})\right|} \sum_{(e,f) \subset p_e} \frac{1}{|L(vp_{sn})|} \left(1 - \frac{\left|K'_L(e,f)\right|}{\left|K_{e,f}\right|}\right) \tag{4}$$

Here,
$|L(vp_{sn})|$ = Number of connections in vp_{sn}.
$\left|K'_{vp_{sn}}\right|$ = Number of ks of vertex of path vp_{sn} not acquired by the intruder.
$\left|K'_L(e,f)\right|$ = Number of keys not obtained by intruder.
$\left|K_{e,f}\right|$ = Number of shared Keys by vertex V_e and V_f.

$\left|K_p(vp_{sn})\right|$ = Number of shared keys of whole connections.

The $\left|K_L'(e,f)\right|$(Eq. 5), $\left|K_{vp_{sn}}'\right|$ (Eq. 6), and $\left|K_p(vp_{sn})\right|$ (Eq. 7) values are calculated as below:

$$K_L'(e,f) = K_{e,f} - K^{cm} \tag{5}$$

$$K_{vp_{sn}}' = \bigcup_{(e,f)\subset p_{sn}} K_L'(e,f) \tag{6}$$

$$K_p(vp_{sn}) = \bigcup_{(e,f)\subset p_{sn}} K_{e,f} \tag{7}$$

Finally, we combined the $DP_v(e)$ and $IP_v(e)$ to obtain Vertex Participation (VP) through Eq. 8.

$$VP(e) = \delta \times DP_v(e) + (1 - \delta) \times IP_v(e) \tag{8}$$

Here,

$VP(e)$ = Vertex Participation value of vertex e.

δ = constant in (0, 1) denoted the consequence of direct and indirect vertex participation. The maximum value of δ is obtained greatest destructiveness of WSN. The VP value is also superior for greatest destructiveness of WSN.

2.2 *Black Hole Optimization Algorithm (BHOA)*

After obtaining the Eq. 8 for each and every vertex in WSN, BHOA is utilized to evaluate the maximum VP values for providing best vertices V^{cm} from available vertices. An area of space–time (x, y, t) with sturdy and influential gravitational field is known as black hole in which not anything can get away from it. An event horizon is a mathematically explained region about a black hole that results in the point of no return. If something goes near to the horizon, then it would be fascinated into the black hole and eternally vanish. The mass, charge, and momentum are three general characteristics of black hole. The momentum and charge are not present in basic black hole but black hole has larger mass with maximum gravitational force.

2.2.1 Basic Preliminaries

Black Hole. The optimal aspirant among the entire aspirants at every repetition is nominated as a black hole in black hole optimization.

Stars. All the aspirants are known as stars excluding the black hole. The black hole is not formed casually and it is a genuine population aspirant.

Movement. All aspirants are stimulated near to the black hole on the basis of their present position and an arbitrary number.

1. BHOA initiates with primary aspirant solutions population to an optimization predicament and an evaluated objective function for optimization problem.
2. The optimal aspirant is nominated as black hole and other aspirants are normal stars in every repetition. The black hole initiates to pull the stars in the region of it after the initial stage.
3. Black hole absorbs the highly nearer stars and new aspirant solution (star) is arbitrary obtained for starting the new search after locating the new star to the search space.

2.2.2 Fitness Value Evaluation

- Initially, the aspirant solutions (stars) population $P(X) = \{X_1^t, X_2^t, X_3^t, \ldots, X_N^t\}$ is arbitrarily obtained and located in search space for optimization.
- Evaluate the entire population fitness by using Eqs. 9 and 10.

$$F_a = \sum_{a=1}^{PS} E(P(t)) \quad (9)$$

$$F_{BH} = \sum_{a=1}^{PS} E(P(t)) \quad (10)$$

Here,
F_a = ath star fitness value.
F_{BH} = Black hole fitness value.
$E(P(t))$ = Evaluation functions for population at time t.
PS = Population Size.

The population is evaluated and the optimal aspirant with optimal fitness value F_a is nominated as black hole and other are normal stars. Black hole starts enthralling the highly nearer stars and other stars begin movement near to the black hole after the initial phase.

Black Hole Fascination Rate of Stars. The black hole begins fascinating the stars in the region of it and other stars begin movement near to the black hole. The black hole fascination rate is evaluated by using Eq. 11.

$$X_a(t+1) = X_a(t) + \text{rand} \times (X_{BH} - X_a(t)) \tag{11}$$

Here,
$a = 1, 2, 3, \ldots N$.
$X_a(t) =$ ath star position at repetition t.
$X_a(t+1) =$ ath star position at repetition (t + 1).
X_{BH} = Black hole position in search space.
rand = random number within (0, 1).
N = Number of aspirant solutions (Stars).

A star may attain a minimum cost position than black hole though moving near to the black hole. Therefore, black hole shifts to that new star position and vice versa. The process will be continued with new black hole, and other stars begin movement near to that new black hole position.

Possibility of the Event Horizon Passage through moving Stars. The possibility of the event horizon passage moving stars is utilized to obtain more optimal value for the optimization problem. Each aspirant solution (star) that passages the black hole's event horizon will be fellated by black hole and an aspirant dies by blowing through black hole each slice of time. Therefore, new aspirant is populated and dispersed arbitrarily in search space to provide new search in population with constant aspirant solutions. The next repetition performs later than the movement of all stars. The horizon radius (R) is evaluated by using Eq. 12.

$$R = \frac{F_{BH}}{\sum_{a=1}^{N} F_a} \tag{12}$$

Here, F_a, F_{BH}, and N have been previously explained. If distance between an aspirant solution (star) and optimal aspirant (black hole) is found smaller than R, then aspirant is shrunken and a new aspirant is formed and dispersed arbitrarily in the search space.

```
Black Hole Optimization Algorithm (BHOA)
Input:  The  Stars  Population  with  arbitrary  positions
located in search space for optimization.
Output: the optimal aspirant (black hole)with fitness value
```

```
START
Initialize the stars population with arbitrary positions
P(X) = {X_1^t, X_2^t, X_3^t, ..................., X_N^t}

WHILE (until the end situation) DO
      FOR a = 1 to N DO
            Calculate the population fitness (eq. 9 and
            10)
            Nominate the black hole (optimal aspirant)
            with maximum fitness value.
            Modify every star position by utilizing eq.
            11.
            IF (A star may attain a minimum cost position
                  than black hole)
            THEN
                  Black hole shifts to that new star
                  position and vice versa.
            IF (an aspirant (star) passages the event
                  horizon)
                  Calculate the horizon radius (R) by
                  utilizing eq. 12.
                  IF (Distance between an aspirant
                        solution (star) and optimal
                        aspirant (black hole) is found
                        smaller than R)
                  THEN
                        Aspirant is shrunken and a new
                        aspirant is formed and dispersed
                        arbitrarily in the search space.
                        Substitute it with a new aspirant
                        in an arbitrary position in search
                        space.
            ELSE
                  BREAK;
      END FOR;
END WHILE;
STOP
```

3 Results and Analysis

The efficiency of BHOA is analyzed and compared with MA, OGA, MREA, GA, and FFOA in terms of several circumstances shown in Table 1.

Table 1 Circumstances of simulation

Circumstances	Values
Number of vertices (V_{sn})	250
Network region	100 m × 100 m
Source vertices (S^v)	20
Area of transmission (T^a)	25
Destination vertices (D^v)	7
Complete keys (K_{sn})	250
Population size (PS)	250
Iterations (N)	250

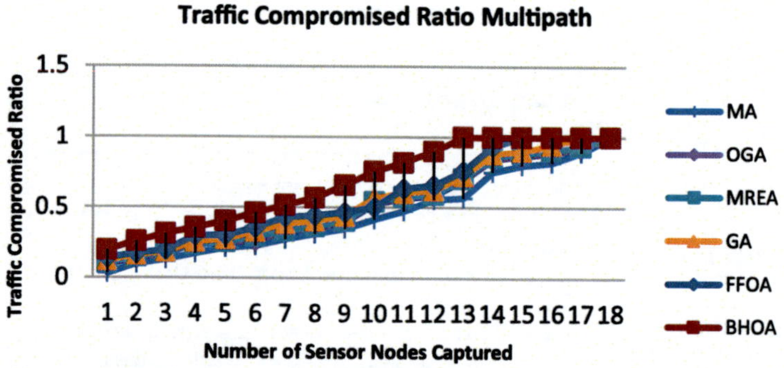

Fig. 1 Traffic compromised ratio (multipath routing)

3.1 Traffic Compromised Ratio

Traffic Compromised Ratio is defined as the proportion of captured paths to the all paths in WSN for single and multipath routing. Figure 1 elaborates the better quality of attacking ability of BHOA in terms of traffic compromised ratio with compromising least sensor nodes to provide 9.09 and 13.33% superior performance against the FFOA, 23.08 and 23.53% superior performance against GA, 28.57 and 27.78% superior performance against MREA, OGA, and MA for single and multipath routing, respectively, to capture complete WSN.

3.2 Power Consumption Cost

The Power Consumption Cost is the cost of power consumption for electing the sensor nodes to capture for co-operating the complete WSN. Figure 2 elaborates the greater efficiency of BHOA in terms of power consumption cost by means of number of keys to capture least sensor nodes to provide 20.06 and 19.47% superior

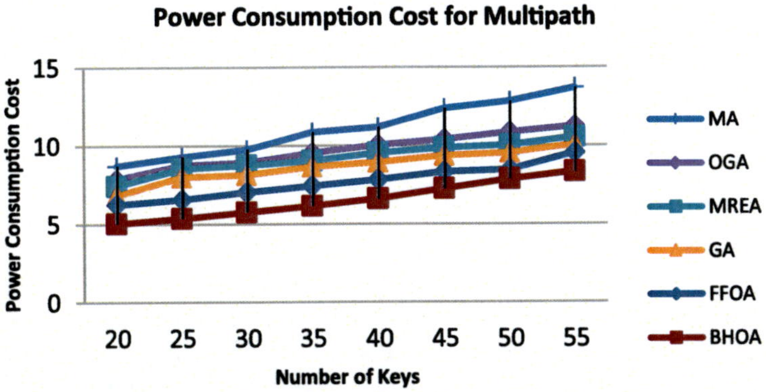

Fig. 2 Power consumption cost (multipath routing)

performance against the FFOA, 25.16 and 26.20% superior performance against GA, 33.18 and 32.42% superior performance against MREA, 40.65 and 36.09% superior performance against OGA, and 53.75 and 42.26% superior performance against MA for single and multipath routing, respectively, to capture complete WSN.

3.3 *Attacking Time*

The Attacking Time is described as the number of attacking rounds to co-operate the complete WSN by capturing the minimum sensor nodes. It is totally based on traffic compromised ratio. Hence, Fig. 3 elaborates the greater performance of BHOA in terms of attacking time by means of number of keys to capture least sensor nodes to provide 9.09 and 13.33% improved performance against the FFOA, 23.08 and

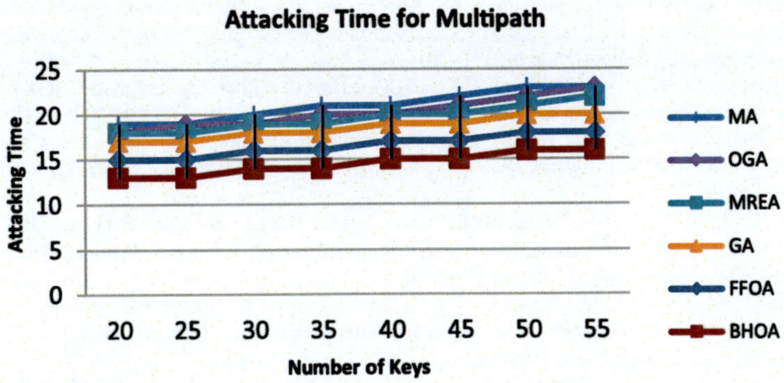

Fig. 3 Attacking time (multipath routing)

23.53% improved performance against GA, 28.57 and 27.78% improved performance against MREA, OGA, and MA for single and multipath routing, respectively, to capture complete WSN.

4 Conclusion

Wireless Sensor Network (WSN) is a combination of numerous research fields like communication, routing, protection, and attacks. The dynamic behavior of WSN makes it highly vulnerable to node capture attack in which an attacker captures some nodes to compromise whole WSN for stealing the important data, routing method, and keys. To enhance the protection of WSN, a Black Hole Optimization Algorithm (BHOA) is introduced in most feasible node capture attack to determine the optimal nodes having advanced opportunity of attack. The BHOA is applied on a function Vertex Participation, and results represent the superior efficiency of BHOA against MA, OGA, MREA, GA, and FFOA on the basis of traffic compromised ratio, power consumption cost, attacking time on MATLAB 2019a environment.

References

1. Xiong, C., Li, S., Liu, L., Li, R., & Jin, Y., (2019). A hybrid key pre-distribution scheme for wireless sensor networks. In *IOP Conf. Series, Journal of Physics* (pp. 1–18). https://doi.org/10.1088/1742-6596/1229/1/012066.
2. Shaukat, H. R., Hashim, F., Shaukat, M. A., & Alezabi, K. A. (2020). Hybrid multi-level detection and mitigation of clone attacks in mobile wireless sensor network (MWSN). *Sensors, MDPI, 20*, 1–23.
3. Chowdary, K., & Satyanarayana, K. V. V. (2017). Malicious node detection and reconstruction of network in sensor actor network. *Journal of Theoretical and Applied Information Technology, 95*(3), 582–591.
4. Ehdaie, M., Alexiou, N., Ahmadian, M., Aref, M. R., & Papadimitratos, P. (2017). Mitigating node capture attack in random key distribution schemes through key deletion. *Journal of Communication Engineering, 6*(2), 1–10.
5. Alshammari, M. R., & Elleithy, K. M. (2018). Efficient and secure key distribution protocol for wireless sensor networks. *Sensors, MDPI, 18*, 1–25. https://doi.org/10.3390/s18103569.
6. Ahlawat, P., & Dave, M. (2018). An attack resistant key predistribution scheme for wireless sensor Networks. *Journal of King Saud University—Computer and Information Sciences, Elsevier*, 1–13.
7. Lin, C., & Wu, G. (2013). Enhancing the attacking efficiency of the node capture attack in WSN: A matrix approach. *The Journal of Supercomputing, Springer Science & Business Media*, 1–19. https://doi.org/10.1007/s11227-013-0965-0.
8. Lin, C., Wu, G., Yu, C. W., & Yao, L. (2015). Maximizing destructiveness of node capture attack in wireless sensor networks. *The Journal of Supercomputing, Springer Science & Business Media, 71*, 3181–3212. https://doi.org/10.1007/s11227-015-1435-7.
9. Lin, C., Qiu, T., Obaidat, M. S., Yu, C. W., Yao, L., & Wu, G. (2016). MREA: A minimum resource expenditure node capture attack in wireless sensor networks. *Security and Communication Networks, Wiley Online Library, 9*, 5502–5517. https://doi.org/10.1002/sec.1713.

10. Bhatt, R., Maheshwary, P., Shukla, P., Shukla, P., Shrivastava, M., & Changlani, S. (2020). Implementation of fruit fly optimization algorithm (FFOA) to escalate the attacking efficiency of node capture attack in wireless sensor networks (WSN). *Computer Communications, Elsevier, 149*, 134–145.
11. Shukla, P. K., Goyal, S., Wadhvani, R., Rizvi, M. A., Sharma, P., & Tantubay, N. (2015). Finding robust assailant using optimization functions (FiRAO-PG) in wireless sensor network. *Hindawi Publishing Corporation, Mathematical Problems in Engineering*, 1–8. https://doi.org/10.1155/2015/594345.
12. Molina, D., Poyatos, J., Ser, J. D., Garcia, S., Hussain A., & Herrera, F. (2020). Comprehensive taxonomies of nature- and bio-inspired optimization: inspiration versus algorithmic behavior. *Critical Analysis and Recommendations. cs.AI,* pp. 1–76. arXiv:2002.08136v2.

A Systematic Analysis of the Human Activity Recognition Systems for Video Surveillance

Sonika Jindal, Monika Sachdeva, and Alok Kumar Singh Kushwaha

Abstract In recent years, human activity recognition has become a prominent research area in numerous fields such as healthcare, smart home activity analysis, suspicious activity recognition, robotics, surveillance, and security. The focus of the current research work is the analysis of human activity recognition systems for video surveillance. The human activity recognition system involves the detection of normal as well as abnormal activities. The recognition of human activities is still considered a challenging issue despite the contributions of numerous researchers. The erratic human behavior and complexities of the video datasets create numerous challenges to precisely observe the human activities with significant performance. The analysis of the human activity detection systems for video surveillance is conducted on the basis of state-of-art contributions by different researchers in the field. The paper also describes the taxonomy of human activity detection. It ends with a discussion of the challenging issues in the field along with the concluding remarks.

Keywords Human activity recognition · Video surveillance · Security surveillance · Human computer interaction · Visual surveillance

1 Introduction

Human activity is the movement of the human body parts and the recognition of these activities belong to the development of the autonomous system as the application of the human–computer interaction. Human activities can be recognized as vision-based devices or sensor-based devices. The sensor-based devices can be contact or

S. Jindal (✉) · M. Sachdeva
Department of Computer Science & Engineering, IKG Punjab Technical University, Jalandhar, India
e-mail: sonikajindal@sbsstc.ac.in

M. Sachdeva
e-mail: monika@ptu.ac.in

A. K. S. Kushwaha
Department of Computer Science & Engineering, Guru GhasidasVishwavidyalaya, Bilaspur, India

P. Nayak et al. (eds.), *IoT and Analytics for Sensor Networks*, Lecture Notes in Networks and Systems 244, https://doi.org/10.1007/978-981-16-2919-8_31

contactless gadgets with any kind of special sensor or the inbuilt sensors available in mobile phones or other advance electronic gadgets. The vision-based devices are the contactless devices in which a camera is considered as the major source of information collector. The captured data can be processed with the help of computer vision techniques to recognize human activities.

The focused section of the current research work is the recognition of human activities with vision-based video data. The video data is useful for the recognition of normal as well as abnormal human activities. The taxonomy of human activity recognition is illustrated in Fig. 1. The normal activities involve the daily routine benign activities like sleeping, walking, playing, standing, sitting, etc. that can be considered as the smart home activities. The activities other than the normal benign activities are referred to the abnormal activities such as health issues (like sudden falling, chest pain, vomit, faint, etc.) of elder people, abnormal crowd activities (like running crowd, punching, slap, etc.), or any kind of suspicious activities (like the appearance of weapon at public places, leftover baggage at public places, theft, etc.).

The surveillance with video data is the major source of visual data for human activity recognition. The video data analysis can be considered as the process with the modules of object recognition, human action recognition, and the classification of the actions. These surveillance systems serve as the protective element for the need of the future. For instance, the aging population of the world is estimated to increase from 0.461 billion to 2 billion by the year 2050 [1]. But the extraction of video data from all the private and public places can lead to the processing of huge data. The manual loop up for the daily activities is a tedious task for a human and the requirement of huge manpower with the expertise to analyze the live data. Moreover, the storage of such huge data can also increase the financial burden. This makes to develop the autonomous video surveillance system that can lesser down the storage

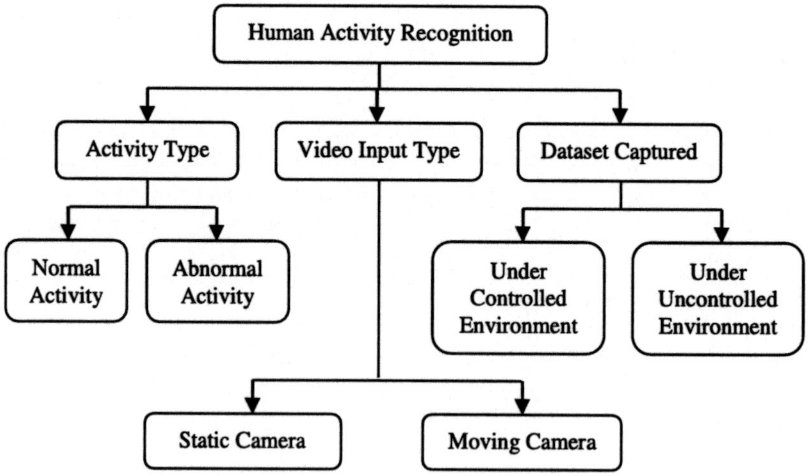

Fig. 1 Taxonomy of human activity recognition

cost, human errors, and huge manpower expertise. The autonomous system should be able to perform the task with ease with the ability to analyze the live abnormal activities so that the immediate problem solution can be determined. It is also vital to perform the object recognition [2] to determine the suspicious activities.

The present work discusses the different state-of-art techniques of human activity recognition systems for video surveillance. These research contributions are described in Sect. 2. The analysis of the human activity recognition systems is conducted on the basis of discussed research contributions presented in Sect. 3. The analysis is described with the discussion of the significant human activity recognition techniques for video surveillance and challenging issues for the same. The paper ends with the concluding remarks of the work.

2 Human Activity Recognition Systems for Video Surveillance

The section describes the different human activity recognition systems for video surveillance. The growing usability of autonomous human activity recognition in different fields has drawn the attention of researchers. The quality contributions of the researchers from the year 2011 to 2020 are discussed with the focused region of video surveillance. Moreover, the findings from the different research contributions are also highlighted in Table 1.

Zhang et al. [3] have employed the improved variant of the Gaussian mixed model that adds the attributes of k-means clustering along with the envelope shape presentation to handle the varying viewpoints in different video backgrounds. The average recognition of 93.23% was evaluated by authors. Gao et al. [4] have presented the network transmission based (NTB) algorithm with a focus on the recognition of abnormal activities. The experimentation was conducted on the recording data obtained from two manually settled cameras and attained the efficient results. Shieh and Huang [5] have focused to care the aging people with video surveillance. The authors have determined the falling activities using the pattern recognition based autonomous falling detection algorithm. The system throughput was improved with the software pipelining system. Chathuramali and Rodrigo [6] have addressed the problems of huge feature vector size and longer training time by adapting the support vector machine classifier. The classifier was implemented for the dataset of UIUC1 and Weizmann and achieved the comparing results. The training time of SVM was a bit higher than the other techniques for the case of fewer data samples. Gupta et al. [7] have conducted human activity recognition using the Gait pattern recognition model. The method utilizes the real-life scenario based on the Hu-moments [8] to evaluate human activities. The average recognition rate of 91.36% and 95.01% was evaluated for the Weizmann and KTH datasets, respectively.

Further, Bengalur [9] have worked for the recognition of 3D joint structures model for the recognition of human activities. The model utilizes the depth feature along

Table 1 Findings from the human activity recognition systems for video surveillance

Author and year	Technique	Dataset	Activity type	Results (%)	Remarks
Zhang et al. [3]	Gaussian mixed model	CASIA Activity analysis	Normal activities	Accuracy 93.23%	No state-of-art comparison
Gao et al. [4]	Network transmission algorithm	Self prepared dataset	Normal and abnormal	Equal Error rate 11.4%	Need to reduce error rate
Shieh and Huang [5]	Autonomous falling detection system	Self prepared dataset	Falling and non-falling activities	Accuracy 93% non-falling & 90% Falling	Possibility of extension for more activities and performance improvement
Chathuramali and Rodrigo [6]	Support vector machine	UIUC1 and Weizmann datasets	Normal activities	average recognition rate 99.32%	SVM classifier takes more time for training the sample
Gupta et al. [7]	Gait pattern recognition	KTH and Weizmann actions	Normal activities	Average recognit-ion rate 93.19%	No comparison with state-of-art
Bengalur [9]	Support vector machine	Self prepared dataset of 650 videos	13 Normal activities	Accuracy 89%	No evaluation on standard benchmark datasets
Kaur and Singh [10]	Kalman filter	Weizmann dataset	10 Normal activities	Not available	No performance evaluation
Htike et al. [11]	KNN, SOM, MLP, FCM, and KMA	Human posture dataset	Normal human postures	MLP accuracy 97% for 5 & 98% SOM for 2 posture	No Consideration of standard benchmark dataset
Taha et al. [12]	Hidden Markov models	Cornell CAD-60 and CAD-120	Normal activities	CAD-60 82.3% & CAD120 94.4%	The work needs improvement for special activities like eating, drinking, etc
Patwardhan [13]	Dynamic probabilistic networks	Self prepared dataset	Normal activities	Average accuracy 75.5%	DPN lacks the HMM for the walking activity

(continued)

Table 1 (continued)

Author and year	Technique	Dataset	Activity type	Results (%)	Remarks
Kamal et al. [14]	Modified Hidden Markov model (MHMM)	IM-daily depth activity and MSR daily 3D activity	Normal daily activities	IM-Daily accuracy 68.3%, MSR 91.3%	Accuracy of the MHMM model was only 68.3% that can be improved in real life evaluation
Khemchandani and Sharma [15]	RLSTW-SVM approach	IXMAS, UMD, UIUC1, & Weizmann	Normal Activities	Average Accuracy 93.76%	The work can be extended for the interaction of multiple people
Ijjina and Chalavadi [16]	Convolutional neural network	Weizmann, SBU Kinect interaction, NATOPS gesture, & MIVIA action	Normal activities	Average accuracy 92.73%	Work can be extended for group activities & categories of human behaviour recognition
Zhao et al. [17]	Region-based mixture model	Weizmann and UT-tower datasets	Normal activities	Average accuracy 98.99	It can be made autonomous for the adjustment of regions in the RMM model
Kamthe and Patil [18]	Semantic based approach	CAVIAR (PETS 2004) and PETS 2006	Suspicious activities	Average accuracy 94.5%	Results evaluated only for two activities
Bouachir et al. [19]	NB, LSVM, LDA, and RBF-SVM	Self prepared dataset	Suicide attempt activity	100% Accuracy with NB & RBF-SVM	The occlusion problem affects the reliability for the detection of the 3D skeleton
Singh et al. [21]	Hidden Markov model	WVU multi-view, MSR view-point, i3D Post, and KTH action	Normal activities	Average accuracy 99.66%	It can be extended for the recognition of group activities recognition

(continued)

Table 1 (continued)

Author and year	Technique	Dataset	Activity type	Results (%)	Remarks
Kong et al. [22]	Three stream convolutional neural network	UR Fall detection and multiple camera Fall	Falling activities	Average recall 99.4%	Method fails for bad quality videos. The hardware requirement is also expensive
Jaouedi et al. [23]	Gated recurrent neural network	KTH, UCF101, and UCF Sports Action	Normal and sports activities	Average accuracy 91.54%	Accuracy needs to improve for the UCF101 and UCF sports datasets
Silva and Marana [24]	Bag of poses approach	KTH and Weizmann datasets	Normal activities	Average accuracy 97.51%	It can be extended for the recognition of special activities and group based interactions

with the support vector machine for skeleton attribute analysis. Kaur and Singh [10] used the Kalman filter to track human activities. The overall process involved the steps of background analysis by calculating the mean of 'N' frames, human detection by subtracting the background, and final human activity recognition. The visuals of the work were significant, but the numerical performance analysis was not conducted by authors. Htike et al. [11] have explored different classifiers for the recognition of human activity recorded with a single camera. The considered classifiers were k-nearest neighbor (KNN), self-organizing maps (SOM), multi-layer perceptron feed-forward neural network (MLP), fuzzy c-means (FCM), and k-means algorithm (KMA). Among the mentioned classifiers, multiplayer perceptron was evaluated as a significant classifier due to higher recognition accuracy. Taha et al. [12] have employed the Hidden Markov Models (HMM) for the recognition of human activity. The experimentation was conducted for the 3D position with depth parameter on the basis of skeleton structure. Patwardhan [13] have investigated the dynamic probabilistic networks (DPN) for normal activity recognition. The method was evaluated for the basic activities of walking, sitting, standing, and walking whose performance comparison was conducted with HMM. The author indicated more accurate results in the indoor light controlled environment than the outdoor environment. Kamal et al. [14] have presented the modified HMM for the recognition of the 3D human body and their respective activities. The feature set of the system was enhanced by fusing the temporal joint features and spatial depth shape attributes. The HMM model was modified to acquire the moving body features along with the static human body features.

Furthermore, Khemchandani and Sharma [15] have proposed the robust least square twin support vector machine (RLSTW-SVM) for human activity recognition.

The method was proposed to handle the heteroscedastic noise and incorporation of the outlier effect. The optic flow and silhouette were also utilized as the feature descriptors. Ijjina and Chalavadi [16] have used the deep leaning based convolutional neural network along with the investigation of features through depth stream videos and RGB motion streams. The approach was multi-model with the ability to tolerate the noise in a robust manner. Zhao et al. [17] have proposed the Region based Mixture Model (RMM) for human activity recognition from low quality video clips. The method of Layered Elastic Motion Tracking was adapted for the extraction of motion features and body shape with the estimation of raw edges. Kamthe and Patil [18] have focused on the detection of suspicious human activities using the semantic approach. The suspicious activities of loitering at ATM and abandoned bag detection were evaluated with the accuracy of 93% and 96%, respectively. Bouachir et al. [19] have described an intelligent system for the recognition of suicide attempts. The motion and pose features especially the 3D joint features are extracted which were utilized for the classification with different classifiers of Naïve Bayes (NB), Linear Support Vector Machine (LSVM), Linear Discriminant Analysis (LDA), and SVM with radial basis kernel (SVM-RBF). The human face features are also vital to detect the face movement [20]. Among the mentioned classifiers, the SVM-RBF has attained the superior results.

Recently, Singh et al. [21] have presented a multi-view based human activity recognition system with the usage of the hidden Markov model (HMM) as the activity detector. The overall process detects the people by removing the background, extract combinational features (linear binary pattern, optical flow motion, and contour based distance signal features), and determine the activity with the HMM approach. The accuracy of the discussed system was better than the other comparison techniques. Kong et al. [22] have focused to determine the falling activities from the multi-view cameras using the three stream convolutional neural network. The approach considers the first two streams adapt the Silhouettes and motion history images as the input and the third stream considers the dynamic images. In the overall scenario, the method was effective in terms of accuracy but lacks due to inefficient results for the bad representation of video clips. Jaouedi et al. [23] have explored the Gated Recurrent Neural Network for the recognition of human activities. The major focus of the paper was the extraction of significant features using Kalman Filter and Gaussian Mixture Model related to moving body. The significant results were noted for the recognition of normal as well as sports activities. Silva and Marana [24] have employed the Bag of Poses approach along with the Spatio-temporal features for the recognition of 2D human poses. The method adapts the indirect features by converting the line segments into the parameter space.

3 Systematic Analysis and Discussion

The analysis of the human activity recognition systems for video surveillance is conducted on the basis of discussed research contributions. The section describes the overall major focused activities, significant research technique, and challenging issues related to the discussed work.

The usability of different activities in the discussed studies is illustrated in Fig. 2. The activities of falling, suicide attempts, suspicious activities, and other alarming activities are considered in the list of abnormal activities. The daily routine activities and sports activities are taken under the category of normal activities. On the basis of Fig. 2, it can be noticed that there are major research contributions for the recognition of normal activities. There is lesser work available for the detection of abnormal activities.

On the basis of different activities, the authors have selected their respected dataset and used the different techniques for the classification and detection of activities. The significance of different research techniques depends on different scenarios such as the video quality, individual activity or group interaction, availability of feature components, and type of activity. In a comprehensive manner, the deep learning models have attained outperformed recognition accuracy.

Further extending the discussion, there is always some limitation or the possible extension that can be considered to improve the work. The possible directions and limitations of the individual studies are mentioned in the remarks column of Table 1. In a comprehensive manner, the following challenging issues are observed:

- The detection of abnormal activities that demand alarming attention needs to be installed with GPU based higher configured systems that can make it expensive for real time implementation.
- The installation of cameras in real time raise issues related to privacy, usability expertise, acceptance in society, and large scale applicability.
- In the discussed work, most of the contributions can be noticed related to work on the specific datasets for specific activities. The real time implementation of these

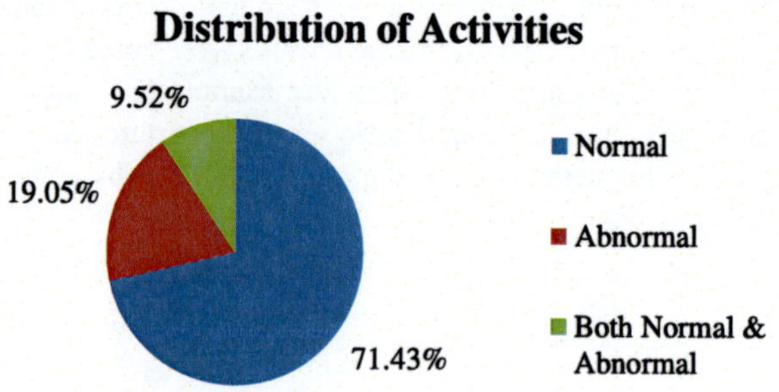

Fig. 2 Distribution of the different human activities in discussed work

systems demands to have higher recognition accuracy regardless of the color, gender, age, size, shape, and other body gadgets.

- The systems should also be resolved for the view-point issues as the human cannot stay focused towards the camera for the activity recognition. This raises the need to install multiple cameras at the same location with the ability of the system to observe the activity from side poses as well.

4 Conclusion

Human activity recognition is one of the significant applications in the digital age to adapt for different applications such as video surveillance, health monitoring, automated security, and human computer interaction. This research paper has focused on video surveillance to monitor different normal and abnormal activities. The contributions of the different researchers related to human activity recognition for video surveillance are thoroughly analyzed and significant outcomes are discussed. In the discussed techniques, the deep learning models are considered as efficient to recognize both the normal and abnormal activities. The different challenging issues determined with the existing techniques are described which can be addressed to implement the human activity recognition for video surveillance in real life.

References

1. Demrozi, F., Pravadelli, G., Bihorac, A., & Rashidi, P. (2020). Human activity recognition using inertial, physiological and environmental sensors: A comprehensive survey. *IEEE Access, 8*, 210816–210836. https://doi.org/10.1109/ACCESS.2020.3037715.
2. Gupta, S., Kumar, M., & Garg, A. (2019). Improved object recognition results using SIFT and ORB feature detector. *Multimedia Tools and Applications, 78*(23), 34157–34171.
3. Zhang, F., Wang, Y., & Zhang, Z. (2011). View-invariant action recognition in surveillance videos. In *Proceeding of the First Asian Conference on Pattern Recognition* (pp. 580–583), Beijing, China.
4. Gao, H., Lin, W., Yang, X., Li, H., Xu, N., Xie, J., & Li, Y. (2011). A new network-based algorithm for multi-camera abnormal activity detection. In *Proceedings of the 2011 IEEE International Symposium of Circuits and Systems (ISCAS)* (pp. 361–364), Rio de Janeiro, Brazil.
5. Shieh, W. Y., & Huang, J. C. (2012). Falling-incident detection and throughput enhancement in a multi-camera video-surveillance system. *Medical engineering & physics, 34*(7), 954–963.
6. Chathuramali, K. M., & Rodrigo, R. (2012). Faster human activity recognition with SVM. In *Proceedings of the International Conference on Advances in ICT for Emerging Regions (ICTer2012)* (pp. 197–203), Colombo, Sri Lanka.
7. Gupta, J. P., Singh, N., Dixit, P., Semwal, V. B., & Dubey, S. R. (2013). Human activity recognition using gait pattern. *International Journal of Computer Vision and Image Processing (IJCVIP), 3*(3), 31–53.
8. Hu, M. K. (1962). Visual pattern recognition by moment invariants. *IRE Transactions on Information Theory, 8*(2), 179–187.

9. Bengalur, M. D. (2013). Human activity recognition using body pose features and support vector machine. In *Proceedings of the 2013 International Conference on Advances in Computing, Communications and Informatics (ICACCI)* (pp. 1970–1975), Mysore, India.
10. Kaur, R., & Singh, S. (2014). Background modelling, detection and tracking of human in video surveillance system. In *Proceedings of the 2014 Innovative Applications of Computational Intelligence on Power, Energy and Controls with their impact on Humanity* (pp. 54–58), Ghaziabad, India.
11. Htike, K. K., Khalifa, O. O., Ramli, H. A. M., & Abushariah, M. A. (2014). Human activity recognition for video surveillance using sequences of postures. In *Proceedings of the Third International Conference on e-Technologies and Networks for Development* (pp. 79–82), Beirut, Lebanon.
12. Taha, A., Zayed, H. H., Khalifa, M. E., & El-Horbaty, E. S. M. (2015). Skeleton-based human activity recognition for video surveillance. *International Journal of Scientific & Engineering Research, 6*(1), 993–1004.
13. Patwardhan, A. S. (2015). Walking, lifting, standing activity recognition using probabilistic networks. *International Research Journal of Engineering and Technology (IRJET), 3*(9), 667–670.
14. Kamal, S., Jalal, A., & Kim, D. (2016). Depth images-based human detection, tracking and activity recognition using spatiotemporal features and modified HMM. *Journal of Electrical Engineering & Technology, 11*(6), 1857–1862.
15. Khemchandani, R., & Sharma, S. (2016). Robust least squares twin support vector machine for human activity recognition. *Applied Soft Computing, 47*, 33–46.
16. Ijjina, E. P., & Chalavadi, K. M. (2017). Human action recognition in RGB-D videos using motion sequence information and deep learning. *Pattern Recognition, 72*, 504–516.
17. Zhao, Y., Di, H., Zhang, J., Lu, Y., Lv, F., & Li, Y. (2017). Region-based mixture models for human action recognition in low-resolution videos. *Neurocomputing, 247*, 1–15.
18. Kamthe, U. M., & Patil, C. G. (2018). Suspicious activity recognition in video surveillance system. In *Proceedings of the 2018 Fourth International Conference on Computing Communication Control and Automation (ICCUBEA)* (pp. 1–6), Pune, India.
19. Bouachir, W., Gouiaa, R., Li, B., & Noumeir, R. (2018). Intelligent video surveillance for real-time detection of suicide attempts. *Pattern Recognition Letters, 110*, 1–7.
20. Gupta, S., Thakur, K., & Kumar, M. (2020). 2D-human face recognition using SIFT and SURF descriptors of face's feature regions. In *The Visual Computer* (pp. 1–10). https://doi.org/10.1007/s00371-020-01814-8.
21. Singh, R., Kushwaha, A. K. S., & Srivastava, R. (2019). Multi-view recognition system for human activity based on multiple features for video surveillance system. *Multimedia Tools and Applications, 78*(12), 17165–17196.
22. Kong, Y., Huang, J., Huang, S., Wei, Z., & Wang, S. (2019). Learning spatiotemporal representations for human fall detection in surveillance video. *Journal of Visual Communication and Image Representation, 59*, 215–230.
23. Jaouedi, N., Boujnah, N., & Bouhlel, M. S. (2020). A new hybrid deep learning model for human action recognition. *Journal of King Saud University-Computer and Information Sciences, 32*(4), 447–453.
24. Silva, M. V. da, & Marana, A. N. (2020). Human action recognition in videos based on spatiotemporal features and bag-of-poses. *Applied Soft Computing, 95*. https://doi.org/10.1016/j.asoc.2020.106513.

Integrating IoT with Blockchain: A Systematic Review

Malvinder Singh Bali, Kamali Gupta, and Swati Malik

Abstract With the advent of IoT, Internet dominance has been extended far and wide, resulting in management of billions of smart devices online. However, all the management frameworks in IoT developed so far are based on centralized models, which have their own set of issues like single point of failure and security constraints. To encounter major issues like this, Blockchain provides an effective alternative to encounter the issues of security and privacy. Blockchain is being a distributed and decentralized ledger framework, when used with IoT helps to encounter a lot of security related constraints. In this paper, a detailed study related to Blockchain has been conducted and how well it works in collaboration with IoT to overcome all the major security and privacy issues in an IoT ecosystem. Many other applications of Blockchain in an IoT environment were also studied, with the intention of exploring all the potential benefits it can provide while working with IoT.

Keywords Blockchain · Decentralization · Smart contract · IoT

1 Introduction

With the advancement of Digital Technology, IoT has fostered various technologies like Cloud Computing and Machine Learning to Data Analytics and Information modelling and thus becoming an integral part of IoT world. All these technologies are dependent on data which is provided by IoT with the help of sensor-based devices. With the passage of time, very smaller to smaller devices are brought under the ambit

M. S. Bali (✉) · K. Gupta · S. Malik
Chitkara University Institute of Engineering and Technology, Chitkara University Rajpura, Patiala, India
e-mail: malvinder.singh@chitkara.edu.in

K. Gupta
e-mail: Kamali.singla@chitkara.edu.in

S. Malik
e-mail: Swati.malik@chitkara.edu.in

P. Nayak et al. (eds.), *IoT and Analytics for Sensor Networks*, Lecture Notes in Networks and Systems 244, https://doi.org/10.1007/978-981-16-2919-8_32

of IoT domain including household devices, health care, industry, precision agriculture and autonomous vehicles. Data collected from various IoT devices contains Critical and Private data, and current IoT architectures are not able to make data reliable and secure [1]. Most of the IoT based architectures are heavily centralized and liable to face failures due to single point which brings Scalability issue and brings lot of issues like Privacy and Security. The introduction of Blockchain brings new opportunity to overcome the challenges faced by IoT where decentralized approach is followed and seen as a long-term solution for the rise of IoT and helps in avoiding centralized failures. Presently Blockchain is seen as the suitable option to provide Secure and Distributed network to IoT [2].

The term Blockchain appeared in article "a cryptographic secured chain of blocks" [3]. Later it was universally introduced by Nakamoto in [4] and brought into implementation in 2008 and 2009 as the main part of Cryptocurrency Bitcoin. With the slow and steady adoption of Blockchain features to decentralized IoT communication and eradicate the need for depending on central trusted servers. Various researches are undergoing through provide reliable, trustworthy, scalable and secure IoT based network using Blockchain. Several benefits of decentralizing IoT with the help of Blockchain are as follows:

- IoT move from centralized server system to Decentralized Blockchain increases fault tolerance and eradicates central point of failure.
- With Blockchain network, users can authenticate data integrity and also the identity of the sending participant.
- IoT data stored in Blockchain are Immutable which ensures accountability.

2 IoT

2.1 Introduction of IoT

It extends internet power beyond computers and smart phones. The concept of IoT in simple terms is to connect all the electronic devices with the internet world. When these things are connected to the internet world, they can send and receive information. This ability makes these things smart. In IoT, objects strongly interconnected with the internet can be categorized as object which gather data and send it, object which receive data and act upon and objects that can do both. The IoT is a system of interconnected digital device, machines, animals or people provided with unique identifiers and the ability to transmit and share data over the network without human to computer or human to human interaction. IoT technology stack is divided into four layers involved in making IoT work. (1). Device hardware: They act as an interface between the real and digital world. From small objects like microphone to big size objects, everything can be practically turned into a connected device by adding sensors or actuators to calculate and gather the required data. (2). Device software: responsible for implementing device communication with cloud, collect

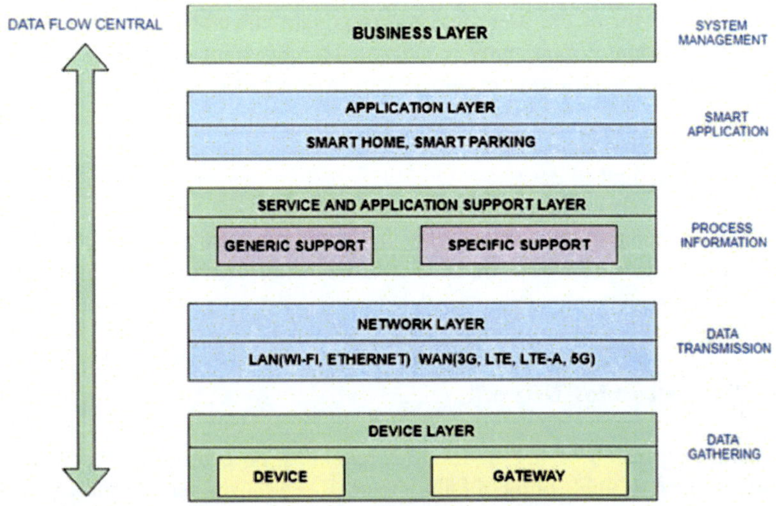

Fig. 1 IoT layered architecture [1]

data and integrate device for performing real time data analysis with IoT network. (3). Communication: which comprises of Cellular and satellite communication and specific protocol used in IoT environment (Zigbee, Zwave, etc.). Choosing relevant solution is one of the major challenges of every IoT stack. (4). Platform: space in which data is gathered, managed, processed, analyzed, and presented in a user friendly way. Several architectures for IoT were present but it never got standardized. Famous IoT architectures which made their presence in the IoT market were Microsoft Azure, RAMI4.0, WS02 and Internet of everything model. The paper highlights an architecture comprised of five layers as shown in Fig. 1. Network layer will provide network connectivity using Wired LANs, WANs, and Radio Satellites. It also facilitates with the feature of AAA (Authentication, Authorization and Accounting). Service and Application Support Layer (SASL) supports two capabilities such as Generic support capabilities and Specific support capabilities provides support for different IoT applications. Business Layer can build various business-related models, graphs, flowcharts and reports by gathering information from application layers [5] to make better decisions.

2.2 *Challenges in IoT*

2.2.1 Data Privacy

Due to diversified integration of IoT based services, lot of data is being generated sending vast majority of information of users to third parties who observe the pattern/behavior of the users making users data vulnerable [6]. To encounter this problem, various policies have been framed for providing privacy but challenge remains to provide solutions to ensure privacy by design.

2.2.2 Management of data

Tremendous amount of data generated by IoT devices is huge and becomes unmanageable. So, to store bulk data, we need scalable infrastructure to efficiently handle growing data [7].

2.2.3 Centralized IoT Network

Continuous expansion of IoT nodes with reliability on central network for data storage will lead to single point of failure issue. If centralized server is down, whole network nodes will be down.

2.2.4 Data Originality Issue

For IoT, it is very tedious to authenticate the originality of the data. It can be altered by anyone without user having the proof of its tampering.

2.2.5 Network Complexity

In IoT, there are difficult network protocols used as per the application. Protocols such as NB-IoT, Sigfox, Lora, NFC, and Bluetooth offer different coverage ranges creating an interoperability issues [7, 8].

2.2.6 Third Party Data Breach

In IoT network, data is uploaded on the cloud server which is mostly handled by the 3rd party. They can also violate the privacy of the user by observing continuous pattern of the user's data [9].

2.2.7 IoT Device Resource Constraints

Most of the devices lack storage, computation and battery power which makes IoT devices vulnerable to different kinds of attacks.

2.2.8 Heterogeneity

Heterogeneity of protocols and IoT data types brings different issues related to incompatibility.

3 About Blockchain Technology

Blockchain is a distributed and decentralized public ledger (Fig. 2).

The answer is very simple. As the name suggests it is literally a chain of blocks as shown in Fig. 3. It consists of many blocks connected to each other in such a way that each block is connected to the next and previous block.

Each block mainly contains three things:

- **Data** The information like sender's address, receiver's address and the amount of transaction is stored in this bit of block.
- **Hash of own block** A unique long identifier consisting on numbers and alphabets which acts as the fingerprint of the block is stored in this bit. For example, "7d86c5626f1516e01182b5ed12e6924c0d2e908cba8ce3cd062f062b8d989 18b"
- **Hash of previous block** Each block has the previous block's hash stored in it. The reason for this is to increase security. If someone tries to tamper with a block, its hash changes and then the hash will not match the hash stored in the previous block and we would get to know which block has been tampered with.

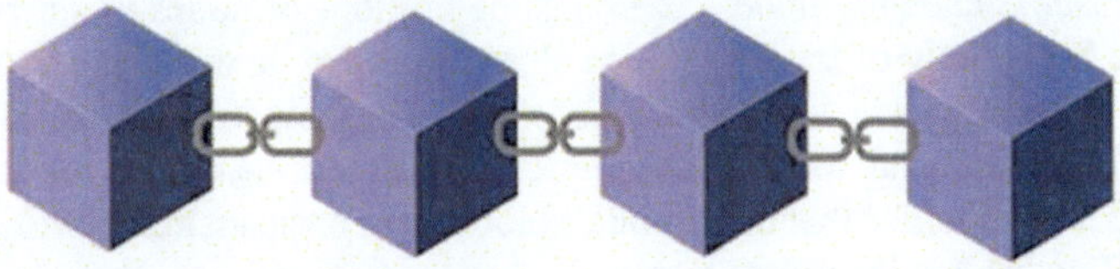

Fig. 2 Simplified blockchain. *Source* Selfmade

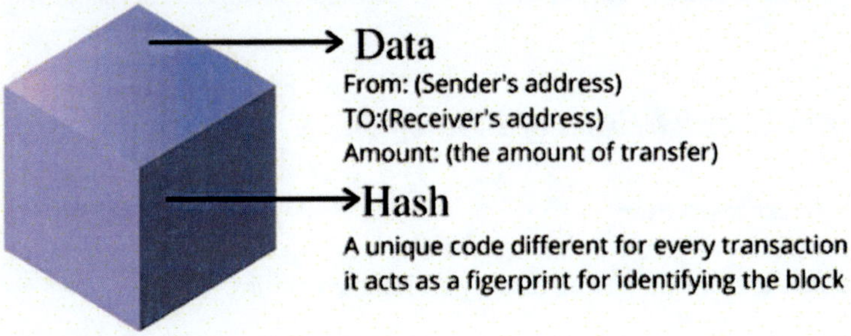

Fig. 3 Structure of a block *Source* Selfmade

3.1 Security Features of Blockchain

3.1.1 SHA256 Hash

The hashes are calculated using the sha256 algorithm which has something called as the avalanche effect. In the avalanche effect if we change even one alphabet in the data the hash changes completely and we get a 100% new hash. The sha256 hash is only one way. We can enter the data and get a hash but not the other way around.

3.1.2 Proof of Work

Nowadays we have a lot of computing power so just using a secure hash is not enough as computers can quickly tamper with the data and recalculate all the hashes of next blocks so fast that it would be impossible to detect that someone has tampered with the data. So, to overcome this flaw proof of work was introduced, this slows down the creation of new blocks and if you tamper with a block you would have to recalculate the proof of work for all other blocks as well which is time consuming.

3.1.3 Decentralization

This is the most important concept in blockchain technology. A blockchain has no owner rather it is distributed in a network of many computers/users, and each computer/user is called a 'node'. There can be millions of nodes in a network and each node has a copy of the blockchain stored in it. So, if we take a hypothetical situation where a hacker logs onto a node's pc and tampers with the blockchain successfully changing all its hashes and proof of work too. As soon as the blockchain is refreshed it will notice that that 1 node's blockchain does not match with the other millions of nodes and revert it to normal.

So, in conclusion, to tamper with the blockchain a hacker has to hack into millions of computers at the same time and change all the hashes and proof of work within seconds which is next to impossible. That is why blockchain is a very secure network.

3.2 Key Characteristics of Blockchain

3.2.1 Decentralization

All the data is stored at and verified by all initiated parties.

3.2.2 Immutability

Data once committed (stored) to the Blockchain cannot be changed or deleted. Once a transaction is entered in the database and the accounts are updated, the records cannot be attached because they are linked to every transaction record that came before them.

3.2.3 Transparency

In Blockchain, users involved in the Transaction process can see the transaction value and their authenticity (node or user) is provided by three plus alphanumeric address. Also, users of the system can reveal their identity in the system or remain silent.

3.2.4 Peer-to-Peer Transmission

Communication takes place between one system to other system without the support of a centralized system. Each node stores and forwards information to all other nodes.

3.2.5 Smart Contract

A smart contract is just a computer program that controls the digital transfer of currency or assets between two or more parties under some given conditions just as in a normal contract, but more securely and 100% fraud-free (Fig. 4). The smart contract states the conditions of the contract and enforces them on its own. They work a lot like food machines; you put the required amount of crypto currency in the machine and you get what you paid for.

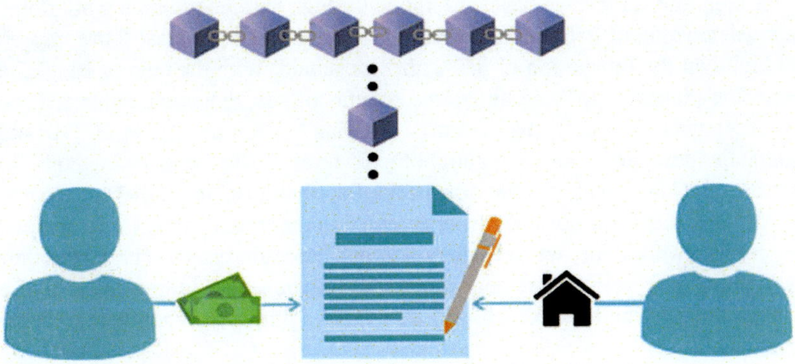

Fig. 4 Smart contract *Source* Selfmade

For example, if someone want to buy a piece of land from a person the standard procedure is that the two parties agree on the price, the land owner gets paid and then they go to a govt. office to get the land ownership changed in the govt. ledger which is just a huge register stored in the govt. archives. And for changing this ownership details, the govt. charges 5–7% of the land value. But on the other hand, if we use a smart contract, we can save this 5% fee. As soon as the buyer pays the smart contract the money gets transferred into the owner's crypto wallet and the land ownership details are now stored in a block, and this block becomes a part of a very large chain and copies itself onto millions of nodes hence securing the data. This is a lot safer than the govt. ledger as it cannot be destroyed due to any natural calamity. See Fig. 2 for illustration of this example.

- **Security of a smart contact**:
  The conditions of a smart contract are 100% transparent, and there are no chances of fraud. You always get what you pay for, and the contract is totally unbiased. It is just a computer code written in a programming language e.g., python, so the contract will always execute and there is no chance of failure.
- **Speed of a smart contract**:
  In the above property example, if we go with the traditional way of transferring ownership, it takes a huge amount of paperwork and travelling to many desks in govt. offices which is very time consuming and difficult. But if we use a smart contract the transaction takes place in a few seconds and the information gets stored in the blockchain immediately and hassle-free. Also, the smart contracts are 100% accurate and have no chances of human error.

4 Literature Review of Integration Between IoT and Blockchain

In our literature survey, we have undergone through various research work done in the amalgamation of IoT and Blockchain. Major research proposals have appeared that highlight the Integration of IoT with Blockchain. With the help of Blockchain Technology features, IoT can utilize the benefits of cost reduction, enhanced security, accelerated data exchange and trust building [7]. Conoscenti et al. [10] did a systematic literature review to highlight the features of Blockchain adopted by IoT. Atzori et al. [11] highlighted the drawback of applying BC in the IoT environment and also presented solutions that have emerged in the recent years. In our review of the previous work done, we have classified our literature research into subsections highlighting the research work done by various researchers to address the challenges in integrating IoT and Blockchain and also various proposals from them to combat the issue.

4.1 Privacy of IoT via Blockchains

4.1.1 Privacy Issues in Centralized IoT

With the introduction of IoT, data, people, virtual environments and different physical objects like wearable sensors, medical devices and smart objects interconnect with each other, resulting in the development of smart environments like, smart cities, smart transport systems, smart agriculture and so on.

However, there is a downside to this technological development also. Numerous challenges like privacy of individuals, security of the processes and the developed system arise. Let us take an example of a smart healthcare application which involves the use of different bio-sensors attached to the patient's body, regularly sending the important data related to the patient like blood pressure, heart rate to the cloud for analysis and evaluation and RFID based tag sensors for user authentication. Data after analysis will be sent to the consulted doctor's smart device and the patient for improvement and feedback.

If we analyze this application intricately, we observe that all the sensitive information regarding location of the patient, medical history and the present medical state of the patient is sent to the cloud using sensors, so there is great need to protect the privacy of the patient, particularly in case of high-profile patient.

The data collected from various IoT sensors can be used for diverse purposes. Certain organizations can predict the user's behavioral pattern and habits as well as the metadata. Customers using IoT based devices put trust on the companies providing smart services and are totally unaware of their data being shared or sold to the third-party.

4.1.2 Decentralization via Blockchain for IoT Privacy

Various ongoing researches in making data decentralized to resolve privacy violation are done and lot of proposed frameworks, architecture and algorithms have been designed. In [11], the author provided a framework called PISCES aiming to provide a strict separation between data provider and data controller. Here providers manage their data privacy and the data controllers are accountable to the privacy and data protocol for IoT. In [12], paper describes PLATIBART, a platform for exchanging IoT Blockchain based application with continuous testing. Using Plaribart, researcher shows how IoT Blockchain based application is developed, tested and then analyzed how effective IoT based application will be in terms of privacy.

5 Convergence of BC and IoT

5.1 Issues Faced in the IoT

5.1.1 Data Privacy

Due to diversified integration of IoT based services, lot of data is being generated sending vast majority of information of users to third parties who observe the pattern/behavior of the users making users data vulnerable [6]. To encounter this problem, various policies have been framed for providing privacy but challenge remains to provide solutions to ensure privacy by design.

5.1.2 Management of Data

Tremendous amount of data generated by IoT devices is huge and becomes unmanageable. So, to store bulk data, we need scalable infrastructure to efficiently handle growing data [13].

5.1.3 Centralized IoT Network

Continuous expansion of IoT nodes with reliability on central network for data storage will lead to single point of failure issue. If centralized server is down, whole network nodes will be down.

5.1.4 Data Originality Issue

In IoT network, it is very difficult to authenticate the originality of the data. It can be altered by anyone without user having the proof of its tampering.

5.1.5 Network Complexity

In IoT, there are difficult network protocols used as per the application. Protocols such as NB-IoT, Sigfox, Lora, NFC, and Bluetooth offer different coverage ranges creating an interoperability issues [10, 14].

5.1.6 Third Party Data Breach

In IoT network, data is uploaded on the cloud server which is mostly handled by the 3rd party. They can also violate the privacy of the user by observing continuous pattern of the user's data [9].

5.1.7 IoT Device Resource Constraints

Most of the devices lack storage, computation and battery power [11] which makes IoT devices vulnerable to different kinds of attacks.

5.1.8 Heterogeneity

Heterogeneity of protocols and IoT data types brings different constraints like privacy and interoperability issues.

5.2 *Benefits of Integrating Blockchain with IoT*

Following the features provided by Blockchain and its potential impact in the current phase, lots of research work is being done to amalgamate the Blockchain technology with IoT to counter the challenges faced in centralized IoT. Below are the benefits offered by BC based IoT framework.

5.2.1 Enhanced Security

Data of IoT can be secured via blockchain as data is stored in the form of Blockchain transactions which are encrypted. Smart contracts in Blockchain technology also help to improve security of IoT. Any data breach will update the users to prompt necessary action thereby improving security [15].

5.2.2 Minimizes Third Party Risks

Decentralized distribution of data removes third party risks as data is stored only at authenticate nodes and cannot be altered without approval of user [16].

5.2.3 Fault Tolerance

Since network of Blockchain based IoT devices is decentralized with point-to-point network where every device attached has the same data. So, failure/fault in one node will not hamper entire network.

6 IoT-Blockchain Applications

6.1 Chain of Things [17]

Chain of Things (CoT) is a Hong Kong-based blockchain + IoT venture production studio involved in a number of ventures in different industry verticals. An integrated blockchain and IoT hardware solution to solve IoT's issues with identity, security, and interoperability.

6.2 My Bit [18]

Provides the first IoT ecosystem in the world. It works on Smart contract system using Ethereum is built by industry veterans in Switzerland.

6.3 Slock.It [19]

Slock is an application combining the technology of Internet of things and Block Chain. This means that you can put advertisement on a blockchain, so the ad is connected to you. The reply to add is also on blockchain, and with internet of things, that transaction of the physical good can be confirmed, without a middleman. Slock is a platform that support all needed to share or sell. Suppose you want to share a room, a bicycle, a car. You find the ad on the blockchain, and you communicate with the owner without needing to give personal data. Only the counterparty, will be involved in all steps to complete a transaction.

6.4 Riddle and Code [20]

It helps various Stakeholders like companies from different spheres, to individuals and institutions to reap the benefits of digital technology of Blockchain and IoT which includes cybersecurity, data trust, avoiding online frauds, etc. The Riddle and

Code amalgamates smart card security features with Bitcoin Technology and IoT. Also, to tackle the projects related to Cryptography, they have hardware and software built in Stacks.

6.5 Modem [21]

Provides real time solutions in Intelligence of Supply chain and Automation making strong ecosystem where features like IoT, Blockchain based Network and Artificial Intelligence can be provided to various Industries.

7 Integration Challenges in Blockchain and IoT

7.1 Scalability Issue

Every time, there is a constant rise in the transaction of Blockchain ledger which might lead to centralization of huge data. Thus, it requires proper management of records to avoid future issues in Blockchain.

7.2 Processing Power and Time

IoT ecosystem has a lot of diversification as compared to other computing networks. IoT network encompasses various devices with varied computing capabilities, some of them lagging the capability of running an identical encryption algorithm at the desired speed.

7.3 Storage

Storage will be tedious. Although in Blockchain there is no need for a centralized server to store device ID's and transactions, the ledger needs to be stored on the nodes themselves and with an increase in time duration, the size of the ledger is also going to increase. Smart devices such as sensors lack this capability due to low storage capacity available.

7.4 Interoperability

The private and public nature of Blockchain will create a problem as their interoperability is also a challenge.

7.5 Legal and Compliance Issues

The absence of any legal and compliance code, due to blockchain being a new territory, poses a difficult threat to be dealt by manufacturers and the service providers in this field. Due to this serious problem, many business domains refrain from using Blockchain technology.

8 Conclusion and Future Work

IoT has revolutionized the digital world by connecting every device with Internet and generating huge amount of data which can be used in expert decision making by observing user's device patterns. Practically, lot of issues like interoperability, scalability, Security and Privacy, Trust and Heterogeneity issues have jeopardized IoT Technology. Also, different Protocols used for machine to machine interaction have made the task of vendors very difficult as interoperability issues will lead to issues like security breach. To overcome all the challenges, Blockchain does provide a solution to overcome these challenges but adoption of regulations by regulatory bodies is key to the Inclusion of Blockchain and IoT as part of Government Infrastructure. The adoption will help citizens, companies and government to adopt twin technology hazzle free. Lastly certain issues have also been highlighted like scalability and storage capacity which will affect Blockchain and IoT. So, research efforts need to be made in this direction to make this twin technology become the next big thing in I.T. world. In the next phase, we will incorporate the features of both the twin technologies and create a secure and Reliable IoT Application using NB-IoT. Also, the tradeoff between Public and Private Blockchain is an area of research where more emphasis will be to select the best Blockchain system for Real time Applications.

References

1. Alaba, F. A., Othman, M., Hashem, I. A. T., & Alotaibi, F. (2017). Internet of Things security: A survey. *Journal of Network and Computer Applications, 88*, 10–28.
2. Huckle, S., Bhattacharya, R., White, M., & Beloff, N. (2016). Internet of Things, blockchain and shared economy applications. *Procedia Computer Science, 98*, 461–466.

3. Haber, S., & Stornetta, W. S. (1990). How to time-stamp a digital document. In *Conference on the Theory and Application of Cryptography* (pp. 437–455). Springer.
4. Nakamoto, S. (2019). *Bitcoin: A peer-to-peer electronic cash system.* Manubot.
5. Kim, T., Kim, D. M., Pratas, N., Popovski, P., & Sung, D. K. (2017). An enhanced access reservation protocol with a partial preamble transmission mechanism in NB-IoT systems. *IEEE Communications Letters, 21*(10), 2270–2273.
6. Zhou, J., Cao, Z., Dong, X., & Vasilakos, A. V. (2017). Security and privacy for cloud-based IoT: Challenges. *IEEE Communications Magazine, 55*(1), 26–33.
7. Gopal, S. (2016). Blockchain for the internet of Things. *Tata Consultancy Services White Paper, TCS Design Services.*
8. Conoscenti, M., Vetro, A., & De Martin, J. C. (2016). Blockchain for the Internet of Things: A systematic literature review. In *2016 IEEE/ACS 13th International Conference of Computer Systems and Applications (AICCSA)* (pp. 1–6). IEEE.
9. Akpakwu, G. A., Silva, B. J., Hancke, G. P., & Abu-Mahfouz, A. M. (2018). A survey on 5G networks for the Internet of Things: Communication technologies and challenges. *IEEE Access, 6*, 3619–3647.
10. Stankovic, J. A. (2014). Research directions for the internet of things. *IEEE Internet of Things Journal, 1*(1), 3–9.
11. Foukia, N., Billard, D., & Solana, E. (2016). PISCES: A framework for privacy by design in IoT. In *2016 14th Annual Conference on Privacy, Security and Trust (PST)* (pp. 706–713). IEEE.
12. Walker, M. A., Dubey, A., Laszka, A., & Schmidt, D. C. (2017). Platibart: A platform for transactive IoT blockchain applications with repeatable testing. In *Proceedings of the 4th Workshop on Middleware and Applications for the Internet of Things* (pp. 17–22).
13. Mohammadi, M., & Al-Fuqaha, A. (2018). Enabling cognitive smart cities using big data and machine learning: Approaches and challenges. *IEEE Communications Magazine, 56*(2), 94–101.
14. Sicari, S., Rizzardi, A., Grieco, L. A., & Coen-Porisini, A. (2015). Security, privacy and trust in Internet of Things: The road ahead. *Computer Networks, 76*, 146–164.
15. Christidis, K., & Devetsikiotis, M. (2016). Blockchains and smart contracts for the internet of things. *IEEE Access, 4*, 2292–2303.
16. http://www.techracers.com/blockchain-key-features
17. Chain of Things. (2017). Retrieved February 1, 2018, from https://www.chainofthings.com/.
18. My Bit. (2017). Retrieved February 1, 2018, from https://mybit.io/.
19. Bahga, A., & Madisetti, V. K. (2016). Blockchain platform for industrial internet of things. *Journal of Software Engineering and Applications, 9*(10), 533–546.
20. Riddle and Code. (2017). Retrieved February 1, 2018, from https://www.riddleandcode.com.
21. Modum. (2017). Retrieved February 1, 2018, from https://modum.io/.
22. Lu, X., Niyato, D., Jiang, H., Kim, D. I., Xiao, Y., & Han, Z. (2018). Ambient backscatter assisted wireless powered communications. *IEEE Wireless Communications, 25*(2), 170–177.

Quality Assisted Spectrum Allocation in Cognitive NOMA Networks

D. Prasanth Varma and K. Annapurna

Abstract Future communications are developed with new communication standards for multi access technologies, where spectrum sharing and Non-orthogonal multiple access (NOMA) are the two latest approaches. To transfer the data from one end to another end through this wireless communication medium interference take place, which affects the total system performance. To minimize the interference in these communications the efficient spectrum sharing technique was developed. Cooperative relaying in NOMA system with spectrum sharing using threshold modeling were proposed in past. To improve the resource allocation in NOMA with cooperative relaying, instantaneous signal to noise ratio were used in obtaining higher outage throughput in a spectrum sharing cognitive radio-NOMA (CR-NOMA) system. In the previous methods constant thresholding is considered in the energy detection model and the quality of signaling is not observed, which results in lower throughput. In this work, a Quality Assisted Spectrum Allocation (QASA) based on loss probability is proposed and a dynamic threshold modeling is suggested for spectrum sensing under dynamic channel condition. Here the Rayleigh and Rician fading models are considered and used the Lagrange mathematical concept to calculate the bit error and to minimize the distortion. The throughput of the proposed work performs well when compared with the conventional method of Simultaneous wireless information and power transfer (SWIPT).

Keywords Cognitive radio · NOMA · Throughput · Dynamic thresholding · Spectrum allocation

1 Introduction

An approach of spectrum sharing for nodes with poorer channel condition is presented in [1]. This approach defines a node as primary with node having poorer channel condition. Here a two slot communication model is presented, which performs the spectrum sharing in the first slot part and perform a data exchange in the second slot.

D. P. Varma (✉) · K. Annapurna
Vignan's Foundation for Science, Technology and Research, Vadlamudi, India

P. Nayak et al. (eds.), *IoT and Analytics for Sensor Networks*, Lecture Notes in Networks and Systems 244, https://doi.org/10.1007/978-981-16-2919-8_33

The present approach is developed with the objective of maximization of network throughput by the optimization of time slot in maximization of network throughput. A NOMA communication based on overlay cooperative cognitive radio with the channel estimation and successive interference cancellation (SCI) is presented in [2]. The approach derives an analytical expression for outage probability computation for the primary and secondary recover as a function of channel error and SCI. The modeling analysis of downlink NOMA network with a independent and non-identical Rayleigh channel with signal to noise ratio (SNR), and data rate is presented in [3]. In [4] the approach of total power over sum-SINR minimization is presented. The analysis of this presented approach defines an accurate approximation of achievable rate and derives the closed form of outage probability. For a near optimal solution to spectrum sensing and sharing, opportunistic interference regulated spectrum utilization is presented in [5]. This approach presented a non-binary coding in spectrum sensing for receiver centric modeling. The model is developed in reference to the interference temperature metric in constraint to spectrum usage. This approach offers an additional usage of spectrum using interference temperature metric in spectrum sharing. The effect of multiple channel interference on the signal propagation in CR-NOMA is presented in [6]. The interference of a PU and SU, Tx-Rx based data rate allocation is presented. [7] outlined a power-dominant multiplexing approach in a CR-NOMA based on time slotting over a real network. A cooperative relaying for the spectrum utilization enhancement in a non-orthogonal multiple access (NOMA) is proposed. A suboptimal power allocation in NOMA network with average rate of coding is presented. The presented approach is developed on independent Rayleigh fading channel interference. [8] outlined an analytical framework for computing the achievable rate by Gauss-Chebyshev integration developed for Rician fading channel. In [9] a full duplex approach with NOMA network is presented to improved network capacity for next generation communication. A secondary user access based on spectrum sharing is outlined in [10]. This approach optimizes the quality-of-service of a PU based on the aggregated rate allocation in a wireless network model. In [11] QoS based coding in consideration to individual user is presented. This approach proposes a CR based NOMA system for energy efficient coding.

In [1] an approach for throughput improvement with optimal power transfer approach is presented. This method outlined the optimal computation of power and information communication in a two slot manner. The presented approach controls the data rate allocation by monitoring the channel gain and power allocation. Wherein throughput is improved, the quality of the received data is unaddressed. In addition to throughput, quality of data exchanged is needed. The impact of QoS on the sharing spectrum has a greater impact on the offered quality of service in the network. In order to achieve the objective of low error resource allocation in a CR-NOMA network, here in this paper, a new approach of resource allocation in CR-NOMA based on observed QoS metric using distortion monitoring parameter is presented. The rest of this paper is outlined in five sections. Section 2 presents the existing approach of joint spectrum resource allocation; Sect. 3 outlines the proposed approach of quality assisted spectrum allocation. Obtained experimental results for the proposed approach is discussed in Sect. 4. The conclusion of this work is presented in Sect. 5.

2 Joint Spectrum Resource Allocation

The block diagram of CR-NOMA operation is shown in Fig. 1 where each frame consists of two subsections. The first subsection is engaged in the spectrum sensing and sharing operation, where a "simultaneous wireless information and power transfer (SWIPT)" is performed. In the second subsection the data exchange takes place.

In spectrum sensing process, each secondary user (SU) measures the energy of the received signal is given by

$$\text{En} = \frac{1}{N}\sum_{k=1}^{N}|R(k)|^2, \tag{1}$$

where En is the signal energy and it is measured from all the signals which are coming from N multiple paths. The computed signal energy is compared with the given threshold in deriving the engagement of channel of PU. The detector performs the comparison as

$$\begin{aligned} \text{En} &< \text{Th};\ H_0, \\ \text{En} &\geq \text{Th};\ H_1. \end{aligned} \tag{2}$$

The spectrum which is sensed S_f to be free is allocated to SU for communication. The threshold limit $I_{\max}$ is defined such that interference should be minimum, and it should be tolerable caused due to SUs. The transmitting power Pt is used for

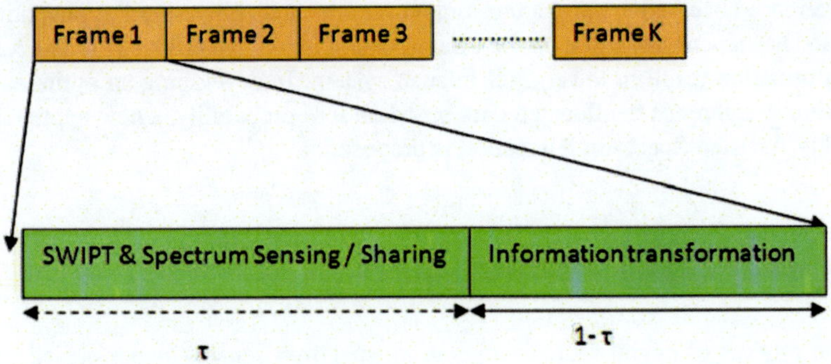

Fig. 1 Structure of CR-NOMA operation

information uplink and a channel gain of Pg is observed for the PU signal. The optimization equation for the rate is given by

$$\max(DR(t)) = P(H_0)(1-t)\left(1 - Q\left(\alpha + \sqrt{tf_s\vartheta}\right)\right)K_0 \tag{3}$$

The optimization function is represented with respect to power transmitted, minimum rate and interference. It is represented by

$$\max DR(t) = (1-t)\sum_{i=1}^{N}\log_2\left(1 + \frac{P_i\vartheta_i}{P_s g_s + \sum_{j=i+1}^{N} P_j\vartheta_j}\right). \tag{4}$$

Subjected to

$$\sum_{i=1}^{N} P_i g_s \leq I_{\max}, \tag{5}$$

$$\sum_{i=1}^{N} P_i \leq P_t \tag{6}$$

$$\log_2\left(1 + \frac{P_i\vartheta_i}{P_s g_s + \sum_{j=i+1}^{N} P_j\vartheta_j}\right) \geq R_i, \tag{7}$$

where the allocation of maximum rate is governed by allocated transmission power P_i and SNR ϑ_i over the given transmission power P_s with a channel gain of g_s. The optimization operation gives an optimal data rate allocation with respect to the interference observed in the channel and power allocated; however, the conventional method has no consideration on the offered QoS with outage probability. The losses of information result in lower QoS in such system. In developing an optimization solution in reference to allocated data based on loss probability, a new approach of Quality Assisted Spectrum Allocation is proposed.

3 Quality Assisted Spectrum Allocation

In the process of cognitive radio operation, spectrum sensing is performed using energy detector. Energy detector is a threshold based correlation of measured signal energy from PU with a given threshold. The decision is made as spectrum free when signal energy is below the threshold limit as given in Eq. (8).

$$\text{En} = \begin{cases} \frac{1}{N}\sum_{i=1}^{N}\left(\frac{R_i}{\sigma}\right)^2 > \text{Thresh}, & (H_1), \\ \frac{1}{N}\sum_{i=1}^{N}\left(\frac{R_i}{\sigma}\right)^2 < \text{Thresh}, & (H_0). \end{cases} \tag{8}$$

This threshold has a major impact in the detection of signals and accuracy in spectrum sensing. In the scenario where channel has a fixed interference effect, a scalar threshold is applicable. However, the probability of estimation under varying channel condition has vacation in signal energy and hence a constant threshold results in false detection. To improve the estimation accuracy, in this paper, dynamic threshold is derived by using the likelihood ratios and considering the signal scaling, shaping and probability of detection.

Probability of estimate $P(R \mid H_0)$ defines the detection probability of the received signal, and estimation of free spectrum is computed by a generalized likelihood ratio derived over a Gaussian distribution. The detection probability of the estimate is developed by the false alarm probability, defined as the estimate probability.

Probability of false alarm P_{Fl} is defined by

$$P_{\text{Fl}} = \text{Prob}\{E > \text{Thresh}|H_0\} \tag{9}$$

The threshold is derived for estimate based on Neyman-Pearson rule is defined by

$$\text{Thresh} = F^{-1}_{\text{Est} \mid H_0}(1 - P_{\text{Fl}}, \theta_0, k_0). \tag{10}$$

Here, $\theta_0$andk_0 are the shaping and scaling parameter, respectively. During the communication when the signal is propagated in medium, it will undergo some fading effects. The Rayleigh fading is defined by

$$f(r) = \frac{r}{\sigma^2}\exp\left\{-\frac{r^2}{2\sigma^2}\right\}, \quad r \geq 0 \tag{11}$$

and the Rician fading is given by

$$k(\text{dB}) = 10\log_{10}\left(\frac{k_d^2}{2\sigma^2}\right) \tag{12}$$

Due to the fading effect the interference takes place and it leads to a packet loss. This loss is measured by using the distortion parameter which is defined as bit error rate. By using the Lagrange method the objective function which is defined as distortion is minimized and updated. The cumulative similarity index is used to calculate the similarity between the signals. The optimization function in reference to similarity index is defined by

$$\min\{J\} = D_{\text{istrSI}} + \lambda_{\text{SI}}.R_a, \tag{13}$$

where J is the Lagrange optimization cost function.

Optimal rate is allocated by considering the distortion, rate requirements and power allocated. The final equation for maximum data rate allocation in addition to the existing constraint is subjected by

$$\begin{aligned} \max(DR) &= \min(\text{Distr}, R_a, P_i) \\ &= (1-t)\sum_{i=1}^{N}\log_2\left(1 + \frac{P_i\vartheta_i}{P_s g_s + \sum\limits_{j=i+1}^{N} P_j\vartheta_j}\right) \geq R_{av}. \end{aligned} \tag{14}$$

4 Experimental Results

The developed approach is compared with the conventional method in terms of throughput. Simulation is carried out by using Matlab software, and the nodes are distributed randomly in the network. Each node has its bandwidth and the power which is required for communication. The network topology is shown in Fig. 2. The

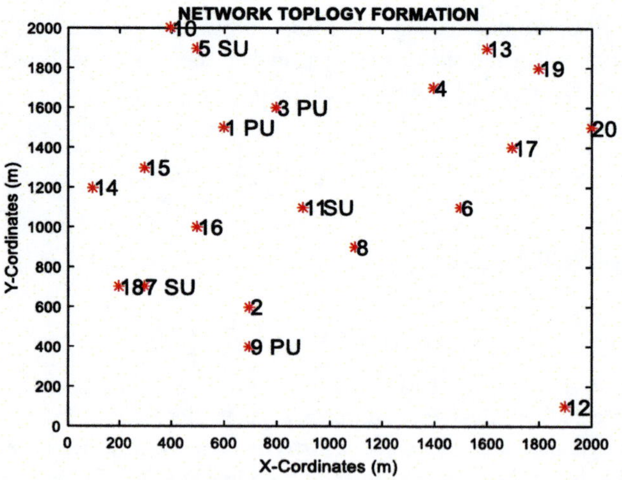

Fig. 2 Network distribution for simulation

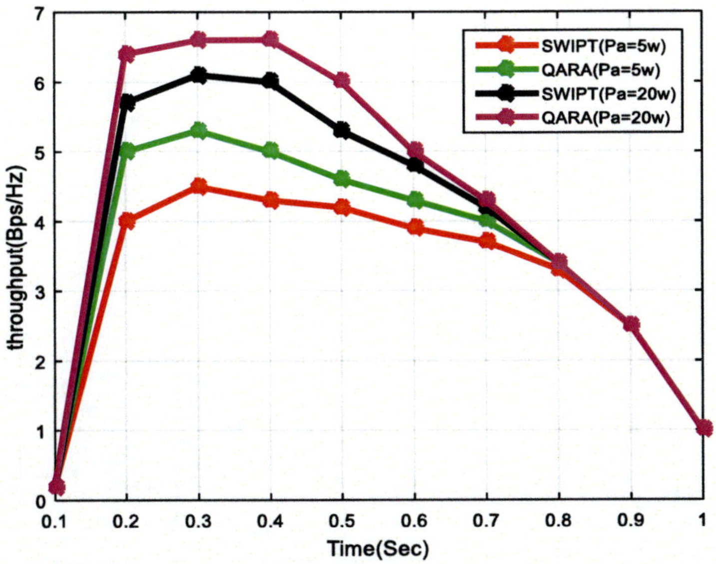

Fig. 3 Throughput with time for varying P_a

SWIPT and QASA are developed for optimal rate allocation. Compilation is carried out for one second. For communication the PU transmission power P_s is ranged from [5, 10, 30] and gain is 4. The P_a is set to [5, 20] W and [10, 30] W and the fading factor is 100. The SNR is [−10:8] dB.

The system throughput for an observation time of 1 s with varying P_a is presented in Fig. 3. The result for the developed approach illustrates an increase in throughput for a initial time period, however with course of time the channel get congested and interference builds. The interference leads to lower data rate allocation hence throughput decreases.

The network throughput with the varying channel gain is presented in Fig. 4. The channel gain increases the power level of the received signal and raises the interference level. The higher interference level builds an overhead in transmission and hence the throughput is decreased.

The throughput of PU with respect to simulation time and by varying P_a is shown in Fig. 5. With an increase in the downlink power, the throughput of PU is decreased. In case of SU with an increase in downlink power the throughput will increase.

The impact of channel gain on the network throughout is presented in Fig. 6 below. Initially the channel gain curve starts at peak point and when the channel gain is increased the throughput started falling down. With an increase in the channel gain, there will be more interference regarding PU which results in lower throughput.

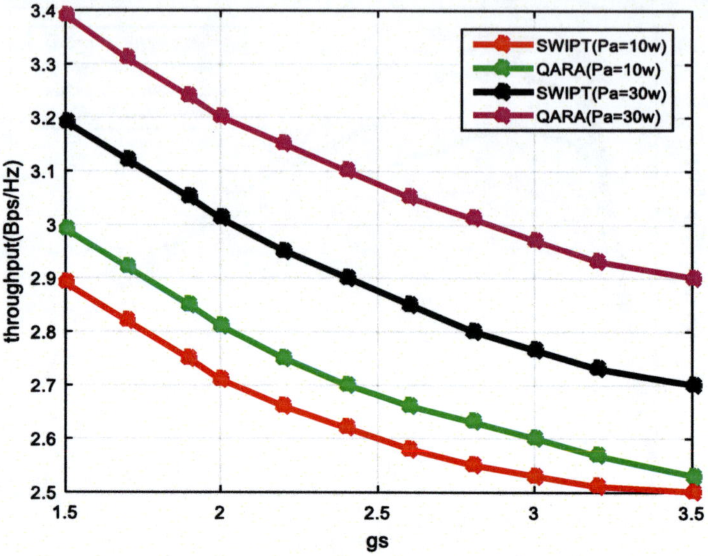

Fig. 4 Network throughput with variation of channel gain g_s

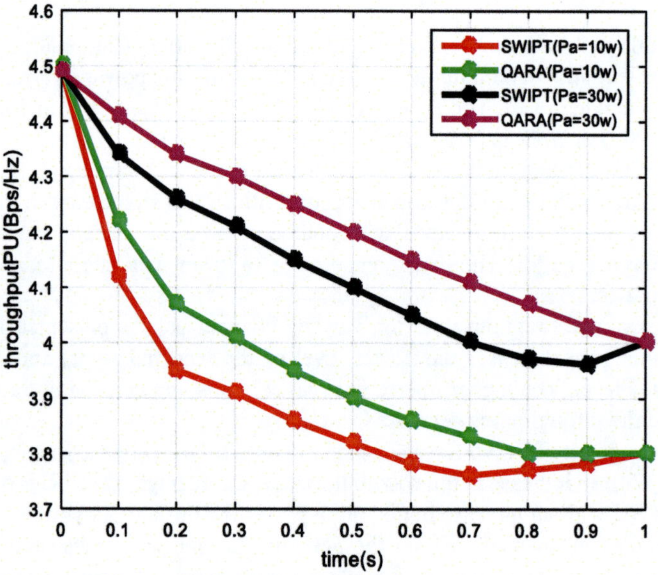

Fig. 5 Throughput PU for different P_a w.r.t. time variation

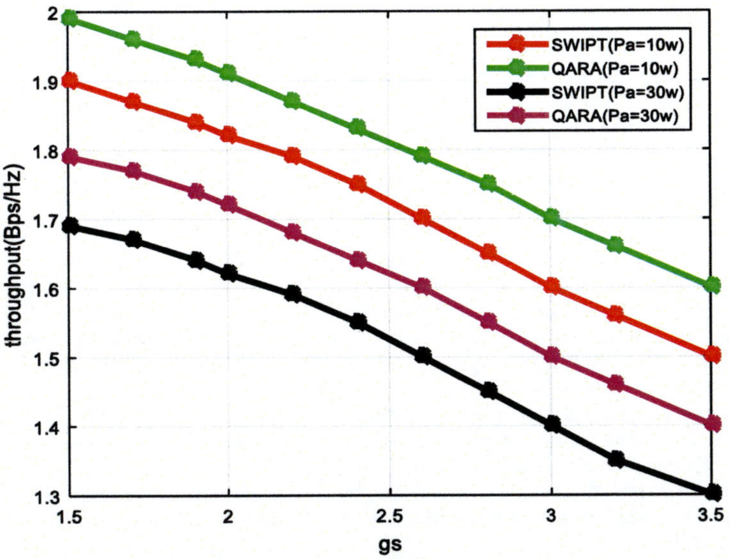

Fig. 6 Throughput over different channel gain with P_a varying

5 Conclusion

The optimum rate is allocated by minimizing the distortion using the developed approach Quality Assisted Spectrum allocation. The distortion is minimized by using the Lagrange optimization cost function and calculated the bit error rate using Lagrange multipliers. Similarity index is used to find the similarity between the transmitted signals and received signals and then using this index the cost function is updated. However in the existing method constant threshold is considered, but as the channels are very dynamic, the dynamic threshold is computed by using the Neyman–Pearson rule. The presence and absence of signal is detected clearly by using this rule with signal scaling and shaping parameters. Simulation is done by varying P_a, P_s, and time, and SNR is considered as $[-10{:}8]$ dB. Here the interference is reduced in this work, and the proposed work outperforms the conventional method.

References

1. Song, Z., Wang, X., Liu, Y., & Zhang, Z. (2019). Joint spectrum resource allocation in NOMA-based cognitive radio network with SWIPT. *IEEE Access, 7*, 89594–89603.
2. Kim, N.S. (2019). Cooperative overlay cognitive radio NOMA network with channel errors and imperfect SIC. *International Journal of Intelligent Engineering and Systems, 12*(5).

3. Zhang, Y., Dong, Y., Wang, L., Liu, J., Peng, Y., & Feng, J. (2019). Outage capacity analysis for cognitive non-orthogonal multiple access downlink transmissions systems in the presence of channel estimation error. *CMC, 60*(1), 379–393.
4. Baidas, M. W., Alsusa, E., & Hamdi, K. A. (2019). Performance analysis and SINR-based power allocation strategies for downlink NOMA networks. *IET Research Journals*, 1–13.
5. Bater, J., Tan, H.-P., Brown, K. N., & Doyle, L. (2007). Modelling interference temperature constraints for spectrum access in cognitive radio networks. In *2007 IEEE Int. Conf. Commun. (ICC)* (pp. 6493–6498).
6. Ghasemi, A., & Sousa, E. S. (2007). Fundamental limits of spectrum-sharing in fading environments. *IEEE Transactions on Wireless Communications, 6*(2), 649–658.
7. Kim, J. B., & Lee, I. H. (2015). Capacity analysis of cooperative relaying systems using non-orthogonal multiple access. *IEEE Communications Letters, 19*(11), 1949–1952.
8. Jiao, R., Dai, L. Zhang, J., MacKenzie, R., & Hao, M. (2017). On the performance of NOMA-based cooperative relaying systems over Rician fading channels. *IEEE Transactions on Vehicular Technolology, 66*(12), 11409–11413.
9. Mohammadi, M., Shi, X., Chalise, B. K., Ding, Z., Suraweera, H. A., Zhong, C., & Thompson, J. S. (2019). Full-duplex non-orthogonal multiple access for next generation wireless systems. *IEEE Communications Magazine, 57*(5), 110–116.
10. Chen, B., Chen, Y., Chen, Y., Cao, Y., Zhao, N., & Ding, Z. (2018). A novel spectrum sharing scheme assisted by secondary NOMA relay. *IEEE Wireless Communications Letters, 7*(5), 732–735.
11. Zhang, Y., Yang, Q., Zheng, T., Wang, H., Ju, Y., & Meng, Y. (2016). Energy efficiency optimization in cognitive radio inspired non-orthogonal multiple access. In *IEEE 27th Annual International Symposium on Personal, Indoor, and Mobile Radio Communications (PIMRC)* (pp. 1–6).

Proficient Dual Secure Multi Keyword Search by Top-K Ranking Based on Synonym Index and DNN in Untrusted Cloud

Rosy Swami and Prodipto Das

Abstract Recent developments in cloud services have increased number of data owners to store their encrypted data in the cloud whereas equal or more data users participate to retrieve data. Secure retrieval of relevant data has become a challenging issue. In this paper, a secure ranking based multi-keyword search using semantic index is being developed. Initially, owner builds an index file by semantic representation of keyword. Security key is provided by Trusted Authority (TA) for decrypting the obtained results at the user side. TA manages dual security processes such as managing secret keys and issuing security devices to the data users. User query reaches proxy server, and it checks whether any frequent keyword matches with given query using Boolean Search. If it does not match, query enters into the main server which stores all document and index files to obtain relevant result using Deep Learning Neural Network. In deep learning neural network, the query is processed with vector space model to retrieve the relevant documents. Finally, user decrypts the relevant results obtained from deep neural network. The experimental result shows that our proposed model provides better performance in terms of recall, ranking privacy, precision, and searching time.

Keywords Cloud service provider · Semantic search · Secure · Multi-keyword search · Trusted authority

1 Introduction

Cloud computing is a new wave in information technology that probes enormous users as well as researchers towards it. In the cloud environment, the users store their data to utilize the flexibility and economic savings provided as well as to handle storage problems. The outsourced data may be in the form of documents, text, voice, multimedia, and so on and at the same time may also contain sensitive data. In general, the sensitive data includes personal identification such as PAN number, passport number, health information, financial information, bank account

R. Swami (✉) · P. Das
Department of Computer Science, Assam University, Silchar, Assam, India

P. Nayak et al. (eds.), *IoT and Analytics for Sensor Networks*, Lecture Notes in Networks and Systems 244, https://doi.org/10.1007/978-981-16-2919-8_34

details, and so on. These sensitive data are vulnerable to several attacks and require confidentiality. Thus in cloud storage, outsourcing of sensitive information to cloud server becomes a risk. To ensure security and privacy of the outsourced sensitive data, it is necessary to encrypt the data before outsourcing. Encryption techniques such as Advanced Encryption Standard (AES) method, Rivest-Shamir-Adleman (RSA) algorithm, Attribute-Based Encryption (ABE) method, and Blowfish algorithms are utilized to ensure security in cloud computing [1]. Comparisons made among these algorithms are based on key size, block size, scalability, encryption and decryption speed, and quality of security. In many cases, researchers have concluded that AES algorithm performs better than other algorithms in which small key size is required for high-level security. Most of the ABE schemes incur high cost of computation [2]. In multi-keyword search, security and privacy plays vital role which can be provided with the consideration of multiple attributes [3].

In cloud storage where multi owners are involved, key management is a complex process [4]. In mass information retrieval from cloud, the term is weighted based on location of the particular term in the document [5]. This method is often classed as N-level vector model (NVM). Keyword frequency is also considered as a main feature in multi-keyword searching process.

The major contributions of this paper are summarized as follows:

- Proxy server is introduced in Cloud Service Provider (CSP) to reduce the searching time, and searching efficiency is increased using Boolean search in the proxy server.
- Main server supports multiple users at a time with the help of Deep learning based Neural Network (DNN), which provides an accurate result.
- Trusted Authority (TA) is employed to provide secure document retrieval for authorized users. TA manages dual security processes as key management and Security Device Issuing (SDI).
- Secure top k ranking is achieved using Euclidean distance calculation, and accuracy of document retrieval is improved.

The rest of the paper is organized as follows: Section 2 surveys previous work regarding multi-keyword search methods and synonym index-based ranking methods. Section 3 presents proposed work and discusses its processes and procedure. Section 4 describes the result and output of proposed system. Section 5 concludes the contributions of the proposed system.

2 Related Work

Personalized search is presented in [6], with the support of Semantic Web Ontology WordNet. User's history is analysed for building user interest model for each user and also scoring mechanism adopted for users to express their interest. Automatic evaluation of keyword priority is realized through the interest model. Personalized Rank Search Encryption (PRSE) is employed with the semantic extension of user's query based on interest model and privacy-preserving. However, building the interest model became complex due to user's dynamic activities.

The authors in [7, 8] have discussed the index structure. Generalized Inverted Index (Ginix) for keyword search, which is efficient structure, is presented in [7]. In Ginix to save storage space all consecutive IDs are in inverted list into intervals. To improve Ginix it is necessary to reorder the documents in datasets using scalable algorithms. Tree-based index structure is used in [8] to facilitate search and updating process for the concern of efficiency. The authors to enhance the security employed dominant function-hidden-inner product encryption by precluding search patterns. But in this scheme revocation process is challengeable. Vector Space Model is used in [9] to construct searchable index to achieve accurate search results, and tree-based index structure is employed to improve efficiency in multi-keyword search. Here index file is secured but has inefficient key management scheme which provides encryption and decryption key for both owner and user.

Authentication is necessary at user side to ensure security of system. In [10], a new Identification Based Authentication protocol based on hierarchical model of identity is presented. It is lightweight and efficient protocol on user side. In this scheme hierarchical model is comprised of root Public Key Generation (PKG) (level-0), data center with sub-PKG (level-1), and bottom level (level-2) with users. Users are provided with unique names for identification. It is a time-consuming process to assign unique name to users since in cloud number of users increase rapidly. In [11], a security device and PKG for user authentication reduces time and also maintain security. The user must have these two factors for authentication. If the security device is lost or stolen, the user is provided with new security device, and lost/stolen device is revoked to prevent encryption and decryption through old device. The server performs some algorithms to revoke the device.

3 Proposed Work

3.1 System Overview

Our work concentrates on searching efficiency and secures ranking process in untrusted cloud server. Our overall architecture is comprised of Cloud Service Provider (CSP), Data Owners (DOs), Data Users (DUs), and Trusted Authority (TA). Figure 1 illustrates our architecture with multiple DOs and multiple DUs. Outsourcing encrypted documents and index file to cloud server is the initial process of DOs. Here multiple DOs are allowed to upload any numbers of documents and single index file. Initially DO separate the documents into keywords and build an index with semantic representation of keywords using WordNet to increase search efficiency. DO calculate Term Frequency (TF), Inverse Document Frequency (IDF), and TF $\times$ IDF values for each keyword and store its weight in index table. Then DO construct all documents into one index file with particular keyword location. After constructing index file DO outsources encrypted document files and index files to cloud server with its secret key and provides the encrypted keys to TA.

Encryption at DO side is performed by using the DO's secret key. Encryption at DO side consists of *index encryption* and *document encryption*. The entire encryption process is shown in Fig. 2. Index encryption is performed by Diffie-Hellman Algorithm (DHA) which encrypts the keywords in index file with their corresponding document. In Document encryption, each document in document set is encrypted by AES algorithm using DO's key. DO provides the encryption keys to TA for key management. AES takes block of data of 128 bits and a key and produces Ciphertext as output.

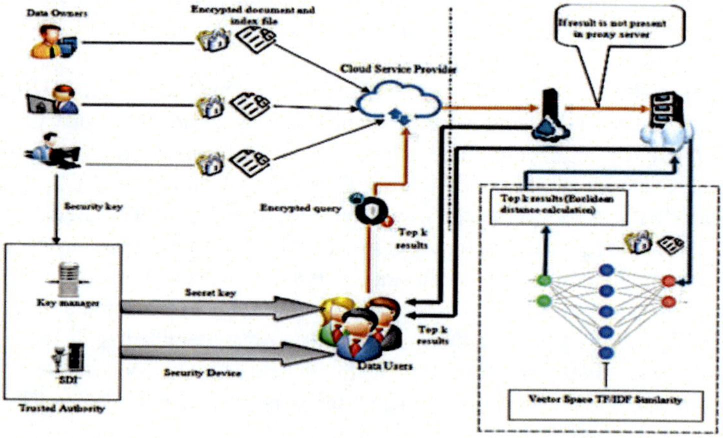

Fig. 1 Overall architecture of our proposed work

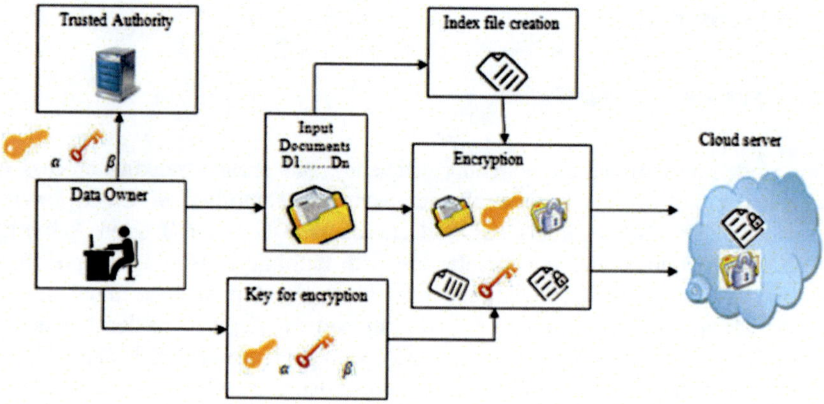

Fig. 2 Encryption process

3.2 Multi-keyword Search

The user query (encrypted multi-keyword) is searched in cloud server. In proposed system, CSP is comprised of main server and proxy server. Hereby we have introduced proxy server for minimizing searching time of user queries in cloud. The relevant result search is first performed in proxy server, if the result is not found for query in proxy, then the query is moved into main server. Proxy server stores the recently searched keywords and top k relevant results for those keywords. User query is arrived at proxy server and searched if there any keyword matches with the query. Consider a query Q with keywords w_1, w_3, and w_4. Let $C = (c_1, c_2, c_3)$ is the recently searched multi-keyword set (candidate set) in proxy server. For Q present in proxy then the candidate set in proxy should satisfy the following condition:

$$c_i = \left\{ \begin{array}{l} 1 \ for \ w_1^\wedge w_3^\wedge w_4 \\ 0 \ for \ otherwise \end{array} \right\} \tag{1}$$

Let $c_1 = (w_1, w_3, w_5)$; $c_2 = (w_1, w_2, w_3)$; $c_3 = (w_1, w_3, w_4)$

Boolean search is performed between Q and each multi-keyword set in C. The top k result of multi-keyword is retrieved to DU when all keywords in query are present in recently searched multi keyword. For above example, the proxy server retrieves the top k result of c_3. The query is moved into main server when it is not able to find any matched keyword in proxy server. In main server, DNN is employed for similarity measurement and to minimize search time.

Main server effectively stores the encrypted documents and index files in the allocated space of each DO. Similarity between query and keyword is processed by DNN. In Fig. 3 queries from multiple users are fed into input layer and in hidden layer similarity between each query and index files are measured by Vector Space Method (VSM).

The DNN allows multiple users at the same time and processes user's queries very effectively and quickly. Input layers process multiple users' queries, then hidden layer performs similarity measurement of each user's query and output layer provides the similarity results. After similarity measurement, the top k relevant results are obtained by calculating distance between keywords based on term location which is provided in index file. In hidden layer similarity between user's query and keywords is calculated by Vector Space TF × IDF model. Consider x_i is input and w_i is weight of input at each layer, then the

Output form hidden layer is

$$z = \sum_{i=0}^{n} w_i x_i \tag{2}$$

Output from output layer is

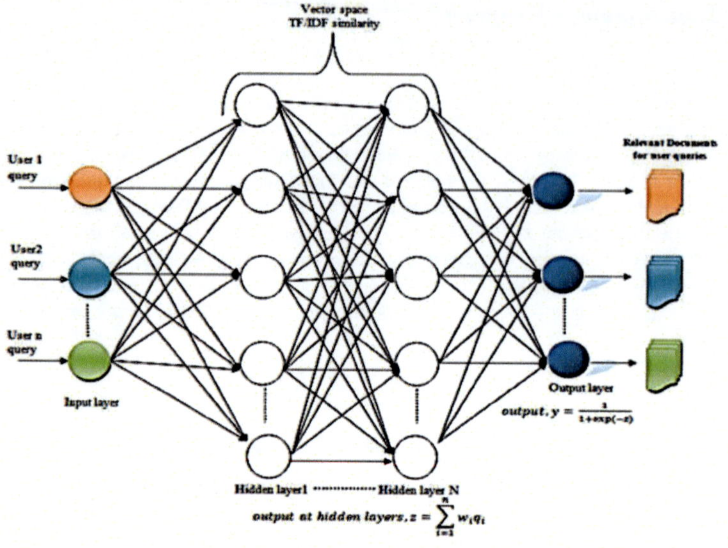

Fig. 3 DNN based keyword search

$$y = \sigma(z) = \frac{1}{1 + \exp(-z)} \tag{3}$$

Our proposed work concentrates on semantic multi-keyword search. Hence in hidden layer, the semantic representation of keywords in query is performed. This increases the relevancy of the results and precision factor. Then similarity measurement takes place in another hidden layer to obtain relevant results for query. At output layer, the relevant documents are obtained from DNN.

Vector Space Method VSM is the most popular way of similarity measurement in keyword search. In VSM, M-dimensional vectors are used to represent documents and queries using each dimension for representing a different keyword. The document vector is constructed using weight of keywords present in that document. The weight of each keyword is given in index file by DO based on TF × IDF value. The document vector for document *D* is

$$\vec{D} = \left(t_{D,w1}, t_{D,w2}, \ldots, t_{D,wn}\right) \tag{4}$$

where $t_{D,w1}$ represents the weight of keyword1 in document *D*. The query vector is denoted as

$$\vec{Q} = (q_1, q_2, \ldots..q_n) \tag{5}$$

The next step after constructing document and query vector is to measure similarity between all the documents and the query. The cosine measure is employed for similarity measurement. In cosine measurement similarity between document $D1$ and query Q is given by

$$Sim(Q, D1) = \cos(\theta) = \frac{\vec{Q}.\overrightarrow{D1}}{\left|\vec{Q}\right| \times \left|\overrightarrow{D1}\right|} \tag{6}$$

When keyword in query $Q1$ lies in document, then Q_i equals t'_{wi}; otherwise it equals zero.

$$\cos\left(\overrightarrow{D1}, \overrightarrow{Q1}\right) = \frac{\sum_{1=1}^{n} t_{D,wi} t'_{wi}}{\sqrt{\sum_{i=1}^{n} \left(t_{D,wi}\right)^2} \sqrt{\sum_{i=1}^{n} \left(t'_{wi}\right)^2}} \tag{7}$$

Top k result retrieval Euclidean distance is calculated between keywords in query and in document to improve retrieval accuracy. The top k relevant result is retrieved by calculating Euclidean distance between keywords in document vector and query vector. Euclidean distance between two keywords is given by

$$d_{EQ} = \sqrt[2]{(q_1 - q_2)^2} \tag{8}$$

$$d_{ED} = \sqrt[2]{(t_{w1} - t_{w2})^2} \tag{9}$$

where

d_{EQ} Distance between keywords in Query.

d_{ED} Distance between keywords in Document.

3.3 Document Retrieval

The initial stage of document retrieval is to generate queries on user side. The DUs encrypt the query using AES algorithm and send the query to CSP with their security device provided Security Device Issuer which is maintained by TA. The security device is a secure random number provided for user identification and the user without the security device will not be allowed to search in cloud server. SDI revokes the security device when DUs lost their security device. When query enters into CSP the multi-keyword search algorithm is performed and top k relevant result is obtained. DU decrypts the retrieved top k documents using secret provided by key manager which is maintained by TA.

4 Experimental Evaluation

4.1 Simulation Setup

Java is used to implement the proposed scheme. The software tool NetBeans 6.9.1, to develop java programs, is installed for our simulation. Wamp server (MySQL 5.1.36) is also installed as server in our work. CloudSim, which offers essential classes for defining data centers, computational resources, and virtual machines, is used for deploying CSP and proxy server. WordNet is a tool for computational linguistics and natural language processing used in our scheme. For simulation, 25–30 numbers of owners, 100–150 numbers of users, 8 numbers of keywords, and 200 numbers of documents are considered.

4.2 Comparative Analysis

In this section, we compared our proposed multi-keyword search system with existing works MRPS method [9], N-level vector model [5], Enhanced-TFIDF method [12], and Space Efficient Semantic search [13].

Effectiveness of Precision Precision is one of the standard measures in keyword search. Precision depends on the accuracy of retrieved documents.

Figure 4 shows that the performance of Enhanced-TFIDF method and MRPS method is significantly changed with respect to number of documents retrieved. These methods are not able to achieve better results with the increasing number of documents. Our proposed scheme using DNN achieves precision of 89% which is significantly higher than previous work.

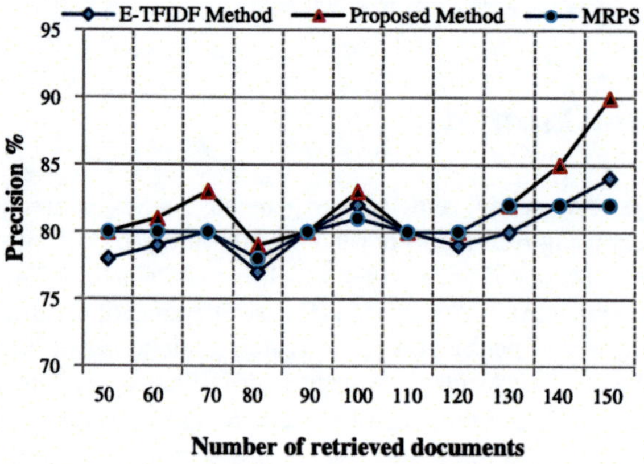

Fig. 4 Comparative results for precision

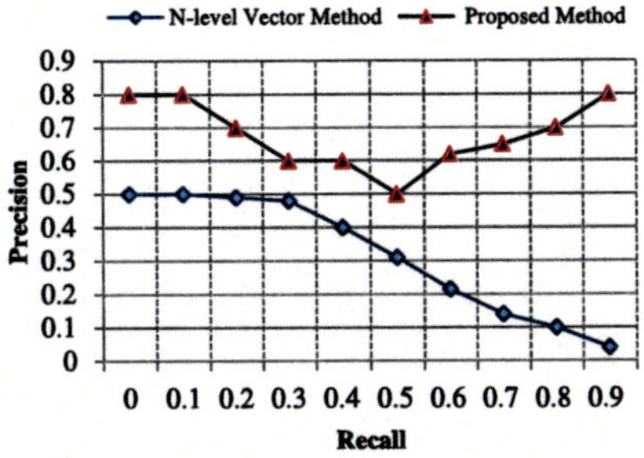

Fig. 5 Comparative analysis for precision-recall

Effectiveness of Recall In information retrieval, recall(sensitivity) is one of the performance metrics which increases the accuracy. It depends on the number of related documents retrieved for query. But the precision and recall are contradictory (i.e.) upon definite threshold the increase of recall causes reduce in precision and increase of precision causes decrease in recall.

Figure 5 shows that our proposed scheme achieves high precision for increased recall compared with previous works. In our work, we have achieved high precision for increased recall compared to N-level vector method. In our work a precision of 0.8 is achieved for increased recall a factor of 0.2.

Effectiveness of Rank Privacy Rank privacy is one of the significant performance metrics of privacy. It depends on rank of each document in the retrieved top k result. Generally, ranking is processed with non-encrypted documents for accurate ranking. But in our proposed work, we have processed ranking over encrypted document to ensure security. Figure 6 shows that E-TFIDF and MRPS methods achieved significantly low-rank privacy. In our work, we provide accurate ranking over encrypted data. The relevant result from DNN is further processed with the calculation of Euclidean distance which provides accurate rank for retrieved documents. Our proposed work achieves rank privacy comparatively higher than previous work. We have achieved rank privacy of 33.5%.

Effectiveness of Search time Search time includes the time taken for index searching time and query processing time. Figure 7a–c show that our proposed scheme reduces

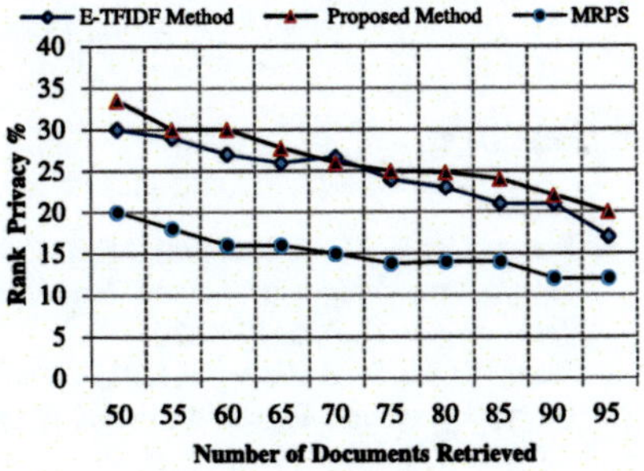

Fig. 6 Comparative results for rank privacy

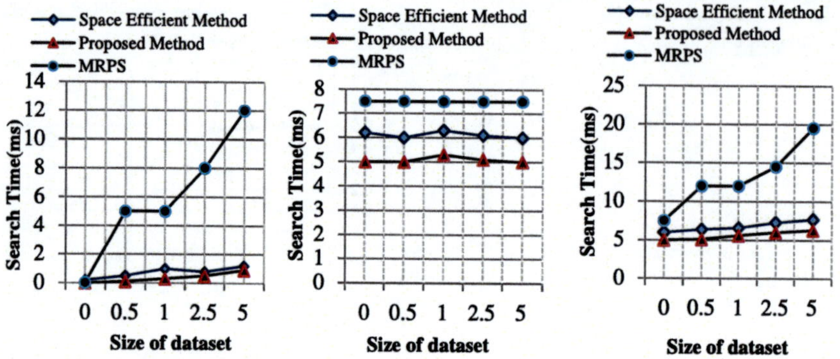

Fig. 7 **a** Comparative results for Hash index search time; **b** Comparative results for query processing time; **c** Comparative results for total search time

the searching time in terms of index searching time, query processing time, total search time due to effective indexing with synonym set. Finally, our proposed scheme achieves high precision, recall, and rank privacy with reduced search time.

5 Conclusion

In this paper, we presented a secure top *k* ranking based multi-keyword search in untrusted cloud. The authorized user can be allowed to search over encrypted data by using security device provided by TA. Proxy server reduces the searching time by storing recently searched keywords with their top *k* results. DNN in main server effectively supports multiple users at a time to search and also gives accurate relevant results using vector space model for similarity calculation in hidden layers. Retrieved relevant documents are again processed by calculating Euclidean distance between keywords and query to rank the retrieved documents. Encryption is performed using AES algorithm. Finally, performance evaluation indicates that our proposed scheme is effective in multi-keyword search and top *k* result retrieval is also effective in precision, recall, ranking privacy, and search time.

References

1. Sonam Darda, M., & Kulkarni, M. K. (2015). Study of Multi-keyword ranked searching and encryption technique over cloud. *International Journal of Computer Science and Information Technologies, 6*(6), 5417–5420.
2. Wang, S., Jia, S., & Zhang, Y. (2019). Verifiable and multi-keyword searchable attribute-based encryption scheme for cloud storage. *IEEE Access, 7*, 50136–50147.

3. Zhang, L., Zhang, Y., & Ma, H. (2018). Privacy-preserving and dynamic multi-attribute conjunctive keyword search over encrypted cloud data. *IEEE Access, 6*, 34214–34225.
4. Zhang, W., Lin, Y., Xiao, S., Wu, J., & Zhou, S. (2016). Privacy preserving ranked multi-keyword search for multiple data owners in cloud computing. *IEEE Transactions on Computers, 65*(5), 1566–1577.
5. Peng, J., Tang, S., Zhang, L., & Liu, R. (2017). Information retrieval of mass encrypted data over multimedia networking with N-level vector model-based relevancy ranking. *Multimedia Tools and Applications, 76*(2), 2569–2589.
6. Fu, Z., Ren, K., Shu, J., Sun, X., & Huang, F. (2016). Enabling personalized search over encrypted outsourced data with efficiency improvement. *IEEE Transactions on Parallel and Distributed Systems, 27*(9), 2546–2559.
7. Wu, H., Li, G., & Zhou, L. (2013). Ginix: Generalized inverted index for keyword search. *Tsinghua Science and Technology, 18*(1), 77–87.
8. Yan, J., Zhang, Y., & Liu, X. (2016). Secure multi-keyword search supporting dynamic update and ranked retrieval. *China Communications, 13*(10), 209–221.
9. Fu, Z., Sun, X., Liu, Q., Zhou, L., & Shu, J. (2015). Achieving efficient cloud search services: Multi-keyword ranked search over encrypted cloud data supporting parallel computing. *IEICE Transactions on Communications, 98*(1), 190–200.
10. Li, H., Dai, Y., Tian, L., & Yang, H. (2009). Identity-based authentication for cloud computing. In M. G. Jaatun, G. Zhao & C. Rong (Eds.), *Cloud Computing. CloudCom*, LNCS 5931 (pp. 157–166.). Beijing, China.
11. Liu, J. K., Liang, K., Susilo, W., Liu, J., & Xiang, Y. (2016). Two-factor data security protection mechanism for cloud storage system. *IEEE Transactions on Computers, 65*(6), 1992–2004.
12. Fu, Z., Sun, X., Linge, N., & Zhou, L. (2014). Achieving effective cloud search services: Multi-keyword ranked search over encrypted cloud data supporting synonym query. *IEEE Transactions on Consumer Electronics, 60*(1), 164–172.
13. Woodworth, J., Salehi, M. A., and Raghavan, V., (2016). S3C: An architecture for space-efficient semantic search over encrypted data in the cloud. In *Proceedings of the 2016 IEEE International Conference on Big Data (Big Data)* (pp. 3722–3731). Washington D.C., USA.

Transfer Learning-Based Detection of Covid-19 Using Chest CT Scan Images

Aryaman Chand, Khushi Chandani, and Monika Arora

Abstract The global pandemic caused by the Severe Acute Respiratory Syndrome Coronavirus 2 (SARS-CoV-2) virus has had historical impact on the world. The virus causes severe respiratory problems and with an R0 of 5.7, spreads at a rapid rate. At the time of writing, there were over 85 million cases and 1.8 million deaths caused by COVID-19. In the proposed methodology, Deep Convolutional Neural Networks (DCNNs) have been trained, with the help of transfer learning, to learn to identify whether a suspected patient is suffering from this disease using their chest CT scan image. Transfer learning technique enables the transfer of knowledge from pre-trained models which have been previously trained on extremely large datasets. Various DCNN models have been applied such as AlexNet, ResNet-18, ResNet-34, ResNet-50, VGG-16, and VGG-19. The DCNNs were evaluated on a set of 2,481 chest CT scan images. Various performance metrics (Accuracy, MCC, Kappa, F1 score, etc.) were calculated for all DCNN models to enable their comparative evaluation. After extensive testing, ResNet50 was found to give the best results in this binary classification task. The highest accuracy achieved was 97.37% and highest kappa was 0.947. Identification of presence of COVID-19 using this method would provide great benefit to society and mankind.

Keywords Convolutional neural network · SARS-CoV-2 · Coronavirus · COVID-19 · Transfer learning

1 Introduction

Amid the ongoing global pandemic caused by the SARS-CoV-2 virus, the world has seen loss of life and property to an extreme degree. The disease caused by this virus, colloquially termed as COVID-19, causes several respiratory problems in humans, including but not limited to dry coughs, chest pain, chest pressure, shortness of breath, and sore throat. In serious cases, this disease can also cause loss of speech and movement [1]. Moreover, this disease spreads at a rapid pace. A recent study

A. Chand (✉) · K. Chandani · M. Arora
Amity University, Noida, Uttar Pradesh, India

P. Nayak et al. (eds.), *IoT and Analytics for Sensor Networks*, Lecture Notes in Networks and Systems 244, https://doi.org/10.1007/978-981-16-2919-8_35

shows that COVID-19 has a basic reproductive number (R0) of 5.7, which means that one infected carrier of this disease can potentially transmit to an average of 5.7 people [2]. The onset of this pandemic has prompted governments all over the world to impose lockdowns and travel bans. However, the spread has been widespread and so extensive that small, remote island nations in Oceania have reported cases as well. At the time of writing, there are over 85 million cases and 1.8 million deaths due to this disease [3]. Hence, healthcare systems worldwide are overburdened and there is a dearth of doctors, nurses, and medical support staff. Moreover, conventional testing methods of COVID-19 diagnosis have issues of their own. The molecular tests (PCR tests/nucleic acid tests/viral RNA tests) are unreliable with a false negative rate of as high as 37% [4]. A professional lab and a few days are required to generate a result from this type of test. Moreover, since the outbreak of COVID-19, medical laboratories are operating at capacity, thus increasing the time taken to generate a result from a normal two days up to ten days in several places [5, 6]. They are also a challenge to mass produce since the testing process requires additional reagents. Antigen tests provide faster results but are even more unreliable, with a reported false negative rate as high as 50% [7].

Therefore, considering these factors, there raises a case for a COVID-19 testing mechanism that offers quick and reliable results. In this research, we explore the possibility of a testing mechanism that is based on the chest CT scan images of the suspected patient. Studies show that an infected patients' chest shows changes in a CT scan even before the onset of symptoms [8]. The infrastructure to get a CT scan done is also widespread and prevalent all over the world, and costs are moderate when compared to conventional testing methods.

Although there has been some research in this space, certain design and execution challenges were identified. Related work in medical applications has been done to identify COVID-19. Transfer learning-based detection of COVID-19 was presented by [9]. They achieved a maximum accuracy of 87%. Deep learning based segmentation system to quantify lung infection areas was presented by [10] and obtained a maximum accuracy of 91%. An automatic DL-based COVID-19 diagnostic system was proposed by [11] and achieved a maximum accuracy of 85%. A DL-based technique to quantify the severity of COVID-19 using segmentation was presented by [12], achieving a maximum accuracy of 94.8%.

In this paper, a model based on transfer learning of state-of-the-art Deep Convolutional Neural Network (DCNN) models is proposed. Transfer learning enables the transfer of knowledge from pre-trained networks which have been trained on massive image datasets, such as ImageNet. The idea behind transfer learning is to transfer the weights of the layers of the DCNN and apply these layers to a new classification task. The detection of COVID-19 from chest CT scan images was casted as a binary classification problem, with the dataset used containing chest CT scans of healthy patients and those suffering from COVID-19. The main contribution of this research was to utilize several off-the-shelf and pre-trained DCNN models via transfer learning approach. Another major contribution of this research was to find the right image transformation techniques and tune the model hyperparameters—variables such as batch size, learning rate, and number of epochs—to obtain better classification results.

The chapter scheme is as follows. Section 2 contains details about the proposed methodology. Section 3 discusses the results. Section 4 concludes the research.

2 Methodology

In the proposed methodology, the goal is to leverage transfer learning technique and apply the pre-trained models in the binary classification task of identifying the presence of COVID-19 in suspected patients. By using these pre-trained models, their demonstrated capabilities in image classification can be transferred to the task at hand [13]. This can be done by training the models without updating the weights of the weighted layers. The training of all of the models was done using the fastai library in Python. The fastai library automatically removes the classification head of the pre-trained DCNN model and replaces it with a new head suitable for the database. It also sets automatic defaults for the learning rate, optimizer, etc. [14]. The execution of the code was done in Google Colaboratory environment due to the need of high-powered GPUs for training DCNNs. The hyperparameters of each of the DCNNs were tuned extensively to improve the results obtained. Hyperparameters include variables such as number of epochs, batch size and learning rate. These variables help control the learning process of the DCNNs. The dataset used to train and evaluate the models consists of 2,481 images of chest CT scans. The dataset has been described in detail in Sect. 2.1. To increase the robustness and integrity of the DCNN models, data augmentation technique was used. Data augmentation virtually increases the size of the dataset by applying certain transformations to each image of the dataset in order to create several copies of it. In theory, this technique should improve the classification accuracy of the DCNN models since they have more data to train on. Various permutation and combinations of different data augmentation techniques were explored in order to obtain the best possible results. Data augmentation has been further elucidated in Sect. 2.2. The several DCNN models used in this research have been explained in detail in Sect. 2.3. After training all the DCNN models, several performance metrics were calculated in order to enable the comparative analysis of the several models. Figure 1 shows the block diagram of the methodology proposed.

2.1 Data Description

The images were collected by researchers in Brazil to encourage the research and development of deep learning based methods of detection of COVID-19. CT scan images of real patients were collected from hospitals in Sao Paulo, Brazil [15]. There

Table 1 Details of the dataset with sample image of both classes

Class name	Number of images	Sample image
COVID-19	1252	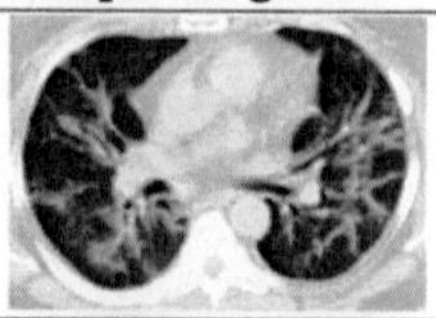
Non-COVID-19	1229	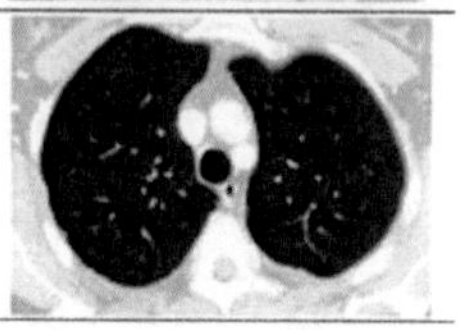

are a total of 2,481 images in the dataset, split into two classes—COVID-19 and non-COVID-19. There are 1252 chest CT scans of patients suffering from COVID-19 and 1229 chest CT scan images of healthy patients. All of the CT scan images are in the transverse anatomical plane; hence, there was little need for preprocessing in that regard. Sample images of infected and healthy patients have been shown in Table 1. Since the DCNNs were trained and evaluated on fastai, there was no need to specify a separate test dataset. The validation split was set to twenty percent for all DCNN models.

2.2 Data Augmentation

Data augmentation is the process of manipulating the images of the dataset with certain image transformations such that several copies of each image are created and the size of the dataset is artificially increased. The copies of each image of the dataset contain the same contextual information from the human perspective. However, the copies are not identical down to the pixel levels; hence, they act as more training data for the DCNN model to train on—since the models are fed with several variations of the same image. There are several transformations available in the fastai library. However, it is also critical to apply these transformations with caution so as to retain the important data within the images of the dataset. The following transformations were found to work best with the dataset—rotation, horizontal flip, lighting, zoom and warp. These transformations were applied to a random image from the dataset, and are shown in Fig. 2.

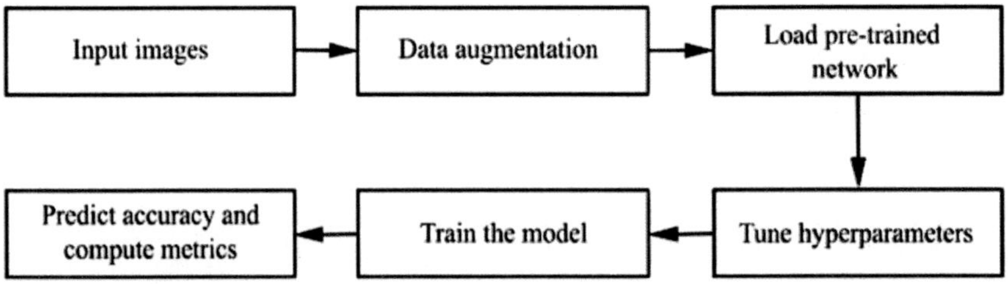

Fig. 1 Block diagram of proposed methodology

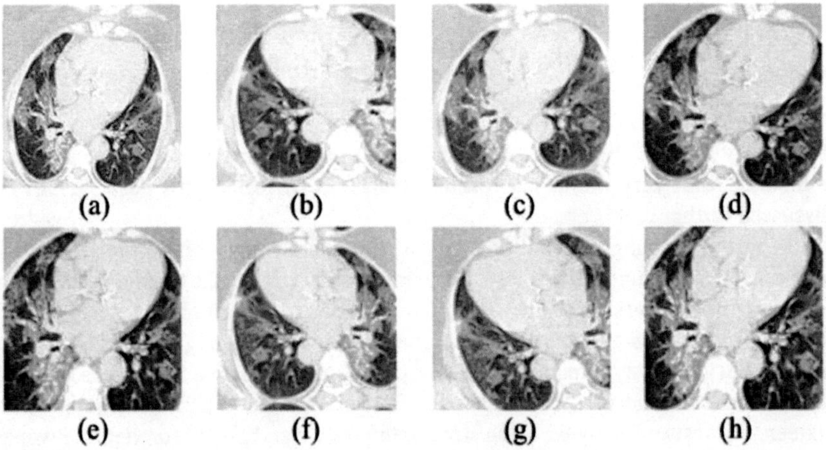

Fig. 2 **a** Original image **b** horizontal flip **c** warp and lighting **d** zoom, rotation and warp **e** zoom and lighting **f** horizontal flip, rotation and warp **g** horizontal flip and warp **h** zoom and lighting

2.3 Model Architectures

The DCNN models used in this research were VGG16, VGG19, AlexNet, Resnet18, ResNet34, and ResNet50

AlexNet

The architecture of AlexNet DCNN comprises of eight layers, including five convolutional layers and three fully connected layers. AlexNet was the first DCNN to make use of the ReLU activation function. There are 60 million parameters in this architecture. It utilizes data augmentation and dropout to counteract overfitting [16].

VGG

VGGnets created by the Visual Geometry Group at Oxford University were derived from the AlexNet architecture. They utilize smaller 3×3 receptive fields and a stride of one to preserve spatial resolution. They use the ReLU activation function, akin to

AlexNet. Various versions of VGGnets exist, but the two best performing models, VGG16 and VGG19, with sixteen and nineteen layers, respectively, were used in this research [17].

ResNet

ResNets are fundamentally different to linear DCNN architectures such as AlexNet and VGG. ResNets utilize the residual block and rely on micro architecture modules. The identity mapping, or the skip connection, feeds the output of the previous layer to the layers ahead, ensuring that the performance of subsequent layers does not fall. This concept enabled much deeper models to be trained [18].

3 Results

The binary classification task of identifying COVID-19 from chest CT scans was completed with very promising results. ResNet50 performed the best overall, as discussed further in this chapter.

Using the learning rate finder graph of the fastai library, the optimal starting learning rate was found out to improve the results and optimize the training time. From Fig. 3, a learning rate slice of ($1e^{-3}$, $7e^{-2}$) was chosen for the training of ResNet50. In this range, the loss has a steep negative slope, ideal for the training process. Extensive testing with various hyperparameters and data augmentation transformations resulted in some notable and predictable results. The optimal batch size was found to be sixteen, and tests with higher batch sizes of thirty-two and sixty-four yielding worse results. It was observed that around seven to eight epochs were needed to converge to local minima, and tests with more training epochs did not yield any better results. Therefore, for the final results obtained for each model, the number of training epochs was set to eight. Further, minimal tweaking of the learning rate was required since

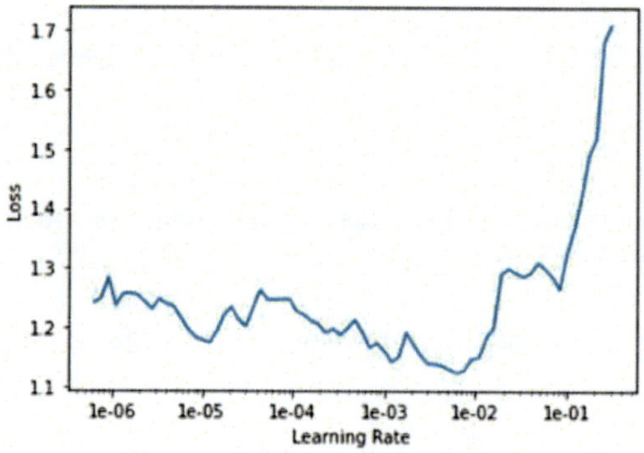

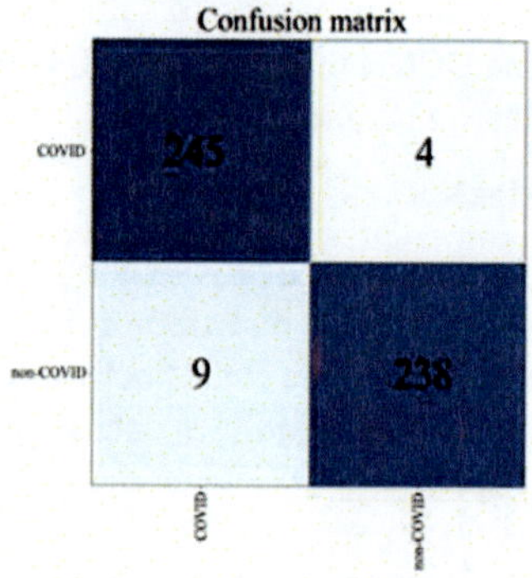

Fig. 3 Learning rate plot and confusion matrix for ResNet152

the in-built learning rate finder of the fastai library provided a good head start and the learner object automatically updated it during training.

Twenty percent of the dataset was reserved for validation, while eighty percent was used for training. The DCNN's predictions on the validation images were represented in the form of a confusion matrix, as shown in Fig. 3. The confusion matrix is a 2 $\times$ 2 matrix that represents the true positives (infected patients that were classified as having the disease), true negatives (healthy patients that were classified as not having the disease), false positives (healthy patients that were classified as having the disease), and false negatives (infected patients that were classified as not having the disease). Minimizing false negatives is one of the most critical targets of this research since conventional testing methods of COVID-19 have a high rate of false negatives.

Performance metrics were calculated using the confusion matrix obtained for each DCNN model. Firstly, basic metrics such as accuracy, specificity, and recall were

calculated (Eqs. 1–3). Recall highlights the probability of detection of the disease and specificity highlights the selectivity of rejection.

$$\text{Accuracy} = \frac{TN + TP}{TN + FN + TP + FP} \tag{1}$$

$$\text{Recall} = \frac{TP}{TP + FN} \tag{2}$$

$$\text{Specificity} = \frac{TN}{TN + FP} \tag{3}$$

Further evaluation needs to be done in order to compare the DCNN models thoroughly since the metrics calculated above are only true indicators for a balanced dataset. Since the dataset used in this research is unbalanced, true comparison between the DCNN models requires more metrics. These metrics are precision, F1 score, false negative rate (FMR), error, MCC and Cohen's Kappa index (Eqs. 4–9). Precision represents the positive predictive value. The F1 score, or the Dice coefficient, is the harmonic mean of precision and recall. FNR is perhaps one of the most critical performance metrics for such a classification task since a false negative on the test has serious public health risks associated with it. Moreover, conventional testing methods of COVID-19 have high tendencies to return false negative results, hence minimizing this metric would prove the efficacy of this model over molecular and antigen tests. MCC and Kappa index are two metrics that are completely agnostic to the size of each class and hence work best for unbalanced datasets such as the one used in this research. The value of MCC lies between -1 and $+1$, with a higher score being more desirable. Cohen's Kappa index is a measure of the performance delta between the trained classifier and a random classifier that makes random predictions according to the frequency of occurrence of each class.

$$\text{Precision} = \frac{TP}{TP + FP} \tag{4}$$

Table 2 Accuracy results of all models

Model	Accuracy (%)	Specificity	Recall	Precision	F1 score	FNR	Error	MCC	Kappa
AlexNet	87.50	0.928	0.821	0.918	0.866	0.178	0.125	0.753	0.750
VGG16	92.94	0.949	0.907	0.943	0.925	0.092	0.07	0.859	0.851
VGG19	93.75	0.933	0.941	0.930	0.935	0.058	0.062	0.874	0.875
ResNet18	92.13	0.928	0.914	0.925	0.919	0.085	0.078	0.842	0.842
ResNet34	96.12	0.973	0.951	0.976	0.963	0.048	0.038	0.923	0.923
ResNet50	97.37	0.983	0.983	0.964	0.964	0.016	0.026	0.947	0.947

$$\text{F1 score} = \frac{2 \times Recall \times Precision}{Recall + Precision} \tag{5}$$

$$\text{False Positive Rate (FNR)} = \frac{FN}{TP + FN} \tag{6}$$

$$\text{Error} = \frac{FP + FN}{TN + FN + TP + FP} \tag{7}$$

$$\text{MCC} = \frac{TP \times TN - FP \times FN}{\sqrt{(TN + FP)(TN + FN)(TP + FP)(TP + FN)}} \tag{8}$$

$$\text{Kappa} = \frac{Accuracy - random\ Accuracy}{1 - random\ Accuracy} \tag{9}$$

The highest classification accuracy was given by ResNet50. ResNet50 also provided the best MCC and Kappa index scores, useful for evaluating DCNN models on an unbalanced dataset. Further, a false negative rate of just 1.6% was achieved, which is a massive improvement over conventional tests such as molecular and antigen tests, which have a reported false negative rate of as high as 50%. The final results with all performance metrics have been depicted in Table 2.

4 Conclusion

Limited work has been done on the detection of COVID-19 through the CT scan image of the patient. In this research, the prospect of using DCNN models via transfer learning approach to detect COVID-19 was explored. After extensive testing and computing performance metrics, very promising results were obtained in this binary classification task. The proposed model was able to successfully identify the presence or absence of the disease with a maximum accuracy of 97.37%. Further, false negatives were at an exceptionally low 1.6%. These results could be further improved with a larger dataset for the DCNN models to train on. However, the proposed model contributes to a possibility of a quick, cost effective and most importantly, accurate diagnosis of COVID-19 using chest CT scan images. Practical use of this model would provide great benefit to society and mankind.

References

1. World Health Organization. (2020). Coronavirus Disease (COVID-19), https://www.who.int/emergencies/diseases/novel-coronavirus-2019/question-and-answers-hub/q-a-detail/q-a-coronaviruses#:~:text=symptoms.
2. Sanche, S., Lin, Y., Xu, C., Romero-Severson, E., Hengartner, N., & Ke, R. (2020). High contagiousness and rapid spread of severe acute respiratory syndrome coronavirus 2. *Emerging Infectious Diseases, 26*(7), 1470–1477. https://doi.org/10.3201/eid2607.200282.
3. Wikipedia. (2020). COVID-19 pandemic data, https://en.wikipedia.org/wiki/Template:COVID-19_pandemic_data.
4. Interpreting a covid-19 test result, 2020. *BMJ, 369*, m1808. https://doi.org/10.1136/bmj.m1808..
5. NBC News. (2020). Frustration over long coronavirus test result wait times grows in Florida, https://www.nbcnews.com/news/latino/frustration-over-long-coronavirus-test-result-wait-times-grows-florida-n1234493.
6. Times of India. (2020). Delhi: Up to 10-day waiting time for Covid-19 test at private labs, https://timesofindia.indiatimes.com/city/delhi/delhi-up-to-10-day-waiting-time-for-covid-19-test-at-private-labs/articleshow/76521078.cms#:~:text=June%2023%2C%202020-Delhi%3A%20Up%20to%2010%2Dday%20waiting%20time%20for%20Covid%2D,19%20test%20at%20private%20labs.
7. Science. (2020). Coronavirus antigen tests: quick and cheap, but too often wrong? https://www.sciencemag.org/news/2020/05/coronavirus-antigen-tests-quick-and-cheap-too-often-wrong.
8. A familial cluster of pneumonia associated with the 2019 novel coronavirus indicating person-to-person transmission: a study of a family cluster. *The Lancet, 395*(10223), 514–523.
9. Sarker, L., Islam, M., Hannan, T., & Ahmed, Z. (2020). COVID-DenseNet: A deep learning architecture to detect COVID-19 from chest radiology images. preprint. https://doi.org/10.20944/preprints202005.0151.v1.
10. Shan, F., Gao, Y., Wang, J., Shi, W., Shi, N., Han, M., Xue, Z., Shen, D., & Shi, Y. (2020). Lung infection quantification of COVID-19 in CT images with deep learning. arXiv Preprint arXiv:2003.04655.
11. Wang, S., Zha, Y., Li, W., Wu, Q., Li, X., Niu, M., Wang, M., Qiu, X., Li, H., Yu, H., & Gong, W. (2020). A fully automatic deep learning system for COVID-19 diagnostic and prognostic analysis. *European Respiratory Journal*. https://doi.org/10.1183/13993003.00775-2020.
12. Gozes, O., Frid-Adar, M., Greenspan, H., Browning, P. D., Zhang, H., Ji, W., Bernheim, A., & Siegel, E. (2020). Rapid AI development cycle for the Coronavirus (COVID-19) pandemic: Initial results for automated detection & patient monitoring using deep learning CT image analysis. arXiv Preprint arXiv:2003.05037.
13. Pan, S. J., & Yang, Q. (2009). A survey on transfer learning. *IEEE Transactions on Knowledge and Data Engineering, 22*(10), 1345–1359.
14. Howard, J., & Gugger, S. (2020). Fastai: A layered api for deep learning, Information *11*, 2.
15. Kaggle. (2020). SARS-COV-2 Ct-Scan Dataset, https://www.kaggle.com/plameneduardo/sarscov2-ctscan-dataset.
16. Krizhevsky, A., Sutskever, I., Hinton, G. E. (2017). ImageNet classification with deep convolutional neural networks. *Communications of the ACM, 60*(6), 84–90. https://doi.org/10.1145/3065386. ISSN 0001-0782.

17. Simonyan, K., & Zisserman, A. (2014). Very deep convolutional networks for large-scale image recognition. In *Proceedings of International Conference on Learning Representations*, http://arxiv.org/abs/1409.1556.
18. He, K., Zhang, X., Ren, S., & Sun, J. (2016). *Deep residual learning for image recognition* (pp. 770–778), https://doi.org/10.1109/CVPR.2016.90.

A Hybridized Machine Learning Model for Optimal Feature Selection and Attack Detection in Cloud SaaS Framework

Reddy Saisindhutheja and Gopal K. Shyam

Abstract Cloud computing offers several profitable services as and when required to the customer, and so it is growing up as a forthcoming drift in the IT sector. Software-as-a-service (SaaS) is one of the outstanding and fastest-growing fields in cloud computing history. It is a license to acquire cloud applications via the Internet. Out of all this curiosity, security is found as one of the key issues that delay the growth of SaaS. The motivation of this research work is to provide security for SaaS by handling massive amounts of data. At first, feature selection is accomplished by Oppositional Crow Search Algorithm (OCSA). These nominated features are sent for detecting the attacks via Deep Belief Network (DBN). The main objective is to introduce an innovative, secure framework for SaaS by performing attack detection when there is enormous traffic. The proposed and conventional models are evaluated using a benchmark dataset. Results prove that the proposed OCSA + DBN outperforms the other existing methods with respect to precision, sensitivity which are positive measures and False Positive Rate (FPR) and False Detection Rate (FDR); the negative measures. Moreover, the proposed work performs better with the existing works with a performance indicator of 3% for all the metrics.

Keywords Cloud computing · SaaS · Attack detection · Crow search algorithm · Deep belief network

1 Introduction

Over the years, digital appliances play a major role in human life [1]. The sudden increase in the world-wide internet usage needs a new strategy to manage the size, range, and availability of data [2, 3]. Cloud computing is an upcoming expertise that is substantial in delivering on-demand, scalable and consistent services to its customers

R. Saisindhutheja (✉)
Sreenidhi Institute of Science & Technology, Hyderabad 501301, Telangana, India

R. Saisindhutheja · G. K. Shyam
REVA University, Bengaluru 560064, Karnataka, India
e-mail: gopalkrishnashyam@reva.edu.in

P. Nayak et al. (eds.), *IoT and Analytics for Sensor Networks*, Lecture Notes in Networks and Systems 244, https://doi.org/10.1007/978-981-16-2919-8_36

at low cost [4]. Besides, it offers a simple strategy to use servers, databases, software, and analytics over the Internet and accomplishes all the application services [5–7].

The service delivery models of cloud are Infrastructure-as-a-service (IaaS), Platform-as-a-Service (PaaS), and Software-as-a-Service (SaaS). SaaS is the top most layer of cloud which is present on PaaS and IaaS. As the characteristics obtained, it has a chance of acquiring the threats as well. Hence, SaaS is highly vulnerable to attacks. Attackers may harm the systems with the most common attacks like data theft, DDoS attack, illegal access, etc. [8–11].

In contempt of this, machine learning algorithms are effectively used for attack detection. Out of these models, deep-learning resembles the most useful method in finding the attack detection. The key contributions of this work are as follows: (i) Selecting the best features using OCSA algorithm, (ii) Performing the attack detection via DBN, and (iii) Comparing the results of proposed work with existing methods. The SaaS providers using this framework will design and develop the detection phase, which fits best in finding various threats.

The remaining part of the paper is organized as follows: Sect. 2 illustrates the literature review. The proposed attack detection system is demonstrated in Sect. 3. The results acquired with the proposed model are discussed in Sect. 4. At the end, conclusions and future work are given in Sect. 5.

2 Literature Review

In [12], the authors have developed a new architecture for detecting DDoS attacks in the cloud. This architecture is accomplished by coupling the "sparse AE for feature extraction and DNN for classification". Further, by using a relevant tuning model, the parameters are fine-tuned for AE and DNN. As a result, the overfitting problem is resolved with the least error.

A new security system is proposed in [13, 14] to handle the activated DDoS attack by unauthorized IoT servers. By considering SDN with cloud, the DDoS attack is mitigated on the IoT servers. Results attained with high accuracy in detecting DDoS attacks.

The DDoS attack detection and prevention system was introduced in [15] using ICRPU and feature selection. To examine the traffic, Hellinger distance function was used. Further, the packets were distinguished as valid groups and extracted features regarding DDoS.

A new method formulated on the study of network flow in the cloud-based services is introduced in [16] for FRC attack detection and prevention. Results of the proposed work are very effective with low overhead and high accuracy.

CRESOM-SDNMS was developed in [17] to counterattack the DDoS attacks in the cloud scenario. After initialization, the improvised topology preservation is used in the SOM-based categorization process to remove the vector quantization problems. The results show that the proposed work had minimized the FPR at DDoS prevention.

The challenges that identified from the literature review are poor attack detection rate, low classification accuracy, redundant, huge and data with high dimensionality which are not managed properly. The proposed work overcomes all these problems.

3 Proposed Attack Detection Framework

The proposed work aims to design and develop a secure framework for detecting the attacks which is given in Fig. 1. Following are the steps used to develop this framework.

Step 1: The input data $Input^{Data}$ is subjected to RSA [18] for carrying out encryption, and the encrypted data after evaluation $Encrypt^{Data}$ is decrypted by mapping $Encrypt^{Data}$ to 9 RSU's. The applications and RSU count will not differ. After decryption the resultant is $Decrypt^{Data}$.

Step 2: As $Decrypt^{Data}$ consists of more features, the training phase might be very complex. Hence, the features Fn are selected from $Decrypt^{Data}$. OCSA is used to accomplish feature selection.

Step 3: As soon as the optimal features Fn are chosen, they are given for DBN to detect the attacks, in which it predicts whether the feature selected is attack free or not.

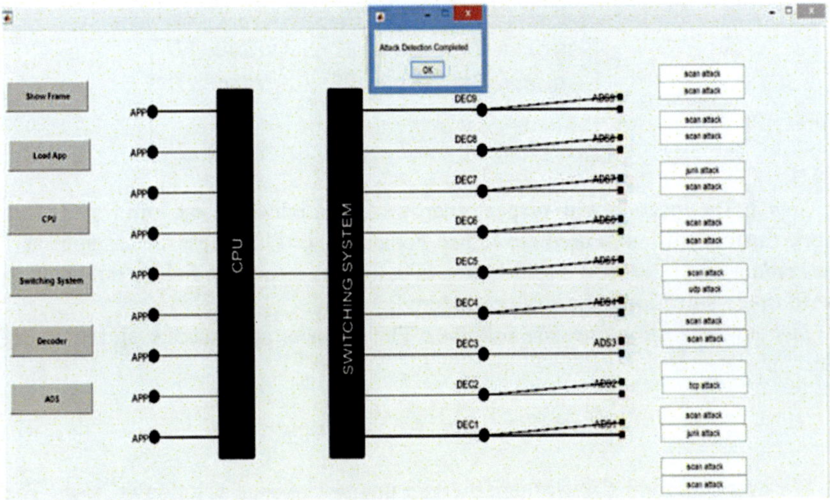

Fig. 1 The architecture of the proposed work

3.1 Feature Selection Using OCSA

The selection of the appropriate features of $Decrypt^{Data}$ improves the capability of the system potential to detect the attacks. In the presented work, the relevant attributes are selected by discarding the unnecessary and duplicate attributes. To complete this target OCSA [30] is used which is combination of Crow Search Algorithm (CSA) and Opposition Based Learning (OBL).

CSA is an advanced meta-heuristic method. It depends on the social aspects and intellectual approach of crows [19–21]. The advantage of CSA is motivated starting with the practice of crow redeeming good enough food in a mysterious location and restoring it at particular time. To safeguard its stuff, it conceals and misguides others away from its actual spot. The major asset of CSA it has only two parameters: flight length and awareness probability, to fix optimization problems.

OBL is a competent method to break the calculation in certain problems of optimization [22, 23]. This method is encouraged by the opposite association with the real-world objects. Without using the notion of counteraction, it would be tough to describe couple of entities like north–south, day-night, east–west, etc., in the existing world. Followings are the steps adapted in OCSA:

Step 1: Pope represents the total quantity of population of the crows (search agents), Flight length $Flight^{length}$, maximum quantity iteration *(Ki)* and $aware^{porb}$ as awareness probability are set. Moreover, crows (i.e., columns of the dataset) are initiated via Eq. (1).

$$Crow_i = \begin{bmatrix} Attribute\ 11\ Attribute\ 12\ ...\ Attribute\ 1D \\ Attribute\ 21\ Attribute\ 22\ ...\ Attribute\ 2D \\ ... \quad ... \quad ... \quad ... \\ Attribute\ n1\ Attribute\ n2\ ...\ Attribute\ nD \end{bmatrix} \tag{1}$$

Step 2: The location with respect to crows are initialized in random way. In this work, the total attributes used are 115 so $Population = 115$. These are assigned as 1 or 0 randomly. If the crow's *i*th location is 0, then its respective field is not opted to perform classification, otherwise considered.

Step 3 (Generating opposite solution): The opposite position is evaluated using Eq. (2).

$$Attribute_i' = a_i + b_i - Attribute_i \tag{2}$$

For every solution $Crow_i$, there exists a unique opposite solution $OCrow_i$. The evaluation of the opposite location is $OCrow_i$ (*Attribute′*) is denoted using Eq. (3). Here, $i \in 1, 2, \ldots, m$.

$$OCrow_i = \begin{bmatrix} Attribute\ 11'\ Attribute\ 12'\ \ldots\ Attribute\ 1D' \\ Attribute\ 21'\ Attribute\ 22'\ \ldots\ Attribute\ 2D' \\ \ldots \quad \ldots \quad \ldots \quad \ldots \\ Attribute\ n1'\ Attribute\ n2'\ \ldots\ Attribute\ nD' \end{bmatrix} \tag{3}$$

Step 4: The memory of crows is initialized and noted. Further, the count is given as $Ki = 1$.

Step 5: Do for i = *1 to n*, a crow is chosen arbitrarily to track *n*th crow.

Step 6: R_m, the random number is produced and position is updated using Eq. (4) if $R_m > = aware^{porb(t)}$. The position update of thief crow m that follows the owner crow *n*.

$$S_m^{t+1} = S_m^t + R_m \times F_m^{length(t)} \times \left(B_n^t - S_m^t\right) \tag{4}$$

Here, B_n^t represents the food source of owner crow *m* and R_m is a random number in the interval [0, 1]. At the iteration *t*, the flight length *x* is denoted as $F_m^{length(t)}$ In case, if $R_n < aware^{prob(t)}$, then revise the random position.

Step 7: By assessing the updated location of crows their memories are updated. Further results are seen and best position is opted. Out of 115 attributes, OCSA selects 36 features which are further given to DBN for detecting the attacks.

3.2 DBN Based Attack Detection

DBN is designed and developed by smolensky in 1986 with multiple layers, in which each layer constitutes visible and hidden neurons which are completely interrelated. The neuron's result is based on probability in Boltzmann machine which is denoted as *outcome*. The *outcome* in Eq. (5) is given with a probability of $prob(\tau)$ in Eq. (6). τ denotes pseudo-temperature of the parameter and used for governing the noise levels.

$$out = \begin{Bmatrix} 0 & with\ probability\ of\ 1 - prob(\tau) \\ 1 & with\ probability\ of\ prob(\tau) \end{Bmatrix} \tag{5}$$

$$prob(\tau) = \frac{1}{1 + e^{\frac{-\tau}{T}}} \tag{6}$$

The energy state which is configured is denoted as *State*, E^{Fun} which is the energy function of the Boltzmann machine is given in Eq. (7). $Wt_{A;B}$ is the weight function between the neurons *A* and B, and φ_A is their biases.

$$E^{Fun}(state) = -\sum_{A<B} State_A State_B wt_{A,B} - \sum_A \varphi_A State_A \tag{7}$$

The energy differential function is evaluated with ΔE^{Fun}. In the Restricted Boltzmann Machine (RBM), for visible and hidden neurons *VIS, HID*, the updated energy function attained using Eqs. (8)–(10).

$$E^{Fun}\left(\vec{VIS}, \vec{HID}\right) = \sum_{A,B} wt_{A,B} VIS_A HID_B - \sum_A VIS_A P_A - \sum_B HID_B Q_B \quad (8)$$

$$E^{Fun}\left(VIS_A, \vec{HID}\right) = \sum_B wt_{A,B} HID_B + P_A \quad (9)$$

$$E^{Fun}\left(\vec{VIS}, HID_B\right) = \sum_A wt_{A,B} VIS_A + Q_B \quad (10)$$

Unsupervised training is performed in RBM. Training set is represented as *Train*, where the rows of matrix denote $\vec{VIS}$. The highest probability is obtained with assigning the weights with $Weight_m$ which is given in Eq. (11).

$$Weight_m = \max_{Weight} \prod_{\vec{VIS} \in Train} prob\left(\vec{VIS}\right) \quad (11)$$

Using Eq. (12), probability is given to every *(VIS, HID)* pair. *Z* is partition function which is mathematically given in Eq. (13).

$$prob\left(\vec{VIS}, \vec{HID}\right) = \frac{1}{Z} e^{-E^{Fun}\left(\vec{VIS}, \vec{HID}\right)} \quad (12)$$

$$Z = \sum\nolimits_{\vec{VIS}, \vec{HID}} e^{-E^{Fun}\left(\vec{VIS}, \vec{HID}\right)} \quad (13)$$

The error is calculated using Root Mean Square Error (RMSE) as per Eq. (14), and it is the variation of predicted and actual values. Objective function regarding fitness is evaluated in Eq. (15).

$$RMSE = Predicted - Actual \quad (14)$$

$$Obj = Min(RMES) \quad (15)$$

The Contrastive Divergence (CD) is used to initialize visible states systematically. The updated weights are given in Eq. (16).

$$wt_{VIS,HID} = \Delta wt_{VIS,HID} + wt_{VIS,HID}. \quad (16)$$

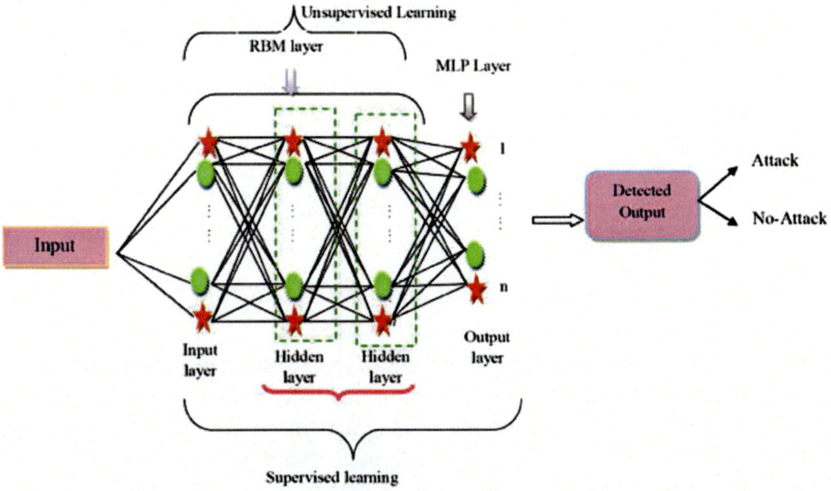

Fig. 2 Architecture of DBN based attack detection

The existence of the attack in the network is determined by DBN. The architecture of attack detection via DBN is given in Fig. 2.

4 Experimental Results

4.1 Execution Setup

The proposed framework was implemented in the MATLAB 2019. The dataset is downloaded from [24]. The performance analysis regarding the optimal feature selection is evaluated for the proposed and other existing works like PSO [25], GA [26], WOA [27], FF [28] and ROA [29]. Additionally, the performance of the classifiers is evaluated using NN [30], SVM [31], and CNN [32]. Evaluation is performed with respect to specificity, precision; the positive measures and FDR, FPR; the negative measures.

4.2 Performance Metrics

Following are the performance metrics used for evaluating the proposed work.

Precision: The proportion of standard data and the total quantity of standard and attack data. Calculated as TP/(TP + FP).

Specificity: The ratio of original positive values that are identified as positive. Calculated as TN/(TN + FP).

False Positive Rate (FPR): Possibility of incorrectly ignoring the null hypothesis. Calculated as FP/(FP + TN).

False Detection Rate (FDR): Actual ratio of type I errors. Calculated as FP/(TP + FP).

4.3 Performance Analysis with Respect to Feature Selection

Analysis using positive measures: The obtained results using positive measures are given in Fig. 3a, b. The accuracy is maximum for OCSA at all the learning percentages. It is observed that the precision and specificity metric recorded an outstanding performance with 98% when compared with other traditional models.

Analysis using Negative measures: The outcomes of FPR and FDR with OCSA and other conventional works are given in Fig. 4c, d.

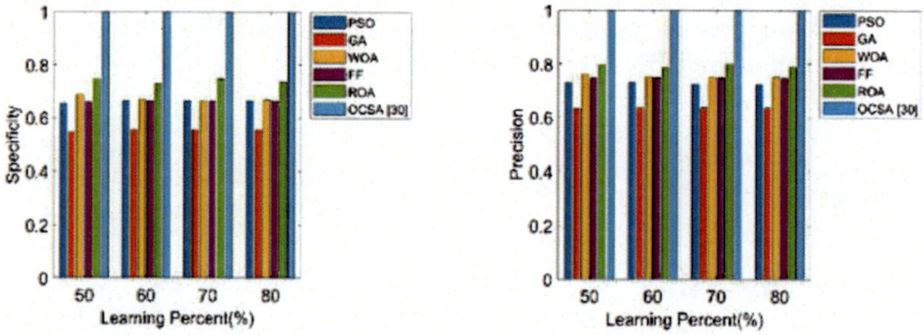

(a) Specificity (b) Precision

Fig. 3 Evaluation of the proposed and conventional algorithms with positive metrics

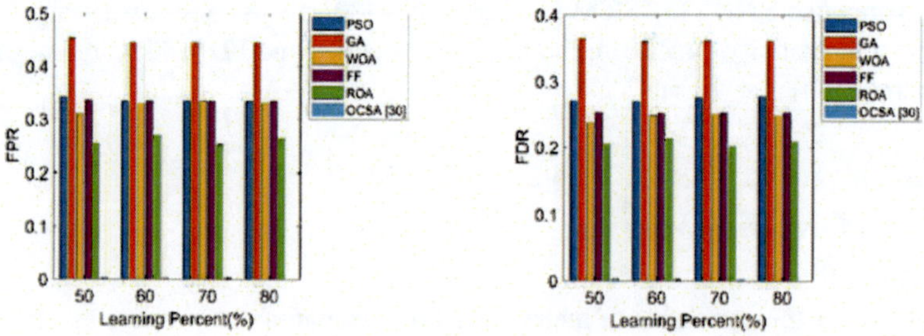

(a) FPR (b) FNR

Fig. 4 Evaluation of the proposed and conventional algorithms with negative metrics

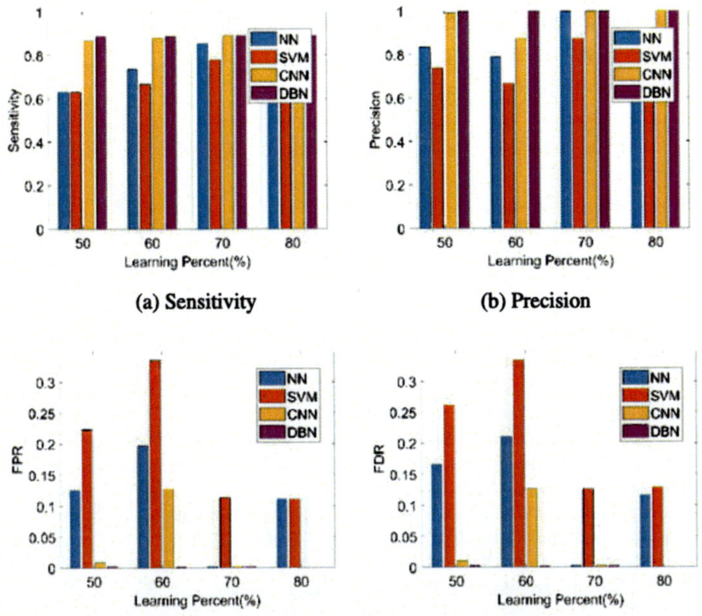

(a) Sensitivity

(b) Precision

(c) FPR

(d) FDR

Fig. 5 Evaluation of the proposed and Traditional classifiers

4.4 Performance Analysis of the Classifier

The performance of the proposed work with DBN is computed with NN, SVM and CNN. The results obtained using the positive measures is given in Fig. 5a, b. On the other hand, evaluations of FPR and FDR are given in Fig. 5c, d. Results prove that the OCSA scores the low values for erroraneous measures.

5 Conclusions and Future Work

This work intends to develop a secure framework for SaaS. As a benchmark, the N_BaIoT dataset is utilized. This dataset contains nine different applications, and each application comprises 115 features. It is subjected to OCSA to perform optimal feature selection. OCSA selects 36 best features from the dataset. After feature selection, the selected features were sent to DBN for detecting the attacks. The proposed OCSA and DBN based attack detection system outperformed with precision, sensitivity, FPR, and FDR over other conventional models and classifiers.

We aim to focus our future work on executing the entire work in the cloud environment. Also, perform cross validation, which states how accurate the proposed model is. Further, aiming at real-time identification of threats and explore the various attacks in the multi-cloud environment.

References

1. Ashok Kumar, C., & Vimala, R. (2020). Load balancing in cloud environment exploiting hybridization of chicken swarm and enhanced raven roosting optimization algorithm. *Multimedia Research, 3*(1), 45–55.
2. Alkadi, O., Moustafa, N., & Turnbull, B. (2020). A review of intrusion detection and block chain applications in the cloud: Approaches, challenges and solutions. *IEEE Access, 8*, 104893–104917.
3. Dong, S., Abbas, K., & Jain, R. (2019). A survey on Distributed Denial of Service (DDoS) attacks in SDN and cloud computing environments. *IEEE Access, 7*, 80813–80828.
4. Vhatkar Kapil Netaji, B. G. P. (2020). Optimal container resource allocation using hybrid SA-MFO algorithm in cloud architecture. *Multimedia Research, 3*(1), 11–20.
5. Wang, W., Du, X., Shan, D., Qin, R., & Wang, N. (2020). Cloud intrusion detection method based on stacked contractive auto-encoder and support vector machine. *IEEE Transactions on Cloud Computing*. https://doi.org/10.1109/TCC.2020.3001017
6. Ravi, N., & Shalinie, S. M. (2020). Learning-driven detection and mitigation of DDoS attack in IoT via SDN cloud architecture. *Internet of Things Journal, 7*(4), 3559–3570.
7. Tian, Z., Luo, C., Qiu, J., Du, X., & Guizani, M. (2020). A distributed deep learning system for web attack detection on edge devices. *IEEE Transactions on Industrial Informatics, 16*(3), 1963–1971.
8. Mishra, P., Varadharajan, V., Pilli, E. S., & Tupakula, U. (2020). VMGuard: A vmi-based security architecture for intrusion detection in cloud environment. *IEEE Transactions on Cloud Computing, 8*(3), 957–971.
9. Virupakshar, K. B., Manjunath Asundi, D. G., & Narayan. (2020). Distributed Denial of Service (DDoS) attacks detection system for open stack-based private cloud. *Proceedings of the Procedia Computer Science, Elsevier, 167*, 2297–2307.
10. Garre, J. T. M., Perez, M. G., & Ruiz-Martinez, A. (2020). A novel Machine Learning-based approach for the detection of SSH botnet infection. *Future Generation Computer Systems, 115*, 387–396.
11. Ninu Preetha, N. S., Brammya, G., Ramya, R., Praveena, S., Binu, D., & Rajakumar, B. R. (2018). Grey wolf optimisation based feature selection and classification for facial emotion recognition. *IET Biometrics, 7*(5), 490–499.
12. Bhardwaj, A., Mangat, V., & Vig, R. (2020). Hyperband tuned deep neural network with well posed stacked sparse autoencoder for detection of DDoS attacks in cloud. *IEEE Access, 8*, 181916–181929.
13. Ravi, N., & Shalinie, S. M. (2020). Learning-driven detection and mitigation of DDoS attack in iot via SDN cloud architecture. *IEEE Internet of Things Journal, 7*(4), 3559–3570.
14. Pillutla, H., & Arjunan, A. (2019). Fuzzy self organizing maps-based DDoS mitigation mechanism for software defined networking in cloud computing. *Journal of Ambient Intelligence and Humanized Computing, 10*, 1547–1559.
15. Bharot, N., Verma, P., Sharma, S., & Suraparaju, V. (2018). Distributed denial-of-service attack detection and mitigation using feature selection and intensive care request processing unit. *Arabian Journal for Science and Engineering, 43*, 959–967.
16. Bhushan, K., & Gupta, B. B. (2019). Network flow analysis for detection and mitigation of Fraudulent Resource Consumption (FRC) attacks in multimedia cloud computing. *Multimedia Tools and Applications, 78*, 4267–4298.
17. Harikrishna, P., & Amuthan, A. (2020) SDN-based DDoS attack mitigation scheme using convolution recursively enhanced self organizing maps *45*(104). https://doi.org/10.1007/s12046-020-01353-x.

18. Zhang, H., et al. (2020). Efficient and secure outsourcing scheme for RSA decryption in internet of things. *IEEE Internet of Things Journal, 7*(8), 6868–6881.
19. Zamani, H., Nadimi-Shahraki, M. H., & Gandomi, A. H. (2019). *CCSA: Conscious neighborhood based crow search algorithm for solving global optimization problems. Applied Soft Computing*. Elsevier.
20. Rizk-Allah, R. M., Hassanien, A. E., & Bhattacharyya, S. (2018). Chaotic crow search algorithm for fractional optimization problems. *Applied Soft Computing, 71*, 1161–1175. Elsevier.
21. Sayed, G. I., Hassanien, A. E., & Azar, A. T. (2019). Feature selection via a novel chaotic crow search algorithm. *Neural Computing and Applications, 31*, 171–188. Springer.
22. Mahdavia, S., Rahnamayana, S., & Debb, K. (2018). Opposition based learning: A literature review. *Swarm and Evolutionary Computation, 39*, 1–23. Elsevier.
23. Ventresca, M., Rahnamayan, S., & Tizhoosh, H. R. (2010). A note on opposition versus randomness in soft computing techniques. *Applied Soft Computing, Elsevier, 10*(3), 956–957.
24. UCI Machine learning Repository, https://archive.ics.uci.edu/ml/datasets/detection_of_IoT_botnet_attacks_N_BaIoT.
25. Tanweer, M. R., Suresh, S., & Sundararajan, N. (2015). Self regulating particle swarm optimization algorithm. *Information Sciences, 294*, 182–202.
26. McCall, J. (2005). Genetic algorithms for modelling and optimization. *Journal of Computational and Applied Mathematics, 184*(1), 205–222.
27. Mirjalili, S., & Lewisa, A. (2016). The whale optimization algorithm. *Advances in Engineering Software, 95*, 51–67.
28. Iztok Fister, I., Xin-SheYang & JanezBrest. (2013). A comprehensive review of firefly algorithms. *Swarm and Evolutionary Computation, 13*, 34–46.
29. Sai Sindhu Theja, R., & Shyam, G. K. (2020). An efficient metaheuristic algorithm based feature selection and recurrent neural network for DoS attack detection in cloud computing environment. *Applied Soft Computing Journal, 100*, 1–11. Elsevier.
30. Binu & Kariyappa. (2018). RideNN: A new rider optimization algorithm-based neural network for fault diagnosis in analog circuits. *IEEE Transactions on Instrumentation and Measurement*, 1–25. IEEE.
31. Gangappa, M., Kiran Mai, C., & Sammulal, P. (2019). Enhanced crow search optimization algorithm and hybrid NN-CNN classifiers for classification of land cover images. *Multimedia Research, 2*(3), 12–22.
32. Thomas, R., & Rangachar, M. J. S. (2019). Fractional rider and multi-kernel-based spherical SVM for low resolution face recognition. *Multimedia Research, 2*(2), 35–43.

English Master AMMU: Advanced Spoken English Chatbot

A. N. Gayathri and V. Viji Rajendran

Abstract Everyone needs to express their ideas, thoughts, and emotions. For a professional or a student, they have to express or communicate their thoughts effectively. To make communication effective, it is important to understand the English language. As English is an international language, all had to have fluent English-speaking skills. Making improvements in spoken English can be done in various ways. Talking to friends or family members in English, practicing phrases and sentences, etc. are some of them. But most people feel some awkwardness talking in front of a crowd or even with friends. Here comes the helper Assistant, Miss AMMU Teacher. Miss AMMU Teacher is the advanced version of AMMU spoken English chatbot. In this work, a chatbot called AMMU, Automatic Mega-agent Managed User guide is built for helping this purpose. AMMU will communicate with the user and will help to improve the spoken English capability of users.

Keywords Chatbot · Dialogflow · Integromat · Google assistant · Bot

1 Introduction

Communication is a very basic need of every individual. A healthy and effective communication makes a relationship or a professional perform better. English may not be the most widely used language in the world, but it is the official language of the world's 53 countries, and nearly 400 million people speak it. If you wish to speak to someone from another country, there is a good chance that you both speak English. As a second language, so many more people are now spending time to learn English. Several nations incorporate English in their school syllabus and children acquire English knowledge at their younger age.

The pandemic outbreak of COVID-19 (coronavirus) [1] demands us to stay home for safety reasons. Children are the most affected by these scenarios. All schools have online classes, and that is the new normal way of teaching. Most of the students face many issues while attending these kinds of classes. And for the professionals also,

A. N. Gayathri (✉) · V. Viji Rajendran
NSS College of Engineering, Palakkad, India

P. Nayak et al. (eds.), *IoT and Analytics for Sensor Networks*, Lecture Notes in Networks and Systems 244, https://doi.org/10.1007/978-981-16-2919-8_37

they have to stay home and do their work and acquire their knowledge through online platforms. As the situation changes, technologies have to change. Here professionals and children both are referred to as students as they both need knowledge. Practicing English with friends and family is a good idea. But most people feel awkward while speaking like this and are afraid of making mistakes. Spoken English courses and classes are very costly and are not affordable for ordinary people.

Here comes the importance of Chatbots. A chatbot is an assistant that can communicate with the human through voice or text, interpret the human queries and make a feeling to the user as they are interacting with a real human. Chatbots, in the near future, will witness a high boom in the field of education [2] because the current COVID situation made all students study through multimedia and telecommunication media. Chatbots for the education system can be developed in various ways. Before going on to the development of English Master AMMU, the following section discusses the most famous and developer-friendly chatbot development frameworks from renowned companies that help to make efficient chatbots. A variety of chatbot development platforms are available for developers. After experimenting with many of them, Google Dialogflow seems to be developer-friendly and is selected for building English Master AMMU. The experimented chatbot development platforms are discussed in the next section.

2 Chatbot Development Framework

There are a variety of chatbot building platforms that help the developer to easily start building chatbots from scratch. Some of them helped the developer to create a chatbot completely without any coding. Also, there are so many chatbot building platforms that help in building chatbots with some code and with some inbuilt features. The chatbot building frameworks from renowned companies are IBM Watson Assistant from IBM, Azure Cognitive services from Microsoft Azure, and Google Dialogflow from Google. The following subsections discuss each framework separately.

2.1 IBM Watson Assistant

IBM Watson Assistant is offered by IBM as a part of the IBM Watson suite. Watson assistant is mainly an AI-powered engine that is capable of providing several NLP services. Watson Assistant incorporates with several cognitive services provided by the IBM Watson platform to enhance the conversation through spell checks, a tone analyzer, and other services. The chatbot that was built using Watson assistant has the ability to know when to search for an answer from a knowledge base when to ask for clarity, etc. [3].

2.2 *Azure Cognitive Services*

Cognitive services are actually a group of REST APIs (Application Programming Interfaces) [4]. Azure cognitive services, a part of Azure and Microsoft public cloud, guarantees the most secure, highly available, and smooth platform for chatbot building. Using Azure Cognitive Services will help the developer force machine learning capabilities in the developer's application, with Microsoft expertise in clearing and training the data, with no load of developing complicated machine learning algorithms. Developers can use the Azure cognitive service QnA maker for making simple Frequently Asked Question (FAQ) bots or question–answer bot. Also, Microsoft Azure provides the web App Bot service, which is a union of web app and bot service [5].

QnA maker is actually an important service provided by Microsoft Azure. The purpose of the QnA maker is to help to enhance the user service with question and answer capability that can be extracted or that can be gathered from semi-structured or structured contents like FAQ, product manuals, etc., that all are available in Uniform Resource Locator (URLs), documents and so on [6].

In Microsoft Azure, it is possible to create a bot into C# or Node.js. The Bot framework provides the core component for creating bots [7]. The main components of it are the Bot connector, Bot framework SDK, and the Bot directory. The main functions of a bot connector are routing the messages, managing the states of the bot, bot registration management, directory tracking, etc. Bot Framework Software Development Kit (SDK) V4 is an open-source SDK that enables the developers to model and builds sophisticated conversations using their favorite programming language [8]. The bot directory is where the user will be able to find bots.

2.3 *Google Dialogflow*

Google Dialogflow is a tool that helps to build smart conversational agents offered by Google. Dialogflow helps the developer to design conversational interfaces and integrate them with other applications. Dialogflow incorporates Google's different AI experiences like learning expertise, search capabilities, speech recognition, and of course the NLU [9]. The developer can create free agents (bots) and make it efficient by creating several intents (use case verb or action). If the bot wants to give dynamic responses, fulfillment features of Dialogflow are used [10].

3 System Design

3.1 System Architecture

The proposed system aims to build a chatbot that helps the user to improve their spoken English skills through simple activities, learning lessons, practice sections and also correcting the user's grammar. Figure 1 shows the diagram for the proposed system.

The proposed chatbot is implemented to be deployed in Google assistant [11]. That is, the user interaction medium for this chatbot system is Google assistant. A user-friendly Google assistant platform is chosen for this work due to the fact that it can be easily accessible for the user, without making any account or linking any Google accounts. The user interacts with the chatbot through Google Assistant, either by means of text messages or by means of voice messages. Then, the Google assistant application sends this message to the Dialogflow, which is the main part of the Chatbot. It is possible to make multiple agents (bots) in for the same chatbot. The agent will categorize the user utterances or input and make a match to the relevant intent. After that, corresponding responses will be sent back to the user. This agent works well if the user only needs static responses from the agent. If the building agent needs more dynamic responses, the Fulfillment feature of Dialogflow can be used. That is, in some cases, it is needed to extract information or data from external databases or services, and the fulfillment facility of Dialogflow is used. In this work, a tool called Integromat [12] is used for the realization of fulfillment in Dialogflow.

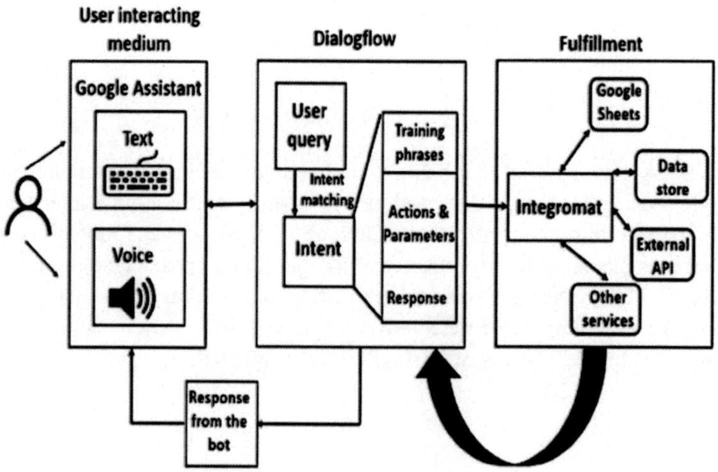

Fig. 1 Proposed system architecture

3.2 Implementation

Initial Version AMMU: English Master AMMU is the advanced version of the Spoken English chatbot AMMU (Automatic Mega-agent Managed User guide). The initial version is a Google Assistant-based chatbot that makes some conversation with the user like taking permission from the user to ask their details, making them comfortable, and will make a friendly relation with the user. After that, it will give small practice activity to the user which helps them to gain confidence with speaking English.

English Master AMMU design: English Master AMMU is designed to be a Master, that is a friendly master, who makes friendly chats with the user and also gives some activities, practice sections, and study materials. AMMU will collect user data for understanding user improvements and for further analysis. The tool Integromat is used for the external API connection, data store integration, and integration of Google sheets for collecting user data.

English Master AMMU has two agents and one mega agent. Using the feature Mega agent, multiple Dialogflow agents, called sub-agents, can be combined into a single agent. The first agent contains friendly conversations with the user. It includes the agent asking the user for their details like name, place, hobbies, etc. These details are collected to the Google sheets for further analysis. Apart from these, this agent also includes some practice activities and learning lessons. The first agent has more than 20 intents for giving the above-mentioned features. The second agent's name is Jokes. It is intended for saying jokes when the user asks. It makes relationships with the user easier.

Development: As mentioned above, the majority of the development of English Master AMMU is done using Google Dialogflow. The main steps for the development of AMMU are as follows:

- Create an account in Dialogflow and create the first agent, here named it as EnglishBot.
- 20 intents are built for this agent.

Fig. 2 Different intents in EnglishBot agent

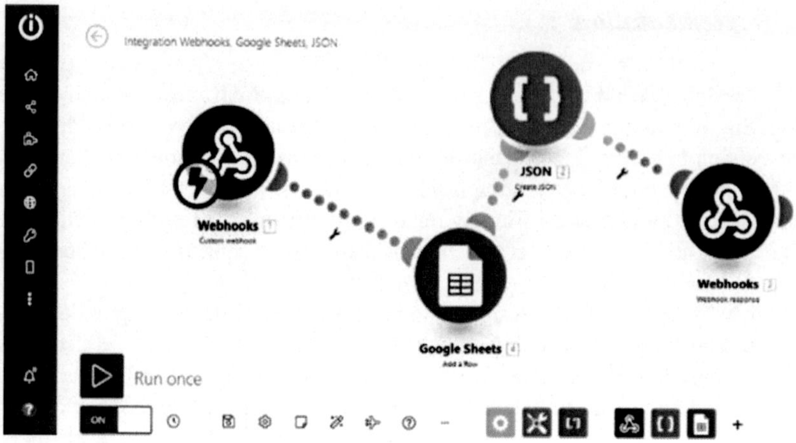

Fig. 3 Google sheet integration in Integromat

- For each intent, give appropriate training phrases and responses. Coding for voice, image, audio, and video are all done in JavaScript or JSON language. Figure 2 gives the intents that are used in EnglishBot agents.
- During the creation of each intent, necessary entities are created. The System entities like @sys.given-name and @sys.location are used for extracting user name and place. Further, necessary entities are created.
- The second agent Jokes is created.
- A mega agent named MegaEnglishBot is created, and the first and second agents are added to it.
- Integromat account is created for fulfillment implementation.
- A scenario is created in Integromat.
- Google sheet is integrated into the Dialogflow chatbot through the webhook in Integromat. Figure 3 shows integration.
- While setting up the Integromat, Integromat will produce a webhook URL. Copy and paste that URL to the Dialogflow webhook fulfillment for all the agents and mega agents.
- Also, the fulfillment should be enabled for the appropriate intents whose data are to be stored. Here, the name, place, and activity intent's fulfillment are enabled.

Testing of English Master AMMU: The last but frequently needed step for building a chatbot is testing. Chatbot AMMU is frequently tested using the "try it now" panel in the Dialogflow interface. After initial testing is done, the created agents are combined with the mega agent. Then the mega agent is integrated into the Google assistant. This will lead to the Action on the Google console page. There is a testing tab, where the fully developed chatbot can be tested. AMMU is perfectly working without errors. At last, the current version is deployed on to the Google assistant. Version 6, now available for users, has been approved by Google.

4 Experimental Results

This section discusses the experimental results that have been obtained after the testing and integration of English Master AMMU with the Google Assistant. The comparison of the current application and the English master AMMU is also discussed here.

English master AMMU takes five versions to get improved and get approved by Google. Version 6 of AMMU is a teaching assistant that initially makes a friendly conversation with the user and after that gives users some teaching or learning lessons and practice activities. Figure 4 shows the welcome UI of English Master AMMU. The current version is perfect and works with dynamic and static user utterances. Figure 5 shows the conversation of AMMU with the user.

After the initial friendly conversation and the first simple activity, the current phase 1 version of AMMU gives the user a provision for studying the basics of English grammar. Then, AMMU asks the user to continue with studying the basics of grammar and practice the activity or not. If the user selects "yes", they will be provided four options for study purposes. If the user selects "No" for the previous question, AMMU will end the conversation. In future versions, AMMU will give some more activities like vocabulary practice and quizzes when the user is not interested in further studying. The study options that AMMU provided are as follows:

- Basic terms in English
- Everyday sentences
- English grammar tenses
- Some helpful phrases.

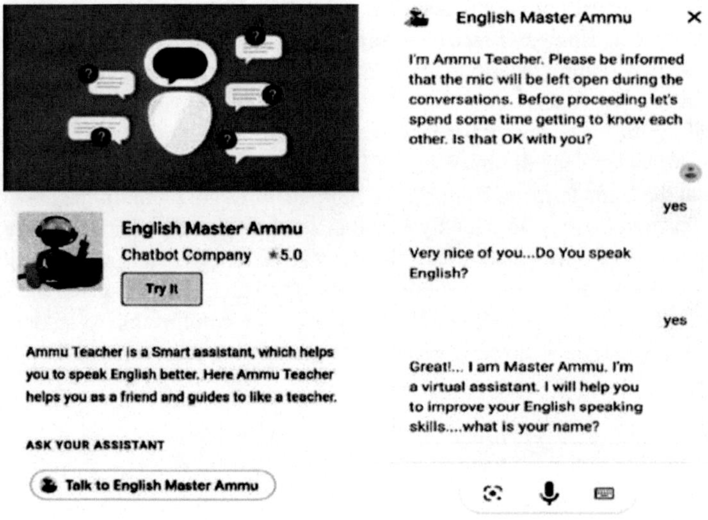

Fig. 4 Welcome page of English master AMMU

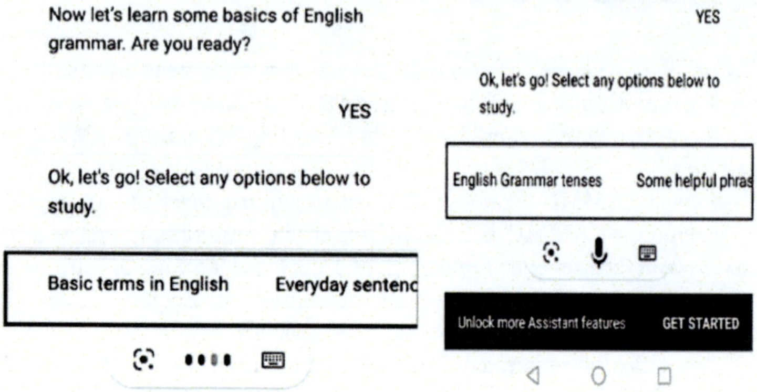

Fig. 5 Working demonstration of latest chatbot release

Figure 5 shows the above-mentioned options that are provided to users. Upon selecting each option, the user is taken to the next page, where each page contains the detailed material about the given main options. Each option has sub-options like the basic terms that given here are nouns, pronouns, verbs, adverbs, adjectives, and prepositions. Additional sub-options are there for deep knowledge.

There are some similar chatbots available in Google Assistant, named English Teacher and Learn English club, both being used to improve English. The comparison between them is shown in Table 1. Learn English Club [13] and English Teacher [14] are the two chatbots that have the same functionality as AMMU. Basically, AMMU, Learn English Club, and English Teacher are integrated with Google Assistant. AMMU will support both text-based and voice-based communication. As Google Assistant chatbot, English Teacher and the Learn English Club also support text and voice-based communication. These three are also available as an open-source facility.

AMMU initially makes a friendly conversation with the user. This creates a pleasant atmosphere for the user. Learn English Club is just giving some knowledge about the basic terms in English. No more features are there. In the case of the English Teacher, there is no friendly conversation between the user and agent. The agent gives some learning activities from the start itself. The learning activities are just one or two practice activities, but on the other hand, AMMU gives the user a very pleasant relationship and then makes them more comfortable to study. There are both study materials and practice activities in English Master AMMU.

Table 1 Comparison on different chatbots

AMMU	Learn english club	English teacher
Google assistant chatbot	Google Assistant chatbot	Google assistant chatbot
Both text and voice-based	Both text and Voice-based	Both text and Voice-based
Open-source	Open-source	Open-source
Makes a friendly conversation to make the user comfortable before start learning	Give basic terms of English grammar. Nothing more	No friendly conversation with the user. Gives some learning activities from the start itself
Focused on correcting user's grammar and vocabularies	No more features to expose	Not trying to correct grammar

5 Conclusion

Chatbots are an intelligent way of communicating with the user. They can make an efficient relationship with the user by interacting like a human. In this paper, an advanced version of Spoken English Chatbot AMMU is discussed. This is capable of making users friendlier with the usage of the language English and makes them more confident in speaking English. Apart from this, practice sections are very helpful for the user for improving their speaking ability, and also learning sections make their grammar parts very strong. Currently, the latest version is available for users in Google assistant in the name "English Master AMMU".

In the next version of AMMU, a data store will be added for data analyzing and Integromat will be utilized for giving the users their improvements on their English learning. The external API connection is also made possible through the help of Integromat and make use of Dialogflow NLU (Natural Language Understanding) for adding grammar correcting feature to the AMMU.

References

1. Singhal, T. (2020). A review of coronavirus disease-2019 (COVID-19). *The Indian Journal of Pediatrics, 87*(4), 281–286.
2. Smutny, P., & Schreiberova, P. (2020). Chatbots for learning: A review of educational chatbots for the Facebook Messenger. *Computers & Education, 151*, 103862.
3. Sabharwal, N., Barua, S., Anand, N., & Aggarwal, P. (2020). Building your first bot using Watson assistant. In *Developing cognitive bots using the IBM Watson engine* (pp. 47–102). Berkeley, CA: Apress.
4. Ed-Douibi, H., Daniel, G., & Cabot, J. (2020, June). OpenAPI Bot: A chatbot to help you understand REST APIs. In *International Conference on Web Engineering* (pp. 538–542). Cham: Springer.
5. Shaikh, K. (2019). Eagle-eye view of azure cognitive services. In *Developing bots with QnA maker service* (pp. 1–29). Berkeley, CA: Apress.
6. Shaikh, K. (2019). The what, when, why, and how of the QnA maker service. In *Developing bots with qna maker service* (pp. 31–59). Berkeley, CA: Apress.

7. Etaati, L. (2019). Bot framework. In *Machine learning with microsoft technologies* (pp. 335–353). Berkeley, CA: Apress.
8. Sabharwal, N., Barua, S., Anand, N., & Aggarwal, P. (2020). Bot frameworks. In *Developing cognitive bots using the IBM Watson engine* (pp. 39–46). Berkeley, CA: Apress.
9. Sabharwal, N., & Agrawal, A. (2020). Introduction to Google dialogflow. In *Cognitive virtual assistants using google dialogflow* (pp. 13–54). Berkeley, CA: Apress.
10. Sabharwal, N., & Agrawal, A. (2020). Advanced concepts in google dialogflow. In *Cognitive virtual assistants using google dialogflow* (pp. 55–117). Berkeley, CA: Apress.
11. Tulshan, A. S., & Dhage, S. N. (2018). Survey on virtual assistant: Google assistant, siri, cortana, alexa. In *International Symposium on Signal Processing and Intelligent Recognition Systems* (pp. 190–201). Singapore: Springer.
12. Hughes, K., Lecky-Thompson, J., Ammon, M., & Murphy, H. (2017). Track the impact of your publications.
13. English Learn Club, https://assistant.google.com/services/a/uid/0000002a507e2159?hl=en_in&source=web.
14. English Teacher, https://assistant.google.com/services/a/uid/00000082469a1382?hl=en_in&source=web.

Investigation of CNN-Based Acoustic Modeling for Continuous Hindi Speech Recognition

Tripti Choudhary, Atul Bansal, and Vishal Goyal

Abstract Recently, Convolutional Neural Network (CNN) gains more popularity over hybrid Deep Neural Network (DNN) and Hidden Markov Model (HMM) based acoustic models. CNN has the ability to deal with speech signals and it makes appropriate choice for the Automatic Speech Recognition (ASR) system. The sparse connectivity, weight sharing, and pooling allow CNN to handle a slight position shift in the frequency domain. This property helps to manage speaker and environment variations. CNN works well for speech recognition, but it was not appropriately examined for the Hindi speech recognition system. The activation functions and optimization techniques play a vital role in CNN to achieve high accuracy. In this work, we investigate the impact of various activation functions and optimization techniques in the Hindi-ASR system. All the experiments were performed on the Hindi speech dataset developed by TIFR, with the help of the Kaldi and Pytorch-Kaldi toolkit. The experiment results show that the ELU activation function with Rmsprop optimization techniques gives the best Word Error Rate (WER) 14.56%.

Keywords Automatic speech recognition · GMM · FBANK · CNN

1 Introduction

ASR is the process of converting human speech into text sequences. The speech recognition field has dramatically benefited from DNN as it estimates the millions of parameters without overfitting [1]. The high-performance gain with less need for engineered features (i.e., MFCC, PLP) makes it the foremost favorite choice for Large Vocabulary Continuous Speech Recognition (LVCSR) system. DNN based acoustic models replace the statistical models (HMM, GMM) in the field of ASR [2]. DNNs are good enough to perform class-based discrimination [3]; on the other hand, it suffers from some serious drawbacks. First, due to the feed-forward fully connected nature, it cannot capture the structural locality of feature space [4], Second, DNN's cannot persist the past information. Finally, DNN has suffered from a vanishing

T. Choudhary (✉) · A. Bansal · V. Goyal
GLA University, Mathura, India

P. Nayak et al. (eds.), *IoT and Analytics for Sensor Networks*, Lecture Notes in Networks and Systems 244, https://doi.org/10.1007/978-981-16-2919-8_38

gradient problem [5]. Except above, the fully connected nature of DNN introduces the millions of learnable parameters that increase the computational load of the ASR system. Generally, neural networks need a huge amount of data to train a large number of parameters. Recently, more emphasis has been given to those models that require fewer parameters. These models have become very popular in low resource languages and show high performance. In this work, we have only a few hours of data, and CNN requires fewer parameters compared to other models. In this work, we choose CNN based acoustic model to test the Hindi ASR performance.

CNN [6] is a modification of neural network (NN). CNN can efficiently seize the structural locality from the feature space [7]. CNN has the capability to capture the position invariant features. CNN is not good enough to handle variable-length inputs. Basically, CNN consist of three layers, these are pooling layer, convolutional layer and Fully Connected (FC) layers. In convolutional layers the filter is glide across the width and height of the input and dot product of input and filter are count at every spatial position. The pooling layer is responsible for reducing the dimensionality of the feature map without losing any useful information. This layer minimizes the computational complexity of the system and control the matter of overfitting. The last layers of CNN are FC layer in this input in the form of a vector. In FC the output of previous layer is compress and form a single vector.

The overall performance of any CNN based acoustic model highly depends on the type of activation function and optimization techniques. The optimization problem becomes more challenging as the number of parameters increases in limited resource conditions. We have tested three different optimization techniques to investigate their role in low or limited resource conditions in this work. This work aims to study the impact of activation function and optimization techniques on limited resource Hindi speech data-set. In this work, we investigate the Word Error Rate (WER) of the Hindi ASR based on five activation functions (Sigmoid, tanh, relu, leaky-relu, and elu) and three optimization techniques (Adam, SGD, and Rmsprop). We have been used the Hindi speech data-set [8] for this work. In comparison to the other European languages, a small amount of work has been done for the Hindi language [4, 9–16]. The major difficulty for Hindi ASR is the lack of speech data available for research. All the hyper-parameters and CNN architecture remained the same for all experiments.

The remaining paper is sorted out as follows: In Sect. 2, CNN-based acoustic modeling for the Hindi language was discussed. In Sect. 3, speech corpus details were explained. Sect. 4 covers the results and experimental setup details. In Sect. 5, the conclusion is discussed.

2 CNN-Based Acoustic Modeling for Hindi Language

The architecture of CNN acoustic modeling is illustrated in Figure 1. The convolutional layers are the main building blocks of any CNN architecture, in which a small size of filters was applied to the input to generate feature maps. 40-FBANK features

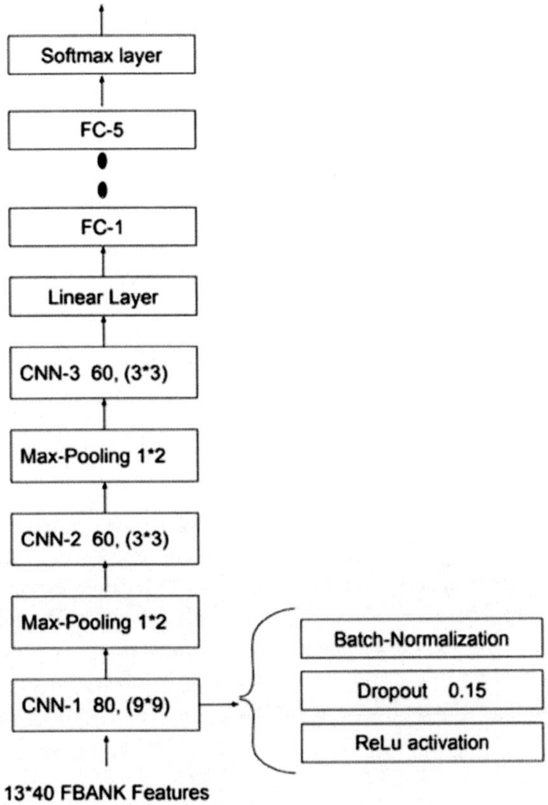

Fig. 1 CNN based acoustic modeling

were used as an input to the CNN architecture throughout this work. We have applied the filter of size 9*9 to the first convolutional layer to get the output feature map. Eighty feature maps were used in the first convolutional layer. The dimensions of the feature maps are reduced by applying the pooling layer known as max pooling layer with a filter size of 1*2. The pooling was only applied in the frequency domain. The input context of 13 frames was used in this work.

Max pooling layer contains the most prominent features of the previous feature map. The final dimension has been reduced to 1*5. Finally, a fully connected layer was used. In this work, we used the FC layer of size five, in which the last layer used a softmax activation function to generate the posterior probabilities. The other details are mentioned in Table 1.

Table 1 CNN architecture configurations

Layers	#Layers	Feature map size	Filter size	#Neurons
Convolution	3	80, 60, 60	9, 3, 3	–
Pooling	3	80, 60	2, 2	–
Fully connected	3	–	–	1024

Table 2 Hindi corpus details

SN	Dataset	#Sentences	Duration
1	Training	800	2.1 h
2	Development	100	10 Min
3	Test	100	10 Min

3 Dataset Details

In this work, we used the Hindi speech dataset created by TIFR, Mumbai, in 2000 [8]. This dataset contains 100 speaker utterances. 10 sentences were recorded by each speaker in an isolated room with two microphones on 16 kHz sampling frequency. This dataset is small in approximately 2.5 hours. We split this dataset into three parts, training set, development set, and test set. We used 80 speakers for training and 10 speakers for development, remaining speakers for the testing. Further details of this dataset can be found in Table 2.

4 Experimental Setup and Results

In this work, 40 log-Mel Filterbank features with a 13-frame context window were used as an input. To extract the features, we used the Kaldi [17] toolkit. The CNN based acoustic modeling was done by the Pytorch-Kaldi [18] toolkit. The speech features and alignments generated via the Kaldi toolkit were used as an input for the Pytorch-Kaldi toolkit. We used the mini batch of size eight. The various optimization techniques were used. The training has been done for 24 epochs. The system performance was measured, after each epoch on the dev set. We have chosen the threshold value as 0.001 for updating the learning rate. We used the initial learning rate of 0.08. Except these, we use the dropout of value 0.2 in all layers for all experiments. The pronunciation dictionary of the Hindi language contains 2803 words made up of 68 Hindi phones. In this work, only closed vocabulary experiments have been done for investigation. For language modeling, the different text was collected from various sources to train the language model. The SRILM [19] toolbox was utilized to prepare the language model. For language modeling, only the labeled data of tarin set was used.

4.1 Experiments with Baseline System

In this section, we prepared the baseline system using the Kaldi toolkit. The alignments were generated by a tri-phone based GMM system was used to train the CNN based acoustic model. The Pytorch-kaldi toolkit was utilized in training of CNN based acoustic modeling. The Sigmoid activation function used to train all the acoustic models is presented in this section. The Stochastic Gradient Descent (SGD) algorithm was used in optimization. Table 3 reports the ASR system performance using different acoustic modeling. First, GMM-HMM-based acoustic modeling took place using 39 MFCC features. We that found the tri-phone-based acoustic model improves the ASR system performance. For DNN based acoustic modeling, 40-dimensional FBANK features were used. We used five-layer architecture with 1024 hidden neurons for this purpose. We used the sigmoid activation function in hidden layers and the softmax activation function in the output layer. DNN works better compare to the conventional GMM-HMM model. Finally, CNN based acoustic model takes place. For CNN, we used 40 FBANK features and 13 context window sizes as an input to the CNN model. All other configurations remained the same as describes in the previous section. Table 3 reports the best WER Of 23.04% was found with CNN based acoustic model. Figure 2 demonstrates the evaluation of loss and accuracy of CNN based acoustic modeling for above mention setup.

Table 3 Baseline system performance evaluation

Acoustic models	Features	Language model	WER(%)
Monophone-GMM	MFCC	tri-gram	30.20
Triphone-GMM	MFCC	tri-gram	28.85
DNN	FBANK	tri-gram	25.50
CNN	FBANK	tri-gram	23.06

Fig. 2 Baseline CNN based acoustic modeling, (left) accuracy evaluation, (right) loss function evaluation

Table 4 Experiments with different activation function and optimization techniques

Optimization	Sigmoid	Tanh	ReLu	Leaky-ReLu	'Elu
SGD	23.06	16.48	15.36	15.88	14.56
Adam	25.08	17.65	16.88	17.44	15.66
Rmsprop	24.66	16.96	16.04	16.88	15.04

4.2 Experiments with Different Activation Function and Optimization Techniques

In this study, we explore the role of three optimization techniques (SGD, Rmsprop, Adam). The impact of five activation function on accuracy was measured. We found that Sigmoid activation function performs very poorly compared to the other activation functions. Tanh activation function works slightly better than the Sigmoid activation function. Except for Sigmoid and Tanh, we tested ReLu, Leaky-ReLu, and Elu activation function. In the same series, we tested our model with different optimization techniques. We found SGD optimization works better compared to the other two optimization techniques. We found elu activation function with SGD optimization technique gives best WER of 14.56%. All other configurations were remained same throughout all the experiments (Table 4).

5 Conclusion

Training of neural network acoustic models on limited resource data mostly suffers from generalization issues. A careful selection of the acoustic model and fine-tuning will address the issue. In this paper, the role of activation functions and optimization techniques were investigated on limited resource Hindi dataset. Overall, 9% relative reduction on WER was measured using elu activation function for Hindi speech recognition. In optimization techniques, SGD technique performs well in comparison to Rmsprop and adam optimization. This work can be investigated to explore the different hyperparameters of the neural networks. The accomplished WERs were still moderately high however inferred the right way for subsequent work.

References

1. Dahl, G. E., Yu, D., Deng, L., & Acero, A. (2011). Context-dependent pre-trained deep neural networks for large-vocabulary speech recognition. *IEEE Transactions on audio, speech, and language processing, 20*(1), 30–42.
2. Hinton, G., Deng, L., Yu, D., Dahl, G. E., Mohamed, A. R., Jaitly, N., Senior, A., Vanhoucke, V., Nguyen, P., Sainath, T. N., & Kingsbury, B. (2012). Deep neural networks for acoustic modeling in speech recognition: The shared views of four research groups. *IEEE Signal Processing Magazine, 29*(6), 82–97.
3. Mohamed, A. R., Dahl, G., Hinton, G. (2009) Deep belief networks for phone recognition. In Proceedings of Nips workshop on deep learning for speech recognition and related applications. vol. 1(9) (p. 39)
4. Passricha, V., & Aggarwal, R. K. (2019). A hybrid of deep CNN and bidirectional LSTM for automatic speech recognition. *Journal of Intelligent Systems, 29*(1), 1261–1274.
5. Glorot, X., Bengio, Y. (2010) Understanding the difficulty of training deep feedforward neural networks. In *Proceedings of the thirteenth international conference on artificial intelligence and statistics* (pp. 249–256).
6. Rumellhart, D. E. (1986). Learning internal representations by error propagation. *Parallel Distributed Processing, 1*, 318–362.
7. LeCun, Y., Bottou, L., Bengio, Y., & Haffner, P. (1998). Gradient-based learning applied to document recognition. *Proceedings of the IEEE, 86*(11), 2278–2324.
8. Samudravijaya, K., Rao, P. V., Agrawal, S. S. (2000) *Hindi speech database*. In *Sixth International Conference on Spoken Language Processing*
9. Sahu, P., Dua, M., Kumar, A. (2018) Challenges and issues in adopting speech recognition. In *Speech and Language Processing for Human-Machine Communications* (pp. 209–215). Singapore: Springer.
10. Kuamr, A., Dua, M., Choudhary, A. (2014) Implementation and performance evaluation of continuous Hindi speech recognition. In *2014 International Conference on Electronics and Communication Systems (ICECS)* (pp. 1–5). India: IEEE.
11. Kumar, A., Aggarwal, R. K. (2020) A time delay neural network acoustic modeling for Hindi speech recognition. In *Advances in Data and Information Sciences* (pp. 425–432). Singapore: Springer.
12. Dua, M., Aggarwal, R. K., & Biswas, M. (2019). Discriminatively trained continuous Hindi speech recognition system using interpolated recurrent neural network language modeling. *Neural Computing and Applications, 31*(10), 6747–6755.
13. Dua, M., Aggarwal, R. K., & Biswas, M. (2019). GFCC based discriminatively trained noise robust continuous ASR system for Hindi language. *Journal of Ambient Intelligence and Humanized Computing, 10*(6), 2301–2314.
14. Kumar, A., Aggarwal, R. K. (2020) A hybrid CNN-LiGRU acoustic modeling using raw waveform sincnet for Hindi ASR. *Computer Science, 21*(4).
15. Kumar, A., Aggarwal, R. K. (2020) Hindi speech recognition using time delay neural network acoustic modeling with i-vector adaptation. *International Journal of Speech Technology*, 1–12.
16. Kumar, A., & Aggarwal, R. K. (2020). Discriminatively trained continuous Hindi speech recognition using integrated acoustic features and recurrent neural network language modeling. *Journal of Intelligent Systems, 30*(1), 165–179.
17. Povey, D., Ghoshal, A., Boulianne, G., Burget, L., Glembek, O., Goel, N., Hannemann, M., Motlicek, P., Qian, Y., Schwarz, P., Silovsky, J. (2011) The Kaldi speech recognition toolkit. In *IEEE 2011 Workshop on Automatic Speech Recognition and Understanding 2011 (No. CONF). IEEE Signal Processing Society*
18. Ravanelli, M., Parcollet, T., Bengio, Y. (2019) The pytorch-kaldi speech recognition toolkit. In *ICASSP 2019–2019 IEEE International Conference on Acoustics, Speech and Signal Processing (ICASSP)* (pp. 6465–6469). IEEE.
19. Stolcke, A. (2002) SRILM-an extensible language modeling toolkit. In *Seventh international conference on spoken language processing.*

Energy Conserving Techniques of Data Mining for Wireless Sensor Networks—A Review

Pragati Patil Bedekar, Atul Raut, and Abhimanyu Dutonde

Abstract Now a day the application of Data mining in many areas has been tremendously increased. Data management and its processing is becoming the active research area for research community. In this paper the major concern is to study and analyse the impact and performance of various data mining techniques over sensor networks (wireless), as the characteristics of sensor nodes and its wireless nature are the primary concerns. The volume and rate for data generation is huge, which is variable in nature; therefore, it is very much essential to design and implement data mining methods for Wireless Sensor Network. This paper mainly focused on to the comprehensive survey of existing data mining techniques, wherein the various limitations and its probable solutions are highlighted in detailed. A transition of traditional mining techniques to newly introduced research is also analysed in this paper. Detailed working and description of compression algorithm has been stated in this paper.

Keywords Data mining · Energy conservation · System efficiency · Wireless sensor network

1 Introduction

To extract useful or meaningful information from the large databases; the concept of data mining is used. Therefore, it relates to the meaningful or relevant information from the warehouse.

P. P. Bedekar (✉)
Department of Computer Science and Engineering, Abha Gaikwad-Patil College of Engineering, Nagpur, India

A. Raut
Department of Information Technology, Jawaharlal Darda Institute of Engineering and Technology, Yavatmal, India

A. Dutonde
Department of Information Technology, Tulsiramji Gaikwad-Patil College of Engineering and Technology, Nagpur, India

P. Nayak et al. (eds.), *IoT and Analytics for Sensor Networks*, Lecture Notes in Networks and Systems 244, https://doi.org/10.1007/978-981-16-2919-8_39

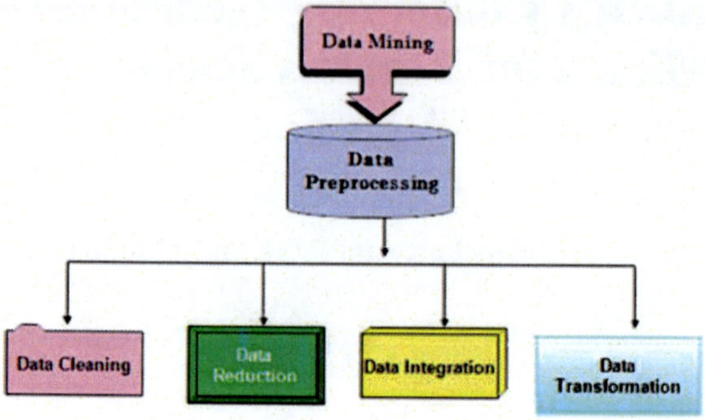

Fig. 1 Data processing techniques of data mining

One of the broader areas of Data Mining is Data Pre-processing which includes techniques like

- **Data Cleaning** This technique is used to remove noise and correct inconsistencies in data.
- **Data Integration** This technique is used to combine data from various sources into single warehouse.
- **Data Reduction** Data reduction can be observed by Data aggregation, clustering methods.
- **Data Transformation** Changing format, values, data (transforming data from one form to another) (Fig. 1).

Reliable and real time monitoring plays a vital role in many applications like field monitoring, health and habitant monitoring, tracking of objects, analysis of data etc. Practical monitoring of above mentioned applications is very impotent because maximum time it deals with the fast rate of dynamic and huge amount of data streams which can be geographically distributed all over network. This huge amount of variable data is then analysed, processed and transformed using mining techniques. In this paper existing and revised data mining techniques has been studied.

2 Overview of Methods Used

Abdelmoghith and Mouftah [1] The proposed work is focused on the compression of data which is being sent between sink and sensor node. The mathematical results proved that the data reduction is up to 70% and energy saving is up to 37%. The data compression algorithm, S-LZW, LEC, Huffman, ND-Encode, RQT etc., has been studied and analyzed, but the proposed MBC has been proved to be the energy efficient scheme. Mamurjon and Ahn [2] The authors have discussed about Data Collection methods in WSN. Generally gathering sensed data from each node is a tedious job. In general, base node starts from node nearby base station, wherein it has travel all over the network for collecting data from each node, return back and upload the data to base node. This entire process is energy and time consuming. Hence the author has introduced the concept of mobile sink, where in the mobile sink used to visit every cluster in the network, return back and send sensed information to sink node. The introduced method has decreased energy consumption of nodes. Mudgule et al. [3] have done review on different data reduction techniques like adaptive filter, tree based method, cluster technique, data prediction method, etc. It has been observed that to reduce energy consumption the data reduction technique of Data Mining is important. Data reduction is used to remove unnecessary data while transmission. In this technique, it reduces the repeated data, its processing and analyzed it before transmission and hence as the amount of data reduce it leads to energy preservation in WSN.

3 Classification of Compression Techniques

Two algorithms are discussed in this paper that is compression and reconstruction algorithm [4]. Compression algorithm takes X as input and generates a compressed Xc whereas reconstruction algorithm executes on compressed representation Xc and gives output as the reconstruction Y.

Based on requirement of reconstruction, a data compression technique has been classified into two broader classes as shown in Fig. 2.

1. Lossless compression scheme, reconstruction of original data from compressed data, in which Y is identical to X and
2. Lossy compression scheme, provide higher compression than aforesaid compression but allow Y to be different from X.

The ultimate aim is to reduce energy consumption of the Network, which can be achieved by using compression techniques. It deals with conversion of data strings of characters into another representation which may have same data in small length as much as possible. The main objective is efficient coding and minimize bandwidth required for transmission.

The main objective of this review is to show performance of different lossless and lossy compression techniques and their comparative study. This technique helps to

Fig. 2 Working of compression technique

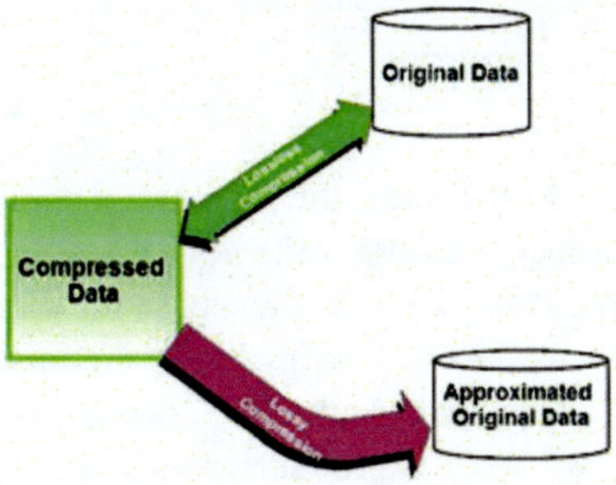

save battery power and data storage capacity or transference capacity. It has been studied that more energy is utilized in transmission rather than the processing of information therefore the sending compressed data can work out at some extent. Following is the review on different algorithms/schemes used for data compression.

3.1 String Based Compression Techniques

LZW is a general technique used for compression of data. This technique is a universal lossless data compression algorithm. It is also called as dictionary based algorithm [5–10]. In the year 1977, A. Lempel and J. Ziv published this algorithm.

Welch [11], Singh and Singh [12], Saidani et al. [13] The LZW algorithm: it never send character, and it tries to send codes for already known strings. Each time a new code is sent, a new code is attached to the dictionary. Working of algorithm is as follow:

a. Reading of incoming text (no analysis will be performed here)
b. Check and verify whether the string has been read before or not
c. If no in step b, then read string and add it to the dictionary
d. Compression of text will be done if the new code is written instead of string.

The start dictionary contains standard character set, that is, ASCII Codes. Two control codes namely 256 and 257 are attached to this dictionary. The 256 is used as an EOF (end of file) character, and the 257 is used as an EOD (end of dictionary) character. Example,

Let us consider the string c a b b a c a b b a.

Index	Dictionary
1	A
2	B
3	C
4	Ca
5	Ab
6	Bb
7	Ba
8	Ac
9	Cab
10	Bba

Step	X	y	x+y	Code	
1		c	C		
2	C	a	Ca	3	
3	A	b	Ab	1	
4	B	b	Bb	2	
5	B	a	Ba	2	
6	A	c	Ac	1	
7	C	a	Ca		String Matched
8	Ca	b	Cab	4	
9	B	b	Bb		
10	Bb	a	Bba	6	String Matched
11	A		A	1	

At the start, the dictionary contains all possible individual characters and x is empty; y contains the next character in the char stream. Now, let us find out whether the string x + y is present in the dictionary or not. If it is not found in the dictionary then add x + y string to the dictionary.

Hence after applying concept of LZW algorithm over the above text the code 3 1 2 2 1 4 6 1 is produced as output.

S-LZW here **S** stands for Sensor. S-LZW is a dictionary/string based lossless compression algorithm which is used to support data compression in WSN. It is stated that to transmit plain text data, more energy is required as compared to compressed text. Compressed form data gives more reliability comparatively. It has been observed that in transmission process of data, packet loss occur in WSN at more rate; hence, S-LZW algorithm follows the strategy to divide the incoming data stream into small and independent blocks as shown in the figure. While doing this the algorithm can assure that if packet loss happened in transmission process then it only affect the packet loss following packets in its own block and therefore rest of blocks will be secure enough. With reference to [8] as resources are limited in WSN, therefore a dictionary of size 512 entries is adopted to fit in small memory of a sensor node. At the same time, the algorithm said to compressed data into the block of 528 Bytes as shown in the figure which helps to improve performance. Now the next step is to divide the compressed data into many blocks of independent packets (Figs. 3 and 4).

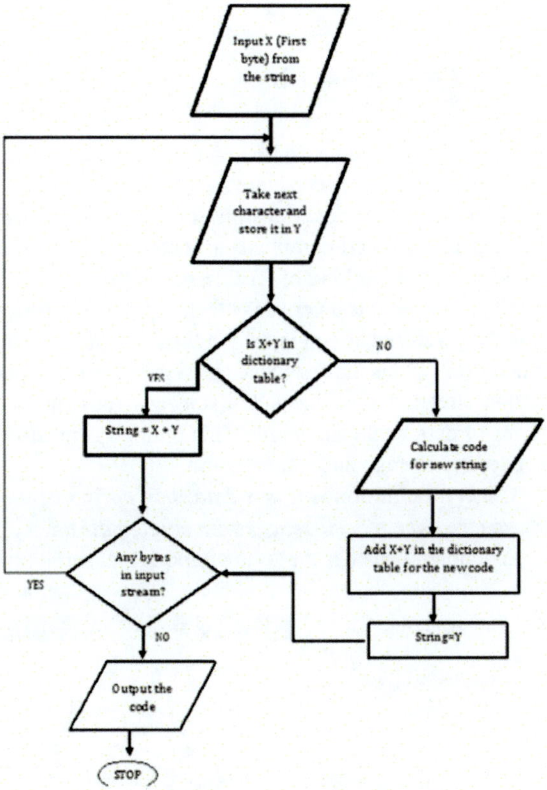

Fig. 3 Flowchart for LZW algorithm

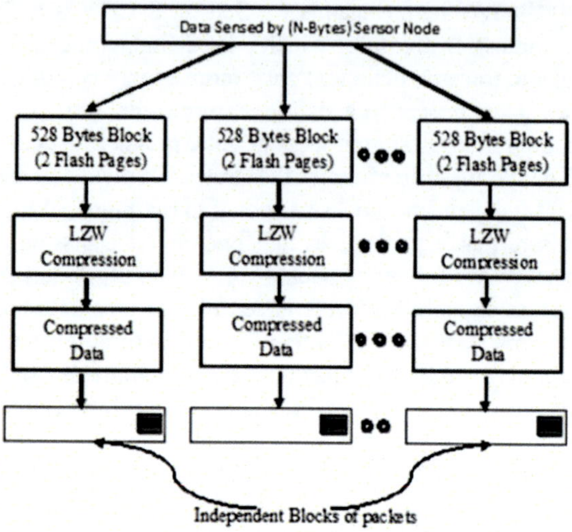

Fig. 4 Working of S-LZW algorithm

3.2 *RLE and K-RLE*

Run Length Encoding (RLE). As mentioned in S-LZW algorithm, transmission and reception of data consume more power. In order to reduce overhead of energy consumption while data transmission which is also called as in-network processing technique, enhanced compression techniques are used—RLE is one of them. First we will see the working of RLE in detail [9].

RLE (lossless) is a very easy technique which is used to compressed digital data. The data is stored as single data in the form of Value and Count. It is more benefited when data carries the sequences of same data value occurrences in the original runs or data stream; hence, instead of storing same original run, the data will be stored in terms of data value and count. This method is needed to reduce the amount of data needed for storage and transmission.

General definition is *d is a data item occurs at n consecutive times in the input stream, replace n occurrences with single pair nd.*

Example: Consider the following original run length.

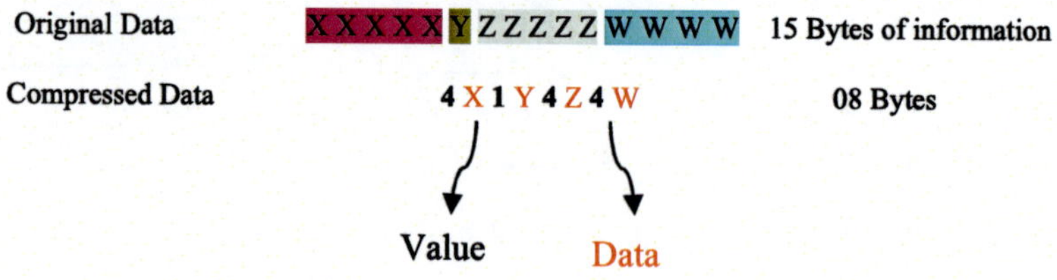

RLE is used where we expect the consecutive sequence of same runs. In the given example, similar occurrences of X, Y, Z and W are counted and compressed as Count followed by data or vice versa.

- With this example we can see the clear difference of using LZW and RLE.

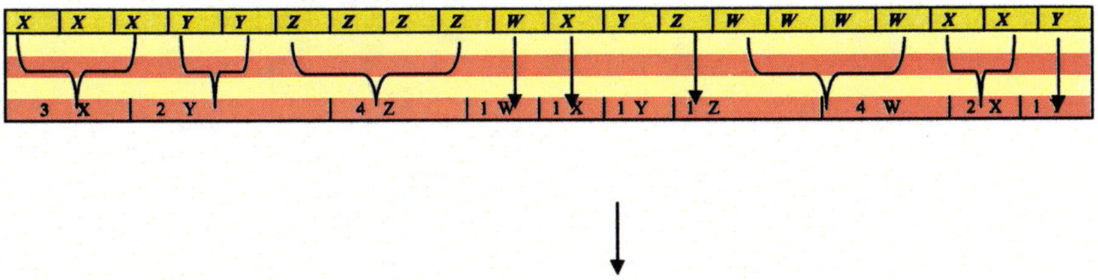

K-RLE means RLE with K-Precisions. In this technique if data items d, d + K, or d-K occurs n consecutive times in the input stream then replace the *n* occurrences with the single pair *nd* [8].

3.3 Coding by Ordering

Coding by ordering is a method used to minimize amount of data need to send for further communication. The main objective is to increase network lifetime by using two schemes mainly a. Data Aggregation b. Data Compression [7].

Step 1. The controller breaks down the space into different regions also known as cuboids and transmit the interested data packets to each region.

Step 2. After step 1, maximum nodes transmit sensed data to the controller after the specified intervals.

Step 3. As almost all nodes send data at the specified time, then it will become easier to combine all the readings into a single packet.

Step 4. One data packet with only one adder will travel from the region to the controller (Fig. 5).

Data Funneling (routing algorithm) has been introduced [7] to reduce the communication overhead while sending the set of sensed data by the number of sensor nodes to the single destination called as sink node. The proposed routing algorithm not only benefited in energy conservation but also used to reduce the data packet collision in the network.

For the compression of data 'Coding by Ordering' scheme has been used which losslessly compressed data encode learning information in the sequence or order.

The motes (nodes) disbursed in large region will send data to the border node. Border node is point where all data from the region (send by n nodes) will be collected. In this case sensor nodes will compute the communication cost for border node only instead of sink (controller) node. As border nodes needs to communicate with controller so it will compute its communication cost.

Fig. 5 Coding by order

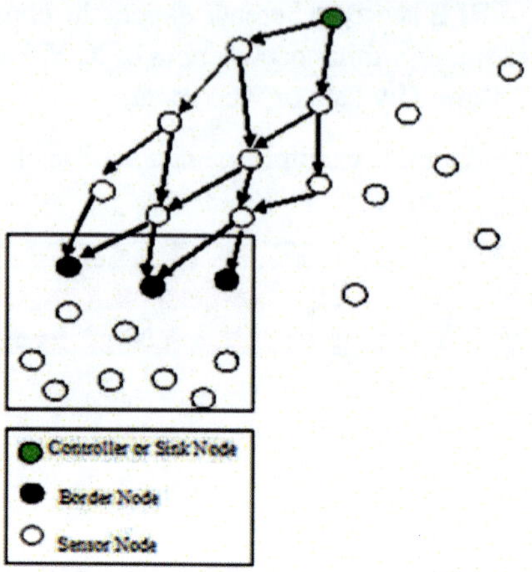

Border node is the responsible one to place each node's packet into a big super-packet containing sensed information from environment then send only one big-packet to control node. The packet (node) contains the NODE ID and node ID contains its position and payload. The Border node contains the nodes ID convey belonging of payload assign to which particular node. The authority to choose ordering of packets to super packets is managed by border node, and it will not affect application as the all data reached at controller at the same time.

Example: Assume that there are four nodes in the region with ID's 1, 2, 3, and 4. Each of these sensor ranging from 1 to 4 sensed some number values let us say {0, …, 5}. It is stated earlier that the border node has freedom, in the below example the data packets of node 4 has been suppressed by border node, and now it will select possible permutations combination for the remaining 3 nodes that is 3! = 6 which is shown in below table [7] (Table 1).

Table 1 Mapping from permutations to integers

Permutations	Integer
123	0
132	1
213	2
231	3
312	4
321	5

3.4 Pipelined In-Network Compression

The authors' motivation behind this compression technique is to save energy required for the communication in wireless network. Thereby reducing the repetitive data stream the said objective has been achieved by Pipelined In-Network compression method.

1. Sensed data by node (mote) is buffered at own node itself
2. Delay for some appropriate time satisfying End-to-End latency-bound
3. Remove same or repeated data from the buffered data.

3.5 Distributed Compression

Distributed Compression as the name suggests it is used to do compression in a distributed manner. Here without the communication between n motes or sensor nodes the data redundancies can be removed in a distributed manner.

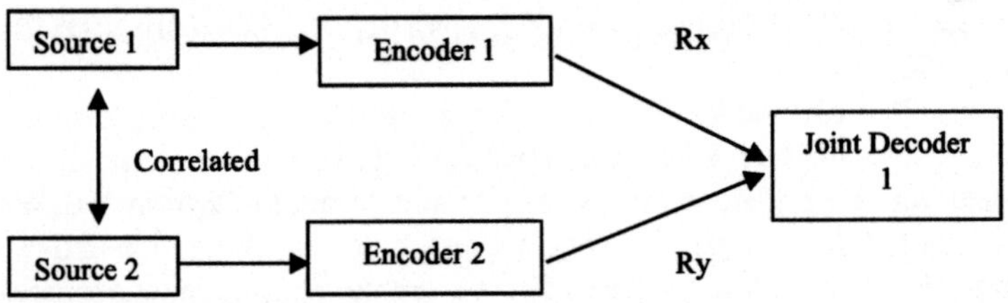

As it is clear from the above figure, information can jointly compress correlated observations, even with no communication between nodes.

- Source coding with side information-discrete alphabets

X and Y are two data sets likely equal with length of three binary data. Hamming Distance between X and Y is 1 at most.

X = [0 1 0], Y = [0 1 0], [0 1 1], [0 0 0], [1 1 0].

X can be compressed to two bits. If both decoder and encoder know Y, what will happen to the compression rate of X? Then if Y is known to the decoder.

According to this method, since the Decoder know Y and X is only one Hamming Distance apart from Y, it is not sufficient to differentiate X = 111 from X = 0000 [6].

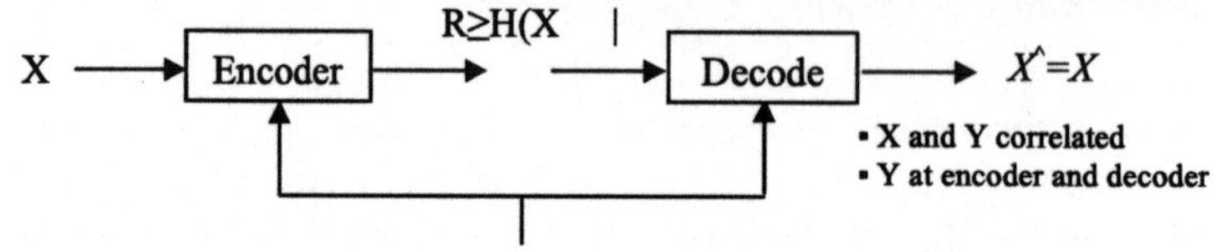

X^=E[X | Y, X ϵ Γi]- estimate given decode Y ut and side information.

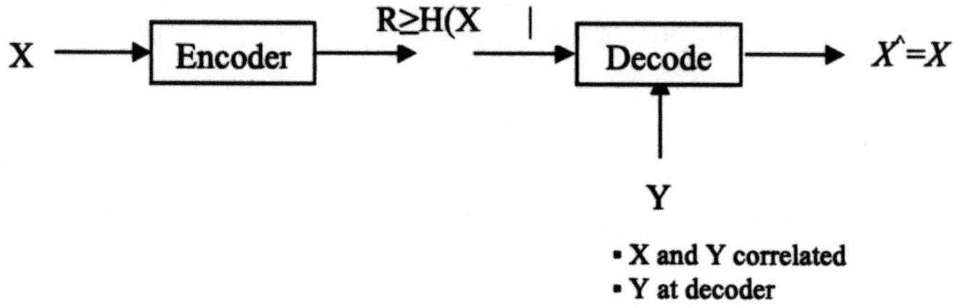

For same reason, X = 001 and 110, X = 010 and 101, and X = 011 and 110 do not need to distinguished from each other. These set of two X values are grouped as four coset and assigned four different binary index numbers:

coset 1 = (000, 111): 00, coset 2 = (001, 110): 01, coset 3 = (010, 101): 10, coset 4 = (011, 100): 11.

If input is X = 010 and Y = 110, Decoder received Y = 110 as a side information from Y and X = 10 as a partial information from X. Then at the Decoder, X = 010 is selected from coset 2 since 110 has a hamming distance of 2 from 110. Whether X know Y or not, X can still reduce or compress three bits information into two bits.

This method is particularly used in discrete sources and continuous sources. The same method useful for lossless and lossy compression schemes.

4 Conclusion

The major concern to study compression techniques is the constrained of battery of sensor nodes. However, many researchers has proposed the use of compression techniques of data mining to extend the life of sensor networks. The studies stated that comparatively lossless data compression methods are easy to implement and understand, also gives better compression efficiency. This paper talks about the comprehensive survey of various compression algorithms and its detailed working. The compression algorithm of both categories lossless and lossy has been studied and analysed here. The main objective to study different types of compression algorithm is to understand how energy can be saved with the help of it. Data gathering, its processing, aggregation and compression have become the interested field of research.

With the use of basic data mining techniques, in future the paper will focus on various types of aggregation methods, fusion techniques, funnelling techniques, storage, analysis, and knowledge extraction, and compression of huge volume of data is also the key concern in future.

References

1. Abdelmoghith, E. M., & Mouftah, H. T. (2013). A data mining approach to energy efficiency in wireless sensor networks. In *International Symposium on Personal, Indoor and Mobile Radio Communication: Mobile and Wireless Networks* (pp. 2612–2626).
2. Mamurjon, D., & Ahn, B. (2013). A novel data gathering method for large wireless sensor networks. In *ELSEVIER-International Conference on Electronic Engineering and Computer Science. IERI Procedia, 4*, 288–294.
3. Mudgule, C. B., Nagaraj, U., & Ganjewar, P. D. (2014). Data compression in wireless sensor network: A survey. *International Journal of Innovative Research in Computer and Communication Engineering, 2*(11), 60–63.
4. Preethi, A., & Ramakrishnan, A. (2015). A survey on data compression techniques in wireless sensor network. *International Journal of Electrical and Computer Engineering, 2*(3).
5. Vidhyapriya, R., & Vanathi, P. (2009). Energy efficient data compression in wireless sensor networks. *The International Arab Journal of Information Technology, 6*(3), 297–303.
6. Sadler, C. M., & Martonosi, M. (2006). Data compression algorithms for energy—constrained devices in delay tolerant networks (pp. 265–278). ACM
7. Kimura, N., & Latifi, S. (2005). A survey on data compression in wireless sensor networks. In *Proceedings of the International Conference on Information Technology: Coding and Computing* (pp. 8–13).
8. Alwadi, M., & Chetty, G. (2015). Sensor selection scheme in temperature wireless sensor network. *International Journal of Wireless & Mobile Networks (IJWMN), 7*(3), 47–53.
9. Kusuma, J., Doherty, L., & Ramchandran, K. (2001). Distributed compression for sensor networks. In *Proceedings of the IEEE Conference on Image Processing, Thessalonki, Greece, October 2001* (pp. 82–85).
10. Capo-Chichi, E. P., & Friedt, J.-M. (2009). K-RLE: A new data compression algorithm for wireless sensor network. In *Third International Conference on Sensor Technologies and Applications* (pp. 502–507).
11. Welch, T. A. (1984). A technique for high-performance data compression. IEEE, 8–18.
12. Singh, N., & Singh, H. (2019). Energy efficient NK-RLE data compression scheme. *International Journal of Engineering and Advanced Technology (IJEAT), 8*(6), 4014–4021.
13. Saidani, A., Jianwen, X., & Mansouri, D. (2020). A lossless compression approach based on delta encoding and T-RLE in WSNs. *Hindawi Wireless Communications and Mobile Computing*, 2–10.

Iot Based Healthcare System for Patient Monitoring

S. Saravanan, M. Kalaiyarasi, K. Karunanithi, S. Karthi, S. Pragaspathy, and Kalyan Sagar Kadali

Abstract The Internet of Things gadgets can acquire and transmit the data straightforwardly with different gadgets through the cloud, giving a gigantic measure of data to be assembled, stored and investigated for data-analytics processes. The aim of this paper is to enhance the patient's quality of life by accessing real time visibility of the patient's condition, through measuring the physiological parameters like systolic, diastolic, pulse rate and body temperature values. The key idea is to administer care to the patients by constantly monitoring the medical parameters include blood pressure, pulse rate and body temperature without the need for the patient to move from facility to facility for constant supervision of their health. Data gathered from the blood pressure sensor and temperature sensor is analyzed and stored in the cloud, which can be monitored by the caregivers of the patient from any location and respond appropriately, based on the alert received.

Keywords Internet of things · Healthcare system · Blood pressure · Temperature · Arduino UNO

1 Introduction

The Internet of things (IoT) is the inter-connecting of different gadgets in order to communicate and share data with other networks and computers. It is utilized in

S. Saravanan
B V Raju Institute of Technology, Narsapur, India
e-mail: Saravanan.s@bvrit.ac.in

M. Kalaiyarasi (✉) · S. Karthi
V.S.B. Engineering College, Karur, India

K. Karunanithi
Vel Tech Rangarajan Dr. Sagunthala R&D Institute of Science and Technology, Tamil Nadu, India

S. Pragaspathy
Vishnu Institute of Technology, Bhimavaram, India

K. S. Kadali
Shri Vishnu Engineering College for Women, Bhimavaram, India

P. Nayak et al. (eds.), *IoT and Analytics for Sensor Networks*, Lecture Notes in Networks and Systems 244, https://doi.org/10.1007/978-981-16-2919-8_40

different areas such as e-health [1], Cultural Heritage [2–6] not failing to remember legitimate area [7–9], Public organization domain [10–13], and Humanitarian Help and Disaster Relief [14–16], yet in addition home computerization, self-ruling and associated vehicles [17], and wearable innovation.

The IoT devices are developed with different types of sensors, actuators, processors and transceivers [18]. With the increasing utilization of sensors by clinical gadgets, remote and constant observing of a patient's heath is getting conceivable. This organization of sensors, actuators and other portable specialized gadgets, alluded to as Internet of Things for Medical Devices (IoT-MD), is ready to revolutionize the functioning of the medical care industry. Associated health technologies have acquired conspicuousness because of the broad event of ongoing illnesses and the significant requirement to control medical care expenses of a maturing populace. An associated medical care climate advances the speedy progression of data and empowers easy access to it. Improved home care facilities and customary health updates to clinicians diminish the odds of repetitive or improper consideration, improve tolerant consideration and security, and decrease generally expenses of care. It can likewise be utilized to follow way of life infections, for example, hypertension, diabetics and asthma which need ceaseless observing.

Technological giant Cisco predicts that 25 billion gadgets will be connected with IoT by 2015 and 50 billion by 2020. The IoT is another innovative, and there is still no steady definition for it [19]. IoT can offer incredible guarantee in the field of medical services. It may be utilized broadly to improve admittance to nature of mind and in addition give mindfulness, backing and help to patients in distant areas. IoT can likewise give help to patient's ongoing sicknesses. IoT can offer incredible guarantee in the field of medical services. It may be utilized broadly to improve admittance to nature of mind and in addition give mindfulness, backing and help to patients in distant areas. IoT can likewise give help to patient's ongoing sicknesses. Auto-ID Center represents the joint effort among industry and private area to deal with new innovation for following merchandise all around the world. Presently Auto-ID Labs are driving scholarly examination network on IoT.

A remote versatile multi-boundary gadget was proposed to procure physiological signals and send them to a nearby worker through Bluetooth remote innovation. Four sorts of screen units were intended to impart by means of the WiFi remote innovation, including a neighborhood screen unit, a control place, cell phones and a web page. The utilization of different screen units is intending to meet distinctive clinical prerequisites for various clinical work force. This framework was exhibited to advance the portability and adaptability for both the patients and the clinical faculty, which further improves the nature of medical services [20].

ECG instruments calculate the electrical function of the heart to assess heart problems. ECG signal consistency affects when a patient has any cardiac problems. Be that as it may, the accessibility of ECG gadgets is causing ECG conclusion moderate and troublesome. The issues found in many medical clinics are that nonstop observing of indispensable parameters are accomplished for ICU patients, yet the screens are nearby to the room in which the patient is conceded. Doctor needs to consult the patient regularly and asses his/her situation by dissecting the deliberate criteria, e.g., temperature, circulatory strain, beat oximeter, E.C.G. and more [21]. Latest and progressed installed gadgets have demonstrated their capacity to analyze health boundaries which are significant and to be persistently checked in clinical fields. Advancement sheets in clinical and non-clinical apparatuses have arisen with quicker and less expensive organization measure.

The IoT-MD gives an environment where a patient's significant information gets communicated by clinical gadgets through a passage onto secure cloud based stages where it is stored and dissected. It helps to store information for a large number of patients and perform investigation continuously, eventually advancing a proof based medication framework. In this paper, IoT based healthcare system for patient monitoring has been proposed.

This paper is further arranged in the following manner. Sect. 2 gives the objective of the proposed healthcare system. Sect. 3 presents the proposed methodology. Implementation steps and results are discussed in Sect. 4, and the conclusion of this paper has been given in Sect. 5.

2 Objectives

- To create IoT-driven framework for making it conceivable to drastically lessen costs and improve health by expanding the accessibility and quality of care.
- To create an embedded innovation for use all through IoT-driven medical healthcare system, including:
- Sensor that gather quiet information.
- Microcontrollers that process analyze and remotely communicate the information.
- To create Healthcare-explicit passages through which sensor information is further dissected and transmitted to the cloud.

3 Proposed System

Wellbeing and health management is one of the most encouraging application zone of IoT innovation. Remotely observing of patients prompts more viable and ideal treatment, prompting better administration of health. Likewise, patients are engaged by getting more perceivability into their genuine ailments, empowering them to assume a functioning part in controlling and impacting their treatment. The proposed method presents the idea of solving medical problems utilizing most recent innovation, Internet of Things.

The block diagram of smart real time health care monitoring and alert system using IoT has been given in Fig. 1. It presents the smart medical care framework utilizing IoT which is intended to give Quality Health Care to everyone. Through this proposed system, patient's significant needy parameters such as blood pressure

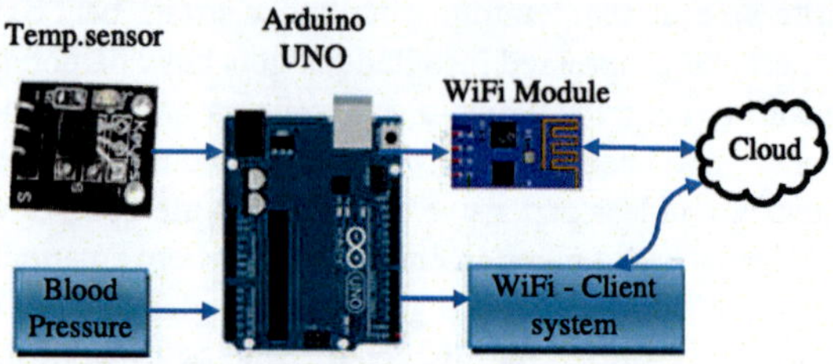

Fig.1 Monitoring and alert system

(BP) and temperature can be measured progressively. The workflow of the proposed system has been given in Fig. 2. It has been explained as follows:

Step:1 Initially, connect the blood pressure sensor and temperature sensor to the Arduino UNO.

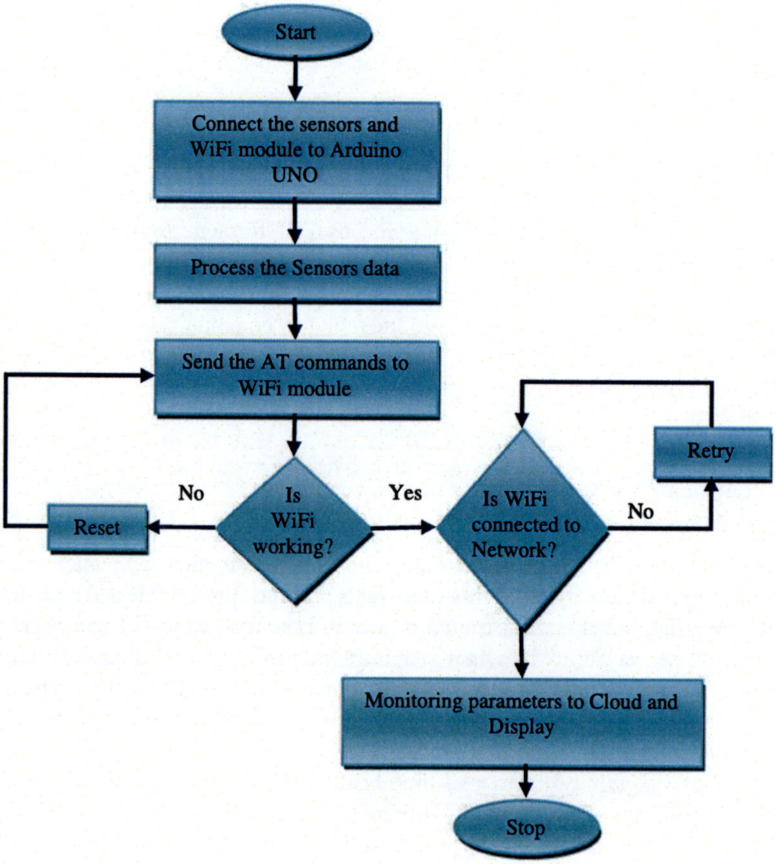

Fig. 2 Flowdiagram

Step:2 Process the sensor data using Arduino UNO with the help of Arduino programming.

Step:3 Establish TCP network connection to the WiFi module.

Step:4 Check whether the WiFi module is connected to the network or not?

Step:5 Once the WiFi connects send the sensor data to Cloud. If not go to step 3.

Step:6 The stored data can be viewed using Android app on the phone.

Step:7 A tweet alert has been sent whenever the data falls below or above the predefined value.

4 Hardware Implementation

The health parameters, for example, BP, beat rate and temperature level are estimated from the pulse and temperature sensors separately. The gathered medical information is stored in Ardunio UNO through interfacing. This information is then sent to the ESP8266. The transmitted data is also transferred to the cloud. Here the pulse sensor is connected to the UART pins (0 and 1) of the ATmega328. Temperature sensor is associated with the pin 4 of Arduino UNO and Esp8266 WiFi module sequential pins (Tx&Rx) to the product sequential pins (8 and 9) of Arduino UNO. The arrangements of the proposed system has been given in Fig. 3. The measured values are shown in Fig. 4.

The measured data has been transmitted to IoT based cloud administration referred as ThingSpeak using the ESP8266 WiFi shield. Since ThingSpeak web service works in collaboration with MATH works, the transmitted data are stored in the graphical format. The stored values of Systolic (SC), Diastolic (DC), Pulse Rate (PR) and Temperature (T) on ThingSpeak have been shown in Fig. 5.

From the analysis of the above figure, the following parameters are obtained.

SC:92mmhg, DC:60mmhg.

Fig. 3 Real time implementation

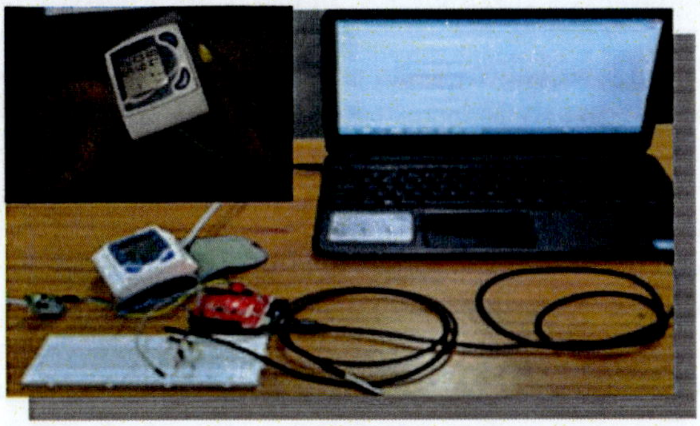

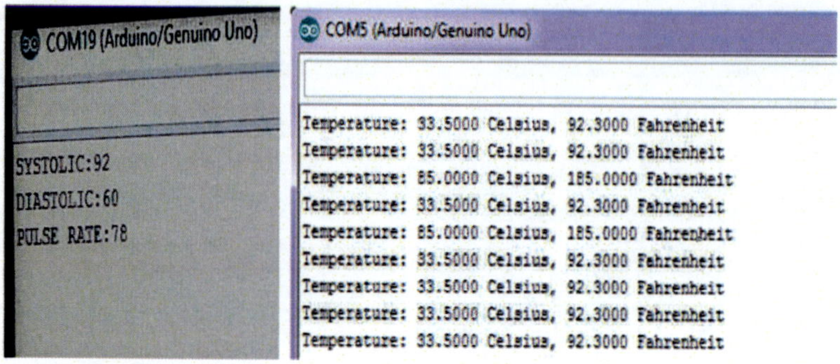

Fig. 4 Pressure and temperature senor values on serial monitor

PR:78 bpm, T:33 °C.

The preserved data in the ThingSpeak cloud administration could be recovered from mobile application. It empowers individual to visualize their ThingSpeak directs in a simple manner by entering the channel ID. It has been appeared in Fig. 6. It shows the different project title. By clicking on the project title, it receives the values from thing speak server and displays the stored (sensors) values in graphical format.

The systolic, diastolic, body temperature and pulse values on Thing View app have been presented in Fig. 7. Based on the principles of the facts reviewed, the psychiatrist identified the illness and then estimated the state of the patient's wellbeing. The main advantage of this method is that it is possible to access vital parameters of a particular person from anywhere in the world with the availability of Internet, IoT health care device and android mobile.

5 Conclusion

The IoT is becoming feasible solution in the field of health surveillance. Using remote patient surveillance to enable an individual's wellbeing to be tracked and help a doctor to identify signs of disease. In this paper, IoT based healthcare system for patient monitoring has been proposed. Here, different types of sensors have been utilized to monitor the patient health parameters. They sense internal heat level and heartbeat rate separately as indicated by their capacity and afterward with the assistance of microcontroller then the data has been stored and processed further. These sensor values are then transmitted via wireless communication to a medical server. These data are subsequently received on an approved IoT framework smart phone. Based on the principles of the facts he reviewed, the psychiatrist identified the illness and then estimated the state of the patient's wellbeing. In future, this framework can be actualized with insignificant advancement board with promptly fitting and play IoT and fundamental parameters can be observed in a pervasive strategy.

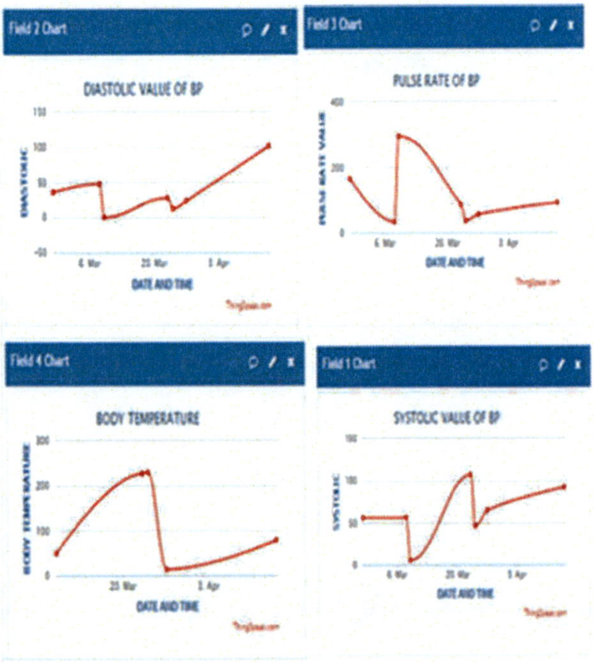

Fig. 5 SC, DC, PR and T values on ThingSpeak

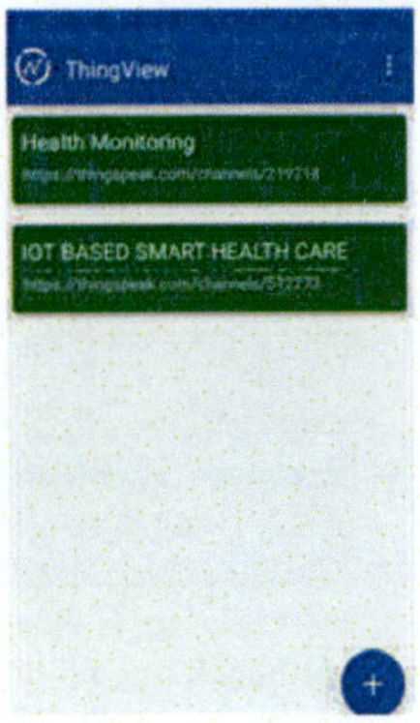

Fig. 6 ThingView app window after addition of channels

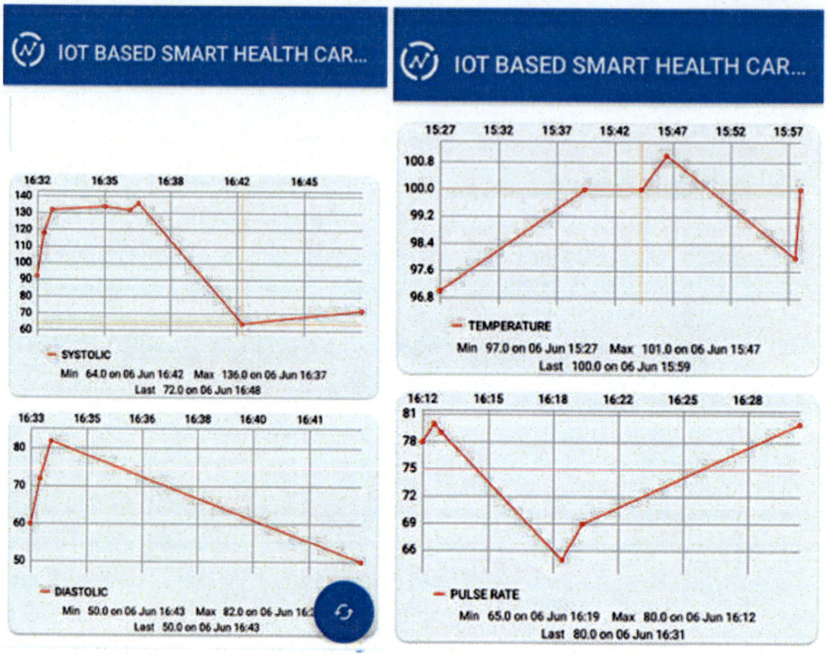

Fig. 7 Systolic and diastolic, body temperature and pulse values

References

1. Guadagni, Y. et al. (2017). RISK: a random optimization interactivesystem based on kernel learning for predicting breast cancerdisease progression. In *Proceedings of the 5th International Work-Conference, IWBBIO 2017,* April 26–28 (pp. 189–196). Spain, Granada.
2. Ohta, S., Nakamoto, H., Shinagawa, Y., & Tanikawa, T. (2002). A Health monitoring system for eldery people living alone. *Journal of Telemedicine and Telecare, 8*(3), 151–156.
3. Pennacchiotti, M., & Zanzotto, F. M. (2008). *Natural language processingacross time: An empirical investigation on Italian* (pp. 371–382). Berlin, Heidelberg: Springer.
4. Beccaceci, R., Fallucchi, F., Giannone,C. F., Spagnoulo, F., & Zanzotto, F. M. (2009). Education with 'living artworks'in museums. In *Proceedings of the 1st International Conference on ComputerSupported Education,* (pp. 1–5).
5. Arcidiacono, G., De Luca, E.W., Fallucchi, F., & Pieroni, A., **2016**. The use of lean six sigma methodology in digital curation. In *Proceedings of the CEUR Workshop,* (pp.1–7).
6. Ambrogio, A. D., Gaudio, P., Gelfusa, M., Luglio, M., Malizia, A., Roseti, C., & Marsella, S. (2017). Use of integrated technologies for fire monitoring and first alert. In *Proceedings of the International Conference of Information andCommunication Technologies, AICT 2016* (pp. 1 5).
7. Pazienza, M. T., Scarpato, N., & Stellato, A. (2009). STIA: Experience of semantic annotation in Jurisprudence domain. *Artificial Intelligence and Applications, 205,* 156–161.
8. Bianchi, M., Draoli, M., Gambosi, G., Pazienza, M. T., Scarpato, N., & Stellato, A. (2009). ICT tools for the discovery of semantic relations in legal documents. In *Proceedings of the CEUR Workshop* (p. 582).

9. Boella, G., Di Caro, L., Humphreys, L., Robaldo, L., Rossi, P., & van der Torre, L. (2016). Eunomos, a legal document and knowledgemanagement system for the Web to provide relevant, reliable and upto-date information on the law. *Artificial Intelligence Law, 24*(3), 245–283.
10. Morabito, V. (2015). Big data and analytics for government innovation. In *Big data and analytics* (pp. 23–45).
11. Zanella, A., et al. (2014). Internet of Things for Smart Cities. *IEEE Internet ofthings, 1*, 22–32.
12. Fallucchi, F., Alfonsi, E., Ligi, A., & Tarquini, M. (2014). Ontology-driven public administration web hosting monitoring system. IJTRE, 8842.
13. Bianchi, M., Draoli, M., Fallucchi, F., & Ligi, A. (2014). Service level agreement constraints into processes for document classification. In *Proceedings of the 16th International Conference onEnterprise Information Systems* (pp. 12–18).
14. Zhang, D., Zhou, L., & NunamakerJ, F. (2002). A Knowledgemanagement framework for the support of decision making inhumanitarian assistance/disaster relief. *Knowledge and Information systems., 4*(3), 370–385.
15. Fallucchi, F., Tarquini, M., & De Luca, E.W. (2016). Knowledge management for the support of logistics during humanitarianassistance and disaster relief (HADR). *Lecture Notes in Business Information Processing* (pp. 1–7).
16. Fallucchi, F., Tarquini, M. and De Luca, E.W. (2016). Supporting humanitarian logistics with intelligent applications for disastermanagement. In *INTELLI 2016 : The Fifth International Conference on Intelligent Systems and Applications* (pp.51–56).
17. Pieroni, A., Scarpato, N., & Brilli, M. (2018). Industry 4. 0 revolution inautonomous and connected vehicle a non-conventional approach tomanage big data. *Journlal of Theoretical and Applied Information echnology, 96*(1), 10–18.
18. Colakovi, A., & Hadžialic, M. (2018). Internet of Things (IoT): a review of enablingtechnologies, challenges and open research issues. *Computer Networks, 144*, 17–39.
19. Wang, C., Daneshmand, M., Dohler, M., Mao, X., Hu, R. Q., & Wang, H. (2013). Guest editorial–special issue on internet ofthings (IoT): architecture, protocols and services. *IEEE Sensors Journal, 13*(10), 3505–3510.
20. Hao, Y., & Foster, J. (2008). Wireless sensor networks for health monitoring applications. *Physiological Measurements, 29*(11), R27–R56.
21. Gowthami, P., & Sathishkumar, P. (2016). An android based patient monitoring system. *International Journal of Innovative Research in Advanced Engineering, 4*(3), 54–57.

Detection and Classification of Intracranial Brain Hemorrhage

K. V. Sharada, Vempaty Prashanthi, and Srinivas Kanakala

Abstract Computer

-aided diagnosis systems (CAD), as their name suggests, utilize computers to assist doctors to obtain a quick and correct diagnosis. They focused on several scholars as they are built upon the concept of processing and examining pictures of various parts of the individual body meant for a fast and correct outcome. CAD systems are generally area specific because they are augmented for some certain kinds of infections, various parts of the individual body, diagnosis methods, etc. They analyze dissimilar types of inputs given, for example, signs, test center, result, health pictures, etc. varying on their territory. One of the maximum common kind of diagnosis depends on medical pictures. Our approach is to develop a model to identify either a brain hemorrhage is present or not in Computed Topography (CT) scan of the brain and also identify the kind of hemorrhage. The process of detecting and identifying hemorrhage contains many steps like image pre-processing, segmentation of image, extracting the features, and classifying the images.

Keywords CT scan · Brain haemorrhage · Image processing · Image segmentation

1 Introduction

Brain hemorrhage implies blood loss within brain. In brief, one quite stroke which causes draining round the tissues by an artery within the cerebrum is observed as brain hemorrhage. Draining can happen in between the cerebrum and also the layers that cover it. The irritation that is caused by the blood from trauma leads to increase of pressure on brain tissues; this results in reduced percentage of oxygen from reaching the brain cells. This brain hemorrhage is observed as medical emergency which needs prompt treatment. There are several factors which might cause or result in

K. V. Sharada · V. Prashanthi (✉)
Gokaraju Rangaraju Institute of Engineering and Technology, Hyderabad, India

S. Kanakala
VNR Vignana Jyothi Institute of Engineering and Technology, Hyderabad, India

P. Nayak et al. (eds.), *IoT and Analytics for Sensor Networks*, Lecture Notes in Networks and Systems 244, https://doi.org/10.1007/978-981-16-2919-8_41

cerebral hemorrhage. These components include head injury, blood vessel anomalies, liver disease, brain tumour, extreme high blood pressure, bleeding disorders, and utilization of illegal medications. Anomalies within the blood vessels are the most explanation in the majority of the intracerebral hemorrhages that out of nowhere happen in youngsters, anyway there could be other potential causes which incorporate blood sicknesses, cerebrum tumors, septicemia, or the work of liquor or illegal medications. Many people who experience a hemorrhage have side effects as though they are having a stroke and can create weakness on one side of their body or a feeling of numbness. Sometimes brain hemorrhage can cause a scope of various indications like sudden, serious migraine, difficulty in gulping, vision problems, loss of coordination with the body, confusion or trouble in understanding, difficulty in talking or stammering discourse, seizures, torpidity or stupor.

Brain hemorrhage often leads to several complications. Because of the draining nerve cells one cannot speak with different pieces of the body and in this manner stops ordinary working. Additionally, there are scarcely any basic issues that emerge after a brain discharge which incorporates development, discourse, or memory issues. A few difficulties might be lasting relying upon the area of drain and the harm that happens. These intricacies may incorporate—loss of motion, vision misfortune, and decreased capacity to talk or get words, disarray, or memory misfortune. These complications made brain hemorrhage an emergency condition that requires immediate treatment. Diagnosing a brain drain could be troublesome as certain individuals do not give any physical indications. Specialists need to do take the assistance of a CT scan or MRI scan so as to locate the specific area of the seeping in the mind. CT pictures are known to claim numerous points of interest over MRI. Along these lines, the standard of CT pictures is sufficiently high to precisely analyse Intracranial Brain Hemorrhage. PC supported determination frameworks (CAD), as their name proposes, use PCs to assist specialists with arriving at a quick and exact conclusion. They have been the focal point of numerous scientists since they depend on preparing and examining pictures of various pieces of the human body for brisk and exact outcomes. Computer aided design frameworks are normally area explicit as they are advanced for certain particular sorts of infections, portions of the body, analysis techniques, and so on. They investigate various assortments of information, for example, manifestations, research facility test results, clinical pictures, and so on depending on their space. One of the most generally perceived sorts of determination is the one that depends on clinical pictures. Such frameworks are helpful on the grounds that they can be incorporated with the product of the clinical imaging machine so as to deliver a brisk and precise finding. Then again, they can be trying since they consolidate the components of man-made reasoning and advanced picture preparing (Fig. 1).

The rest of the paper is organized as follows: In Session 2 Literature survey is discussed, in Session 3 proposed system is described, Session 4 describes the implementation, Session 5 explains the experimentation and results and finally in Session 6 conclusion is given of our main proceedings series.

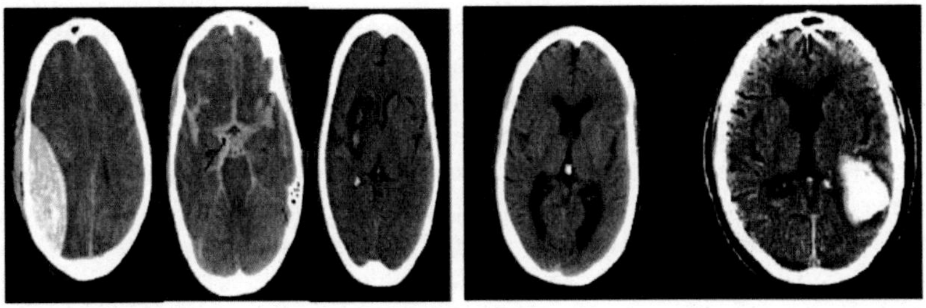

Fig. 1 CT scan images of five types of brain hemorrhages

2 Literature Survey

The paper [1] mainly discussed the pre-segmentation process. In this the techniques of dividing the input image into four quadrants using the method of splitting and merging are used. This is called as pre-segmentation of imaging. The pre-segmentation can be done with the following methods: (1) Thresholding: in thresholding the pixels of the input image is given certain threshold values and those pixels are mapped to related areas. (2) Region growing technique. (3) Supervised segmentation methods and unsupervised segmentation methods. Process of segmentation is as follows: as a first step the image is split into quadrants. Then histogram and pixel values are computed for each quadrant separately. The succeeding process for the above step is comparing each histogram value and pixel value with the predefined threshold criteria (or) histogram, pixel values. The final step is taking only the region of abnormality that is detected and feeding this as input to the segmentation process.

In [2] the method includes a combination of the machine with knowledge discovery techniques. During a CT scan the Intracranial pixel depth is noted. Based on the depth the pixel intensities are normalized using depth-dependent gray level normalization and the dense area of the intracranial region is segmented with the process of region growing and multi-resolution thresholding. The succeeding step is the construction of the decision tree by applying the algorithm. The first step in the method is pre-processing, and it involves the separation of pixels containing the extracranial part and the skull (intracranial pixels). This is used for recognition, measurement, and classification of a hematoma. In the next step the optimal threshold is chosen for hematoma segmentation by human experts and this differs from case to case which leads to instability. In the succeeding step the identification of the largest area containing Intracranial hematoma is done by considering the largest connected hyperdense region and excluding other smaller hyperdense regions. The final step is the construction of a decision tree using algorithm. Limitations: there is difficulty in recognizing subdural and epidural hematomas.

In [3] there is much focus on Cerebral Ischemia and had been said that if cerebral ischemia developed once there might not be a successful or complete cure for the abnormality. The method involved is based on the average thickness of the blood layer and based on thickness, and it is divided into grades such as good, recovery, dead scaling them as 1, 2, and 3. Computed Tomography (CT) scanned images are given as input to the process. It is described in this process that a thick localized subarachnoid layer of blood or diffusion of blood might lead to delayed cerebral ischemia. This method only focuses on Cerebral Ischemia.

In [4] there is an automatic classification of images into two classes based on features. The two classes are abnormal and normal. This classification involves four phases. Pre-process includes the separation of pixels containing the extracranial part and intracranial part. In feature extraction the original dataset is restricted, and these extracted features serve as training data and saved into feature library. Next, they applied the SVM classifier and KNN classifier. While performing this method it was observed that the K-Nearest Neighbors classifier yielded better results in comparison with the SVM classifier. The postprocessing step is performed after image is classified as abnormal. In the post-processing step the image of the skull is removed and then the abnormal region is extracted. Generally, for removing skull we use brightest pixel cells as the brain matter appears in grey color. This approach categorizes only two classes (abnormal and normal) and gives generic results without any specific label.

In [5, 6] the cerebral microbleed is visualized by using susceptibility-weighted imaging (SWI). Here depending on susceptibility-weighted imaging they constructed rank based average pooling to identify cerebral microbleed. For detection of cerebral microbleed, Convolutional Neural Network is used which contains multiple layers. ReLU layer is used to map the convolution layer with SoftMax activation, tangent activation and Rectified linear unit activation. It is more effective than the convolution layer. So, it is more popular. The convolutional layer may have a greater number of elements in the feature set which will lead to dimension disaster termed as overfitting. In this situation pooling plays a major role that replaces the clusters of related elements in the feature set with statistic summary value. This is done using a pooling function. The last advance is rank-based normal pooling which incorporates a normal of non-zero negative actuations. Rank-based normal pooling can beat the issue of loss of helpful data which is brought about by normal pooling and most extreme pooling. This is a very complex method and has a computational burden.

In [7] the method proposed is a densely connected neural network which is called densenet. To detect cerebral microbleeds the algorithm used is densenet. A sliding window is utilized to cover the arrangement of unique pictures from left to right and through and through. The objective worth is chosen dependent on the focal pixel of the sub models. Later the cost matrix will be employed, based on the comparison between the entries in the cost matrix and target value final abnormality region is detected. Following are the steps included in this algorithm: (1) Traditional Convolutional Neural Network: the layer age of straight enactment, the capacity is done by means of convolutions [8]. The following layer, for example, ReLU layer, is utilized to outline convolution layer with SoftMax enactment, digression actuation, Rectified straight unit initiation, and so on. It is more effective than the convolution layer. So, it is more popular. (2) Densenet: It is used to establish connections between layers. These connections are used for feature maps. The layers that are between the blocks are called transition layers. Each layer receives information from its previous layer. (3) Transfer learning: it is based on fully convolutional multiscale residual densenets. It considers the labeled samples to get high classification accuracy. It is meant to retrain the later layers of densenet. The usage of this transfer learning leads to an increase in accuracy rate.

In [9, 10] a complete CNN is trained with computed tomography (CT) scans. The algorithm resulted in high accuracy for the detection of acute Intracranial Hemorrhage. This algorithm has shown more accurate results than the measurements that are calculated by 2–4 radiologists.

The method used in [11, 12] is 3D quantitative analysis which performs 3D measurements of the parameters of the Intracerebral Brain Hemorrhage region based on computed tomography (CT) images. This resultant data is correlated with patient mortality. The image segmentation in this is done using a clustering algorithm whose strategy is based on fuzzy c-means which minimizes the objective function that represents the distance of feature vectors from the cluster centers. Based on the features of the clusters rule-based labelling is done whose components include fact-list, knowledgebase (or) rule-base, and inference engine. The labelling is done on the area, the color of the region. This is taken as input for the final step. This method not always perform correct segmentation.

In [13–15] the first step is pre-processing which includes division of the image into four quadrants called segmentation. Pre-processing is done using a tracking algorithm to separate skull images from the Gray dura matter in the CT scan image. Among these four regions the most vulnerable region which has the possibility of abnormality is considered and remaining regions are excluded. The next step is a grouping of all the identical homogenous regions to point out the abnormality. Following are the segmentation methods that are used: (1) Thresholding: it compares the pixel values with the threshold values. (2) Region growing techniques: it extracts useful or connected regions. (3) Supervised and unsupervised method: in this method there is the segmentation of the image in the training stage. Compared to supervised, unsupervised is efficient and less error sensitive. This method is not fully automatic. It requires a lot of human intervention.

3 Proposed Method

The proposed method includes CT scan images as the input datasets. Initially CT Scan images are converted into the format that a network/model can take as input. Images are converted into jpeg format. Later the images are processed so as to analyze the images accurately. Otsu's strategy is utilized with the end goal of segmentation. In general, it is a method to segment an advanced picture into numerous regions bolstered a few rules like arrangements of pixels, and so forth. Noise and unwanted pixels from the image are removed such as the skull part of the image which is of high intensity. The objective of the segment is to disentangle a picture to be increasingly significant and simpler to examine. Numerous methodologies for division exist, for example, thresholding and grouping. This stage sections the cerebrum picture into a few locales with the goal that we can segregate the ROI (the drain district). The result of this stage is regularly used to distinguish the presence of hemorrhages with most extreme exactness. Later classification of images is done using the Weka tool. Otsu's method of segmentation also minimizes the within-class variance of the system by finding a threshold value automatically based upon the pixel values, probability values of a segment. Morphological operations and region growing techniques can be used to obtain further improvements on the segmented image and finally most filtered part of ROI is obtained and features like the area of ROI, the axis of ROI, the circumference of ROI, distance between skull and ROI are extracted, and they are fed into a neural network through a feature vector pattern. For the extraction of features accurately Region props tool of MATLAB is used. For the purpose of classification between various types of hemorrhages the shape of ROI plays a major role generally for Epidural and Intraparenchymal hemorrhages, there is a convex hull that is present in the region of Interest whereas for subdural hemorrhage it is concave in shape. The Region surrounding the ROI is called Bounding Box, which also plays an important role to extract features of region of interest (ROI). The extracted features from ROI will be stored as input to train the model. The kind of hemorrhage is then recognized based on the neural network which is trained. Once the set of input and output images are saved at that instance then there will be a saved network which is called a Network file. Later working of a network file is determined using a training percentage method.

3.1 Preprocessing

Preprocessing is the beginning step of the method that involves removal of high intensity pixel part of the image such as the skull part and removal of noise. Later the image is segmented into parts so that it becomes easy for the purpose of analyzing and it is done by the OTSU'S method which has 100% of accuracy and clustering of segments with similar properties is made so that identification abnormalities become easy.

3.2 Segmentation

Segmentation step is continued by the feature extraction step which is main step to group segments with similar properties together and then later feed into a neural network. The pixel intensity of the brightest part of the skull is about 250 pixels, and it must be removed in order to identify the part that has abnormality. It is done through MATLAB. Image based cad system is used for the purpose of image segmentation. Later based on the properties of segmented part the type of hemorrhage is detected using a classifier algorithm (Type detection algorithm).

Type Detection Algorithm (Segmentation)

```
1: BTD (Ss, p, q, Cc, W) ◁ Ss CT scan Data set. n count of feature. C classifier
W is vector.
2: Dd ← BUILDDATASET1 (Ss, p, q)
3: for c ∈ Cc do
4: Acu ← TESTINCLASSIFIER (Dd, c, wc)
5: end for
6: return c Acu
7: end
8: BUILDDATASET1 (Ss, p, q) ◁ Ss Scan Dataset. n count of feature
9: Assign Dd to empty
10: for j ← 0,9 do
11: tt ← EXTRACINTFEATURE (j, n)
12: add tt to the end of Dd
13: end for
14: return Dd
15: end
16: EXTRACTINFEATURE (j, n) ◁ i brains CT scan
17: j is ≥ 251, j = 0; Apply steps 18 to 23
18: perform Otsu j
19: morphological j
20: regions growing j
21: region props
22: classifier
23: return vi
24: end
25: TESTINCLASSIFIER (Dd, c, wc)
26: partition Dd into 10 subset: Dd0..Dd9
27: for k ← 0,9 do
28: Ta ← Dk
29: Tb ← Sk = 0,...,9
30: calculate accuracy
31: return avg of Aj's for j = 0 to 9
32: end for
33: end procedure.
```

3.3 Morphological Techniques

The followings are the morphological techniques used Opening-by-remaking, shutting by-remaking, Supplement picture, Compute territorial maxima, Superimpose the picture, Process Background Markers, Watershed Transformation and the Segmentation, Imagine the Result, Evacuate Background Noise. Morphological operations basically remove the background and then dilates the inner segment region. An image

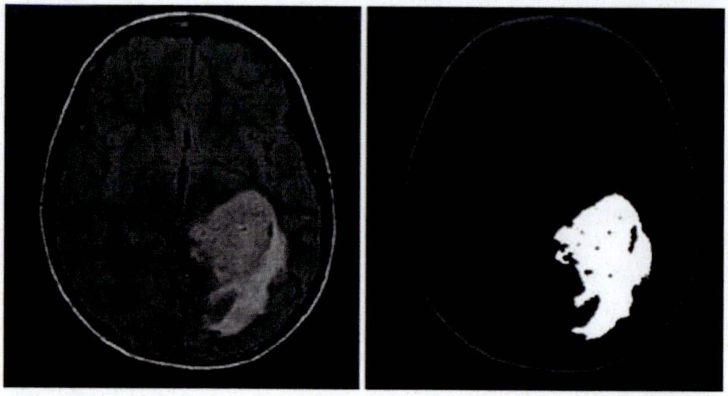

Fig. 2 Haemorrhagic region after applying segmentation and pre-processing techniques

containing potential regions with suspicious mass is the result of morphological operations.

3.4 Obtaining Region of Interest

Figure 2 highlighted images that indicate the hemorrhagic region after applying segmentation and pre-processing techniques. Region growing starts from a single point and then expand from that point by finding the region that is similar to that of the seed region. At the end of this step 'Region of Interest' is obtained which is utilized as contribution for 'Neural Network' and furthermore as training set. ROI is very crucial for the success of the proposed system. Resultant image after applying region growing techniques will be blank for normal brain image, whereas there will be some portion of image existing for brain image in abnormal condition after application of region growing technique.

4 Experimentation and Results

Experiments were done by changing the parameters given to the model: changing the way the data was split was tried by toggling the value of the shuffle argument—True or False as shown in Fig. 3. The change has not got any specific modifications to the output of the model. This has happened as little number of samples was utilized for performing segmentation. Next by changing the image size to 256X256 and the batch size to 64, the training process a lot more slower compared to when the values were set to 128 $\times$ 128 and 32 which trained the model approximately in 25 min. The training loss for the 256 $\times$ 256 images started learning later because the batch

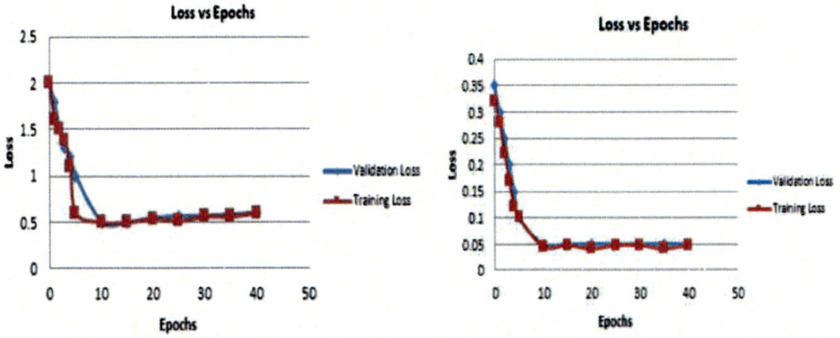

Fig. 3 (a) Loss vs Epoch: for 256X256 image size (b) Loss vs Epoch: for 128X128 image size

size was doubled and further we think it did not reach it. The model trained on 50 epochs but could have been appropriately trained for less number of epochs. This experiment mainly focused on segmentation of the data and localizing the affected area.

5 Conclusion

Automated frameworks for grouping medical images have increased a phenomenal degree of consideration recently. It has considerable impact in recognizing the existence of cerebral hemorrhage (the paired characterization issue) and in the event that it exists classification of hemorrhage is done and also problem of multiclass is also resolved. Although there are some algorithms which detect the hemorrhage efficiently and produce reasonable results, they are having few limitations like few algorithms find it difficult in recognizing subdural and epidural hematomas, few of them cannot detect subarachnoid hemorrhage, few consume more power and some are not suitable for large datasets, having longer computation time. Main cause for these limitations is existence of standardized procedures in less numbers. Since the diagnosis of Brain Hemorrhage is the very complicative and sensitive task, accuracy and reliability are given much priority. The investigations indicate that later pre-handling CT Scans, the double arrangement issue untraveled with 100% exactness. Additionally, the actualized framework accomplished over 92% exactness for the order issue of deciding the discharge type utilizing convolutional neural systems as a classifier. The outcomes are truly promising and more elevated levels of exactness for the characterization issue will be accomplished by getting a vastly improved dataset with high-level goals pictures held legitimately from the CT scanner. Also, unique element extraction and highlight choice procedures could be utilized to improve the presentation of the framework.

References

1. Li, Y., Hu, Q., Wu, J., & Chen, Z. (2009). A hybrid approach to detection of brain hemorrhage candidates from clinical head ct scans. In *2009 Sixth International Conference on Fuzzy Systems and Knowledge Discovery*. vol. 1. IEEE.
2. Liao, CC., Xiao, F., Wong, J. M., & Chiang, I. J. (2008). A knowledge discovery approach to diagnosing intracranial hematomas on brain CT: recognition, measurement and classification. In *International conference on medical biometrics*. Berlin, Heidelberg, Springer.
3. Bhadauria, H. S., & Dewal, M. L. (2014). Intracranial hemorrhage detection using spatial fuzzy c-mean and region-based active contour on brain CT imaging. *Signal, Image and Video Processing, 8*(2), 357–364.
4. Mohsen, F., Pomonis, S., & Illingworth, R. (1984). Prediction of delayed cerebral ischaemia after subarachnoid haemorrhage by computed tomography. *Journal of Neurology, Neurosurgery & Psychiatry, 47*(11), 1197–1202.
5. Kyaw, M. M. (2013). Pre-segmentation for the computer aided diagnosis system. *International Journal of Computer Science & Information Technology, 5*(1), 79.
6. Dou, Q., Chen, H., Yu, L., Zhao, L., Qin, J., Wang, D., & Heng, P. A. (2016). Automatic detection of cerebral microbleeds from MR images via 3D convolutional neural networks. *IEEE transactions on medical imaging, 35*(5), 1182–1195.
7. Ramteke, R. J., & Monali, K. Y. (2012). Automatic medical image classification and abnormality detection k-nearest neighbour. *International Journal of Advanced Computer Research* 2(4), 190–196.
8. Wang, S., Jiang, Y., Hou, X., Cheng, H., & Du, S. (2017). Cerebral micro-bleed detection based on the convolution neural network with rank based average pooling. *IEEE Access, 5*, 16576–16583.
9. Kuo, W., Häne, C., Mukherjee, P., Malik, J., & Yuh, E. L. (2019). Expert-level detection of acute intracranial hemorrhage on head computed tomography using deep learning. *Proceedings of the National Academy of Sciences, 116*(45), 22737–22745
10. Loncaric, S., Dhawan, A. P., Cosic, D., Kovacevic, D., Broderick, J., & Brott, T. (1999). Quantitative intracerebral brain hemorrhage analysis. In *Medical Imaging 1999: Image Processing* (Vol. 3661). International Society for Optics and Photonics.
11. Kyaw, M. M. (2013). Computer-Aided Detection system for Hemorrhage contained region. *International Journal of Computational Science and Information Technology*, 11–16.
12. Magoulas, G. D., & Prentza, A. (1999). Machine learning in medical applications. In *Advanced course on artificial intelligence*. Berlin, Heidelberg, Springer.
13. Prashanthi, V., & Srinivas, K. (2020). Plant disease detection using convolutional neural networks. *International Journal of Advanced Trends in Computer Science and Engineering, 9*(3), 2632–2637.
14. Hayward, R. D. (1977). Subarachnoid haemorrhage of unknown aetiology: A clinical and radiological study of 51 cases. *Journal of Neurology, Neurosurgery & Psychiatry, 40*(9), 926–931.
15. Prashanthi, V., & Srinivas, K., Generating analytics from web log. *International Journal of Engineering and Advanced Technology, 9*(4), 161–165.

Implementation of Efficient Technique to Conduct DDoS Attack Using Client–Server Paradigm

Seema Rani and Ritu Nagpal

Abstract In modern times, every human being relies upon the internet for fulfilling their hefty needs as the internet offers a vast amount of information to users, so its availability to users is indispensable. Major objectives of security are availability, integrity, and confidentiality. DDoS (Distributed Denial of Service) is a universal cyber-attack that is a major intimidation for cyberspace. DDoS attack slows down network availability by overflowing illegal traffic over network bandwidth. Day by day, attackers improve upon their strategies by using new technologies and techniques. In this paper, a DDoS attack is proposed using python script. We focus on the volumetric DDoS attack effect on the performance of the server ultimately shutting it down. A DDoS attack is undertaken using a python script on the server in which multiple clients send multiple fake requests to the server to slow down the services/performance of the server.

Keywords Distributed denial of service (DDoS) · Denial of service (DoS) · Client · Server · Security · Volumetric DDoS attack · Zombies

1 Introduction

Cybersecurity is the most significant and crucial technology across the world, and social media, the internet, net banking, digital media create new challenges for cybersecurity [1]. Security industry experts, researchers, and related organization personnel are seen analyzing the cyber-attacks across the world and develop different techniques to prevent the cyber-attacks [1].

- Cyber-attack is a sort of outrageous action that exploits computer systems, computer networks, personal devices, and information systems.
- Attacks are being detected every moment, DoS and DDoS have grown over the internet like parasites [2].
- Government, private organizations, banks, schools, hospitals, and people across the world face financial crises due to cyber-attacks [1, 2].

S. Rani (✉) · R. Nagpal
Computer Science & Engineering, GJUS&T, Hisar, India

P. Nayak et al. (eds.), *IoT and Analytics for Sensor Networks*, Lecture Notes in Networks and Systems 244, https://doi.org/10.1007/978-981-16-2919-8_42

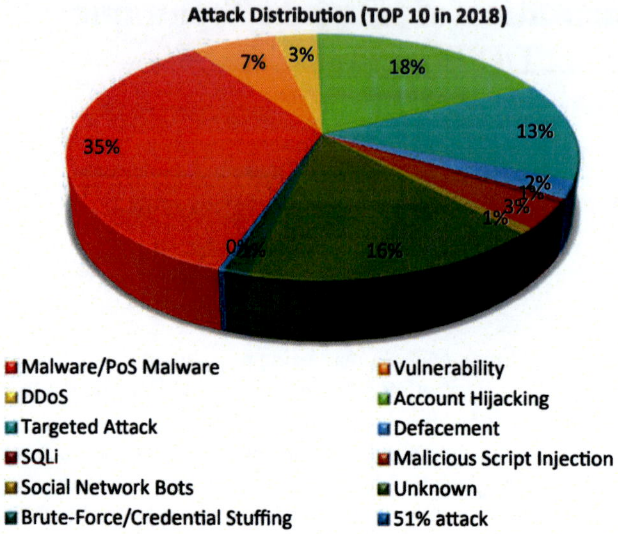

Fig. 1 Top internet attack distribution data

- The primary concern of the internet is security against repudiation, integrity, confidentiality, and availability of information.
- Distributed Denial of Service attack obstruct the legitimate user to access the network services and allows an attacker to access all leverage of services [3].
- It can be conducted by using a single machine or multiple machines named zombies [4].
- DDoS is a distributed version of DoS which is faster and difficult to detect and adverse than DoS [5, 6].

From the Fig. 1, DDoS is a measured attack used by cyber attackers to degrade the performance of the server. In this paper, efficient technique is used to attack the server which is described in the further sections.

2 DDoS: A Major Threat to the Isps

The effect of a fruitful DDoS assault on an ISP is boundless [7, 8]. Site execution is seriously undermined, bringing about baffled clients and different clients. Administration level arrangements (SLAs) [9] are disregarded, bringing about expensive help credits. The developing reliance on the Internet has the effect of fruitful DDoS assaults [10]. DDoS on ISPs [11] brings about the accompanying:

- Lost income
- Lost profitability

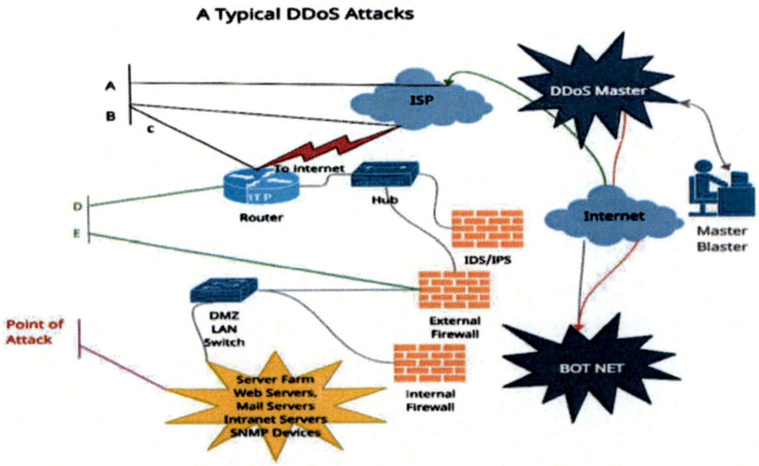

Fig. 2 DoS network

- Increased IT costs
- Mitigation costs
- Loss of clients
- Figure 2 shows how DDoS assault is extended on ISP.

Point A: This is the passage purpose of ISP.

Point B: This is the leave purpose of ISP.

Point C: This is the passage highlight your organization.

Point D and E: This is where Anti-DDoS or Firewalls or your IPS/IDS frameworks dwell.

From the above chart, it is very apparent that DDoS may assault a solitary point in your framework, yet the repercussions are felt from Point B onwards and can be ruined at Point B itself.

The graph in Fig. 3 shows the percentage of worldwide DDoS attack traffic between November 2017 and April 2018, sorted by originating countries. It is shown that during that period, 30% of DDoS attack traffic originated from the United States [12] (Table 1).

3 History of DDoS Attacks

Chowdhury [1], **in this paper**, discuss the recently occurred cyber-attacks and evaluate the loss of the economy due to increasing cyber-attacks. Denial of service (DoS), SQL injection, and decrypting RSA with Obsolete and Weakened Encryption (DROWN) attack have been discussed. The impact of these attacks on the business, organization, and individual has been defined. Other strategies are software patching,

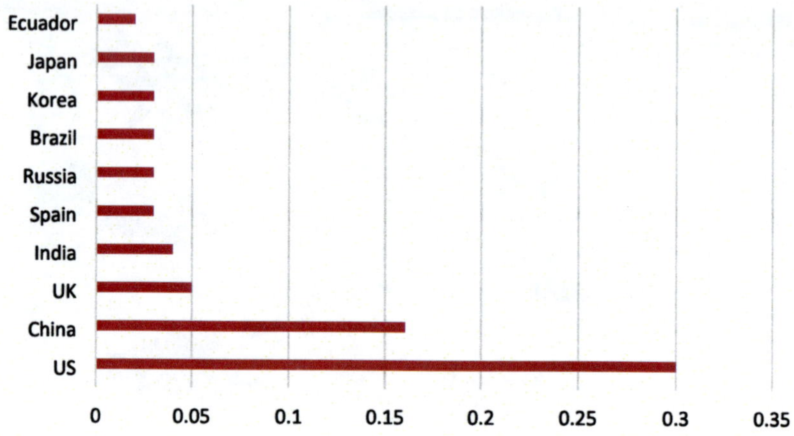

Fig. 3 Chart of various countries originating the DDoS attack

Table 1 Chart of DDoS attack of various countries

Countries	US	China	UK	India	Spain	Russia	Brazil	Korea	Japan	Ecuador
Total (100%)	30%	16%	5%	4%	3%	3%	3%	3%	3%	2%

the digital immune system by minimizing administration privilege using avoidance and detection system, using a firewall, and log file monitoring.

Mirkovic and Reiher [2] **have** discussed DDoS attack detection and mitigation taxonomy in this paper. The first DDoS attack is conducted using three-phase recruit, infect, and exploit. It proposed a degree of automation manually, semi-automatic, automatic. Describes various host scanning and vulnerability scanning strategies. Source address validity that exploited the weakness to deny service, attack rate dynamics, impact on the victim are explained in detail.

Purwanto et al. [3], **in** this paper, show how to mitigate and detect flooding DDoS attack. The researcher conducts a flooding attack and detects it by using signature-based and anomaly-based detection techniques. Statics detection, information theory, forecasting, soft computing, signal processing are categories of the anomaly that are discussed.

Sachdeva et al. [4], **in** this paper, analyze the impact of DDoS attack and defense. NS-2 simulation is used to evaluate DDoS attack conducted at internet-based web application. Quantitatively measure the performance using throughput, response rate, number of requests service to the server with or without attacks.

Li et al. [5], **in** this paper, discuss the volumetric DDoS attack in IoT (Internet of Things) devices. A DDoS detection method is proposed in three levels of the sliding window, single direction packet filter, and quintile deviation algorithm. An entropy calculator is used for the calculation of optimized slide window time which has a variation updating method and value counter dictionary and counter matrix.

4 DDoS Client/Server Attack

DDoS attacks jam resources at server making them unavailable to the users [13], slowing down the network [14], and making websites inaccessible [15]. We are sending a huge amount of traffic to the server to stop the response from the server. A server-side script response to the client causes a DDoS attack, many requests are forwarded to the server, so that the server stops responding [16–18]. To prevent the attack and allow only authorized users to access the server, a certain amount of time after that, the server is automatically closed.

The DDoS attack is executed using client scripts in which multiple clients send huge miscellaneous data per second and make the server stop working. During the DDoS attack, the server was on the network having IP address 192.56.101.1 on port 23. After the server started, client machines start their execution one by one sending data to the server. When clients send data to the server, it works properly, but as the number of users increases, server performance decreases gradually. Up to the eighth machine, server works and receives data, but at the ninth client machine server completely stopped working, and service was unreachable to the clients during a willful attack against the server and a denial of service attack takes place.

5 Implementation of Techniques to Generate DDoS Attack

5.1 Number of Attacks Sent

We generate an attack, the number of attack packets was sent to the victim while generating a DDoS attack. Denoted the time at which attack packets were sent to the server from different zombie [19, 20] machines and attack data saved into the text file for counting the number of attacks sent to the server. It is simply stored in the text file, and by using these files, we got the number of attacks sent to the server by counting the number of lines in the data file. Let us denote the number of attack packets sent to the server as Psuccess.

For experimental purposes, data is collected by sending packets to the server at the start time Tx by all the machines, and after the error time Ty, machines get an error message from the server and the server stops servicing them. Based upon the gathered information, we measure the performance of servers and clients. Approximately 60 packets per second are delivered to the server, and the number of attacks per second is evaluated using three attributes, start time of packet sent, error time, and a number of packet successful packet, after the total attack at the instant is computed using attack per second (Y).

5.2 Attacks Per Second

Attacks per second are evaluated using start time (Tx), error time (Ty), and number of attack packets sent (Psuccess).

$$\text{Start time of packet sent} = \text{Tx}, \ \text{Error time} = \text{Ty}$$
$$\text{Number of a packet sent} = \text{Psuccess}$$
$$Attack\ per\ second\ (y) = Psuccess/Ty - Tx$$

Using the above formula, attacks per second for the first client are calculated as follows: start time is 19:28:38 and error time is 19:29:43; thus, the difference between Ty – Tx is 65 seconds and the number of packets Psuccess is 4152. The value of attacks per second Y calculated is 4152/65 = 63.88s, and so on.

5.3 Total attacks at instant

The value of total attacks at instant is computed using the value of attacks per second (Y), i.e., Yn, Yn+1, Yn+2, where n is number of client machine.

$$\sum_{i=1}^{n} Y_n = X_i$$

where X = Total attacks at instant.

For the first machine, total attack packets/sec is 63.88, and for the second machine is obtained by adding the first and second machine data attacks per second, and for the third machine by adding machine first to third machine attacks per second, and so on.

5.4 Server Performance (Z)

At last, using the computed value of total attacks at instant and attacks per second, server performance is evaluated.

$$\text{X} = \text{Total attacks at instant} \quad \text{Y} = \text{Attacks per Second} \quad \text{Z} = \text{Server Performance}$$
$$\text{Z} = Maximum\ Attack\ per\ Second/Total\ Attack\ Instant * 100$$
$$\text{Z} = \frac{Y_i}{X_{initial}} * 100 \quad \text{where } i = 1, 2, 3 \ldots n$$

Table 2 Attack traffic from clients to server

Client machine	Start time	Error time	Number of attacks sent	Attacks per second	Total attacks at instant (Server)	Server performance
Machine#01	19:28:38	19:29:43	4152	63.88	63.88	100
Machine#02	19:28:49	19:29:43	3404	63.03703704	126.92	98.6851495
Machine#03	19:29:01	19:29:43	2652	63.14285714	190.06	98.850812
Machine#04	19:29:06	19:29:43	2329	62.94594595	253.01	98.54254544
Machine#05	19:29:14	19:29:43	1806	62.27586207	315.28	97.49352203
Machine#06	19:29:22	19:29:43	1302	62	377.28	97.06165703
Machine#07	19:29:33	19:29:43	604	60.4	437.68	94.55684008
Machine#08	19:29:40	19:29:43	152	50.66666667	488.35	79.3192036
Machine#09	19:29:47	19:29:47	0	0	0.00	0
Attacks faced by server at the time of failure						488.35

Server performance is calculated by dividing the attacks per second by the maximum value of total attacks at an instant and multiply by 100. For example, for the first machine divide 63.88 by 63.88 and multiply by 100, second machine divide 63.037 by 63.88 and multiply by 100, and so on. The attacks faced by the server at the time of failure are 488.35. Table 2 shows the details of attacks by client machines.

6 Results of DDoS Attack Using Client/Server Paradigm

To measure the server performance, a graph was plotted between the server performance and the number of attacks per second. The graph below shows that server performance decreases gradually as the number of machines attacking the server increase, and at particular instant of time server completely stops working. The attacks faced by the server at the time of failure were 488 (Fig. 4).

The volumetric attacks [21] were launched to the server by sending a huge number of packets from multiple machines to the server [22], and this slows down its functioning and finally stops its services. Around 480 packets per second were sent to the server and made it stop working. The individual performance evaluation of each client machine was stored in a dataset as shown in Table 3.

Green, red, and yellow are used to represent different states of the machines. Green shows that the machine is up, red shows the machine is down, and yellow indicates that the machine has yet not started. At 15:48:25 machine #1 and server were running, at 15:48:32 machine #1, machine #2, and server were running. As time increased, at 15:49:02 eight machines were running and the server was down, and

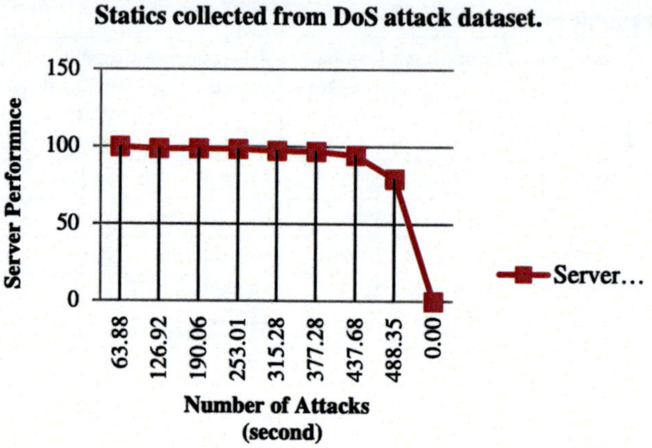

Fig. 4 Statics collected from DoS attack dataset

at 15:49:06 all machines were down. Based on this data, the performance evaluation graph was drawn and is shown in Fig. 5.

The light green line in the graph indicates the server, and the rest of the colored lines show machines, at time 15:49:02 server was down and after few microseconds, all machines were down, at 15:49:04 except machine #8 all machines were down, after two microseconds machine #8 also stopped working.

7 Conclusion and Future Work

In this paper, we computed the server performance that goes on decreasing when a number of zombie machines made bogus requests to the server and after a stipulated duration server completely stopped working. Our results will help in future courses of action to overcome the problem of rapidly increasing DDoS attacks. No doubt that attackers use advanced technologies and techniques to attack the machines, but as security practitioners, we need to think like attackers. To mitigate DDoS attacks, we can reduce the burden on the servers caused by attackers by load balancing and backups. As resources are limited, we must allow access only to legitimate users. We can grant time limits to the clients and after that duration, user will have to make the request again to the server for access to its resources. We can integrate this technique for the mitigation of DDoS attacks.

Table 3 Individual performance evaluation

Machine/Time	15:48:25	15:48:32	15:48:37	15:48:42	15:48:47	15:48:53	15:48:58	15:49:02	15:49:04	15:49:06
Machine #1	1	1	1	1	1	1	1	1	0	0
Machine #2		2	2	2	2	2	2	2	0	0
Machine #3			3	3	3	3	3	3	0	0
Machine #4				4	4	4	4	4	0	0
Machine #5					5	5	5	5	0	0
Machine #6						6	6	6	0	0
Machine #7							7	7	0	0
Machine #8								8	8	0
Server	10	10	10	10	10	10	10	0	0	0

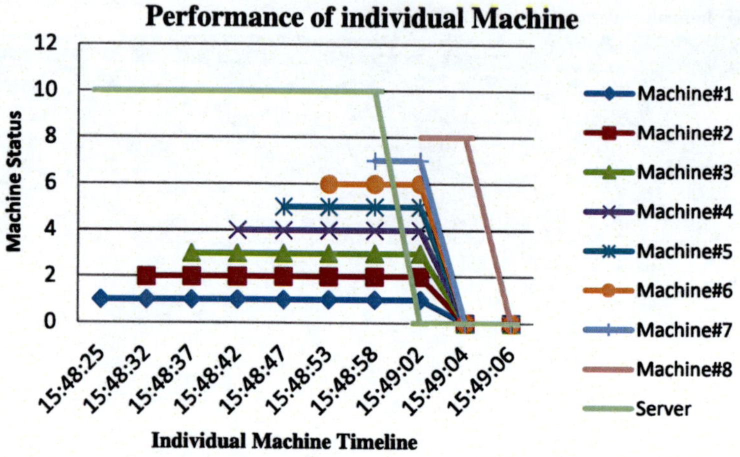

Fig. 5 Performance of individual machine

References

1. Chowdhury, A. (2016). In L. Batten & G. Li (Eds.), *Applications and techniques in information security* (Vol. 651, pp. 54–65).
2. Mirkovic, J., & Reiher, P. (2004). A taxonomy of DDoS attack and DDoS defense mechanisms. *ACM SIGCOMM Computer Communication Review, 34*(2), 39.
3. Purwanto, Y., Kuspriyanto, Hendrawan, & Rahardjo, B. (2014). Traffic anomaly detection in DDoS flooding attack. In *8th International Conference on Telecommunication Systems Services and Applications* (pp. 1–6). Kuta, Bali, Indonesia: IEEE.
4. Sachdeva, M., Kumar, K., Singh, G., & Singh, K. (2009). Performance analysis of web service under DDoS attacks. In *IEEE International Advance Computing Conference* (pp. 1002–1007). Patiala, India: IEEE.
5. Li, J., Liu, M., Xue, Z., Fan, X., & He, X. (2020). RTVD: A real-time volumetric detection scheme for DDoS in the internet of things. *IEEE Access, 8*, 36191–36201.
6. Yihunie, F., Abdelfattah, E., & Odeh, A. (2018). Analysis of ping of death DoS and DDoS attacks. In *IEEE Long Island Systems, Applications and Technology Conference* (pp 1–4). Farmingdale, NY: IEEE.
7. Jamshed, M., & Brustoloni, J. (2010). In-network server-directed client authentication and packet classification. In *IEEE Local Computer Network Conference* (pp. 328–331). Denver, CO, USA: IEEE.
8. Guleria, A., Kalra, E., & Gupta, K. (2019). Detection and prevention of DoS attacks on network systems. In *International Conference on Machine Learning, Big Data, Cloud and Parallel Computing* (pp. 544–548). Faridabad, India: IEEE.
9. Bajpai, D., Vardhan, M., Gupta, S., Kumar, R., & Kushwaha, D. S. (2012). In N. Meghanathan, D. Nagamalai, & N. Chaki (Eds.), *Advances in computing and information technology* (Vol. 176, pp. 719–728).
10. Tupakula, U. K., & Varadharajan, V. (2003). Counteracting DDoS attacks in multiple ISP domains using routing arbiter architecture. In *The 11th IEEE International Conference on Networks* (pp 455–460). Sydney, Australia: IEEE.
11. Kumar, K., Joshi, R. C., & Singh, K. (2007). In T. Sobh, K. Elleithy, A. Mahmood, & M. Karim (Eds.), *Innovative algorithms and techniques in automation, industrial electronics and telecommunications* (pp. 235–240).

12. Akamai (https://www.akamai.com/us/en/multimedia/documents/state-of-the-internet/akamai-q2-2016-state-of-the-internet-security-report.pdf)
13. Choi, J., Choi, C., Ko, B., & Kim, P. (2014). A method of DDoS attack detection using HTTP packet pattern and rule engine in cloud computing environment. *Soft Computing, 18*(9), 1697–1703.
14. Zhijun, W., Wenjing, L., Liang, L., & Meng, Y. (2020). Low-rate DoS attacks, detection, defense, and challenges: A survey. *IEEE Access, 8*, 43920–43943.
15. Carl, G., Kesidis, G., Brooks, R. R., & Rai, S. (2006). Denial-of-service attack-detection techniques. *IEEE Internet Computing, 10*(1), 82–89.
16. Tan, L., Pan, Y., Wu, J., Zhou, J., Jiang, H., & Deng, Y. (2020). A new framework for DDoS attack detection and defense in SDN environment. *IEEE Access, 8*, 161908–161919.
17. Raj, A. G. R., Sunitha, R., & Prasad, H. B. (2020). Mitigating DDoS flooding attacks with dynamic path identifiers in wireless network. In *Second International Conference on Inventive Research in Computing Applications* Coimbatore (pp. 869–874). Coimbatore, India: IEEE.
18. Agrawal, N., & Tapaswi, S. (2019). Defense mechanisms against DDoS attacks in a cloud computing environment: State-of-the-art and research challenges. *IEEE Communications Surveys & Tutorials, 21*(4), 3769–3795.
19. Munshi, A., Alqarni, N. A., & Abdullah Almalki, N. (2020). DDoS attack on IOT devices. In *2020 3rd International Conference on Computer Applications & Information Security* (pp. 1–5). Riyadh, Saudi Arabia: IEEE.
20. Udhayan, J., Hamsapriya, T., & Vasanthi, N. A. (2012). DDoS attack detection through flow analysis and traffic modeling. In V. V. Das, E. Ariwa, & S. B. Rahayu (Eds.), *Signal processing and information technology* (Vol. 62, pp. 89–94).
21. Salim, M. M., Rathore, S., & Park, J. H. (2020). Distributed denial of service attacks and its defenses in IoT: A survey. *The Journal of Supercomputing, 76*(7), 5320–5363.
22. Bhuyan, M. H., Bhattacharyya, D. K., & Kalita, J. K. (2014). Network anomaly detection: Methods, systems and tools. *IEEE Communications Surveys & Tutorials, 16*(1), 303–336.

Design and Development of Retrieval-Based Chatbot Using Sentence Similarity

Haritha Akkineni, P. V. S. Lakshmi, and Lasya Sarada

Abstract Chatbots or the well-known automated conversational agents have become a raging trend among all the sectors of businesses as a result of the rapid transition happening towards automation in processes. They are already being used extensively and will spread their wings to newer horizons shortly. The basic model of Chatbots is to interact with the user to answer their questions using various modes like text messages, voice replies, or any other predefined suitable interface. This paper discusses the development of a Chatbot for the college, Prasad V Potluri Siddhartha Institute of technology, to answer various questions related to the college like the facilities, procedures, policies, etc. This is a web-based software application implemented using Flask framework. This model is designed to capture text inputs from the user through a console and outputs the response in text format using machine learning concepts. A retrieval approach is implemented to process the input and to respond with an appropriate answer using logic adapters. The performance of this model is analyzed using a questionnaire that uses various parameters like performance, humanity, effect, and accessibility. This paper presents the overall approach used to design the Chatbot and compares the web application as-is study with the to-be website when the Chatbot is incorporated. The web application along with the Chatbot showed a 20% improvement in performance and 5% increase in accessibility by analyzing the performance metrics.

Keywords Chatbot · Machine learning · Retrieval approach · Flask framework

1 Introduction

The usage of Chatbots has skyrocketed in the recent times. They have a strong footing in the customer support industry which involves text-based dealings or support. A chatbot is basically a term for a conversational agent which in actual is a computer system software that can take natural language input and returns a chatty response instantaneously [1]. As observed, chatbots have been in great demand over the past

H. Akkineni (✉) · P. V. S. Lakshmi · L. Sarada
Prasad V Potluri Siddhartha Institute of Technology, Vijayawada, India

P. Nayak et al. (eds.), *IoT and Analytics for Sensor Networks*, Lecture Notes in Networks and Systems 244, https://doi.org/10.1007/978-981-16-2919-8_43

few years. The number of internet users is increasing year-by-year. Proportionately, we can also observe a significant rise in the e-commerce sector which in turn demands increased customer support services for these online digital platforms. A delay of even 5 minutes in providing the customer support could reduce the chances of reaching out and selling to a customer [2]. A Chatbot of any particular organization is designed to perform certain tasks pertaining to their requirements. In reality, if the chatbots do not exist, the chances of getting the information across to each and every person of a particular organization (college in this case) are very meek. A Chatbot completes almost 25–30% of the tasks that are done by the organization to the end-user.

The College Enquiry Chatbot uses Machine Learning (ML) algorithms. The chatbot provides simple and trouble-free query clarification platform to students by addressing their queries in text format. The chatbot system makes it easy for the student to clarify his/her queries in lesser time. It is used to perform tasks like answering the questions related to the organization, supply the relevant links for the questions encountered by the user (if necessary), etc.

Basically, the initial chatbot named "ELIZA" was designed by Joseph Weizenbaum. With the innovation of the existing Chatbots, there came the evolution of the Chatbot technology. Chatbots can be implemented based on two approaches; Rule-Based Approach and Self-Learning Approach. In rule-based approach, there are some specific rules to be followed when training the bot. Here, the chatbot cannot answer complex questions triggered by the user. In self-learning approach, the chatbot follows the machine learning approach. This approach can further be divided into two models; Retrieval-Based Model (Chatbot retrieves the best response from the collection of responses) and Generative Model (Chatbot answers the question asked by the user from the set of answers). Chatbots that implement generative models are termed as intelligent chatbots.

The main sequence of flow for a functional chatbot involves three stages: (i) to get natural language user input, (ii) generate most appropriate automated response, and (iii) returning prudent natural language output. This College Enquiry Chatbot is developed using the Python Flask framework, where the front-end is written in HTML, CSS, and internal JavaScript. To implement the working of College Enquiry Chatbot, Chatterbot Library is used. Chatterbot is language independent and makes it easy to design a product that takes part in conversation. The inspiration to build this chatbot came when going through our college's website. This Chatbot follows the generative approach, which answers the user from a predefined set of answers. This Chatbot functions similar to the initial chatbot "ELIZA." Section 2 discusses the motivation behind developing this chatbot. Section 3 talks about the related work done in this area. Section 4 is about the proposed approach and Sect. 5 shows the implementation part and results. Section 6 is the conclusion part.

2 Motivation

For every student, to get the information related to the fee structure, due fees, events that are being organized, etc., is a lengthy and difficult process most of the times. Some of the events that are being organized by one department are not known to the other departments due to lack of effective communication channels. Sometimes, the changes in the fee structure are often not known to many students. We are living in a world where people are striving towards increased automation rather than manual intervention for repetitive tasks to arrive at faster and simpler solutions for the complex problems. This is the main reason behind building the College enquiry Chatbot where the user can ask the queries and a bot responds within no time. It reduces the users' time and can be easily available to any user with a mobile phone as well. With this Chatbot, the student no longer needs to take trips to notice boards or contact his/her friends or faculty to get timely information. This Chatbot in turn reduces the work of the organization. The user can get answers for any questions related to the college like address, admission process, and subsequent procedures which may be very difficult to find out for many students due to the pandemic situations.

3 Related Work

Harsh Pawar et al. developed College Enquiry Chatbot using Knowledge Database. This System is built using LUIS.ai, Microsoft Bot builder, and MongoDB for database. The user can interact with this system through web interface, in which the web application is connected to the bot through bot connector. This system will be a chatbot that responds to the questions by the user. In this system, they used LUIS.ai for training language model, which is used to identify the intention of the user and fetches the adequate response using pattern matching algorithm [3].

Nidhi Mishra et al. developed Dr. Vdoc, a medical Chatbot in 2018 which acts as a virtual doctor. The interaction can take place between a patient and a virtual chatbot, and this is made possible using natural language processing and pattern matching algorithms. It is more useful to provide awareness on the disease in case of unavailability of the doctors. The results showed that Dr. Vdoc is capable of returning the responses with 80% accuracy [4].

Pooja Prashanth et al. designed a college enquiry chatting system using knowledgeable database. In education system, few works can be very lengthy, time consuming, and require extra manpower. But in today's world, we see that almost everything is turning to digital format. This chatbot updates the students with college's cultural activities. It is based on client server architecture. All the information is kept in the database on the central server. By installing application in the smart phone, a user can access the information from the database. This system is developed with the combined technologies of Artificial Intelligence, Knowledgeable Database, and Virtual Assistance, such that the chatbot can make conversations between humans and machine [5].

Sagar Pawar et al. developed a web-based college enquiry chatbot with results. The system uses bigram and some sentence similarity algorithms to provide answers to student or any other user queries. If the response is invalid or irrelevant to the user query, then user can mark the response as invalid. This is sent to the admin. The admin will be able to delete or modify the data [6].

Jayesh Gangrade et al. designed a Review on College Enquiry Chatbot in 2019. An intelligent voice recognition chatbot is designed and developed which replies to the user queries in a graphical user interface, which gives them a feel of talking to a person. In this system, the user must login into the web application to ask queries. This system also contains a notice board where the important notices are placed in the form of documents or portable document formats (pdf). Here, important keywords are fetched from the query, and those keywords are searched in the knowledge database using keyword matching algorithms. If the answer is not found, then the system displays the following—"Answer to this query is not available at the moment, please revert after some time" [7].

4 Proposed Approach

The proposed system is designed to solve the user queries in no time without the user approaching the college or contacting any other person. A retrieval-based model uses an archive of predefined responses and performs some analysis to pick a desired response based on the given input and text [8]. This analysis can be a simple expression match, or Machine Learning classifiers that work on the principle directed towards flows of graphs. The systems that are designed using this model choose an appropriate response from a fixed set instead of returning a new text. The chatbot is trained based on existing information in the database to provide best possible responses from the predefined responses. In case of retrieval-based chatbot [9], the selection of appropriate responses can be determined based on techniques like machine learning, deep learning, and keywords matching. This model cannot generate new output and can only provide predefined responses. Retrieval-based chatbot design is shown in Fig. 1.

User Query: The students can approach the Chatbot to resolve their queries related to the College. This Chatbot receives the queries that are of any type (Greetings, College related queries, General questions, etc.) by the user in the form of text. The user inputs his queries in the textbox.

Tokenizing the words: Tokenization is a process of breaking a sentence or the sequence into pieces called tokens, perhaps eliminating the special characters like punctuation marks. For example, if the input is "Hello, How are you?," then the tokens obtained after tokenization are "Hello," "How," "are," and "you." After tokenizing,

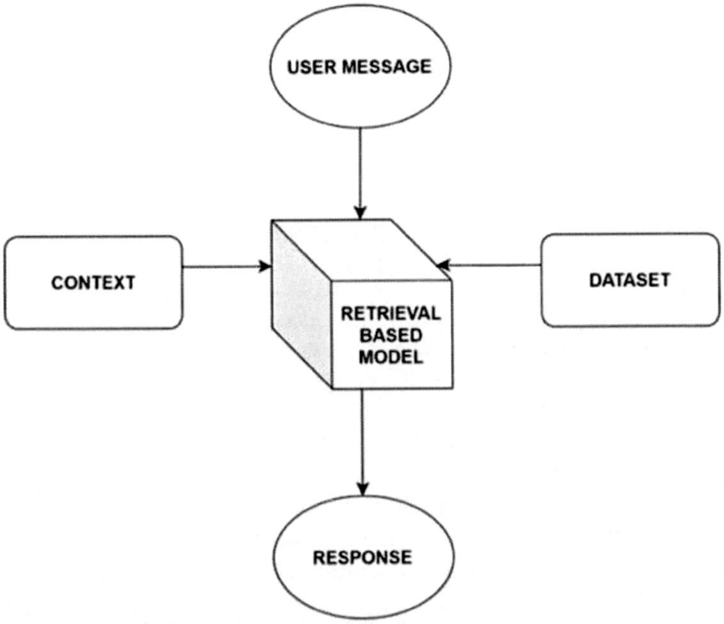

Fig. 1 Retrieval-based chatbot design

the tokens are processed by normalizer and matcher. Tables 1 and 2 depict how the normalizer and matcher process the tokens.

A normalizer processes the tokens by matching them with the patterns stored in the dataset. It matches the keywords of the input sentence with the patterns of the dataset and stores the matched keywords. A Matcher is used to find out the appropriate output

Table 1 Processing of the normalizer

S. No	Input	Output of the normalizer
1	I want	I want
2	to talk	to talk
3	in Hindi	in Hindi

Table 2 Processing of matcher

S. No	Input sentence	Pattern in database	Similarity threshold	Output sentence
1	I code in python	I prefer to code in python	0.5	The output will be "I prefer to code in python"
		I like coding in python	0.33	

for the user query from the existing dataset. The Matcher checks by converting text into ASCII and checks the sensitivity of the characters (lower/upper case).

Retrieving the matched response: Upon finding a sentence similar to the query from the predefined dataset, the chatbot retrieves an appropriate response from the same [10].

Calculating the similarity: The similarity threshold is calculated for the input query, out of many sentences of the dataset that match with the input. The higher threshold valued sentence is sent to the user.

For example, Input—"I code in python" and the dataset has two related sentences, Sentence 1—"I love to code in python" and Sentence 2—"I code in python language."

The similarity is checked by calculating the threshold of the input query with each of the two sentences in the dataset. The similarity threshold of the "Sentence 2" is higher than that of the "Sentence 1." So, the appropriate answer for the "Sentence 2" is returned to the user.

The intersection between two sentences, one which is the user query and the other is the pattern in database, gives the sentence similarity score. It is represented by S1 ∩ S2 and S2 ∩ S1, where S1 and S2 are two sentences.

The formula to calculate sentence similarity score is shown in Eq. (1).

$$Count(S1 \cap S2) \cup Count(S2 \cap S1)/Count(S1) \cup Count(S2) \tag{1}$$

Sample user input statement

S1 = "I code in python." and the pattern in the database as: S2 = "I prefer to code in python."

Using the similarity algorithm, we get:

S1 = {I code, code in, in python} Count (S1) = 3

S2 = {I prefer, prefer to, to code, code in, in python} Count (S2) = 5

Now, S1 ∩ S2 = 2 and S2 ∩ S1 = 2.

Calculated score for Sentence Similarity

Let a = Count (S1 ∩ S2) v Count (S1 ∩ S2) = 2 ∪ 2 = 4.

Let b = Count (S1) ∪ Count (S2) = 3 ∪ 5 = 8.

Similarity score is a/b = 0.5.

This procedure is adopted to calculate the sentence similarity score for each pattern stored in the database. The output for the users input query is displayed in the designed College enquiry Chatbot web app.

The proposed system design is shown in Fig. 2 which depicts the procedure of how the system is developed.

Fig. 2 College enquiry chatbot system design

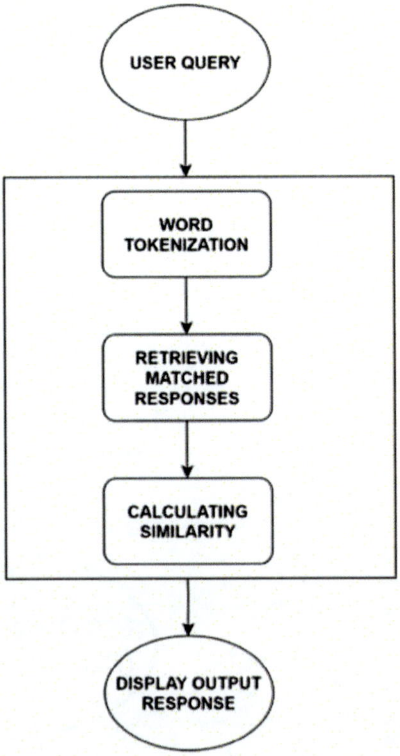

5 Results and Discussion

This system follows three main steps as shown in Fig. 3, i.e., Get Input, Process Input, and Return Response.

Get Input: In the first step, the user asks the Chatbot a question from any source (sources can be screen, speech recognition, or API). In this project, the user uses a console to converse with the Chatbot [11]. The initial message is displayed as "Hi, Welcome to PVPSIT Chatbot. I am at your service. You can ask me questions!!"

Process Input: The input given by the user is processed by the logic adapters. The logic adapters are those which are used to decide the logic for the Chatterbot to select a response for the input asked by the user. Any number of logic adapters can be used in the project. There are two logic adapters used in this paper; "Mathematical Evaluation and "Best Match." The bot gives back the response of the logic adapter that has higher confidence value. If any two logic adapters have the same threshold/confidence, then the first logic adapters' response is returned to the user. One can set the parameters for the logic adapters like setting the default response, threshold, paths, etc. The "Best Match" logic adapter responds the user with the confidence level of 0.70, and when the bot does not find the best match for the input, it gives the default response

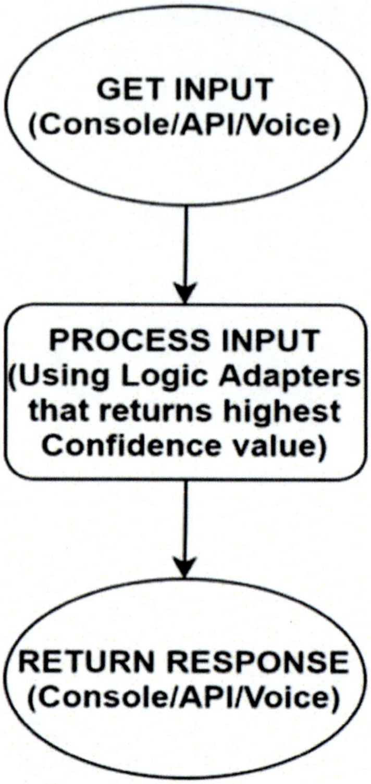

Fig. 3 Flow chart diagram of college enquiry chatbot

saying "I am sorry, but I do not understand. I am still learning." Figure 4 gives the responses after processing the input data by the Chatbot.

Return Response: The bot returns the response through the console. The bot in this project is trained with the data related to the college and English language. This Chatbot can also respond to the input of various kinds like mathematical calculations, sports, emotions, and simple conversations. Figure 5 depicts the logic of how training is performed.

A questionnaire was constituted of various questions related to performance levels, humanity, effect, and accessibility, and was distributed among the students of the department. A sample of 240 responses was taken and analyzed, and the results conclude that the performance of the website improved by 20% when the chatbot is included. Considering humanity and affection, when compared to the existing one, it reached an average level and a 5% improvement over the previous version. The statistics are shown in Table 3.

Performance, Humanity, Effect, and Accessibility were the parameters that were considered to analyze the performance of the bot. A questionnaire was used for the comparative analysis [12]. This helps in designing more effective bots. To assess

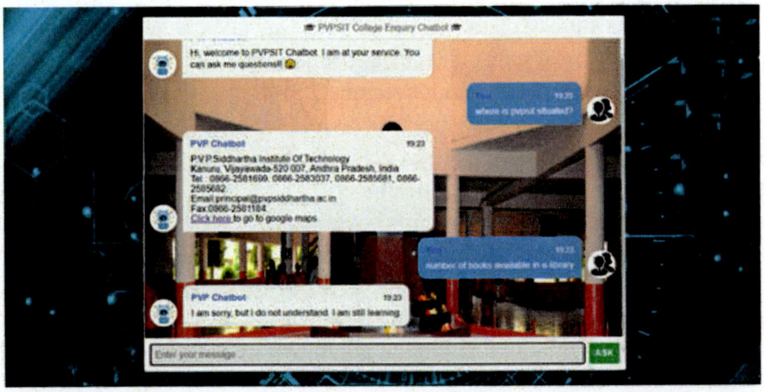

Fig. 4 Responses after processing the input data by the chatbot

```
# Prepare the training data
with open('training_data/personal_ques.txt','r') as fd:
    conv = []
    last = ''
    for line in fd:
        # Remove newline and whitespaces only from the end of the line
        line = line.rstrip()
        if line.startswith(' '):
            last += line
        else:
            if last:
                conv.append(last)
            last = line
# Add the final string... last will be '' only if the file is empty
if last:
    conv.append(last)

# Training with Personal Ques & Ans
trainer=ListTrainer(chatbot)
trainer.train(conv)

# Training with English Corpus Data
trainer_corpus = ChatterBotCorpusTrainer(chatbot)
trainer_corpus.train(
    'chatterbot.corpus.english'
)
```

Fig. 5 Training the chatbot

Table 3 ChatBot quality assessment

Category	Quality attribute	Metric	Old system	New system
Performance	Robustness to unexpected input	% of success	60%	80%
Humanity	Able to maintain discussion	Low to high	Low	Low
Affect	Provides greetings	Low to high	Low	Average
Accessibility	Can detect meaning and intent	% of success	70%	75%

the performance questions related to robustness, an unexpected input was questioned on. After analysis, it was evident that respondents stated that there is a clear improvement of 20% over the previous version. Coming to the category of humanity, questions based on maintaining discussion was assessed, and there was a significant improvement from average to high [13]. Coming to the effect, questions on providing greetings were assessed. There is no specific change noted in both the versions [14]. Coming to the accessibility, there was improvement by 5% which is not quite noteworthy. This paper presents the overall approach used to design the chatbot, and compares the web application as-is study with the to-be website when the chatbot is incorporated.

6 Conclusion

In this paper, a machine learning approach for College Enquiry Chatbot has been proposed. The responses of the Chatbot are obtained with the help of a "Retrieval-Based" approach. This Chatbot can perform almost 30% of the entire task of the college. This makes the work easy and provides results within seconds. This proposed chatbot is designed in python using "Flask Framework." The chatbot mainly follows three steps: Getting input queries from the user, processing the input, and returning the responses to the user. The methodology of the Chatbot is that it takes the input from the user, tokenizes the input, retrieves the matched response, calculates the similarity score, and then returns the appropriate response to the user. The results of to-be website when Chatbot is incorporated and the as-is website are compared based on the performance, humanity, effect, and accessibility. Of all those tests, the website that incorporates the chatbot shows better performance, humanity, effect, and accessibility. Thus, implementing chatbot will reduce the work of the organization. Some of the limitations of the chatbot are domain knowledge, not possessing the personality.

References

1. Schachner, T., Keller, R., & von Wangenheim, F. (2020). Artificial intelligence-based conversational agents for chronic conditions: Systematic literature review. *Journal of Medical Internet Research, 22*(9), e20701. https://doi.org/10.2196/20701
2. Xu, A., Liu, Z., Guo, Y., Sinha, V., & Akkiraju, R. (2017). A new chatbot for customer service on social media. In *Proceedings of the 2017 CHI Conference on Human Factors in Computing Systems—CHI '17* (pp. 3506–3510).
3. Pawar, H., Prabhu, P., Yadav, A., Mendonca, V., & Lemos, J. (2018). College enquiry chatbot using knowledge in database. *International Journal for Research in Applied Science & Engineering Technology (IJRASET)*. ISSN: 2321-9653.
4. Mishra, S. K., Bharti, D., & Mishra, N. (2017), Dr. Vdoc: A medical chatbot that acts as a virtual doctor. *Research & Reviews: Journal of Medical Science and Technology, 6*(3).

5. Prashant, B. P., Anil, M. S., Dilip, K. M. (2018). Online chatting system for college enquiry using knowledgeable database. *International Journal of Engineering Research in Electronics and Communication Engineering, 5*.
6. Pawar, S., Rane, O., Wankhade, O., Mehta, P. (2018). A web based college enquiry chatbot with results. *International Journal of Innovative Research in Science, Engineering and Technology, 7*(4).
7. Gangrade, J., Surme, S. S., Somu, S., Raskonda, S., Gupta, P. (2019). A review on college enquiry chatbot. *International Journal of Engineering Science and Computing, 9*(3).
8. Vamsi, G. K., Rasool, A., & Hajela, G. (2020). Chatbot: A deep neural network based human to machine conversation model. In *11th International Conference on Computing, Communication and Networking Technologies (ICCCNT)* (pp. 1–7). Kharagpur, India. https://doi.org/10.1109/ICCCNT49239.2020.9225395
9. Wu, Li, Z., Wu, W., & Zhou, M. (2018). Response selection with topic clues for retrieval-based chatbots. *Neurocomputing, 316*, 251–261.
10. Setiaji, B., & Wibowo, F. W. (2016). Chatbot using a knowledge in database: Human-to-machine conversation modeling. In *7th International Conference on Intelligent Systems, Modelling and Simulation (ISMS)* (pp. 72–77). Bangkok. https://doi.org/10.1109/ISMS.2016.53
11. Khan, R., & Das, A. (2018). Introduction to chatbots. In *Build better chatbots* (pp. 1–11). Berkeley, CA: Apress.
12. Maroengsit, W., Piyakulpinyo, T., Phonyiam, K., Pongnumkul, S., Chaovalit, P., & Theeramunkong, T. (2019). A survey on evaluation methods for chatbots. In: *Proceedings of the 2019 7th International Conference on Information and Education Technology (ICIET 2019)* (pp. 111–119). New York, NY, USA: Association for Computing Machinery. https://doi.org/10.1145/3323771.3323824
13. Miner, A. S., Milstein, A., Schueller, S., Hegde, R., Mangurian, C., & Linos, E. (2016). Smartphone-based conversational agents and responses to questions about mental health, interpersonal violence, and physical health. *JAMA Internal Medicine, 176*(5), 619–625.
14. Seeger, A. M., & Heinzl, A. (2017). Human versus machine: Contingency factors of anthropomorphism as a trust-inducing design strategy for conversational agents. In *Lecture notes in information systems and organization* (Vol. 25, pp. 129–139). Cham: Springer.

A Pragmatic Study on Movie Recommender Systems Using Hybrid Collaborative Filtering

Akhil M. Nair and N. Preethi

Abstract The Movie Recommendation System (MRS) is part of a comprehensive class of recommendation systems, which categorizes information to predict user preferences. The sum of movies is increasing tremendously day by day, and a reliable recommender system should be developed to increase the user satisfaction. Most of the approaches are made to prevent cold-start, first-rater drawbacks, and gray sheep user problems, nevertheless, in order to recommend the related items, various methods are available in the literature. Firstly, content-based method has some drawbacks like data of similar user could not be achieved, and what category of these items the user likes or dislikes are also not known. Secondly, this paper discusses about collaborative filtering to find both user and item attributes that have been considered. Since there exist some issues pictured with collaborative filtering, so this paper further aims into hybrid collaborative filtering and deep learning with KNN algorithm of ratings of top K-nearest neighbors.

Keywords Collaborative filtering · Deep learning · Cold-start · First-rater · Gray sheep · KNN algorithm

1 Introduction

Nowadays, every online site provides Recommendation System for gaining users' interest in their websites or applications [1]. It is a very big challenge for the users to search for their kind of movies in this huge catalog. Recommendation Systems help in saving time and energy for both users and owners by creating codes to ensure user interest is always prioritized and they are offered similar products or items. These recommendation systems analyze our shopping, watching, listening patterns, and make predictions based on our past activities. The most basic model

A. M. Nair (✉) · N. Preethi
CHRIST (Deemed To Be University), Bengaluru, India
e-mail: akhil.nair@christuniversity.in

N. Preethi
e-mail: preethi.n@christuniversity.in

P. Nayak et al. (eds.), *IoT and Analytics for Sensor Networks*, Lecture Notes in Networks and Systems 244, https://doi.org/10.1007/978-981-16-2919-8_44

of the recommendation system is collaborative filtering which tends to assume that users may like items that are liked by similar users. This development has led to the collection of a humungous amount of data that may be difficult to be dealt with individually, so groups of similar interests and users are constructed in advance, to tackle individually [2]. This method of collaborative filtering helps to solve the problems and drawbacks of content-based recommendations.

2 Literature Review

The collaborative filtering method uses implicit or explicit feedback from the users [1]. Collaborative filtering involves two techniques: the memory-based approach and the model-based approach [3]. The memory-based approach finds user similarity using cosine similarity or Pearson correlation coefficient and takes weighted average ratings of users, whereas in the model-based approach machine learning is used to find user ratings of unrated item instances using Principal Component Analysis (PCA), Singular Value Decomposition (SVD), Neural Network, and Deep-Matrix Factorization [4–6]. In this paper, we are using a memory-based approach which helps in the easy creation and elucidation of results. Figure 1 shows the architecture of a collaborative filtering-based recommender system. The system assumes a user to be x. It creates a set N of other similar users whose preferences match with the user x. Finally, the system predicts the x's rating based on the ratings provided by x's, similar users. Figure 1 shows a basic architecture of such recommender system.

The memory-based collaborative filtering techniques can be further classified into two: user-item filtering and item-item filtering [7]. In user-item filtering, a target user is chosen, and users similar to the target user are discovered based on the user preferences and feedback. Item-item filtering takes a target item and user ratings of

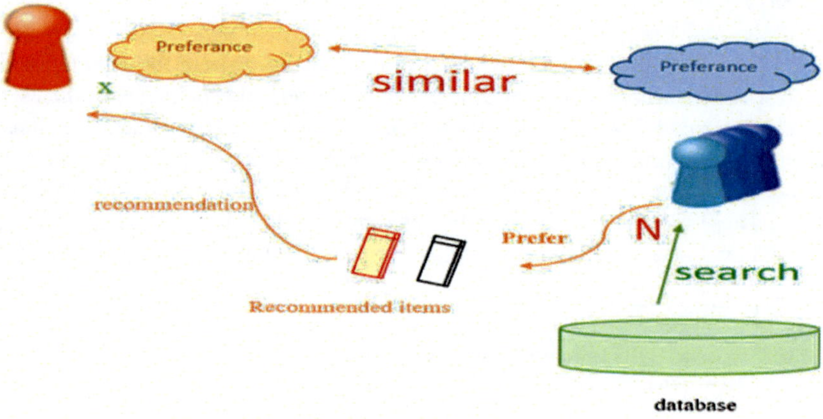

Fig. 1 Architecture of collaborative filtering system

that item are grouped together. There are certain disadvantages in the collaborative filtering method like Cold-Start, Sparsity Problems, First-Rater, Popularity Bias, etc. The complexity in finding K-most similar users or items is high and is given by O(|U|), Where |U| is the size of the utility matrix [8, 9]. To overcome these disadvantages in CF, hybrid CF-method and Deep Learning are considered in this paper.

To overcome the limitations of CF, a global baseline is combined with CF [6]. The negative predictions achieved by the system are corrected by averaging the distance from the baseline predictor. It also helps to extend collaborative filtering to large user bases. Since the difference between the baseline estimate and the rating matrix is evaluated and thus similarities can be applied to a normalized matrix [10]. While the ratings are limited to be non-negative, it is noted that certain ratings average together and compute the predictions to be negative. The baseline estimator for the weighted averaged is given by Eq. 1.

$$r_{xi} = b_{xi} + \frac{\sum_{j\varepsilon N(i:x)} S_{ij}*\left(r_{xj} - b_{xj}\right)}{\sum_{j\varepsilon N(i:x} S_{ij}} \tag{1}$$

where $b_{xi} = \mu + b_x + b_i$
$b_x = rating\ deviation\ of\ user\ x = (avg \cdot rating\ of\ user) - \mu$
$\mu = overall\ mean\ movie\ rating$
$b_i = rating\ deviation\ of\ movie.$

Equation 1 specifies the baseline estimator for the prediction. To start with similarity s_{ij} of items i. This provides the most similar items I that are rated by the user x. The rating r_{xi} is then estimated as the weighted average. The predictions are evaluated using Root-Mean-Square-Error (RMSE).

Evaluating predictions: Root-Mean-Square-Error (RMSE) is calculated by Eq. 2.

$$\sqrt{\frac{\sum_{(x,i)\in T}\left(r_{xi} - r_{xi}^{*}\right)^2}{N}} \tag{2}$$

where $N = |T|$, r_{xi} is the predicted rating.

3 Proposed Methodology

This study uses the MovieLens10M dataset obtained from the movie lens database (https://grouplens.org/datasets/movielens/latest/). We exercised the two original files, namely, movies.csv and ratings.csv with records containing 9762 rows and 100,836 rows, respectively. The dataset consists of the ids of users, movies, and ratings of the movies provided by the users. The study evaluates the results of the proposed method with that of state-of-the-art collaborative filtering approaches. A cosine similarity metric is applied for one dimension, which is "ratings" to find

Table 1 Meta data about movie data set

S. No	Parameter	Numbers of instances
1	Users	610
2	Movies	9742
3	Ratings	1,00,836
4	Rating scale	5 Star (0.5 stars to 5 Star)

the similarity among the users. K-nearest neighbour (KNN) finds the n nearest and similar users as per the preferences and feedback. KNN builds a neighborhood of K similar users concerning the target users rendering to a similarity measure which is evaluated from the user ratings and other parameters. Item-item-based system is tested with KNN using both cosine and Pearson similarity measures. The value of k for the number of neighbors was set to 20 although the results were quite similar for both higher and lower values. The study was conducted on two types of recommendation algorithms that are collaborative filtering and neural networks using KNN. The proposed method is evaluated using the metrics shown in Table 1.

Neural networks are considered to be black-box algorithms, and the interpretation of their weights is done in an insightful way without the application of specific visualization techniques. The ratings and its frequency of occurrences is shown in Fig. 2.

Table 2 displays the KNN Methods for the number of movies which contains training set 6819 and testing set 2923 for k = 1, k = 3, and neural network (NN). The Accuracy, Recall, Precision, and F-measures are shown for various methods. For the size K = 1, the accuracy is 86.25%, Recall is 0.872, Precision is 0.853, and F-measures is 0.873. When k = 2, the accuracy is 91.2%, Recall is 0.926, Precision

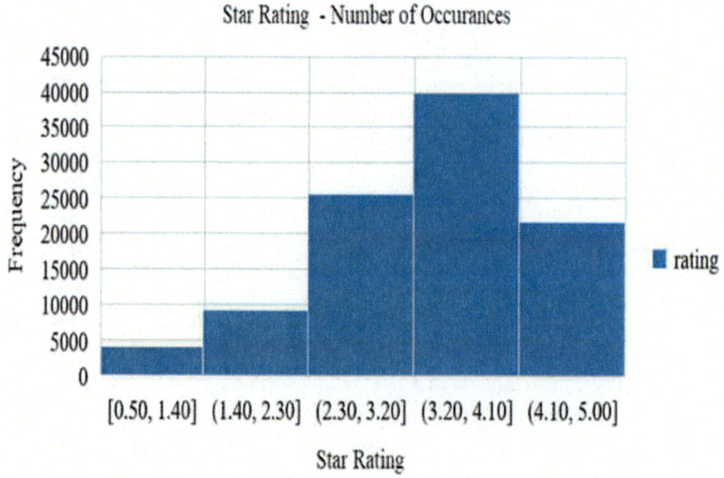

Fig. 2 Star rating—frequency of occurrences

Table 2 Performance of NN, KNN K = 1, and K = 3 classifiers

Method	Data set—number of movies		Accuracy (%)	Recall	Precision	F-Measure
	training set	Testing Set				
KNN K = 1	6819	2923	86.25	0.872	0.853	0.873
KNN K = 3	6819	2923	91.2	0.926	0.932	0.912
NN	6819	2923	94.6	0.943	0.932	0.964

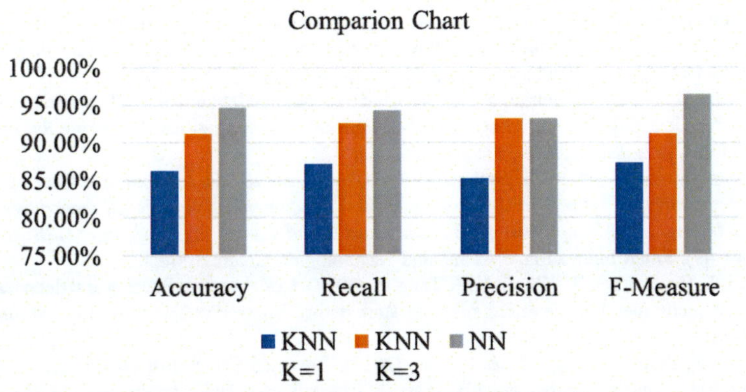

Fig. 3 Assessment graph for the NN, KNN K = 1, and 3 classifiers

is 0.932, and F-measures is 0.912, whereas in a neural network, the accuracy level is 94.6%, Recall is 0.943, Precision is 0.932, and F-measures is 0.964 which gives higher accuracy when compared with normal KNN method. The Accuracy, Recall, Precision, and F-measures values for the KNN and NN methods are shown clearly in the following Table 2 and Fig. 3.

4 Conclusion

Movie recommendation is one of the topics that is increasing research interest. Nevertheless, to convey relevant recommendations, it is required to deal with various recommendation methods to extract the information from movies. This study proposes a hybrid recommendation model which integrates collaborative filtering with deep learning to precise the cold-start problem, sparsity problems, first-rater, and popularity bias of collaborative filtering recommendation models. The proposed study evaluates the model with respect to RMSE on a movie dataset. We have implemented both the models with the specified dataset. It is clear from the results that proposed model outperforms the simple model, which uses the average rating of users for rating prediction of various items. Also, the hybrid model outperforms the global baseline model for rating prediction of items with few ratings. The proposed model shows that the hybrid model is effective on rating predictions. A hybrid recommendation technique will eventually be adopted by the recommender systems community.

References

1. Sánchez-Moreno, D., Gil González, A., Muñoz Vicente, M., López Batista, V., & Moreno García, M. (2016). A collaborative filtering method for music recommendation using playing coefficients for artists and users. *Expert Systems with Applications, 66*, 234–244.
2. Arnold, A. N., & Vairamuthu S. (2019). Music recommendation using collaborative filtering and deep learning. *International Journal of Innovative Technology and Exploring Engineering* (IJITEE), *8*(7). ISSN: 2278-3075
3. Suganeshwari, G., & Syed Ibrahim S. P. (2016). A survey on collaborative filtering based recommendation system. In V. Vijayakumar & Neelanarayanan V. (Eds.), *Proceedings of the 3rd International Symposium on Big Data and Cloud Computing Challenges (ISBCC—16')* (Vol. 49). Smart Innovation, Systems and Technologies. Cham: Springer.
4. Wei, J., He, J., Chen, K., Zhou, Y., & Tang, Z. (2017). Collaborative filtering and deep learning based recommendation system for cold start items. *Expert Systems with Applications, 69*, 29–39.
5. Park, Y., Park, S., Jung, W., & Lee, S. (2015). Reversed CF: A fast collaborative filtering algorithm using a k-nearest neighbour graph. *Expert Systems with Applications, 42*(8), 4022–4028.
6. Sunitha, M., & Adilakshmi, T. (2018). Music recommendation system with user-based and item-based collaborative filtering technique. In G. Perez, K. Mishra, S. Tiwari, & M. Trivedi (Eds.), *Networking Communication and Data Knowledge Engineering* (Vol. 3). Lecture Notes on Data Engineering and Communications Technologies. Singapore: Springer.
7. Segaran, T. (2007). *Programming collective intelligence: Building smart web 2.0 applications.* O'Reilly Media.
8. Herrada, O. C. (2009). *Music recommendation and discovery in the long tail.* Ph.D. thesis, Universitat Pompeu Fabra.
9. Burke, R. (2007). *Hybrid web recommender systems.* The Adaptive Web (pp. 377–408). Springer.
10. Piotte, M., & Chabbert, M. (2009, August). *The pragmatic theory solution to the Netflix Grand Prize.* Netflix Prize Report. Retrieved February 5, 2017, from http://www.netflixprize.com/assets/GrandPrize2009_BPC_PragmaticTheory.pdf

An Anatomization of FPGA-Based Neural Networks

Anvit Negi, Devansh Saxena, Kunal, and Kriti Suneja

Abstract Ongoing advancements in the improvement of multilayer convolutional neural organizations have brought about upgrades in the precision of important recognition jobs, for example, huge category picture classification and cutting-edge automated recognition of speech. Custom hardware accelerators are crucial in improving their performance, given the large computational demands of Convolution Neural Networks (CNN). The Field-Programmable Gate Arrays (FPGAs) reconfigurability, computational abilities, and high energy efficacy makes it a propitious CNN hardware acceleration tool. CNN have demonstrated their value in picture identification and recognition applications; nonetheless, they require high CPU use and memory transmission capacity tasks that cause general CPUs to neglect to accomplish wanted execution levels. Consequently, to increase the throughput of CNNs, hardware accelerators using Application-Specific Integrated Circuits (ASICs), FPGAs, and Graphic Processing Units (GPUs) have been employed to improve CNN performance. To bring out their synonymity and dissimilarity, we group the works into many groups. Thus, it is anticipated that this review will lead to the upcoming development of successful hardware accelerators and be beneficial to researchers in deep learning.

Keywords FPGA · ASIC · Deep learning · Neural net

1 Introduction

In the past decade, artificial intelligence (AI) and machine learning (ML) instruments have gained considerable prominence due to advancements in computational structure which consists of area, power, and effectiveness. A range of programs have started using AI algorithms to improve overall efficiency than conventional methods. These applications include image processing [1], for example, face identification,

A. Negi · D. Saxena (✉) · Kunal · K. Suneja
Department of Electronics and Communication Engineering, Delhi Technological University, New Delhi, India

K. Suneja
e-mail: kritisuneja@dtu.ac.in

P. Nayak et al. (eds.), *IoT and Analytics for Sensor Networks*, Lecture Notes in Networks and Systems 244, https://doi.org/10.1007/978-981-16-2919-8_45

banking and statistical surveying [2], mechanical arms in the robotized producing area [3], medical services applications [4], database administration and management [5], and face checking and investigation security applications [6].

The explosive development of big data over the past decade has inspired revolutionary approaches to obtain data from various sensors such as photos and voice samples. In reality, these Convolution Neural Networks (CNNs) [7] now are conceived as the standard method among the proposed methods by delivering "human-like" accuracy in various computer vision-related applications, such as classification [8], detection, segmentation [9], and speech recognition [10].

As CNNs need up to 38 GOP/s to identify a single frame [11], this output is obtained at the expense of a high computational price. Hence, to accelerate their execution, dedicated hardware is required. The most frequently used platform for implementing CNNs is Graphics Processing Units (GPUs), as they provide the highest performance with respect to computational throughput, hitting 11 TFLOP/s [12].

Nonetheless, FPGA systems are considered to be better energy efficient in terms of power consumption (vs GPUs). Thereby, several FPGA-based CNN accelerators were suggested, targeting both data centers for High Performance Computing (HPC) [13] and embedded applications [14].

Although GPU implementations have shown exceptional computational efficiency, for two reasons, CNN acceleration is heading momentarily towards FPGAs. First, recent advances in FPGA technology have brought about FPGA performance with a recorded performance of 9.2 TFLOP/s for the latter in striking distance to GPUs. Second, recent CNN production patterns are increasing the sparsity of CNNs and using extremely compact types of data.

The remaining paper is structured in the following way: Literature Review, summarizing the use of AI algorithms in the past decade. Explanation of CNN and challenges of FPGA-based implementation. A detailed analysis of FPGA, GPU, and ASIC-based implementation of AI networks which compares between the power, area performance, and efficiency parameters. Finally, we look into the optimizations available and scope of future research.

2 Literature Review

Lately, the application of AI algorithm as a replacement of conventional algorithms have become popular. With the advent of ML and AI, there is a worry to basically address computational cost and power-burning through AI calculations. This leads to the necessity for particular equipment with high capacity to perform AI estimations with large scope issues [15]. The aim of all the research is to achieve better and more capable handling of AI algorithmic calculations. Numerous researches have been conducted over the span of the latest decade discussing hardware and programming advancements and execution strategies in this field [1, 16–21].

Table 1 A comprehensive compendium of survey papers

References	Publication year	Description and scope
[16]	2010	This research addresses all the main architecture methods and models for the 1990–2010 "hardware neural network." They exemplify numerous HNN implementations in a variety of "ANN models" like CNN, neural bursts, and so on. Digital, analog, composite, neuromorphic, "FPGA," or optical implementations are used in these HNN
[19]	2013	The scope of the research work was if developers were designing customized versions or were using pre-tailored free platforms for the development. The "artificial neural networks (ANN)" software framework table was also included in the work
[1]	2017	The paper provides a brief analysis of past work on the main approaches of FPGA-based neural inference networks
[21]	2018	This survey paper gives insight about the Moore's law in conjunction with the increasing use of CPUs and GPUs. Major applications of AI are also discussed
[22]	2020	This paper summarizes AI neural network's hardware implementation, with emphasis on all three-hardware equipment: ASIC, GPU, and FPGA. A comprehensive flatting strategy has been used to pick the research articles to address issues of science

Here in Table 1, we discuss corresponding survey articles published roughly between 2009 and 2019 on the implementation of AI algorithms, more precisely neural nets for hardware. Few studies have covered hardware implementations of artificial neural networks [16, 19, 20]. While other studies have centered on FPGA-based deep learning neural network accelerators [1, 17, 18]. In [19], the FPGA implementation of coevolutionary neural networks was the subject of the authors (CNNs). The survey addressed GPU implementations in [21].

This examination distinguishes and addresses various papers. These papers were evaluated as per the hardware used. The bulk of the documents include FPGA-based implementations. With its high adaptability and solidness, the FPGA is viewed as a promising option for the use of these calculations. Likewise, recent FPGA design improvements have brought about making it more accessible because of which profound learning research has procured significant consideration [23].

One of the powerful "application-specific integrated circuit" (ASICs) is also used for AI algorithm implementation. ASICs are personalized chips that are conceived for a particular purpose. They save high-power and are speed-efficient, consume less silicon area, making them ideal solutions for accelerating AI algorithms [24]. To accelerate AI algorithms, graphical processing units (GPUs) are also used to pace up algorithms to hundreds of times the initial speed [25]. GPUs are made to carry out scalar and parallel intensive computing [26]. Unlike multicore CPUs, when accessing DRAM memories, GPUs try not to depend on shrouded latencies utilizing large cache memories [27]. Such highlights have made GPUs increasingly more useful for AI acceleration.

Using small single-board processors, some AI algorithms are applied. For example, raspberry pi. As a result of their little size and low force prerequisites, single-board PCs are viewed as a decision for AI usage.

3 Overview of CNN

One of the classic profound learning networks is the deep convolutional neural network. They are widely used in these areas for the continued development of deep learning technologies, machine vision, and language recognition [28]. Earlier studies have indicated that the calculation of the cutting-edge CNNs is overwhelmed by the convolutional layers [29, 30] formation.

It contains numerous falling layers, pooled layers, and fully-connected complex layers. The convolutionary neural network distinguishes the image as the input, and obtains the outcome across several "convolutional layers, pooling layers and associated layers."

3.1 Model of Convolutional Layer

The input f^{in} and convolution kernel composed of weights w_{ij}, comprises the convolution layer. The sampled function is set by balancing results to get the output $f^{out.}$ [30].

$$f_i^{out} = \sum_{i=1}^{n_{in}} f_i^{in} * w_{i,j} + b_i, 1 \le i \le n_{out} \tag{1}$$

3.2 Model of Pooling Layer

The pooling layer typically uses the maximum scan or core scan to minimize the size of the input matrix. The activity will viably lessen the following layer's data processing ability while forestalling the loss of characteristic information.

3.3 *Full Connection Layer*

The layer changes over the input to straight space, hence, changing over the input to a direct space. Output is received.

$$f^{out} = \sum_{j=1}^{n} f_j^{in} * w_{i,j} + b \tag{2}$$

3.4 *Activation Layer*

A nonlinear change of the input is presented by the activation function excitation, and the output after each layer is regularly handled. Some common functions include nonlinear (Relu), trigonometric function (Tanh), shock response (Sigmod), etc.

4 Challenges in FPGA

4.1 *Tradeoff Between RTL and HLS Approaches*

Via high-level abstractions, the HLS-based design approach enables fast growth. This demonstrates the conventional plan utilizing the OpenCL-based methodology [31]. Key features including partitioning, automatic pipelining of the CNN model, etc., can also be supported but for HW performance, however, the resulting design cannot be optimized.

4.2 *Necessity of Diligent Design*

Considering the particular characteristics of both FPGAs and Convolutional Neural Net, the optimization of throughput demands careful design. In general, two architectures for the same use of resources may have significantly different performances [30],and thus, the use of resources may not be a reliable overall performance measure. There are vastly different criteria for separate CNN layers. In addition, the use of FPGA relies heavily on the burst duration of memory access [32], so the access pattern of CNN memory must be cautiously configured. Furthermore, the compute-throughput ought to be adjusted with the memory or the memory can end up being the bottleneck [19].

4.3 FPGAs Over GPUs/ASICs/CPUs

Although the computer software environments are already mature for developing convolutional Neural Net designs for CPUs or GPUs, those for FPGAs are even now nascent. Furthermore, despite the HLS software, it can take several months for an experienced HW designer to implement the CNN model on FPGAs [33]. Therefore, in general, modeling efforts for GPUs is quite less when compared to FPGAs. Accounting for rapid changes growth related to neural nets, it may not be feasible to re-architect accelerators based on FPGA with every upcoming neural net algorithm. Therefore, the most recent NN algorithm cannot efficiently model an FPGA-based architecture. In addition, a Field-Programmable Gate Array (FPGA) modeling demands greater resource overhead than an ASIC design because of reconfiguration of logic blocks and switches. These factors give these custom boards a competitive disadvantage over other HW acceleration computing platforms for NNs.

4.4 A Comparative Study Between FPGA, ASIC, and GPU

Here, we bring the execution technique, GPUs/FPGAs/ASICs, in relation to regular merit figures. A distinction between the three methods is seen in Table 2. Since the ASIC specification is unaltered post-production, it becomes unsuitable for application where subsequent to installation, the model needs to be revised. ASIC is better used where low power and area are the goal. In comparison to ASICs, FPGAs are available for post-implementation upgrades. It is worth mentioning that even though the ultimate objective is ASIC execution, FPGAs are also used for prototyping and validation [22].

GPUs provide a solution at the software level, while FPGAs and ASICs have a solution at the hardware level. FPGAs and ASICs thus have greater stability compared to GPUs during the deployment process. As a result, the implementation of the GPU

Table 2 A comparison between the aforesaid

	FPGA (Hardware)	ASIC (Hardware)	GPU (Software)
Skills required	Verilog	Verilog	High level language
Implementation level	Medium	High	Medium
Versatility	High	High	Low
Post implementation change	High	Very low	High
Overall expenditure	Medium	High	Low
Area consumption	High	Low	High
Performance	Medium	High	Low/Medium
Power consumption	Medium	Low	High

is constrained by the current underlying hardware. Thus, in some situations, FPGAs can be quicker than GPUs.

5 A Compendium of Numerous Hardware Implementations

Implementation details of neural nets on various FPGA boards are shown in Table 3. Loop optimizations were first investigated in [30] to extract an FPGA-based CNN accelerator. This design was modeled utilizing HLS tools, therefore, it relies on the arithmetic of 32 floating points. Works in [34, 35] pursue the same unrolling scheme. In addition, [35] design has 16 bits of fixed-point arithmetic and RTL design, which results in an increase of performance by approximately two times. In recent works [36], the same unrolling and tiling scheme is used where authors report an improvement of x13.4 over their original works [30]. Consequently, unrolling and tiling loops can be proficient as it were for devices with large computational capabilities (i.e., DSP). This is often illustrated in works of Rahman et al. [34] which improves speed by 1.2 times over [30].

Conversely, all modeling factors looking for ideal loop unroll are fully explored in the works of Ma et al. [37, 40, 41]. More specifically, researchers show that using unroll function for single input arithmetic, the input FMs and weights are optimally reused [38, 39, 42–46].

6 Conclusion and Future Work

In this paper, a comprehensive study related to development and deployment of FPGA in CNN accelerators has been provided. To highlight their similarities and distinctions, we categorized the works based on many parameters. We outlined the influential avenues for study and summarized the key themes of numerous works. Usage of more than one FPGA is important for computational acceleration of massive and extensive Neural Net models, considering the limited hardware resources available on an FPGA board. This allows the architecture to be partitioned across several FPGAs. Although a software for automatic partitioning will not be readily accessible, the manual approach to partitioning is vulnerable to error and unscalable. In addition, random splitting will increase the complexity of interconnections in a manner that the I/O pin count cannot complete the requirement of number of connections.

Lately, researchers have also reconnoitered models of the spike neural network (SNN) that model the biological neuron more closely [47]. This study concentrates on (ANN) which is an acronym for artificial neural network. SNNs and ANNs vary considerably. As a result, their training rules and network architectural structure

Table 3 Specifications of various hardware implementations

	Neural net	FPGA board	Frequency (MHz)	Performance (GOPs)	Consumed power (W)	LUT (K)	DSP	Memory (MB)
[30]	AlexNet-C	Virtex7 VX485T	100	62	19	186	2240	19
[14]	VGG16SVD-F	Zynq Z7045	150	137	10	183	780	18
[31]	AlexNet-C	Stratix5 GSD8	120	187	34	138	635	18
	AlexNet-F			72		272	752	30
	VGG16-F			118		524	1963	52
[34]	AlexNet-C	Virtex7 VX485T	100	75		28	2695	20
[36]	AlexNet-F	Virtex7 VX690T	150	826	126		14,400	
	VGG16-F			1280	160		21,600	
[35]	NIN-F	Stratix5 GXA7	100	114	20	224	256	47
	AlexNet-F			134	19	242	256	31
[40]	AlexNet-F	Virtex7 VX690T	156	566	30	274	2144	35
[41]	AlexNet-C	Virtex7 VX690T	100	62		273	2401	20
[37]	VGG16-F	Arria10 GX1150	150	645	50	322	1518	38
[42]	AlexNet-F	Arria10 GT1150	240	360		700	1290	47
	VGG-F		222	460		708	1340	49
	VGG-F		232	117		626	1500	33
[43]	AlexNet-C	Cyclone5 SEM	100	12		22	28	1
		Virtex7 VX485T	100	445			2800	
[39]	NiN	Stratix5 GXA7	150	283		453	256	30
	VGG16-F			352		424	256	44
	ResNet-50			251		347	256	39
	NiN	Arria10 GX1150	200	588		320	1518	31
	VGG16-F			720		263	1518	45
	ResNet-50			619		437	1518	39
[38]	AlexNet-F	Virtex7 VX690T	100	446	25	207	2872	37
	VGG16SVD-F			473	26	224	2950	47

vary drastically. Large-scale SNN modeling onto SOC/FPGA boards will offer an exhilarating and remunerating challenge for IT designers later on [45].

References

1. Sze, V., Chen, Y. H., Yang, T. J., & Emer, J. S. (2017). Efficient processing of deep neural networks: A tutorial and survey. *Proceedings of the IEEE.*
2. Pau, L. F. (1991). Artificial intelligence and financial services. *IEEE Transactions on Knowledge and Data Engineering.*
3. Yao, X., Zhou, J., Zhang, J., & Boer, C. R. (2017). From intelligent manufacturing to smart manufacturing for industry 4.0 driven by next generation artificial intelligence and further on. In *Proceedings—2017 5th International Conference on Enterprise Systems ES.*
4. Bishnoi, L., & Narayan Singh, S. (2018). Artificial intelligence techniques used in medical sciences: A review. In *Proceedings of 8th International Conference on Cloud Computing, Data Science & Engineering (Confluence).*
5. Parker, D. S. (1989). Integrating AI and DBMS through stream processing.
6. Fraley, J. B., & Cannady, J. (2017). The promise of machine learning in cybersecurity. In *Conference of Proceedings—IEEE SOUTHEASTCON.*
7. Lecun, Y., Bengio, Y., & Hinton, G. (2015). Deep learning. *Nature.*
8. Russakovsky, O., Deng, J., Su, H., Krause, J., Satheesh, S., Ma, S., et al. (2015). ImageNet large scale visual recognition challenge. *International Journal of Computer Vision.*
9. Long, J., Shelhamer, E., & Darrell, T. (2015). Fully convolutional networks for semantic segmentation. In *Proceedings of the IEEE Conference on Computer Vision and Pattern Recognition.*
10. Zhang, Y., Pezeshki, M., Brakel, P., Zhang, S., Laurent, C., Bengio, Y., et al. (2016). Towards end-to-end speech recognition with deep convolutional neural networks. In *Proceedings of Annual Conference of the International Speech Communication Association, INTERSPEECH.*
11. Simonyan, K., & Zisserman, A. (2015). Very deep convolutional networks for large-scale image recognition. In *3rd International Conference on Learning Representations, ICLR 2015—Conference Track Proceedings.*
12. Nurvitadhi, E., Venkatesh, G., Sim, J., Marr, D., Huang, R., Ong, J. G. H., et al. (2017). Can FPGAs beat GPUs in accelerating next-generation deep neural networks? In *FPGA 2017—Proceedings 2017 ACM/SIGDA International Symposium on Field-Programmable Gate Arrays.*
13. Ovtcharov, K., Ruwase, O., Kim, J., Fowers, J., Strauss, K., & Chung, E. S. (2015). Accelerating deep convolutional neural networks using specialized hardware. *Microsoft Research Whitepaper.*
14. Qiu, J., Wang, J., Yao, S., Guo, K., Li, B., Zhou, E., et al. (2016). Going deeper with embedded FPGA platform for convolutional neural network. In *FPGA 2016—Proceedings of the 2016 ACM/SIGDA International Symposium on Field-Programmable Gate Arrays.*
15. Rigos, S., Mariatos, V., & Voros. N. (2012). A hardware acceleration unit for face detection. In *2012 Mediterranean Conference on Embedded Computing.*
16. Misra, J., & Saha. I. (2010). Artificial neural networks in hardware: A survey of two decades of progress. *Neurocomputing.*
17. Baji, T. (2018). Evolution of the GPU device widely used in AI and massive parallel processing. In *2018 IEEE Electron Devices Technology and Manufacturing Conference EDTM 2018—Proceedings.*
18. Shawahna, A., Sait, S. M., & El-Maleh, A. (2019). FPGA-based accelerators of deep learning networks for learning and classification: A review.
19. Mittal, S. (2020). A survey of FPGA-based accelerators for convolutional neural networks. *Neural Computing & Applications.*

20. Guo, K., Zeng, S., Yu, J., Wang, Y., & Yang, H. (2017). [DL] A survey of FPGA-based neural network inference accelerator.
21. Blaiech, A. G., Ben Khalifa, K., Valderrama, C., Fernandes, M. A. C., & Bedoui, M. H. (2019). A survey and taxonomy of FPGA-based deep learning accelerators. *The Journal of Systems Architecture.*
22. Talib, M. A., Majzoub, S., Nasir, Q., & Jamal, D. (2020) A systematic literature review on hardware implementation of artificial intelligence algorithms. *The Journal of Supercomputing.*
23. Schneider, S., Taylor, G. W., Linquist, S., & Kremer, S. C. (2019). Past, present and future approaches using computer vision for animal re-identification from camera trap data. *Methods in Ecology and Evolution.*
24. Faraone, J., Gambardella, G., Fraser, N., Blott, M., Leong. P., & Boland, D. (2018). Customizing low-precision deep neural networks for FPGAs. In *Proceedings—2018 International Conference on Field Programmable Logic and Applications FPL.*
25. Cheng, K. T., & Wang, Y. C. (2011). Using mobile GPU for general-purpose computing a case study of face recognition on smartphones. In *Proceedings of 2011 International Symposium on VLSI Design, Automation and Test VLSI-DAT 2011.*
26. Ouerhani, Y., Jridi, M., & AlFalou, A. (2010). Fast face recognition approach using a graphical processing unit "GPU." In *2010 IEEE International Conference on Imaging Systems and Techniques IST 2010—Proceedings.*
27. Li, E., Wang, B., Yang, L., Peng, Y. T., Du, Y., Zhang, Y., et al. (2012). GPU and CPU cooperative acceleration for face detection on modern processors. In *Proceedings—IEEE International Conference on Multimedia and Expo.*
28. Lu, L., Liang, Y., Xiao, Q., & Yan, S. (2017). Evaluating fast algorithms for convolutional neural networks on FPGAs. In *Proceeding—IEEE 25th Annual International Symposium on Field-Programmable Custom Computing Machines FCCM 2017.*
29. Szegedy, C., Liu, W., Jia, Y., Sermanet, P., Reed, S., Anguelov, D., et al. (2015). Going deeper with convolutions. In *Proceedings of the IEEE Conference on Computer Vision and Pattern Recognition.*
30. Zhang, C., Li, P., Sun, G., Guan, Y., Xiao, B., & Cong, J. Optimizing FPGA-based accelerator design for deep convolutional neural networks. In *FPGA 2015—2015 ACM/SIGDA International Symposium on Field-Programmable Gate Arrays.*
31. Suda, N., Chandra, V., Dasika, G., Mohanty, A., Ma, Y., Vrudhula, S., et al. (2016). Throughput-optimized openCL-based FPGA accelerator for large-scale convolutional neural networks. In *FPGA 2016—Proceedings of the 2016 ACM/SIGDA International Symposium on Field-Programmable Gate Arrays.*
32. Zhang, C., Fang, Z., Zhou, P., Pan, P., & Cong, J. (2016). Caffeine: Towards uniformed representation and acceleration for deep convolutional neural networks. In *IEEE/ACM International Conference on Computer-Aided Design Digital Technical Paper ICCAD.*
33. Guan, Y., Liang, H., Xu, N., Wang, W., Shi, S., Chen, X., et al. (2017). FP-DNN: An automated framework for mapping deep neural networks onto FPGAs with RTL-HLS hybrid templates. In *Proceedings—IEEE 25th Annual International Symposium on Field-Programmable Custom Computing Machines FCCM 2017.*
34. Rahman, A., Lee, J., & Choi, K. (2016). Efficient FPGA acceleration of convolutional neural networks using logical-3D compute array. In *Proceedings of 2016 Design, Automation & Test in Europe Conference & Exhibition DATE 2016.*
35. Ma, Y., Suda, N., Cao, Y., Seo, J. S., & Vrudhula, S. (2016). Scalable and modularized RTL compilation of Convolutional Neural Networks onto FPGA. In *FPL 2016—26th International Conference on Field-Programmable Logic and Applications.*
36. Zhang, C., Wu, D., Sun, J., Sun, G., Luo, G., & Cong. J. (2016). Energy-efficient CNN implementation on a deeply pipelined FPGA cluster. In *Proceedings of International Symposium on Low Power Electronics and Design.*
37. Ma, Y., Cao, Y., Vrudhula, S., & Seo, J. S. (2017). Optimizing loop operation and dataflow in FPGA acceleration of deep convolutional neural networks. In *FPGA 2017—Proceedings of the 2017 ACM/SIGDA International Symposium on Field-Programmable Gate Arrays.*

38. Liu, Z., Dou, Y., Jiang, J., Xu, J., Li, S., Zhou, Y., et al. (2017). Throughput-optimized FPGA accelerator for deep convolutional neural networks. *ACM Transactions on Reconfigurable Technology and Systems*.
39. Ma, Y., Cao, Y., Vrudhula, S., & Seo, J. S. An automatic RTL compiler for high-throughput FPGA implementation of diverse deep convolutional neural networks. In *2017 27th International Conference on Field-Programmable Logic and Applications FPL*.
40. Li, H., Fan, X., Jiao, L., Cao, W., Zhou. X., & Wang. L. (2016). A high performance FPGA-based accelerator for large-scale convolutional neural networks. In *FPL 2016—26th International Conference on Field-Programmable Logic and Applications*.
41. Alwani, M., Chen, H., Ferdman, M., & Milder, P. (2016). Fused-layer CNN accelerators. In *2016 49th Annual IEEE/ACM International Symposium on Microarchitecture (MICRO)*.
42. Wei, X., Yu, C. H., Zhang, P., Chen, Y., Wang, Y., Hu, H., et al. (2017). Automated systolic array architecture synthesis for high throughput CNN inference on FPGAs. In *Proceedings of the 54th Annual Design Automation Conference 2017*.
43. Motamedi, M., Gysel, P., & Ghiasi, S. (2017). PLACID: A platform for FPGA-based accelerator creation for DCNNs. *ACM Transactions on Multimedia Computing, Communications, and Applications*.
44. Ma, Y., Kim, M., Cao, Y., Vrudhula, S., & Seo, J. S. (2017). End-to-end scalable FPGA accelerator for deep residual networks. In *Proceedings—IEEE International Symposium on Circuits and Systems*.
45. Maguire, L. P., McGinnity, T. M., Glackin, B., Ghani, A., Belatreche, A., & Harkin, J. (2007). Challenges for large-scale implementations of spiking neural networks on FPGAs. *Neurocomputing*.
46. Negi, A., Saxena, D., & Suneja, K. (2020). High level synthesis of chaos based text encryption using modified Hill Cipher algorithm (pp. 3–7).
47. Thapa, S., Adhikari, S., Naseem, U., Singh, P., Bharathy, G., & Prasad, M. (2020). Detecting Alzheimer's disease by exploiting linguistic information from Nepali transcript. *Communication in Computer and Information Science*.

Author Index

P. Nayak et al. (eds.), *IoT and Analytics for Sensor Networks*, Lecture Notes in Networks and Systems 244, https://doi.org/10.1007/978-981-16-2919-8